Springer Series in Computational Mathematics

Volume 62

This series publishes high-level monographs in numerical analysis and computational mathematics. It includes theoretical contributions as well as application-oriented and software-related volumes. A list of titles and more information about the series can be found below.

Michael Griebel • Peter Oswald

Hilbert Space Splittings and Iterative Methods

Springer

Michael Griebel
Institut für Numerische Simulation
Universität Bonn
Bonn, Nordrhein-Westfalen, Germany

Fraunhofer Institute for Algorithms
and Scientific Computing (SCAI)
Schloss Birlinghoven
Sankt Augustin, Germany

Peter Oswald
Institut für Numerische Simulation
Universität Bonn
Bonn, Nordrhein-Westfalen, Germany

ISSN 0179-3632 ISSN 2198-3712 (electronic)
Springer Series in Computational Mathematics
ISBN 978-3-031-74369-6 ISBN 978-3-031-74370-2 (eBook)
https://doi.org/10.1007/978-3-031-74370-2

Mathematics Subject Classification (2020): 65J10, 65D15, 46E36, 65N22, 65N12, 65N55, 65T60, 65F10, 65F08

This Springer imprint is published by the registered company Springer Nature Switzerland AG
The registered company address is: Gewerbestrasse 11, 6330 Cham, Switzerland

Preface

This is a book on the theory of so-called Schwarz methods for solving variational problems in a Hilbert space V arising from linear equations and their associated quadratic minimization problems. Schwarz methods are based on the construction of a sequence of approximate solutions by solving auxiliary variational problems on a set of (smaller, finite-dimensional) Hilbert spaces V_i in a certain order, combining them, and using the combined approximations in an iterative method within a proper update procedure. The spaces V_i form a so-called space splitting for V, and we highlight which of its properties influence the convergence behavior of Schwarz methods based on it. This includes a detailed discussion of traditional deterministic and more recent greedy and stochastic orderings in the subproblem solution process and the investigation of accelerated methods. To illustrate the abstract theory, we discuss the numerical linear algebra analogs of the iterative methods studied in this book. We explain in an exemplary way its standard application to the convergence theory of multilevel and domain decomposition methods for solving PDE problems, but also mention links to optimization theory and online learning algorithms.

The selection of material for this book largely reflects the authors' research collaboration over the past 30 years, which began in 1993, initially focusing on finite element PDE solvers and preconditioning. The present book also includes some more recent developments in this and related areas of numerical analysis, approximation theory, optimization theory, and scientific computing, and could serve as the basis for a one- or two-semester course for M.S. and Ph.D. students specializing in numerical analysis and scientific computing.

We thank our families and our host institutions for their continued support. We also thank our unnamed mentors and colleagues whose advice and contributions to the subject matter of this book have inspired our own work over the years. We are especially grateful for the support of the *Hausdorff Center for Mathematics* in Bonn, funded by the Deutsche Forschungsgemeinschaft (DFG) under the German Excellence Strategy EXC-2047/1-390685813, and the Collaborative Research Centre 1060 *The Mathematics of Emergent Effects*.

Bonn, Dresden,
April 2024

Michael Griebel
Peter Oswald

The original version of the book has been revised. A correction to this book can be found at https://doi.org/10.1007/978-3-031-74370-2_8

Contents

Chapter 1
Introduction

Abstract The introduction explains the main topics of this book in a non-technical way. It also provides the motivation for the idea of space splitting for variational problems. Finally, the goals of the book, its structure, and the contents of the remaining chapters of this book are reviewed in more detail.

The theme of this book is the convergence theory of a class of iterative approximation methods for variational problems based on the divide-and-conquer approach. Roughly speaking, we deal with situations where a given problem is decomposed into a set of smaller problems that are cheaper to solve. The solutions of these subproblems are then combined in an appropriate way to obtain an approximation to the solution of the original problem. In special cases, this approach can actually lead to the exact solution of the original problem at lower cost. Classical examples are sorting algorithms such as Mergesort [109, 163] or the fast Fourier transform [28, 48]. In general, however, it only leads to a (hopefully) good approximation to the solution of the original problem, which is then used in an iterative method within a proper update procedure or defect correction step. After a sufficient number of iterations, the full solution is finally obtained with some prescribed accuracy, provided that the iteration is convergent.

The roots of what we call in this book Schwarz iteration and approximation methods are old and go back to classical methods such as the Gauss-Seidel method for solving linear systems of equations in numerical linear algebra or the von Neumann-Halperin alternating projection method for finding the closest point in the intersection of affine subspaces in optimization theory. In their modern form, their study began in the late 1980s and early 1990s to provide tools for the theoretical foundation of multigrid and domain decomposition solvers for large-scale discretizations of elliptic and parabolic partial differential equations (PDEs). In addition, the term Schwarz iterative methods was coined, with historical reference to the Schwarz alternating method of 1870 for proving the existence of solutions to the Dirichlet problem. We refer to [180, 190, 194] where the basic theory of Schwarz iterative methods is presented with emphasis on applications to PDEs. Developments in this

M. Griebel, P. Oswald, *Hilbert Space Splittings and Iterative Methods*, Springer Series in Computational Mathematics 62, https://doi.org/10.1007/978-3-031-74370-2_1

area of scientific computing at that time were driven in part by the availability of massively parallel and distributed computing platforms, for which decomposition into subproblems was an attractive feature because the subproblem solutions could now be computed in parallel by the many processors. But even on single-processor machines, there are advantages when the whole problem does not fit in RAM, but the subproblems do.

How to split a given problem into subproblems is often (but not always) induced in an obvious way by the problem itself. For example, naive sorting of a list of N real numbers has a worst-case complexity of $\approx N^2$ comparisons, but sorting of two lists of $\approx N/2$ numbers and then merging the two ordered lists requires only $\approx 2N^2/4 + N \approx N^2/2$ comparisons. This suggests to split the original list of N numbers into two lists of $\approx N/2$ numbers, and then into 4 lists of $\approx N/4$ numbers, and so on, until very short lists of a few numbers are obtained. The short lists are ordered in a naive way and then recursively merged into longer ordered lists. The final ordered list of N numbers is then obtained by a total of $O(N \log N)$ comparisons. This is what Mergesort does. The fast Fourier transform (FFT) applied to a vector in $\mathbb{C}^N$ with $N = 2^m$ entries proceeds in a similar way by recursively reducing the underlying trigonometric interpolation problem into 2^k similar trigonometric interpolation problems with vectors of length 2^{m-k} in a forward loop $k = 1, \ldots, m$, which results at the end in $N = 2^m$ trivial interpolation problems for $k = m$. The solutions of the latter are then pairwise combined in a backward loop $k = m, \ldots, 1$ to finally obtain the solution of the interpolation problem for the original vector of length N. This corresponds to a factorization of the underlying dense $N \times N$ interpolation matrix into $m = \log_2 N$ sparse matrices, each with $O(N)$ entries, and explains the reduced $O(N \log_2 N)$ operation count of the FFT.

Recursive splittings are typical in many areas. In particular, they lead to very efficient multilevel and multigrid methods for solving sparse systems of linear equations of large dimension N resulting from finite element or wavelet discretizations of PDEs, an application that motivated the development of the theory of Schwarz iterative methods and is used throughout this book for illustration. In contrast to the previous two examples, where the splitting into subproblems led to the exact solution of the original problem, here the combination of the subproblem solutions yields only an approximation to the solution of this system, which is then improved iteratively. Each update step typically takes $O(N)$ operations, and $O(|\log \epsilon|)$ iteration steps are needed to obtain an error bound $\leq \epsilon$ for given $\epsilon > 0$.

In other applications, the splitting is obtained in a more direct way. Some examples are given next: The typical feasibility problem in linear programming consists in finding a vector $x \in \mathbb{R}^m$ that satisfies a set of linear inequalities $a_i^T x \leq b_i$, where $a_i \in \mathbb{R}^m$ and $b_i \in \mathbb{R}$, $i = 1, \ldots, n$, are given. In other words, the feasible domain is described by n affine half-spaces $X_i \subset \mathbb{R}^m$ and one looks for a point x in their intersection $X := \cap_{i=1}^n X_i$. A simple idea is to take an arbitrary $x_0 \in \mathbb{R}^m$ and, if it is not already in the feasible domain X, compute its projection onto X, i.e., solve the quadratic minimization problem

$$Q_X x_0 := \operatorname{argmin}(\|x - x_0\|_2^2 : x \in X). \tag{1.1}$$

Here, the splitting idea is to replace (1.1) by n simpler, explicitly solvable minimization problems

$$Q_{X_i}y := \operatorname{argmin}(\|y - x_0\|_2^2 : x \in X_i),$$

and get approximations to $Q_X x_0$ by applying them iteratively. There are many different ways to implement this idea, see [37, 65, 77], where the subject is treated for arbitrary Hilbert spaces. Their analysis is facilitated by the product space formalism of [145] and its generalizations for (subgradient) projection methods. We give two basic versions, which turn out to be partial instances of what will be called multiplicative and additive Schwarz iterations in later chapters. The von Neumann-Halperin alternating projection method applied to the situation above is given by setting $x^{(0)} := x_0$ and, for $k = 0, 1, \ldots$, executing the inner loop

$$x^{(kn+i)} = Q_{X_i}x^{(kn+i-1)}, \qquad i = 1, 2, \ldots, n, \tag{1.2}$$

until a termination criterion is reached. In the method of simultaneous projections, one again sets $x^{(0)} := x_0$ at the beginning and then iterates according to

$$x^{(m+1)} = \frac{1}{n}\sum_{i=1}^{n} Q_{X_i}x^{(m)}, \qquad m = 0, 1, \ldots. \tag{1.3}$$

It can be proved that both processes geometrically converge to $Q_X x_0$, but with different speed. For a fair comparison, it is natural to compare the error after executing k inner loops in (1.2) with the error of $x^{(k)}$ obtained by (1.3).

As another example, consider an unconstrained convex minimization problem

$$J(x) \longrightarrow \min, \qquad x \in \mathbb{R}^N, \tag{1.4}$$

with differentiable convex $J : \mathbb{R}^N \longrightarrow \mathbb{R}$ such that $J(x) \longrightarrow \infty$ as $\|x\|_2 \longrightarrow \infty$. In block-coordinate descent methods, which have become popular in large-scale distributed optimization in recent years, the index set $\{1, \ldots, N\}$ of the state vector x is partitioned into n mutually disjoint non-empty subsets of size N_i, $i = 1, \ldots, n$, which induces a natural decomposition

$$x = \sum_{i=1}^{n} x_i, \qquad x_i \in \mathbb{R}^{N_i},\ i = 1, \ldots, n, \quad N = \sum_{i=1}^{n} N_i.$$

Here, the splitting idea is implemented by replacing the minimization problem (1.4) on $\mathbb{R}^N$ by n partial minimization problems with respect to $\mathbb{R}^{N_i}$ and iteratively combining their solutions. In practice, given a current iterate

$$x^{(m)} = \sum_{i=1}^{n} x_i^{(m)},$$

one often (approximately) solves one or more of the partial minimization problems

$$J_i(x_i; x^{(m)}) := J(x_i + x^{(m)}) \longrightarrow \min, \qquad x_i \in \mathbb{R}^{N_i}, \quad i = 1, \ldots, n, \qquad (1.5)$$

and sets $x^{(m+1)} = x^{(m)} + z^{(m)}$, where $z^{(m)}$ is some linear combination of the obtained approximations $z_i^{(m)}$ to the solution of the subproblems (1.5). The partitioning of the index set is usually determined by the particular structure of $J(u)$ and aims at making the solution of the subproblems as efficient as possible, while guaranteeing a sufficiently fast convergence of the iterative process to the set of minimizers of (1.4). Further examples of natural splittings in the context of SOR-type iterative methods for systems of linear equations and PDE discretizations are given in Section 1.1.

At this point, it is important to clarify that for arbitrarily given problems, there is no canonical method for deriving a splitting into subproblems, combining the solutions of the subproblems, and devising an iterative scheme to solve the original problem using them. Obviously, we are only interested in splittings such that the solutions of the subproblems and their combination are cheap to compute, and the iteration converges sufficiently fast. Even for simple models of measuring computational cost, a theory of optimality of splitting approaches can be expected only for certain classes of structured problems. In this book, we do not venture in that direction, except to add a few comments at some points on the aspect of computational cost and complexity. Rather, we focus our attention on the class of linear operator equations $Au = f$ with bounded symmetric and coercive operators $A = A' : V \longrightarrow V'$ and solution u in a separable Hilbert space V and splittings into similar problems on (usually finite-dimensional) Hilbert spaces V_i, $i \in \Omega$, where the index set Ω is either finite or countable. Equivalently, these problems can be cast as variational problems with symmetric and coercive bilinear forms $a(\cdot,\cdot)$ on V and $a_i(\cdot,\cdot)$ on V_i, generated by the operators A and A_i, respectively. From an optimization point of view, this refers to the class of unconstrained quadratic minimization problems on a Hilbert space. Although this class is very specific, it includes many problems of central importance in scientific computing, such as boundary value problems for elliptic PDEs, large sparse linear systems with symmetric positive definite matrices, inverse problems, kernel regression, and others. Such Hilbert space splittings of $(V; a)$ into $(V_i; a_i)$, $i \in \Omega$, are described in more detail below and studied in Chapter 2.

In this book we always start from a given space splitting, i.e., we do not discuss its construction or optimization, except for a few examples. Instead, we focus on the asymptotic convergence theory of a variety of iterative methods based on the given space splitting, including randomized, greedy, and accelerated schemes that have gained more attention in recent years. It turns out that the stability of the space splitting and the associated spectral properties of the so-called Schwarz operator play a central role in this convergence theory. While the case of finite Ω (i.e., decompositions into finitely many subproblems) is more common and covers various special iteration methods in finite-dimensional spaces, the case of countable Ω is also useful for several reasons. On the one hand, it is related to successive approximation schemes in infinite-dimensional separable Hilbert spaces, and in particular to adaptive methods that construct approximate solutions of operator equations on the fly, individually adapted to the problem instance under consideration. On the other hand, from the infinite-dimensional case and splittings with countable Ω, one gains insights for the finite-dimensional case and finite splittings with index sets $\Omega' \subset \Omega$, and one often gets convergence guarantees that are uniform with respect to Ω'. A typical situation is multilevel finite element or wavelet methods based on nested hierarchies of subspaces $V_0 \subset V_1 \subset \cdots \subset V_j \subset \cdots \subset V$.

The following section contains more details on the space splitting concept in a Hilbert space, which is the basis for the construction and study of the Schwarz iteration and approximation methods in this book. We also provide three additional examples to illustrate it. Section 1.2 contains some further comments on what to expect and what not to expect from this book, followed by a brief overview of the subsequent chapters and some explanations of notational conventions.

1.1 Schwarz methods: Generalities and examples

1.1.1 Hilbert space framework

In this introductory chapter, we will simplify things a bit and leave out some technical details. We refer to later chapters for a complete treatment. Let V be a separable real Hilbert space equipped with a designated scalar product $a(\cdot,\cdot)$ and norm $\|u\| := a(u,u)^{1/2}$. This includes finite-dimensional spaces such as $V = \mathbb{R}^n$ with a scalar product of the form $a(x,y) := (Ax,y)_2 = y^T Ax$ induced by a symmetric positive definite (spd) matrix $A \in \mathbb{R}^{n\times n}$, but it also includes infinite-dimensional Sobolev spaces $V = H^s(G)$. The latter are essential for the solution theory of strongly elliptic PDEs on domains $G \subset \mathbb{R}^d$, where $a(\cdot,\cdot)$ is typically induced by the elliptic operator. In addition, Sobolev spaces appear as smoothness spaces in several other applications. For example, they are used in regularization methods for inverse problems and as reproducing kernel Hilbert spaces in machine learning. Furthermore, for $s = 0$ we have the Lebesgue space $V = L^2(G)$ of square-integrable real-valued functions with domain G, which is often the Hilbert space of choice in signal processing and data analysis. Note that the designated scalar product $a(\cdot,\cdot)$ in V may differ from the canonical scalar product $(\cdot,\cdot)_V$ with associated norm $\|u\|_V := (u,u)_V^{1/2}$, but it should at least be equivalent in the sense that $c\|u\|_V^2 \le a(u,u) \le C\|u\|_V^2$ holds with constants $0 < c \le C < \infty$, independent of $u \in V$.

The goal is to solve the variational problem

$$a(u,v) = f(v) \qquad \forall\, v \in V, \tag{1.6}$$

where f is a given linear functional in the dual space V'. The solution u of this problem exists and is unique. The variational problem (1.6) can be rewritten as a linear operator equation

$$Au = f \tag{1.7}$$

in V' with bounded symmetric and coercive operator $A = A' : V \to V'$ by defining $Au \in V'$ as the linear functional $a(u,\cdot)$. Conversely, (1.7) can be used to define the scalar product $a(\cdot,\cdot)$ and the variational problem (1.6). Equivalently, $u \in V$ solves the quadratic minimization problem

$$J(v) := \frac{1}{2}a(v,v) - f(v) \to \min, \qquad v \in V.$$

The latter formulation provides the link to unconstrained convex optimization theory.

Next, for each $i \in \Omega$, we fix an auxiliary Hilbert space V_i with scalar product $a_i(\cdot,\cdot)$ and norm $\|v_i\|_i := a_i(v_i,v_i)^{1/2}$, $v_i \in V_i$, where $\Omega = \{1,\dots,n\}$ is finite. The case of countable $\Omega = \mathbb{N}$ is similar, an example is given below. For maximum flexibility, the V_i need not be subspaces of V, nor must the scalar products in V_i be inherited from the scalar product in V. These auxiliary Hilbert spaces are used to define the subproblems in the iterative methods for solving (1.6) described below. To connect the subproblems on V_i and the problem on V, we require the existence of bounded linear operators $R_i : V_i \to V$, and we define their duals $R_i^* : V \to V_i$ by finding the solutions $R_i^* u := u_i \in V_i$ of the variational problems

$$a_i(u_i, v_i) = a(u, R_i v_i) \qquad \forall\, v_i \in V_i, \qquad i \in \Omega, \tag{1.8}$$

for each $u \in V$.

Based on the auxiliary problems (1.8), we now construct a family of iterative solvers for the original problem (1.6) in the following way: Given the iterates $u^{(k)}$ with $k \le m$, we find the new iterate $u^{(m+1)}$ by solving the subproblems of the form

$$a_i(d_i^{(m)}, v_i) = a(u - w^{(m)}, R_i v_i) = f(R_i v_i) - a(w^{(m)}, R_i v_i) \qquad \forall\, v_i \in V_i, \tag{1.9}$$

for all i from some (finite) subset $I_m \subset \Omega$. Here, $w^{(m)}$ is a specified linear combination of the iterates $u^{(k)}$ with $k \le m$. The formulas in (1.8) and (1.9) show that

$$d_i^{(m)} = R_i^*(u - w^{(m)})$$

can be computed without knowing the unknown solution u of (1.6). With these $d_i^{(m)} \in V_i$ at hand, we define the next iterate as

$$u^{(m+1)} = v^{(m)} + \sum_{i \in I_m} \xi_i^{(m)} R_i d_i^{(m)},$$

where $v^{(m)}$ is another specified linear combination of the iterates $u^{(k)}$ with $k \le m$ (possibly different from $w^{(m)}$), and $\xi_i^{(m)} > 0$ are so-called relaxation or step-size parameters. In operator notation, the update formula has the form

$$u^{(m+1)} = v^{(m)} + \sum_{i \in I_m} \xi_i^{(m)} P_i(u - w^{(m)}), \tag{1.10}$$

where

$$P_i := R_i R_i^* : V \to V$$

are symmetric positive (but not necessarily positive definite) operators, which play an important role in the convergence theory of the above methods. Iterations with (1.10) as update formula appear in the literature under various names (subspace correction, Schwarz iteration, or alternating direction method). In this book, we use the terminology *Schwarz iterative methods*.

Due to the freedom of choosing the linear combinations $w^{(m)}$, $v^{(m)}$, the index set $I_m \subset \Omega$, and the relaxation parameters $\xi_i^{(m)}$, the approach in (1.10) covers a whole zoo of concrete iterations. In practice, it is sufficient to deal with one- and two-step methods. In this introduction, we discuss only one-step methods for which $w^{(m)} = v^{(m)} = u^{(m)}$, and focus on the two extremes for the choice of the index set I_m, namely $I_m = \{i_m\}$ (solving a single subproblem per iteration step) and $I_m = \Omega$. As far as the relaxation parameters are concerned, we consider them to be constant throughout the iteration process. More general situations are treated in Chapter 4.

With these specifications, in the case of $I_m = \{i_m\}$, the update formula (1.10) is simplified to

$$u^{(m+1)} = u^{(m)} + \omega\rho_{i_m}P_{i_m}(u - u^{(m)}), \qquad m = 0, 1, \ldots. \tag{1.11}$$

We start with an arbitrary $u^{(0)} \in V$. The weights $\rho_{i_m} > 0$ are useful to account for differences in the relative scaling of the subproblem contributions $P_{i_m}(u - u^{(m)}) = R_{i_m}d_{i_m}^{(m)}$, while $\omega > 0$ is a common step-size parameter. These parameters can be used to optimize the convergence speed of this non-stationary iteration. The associated error propagation operator after m steps has the product form

$$(I - \omega\rho_{i_{m-1}}P_{i_{m-1}})\cdots(I - \omega\rho_{i_1}P_{i_1})(I - \omega\rho_{i_0}P_{i_0}),$$

whose analysis in Chapter 4 depends on the way the sequence $(i_m)_{m\geq 0}$ is chosen and is in some cases non-trivial, especially if the individual P_i do not commute. Schemes like (1.11) belong to the class of so-called multiplicative Schwarz methods. The von Neumann-Halperin alternating projection method for the above feasibility problem is of this type.

The case $I_m = \Omega$ results in the update rule

$$u^{(m+1)} = u^{(m)} + \omega\sum_{i\in\Omega}\rho_i P_i(u - u^{(m)}) = u^{(m)} + \omega P_\rho(u - u^{(m)}), \tag{1.12}$$

with $m = 0, 1, \ldots$, where

$$P_\rho := \sum_{i\in\Omega}\rho_i P_i \;:\; V \to V$$

is supposed to be symmetric and coercive. The method of simultaneous projections for the feasibility problem is similar to this type of update formula. The stationary iteration (1.12), called additive Schwarz iteration, is much easier to analyze since the error propagation operator after m steps is simply

$$(I - \omega P_\rho)^m,$$

which leads to convergence estimates in a straightforward manner. As in classical numerical linear algebra, the condition number $\kappa(P_\rho)$ determines the convergence speed of the iteration (1.12). More sophisticated two-step Schwarz methods reduce this to a dependence of the convergence speed on $\sqrt{\kappa(P_\rho)}$. This points to another aspect of the space splitting approach: In some cases, the subproblems can be chosen such that even for operator equations with ill-conditioned A, the operator P_ρ has a small to moderate condition number. We will comment on this preconditioning effect in connection with the examples in Subsection 1.1.3 and 1.1.4. Operators like P_ρ also play an important role in the analysis of multiplicative Schwarz iterations and approximation methods, especially of their randomized versions. We note here that the randomness of the choice of the index set I_m in the update formula of a Schwarz method (or of i_m in the iteration (1.11)) may be the result of random failures of processors in an unreliable distributed computing network, or it may be artificially imposed to improve the average performance of the method when no preferable deterministic ordering is available.

Before discussing some differences between Schwarz iterative methods (finite Ω) and Schwarz approximation methods (countable Ω) in Subsection 1.1.5, we consider some typical examples in the next three subsections to illustrate the above abstract scheme.

1.1.2 Example: SOR-type methods

We begin with our first example, which is related to the classical theory of a variety of successive overrelaxation (SOR) methods in numerical linear algebra. It is repeated as Example 1 in Subsection 2.1.2. Consider the problem of solving the system

$$Ax = b$$

of linear equations, where $A \in \mathbb{R}^{n\times n}$ is symmetric positive definite, $b \in \mathbb{R}^n$, and $x \in \mathbb{R}^n$ denotes its unique solution. To solve it numerically, a variety of direct and iterative methods exists, where the preference depends on the structure of A, its dimension n, and the available computer hardware. In particular, sparse A with large n often require iterative methods, since A^{-1} typically does not have a sparse factorization (as in the case of the FFT) and the fill-in prevents us from using direct methods.

A classical iteration idea, developed in the pre-computer era for hand calculations, is to update in each step only the i-th entry of the iteration vector x^{old} by solving the corresponding i-th equation. This gives $x_k^{new} = x_k^{old}$ for $k \neq i$ and

$$x_i^{new} = x_i^{old} + a_{ii}^{-1}(b_i - \sum_{k=1}^{n} a_{ik}x_k^{old}).$$

In each step, a different i is used, typically in cyclic order or in a greedy fashion, i.e., by choosing the index i with the largest residual entry. This is the Gauss-Seidel method. It was mentioned by Gauss in a letter of 1823, also explaining the greedy variant for a small 4×4 system, and was later reintroduced by Seidel [164]. By adding a relaxation parameter $\omega \in (0, 2)$ before a_{ii}^{-1}, one obtains the SOR-method. Alternatively, the weighted Jacobi iteration of 1845 updates all entries at once by a similar formula, namely

$$x_i^{new} = x_i^{old} + \omega a_{ii}^{-1}(b_i - \sum_{k=1}^{n} a_{ik} x_k^{old}), \qquad i = 1, \dots, n,$$

where ω is a properly chosen step-size parameter. For a nice description of the historical background of these iterative methods, see [154].

Next, we will explain how these methods fit into the framework of multiplicative and additive Schwarz methods. To do this, we set $V = \mathbb{R}^n$ and

$$a(x, y) = (Ax, y)_2 = y^T Ax.$$

Furthermore we set $V_i = \mathbb{R}$ and

$$a_i(x_i, y_i) = x_i y_i, \qquad x_i, y_i \in \mathbb{R},$$

where $i \in \Omega = \{1, \dots, n\}$. Finally, we define the operators $R_i : V_i \to V$ by $R_i x_i = (0, \dots, x_i, \dots, 0)^T = x_i e_i$, which is the extension-by-zero of x_i to a vector in $\mathbb{R}^n$ with x_i as the i-th entry, where e_i denotes the i-th unit coordinate vector, and we set $\rho_i = a_{ii}^{-1}$. Obviously, by definition of R_i^* and P_i we have

$$R_i^* x = \sum_{k=1}^{n} a_{ik} x_k, \qquad P_i x = \left(\sum_{k=1}^{n} a_{ik} x_k \right) e_i.$$

Substituted into the equations (1.11) and (1.12), this gives exactly the SOR- and weighted Jacobi iterations with weights $\rho_i = a_{ii}^{-1}$ and common relaxation parameter $\omega > 0$, respectively.

As detailed in Section 5.1, by applying the convergence estimates for general Schwarz iterative methods from Chapter 4, one recovers the well-known classical convergence results for the class of linear systems with spd matrix A. In particular, we have convergence of the SOR-method for $0 < \omega < 2$ and of the weighted Jacobi method for $0 < \omega < 2/\lambda_{\max}(D^{-1}A)$, where D is the diagonal part of A. The bounds on the rate of convergence depend critically on the spectral condition number $\kappa(D^{-1/2}AD^{-1/2})$. This is not surprising, since the proofs for general space splittings mimic the proofs from classical numerical algebra, and the associated Schwarz operator P_ρ is given by

$$P_\rho x = \sum_{i=1}^{n} \rho_i P_i x = \sum_{i=1}^{n} a_{ii}^{-1} \left(\sum_{k=1}^{n} a_{ik} x_k \right) e_i = D^{-1} Ax.$$

Therefore, no significant preconditioning effect is expected from this splitting.

Throughout the book, the model case of solving sets of linear equations is used repeatedly to illustrate the general theory and to check whether it reproduces classical results. For example, Section 5.2 is devoted to variants of the Kaczmarz method for solving arbitrary linear systems $Ax = b$ with $n' \times n$ matrix and least-squares problems. Here, a different canonical space splitting is used and the convergence estimates depend on upper and lower bounds on the positive eigenvalues of $A^T A$, see Example 2 in Subsection 2.1.2.

We also cover randomized and greedy versions of Schwarz iteration methods and their accelerated versions, for which the consequences for iterative methods for linear systems are less well-known. Many of these can only be found in journal articles published in the last decades. As an example, consider the iteration

$$x_i^{(m+1)} = x_i^{(m)} + \begin{cases} a_{ii}^{-1}(b_i - \sum_{k=1}^n a_{ik}x_k^{(m)}), & i = i_m, \\ 0, & i \in \Omega,\ i \neq i_m, \end{cases} \qquad m \geq 0,$$

where $x^{(0)}$ is given at the beginning. This is exactly the multiplicative Schwarz iteration with update formula (1.11) and $\omega = 1$ when using the above space splitting for dealing with linear systems $Ax = b$ and $n \times n$ spd matrices A. In the randomized version, if i_m is an i.i.d. index chosen from Ω with an underlying probability distribution given by

$$\mathbb{P}(i_m = i) = \frac{a_{ii}}{a_{11} + \cdots + a_{nn}}, \qquad i = 1, \ldots, n,$$

then, in expectation, the error $e^{(m)} = x - x^{(m)}$ satisfies

$$\mathbb{E}(\|e^{(m)}\|^2) \leq (1 - \mu_R)^m \|e^{(0)}\|^2, \quad \mu_R := \frac{\lambda_{\min}(A)}{a_{11} + \cdots + a_{nn}}, \qquad m \geq 1.$$

This bound was originally established in [174] and sparked renewed interest in randomized iterative methods. In fact, this bound is at least as good as, and in some cases better than, the best estimate for the SOR method with fixed cyclic ordering. It is also compatible with the convergence bounds for the weighted Jacobi method.

In the greedy version, the next index i_m is chosen so that the scaled residual quantity

$$\Delta_i^{(m)} := a_{ii}^{-1}(b_i - \sum_{k=1}^n a_{ik}x_k^{(m)})^2$$

is maximized for i_m. Then one can prove that

$$\|e^{(m)}\|^2 \leq (1 - \mu_G)^m \|e^{(0)}\|^2, \qquad m \geq 1,$$

with some constant $\mu_G \geq \mu_R$, which depends on certain metric properties of the matrix $D^{-1}A$. Although this is only an upper bound on the convergence rate, it is consistent with the fact often observed in numerical practice that greedy versions are faster than randomized and deterministic versions of iterative methods.

1.1.3 Example: Domain decomposition

To demonstrate the idea of domain decomposition, we use a simple test problem, namely the linear system $Ax = b$ arising from a standard 5-point finite difference discretization of the Poisson equation

$$-\Delta u = f \quad \text{in } G, \qquad u = 0 \quad \text{on } \Gamma, \tag{1.13}$$

with homogeneous Dirichlet boundary data with respect to a uniform grid on the rectangle $G = [-1, 1] \times [0, 1]$ in the plane. Here Γ denotes the boundary of G where the zero Dirichlet boundary condition is imposed. We denote the mesh-size of the finite difference grid by $h = N^{-1}$, with sufficiently large $N > 2$. If the entries of the grid vector x are denoted as usual by $x_{j,k}$, i.e., $x_{j,k}$ corresponds to the grid point $P_{j,k} := (jh, kh)$, then the system $Ax = b$ is equivalent to the set of linear equations

$$4x_{j,k} - x_{j-1,k} - x_{j+1,k} - x_{j,k-1} - x_{j,k+1} = h^2 f_{j,k}, \qquad |j| < N,\ 0 < k < N,$$

where $f_{j,k} = f(P_{j,k})$ and $x_{j,k} = 0$ for $P_{j,k} \in \Gamma$, i.e., when $j = -N$ or $j = N$ and $k = 0$ or $k = N$. It is well-known that A is symmetric positive definite and that its condition number $\kappa(A)$ scales as $h^{-2} = N^2$ when $h = N^{-1} \to 0$. Furthermore, the bilinear form $a(\cdot, \cdot)$ induced by A as in the previous example has an explicit expression given by

$$a(x, x) = x^T A x = \sum_{j=-N}^{N-1} \sum_{k=1}^{N-1} (x_{j+1,k} - x_{j,k})^2 + \sum_{j=-(N-1)}^{N-1} \sum_{k=0}^{N-1} (x_{j,k+1} - x_{j,k})^2.$$

Because of this very specific structure, there are efficient direct methods for solving linear systems with this A (e.g., using the two-dimensional FFT), but our goal is to demonstrate the domain decomposition approach in the simplest case of splitting the rectangular domain G into two overlapping domains $G_1' = [-1, \ell h] \times [0, 1]$ and $G_2' = [-\ell h, 1] \times [0, 1]$, which overlap in the strip $G_1' \cap G_2' = [-\ell h, \ell h] \times [0, 1]$. Here we assume $0 < \ell < N/2$. For $h = 1/8$ and $\ell = 2$, this situation is shown in Figure 1.1. The case $\ell = 0$ is also important, it concerns non-overlapping domain decomposition methods, which we consider in Subsection 7.2.1 in the finite element context.

The idea now is to reduce the solution of $Ax = b$ to similar problems with respect to the subdomains G_i' using the space splitting approach. Let $n = 2$, i.e., $i \in \Omega = \{1, 2\}$, and define V and V_i as the spaces of all grid vectors x and x^i associated with the grid points in the interior of G and G_i', respectively. By convention, grid values $x_{j,k}$ and $x^i_{j,k}$ associated with grid points on Γ appearing in the formulas below are always fixed to zero. The matrices A_i are identical to submatrices of A, and the bilinear forms $a_i(\cdot, \cdot)$ are obtained from the above explicit expression for $a(\cdot, \cdot)$ by restricting the summation with respect to j to the ranges $j = -(N-1), \ldots, \ell - 1$ and $j = -(\ell - 1), \ldots, N - 1$, respectively. For R_i we again take the natural extension-by-zero operators. Then, in matrix notation, the operators $R_i^* : V \to V_i$ have the form

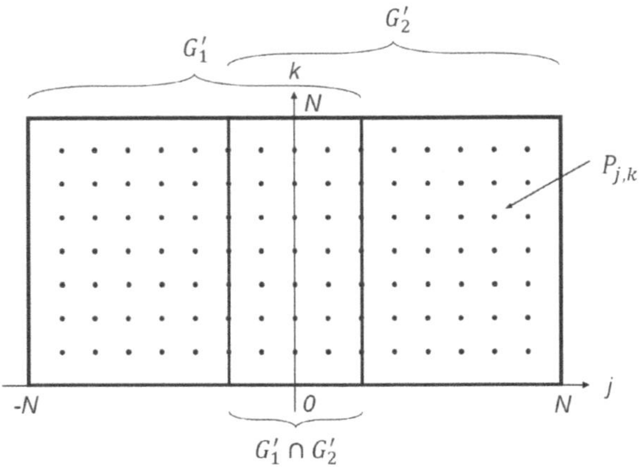

Fig. 1.1 Decomposition of the domain G into two overlapping subdomains G_1' and G_2' with overlap region $G_1' \cap G_2'$ for $N = 8$, $h = 1/8$, and $\ell = 2$.

$A_i^{-1} R_i^T A$, since

$$a_i(x^i, y^i) = (y^i)^T A_i x^i = a(x, R_i y^i) = (y^i)^T R_i^T A x, \qquad \forall\, y^i \in V_i,$$

implies the formula $x^i = A_i^{-1} R_i^T A x$ for the solution of the subproblem. Thus,

$$P_i = R_i A_i^{-1} R_i^T A$$

and

$$P = P_1 + P_2 = \underbrace{(R_1 A_1^{-1} R_1^T + R_2 A_2^{-1} R_2^T)}_{C:=} A, \tag{1.14}$$

where we have set $\rho_1 = \rho_2 = 1$ and omitted the subscript ρ in the notation.

This is the Schwarz operator, which governs the additive Schwarz iteration. Each step of (1.12) with the above P_i would require a residual computation $r = b - Ax^{(m)}$ with the current $x^{(m)} \in V$, whose restrictions r^i to V_i are then used as right-hand sides of the linear systems with matrix A_i, $i = 1, 2$. These two systems can be solved in parallel, and their solutions, after extension-by-zero to vectors in V and multiplication by ω, must be added to $x^{(m)}$ to obtain $x^{(m+1)}$. The parallelization potential is formally even greater if the domain G is split into a larger number of smaller domains. This results in smaller matrices A_i and faster runtime, provided enough processors are available. Note, however, that the overall efficiency also depends on the communication costs and the convergence speed of the additive Schwarz method, i.e., on the spectral radius of $I - \omega P$.

This brings us to another aspect worth mentioning. It turns out that the symmetric matrix C in (1.14) acts as a preconditioner for A. More precisely, $P = CA$ is spd with respect to the scalar product $a(\cdot, \cdot)$ and $\kappa(P) \approx (\ell h)^{-1}$ is inversely proportional to the width of the overlap region. This compares favorably with $\kappa(A) \approx h^{-2}$, and also shows that we can even get $\kappa(P) = \mathrm{O}(1)$ if the overlap is generous, i.e., $\ell \geq cN$ for some fixed $c > 0$. For the case of many overlapping domains and finite element discretizations, results of similar flavor are discussed in Subsection 7.2.2.

Estimates for the condition number of Schwarz operators can be derived from a general result about space splittings proved in Theorem 2.6 in Section 2.2, namely

$$\kappa(P_\rho) \leq C_1 \cdot C_2. \tag{1.15}$$

Here the constants C_1, C_2 are such that, on the one hand,

$$a(\sum_{i\in\Omega} \rho_i R_i v_i, \sum_{i\in\Omega} \rho_i R_i v_i) \leq C_1 \sum_{i\in\Omega} \rho_i a_i(v_i, v_i) \qquad \forall\, v_i \in V_i, \quad i \in \Omega, \tag{1.16}$$

and, on the other hand, for each $u \in V$ there exists a representation $u = \sum_{i\in\Omega} \rho_i R_i v_i$ with certain $v_i \in V_i$, $i \in \Omega$, which satisfies

$$\sum_{i\in\Omega} \rho_i a_i(v_i, v_i) \leq C_2 a(u, u). \tag{1.17}$$

The two inequalities (1.16) and (1.17) are in a sense inverse to each other, and (1.15) is tight: The infimum of the right-hand side in (1.15) with respect to the constants C_1 and C_2 is equal to $\kappa(P_\rho)$. It is called the condition number of the underlying space splitting for given weights $\rho_i > 0$. In Chapter 2 we provide a more abstract framework to study the inequalities (1.16) and (1.17). There we consider the operators $R_\rho : \tilde{V}_\rho \to V$ defined by

$$R_\rho \tilde{v} = \sum_{i\in\Omega} \rho_i R_i v_i, \qquad \tilde{v} = (v_i)_{i\in\Omega} \in \tilde{V}_\rho, \quad v_i \in V_i,\ i \in \Omega,$$

and their adjoints $R_\rho^* : V \to \tilde{V}_\rho$. This results in the operators $P_\rho = R_\rho R_\rho^* : V \to V$ and $\tilde{P}_\rho = R_\rho^* R_\rho : \tilde{V}_\rho \to \tilde{V}_\rho$, where $\tilde{V}_\rho$ is the weighted Hilbert sum of the spaces $(V_i; a_i)$, namely

$$\tilde{V}_\rho = \{\tilde{v} : \|\tilde{v}\|^2_{\tilde{V}_\rho} = \tilde{a}_\rho(\tilde{v}, \tilde{v}) := \sum_{i\in\Omega} \rho_i a_i(v_i, v_i) < \infty\}.$$

Both finite and countable index sets Ω are allowed. This abstraction also provides links to frame and operator theory.

In our simple domain decomposition example, (1.16) holds with $C_1 = 2$, since

$$a(\sum_{i=1}^{2} R_i x^i, \sum_{i=1}^{2} R_i x^i) \leq 2 \sum_{i=1}^{2} a(R_i x^i, R_i x^i) = 2 \sum_{i=1}^{2} a_i(x^i, x^i).$$

A suitable representation $x = R_1 x^1 + R_2 x^2$ to prove the inequality (1.17) with $C_2 = \mathrm{O}((\ell h)^{-1})$ is given by setting

$$x^1_{j,k} = \begin{cases} x_{j,k}, & -N \le j \le -\ell, \\ (1 - j/\ell)x_{j,k}/2, & -\ell < j < \ell, \end{cases} \qquad k = 1, \dots, N-1,$$

and $x^2 = R_2^T(x - R_1 x^1)$. One can easily check that

$$a_1(x^1, x^1) \le 2a(x,x) + \frac{1}{2\ell^2} \sum_{j=-\ell}^{\ell-1} \sum_{k=1}^{N-1} (x_{j,k})^2,$$

similarly for $i = 2$. A bound on the last term follows by observing that

$$(x_{j,k})^2 = \left(\sum_{l=j}^{N-1} (x_{l+1,k} - x_{l,k}) \right)^2 \le N \sum_{l=0}^{N-1} (x_{l+1,k} - x_{l,k})^2$$

holds for $j \ge 0$ since $x_{N,k} = 0$ at the right boundary of G, similarly for $j < 0$ using the left boundary of G. Summing over all $-\ell \le j < \ell$ and $k = 1, \dots, N-1$ and considering the previous estimates, we get (1.17) with constant

$$C_2 = 4 + \frac{N\ell}{2\ell^2} = 4 + \frac{1}{2\ell h}.$$

This proves $\kappa(P) \le 8 + (\ell h)^{-1}$. We leave it to the reader to find suitable lower bounds for C_1 and C_2 which show that indeed $\kappa(P) \approx (\ell h)^{-1}$ when $\ell h \to 0$.

We conclude with two remarks. A representation $G = G_1' \cup G_2'$ with two overlapping domains already appeared in the seminal paper [161] by Schwarz on his alternating method for establishing the Dirichlet principle for the existence of solutions of the Laplace equation on general domains G. In modern language and rephrasing the results in [161], he proved the existence of solutions of the Poisson problem (1.13) in $V = H_0^1(G)$ by a multiplicative Schwarz iteration with $V_i = H_0^1(G_i')$, $i = 1, 2$, provided that the Dirichlet principle holds for the subdomains and that the overlap region $G_1' \cap G_2'$ is sufficiently large. Since for a splitting into two subdomains the only natural order of access to the subproblems is cyclic, i.e., $(i_m)_{m \ge 0} = (1, 2, 1, 2, \dots)$, the name alternating method was used. Schwarz's paper also shows that space splittings and iterations based on them can be useful even in the context of infinite-dimensional Hilbert spaces. For the history of the domain decomposition method, see [78, 180].

As for the case $\ell = 0$ in the example above, we have several possibilities to decompose the space of grid vectors V associated with G. On the one hand, we can leave the definition of V_i as it is and add a third space V_3 of grid vectors associated with grid points $P_{0,k}$, $k = 1, \dots, N-1$. Algebraically, this is a direct decomposition of V into three spaces. Then the main challenge is to construct the bilinear form $a_3(\cdot, \cdot)$ and the operator R_3 in such a way that the new Schwarz operator has a small or moderate condition number and, at the same time, the solution of the linear system with matrix A_3 and the application of R_3 is cheap. This is known in the literature as the interface (or Schur complement) problem. For the case of finite elements and many subdomains, this approach is discussed in Subsection 7.2.1. Alternatively, the grid values $x_{0,k}$ associated with the interface $\tilde{\Gamma} = G_1 \cap G_2$ can be included in one of the V_i (or both). For example, if they are only included in V_1, then the subproblem associated with G_1 corresponds to a finite difference discretization of a Poisson problem on G_1 with unknown Neumann data on $\tilde{\Gamma}$ and zero Dirichlet data on $\Gamma_1 \backslash \tilde{\Gamma}$, while the subproblem on G_2 can be viewed as the discretization of the Poisson equation on G_2 with unknown Dirichlet data on $\tilde{\Gamma}$ and zero Dirichlet data on $\Gamma_2 \backslash \tilde{\Gamma}$. This leads to so-called transmission problems between Neumann and Dirichlet data on $\tilde{\Gamma}$. The latter can be incorporated into the construction of a reasonable space splitting by modifying the operators R_i. This possibility is not considered further in this book, for an introduction see [180, Chapter 2].

1.1.4 Example: Multilevel methods

With this example we introduce an important case of space splittings related to local approximation schemes such as finite elements and wavelets in function spaces, in particular Sobolev spaces. We deal here for the first time with countable $\Omega = \mathbb{Z}_+$ and assume that V is infinite-dimensional and its scalar product $a(\cdot,\cdot)$ is induced by a symmetric and coercive operator $A : V \to V'$. We consider space splittings generated by a ladder of strictly nested finite-dimensional subspaces

$$\{0\} \neq V_0 \subset V_1 \subset \cdots \subset V_j \subset \cdots \subset V,$$

whose union $\cup_{j\ge 0} V_j$ is dense in V. We are interested in methods for solving variational problems

$$a(\hat{u}, \hat{v}) = f(\hat{v}) \qquad \forall\, \hat{v} \in \hat{V} \tag{1.18}$$

for a variety of finite-dimensional subspaces $\hat{V}$ associated with $(V_j)_{j\ge 0}$, hopefully with good bounds on their performance. Note that (1.18) is the Galerkin problem on $\hat{V}$ associated with (1.6), and the unique solvability of (1.18) is guaranteed for all $\hat{V}$. If we now have a stable space splitting for V based on the V_j equipped with appropriate scalar products $a_j(\cdot,\cdot)$, this can indeed lead, under certain conditions, to finite space splittings for many $\hat{V} \subset V$ with uniform convergence guarantees for Schwarz iterative schemes based on them. The underlying theory is developed in Section 3.1, with applications treated in Chapter 7. Here we discuss only a motivating example and do not provide any proofs.

Consider again the Poisson problem (1.13), but this time in its weak form with the Sobolev space $V := H_0^1(G)$ as the solution space and with

$$a(u, v) = \int_G (\nabla u(x), \nabla v(x))_2 \, dx, \qquad \|v\| := \Big(\int_G \|\nabla v(x)\|_2^2 \, dx\Big)^{1/2}$$

as designated scalar product and norm, respectively. The associated Laplace operator $A := -\Delta$ maps $H_0^1(G)$ to its dual $H^{-1}(G)$. For the purpose of discussion, we will restrict ourselves to the square $G = [-1, 1]^2 \subset \mathbb{R}^2$. On such a domain G, the construction of ladders of nested subspaces $(V_j)_{j\ge 0}$ using finite element functions on nested sequences of rectangular and triangular partitions, wavelets, or trigonometric functions is straightforward. We stick to the case of bilinear C^0 Lagrange finite elements on uniform dyadic partitions of G, i.e., the functions in V_j are continuous, vanish on the boundary Γ of G, and coincide with a bilinear function on each dyadic square of side-length 2^{-j} in G. Each function in V_j is uniquely described by its values at $\mathcal{V}_j$, the set of all dyadic points of the form $(i2^{-j}, k2^{-j})$ inside G, which we call level j nodal points (values at level j nodal points on Γ are always silently set to zero). The canonical basis in V_j is given by the set of nodal basis functions $N_{P,j} \in V_j$, $P \in \mathcal{V}_j$, which satisfy $N_{P,j}(P') = 0$ for all level j nodal points $P' \neq P$ and $N_{P,j}(P) = 1$. Figure 1.2 shows the nodal point sets $\mathcal{V}_j$ for $j = 0, 1, 2$ for three successive uniform rectangular partitions on the domain $G = (-1, 1)^2$, and gives the bilinear finite element basis function $N_{P,j}$ for some level j nodal point P.

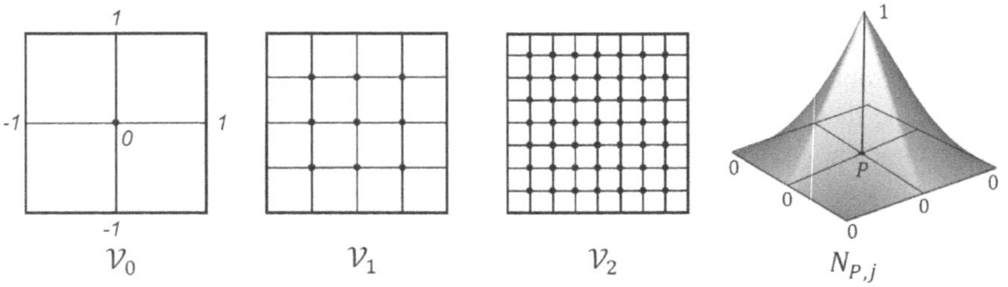

Fig. 1.2 Nodal point sets $\mathcal{V}_j$ for $j = 0, 1, 2$ and the nodal basis function $N_{P,j}$ for some level j nodal point P for the bilinear finite element example.

If one sets $\hat{V} = V_j$, $j \geq 0$, in (1.18), then the solutions u_j of this variational problem converge to the solution u of (1.6), and can be determined by solving sparse linear systems $A_j x_j = b_j$, where $\kappa(A_j) \approx 2^{2j}$ grows exponentially and x_j is the nodal basis vector associated with u_j, i.e.,

$$x_j = (x_{P,j})_{P \in \mathcal{V}_j} : \qquad u_j = \sum_{P \in \mathcal{V}_j} x_{P,j} N_{P,j}.$$

In general, it is not promising to naively split $V = H_0^1(G)$ into V_j by considering the V_j as subspaces of V, i.e., by taking $a_j(\cdot, \cdot) = a(\cdot, \cdot)$ as designated scalar products in V_j and identifying R_j with the natural embedding of V_j in V. The reason for this is twofold. First, the linear systems associated with the subproblems are given by ill-conditioned matrices A_j and are hard to solve (at least for elliptic PDEs with non-constant coefficients and more complicated G). Second, the inequalities (1.16) and (1.17) cannot hold simultaneously for any fixed set of weights $\rho_j > 0$. This means that we cannot expect a well-conditioned P_ρ and thus fast convergence of Schwarz iterative methods. In fact, if (1.16) holds, we must have $\sum_{j \geq 0} \rho_j \leq C_1 < \infty$. To see this, we take an arbitrary $0 \neq u_0 \in V_0$, set $v_j = \rho_j (\sum_{k \leq j} \rho_k)^{-1} u_0 \in V_j$ in

(1.16), and let $j \to \infty$ in the resulting bound. Now, if we take any $0 \neq w_j$ in the orthogonal complement of V_j with respect to V_{j-1} as u in (1.17), then this implies $\rho_j \geq C_2^{-1} > 0$, since $u = \rho_j(\rho_j^{-1}w_j)$. So $v_j = \rho_j^{-1}w_j$ and $v_k = 0$ for $k \neq j$ is the optimal representation that minimizes the left-hand side in (1.17) for this u. But since j is arbitrary here, there is a contradiction.

An alternative approach that circumvents both of the above obstacles can be derived from approximation theory and Besov spaces. This was observed around 1990, but the underlying results are even older. For subspace ladders of finite element spaces, such as our bilinear finite element spaces V_j, we refer to [135, 202] for earlier expositions and references, see also Section 3.1. It turns out that choosing scaled $L^2(G)$ scalar products

$$a_j(u_j, v_j) := 2^{2j}(u_j, v_j)_{L^2}$$

on V_j and setting $\rho_j = 1$ for all $j \geq 0$ leads to a space splitting for which (1.16) and (1.17) hold with finite constants C_1 and C_2. There is also a moderate bound for $\kappa(P_\rho)$. Since $V_j \subset V$ as sets, the natural embeddings still serve as R_j. Due to the choice of the auxiliary scalar products $a_j(\cdot,\cdot)$ on V_j, the subproblems are governed by the well-conditioned $L^2(G)$ Gram matrices associated with the nodal bases in V_j.

These positive results hold for a splitting of the infinite-dimensional space $V = H_0^1(G)$ into countably many finite-dimensional V_j. We need to take further steps before discussing the convergence properties of any iterative methods built from the splitting. One such step follows from the proof of the associated inequalities (1.16) and (1.17), which, roughly speaking, is based on the existence of uniformly L^2 bounded projections $Q_j : L^2(G) \to V_j$, $j \geq 0$. Here we skip some subtleties related to the zero boundary conditions. It can be shown that in our case the validity of (1.16) and (1.17) is equivalent to a two-sided inequality

$$\tilde{C}_1\|u\|_{H_0^1}^2 \leq \|Q_0 u\|_{L^2}^2 + \sum_{j\geq 1} 2^{2j}\|(Q_j - Q_{j-1})u\|_{L^2}^2 \leq \tilde{C}_2\|u\|_{H_0^1}^2 \qquad \forall u \in H_0^1(G),$$

where $\tilde{C}_1 \approx C_2^{-1}$ and $\tilde{C}_2 \approx C_1^{-1}$ are fixed positive constants depending on the chosen sequence of projections. An example of such Q_j are the L^2 orthogonal projections onto V_j, and in Subsection 3.1.3 we deal with a family of so-called quasi-interpolant projections. By restriction, this inequality automatically implies for any finite $J \geq 1$ that

$$\tilde{C}_1\|u_J\|_{H_0^1}^2 \leq \|Q_0 u_J\|_{L^2}^2 + \sum_{j=1}^{J} 2^{2j}\|(Q_j - Q_{j-1})u_J\|_{L^2}^2 \leq \tilde{C}_2\|u_J\|_{H_0^1}^2 \qquad \forall u_J \in V_J.$$

This in turn implies that (1.16) and (1.17) hold for the finite space splitting of V_J (with scalar product induced by a) into the spaces V_j (with the scaled L^2 scalar products a_j), $j = 0, \ldots, J$, with constants C_1 and C_2 independent of J. Therefore, the associated Schwarz operators are uniformly well-conditioned, which carries over to convergence guarantees for Schwarz iterative methods based on these splittings, again uniformly in J. In other words, these finite splittings provide uniform O(1) preconditioning for the linear systems arising from the variational problems (1.18) with $\hat{V} = V_J$. The statement also holds for many other $\hat{V}$, such as spaces that appear in adaptive finite element methods [55]. This is discussed in Subsection 7.1.1.

Second, there is another useful modification that makes these preconditioners even more efficient. It is well-known that finite element nodal bases are L^2 stable. For example, in our two-dimensional example we have

$$2^{2j}\|u_j\|_{L^2}^2 \approx \sum_{P\in\mathcal{V}_j} x_{P,j}^2, \qquad u_j = \sum_{P\in\mathcal{V}_j} x_{P,j}N_{P,j} \in V_j,$$

with constants that are independent of $j \geq 0$. Thus, we can replace the above spaces (V_j, a_i) by their discrete counterparts, namely the spaces of all nodal basis vectors x_j that can be identified with $\mathbb{R}^{n_j}$, where $n_j = (2^{j+1}-1)^2$, equipped with the Euclidean scalar product $(\cdot,\cdot)_2$. This can be done in such a way that the inequalities (1.16) and (1.17) are preserved for the new space splitting up to slight changes in the constants C_1 and C_2. For this purpose, new R_i must be introduced by assigning to each nodal basis vector $x_j = (x_{P,j})_{P\in\mathcal{V}_j} \in \mathbb{R}^{n_j}$ the corresponding $u_j \in V_j \subset V$. The subproblems of the new splitting are trivial, since their matrix representations are now identity matrices. This means that in this particular case, besides the O(1) preconditioning, the resulting preconditioner C_J for the matrix A_J has an extremely simple structure, namely

$$C_J = \sum_{j=0}^{J} R_j R_j^T.$$

It turns out that the matrix-vector multiplication with this C_J can be implemented by a recursion using only $\mathrm{O}(n_J)$ floating-point operations. The associated multiplicative Schwarz iteration, when used in the appropriate cyclic order, is then equivalent to a $V^{(1,0)}$-cycle multigrid method of optimal complexity per iteration. The additive Schwarz method with this splitting was introduced in [22] and is commonly called the BPX-method. For a more complete treatment of multigrid methods using algebraic rather than Hilbert space techniques, we refer to [23, 96, 187], see also [18, 190] for expositions closer to our approach.

Finally, we mention that these results can be interpreted in the language of frame theory: The system

$$\Phi := \{N_{P,j} : P \in \mathcal{V}_j,\ j \geq 0\},$$

consisting of the nodal basis functions of all levels $j \geq 0$, is a frame in $H_0^1(G)$ for the considered G. Recall that frames are redundant systems of elements of a Hilbert space that generalize complete orthogonal systems and Riesz bases but still share some properties of the latter. They have received much attention in applied harmonic analysis. The stable space splittings considered in this book generalize the frame concept by replacing systems of elements of a given Hilbert space V by systems of Hilbert spaces V_i connected to V in a prescribed way via the operators R_i. We refer to Chapter 2 for the connections to frames and to Sections 3.1 and 7.1 for further material on finite element and wavelet multiscale splittings for Sobolev spaces.

1.1.5 Schwarz approximation methods

If V is infinite-dimensional, then space splittings with countably many finite-dimensional V_i and recursions using them, as in the multiplicative Schwarz method, are also common. But in the case of countable Ω, it would be unreasonable to call the resulting computational scheme an iterative method. Instead, we call such schemes Schwarz approximation methods.

It is instructive to consider the synthesis problem for a complete orthonormal system (CONS) $D = \{\phi_i\}_{i\in\Omega}$ in V with index set $\Omega = \mathbb{N}$ as an example. We keep the notation $a(\cdot,\cdot)$ for the scalar product in V. The synthesis problem is to recover $u \in V$ from its sequence of coefficients $c_i = a(u, \phi_i)$, $i \in \Omega$, which belongs to $\ell^2(\Omega)$. In its simplest form it can be interpreted as a multiplicative Schwarz method (1.11) with $\omega = 1$, if we set $V_i = \mathbb{R}$, $a_i(x, y) = xy$, $R_i x = x\phi_i$, with $x, y \in \mathbb{R}$, and $\rho_i = 1$ for all $i \in \Omega$. Then we get $P_i v = a(v, \phi_i)\phi_i$ and (1.11) transforms into the incremental approximation process

$$u^{(m+1)} = u^{(m)} + (c_{i_m} - a(u^{(m)}, \phi_{i_m}))\phi_{i_m}, \qquad m \geq 0. \tag{1.19}$$

If we start with $u^{(0)} = 0$ and use the deterministic sequence $i_m = m+1$, $m \geq 0$, then we get the sequence of partial sums

$$u^{(m)} = S_m := \sum_{i=1}^{m} c_i\phi_i, \qquad m \geq 0,$$

of the orthogonal series $u = \sum_{i\in\Omega} c_i\phi_i$. The greedy version would be to choose i_m as the index with the largest in absolute value residual $r_i^{(m)} := c_i - a(u^{(m)}, \phi_i)$. It gives the sequence of best m-term approximations with terms from the orthogonal series, namely

$$u^{(m)} = S_{m,*} := \sum_{i\in\Omega_{m,*}} c_i\phi_i, \qquad m \geq 0,$$

where the index set $\Omega_{m,*} \subset \Omega$ of cardinality m depends on u and realizes the infimum in

$$\sigma_m(u) := \inf_{\Omega_m\subset\Omega:\ |\Omega_m|\leq m} \|u - \sum_{i\in\Omega_m} c_i\phi_i\|.$$

This corresponds to selecting the terms from the orthogonal series with the m largest in absolute value coefficients c_i. See Section 6.1 for details.

Similar schemes can be devised for Riesz bases, frames, and general so-called dictionaries D. For example, greedy approximation schemes using dictionaries in Hilbert (and Banach) spaces is a much studied topic, see [178, 179]. In Subsection 7.1.4, we review some properties of the adaptive wavelet method, which can be considered as another instance of a greedy multiplicative Schwarz method. Here, D is a wavelet Riesz basis in V and group updates are allowed in (1.11), i.e., the recursion is now of the form

$$u^{(m+1)} = u^{(m)} + \omega \sum_{i \in I_m} \rho_i P_i(u - u^{(m)}),$$

where $I_m \subset \Omega$ is a finite index set chosen according to some optimization criterion.

Randomized Schwarz approximation schemes are also of interest. For example, the $P_i u$ may not be available deterministically, as has been implicitly assumed so far, but may be provided by a random measurement scheme that may or may not contain additional random noise. A model for such a situation is given by a large, unreliable network of sensors indexed by $i \in \Omega$, where the sensors return measurements of quantities such as $a(u, \phi_i)$ in an uncoordinated fashion to a central processing unit responsible for recovering u. An example in the context of online learning algorithms for kernel regression is discussed in Section 6.5.

Obviously, in infinite-dimensional V, the error decay of such incremental approximation processes can be arbitrarily slow, i.e., convergence rates can be expected only for u belonging to certain compact subsets of V. For the greedy version of (1.19), the appropriate subsets are given by u with $(c_i)_{i\in\Omega} \in \ell^p(\Omega)$ for some $0 < p < 2$. In the randomized version, where the i_m are i.i.d. indices with respect to a discrete probability distribution $\mathbb{P}(i_m = i) = p_i$, $i \in \Omega = \mathbb{N}$, the classes of relevance are constructed using the appropriately defined powers P_p^s, $s \in \mathbb{R}$, of the Schwarz operator $P_p = \sum_{i\in\Omega} p_i P_i$. In Chapter 6, we establish convergence rates using similar classes for arbitrary space splittings with finite-dimensional Hilbert spaces $(V_i; a_i)$ under rather minimal conditions. The obtained convergence rates are algebraic, i.e., the bounds on the error $e^{(m)} = u - u^{(m)}$ are of the form $\mathrm{O}((m+1)^{-t})$, where $t > 0$ depends on the chosen compact subset of V.

1.2 Content of this book

In this section, we summarize the main goals of this book. We also briefly indicate what we will not cover. Then we give a short overview of the different chapters of this book and their contents. Finally, we introduce some notational conventions that we have tried to use consistently throughout this book.

1.2.1 Goals

In this book we focus on the asymptotic convergence theory of a class of iterative methods for solving variational problems on a separable Hilbert space V. The main assumption is that the iterative method is built by successively using only solutions of a certain fixed set of subproblems associated with variational problems on a family $\mathbf{V} = (V_i)_{i\in\Omega}$ of typically finite-dimensional Hilbert spaces V_i. As we have tried to convince the reader with the above examples, this assumption is characteristic for many numerical methods used in classical numerical linear algebra as well as in modern scientific computing, with the numerical solution of large-scale PDE discretizations being one of the prominent examples. Since the interest of the authors in this topic started with the intensive development of the theory of multilevel and domain decomposition methods for solving elliptic PDE problems about 30 years ago, we use the name Schwarz methods in reference to the historical publication [161], which explored a splitting into two similar problems for solving Laplace's equation on complicated domains.

We give the abstract theory of such Schwarz methods, discuss their main building blocks, and derive tools for the convergence analysis of the corresponding iterative methods. Depending on the splitting and the specific method used, this leads to algorithms with significantly improved or even optimal convergence properties, allowing fast solutions of large problems. Due to the inherent divide-and-conquer methodology of space splittings, these iterative methods are also interesting for parallel and distributed computing. However, the main focus of this book is on the asymptotic convergence rates of Schwarz methods. In many applications, the underlying problem is formulated for a finite-dimensional V of large dimension and is split into a finite number of subproblems. Thus, the notion of Schwarz iterative methods seems appropriate (in the literature they are also called subspace correction methods). If V is infinite-dimensional and split into an infinite number of subproblems, we use the name Schwarz approximation method.

For a given finite Hilbert space splitting, many different Schwarz iterative methods can be derived. As mentioned above, in the simplest case, where updates are computed simultaneously (similar to Jacobi iterations), we get an additive Schwarz method. If the updates are performed successively (similar to the Gauss-Seidel iteration), we get a multiplicative Schwarz method. The performance of the former is by construction independent of any ordering of subproblems and updates within an iteration cycle, while for the latter it depends on the ordering in which the subproblems are solved and used in the update formula. Here, the ordering can be prescribed in a deterministic way, but also greedy techniques for choosing the next subproblem to update and even random choices can be considered, as we will discuss in the course of this book. The latter two possibilities, i.e., greedy and random versions of Schwarz iterative methods, have been thoroughly investigated in recent years, with most of the results available only in journal and conference papers. Note that many of these publications are based on research on convex optimization problems with large state space, where speedup is expected from distributed and parallel computation. This comment also applies to results on Schwarz approximation methods for infinite-dimensional V and splittings into an infinite number of spaces V_i, where applications to uncertainty quantification, machine learning, and inverse problems motivated the development. We therefore believe that it is timely to present the basic theory of Schwarz methods and their underlying space splittings in one place.

Using our abstract theory of Hilbert space splittings, we will give asymptotic convergence rate estimates for the various Schwarz methods discussed in this book in terms of quantities that depend on certain metric properties of the splitting, on bounds of the spectrum and the condition number of the associated Schwarz operator, on the size of the initial error, and, in some cases, on certain norms of u

and noise characteristics. Therefore, our bounds should be considered as qualitative, since, with a few exceptions, exact knowledge of all quantities entering the estimates is not possible. More quantitative bounds, with tight control of the constants and matching lower bounds, should not be expected by the reader from the basic theory developed in this book. Nevertheless, it is instructive, interesting, and important to derive asymptotic bounds on the convergence rates of the various Schwarz iterative and approximation algorithms through a common basic theory focusing on the underlying space splitting that covers a wide range of practical algorithms.

Note additionally that algorithmic details and implementation concerns of the discussed Schwarz methods are not the focus in this book, and therefore should not be expected by the reader either. Such aspects are only occasionally mentioned in an exemplary way. This also means that a thorough cost analysis and a formal complexity discussion of the different iterative methods are not presented. Furthermore, we do not provide a detailed discussion of the parallelization properties of the different methods, and thus, the reader should not anticipate any analysis of parallel scaling or speedup.

Finally, a few remarks about our bibliography. In a book about such a fundamental concept as splittings of a Hilbert spaces V and, in a sense, standard linear iterative schemes for solving linear equations in V governed by symmetric and coercive operators (or, from an optimization point of view, for solving unconstrained quadratic minimization problems), it is clear that most of the facts and results about the corresponding iterative methods have appeared elsewhere in some equivalent form. This led to our decision to refer only to those sources that we consulted during the writing of the book or that, in our opinion, initiated or significantly influenced the development of the theory presented. In addition, a select number of works have been cited for further reading on topics complementary to the core material. We have also refrained from naming the authors of publications (with the exception of a few historical figures). Overall, any failure on our part to give fair credit to the work of other authors should be held against us, and we ask forgiveness from any colleagues whose work has not been duly acknowledged, whether partially or entirely.

1.2.2 Structure

In the following, we will give a brief overview of the different chapters of this book and their content. The even-numbered chapters are devoted to the abstract theory of Hilbert space splittings and Schwarz methods with finitely and infinitely many spaces V_i, respectively. Here we have tried to give a comprehensive treatment, with full proofs. Nevertheless, we do not claim to have covered all possible versions of Schwarz methods, since the number of publications on such algorithms and their convergence behavior has exploded in the last decades. In this book, we have developed the theory of Hilbert space splittings with an emphasis on the treatment of Schwarz methods, but have not attempted to highlight its historical roots and its continuing importance in operator theory, harmonic analysis and other areas.

The remaining odd-numbered chapters contain supplementary material that provides details on specific applications and is intended to illustrate the basic theory. For example, we discuss linear systems $Ax = b$ with square and rectangular matrices A, where we recover classical and recent convergence results for SOR- and Kaczmarz-type iterations from the abstract theory. A special focus is on multiscale splittings for infinite-dimensional Sobolev spaces and the associated Schwarz methods for their finite-dimensional subspaces, which in the case of finite element spaces V_i are workhorses for practical PDE solvers. Finally, we also mention links to convex optimization and stochastic approximation algorithms. In all these chapters, however, the presentation is less complete. Often, we give only some key results or survey a topic without providing proofs. In particular, we treat only a few applications to PDE solvers in the model situation of a scalar second-order elliptic problem. In the theory of PDE solvers, the space splitting approach is now widely accepted and has been used and optimized in a large number of publications over the last 30 years, also dealing with linear and nonlinear PDE systems, time-dependent PDEs, robustness aspects, and adaptive methods.

In Chapter 2 we consider a separable Hilbert space V with a designated scalar product $a(\cdot,\cdot)$, which we denote as $(V;a)$. Then we introduce the general space splitting

$$(\mathbf{V};\mathbf{a};\mathbf{R};\rho) = (V_i;a_i;R_i;\rho_i)_{i\in\Omega}, \qquad \Omega \subset \mathbb{N},$$

associated with $(V;a)$. Next, we study the relation between $(V;a)$ and a splitting $(\mathbf{V};\mathbf{a};\mathbf{R};\rho)$ on an abstract level. We introduce the Schwarz operator associated with the space splitting and give some illustrative examples. We discuss the stability of a space splitting, introduce the associated Schwarz and Gram equations, and illustrate their preconditioning effect. We then consider operations on space splittings. We focus on scaling, refinement and coarsening, the selection of subsplittings, and tensor product splittings. Finally, we discuss scales of Hilbert spaces associated with space splittings. We deal with variations of the weights ρ and with scales of Hilbert spaces induced by P_ρ.

In Chapter 3 we present some important examples of Hilbert space splittings. They are motivated mainly by their relevance to the solution of large linear systems arising from discretizations of elliptic partial differential equations, where the Hilbert spaces V of interest are Sobolev spaces on domains. First, we consider multiscale splittings as they may arise from multivariate Fourier series expansions, finite element discretizations, and wavelet-type methods, and prove their stability with respect to certain Sobolev spaces. We then consider auxiliary space splittings, both in theory and in applications, which arise from non-matching triangulations, nonconforming finite elements, and also from algebraic multigrid methods. Finally, we discuss paving and sketching as methods, mainly based on probabilistic techniques, to generate space splittings for linear problems in $\mathbb{R}^n$ and $\ell^2(\mathbb{N})$.

In Chapter 4 we review the convergence theory of subspace correction methods for a finite number of subproblems. First, we present some generalities of Schwarz iterative methods. We then focus on iterations with cyclic orderings of the subproblems, analyze their error propagation operator, and derive convergence estimates that depend on the condition number $\kappa(P_\rho)$ of the associated Schwarz operator. Next, we consider certain random orderings, discuss randomization strategies, and provide convergence estimates in terms of the expected squared error. We also analyze greedy strategies and derive their convergence properties. Finally, we discuss improved iterative methods, such as the conjugate gradient method applied to the Schwarz equation and Nesterov-type acceleration, for various ordering strategies. The convergence estimates obtained now depend mainly on $\sqrt{\kappa(P_\rho)}$.

Chapter 5 is devoted to the application of the general theory developed in Chapter 4 to more specific situations. To this end, we resort to linear systems with symmetric positive definite matrices, which correspond to the finite-dimensional situation. We also analyze the Kaczmarz method and least-squares problems, and establish links to block-coordinate descent methods in convex optimization. We then present lower bounds for the convergence rate of Schwarz iterative methods with finite Ω and discuss the sharpness of the upper bounds for the convergence rate of the iterative methods obtained in the previous chapter. Finally, we consider the robustness and sensitivity of Schwarz iterations to errors and explain the link to stochastic approximation algorithms.

In Chapter 6 we focus on the theory of Schwarz methods for infinite-dimensional V using space splittings with countably many spaces V_j. The infinite-dimensional setting brings new aspects and requires the definition of new classes of compact subsets related to the space splitting and the associated Schwarz operator. The new classes make it possible to formulate asymptotic convergence rates for Schwarz methods with greedy and random orderings under minimal assumptions on the space splitting. These methods are now similar to certain successive and adaptive approximation methods, which explains the name change to Schwarz approximation methods. As a preparation, in Section 6.2 we discuss some Krylov subspace iterations with $P \in \mathcal{B}_{pos}(V)$ in V and derive their convergence rates in terms of the spectral spaces V_P^s associated with P. Similar results are known from iterative regularization theory for inverse problems. We then consider randomized Schwarz approximation methods in more detail. Again, the spaces V_P^s are important, but this time with P as the Schwarz operator P_ρ for the space splitting with weights ρ_i determined from the probability distribution underlying the random index selection process. We then study greedy-type Schwarz approximations and discuss their convergence properties. Finally, we apply our theory to the specific setting of online algorithms for kernel regression, which are relevant for machine learning applications [14]. This requires a further extension of the space splitting concept, where now Ω is a compact metric space with a Borel probability measure ρ. We derive convergence rates in the associated reproducing kernel Hilbert space (RKHS) norm by identifying the online learning algorithm with a randomized Schwarz approximation that includes additional noise. For the specific task of retrieving an unknown element in V from noisy measurements of its coefficients with respect to a CONS via an online method, we derive sharp convergence bounds. Finally, we give some remarks on Schwarz approximation schemes for non-symmetric problems.

In Chapter 7 we describe exemplarily how the theory of the previous chapters applies to the design and analysis of solvers for PDE discretizations. Our focus here is on

C^0 Lagrange finite element discretizations of standard second-order boundary value elliptic problems on nested sequences of quasi-uniform partitions into d-dimensional simplicial or rectangular elements. First, we consider multilevel methods. We discuss the additive BPX-type preconditioner for adaptive finite element spaces. We also mention multiplicative versions, point out their connection to multigrid methods, and introduce ideas that lead to optimal adaptive wavelet methods. Next, we consider the domain decomposition method for both overlapping and non-overlapping domain decompositions, again restricted to the situation of nested partition sequences. Finally, we focus on tensor product decomposition methods and sparse grid techniques for higher-dimensional PDE problems.

Each of the Chapters 2-7 has a problem section. Most of the problems are supplementary to the material in the text, and should help the interested reader to better understand some aspects of the theory by attempting to solve them. However, a few of the problems concern technical details needed to complete certain proofs, or require verification of certain statements in the text, or slight generalizations of theorems in the given chapter. In this case, we have provided hints that should be sufficient to solve these problems and complete the corresponding proofs. We believe that the problem sections may also be useful if some parts of the book are used as reading material for an advanced one- or two-semester course in numerical methods or scientific computing.

We conclude the book with an appendix that defines much of the notation and terminology from Hilbert space and linear operator theory that we use freely throughout the book. It also covers variational problems on Hilbert spaces and the associated linear equations. In addition, it discusses various constructions on Hilbert spaces, such as Hilbert sums, Bochner integration, and tensor products. Finally, some facts about Sobolev spaces on Lipschitz domains are given, which are necessary for the treatment of strongly elliptic second-order PDEs and their finite element discretizations.

1.2.3 Notation

In this subsection, we comment on some notational conventions and notation used throughout the book. Further definitions of systematically used notation, especially for concrete Banach and Hilbert function spaces and operators are given in the appendix. The rest is explained in the main text.

We tried to be consistent, but sometimes we had to use the same letter for different things (or use a different letter for the same thing). For example, the letters $f, g, \phi, \psi, \ldots$ are used for functionals (elements of dual spaces) and elements of orthogonal systems and dictionaries, but are also employed to denote functions of various kinds. Furthermore, $X, Y, \ldots$ can be a generic set, or a metric or Banach space, or an (auxiliary, linear) subset or subspace of a Hilbert space. Generic positive constants are denoted by C, c and may change from line to line, sometimes we add a subscript if it helps to distinguish between constants in some formulas. However, the same letter C is used for preconditioners and for cells in partitions that appear in finite element constructions. In formulas involving matrices or operators, I stands for the identity matrix or operator, respectively, without specifying the linear space on which it acts. Subscripted I, such as I_m (a subset of indices of Ω appearing in a generic Schwarz method), I_ρ (a diagonal matrix associated with the weights ρ_i), and I_j (intergrid transfer operators and matrices in multilevel finite element methods) have different meanings, which should be clear from the context. Finally, the letter Q is consistently used for projection operators, while P is reserved for operators appearing in the construction of Schwarz operators (however, P also denotes grid points and nodal points in domain partitions related to the finite element method). Further often used notation is contained in the following list.

- $\mathbb{N}$, $\mathbb{Z}/\mathbb{Z}_+$, $\mathbb{R}/\mathbb{R}_+$, and $\mathbb{C}$ denote the sets of natural numbers (= positive integers), integers/non-negative integers, real/non-negative real numbers, and complex numbers, respectively. $\mathbb{Z}^d/\mathbb{Z}^d_+$ and $\mathbb{R}^n$ stand for the corresponding Cartesian products, with their elements sometimes interpreted as multi-indices, elements of Euclidean space, or column vectors (the latter interpretation is used by default in the context of numerical linear algebra). $\mathbb{P}$ and $\mathbb{E}$ are used for probabilities and expectations.
- The p-norm in $\mathbb{R}^n$ is denoted by $\|\cdot\|_p$, $1 \le p \le \infty$, and the Euclidean scalar product is denoted by $(\cdot,\cdot)_2$. For $\alpha \in \mathbb{Z}^n$ we use $|\alpha|_p$ and $|\alpha| = |\alpha|_1$ instead. Finally, $|X|$ also denotes the cardinality of a finite set X.
- Indexed lists, collections of objects of various types, such as $(V;a)$ (notation for a Hilbert space V with its scalar product a), or $(\mathbf{V};\mathbf{a};\mathbf{R};\rho)$ for space splittings, or $(X;d_X)$ (a metric space X with its metric d_X), are enclosed in parentheses and separated by semicolons. Sequences and indexed families of elements of the same type, such as $x = (x_1, x_2, \ldots) = (x_i)_{i\in\mathbb{N}}$ (elements of $\ell^p(\mathbb{N})$), $(\rho_i)_{i\in\Omega}$ (weight sequences), or $\tilde{v} = (v_i)_{i\in\Omega}$ (elements in $\tilde{V}_\rho$), are also shown in parentheses, with the elements separated by commas. Finally, $\{\cdot\}$ is used for sets. For the sake of simplicity, there may be some tolerable inconsistencies. For example, writing $(\mathbf{V};\mathbf{a};\mathbf{R};\rho) = (V_i;a_i;R_i;\rho_i)_{i\in\Omega}$ is not formally correct, but it is not confusing and can be understood from its context.
- We often use := to temporarily define a variable or object in the proof of a theorem or in an entire section. The Landau symbols $\mathrm{O}(\cdot)$ and $\mathrm{o}(\cdot)$ are defined as usual, and $A \approx B$ means $cA \le B \le CA$ with positive constants, independent of the quantities on which the expressions A and B depend.
- The element 0 is zero, no matter what linear space it belongs to. $\{0\}$ denotes the trivial subspace of any linear space.
- In some places the Fourier transform $\hat{f}(\theta) = \mathcal{F}f(\theta)$ of a (real- or complex-valued) function $f(x)$ defined on $\mathbb{R}^d$ appears without a precise definition. What we use is the formula

$$\hat{f}(\theta) = \mathcal{F}f(\theta) = \int_{\mathbb{R}^d} f(x)e^{-i(x,\theta)_2}\,dx, \qquad \theta \in \mathbb{R}^d,$$

assuming the formula makes sense a.e. on $\mathbb{R}^d$.

Chapter 2
Hilbert space splittings: Abstract theory

Abstract This chapter focuses on the study of the the relation between $(V; a)$, namely a separable Hilbert space V with scalar product $a(\cdot, \cdot)$, and what we call a *space splitting*

$$(\mathbf{V}; \mathbf{a}; \mathbf{R}; \rho) = (V_i; a_i; R_i; \rho_i)_{i\in\Omega}, \qquad \Omega \subset \mathbb{N},$$

associated with $(V; a)$ at an abstract level. The notion of stability, the role of the Schwarz operator associated with a space splitting, and operations on space splittings are discussed.

2.1 Space splittings: Definitions and basic properties

Throughout this book, let V be a separable real Hilbert space with basic scalar product $(\cdot, \cdot)_V$, norm $\|\cdot\|_V$, and duality pairing $\langle \cdot, \cdot \rangle_{V' \times V}$, see the appendix for background material on functional analysis and Hilbert space notation. Let

$$a(u, v) = \langle Au, v \rangle_{V' \times V} \tag{2.1}$$

denote a bounded bilinear form on V induced by a bounded linear operator $A : V \longrightarrow V'$. If this bilinear form is also symmetric and coercive, i.e., if

$$a(u, v) = \langle Au, v \rangle_{V' \times V} = \langle Av, u \rangle_{V' \times V} = a(v, u), \qquad u, v \in V,$$

and if there exists a positive constant c such that

$$a(u, u) \geq c\|u\|_V^2, \qquad u \in V,$$

or, equivalently, if $A = A'$ has a bounded inverse $A^{-1} : V' \longrightarrow V$, then $a(\cdot, \cdot)$ defines a new scalar product on V, and

$$\|u\| := \sqrt{a(u, u)}, \qquad u \in V,$$

M. Griebel, P. Oswald, *Hilbert Space Splittings and Iterative Methods*, Springer Series in Computational Mathematics 62, https://doi.org/10.1007/978-3-031-74370-2_2

is an equivalent Hilbert space norm on V. If this is the case then we use the notation $(V; a)$, and consider V equipped with this new scalar product and the norm induced by it. There are situations where the assumptions on $a(\cdot,\cdot)$ only hold for a proper subspace $U \subset V$, in which case one usually considers $(U; a|_U)$ instead.

Throughout this book we consider at most countable space splittings which, for convenience, we associate with either the finite index set $\Omega := \{1,\dots,n\}$ for some $n \in \mathbb{N}$ (finite space splittings), or the index set $\Omega := \mathbb{N} = \{1,2,\dots\}$ (countable space splittings). If for a given application or example other countable index sets seem more natural, we will use them, but we will state this explicitly. A more general situation, where Ω is a compact metric space with measure, is considered in Section 6.5.

We consider collections

$$(\mathbf{V}; \mathbf{a}; \mathbf{R}; \rho) := (V_i; a_i; R_i; \rho_i)_{i\in\Omega}$$

related to a given separable real Hilbert space $(V; a)$ consisting of

- a family $\mathbf{V} := (V_i)_{i\in\Omega}$ of separable real Hilbert spaces and an associated family of scalar products $\mathbf{a} = (a_i)_{i\in\Omega}$ and norms denoted $\|u_i\|_i = a_i(u_i,u_i)^{1/2}$, $u_i \in V_i$,
- a family $\mathbf{R} := (R_i : V_i \to V)_{i\in\Omega}$ of bounded linear operators, and
- a positive weight sequence $\rho := (\rho_i > 0)_{i\in\Omega}$.

There are special situations where it makes sense to shorten the notation for better readability. The following conventions will be adhered to in all later places:

- When the family $\mathbf{V}$ consists of subspaces of V and the operators R_i are the natural injections, we indicate this by using the notation $(\mathbf{V}; \mathbf{a}; \rho)$, and do not display R_i in the associated formulas.
- If, in addition, the scalar products on these subspaces are inherited from $(V; a)$, i.e., if $a_i(u_i, v_i) = a(u_i, v_i)$, $u_i, v_i \in V_i$, we shorten the notation even further to $(\mathbf{V}; \rho)$, and call it a *subspace splitting*.
- If the weights are fixed at $\rho_i = 1$, we indicate this by writing $(\mathbf{V}; \mathbf{a}; \mathbf{R})$, $(\mathbf{V}; \mathbf{a})$, and $\mathbf{V}$ instead of $(\mathbf{V}; \mathbf{a}; \mathbf{R}; \rho)$, $(\mathbf{V}; \mathbf{a}; \rho)$, and $(\mathbf{V}; \rho)$, respectively. The subscript ρ is then also omitted from further notation.

We call a collection $(\mathbf{V}; \mathbf{a}; \mathbf{R}; \rho)$ related with $(V; a)$ a *space splitting associated with* $(V; a)$ if there exists a constant B for which the inequality

$$\|\sum_{i\in\Omega'} \rho_i R_i v_i\|^2 \le B \sum_{i\in\Omega'} \rho_i \|v_i\|_i^2, \qquad v_i \in V_i, \tag{2.2}$$

holds for all finite index subsets $\Omega' \subset \Omega$. In addition, if the linear subset

$$X_0 := \operatorname{span}(\{\operatorname{ran}(R_i)\}_{i\in\Omega})$$

is dense in V then we call $(\mathbf{V}; \mathbf{a}; \mathbf{R}; \rho)$ a *space splitting for* $(V; a)$.

The condition (2.2) is a non-trivial requirement only for infinite space splittings. In fact, due to the boundedness of the R_i we have

$$\|\sum_{i\in\Omega'} \rho_i R_i v_i\|^2 \le \left(\sum_{i\in\Omega'} \rho_i \|R_i\|_{V_i\to V}\|v_i\|_i\right)^2 \le \left(\sum_{i\in\Omega'} \rho_i \|R_i\|^2_{V_i\to V}\right)\left(\sum_{i\in\Omega'} \rho_i \|v_i\|_i^2\right)$$

for all finite subsets $\Omega' \subset \Omega$. So (2.2) holds with $B = B_0$ if

$$B_0 := \sum_{i\in\Omega} \rho_i \|R_i\|^2_{V_i\to V} < \infty. \tag{2.3}$$

If $\Omega = \mathbb{N}$ this sufficient condition is not necessary, see Subsection 2.1.1. Another sufficient condition based on so-called strengthened Cauchy-Schwarz inequalities is discussed in Section 2.2.

We give several reinterpretations of (2.2). From now on we tacitly assume that $(V; a)$ and the families $\mathbf{V}$, $\mathbf{a}$, $\mathbf{R}$ are temporarily fixed but that ρ can vary (see Subsection 2.3.1 for a discussion of the role of the weight sequence). This is reflected in the notation by the subscript ρ (when $\rho_i = 1$ we drop it).

Let $(\tilde V_\rho; \tilde a_\rho)$ denote the weighted Hilbert sum of the spaces $(V_i; a_i)$ with weight sequence ρ. Recall that

$$\tilde a_\rho(\tilde u, \tilde v) := \sum_{i\in\Omega} \rho_i a_i(u_i, v_i), \qquad \tilde u := (u_i)_{i\in\Omega},\ \tilde v := (v_i)_{i\in\Omega},\ v_i, u_i \in V_i,\ i \in \Omega,$$

is the scalar product in

$$\tilde V_\rho = \{\tilde v :\ \|\tilde v\|^2_{\tilde V_\rho} := \tilde a_\rho(\tilde v, \tilde v) < \infty\}.$$

Let $\tilde V_0 \subset \tilde V_\rho$ denote the set of all finitely supported $\tilde v$. Obviously, $\tilde V_0$ is independent of ρ and dense in $\tilde V_\rho$. We can reformulate the above definitions of space splittings as follows: The collection $(\mathbf{V}; \mathbf{a}; \mathbf{R}; \rho)$ is a space splitting associated with $(V; a)$ if and only if the linear operator $R_\rho : \tilde V_0 \to X_0$ defined by

$$R_\rho \tilde v := \sum_{i\in\Omega} \rho_i R_i v_i, \qquad \tilde v \in \tilde V_0, \tag{2.4}$$

is bounded (considered as operator acting from $\tilde V_\rho$ to V), with norm $\le B$. In addition, if X_0 is dense in V then we have a space splitting for $(V; a)$. In fact, (2.2) is then equivalent to

$$\|u\|^2 = a(u, u) \le B|||u|||^2_\rho, \qquad u \in \mathrm{ran}(R_\rho), \tag{2.5}$$

where

$$|||u|||^2_\rho := \inf_{\tilde v\in \tilde V_0:\, R_\rho \tilde v = u} \|\tilde v\|^2_{\tilde V_\rho} \tag{2.6}$$

defines a semi-norm on $\mathrm{ran}(R_\rho)$ that will play an important role throughout this chapter. Naturally, (2.5) and (2.6) carry over to the unique extension by continuity of $R_\rho : \tilde V_0 \to X_0$ to all of $\tilde V_\rho$. From now on, R_ρ denotes this extension which, as the following statement shows, is also given by the formula (2.4).

Theorem 2.1 *If* $(\mathbf{V}; \mathbf{a}; \mathbf{R}; \rho)$ *is a space splitting associated with* $(V; a)$ *then the series in (2.4) converges unconditionally in* V *for any* $\tilde{v} \in \tilde{V}_\rho$, *and defines a bounded linear operator* $R_\rho : \tilde{V}_\rho \to V$ *with norm satisfying*

$$\|R_\rho\|_{\tilde{V}_\rho \to V} = \min_{(2.2)} \sqrt{B}.$$

Its Hilbert adjoint $R_\rho^* : V \to \tilde{V}_\rho$ *is given by*

$$R_\rho^* u = (R_i^* u)_{i \in \Omega}, \qquad u \in V,$$

where $R_i^* : V \to V_i$ *is defined by*

$$a_i(R_i^* u, v_i) = a(u, R_i v_i) \qquad \forall\, v_i \in V_i, \quad i \in \Omega. \tag{2.7}$$

Also, $\|R_\rho^*\|_{V \to \tilde{V}_\rho} = \|R_\rho\|_{\tilde{V}_\rho \to V}$, *and* R_ρ^* *is injective if and only if* $(\mathbf{V}; \mathbf{a}; \mathbf{R}; \rho)$ *is a space splitting* for $(V; a)$, *i.e., if* $\mathrm{ran}(R_\rho)$ *is dense in* V.

Proof If the splitting is finite then $\tilde{V}_0 = \tilde{V}_\rho$, and there is nothing to prove. Otherwise, consider $\tilde{v} \in \tilde{V}_\rho$, and let $(i_1, i_2, \ldots)$ be an arbitrary index permutation of $\Omega = \mathbb{N}$. Taking $\Omega' = \{i_n, \ldots, i_m\}$ for $n \le m$ in (2.2), we have

$$\|\sum_{j=n}^{m} \rho_{i_j} R_{i_j} v_{i_j}\|^2 \le B \sum_{j=n}^{m} \rho_{i_j} \|v_{i_j}\|_{i_j}^2 \le B \sum_{i=\min_{j \ge n} i_j}^{\infty} \rho_i \|v_i\|_i^2.$$

The right-hand side of this inequality converges to zero since $\min_{j \ge n} i_j \to \infty$ as $n \to \infty$, and $\tilde{v} \in \tilde{V}_\rho$ implies

$$\|\tilde{v}\|_{\tilde{V}_\rho}^2 = \tilde{a}(\tilde{v}, \tilde{v}) = \sum_{i=1}^{\infty} \rho_i \|v_i\|_i^2 < \infty.$$

So,

$$\|\sum_{j=n}^{m} \rho_{i_j} R_{i_j} v_{i_j}\| \to 0, \qquad m \ge n \to \infty,$$

for any index permutation that establishes the unconditional convergence of the series in (2.4) in V for any $\tilde{v} \in \tilde{V}_\rho$.

The formula for the adjoint follows from (2.4) and (2.7), i.e.,

$$a(R\tilde{v}, u) = \sum_{i \in \Omega} \rho_i a(R_i v_i, u) = \sum_{i \in \Omega} \rho_i a_i(v_i, R_i^* u) = \tilde{a}_\rho(\tilde{v}, (R_i^* u)_{i \in \Omega}).$$

The remaining statements follow in a straightforward way from the definition of operator norms and general properties of adjoint operators. □

The condition (2.2) is equivalent to the boundedness of R_ρ^* as an operator from V into $\tilde{V}_\rho$. In fact, (2.2) holds if and only if

$$\|R_\rho^* u\|_{\tilde{V}_\rho}^2 = \tilde{a}_\rho(R_\rho^* u, R_\rho^* u) = \sum_{i\in\Omega} \rho_i a_i(R_i^* u, R_i^* u) \le Ba(u,u) \tag{2.8}$$

for all $u \in V$. Note that the definition of R_ρ^* is formally independent of ρ. This raises the possibility of starting the definition of space splittings with a sequence of bounded linear operators $T_i : V \to V_i$, such that

$$\sum_{i\in\Omega} \rho_i a_i(T_i u, T_i u) \le Ba(u,u), \qquad u \in V, \tag{2.9}$$

for some $B < \infty$, and defining $R_i = T_i^*$ as the adjoint of T_i, $i \in \Omega$. This was actually done with the introduction of g-frames [175], see Section 2.2 for their definition. Our decision to begin with R_i is motivated by applications to preconditioning and iterative methods for operator equations, where the operators R_i are explicitly given, while the operators $T_i = R_i^*$ are defined by solving auxiliary variational problems on the spaces V_i and are only available in implicit form.

2.1.1 The Schwarz operator

Next, we introduce operators associated with space splittings which will play a central role in the following chapters. Let $(\mathbf{V}; \mathbf{a}; \mathbf{R}; \rho)$ be a space splitting associated with $(V; a)$. Then the bounded linear operators $P_\rho := R_\rho R_\rho^* : V \to V$ and $\tilde{P}_\rho := R_\rho^* R_\rho : \tilde{V}_\rho \to \tilde{V}_\rho$ are called the *(additive) Schwarz operator* and the *Gramian operator* of the space splitting, respectively.

It is easy to see that P_ρ can be represented as

$$P_\rho = \sum_{i\in\Omega} \rho_i P_i, \qquad P_i := R_i R_i^* : V \to V, \quad i \in \Omega. \tag{2.10}$$

The symmetric and positive operators P_i will appear in the iterative schemes considered in subsequent chapters.

Theorem 2.2 *The operator P_ρ is symmetric and positive on $(V; a)$, that is,*

$$a(P_\rho u, v) = a(u, P_\rho v), \qquad a(P_\rho u, u) = \tilde{a}_\rho(R_\rho^* u, R_\rho^* u) \ge 0, \qquad u, v \in V.$$

Similarly, $\tilde{P}_\rho$ is symmetric and positive on $(\tilde{V}_\rho; \tilde{a}_\rho)$, that is,

$$\tilde{a}_\rho(\tilde{P}_\rho \tilde{u}, \tilde{v}) = a(\tilde{u}, \tilde{P}_\rho \tilde{v}), \qquad \tilde{a}_\rho(\tilde{P}_\rho \tilde{u}, \tilde{u}) = \|R_\rho \tilde{u}\|^2 \ge 0, \qquad \tilde{u}, \tilde{v} \in \tilde{V}_\rho.$$

We have

$$\|P_\rho\|_V = \|\tilde{P}_\rho\|_{\tilde{V}_\rho} = \inf_{(2.2)} B.$$

Proof The symmetry of P_ρ on $(V; a)$ follows from a simple calculation using the unconditional convergence in (2.10) and (2.7). We have

$$a(P_\rho u, v) = \sum_{i\in\Omega} \rho_i a(R_i R_i^* u, v) = \sum_{i\in\Omega} \rho_i a_i(R_i^* u, R_i^* v) = \tilde{a}_\rho(R_\rho^* u, R_\rho^* v) = a(u, P_\rho v)$$

for all $u, v \in V$. This also shows the positivity of P_ρ. Similarly, symmetry and positivity of $\tilde{P}_\rho$ on $(\tilde{V}_\rho; \tilde{a}_\rho)$ follow from

$$\begin{aligned}\tilde{a}_\rho(\tilde{P}_\rho \tilde{u}, \tilde{v}) &= \sum_{i\in\Omega} \rho_i a_i(R_i^* R_\rho \tilde{u}, v_i) = \sum_{i\in\Omega} \rho_i a(R_\rho \tilde{u}, R_i v_i)\\ &= a(R_\rho \tilde{u}, R_\rho \tilde{v}) = \tilde{a}_\rho(\tilde{u}, \tilde{P}_\rho \tilde{v}).\end{aligned}$$

Note that these formulas imply

$$\begin{aligned}\ker(P_\rho) &= \ker(R_\rho^*) = \operatorname{ran}(R_\rho)^\perp,\\ \ker(\tilde{P}_\rho) &= \ker(R_\rho) = \operatorname{ran}(R_\rho^*)^\perp,\end{aligned} \tag{2.11}$$

which concludes the proof. □

The operators introduced in this section and their quantitative properties will be further studied below. The Schwarz operator P_ρ will play a crucial role in the convergence theory of iterative schemes for the operator equation $Au = f$ generating $(V; a)$. We will call two space splittings associated with the same $(V; a)$ *equivalent* if their Schwarz operators coincide.

2.1.2 Illustrative examples and special cases

Example 1: Linear systems with symmetric positive definite matrix

Let $V = \mathbb{R}^n$ (interpreted as set of column vectors of length n) be equipped with the Euclidean scalar product $(x, y)_2 = \sum_{i=1}^n x_i y_i = y^T x$. Consider a linear system $Ax = b$ with symmetric positive definite (spd) matrix $A = (a_{ij})_{i,j=1,\dots,n}$ and right-hand side $b \in \mathbb{R}^n$. Therefore, $a(x, y) := y^T A x$ defines a spectrally equivalent scalar product on V.

The *standard space splitting* for this $(V; a)$ is as follows: Take $\Omega = \{1, \dots, n\}$, set $V_i = \mathbb{R}$, $a_i(x_i, y_i) = x_i y_i$, and $R_i : x_i \in \mathbb{R} \to (0, \dots, x_i, \dots, 0)^T \in \mathbb{R}^n$ for each $i \in \Omega$. Then, elements $\tilde{x}$ of $\tilde{V}_\rho = \ell_\rho^2(\Omega)$ can be identified with vectors x in $V = \mathbb{R}^n$ via

$$\tilde{x} = (x_i)_{i\in\Omega} \quad \leftrightarrow \quad x = \sum_{i\in\Omega} R_i x_i = (x_1, \dots, x_n)^T,$$

which is tacitly used in the formulas below. Employing the notation $I_\rho := \operatorname{diag}((\rho_i)_{i\in\Omega})$, which represents the diagonal matrix with diagonal entries given by the weight sequence ρ, we compute

$$\tilde{a}_\rho(\tilde{x}, \tilde{y}) = \sum_{i\in\Omega} \rho_i x_i y_i = y^T I_\rho x, \qquad R_\rho \tilde{x} = \sum_{i\in\Omega} \rho_i R_i x_i = I_\rho x.$$

The adjoint operators R_i^*, $i \in \Omega$, are given by $R_i^* x = (Ax)_i$ since

$$a(x, R_i y_i) = (0, \ldots, y_i, \ldots, 0)Ax = y_i(Ax)_i.$$

Thus, $R_\rho^* x = Ax$, and applying the Schwarz operator

$$P_\rho x = I_\rho Ax$$

amounts to multiplication by A followed by diagonal scaling. Moreover, the best possible constant B in (2.2) is given by

$$B = \|P_\rho\|_{V\to V} = \sup_{x\neq 0} \frac{a(P_\rho x, x)}{a(x,x)} = \sup_{x\neq 0} \frac{x^T A I_\rho A x}{x^T A x} = \lambda_{\max}(I_\rho A).$$

A similar calculation shows that the action of $\tilde{P}_\rho$ is given by $\tilde{P}_\rho \tilde{x} = AI_\rho x$. Typical choices for the weight sequence, which corresponds to diagonal rescaling of the matrix A, are $\rho_i = 1$ and $\rho_i = a_{ii}^{-1}$. The standard space splitting is related to the classical Jacobi-Richardson, Gauss-Seidel, and SOR-type iterations.

Example 2: Least-squares problems

Consider now the generally ill-posed problem of solving a linear system $Ax = b$ with a given $n' \times n$ matrix A and a right-hand side $b \in \mathbb{R}^{n'}$. A solution $x \in \mathbb{R}^n$ exists if and only if $b \in \mathrm{ran}(A)$, and it is unique if and only if $\ker(A) = \{0\}$. In the general case, the least-squares solution x^{LS} is defined as the solution with minimal Euclidean norm in R^n of the quadratic minimization problem

$$\|Ax - b\|_2^2 \rightarrow \min .$$

It is well-known that x^{LS} belongs to $\mathrm{ran}(A^T) \subset \mathbb{R}^n$, and satisfies the system of normal equations $A^T A x^{LS} = A^T b$. Thus, one option to formalize the least-squares problem is to set $V = \mathrm{ran}(A^T) \subset \mathbb{R}^n$, and to introduce $a(x, y) = y^T A^T A x$ as the bilinear form on V. Following the previous example, $\rho_i > 0$ for $i \in \Omega = \{1, \ldots, n\}$ and setting $V_i = \mathbb{R}$, $a_i(x_i, y_i) = x_i y_i$, $R_i x_i = Q_V(0, \ldots, x_i, \ldots, 0)$ results in a space splitting for this $(V; a)$. Here, Q_V denotes the ortho-projection onto V with respect to the Euclidean scalar product on $\mathbb{R}^n$. It can be used to investigate distributed versions of the Landweber iteration.

Alternatively, if one chooses $a(x, y) = y^T x$ as the bilinear form on $V = \mathrm{ran}(A^T)$, then the following space splitting for $(V; a)$ is of interest: For $i \in \Omega = \{1, \ldots, n'\}$, let $V_i = \mathbb{R}$, $a_i(x_i, y_i) = x_i y_i$, $R_i x_i = x_i (A_i)^T$, and $\rho_i > 0$. Here, A_i denotes the i-th row vector of A, thus $(A_i)^T \in V$, $\mathrm{ran}(R_i) \subset V$, and $\mathrm{span}(\{\mathrm{ran}(R_i)\}_{i\in\Omega}) = V$. One computes $R_i^* x = (Ax)_i$,

$$P_\rho x = \sum_{i=1}^{n'} \rho_i (Ax)_i (A_i)^T = A^T I_\rho A x, \qquad x \in V \subset \mathbb{R}^n,$$

and

$$\tilde{P}_\rho y = AA^T I_\rho y, \qquad y \in \tilde{V}_\rho = \mathbb{R}^{n'}.$$

Thus, the equation

$$P_\rho x = A^T I_\rho b, \qquad x \in V = \mathrm{ran}(A^T),$$

has x^{LS} as the unique solution. Multiplying by the diagonal $n' \times n'$ matrix I_ρ can be viewed as row scaling in A induced by the parameters $\rho_i > 0$, $i \in \Omega$, and may be used to improve the spectral properties of the system of normal equations. This space splitting is related to the classical Kaczmarz method.

Example 3: Galerkin projection and subspace splittings

A space splitting $(\mathbf{V}; \mathbf{a}; \mathbf{R}; \rho)$ associated with $(V; a)$ is said to be *generated by Galerkin projection* if the auxiliary bilinear forms on V_i are given by

$$a_i(u_i, v_i) = a(R_i u_i, R_i v_i) \qquad \forall\, u_i, v_i \in V_i, \qquad i \in \Omega. \tag{2.12}$$

If (2.12) holds, we use the notation $(\mathbf{V}; \mathbf{R}; \rho)$. If, in addition, $\mathbf{V}$ consists of subspaces $V_i \subset V$ and the R_i are the natural injection operators, then $(\mathbf{V}; \mathbf{R}; \rho)$ is a subspace splitting and is denoted by $(\mathbf{V}; \rho)$.

Space splittings generated by Galerkin projection possess useful properties. In particular, $I - P_i$ is an ortho-projection onto the subspace $\mathrm{ran}(R_i)^\perp$ in V, while P_i is an ortho-projection onto $\mathrm{ran}(R_i)$ if the latter is closed, i.e., if it is a subspace of V itself. We refer to Problem 2.1.

In the special case of a subspace splitting $(\mathbf{V}; \rho)$, the operators P_i coincide with the ortho-projections Q_{V_i} onto V_i. According to (2.8), $(\mathbf{V}; \rho)$ is a subspace splitting associated with $(V; a)$ if and only if

$$\sum_{i \in \Omega} \rho_i \|Q_{V_i} u\|^2 \le B\|u\|^2, \qquad u \in V. \tag{2.13}$$

Moreover,

$$R_\rho \tilde{u} = \sum_{i \in \Omega} \rho_i u_i, \qquad \tilde{u} \in \tilde{V}_\rho.$$

The Schwarz and Gramian operators of a subspace splitting are given by

$$P_\rho u = \sum_{i \in \Omega} \rho_i Q_{V_i} u, \qquad \tilde{P}_\rho \tilde{u} = \left(Q_{V_i} \Big(\sum_{j \in \Omega} \rho_j u_j \Big) \right)_{i \in \Omega},$$

for $u \in V$, $\tilde{u} \in \tilde{V}_\rho$, respectively. Note that, on a theoretical level, any space splitting generated by Galerkin projection can be rewritten as an equivalent subspace splitting, see Part d) of Problem 2.1.

Subspace splittings are related to the theory of frames of subspaces [35, 76], which were later renamed *fusion frames* [36]. Actually, what we call a subspace splitting $(\mathbf{V}; \rho)$ associated with $(V; a)$ is known in [35] as a Bessel sequence of subspaces with respect to $\sqrt{\rho}$, requiring (2.13), and not (2.2), as the definition. Our operators R_ρ, R_ρ^* and P_ρ correspond to what in [35] are called synthesis, analysis and frame operators, respectively.

Space splittings related to multiscale approximation and algorithms, where

$$V_0 \subset V_1 \subset \cdots V_j \subset \cdots \subset V,$$

are of special interest. Another important case is subspace splittings, where each V_i is one-dimensional, and generated by a single non-zero element $\phi_i \in V$. They will be dealt with in more detail in the following example.

Example 4: Space splittings generated by dictionaries

We call any collection $D = (\phi_i)_{i\in\Omega}$ of finitely ($\Omega = \{1, 2, \ldots, n\}$) or countably ($\Omega = \mathbb{N}$) many $0 \neq \phi_i \in V$ a *dictionary* in V. The terminology is well established as a model for various nonlinear approximation processes, e.g., in the theory of greedy algorithms [179]. For a given weight sequence ρ, such a dictionary D generates a particular space splitting associated with $(V; a)$ denoted for short by $(\mathbf{V}_D; \rho)$ as follows: Set $V_i = \mathbb{R}$, $a_i(x_i, y_i) = x_i y_i$, and $R_i x_i = x_i\phi_i$ for all $i \in \Omega$. Then $\tilde{V}_\rho = \ell^2_\rho(\Omega)$, and the series $R_\rho\tilde{x} = \sum_{i\in\Omega} \rho_i x_i \phi_i$ with respect to D transforms coefficient sequences $\tilde{x} \in \ell^2_\rho(\Omega)$ into elements of V. Since

$$a(u, R_i x_i) = a(u, \phi_i)x_i = a_i(a(u, \phi_i), x_i), \qquad u \in V, \quad x_i \in \mathbb{R},$$

we have $R_i^* u = a(u, \phi_i)$, $i \in \Omega$. Thus, according to (2.8) and the definition of space splittings, $(\mathbf{V}_D; \rho)$ becomes a space splitting associated with $(V; a)$ if and only if the linear operator

$$R_\rho^* : u \quad \to \quad (a(u, \phi_i))_{i\in\Omega}$$

mapping $(V; a)$ into $\ell^2_\rho(\Omega)$ is bounded. If, in addition, $\operatorname{span}(D)$ is dense in $(V; a)$, then we get a space splitting for $(V; a)$.

The Schwarz operator acting on V is given by

$$P_\rho u = \sum_{i\in\Omega} \rho_i a(u, \phi_i)\phi_i,$$

while the action of the Gramian operator on $\ell^2_\rho(\Omega)$ can be computed as

$$(\tilde{P}_\rho\tilde{x})_i = \sum_{j\in\Omega} \rho_j a(\phi_i, \phi_j)x_j, \qquad i \in \Omega.$$

A look at the matrix representation of $\tilde{P}_\rho$ in the coordinate basis of $\ell^2_\rho(\Omega)$ explains its name Gramian operator.

This example is related to the study of series representations and incremental approximation processes with respect to dictionaries (and in particular frames, Riesz bases, and orthogonal systems). The standard splitting for linear systems with spd matrix A introduced in Example 1 is a special instance, if we take the unit coordinate vectors in $\mathbb{R}^n$ as ϕ_i.

We note that generating space splittings $(\mathbf{V}_D; \rho)$ from a dictionary D is equivalent to considering subspace splittings of the form $(\mathbf{V}; \theta)$ involving one-dimensional subspaces V_i of V, albeit with a different weight sequence θ. Indeed, if D is given, then one takes the one-dimensional subspaces

$$V_i = \{x_i \phi_i : x_i \in \mathbb{R}\}, \qquad i \in \Omega,$$

of $(V; a)$ generated by the dictionary elements. Since the R_i are natural inclusion operators, it follows that R_i^* is given by

$$R_i^* u = \frac{a(u, \phi_i)}{\|\phi_i\|^2} \phi_i, \qquad i \in \Omega.$$

Consequently, if we define $\theta_i = \|\phi_i\|^2 \rho_i$, then the Schwarz operator associated with the subspace splitting $(\mathbf{V}; \theta)$ will coincide with the above P_ρ, since

$$P_\theta u = \sum_{i \in \Omega} \theta_i R_i^* u = \sum_{i \in \Omega} \rho_i a(u, \phi_i) \phi_i.$$

Conversely, if $(\mathbf{V}; \theta)$ with one-dimensional subspaces $V_i \subset V$ is a subspace splitting associated with $(V; a)$, then it suffices to create the dictionary D by picking for each $i \in \Omega$ an arbitrary $\phi_i = R_i \phi_i' \neq 0$ from $\mathrm{ran}(R_i) \subset V$, and setting $\rho_i = \|\phi_i'\|_i^{-2} \theta_i$.

Finally, for any space splitting $(\mathbf{V}; \mathbf{a}; \mathbf{R}; \theta)$ associated with $(V; a)$, one can find a dictionary D in V and a weight sequence such that the resulting $(\mathbf{V}_D; \rho)$ is equivalent to $(\mathbf{V}; \mathbf{a}; \mathbf{R}; \theta)$, i.e., their Schwarz operators coincide. For instance, one can consider CONS $\{\phi_{ik}\}_{k \in \Omega_i}$ in each $(V_i; a_i)$ and set

$$D = \{R_i \phi_{ik}\}_{k \in \Omega_i,\, i \in \Omega}, \qquad \rho = (\rho_{ik} = \theta_i)_{k \in \Omega_i,\, i \in \Omega}.$$

The details are left as an exercise, see Problem 2.2.

Example 5: A probabilistic scenario

Let $\mathbf{p} = (p_i > 0)_{i \in \Omega}$ correspond to a discrete probability distribution supported on $\Omega \subset \mathbb{N}$, i.e., $\sum_{i \in \Omega} p_i = 1$. Let V_i be a given sequence of subspaces of $(V; a)$, and let $Q_{V_i} : V \to V_i$ be the sequence of associated ortho-projections, $i \in \Omega$. Define, for every $u \in V$, a V-valued random variable $X_u : \Omega \to V$ by setting $X_u(i) = Q_{V_i} u$ with probability p_i. Random variables like X_u arise in the study of randomized versions of the alternating projection method. Then the expectation of X_u is given by

$$\mathbb{E}(X_u) = \sum_{i \in \Omega} p_i Q_{V_i} u = P_{\mathbf{p}} u,$$

and we have a formula for the variance, namely

$$\mathbb{E}(\|X_u - \mathbb{E}(X_u)\|^2) = \sum_{i \in \Omega} \mathbf{p}_i a((Q_{V_i} - P_{\mathbf{p}})u, (Q_{V_i} - P_{\mathbf{p}})u) = a(P_{\mathbf{p}} u, u) - a(P_{\mathbf{p}} u, P_{\mathbf{p}} u).$$

Thus, the investigation of random alternating projection methods naturally involves the Schwarz operator $P_{\mathbf{p}}$ of the subspace splitting $(\mathbf{V};\mathbf{p})$ associated with $(V;a)$. In contrast to the other examples, where the weights ρ_i were more or less arbitrary scaling parameters at our disposal, here the ρ_i are given by an underlying probability measure. Probabilistic aspects related to space splittings will be studied in more detail in later chapters.

2.1.3 Sufficient conditions

To check (2.2) in the case of infinite space splittings, a sufficient condition of independent interest was proposed in [194]. We say that $(\mathbf{V};\mathbf{a};\mathbf{R};\rho)$ satisfies *strengthened Cauchy-Schwarz inequalities* with respect to $(V;a)$ if

$$a(R_i u_i, R_j v_j) \le \gamma_{ij}\|u_i\|_i \|v_j\|_j, \qquad u_i \in V_i, \quad v_j \in V_j, \tag{2.14}$$

holds with constants $\gamma_{ij} = \gamma_{ji}$, $i, j \in \Omega$, such that the symmetric matrix operator

$$\Gamma_\rho x := (\sqrt{\rho_i}\sum_{j\in\Omega}\sqrt{\rho_j}\gamma_{ij}x_j)_{i\in\Omega}, \qquad x = (x_j)_{j\in\Omega} \in \ell^2(\Omega), \tag{2.15}$$

is bounded on $\ell^2(\Omega)$. The condition (2.14) has a natural reformulation in terms of the Gramian operator $\tilde{P}_\rho$, which is given in Problem 2.3.

Theorem 2.3 *If* $(\mathbf{V};\mathbf{a};\mathbf{R};\rho)$ *satisfies strengthened Cauchy-Schwarz inequalities with respect to* $(V;a)$*, then (2.2) holds with*

$$B = B_1 := \|\Gamma_\rho\|_{\ell^2}. \tag{2.16}$$

Proof Given any $\tilde{v} \in \tilde{V}_0$, consider the finitely supported sequence $x = (\sqrt{\rho_j}\|v_j\|_j)_{j\in\Omega}$ in $\ell^2(\Omega)$. Then, with (2.14), we have

$$\begin{aligned} a(R_\rho\tilde{v}, R_\rho\tilde{v}) &= \sum_{i,j\in\Omega}\rho_i\rho_j a(R_i v_i, R_j v_j) \le \sum_{i,j\in\Omega}\gamma_{ij}\rho_i\rho_j\|v_i\|_i\|v_j\|_j \\ &= (\Gamma_\rho x, x)_{\ell^2} \le \|\Gamma_\rho\|_{\ell^2}\|x\|_{\ell^2}^2 = \|\Gamma_\rho\|_{\ell^2}\|\tilde{v}\|_{\tilde{V}_\rho}^2. \end{aligned}$$

This proves the claim. The argument also yields the more general inequality

$$\sum_{i,j\in\Omega}\beta_{ij}\rho_i\rho_j a(R_i v_i, R_j v_j) \le (\sup_{i,j\in\Omega}|\beta_{ij}|)\|\Gamma_\rho\|_{\ell^2(\Omega)}\|\tilde{v}\|_{\tilde{V}_\rho}^2,$$

valid for arbitrary $\tilde{v} \in \tilde{V}_\rho$ and with arbitrary weights β_{ij}. □

If the sufficient condition (2.3) holds for $(\mathbf{V};\mathbf{a};\mathbf{R};\rho)$, then $(\mathbf{V};\mathbf{a};\mathbf{R};\rho)$ satisfies strengthened Cauchy-Schwarz inequalities for $(V;a)$. Indeed, in (2.14) we can always set

$$\gamma_{ij} = \|R_i\|_{V_i \to V} \|R_j\|_{V_j \to V}, \qquad i, j \in \Omega.$$

Then, the resulting operator Γ_ρ is a rank-1 operator $\Gamma_\rho = x_\rho \otimes x_\rho$ induced by $x_\rho = (\sqrt{\rho_i}\|R_i\|_{V_i \to V})_{i \in \Omega}$. Its boundedness follows from (2.3), i.e., we have

$$\|\Gamma_\rho\|_{\ell^2} = \|x_\rho\|_{\ell^2}^2 = \sum_{i \in \Omega} \rho_i \|R_i\|_{V_i \to V}^2 = B_0 < \infty.$$

Thus, (2.3) implies (2.16), which in turn implies (2.2) according to Theorem 2.2. Both implications cannot be reversed. To see this for the first implication, consider the special case of a space splitting $\mathbf{V}_D$ induced by an orthonormal system $D = \{\phi_i\}_{i \in \mathbb{N}}$ in an infinite-dimensional Hilbert space $(V; a)$ with weight sequence $\rho_i = 1$, $i \in \mathbb{N}$. Strengthened Cauchy-Schwarz inequalities are then satisfied with $\gamma_{ii} = 1$, $\gamma_{ij} = 0$ if $i \neq j$, and

$$B_1 = \|\Gamma_\rho\|_{\ell^2} = 1.$$

On the other hand, $B_0 = \sum_{i \in \Omega} \rho_i \|R_i\|_{V_i \to V}^2 = \infty$, and the sufficient condition (2.3) does not hold for this space splitting.

It is worth noting that for subspace splittings with $n = 2$ one has the equality

$$\|P_\rho\|_V = \|\tilde{P}_\rho\|_{\tilde{V}} = \|\Gamma_\rho\|_{\ell^2},$$

see Problem 2.4. In this case, the bound established in Theorem 2.3 is sharp. For space splittings into more than two auxiliary spaces, this is not true. For an example of a space splitting $(\mathbf{V}; \mathbf{a}; \mathbf{R}; \rho)$ for $(V; a)$, which does not satisfy strengthened Cauchy-Schwarz inequalities at all, we refer to Problem 2.5. Finally, we note that the entries of Γ_ρ are related to the notion of angles between subspaces, see Problem 2.6.

2.2 Stability

Roughly speaking, stable space splittings are space splittings for $(V; a)$, for which the Schwarz operator $P_\rho : V \to V$ is invertible on V. The term was first used in connection with preconditioning techniques for PDE solvers, see [135]. Although it is not standard, we will use it throughout this book.

A collection $(\mathbf{V}; \mathbf{a}; \mathbf{R}; \rho)$ is called a *stable space splitting* for $(V; a)$ if it is a space splitting for $(V; a)$, i.e., (2.2) is valid and $\mathrm{ran}(R_\rho)$ is dense in V, and the following holds: There is a constant $A > 0$ such that, for any $u \in \mathrm{ran}(R_\rho)$, there is a $\tilde{u} \in \tilde{V}_\rho$ such that $u = R_\rho \tilde{u}$ and

$$\tilde{a}_\rho(\tilde{u}, \tilde{u}) = \sum_{i \in \Omega} \rho_i a_i(u_i, u_i) \leq \frac{1}{A} a(u, u). \tag{2.17}$$

For finite space splittings, condition (2.17) was noted in [194]. The next theorem offers several equivalent conditions for stability.

Theorem 2.4 *Let* $(\mathbf{V};\mathbf{a};\mathbf{R};\rho)$ *be a space splitting for* $(V;a)$. *Then the following statements are equivalent:*

1) $(\mathbf{V};\mathbf{a};\mathbf{R};\rho)$ *is stable, i.e., (2.17) holds.*
2) An inequality opposite to (2.9) holds with some $A > 0$:

$$|||u|||_\rho^2 \le \frac{1}{A}\|u\|^2, \qquad u \in V. \tag{2.18}$$

3) An inequality opposite to (2.8) holds with some $A > 0$:

$$A\|u\|^2 \le \|R_\rho^* u\|_{\tilde{V}_\rho}^2 = a(P_\rho u, u), \qquad u \in V. \tag{2.19}$$

4) P_ρ *is invertible, and we have with some* $A > 0$

$$\|P_\rho^{-1}\|_V \le \frac{1}{A}. \tag{2.20}$$

5) P_ρ *(and thus* R_ρ*) is surjective, i.e.,* $\mathrm{ran}(P_\rho) = \mathrm{ran}(R_\rho) = V$.

In each of the statements 1)-4) the supremum of the admissible constants A is the same.

Proof Since $(\mathbf{V};\mathbf{a};\mathbf{R};\rho)$ is a space splitting for $(V;a)$, the operators $R_\rho : V_\rho \to V$ and P_ρ are well-defined, and, by (2.11), we have

$$\ker(P_\rho) = \mathrm{ran}(P_\rho)^\perp = \mathrm{ran}(R_\rho)^\perp = \{0\}.$$

So P_ρ is injective and its range is dense in V. Thus, the invertibility of P_ρ with norm bound $\|P_\rho^{-1}\|_V \le \frac{1}{A}$ is equivalent to the inequality

$$\|P_\rho u\| \ge A\|u\|, \qquad u \in V,$$

which in turn is equivalent to (2.19). This shows 3) ⇔ 4). The equivalence of 4) and 5) is a consequence of the open mapping theorem. The equivalence of 1) and 2) is obvious by definition of the semi-norm $|||\cdot|||_\rho$. Also, the implication 4) ⇒ 2) is not difficult. Indeed, set $v = P_\rho^{-1}u$ and $\tilde{u} = R_\rho^* v$. Then $u = R_\rho \tilde{u}$ and, using the Cauchy-Schwarz inequality and (2.20), we obtain (2.18), i.e., we have

$$|||u|||_\rho^2 \le \tilde{a}(\tilde{u},\tilde{u}) = a(P_\rho v, v) = a(u, P_\rho^{-1}u) \le \|u\|\|P_\rho^{-1}u\| \le \frac{1}{A}\|u\|^2.$$

It remains to show 2) ⇒ 3). For arbitrarily given $v \in V$, set $u = P_\rho v$ and $\tilde{u} = R_\rho^* v$. Then

$$|||u|||_\rho^2 = \inf_{\tilde{v}\in\tilde{V}_\rho:\, u=R_\rho\tilde{v}} \tilde{a}_\rho(\tilde{v},\tilde{v}) = \tilde{a}_\rho(\tilde{u},\tilde{u}) = a(P_\rho v, v),$$

since the infimum is given by the ortho-projection of $\tilde{u}$ onto the subspace $\ker(R_\rho)^\perp = \mathrm{clos}(\mathrm{ran}(R_\rho^*))$ in $(\tilde{V}_\rho;\tilde{a}_\rho)$. Obviously, this projection must coincide with $\tilde{u}$ since $\tilde{u} = R_\rho^* v$ belongs to this subspace. Thus, if (2.18) holds, then

$$a(P_\rho v, v) \le \frac{1}{A}\|P_\rho v\|^2 = \frac{1}{A}a(P_\rho v, P_\rho v), \qquad v \in V.$$

This is equivalent to

$$\|P_\rho^{1/2} v\|^2 \le \frac{1}{A}\|P_\rho v\|^2, \qquad v \in V.$$

After renaming $P_\rho^{1/2} v$ as u, this yields

$$A\|u\|^2 \le a(P_\rho u, u), \qquad u \in \operatorname{ran}(P_\rho^{1/2}).$$

By the density of $\operatorname{ran}(P_\rho^{1/2})$ in V, this implies (2.19). This completes the proof of Theorem 2.4. □

Corollary 2.1 *If* $(\mathbf{V}; \mathbf{a}; \mathbf{R}; \rho)$ *is a stable space splitting for* $(V; a)$, *then*

$$|||u|||_\rho = a(P_\rho^{-1} u, u)^{1/2}, \qquad u \in V, \tag{2.21}$$

defines an equivalent norm on $(V; a)$, *i.e., with constants* $0 < A \le B < \infty$, *we have*

$$A|||u|||_\rho^2 \le \|u\|^2 = a(u, u) \le B|||u|||_\rho^2, \qquad u \in V. \tag{2.22}$$

Moreover,

$$|||u|||_{\rho,*} := \|R_\rho^* u\|_{\tilde{V}_\rho} = a(P_\rho u, u)^{1/2}, \qquad u \in V,$$

also defines an equivalent norm on $(V; a)$, *i.e., with constants* $0 < A \le B < \infty$, *we have*

$$A\|u\|^2 \le |||u|||_{\rho,*}^2 \le B\|u\|^2, \qquad u \in V. \tag{2.23}$$

The supremum of the constants A in (2.17), (2.22), (2.23), and the infimum of the constants B in (2.2), (2.22), (2.23), respectively, are the same.

We note that the norm equivalence (2.22) was used in [135, Chapter 4] to define stable space splittings, while the equivalent condition (2.23) was used to define g-frames. To be independent of the properties of the operators R_ρ and P_ρ in the case of infinite Ω, it is sufficient to require the norm equivalence between $\|u\|$ and

$$|||u|||_\rho = \inf_{\tilde{v} \in \tilde{V}_0 :\, u = R_\rho \tilde{v}} \tilde{a}_\rho(\tilde{v}, \tilde{v})^{1/2}$$

on the subset $X_0 = \operatorname{ran}(R_\rho|_{\tilde{V}_0})$ of V, since this avoids any discussion of the meaning of $R_\rho \tilde{v}$ for $\tilde{v} \notin \tilde{V}_0$.

Example 3 and 4 continued

Formally, stable space splittings $(\mathbf{V}; \mathbf{a})$ for $(V; a)$ with subspaces $V_i \subset V$ and weights $\rho_i = 1$ were introduced in [135, Section 4.1] in the context of preconditioning applications. The generalization to the case of arbitrary space splittings was sketched in [135, Section 4.6.1]. In [35], stable subspace splittings $(\mathbf{V}; \rho)$ reappeared under the name *frames of subspaces with respect to* ρ, later renamed *fusion frames*. The definition in [35] required the existence of constants A and B, so that

$$A\|u\|^2 \le \sum_{i\in\Omega} \rho_i \|Q_{V_i} u\|^2 \le B\|u\|^2, \qquad u \in V, \tag{2.24}$$

which is in this particular case nothing but (2.23). We refer to [35, 36] for alternative proofs of most of the above statements for the special case of fusion frames.

Subspace splittings $\mathbf{V}_D$ generated by a dictionary D (i.e., $(\mathbf{V}_D; \rho)$ as described in Example 4 with weights $\rho_i = 1$) are stable if and only if $D = \{\phi_i\}_{i\in\Omega}$ is a *frame* for $(V; a)$. Frame theory has a relatively long history, and frames in a Hilbert space are generalizations of complete orthogonal systems and Riesz bases, but allow redundancy. The original definition of the frame property is a special case of (2.24) (and thus of (2.23)), i.e.,

$$A\|u\|^2 \le \sum_{i\in\Omega} a(u, \phi_i)^2 \le B\|u\|^2, \qquad u \in V.$$

The frame operator $P : V \longrightarrow V$ (since $\rho_i = 1$, we temporarily suppress the subscript ρ) is given by

$$Pu = \sum_{i\in\Omega} a(u, \phi_i)\phi_i, \qquad u \in V. \tag{2.25}$$

We refer to [93] and [39] for more on frames, their history, and their use in harmonic analysis and signal processing. Dual and tight frames, which play a prominent role in these applications, are discussed in Problem 2.7.

A generalization of the fusion frame concept, which is essentially equivalent to our definition of stable splittings, is the introduction of *g-frames* for $(V; a)$ in [175], where g stands for generalized. In contrast to our definition, which starts with the operators $R_i : V_i \longrightarrow V$, the definition of g-frames takes bounded linear operators $T_i : V \longrightarrow V_i$ as the point of departure and postulates a norm equivalence similar to (2.23), namely

$$A\|u\|^2 \le \sum_{i\in\Omega} \rho_i \|T_i u\|_i^2 \le B\|u\|^2, \qquad u \in V,$$

with constants $0 < A \le B < \infty$ as their definition. Setting $R_i = T_i^*$ shows that the definition of g-frames is equivalent to our definition of stable space splittings $(\mathbf{V}; \mathbf{a}; \mathbf{R}; \rho)$ for $(V; a)$. The g-frame concept was subsequently extended further to Hilbert C^*-modules, with applications to operator theory in mind.

2.2.1 Special cases

We now consider two particular settings, namely orthogonal subspace splittings and Riesz splittings.

2.2.1.1 Orthogonal subspace splittings

A collection $\mathbf{W} = (W_i)_{i\in\Omega}$ of subspaces W_i of V is called *orthogonal* (with respect to $(V;a)$) if

$$a(w_i, w_j) = 0, \qquad w_i \in W_i, \quad w_j \in W_j,$$

for all $i, j \in \Omega$, $i \neq j$. Since in this case we have

$$a(R_\rho\tilde{w}, R_\rho\tilde{w}) = \sum_{i\in\Omega} \rho_i^2 a(w_i, w_i)$$

for all finitely supported $\tilde{w} = (w_i)_{i\in\Omega}$, the collection $(\mathbf{W};\rho)$ becomes an *orthogonal subspace splitting associated with* $(V;a)$ if and only if

$$\rho_i \leq B, \qquad i \in \Omega,$$

for some $B < \infty$. If, in addition, the subspaces W_i form a complete orthogonal decomposition of $(V;a)$, i.e., if

$$V = \bigoplus_{i\in\Omega} W_i,$$

then $(\mathbf{W};\rho)$ is an *orthogonal subspace splitting for* $(V;a)$. Finally, such a space splitting is *stable* for $(V;a)$ if the inequality

$$\rho_i \geq A, \qquad i \in \Omega,$$

holds with a constant $A > 0$. The constants A and B play the same role as in (2.17) and (2.2). Note that in this case $R_\rho : \tilde{W}_\rho \to V$ is injective since $Q_{W_i} R_\rho \tilde{w} = \rho_i w_i$ by orthogonality for any $\tilde{v} \in \tilde{W}_\rho$, where $Q_{W_i} : V \to W_i$ are the associated ortho-projections. Thus, the unique $\tilde{w} \in \tilde{V}_\rho$ for which $u = R_\rho\tilde{w}$ is given by $w_i = \rho_i^{-1} Q_{W_i} u$, and

$$a(P_\rho^{-1}u, u) = |||u|||_\rho^2 = \sum_{i\in\Omega} \rho_i^{-1} \|Q_{W_i} u\|^2.$$

The Schwarz operator P_ρ is given by

$$P_\rho = \sum_{i\in\Omega} \rho_i Q_{W_i},$$

and so

$$a(P_\rho u, u) = \sum_{i\in\Omega} \rho_i a(Q_{W_i} u, u) = \sum_{i\in\Omega} \rho_i \|Q_{W_i} u\|^2.$$

The Gramian operator $\tilde{P}_\rho$ reduces to simple diagonal scaling given by

$$\tilde{P}_\rho \tilde{w} = (\rho_i w_i)_{i\in\Omega}, \qquad \tilde{w} \in \tilde{W}_\rho,$$

and

$$a_\rho(\tilde{P}_\rho \tilde{w}, \tilde{w}) = \sum_{i\in\Omega} \rho_i^2 \|w_i\|^2.$$

Orthogonal space splittings are used throughout the book for illustration and to construct counterexamples, since they allow us to do explicit calculations. Another important aspect is that a collection $(\mathbf{W}; \rho)$ consisting of mutually orthogonal subspaces W_i of some Hilbert space $(H; (\cdot,\cdot)_H)$ can be used to obtain (stable) space splittings for other Hilbert spaces, see Section 2.4 and Chapter 3 for examples.

2.2.1.2 Riesz splittings

A stable space splitting $(\mathbf{V}; \mathbf{a}; \mathbf{R}; \rho)$ for $\{V; a\}$ is called a *Riesz splitting for* $(V; a)$ if R_ρ is injective, i.e., if $\ker(R_\rho) = \{0\}$. Since, for stable space splittings, R_ρ is also surjective and bounded, this means that R_ρ^{-1} exists, and

$$|||u|||_\rho = \|R_\rho^{-1} u\|_{\tilde{V}_\rho}.$$

A more explicit definition of Riesz splittings in the spirit of (2.17) and (2.2) is to require the density of X_0 (the span of the ranges of the R_i) in $(V; a)$, and a two-sided inequality

$$A \sum_{i\in\Omega} \rho_i \|v_i\|_i^2 \le \|\sum_{i\in\Omega} \rho_i R_i v_i\|^2 \le B \sum_{i\in\Omega} \rho_i \|v_i\|_i^2, \qquad \tilde{v} \in \tilde{V}_0, \tag{2.26}$$

to hold with certain constants $0 < A \le B < \infty$. Obviously, the Riesz property implies linear independence of the linear sets $\mathrm{ran}(R_i)$ in V. The converse is not true, see Problem 2.8.

Due to the uniqueness of the series representation

$$u = R_\rho \tilde{v} = \sum_{i\in\Omega} \rho_i R_i v_i$$

by elements $\tilde{v} \in \tilde{V}$, Riesz splittings have certain advantages (we will come back to this issue in Section 2.3). In particular, if the ratio B/A in (2.26) is moderate, then, given the series representation of u, its norm $\|u\|$ can be estimated in terms of the norms $\|v_i\|_i$ of the "coefficients" v_i in the series representation $u = R_\rho \tilde{v}$.

2.2.2 Sufficient conditions

If V is finite-dimensional, then a space splitting for $(V; a)$ is automatically stable. Indeed, in this case, the ranges of the operators R_i are finite-dimensional subspaces of V, and their span must also be a subspace of V. Thus, the density of their span implies the surjectivity of R_ρ. Consequently, R_ρ^* must be injective, and P_ρ invertible, compare (2.11).

If V is infinite-dimensional, then the stability of a space splitting for $(V; a)$ is not automatic, i.e., (2.17) and the density of $\mathrm{ran}(R_\rho)$ does not imply (2.2). To see this, consider the collection $(\mathbf{V}_D; \rho)$ generated by an orthonormal system $D = (\phi_i)_{i\in\Omega}$ in $(V; a)$, where we can take $\Omega = \mathbb{N}$. Then $(\mathbf{V}_D; \rho)$ is a space splitting associated with $(V; a)$, that is, it satisfies (2.2), if and only if

$$\sup_{i\in\Omega} \rho_i < \infty.$$

It becomes a space splitting for $(V; a)$ if the orthonormal system D is complete, and satisfies the stability condition (2.17) if additionally

$$\inf_{i\in\Omega} \rho_i > 0.$$

This is because, in this case, $R_\rho^* u = a(u, \phi_i)$ and $P_\rho = \sum_{i\in\Omega} \rho_i a(u, \phi_i)\phi_i$, see Example 4 in Subsection 2.1.2. The example also shows that the conditions (2.17) and (2.2) are independent.

If V is infinite-dimensional, then an unstable space splitting for $(V; a)$ with index set $\Omega = \{1\}$ is given by $(V; a; R)$, where R is a compact symmetric operator $R : V \to V$ with dense range. In fact, in this case, $(\tilde{V}_\rho; a_\rho)$ can be identified with $(V; a)$, R_ρ with R, and P_ρ with R^2, respectively. Since R^2 is compact, P_ρ cannot be invertible, and thus this space splitting cannot be stable. Another example of this kind is described in Problem 2.9.

We do not know of sufficient conditions for the stability of space splittings which are easy to state and at the same time applicable in the general case. The following lemma can sometimes be used for space splittings with $V_i \subset V$, especially for sequences of nested subspaces. It is easy to state, but requires knowledge about properties of ortho-projections onto certain subspaces $W_i \subset V_i$ in $(V; a)$, which form an orthogonal subspace splitting of $(V; a)$ that is only available in special cases. An example is considered in Subsection 3.1.3.

Lemma 2.1 *Let $(\mathbf{V}; \mathbf{a}; \rho)$ be a space splitting for $(V; a)$ with $V_j \subset V$, $j \in \Omega$. Assume that $(V; a)$ admits an orthogonal subspace splitting $\mathbf{W}$ such that $V = \oplus_{j\in\Omega} W_j$ and $W_j \subset V_j$ for all $j \in \Omega$. If, for some fixed constant C_S, we have*

$$\|w_j\|_j^2 \le C_S \rho_j \|w_j\|^2, \qquad w_j \in W_j, \quad j \in \Omega, \tag{2.27}$$

then $(\mathbf{V}; \mathbf{a}; \rho)$ is stable. More generally, the statement also holds if $\mathbf{W}$ is a stable subspace splitting for $(V; a)$, not necessarily an orthogonal one.

Proof The proof is straightforward. Set $w_j = \rho_j^{-1} Q_{W_j} u \in V_j$. Then $u = \sum_j Q_{W_j} u = \sum_j \rho_j w_j$ is an admissible decomposition for u and therefore

$$|||u|||_\rho^2 \le \sum_{j\in\Omega} \rho_j \|w_j\|_j^2 \le C_S \sum_{j\in\Omega} \|Q_{W_j} u\|^2 = C_S \|u\|^2.$$

If the orthogonality of $\mathbf{W}$ in $(V; a)$ is weakened to stability, then only minor changes are needed and an additional constant appears in the last estimation step. □

Note that, as a by-product, we also obtain a more explicit norm equivalence with respect to the ortho-projections Q_{W_j}, namely

$$\|u\|^2 \approx \sum_{j\in\Omega} \rho_j^{-1} \|Q_{W_j} u\|_j^2, \qquad u \in V.$$

In practice, using $(V; a)$ ortho-projections is prohibitive, since this means solving linear equations involving A on the subspaces W_j. Moreover, making the W_j explicit is a problem in itself.

2.2.3 Schwarz equation and preconditioning

Now we turn to operator equations. Suppose $A : V \longrightarrow V'$ is a symmetric, boundedly invertible operator such that V can be equipped with the spectrally equivalent scalar product $a(\cdot, \cdot)$ defined by (2.1). Then

$$Au = f, \qquad f \in V', \tag{2.28}$$

has a unique solution $u \in V$. We consider reformulations of this linear operator equation in terms of stable space splittings for $(V; a)$.

Theorem 2.5 *Let* $(\mathbf{V}; \mathbf{a}; \mathbf{R}; \rho)$ *be a stable space splitting for* $(V; a)$, *and let* $0 < A \leq B < \infty$ *be constants for which (2.23) holds. Define* $g_i \in V_i$, $i \in \Omega$, *as the solution of the variational problems*

$$a_i(g_i, v_i) = \langle f, R_i v_i \rangle_{V' \times V}, \quad v_i \in V_i,$$

for given $f \in V'$. *Then* $\tilde{g} = (g_i)_{i\in\Omega}$ *belongs to* $\mathrm{ran}(R_\rho^*) \subset \tilde{V}_\rho$, *and the following assertions are equivalent:*

a) $u \in V$ *solves the operator equation (2.28), or, equivalently, the variational problem*

$$a(u, v) = \langle f, v \rangle_{V' \times V} \qquad \forall\, v \in V. \tag{2.29}$$

b) $u \in V$ *solves the equation* $R_\rho^* u = \tilde{g}$.

c) $u \in V$ *is a solution of the* Schwarz equation

$$P_\rho u = g := R_\rho \tilde{g}. \tag{2.30}$$

The spectrum of the symmetric operator $P_\rho : V \longrightarrow V$ *lies in the interval* $[A, B]$, *and its spectral condition number satisfies* $\kappa(P_\rho) \leq B/A$.

d) $u \in V$ *is given by* $u = R_\rho \tilde{u}$, *where* $\tilde{u} \in \tilde{V}_\rho$ *is an arbitrary solution of the* Gramian equation

$$\tilde{P}_\rho \tilde{u} = \tilde{g}.$$

Proof a) ⇔ b): Since R_ρ is surjective for a stable splitting, the span of the ranges of all R_i must be dense in V. So u solves (2.29) if and only if

$$a(u, R_i v_i) = \langle f, R_i v_i\rangle_{V'\times V}, \qquad v_i \in V_i, \quad i \in \Omega.$$

By the definition of R_i^* and g_i, this is again equivalent to $R_i^* u = g_i$ for all $i \in \Omega$. But this is equivalent to b), and also shows that $\tilde{g}$ belongs to $\mathrm{ran}(R_\rho^*)$ for any $f \in V'$ (since, by the invertibility of A, there is a $u \in V$ which solves (2.28)).

b) ⇔ c): If u is a solution in b), apply R_ρ to both sides and observe that u is also a solution in c). Conversely, if u is a solution in c), we have $R_\rho(R_\rho^* u - \tilde{g}) = 0$ by linearity. Thus, $R_\rho^* u - \tilde{g} \in \ker(R_\rho)$. But since $\tilde{g} \in \mathrm{ran}(R_\rho^*)$ and $\ker(R_\rho) = \mathrm{ran}(R_\rho^*)^\perp$, we must have $R_\rho^* u - \tilde{g} = 0$. Consequently, u also solves b).

d) ⇔ b) is easy, since R_ρ is surjective. □

The reformulations c) and d) in Theorem 2.5 give the basis for our analysis of Schwarz iterative methods in Chapter 4, which rely on solving auxiliary variational problems on the spaces V_i. Additionally, the spectra of P_ρ and $\tilde{P}_\rho$, which are controlled by the constants A and B in the inequalities (2.17) and (2.2) characterizing stable space splittings, will play a central role in the convergence theory of these methods.

Theorem 2.6 *Let* $(\mathbf{V};\mathbf{a};\mathbf{R};\rho)$ *be a stable space splitting for* $(V;a)$. *Then the spectrum of* P_ρ *satisfies* $\sigma(P_\rho) = \sigma(\tilde{P}_\rho)\backslash\{0\}$ *and is tightly contained in the interval* $[\lambda_{\min}, \lambda_{\max}]$, *where*

$$\lambda_{\min} = \lambda_{\min}(P_\rho) = \sup_{(2.17)} A; \qquad \lambda_{\max} = \lambda_{\max}(P_\rho) = \inf_{(2.2)} B,$$

denote the lower and upper stability bounds of the splitting. Consequently, the condition number of P_ρ *coincides with the condition number* κ *of the space splitting, i.e.,*

$$\kappa := \frac{\lambda_{\max}}{\lambda_{\min}} = \kappa(P_\rho) = \inf_{(2.17),(2.2)} \frac{B}{A}.$$

Proof By property 5) in Theorem 2.4, the restriction of R_ρ to the subspace

$$\tilde{X}_\rho := \tilde{V}_\rho \ominus \ker(R_\rho) \supset \mathrm{ran}(R_\rho^*)$$

is bijective and bounded as a linear map from $\tilde{X}_\rho$ onto V, and thus invertible. Note that then $R_\rho^* : V \to \tilde{X}_\rho$ is also invertible, since $R_\rho^* = R_\rho^{-1} P_\rho$. Consequently, $\tilde{X}_\rho = \mathrm{ran}(R_\rho^*)$. From the invertibility of R_ρ and the identity

$$(P_\rho - \lambda I)R_\rho = R_\rho R_\rho^* R_\rho - \lambda R_\rho = R_\rho(\tilde{P}_\rho - \lambda \tilde{I}), \qquad \lambda \in \mathbb{C},$$

we conclude that $P_\rho - \lambda I$ is invertible on V if and only if $\tilde{P}_\rho - \lambda \tilde{I}$ is invertible on $\tilde{X}_\rho$. This means that the resolvent sets of P_ρ and $\tilde{P}_\rho$ (as operator restricted to $\tilde{X}_\rho$) coincide and implies the formula for the spectrum of P_ρ.

The formulas for $\lambda_{\min}$, $\lambda_{\max}$, and κ follow from Corollary 2.1. □

Clearly, any of the inequalities equivalent to (2.17) and (2.2) can be used to estimate $\lambda_{\min}$ and $\lambda_{\max}$, respectively. These spectral bounds and the condition number κ will enter the convergence estimates of the Schwarz iterative methods based on stable space splittings considered in the following chapters. As a rule, the smaller κ, the faster the convergence. Thus, designing stable space splittings with small κ for a given variational problem (2.29) (or, equivalently, operator equation (2.28)) is a natural goal, called *preconditioning*. The terminology is justified when we consider the operator representations in terms of duality pairings. Recall the definition (2.1) of the scalar product $a(\cdot,\cdot)$ in terms of the operator A. Similarly, introduce the operators $A_i : V_i \to V_i'$ and $R_i' : V' \to V_i'$ by the identities

$$a_i(u_i, v_i) = \langle A_i u_i, v_i\rangle_{V_i'\times V_i}, \qquad \langle R_i' f, v_i\rangle_{V_i'\times V_i} = \langle f, R_i v_i\rangle_{V'\times V},$$

valid for all $f \in V'$ and $u_i, v_i \in V_i, i \in \Omega$, respectively. Since $a_i(\cdot,\cdot)$ is a scalar product on V_i, the operators A_i are invertible. Then one easily computes $R_i^* = A_i^{-1}R_i'A$, and

$$P_\rho = \sum_{i\in\Omega} \rho_i R_i A_i^{-1} R_i' A = C_\rho A, \qquad C_\rho = \sum_{i\in\Omega} \rho_i R_i A_i^{-1} R_i' \;:\; V' \to V. \tag{2.31}$$

If $\kappa = \kappa(P_\rho)$ is moderate, then the operator $C_\rho : V' \to V$ acts as an approximate inverse to A, and is referred to as a *preconditioner* for A that corresponds to the underlying stable space splitting. Therefore, switching from the operator equation (2.28) to the Schwarz equation (2.30) can be seen as a preconditioning step.

Preconditioning of linear operator equations in Hilbert spaces can be theoretically treated by the so-called *fictitious space lemma*, the original formulation [120] of which preceded the introduction of stable space splittings. Indeed, setting $(\tilde V; \tilde a) = (\tilde V_\rho; \tilde a_\rho)$ and $R = R_\rho$ in the following lemma recovers (2.31) as a special case.

Lemma 2.2 *Let $(V; a)$ and $(\tilde V; \tilde a)$ denote two Hilbert spaces equipped with scalar products generated by operators $A : V \to V'$ and $\tilde A : \tilde V \to \tilde V'$, respectively. Suppose $R : \tilde V \to V$ is a bounded and surjective linear operator such that*

$$A|||u|||^2 \le a(u,u) \le B|||u|||^2 \qquad \forall\, u \in V, \qquad |||u|||^2 := \inf_{\tilde v \in \tilde V:\, u = R\tilde v} \tilde a(\tilde v, \tilde v), \tag{2.32}$$

holds with constants $0 < A \le B < \infty$, and define $R' : V' \to \tilde V'$ by duality, i.e.,

$$\langle R' f, \tilde v\rangle_{\tilde V'\times \tilde V} = \langle f, R\tilde v\rangle_{V'\times V} \qquad \forall f \in V',\ \tilde v \in \tilde V.$$

Then the operator

$$P = CA : V \to V, \qquad C := R\tilde A^{-1} R',$$

is symmetric and coercive (thus, invertible) in $(V; a)$ with spectral bounds and condition number given by

$$A \le \lambda_{\min}(P) \le \lambda_{\max}(P) \le B, \qquad \kappa(P) \le B/A.$$

Proof Since by definition of the operators involved and the bilinear forms

$$\begin{aligned} a(Pu, v) &= \langle AR\tilde{A}^{-1}R'Au, v\rangle_{V'\times V} = \langle Av, R\tilde{A}^{-1}R'Au\rangle_{V'\times V} \\ &= \langle R'Av, \tilde{A}^{-1}R'Au\rangle_{\tilde{V}'\times\tilde{V}} = \tilde{a}(\tilde{A}^{-1}R'Au, \tilde{A}^{-1}R'Av), \qquad u, v \in V, \end{aligned}$$

we see that P is symmetric and positive. Setting $v = u$, one has

$$a(Pu, u) = \tilde{a}(\tilde{A}^{-1}R'Au, \tilde{A}^{-1}R'Au) \geq \|\tilde{A}'\|^{-2}_{\tilde{V}\to\tilde{V}'}\|R'Au\|^2_{\tilde{V}'}.$$

So $a(Pu, u) = 0$ is equivalent to $Au \in \ker(R')$. Consequently, P must be injective because $\ker(R') = \{0\}$ due to the assumed surjectivity of R.

Since the closure of $\operatorname{ran}(P)$ is the orthogonal complement of $\ker(P)$, the injectivity of P also shows that $\operatorname{ran}(P)$ must be dense in V. Thus, according to Part b) of the Theorem A.1 in the appendix, P is invertible if $\|Pu\| \geq A\|u\|$ for all $u \in V$ with some constant $A > 0$. To see this with the constant A from (2.32), take any $\tilde{v} \in \tilde{V}$ such that $Pu = R\tilde{v}$ (because of the surjectivity of R, such $\tilde{v}$ exist). Then we have

$$\begin{aligned} a(Pu, u)^2 &= \langle Au, R\tilde{v}\rangle^2_{V'\times V} = \langle R'Av, \tilde{v}\rangle^2_{\tilde{V}'\times\tilde{V}} \\ &= \tilde{a}(\tilde{A}^{-1}R'Au, \tilde{v}) \leq \tilde{a}(\tilde{A}^{-1}R'Au, \tilde{A}^{-1}R'Au)\tilde{a}(\tilde{v}, \tilde{v}) = a(Pu, u)\tilde{a}(\tilde{v}, \tilde{v}), \end{aligned}$$

with equality for the specific choice $\tilde{v} = \tilde{A}^{-1}R'Au$. Taking the infimum with respect to $\tilde{v}$, we get

$$a(Pu, u) = |||Pu|||^2. \tag{2.33}$$

Repeating the same calculations with any $\tilde{v}$ such that $u = R\tilde{v}$, we get

$$a(u, u)^2 \leq a(Pu, u)\tilde{a}(\tilde{v}, \tilde{v}), \qquad u = R\tilde{v}, \quad \tilde{v} \in \tilde{V},$$

and, after taking the infimum with respect to $\tilde{v}$, by the lower bound in (2.32) we get

$$\|u\|^4 = a(u, u)^2 \leq a(Pu, u)|||u|||^2 \leq \frac{1}{A}\|Pu\|\|u\|^3, \qquad u \in V.$$

This is the desired inequality which establishes the existence of P^{-1}.

The statements about the spectrum of P follow from (2.32) and (2.33). Indeed, replace u with $P^{-1}u$ in (2.33). Then

$$\frac{1}{B}a(u, u) \leq |||u|||^2 = a(P^{-1}u, u) \leq \frac{1}{A}a(u, u).$$

This concludes the proof. □

The application of this lemma requires checking the boundedness and surjectivity of R. In the context of space splittings, i.e., when $(\tilde{V}; \tilde{a}) = (\tilde{V}_\rho; \tilde{a}_\rho)$ and $R = R_\rho$, the lemma has been predominantly applied to finite-dimensional spaces V and finite space splittings, where the verification of the boundedness and surjectivity of R_ρ is easy. Otherwise, checking the required properties of R_ρ is as difficult as the proofs for stable space splittings given in this and the previous subsection.

The concept of stable space splittings is attractive for preconditioning purposes for two reasons. On one hand, it comes with a set of equivalent inequalities that allow us to obtain precise estimates of the preconditioning effect in terms of spectral bounds of $P_\rho = C_\rho A$ depending on the components of the space splitting, that is, the choices for the families of auxiliary Hilbert spaces $(V_i; a_i)$, the operators R_i, and the weights ρ_i. On the other hand, the structure of the preconditioner C_ρ looks promising for implementation on modern distributed and parallel computing platforms.

2.3 Operations on Hilbert space splittings

In this section, we consider operations on space splittings that are often used in application scenarios in order to simplify the analysis of iterative methods, or to derive new methods and algorithms from existing ones.

2.3.1 Scaling

The definition of space splittings includes one weight sequence ρ, which influences the relative scaling of the components of the space splitting. However, without essentially changing the components of the space splittings and the associated operators, there are other scaling options. For example, one can scale the operators R_i, or the bilinear forms $a_i(\cdot,\cdot)$ (or what is the same, the operators $A_i : V_i \to V_i'$ inducing them). Let us evaluate the effect of these different scaling options by comparing the resulting Schwarz operators.

Let $(\mathbf{V};\mathbf{a};\mathbf{R};\rho)$ be a space splitting associated with $(V;a)$. Consider now the collection $(\mathbf{V};\bar{\mathbf{a}};\bar{\mathbf{R}};\theta)$, where

$$\bar{R}_i = \xi_i R_i, \qquad \bar{a}_i(\cdot,\cdot) = \eta_i a_i(\cdot,\cdot), \qquad i \in \Omega, \tag{2.34}$$

and ξ, η, θ are three positive sequences with index set Ω. We will tacitly assume that condition (2.2) holds for this splitting as well, i.e., that we can use the associated Schwarz operator

$$\bar{P}_\theta = \sum_{i\in\Omega} \theta_i \bar{R}_i \bar{R}_i^*.$$

By definition of the adjoints R_i^* and $\bar{R}_i^*$, we have

$$\eta_i a_i(\bar{R}_i^* u, v_i) = \bar{a}_i(\bar{R}_i^* u, v_i) = a(u, \bar{R}_i v_i) = \xi_i a(u, R_i v_i) = \xi_i a_i(R_i^* u, v_i),$$

which gives $\bar{R}_i^* = \xi_i \eta_i^{-1} R_i^*$, and

$$\bar{P}_\theta = \sum_{i\in\Omega} \theta_i (\xi_i R_i)(\frac{\xi_i}{\eta_i} R_i^*) = \sum_{i\in\Omega} \frac{\theta_i \xi_i^2}{\eta_i} R_i R_i^*.$$

This leads us to the following conclusion.

Lemma 2.3 *If*

$$\rho_i = \frac{\theta_i \xi_i^2}{\eta_i}, \qquad i \in \Omega,$$

then the collections $(\mathbf{V};\mathbf{a};\mathbf{R};\rho)$ *and* $(\mathbf{V};\bar{\mathbf{a}};\bar{\mathbf{R}};\theta)$ *with* $\bar{\mathbf{a}}$ *and* $\bar{\mathbf{R}}$ *given in (2.34) are equivalent as space splittings associated with* $(V;a)$*, in the sense that they simultaneously satisfy the associated condition (2.2), and their Schwarz operators coincide, that is*

$$P_\rho = \bar{P}_\theta.$$

Note that while the Schwarz operators match, other ingredients like the operator $\bar{R}_\theta$ or the norm in the space $\tilde{\bar{V}}_\theta$ associated with $(\mathbf{V};\bar{\mathbf{a}};\bar{\mathbf{R}};\theta)$ look different. In fact,

$$\bar{R}_\theta = \sum_{i\in\Omega} \theta_i \xi_i R_i, \qquad \tilde{\bar{a}}_\theta(\tilde{u},\tilde{u}) = \sum_{i\in\Omega} \theta_i \eta_i a_i(u_i,u_i).$$

Since the surjectivity of R_ρ and $\bar{R}_\theta$ as well as the stability condition (2.17) can be expressed equivalently in terms of the Schwarz operators (and the latter coincide), Lemma 2.3 holds also in the case of (stable) space splittings for $(V;a)$.

Lemma 2.3 essentially says how the different scaling options interact and compensate for each other. It can be used to obtain convenient properties of the components of a space splitting while leaving the Schwarz operator (and thus the Schwarz equation) unchanged. For example, if one chooses $\xi_i = \rho_i$, $\theta_i = \rho_i^{-1}$ and $\eta_i = 1$ then

$$\bar{R}_\theta \tilde{v} = \sum_{i\in\Omega} R_i v_i, \qquad \bar{P}_\theta = P_\rho, \qquad \tilde{\bar{a}}_\theta(\tilde{u},\tilde{v}) = \sum_{i\in\Omega} \rho_i^{-1} a_i(u_i,v_i). \tag{2.35}$$

This way the operator $\bar{R} := \bar{R}_\theta$ is formally independent of the weight sequence. Instead, $\bar{R}_\theta^*$ now explicitly includes ρ, since $(\bar{R}_\theta^*)_i = \rho_i R_i$. This results in alternative representations for the norm $|||\cdot|||_\rho$ introduced above, namely

$$|||u|||_\rho^2 = \inf_{\tilde{v}:\, u=\bar{R}\tilde{v}} \sum_{i\in\Omega} \rho_i^{-1} \|v_i\|_i^2, \qquad u \in V. \tag{2.36}$$

That this formula is the same as the one in the original definition is also clear by substitution: Without changing the infimum, we can replace $v_i \in V_i$ with $\rho_i v_i \in V_i$ in (2.36), and we are back to (2.6). It is a matter of convenience whether to use (2.6) or (2.36).

Scaling of the R_i is also used to guarantee the uniform boundedness of the operators R_i and R_i^* (if it is not automatically satisfied), i.e.,

$$\sup_{i\in\Omega} \|R_i\|_{V_i\to V} = \sup_{i\in\Omega} \|R_i^*\|_{V\to V_i} \le \Lambda < \infty. \tag{2.37}$$

According to Lemma 2.3, setting $\eta_i = 1$ in (2.34) and taking $\theta_i = \xi_i^{-2}\rho_i$ will not change the Schwarz operator but allows us to control the operator norm of the R_i. For example, by setting $\xi_i = \|R_i\|^{-1}_{V_i\to V}$, we can always achieve unit norm operators $\bar{R}_i$ in the scaled space splitting $(\mathbf{V};\bar{\mathbf{a}};\bar{\mathbf{R}};\theta)$. Finally, scaling parameters in the form of relaxation parameters are implicitly introduced into the iterative methods to improve the convergence estimates and to adjust the spectral properties of the Schwarz operator. We will come back to this aspect in Chapters 4-6.

2.3.2 Refinement and coarsening

Building new (stable) space splittings from available ones by refinement and coarsening is a common procedure. Let $(\mathbf{V};\mathbf{a};\mathbf{R};\rho)$ be a space splitting associated with $(V;a)$, and let

$$(\mathbf{V}_i;\mathbf{a}_i;\mathbf{R}_i;\theta_i) := (V_{ij};a_{ij};R_{ij};\theta_{ij})_{j\in\Omega_i}$$

be space splittings associated with $(V_i;a_i)$, $i\in\Omega$. Here, $\theta_i=(\theta_{ij})_{j\in\Omega_i}$ are positive weight sequences, $(V_{ij};a_{ij})$ are appropriate auxiliary Hilbert spaces, $R_{ij}:V_{ij}\to V_i$ are bounded linear operators, and the Ω_i are at most countable index sets. Then we can consider the collection

$$(\mathbf{V}_{ref};\mathbf{a}_{ref};\mathbf{R}_{ref};\rho_{ref}) = (V_{ij};a_{ij};R_iR_{ij};\rho_i\theta_{ij})_{j\in\Omega_i,\,i\in\Omega}, \tag{2.38}$$

called the *refinement* of $(\mathbf{V};\mathbf{a};\mathbf{R};\rho)$, as the new space splitting associated with $(V;a)$. The new index set Ω_{ref} is the concatenation of all Ω_i, which in turn is at most countable. The following simple sufficient condition provides rough estimates for the spectrum of the Schwarz operator P_{ref} associated with (2.38). To formulate it, denote the constants in the conditions (2.2) for the space splittings associated with $(V;a)$ and $(V_i;a_i)$ by B and B_i, respectively. If the splittings are stable, we use A and A_i for the corresponding constants in (2.17).

Lemma 2.4 *Let the collections $(\mathbf{V};\mathbf{a};\mathbf{R};\rho)$ and $(\mathbf{V}_i;\mathbf{a}_i;\mathbf{R}_i;\theta_i)$ be space splittings associated with $(V;a)$ and $(V_i;a_i)$, $i\in\Omega$, respectively.*
a) If $\bar{B} := \sup_{i\in\Omega} B_i < \infty$, then the refinement (2.38) is also a space splitting associated with $(V;a)$, and (2.2) holds with constant $B_{ref}=B\bar{B}$.
b) If the given collections are space splittings for *$(V;a)$ and $(V_i;a_i)$, respectively, then (2.38) is also a space splitting* for *$(V;a)$.*
c) If the given collections are stable space splittings for $(V;a)$ and $(V_i;a_i)$, respectively, and if, in addition to $\bar{B}<\infty$ in a), we have $\underline{A} := \inf_{i\in\Omega} A_i > 0$, then (2.38) represents a stable space splitting for $(V;a)$, and (2.2) and (2.17) hold with constants $B_{ref}=B\bar{B}$ and $A_{ref}=A\underline{A}$, respectively.

Since these statements are obvious by inspection of (2.2) and (2.17), we omit the argument. Alternatively, one can check that the Schwarz operator P_{ref} associated with the refinement (2.38) is given by

$$P_{ref}u = R_\rho(P_{\theta_i}R_i^*u)_{i\in\Omega}, \qquad u\in V,$$

and use the norm bounds

$$\|R_\rho\|_{\tilde V_\rho\to V} = \|R_\rho^*\|_{\tilde V_\rho\to V} \le \sqrt{B}, \qquad \|P_{\theta_i}\|_{V_i} \le B_i \le \bar B. \qquad i\in\Omega,$$

for the norms of the operators involved. For c), the inverses of R_ρ, R_ρ^* (which exist when considered as acting between V and X_ρ, see the proof of Theorem 2.6), and P_{θ_i} come into play.

Using scaling options as discussed in the previous subsection, e.g. multiplying the weight sequences θ_i by appropriate factors, we would be able to obtain bounds on the condition number of P_{ref} in terms of the condition numbers of the given stable space splittings.

The opposite process, namely replacing a "refined" space splitting of the form (2.38) by the original "coarser" space splitting, is called *coarsening*. We keep the above notation, although the splitting (2.38) is now the starting point, and the process of coarsening consists of identifying a partition $\{\Omega_i\}_{i\in\Omega}$ of Ω_{ref} and appropriate intermediate spaces $(V_i; a_i)$. Now, we only have a statement for the coarsening of stable splittings if the condition $\underline{A} > 0$ holds. Counterexamples can simply be obtained by scaling. For example, let $\mathbf{W}$ be a complete orthogonal subspace decomposition for $(V; a)$ consisting of infinitely many subspaces W_i of dimension $\dim(W_i) > 1$. For each subspace W_i, we can then choose its own complete orthogonal subspace collection $\mathbf{W}_i = (W_{ij})_{j\in\Omega_i}$ consisting of at least two subspaces. Naturally, if we set $\rho_i = i$ and $\theta_{ij} = i^{-1}$, $j \in \Omega_i$, then the "refined" collection $(\mathbf{W}_{ref}; \rho_{ref})$, where

$$\mathbf{W}_{ref} = (W_{ij})_{j\in\Omega_i,\, i\in\Omega}, \qquad \rho_{ref} = (1)_{j\in\Omega_i,\, i\in\Omega},$$

is a stable orthogonal subspace splitting for $(V; a)$ (with $A_{ref} = B_{ref} = 1$). On the other hand, since $\rho_i \to \infty$, its coarsening $(\mathbf{W}; \rho)$ does not satisfy (2.2) and cannot serve as a space splitting for $(V; a)$. The orthogonal subspace splittings $(\mathbf{W}_i; \theta_i)$ for W_i are stable with $A_i = B_i = i^{-1}$. All this easily follows from the material in Subsection 2.2.1.1.

Lemma 2.5 *Assume that* $(\mathbf{V}_i; \mathbf{a}_i; \mathbf{R}_i; \theta_i)$ *are stable space splittings for* $(V_i; a_i)$, $i \in \Omega$, *for which* $\underline{A} > 0$.
a) If (2.38) is a space splitting associated with $(V; a)$, *with constant* B_{ref} *in (2.2), then its coarsening* $(\mathbf{V}; \mathbf{a}; \mathbf{R}; \rho)$ *is also a space splitting for* $(V; a)$, *and (2.2) holds with constant* $B = B_{ref}/\underline{A}$.
b) If (2.38) is a stable space splitting for $(V; a)$, *with constant* A_{ref} *in (2.17), then* $(\mathbf{V}; \mathbf{a}; \mathbf{R}; \rho)$ *is a stable space splitting for* $(V; a)$, *with constant* $A = A_{ref}/\bar B$.

Proof To prove a), we start from (2.2) for (2.38). We have

$$\|\sum_{i\in\Omega}\rho_i R_i(\underbrace{\sum_{j\in\Omega_i}\theta_{ij}R_{ij}v_{ij}}_{u_i:=R_{\theta_i}\tilde v_i})\|^2 \le B_{ref}\sum_{i\in\Omega}\rho_i\sum_{j\in\Omega_i}\theta_{ij}\|v_{ij}\|_{ij}^2.$$

Now temporarily fix $u_i \in \mathrm{ran}(R_{\theta_i})$, and take the infimum with respect to all $\tilde v_i = (v_{ij})_{j\in\Omega_i} \in \tilde V_{\theta_i}$ such that $u_i = R_{\theta_i}\tilde v_i$, $i \in \Omega$. This gives

$$\|\sum_{i\in\Omega}\rho_i R_i u_i\|^2 \le B' \sum_{i\in\Omega}\rho_i |||u_i|||_{\theta_i}^2,$$

where $|||\cdot|||_{\theta_i}$ is the norm associated with the stable space splitting $(\mathbf{V}_i;\mathbf{a}_i;\mathbf{R}_i;\theta_i)$ for $(V_i;a_i)$, $i\in\Omega$. Since the space splittings $(\mathbf{V}_i;\mathbf{a}_i;\mathbf{R}_i;\theta_i)$ are stable for $\{V_i;a_i\}$, we have $\mathrm{ran}(R_{\theta_i}) = V_i$ and

$$|||u_i|||_{\theta_i}^2 \le A_i^{-1}\|u_i\|_i^2 \le \underline{A}^{-1}\|u_i\|_i^2,$$

for all $i\in\Omega$. Substitution gives (2.2) for the coarsening $(\mathbf{V};\mathbf{a};\mathbf{R};\rho)$ with constant $B = B_{ref}/\underline{A}$.

Concerning b), we start with (2.17) for (2.38) and argue similarly: For any $u\in V$, there exists a decomposition

$$u = \sum_{i\in\Omega}\rho_i R_i \underbrace{(\sum_{j\in\Omega_i}\theta_{ij}R_{ij}v_{ij})}_{u_i' := R_{\theta_i}\tilde{v}_i} = \sum_{i\in\Omega}\rho_i R_i u_i$$

such that

$$A_{ref}\sum_{i\in\Omega}\rho_i |||u_i'|||_{\theta_i}^2 \le A_{ref}\sum_{i\in\Omega}\rho_i\sum_{j\in\Omega_i}\theta_{ij}\|v_{ij}\|_{ij}^2 \le \|u\|^2.$$

Since

$$|||u_i'|||_{\theta_i}^2 \ge B_i^{-1}\|u_i'\|_i^2 \ge \bar{B}^{-1}\|u_i'\|_i^2, \qquad i\in\Omega,$$

we obtain (2.17) for $(\mathbf{V};\mathbf{a};\mathbf{R};\rho)$ with the constant $A = A_{ref}/\bar{B}$. □

2.3.3 Selection of subsplittings

What happens to a space splitting $(\mathbf{V};\mathbf{a};\mathbf{R};\rho)$ if we replace some of the V_i with their proper subspaces U_i or leave out some of the components, i.e., if we move to subsplittings

$$(\mathbf{U};\mathbf{a}_\mathrm{U};\mathbf{R}_\mathrm{U};\rho')_{\Omega'} := (U_i; a_i|_{U_i}; R_i|_{U_i};\rho_i)_{i\in\Omega'}$$

for some index subset $\Omega'\subset\Omega$? Here, we have set $\rho' = \rho|_{\Omega'}$. The question has a trivial answer for space splittings $(\mathbf{V};\mathbf{a};\mathbf{R};\rho)_\Omega$ associated with $(V;a)$: Any subsplitting $(\mathbf{U};\mathbf{a}_\mathrm{U};\mathbf{R}_\mathrm{U};\rho')_{\Omega'}$ preserves this property, since condition (2.2) automatically holds for any subsplitting with the same or better constant B.

With the selection of subsplittings, the space $\tilde{V}_\rho$ is transformed into a new space $\tilde{U}_{\rho'}$, which can be identified with the subspace of $\tilde{V}_\rho$ consisting of all $\tilde{v}\in\tilde{V}_\rho$ such that $v_i\in U_i$ if $i\in\Omega'$ and $v_i = 0$ if $i\notin\Omega'$. Consequently, the closure of the range of the new $R_{\rho'} = R_\rho|_{\tilde{U}_{\rho'}}$ may not coincide with V anymore. The natural questions are therefore: If $(\mathbf{V};\mathbf{a};\mathbf{R};\rho)_\Omega$ is a stable space splitting for $(V;a)$, when is its subsplitting $(\mathbf{U};\mathbf{a}_\mathrm{U};\mathbf{R}_\mathrm{U};\rho')_{\Omega'}$ a stable space splitting for the subspace $(U_{\rho'};a)$ of $(V;a)$, where

$U_{\rho'}$ denotes the closure of the span of the ranges $\mathrm{ran}(R_i|_{U_i})$, $i \in \Omega'$, in V? And how does the constant A in the stability condition (2.17) change?

These questions can only be answered satisfactorily in special cases. Obviously, if Ω is finite and V is finite-dimensional, any subsplitting must be stable, but control over A is not possible in general, see the example below. An important special case is Riesz splittings, which were defined in Subsection 2.2.1.2. Due to the equivalent definition of Riesz splittings by the inequalities (2.26), the following statement is obvious:

Lemma 2.6 *If* $(\mathbf{V}; \mathbf{a}; \mathbf{R}; \rho)_\Omega$ *is a Riesz splitting for* $(V; a)$, *i.e., it satisfies (2.26), then any subsplitting* $(\mathbf{U}; \mathbf{a}_\mathrm{U}; \mathbf{R}_\mathrm{U}; \rho')_{\Omega'}$ *is a Riesz splitting for* $(U_{\rho'}; a)$, *with constants* $A \le A' \le B' \le B$ *in the associated condition (2.26), and the condition number can only improve, i.e. we have*

$$\kappa' \le \kappa.$$

In general, subsplittings of stable space splittings may not be stable, and even if they are, the associated constant A' can be arbitrarily smaller than A, and the Schwarz operator for the subsplitting can thus have an arbitrarily large condition number. Simple academic examples illustrate this. The first example demonstrates the possible loss of stability for subsplittings when V is infinite-dimensional. Take $\Omega = \{1, 2\}$, set $V_i = V$, $a_i(\cdot,\cdot) = a(\cdot,\cdot)$, $\rho_i = 1$, $i \in \Omega$. Now take for R_1 the identity operator on V, and take for $R_2 : V \to V$ an injective symmetric compact operator (such an operator cannot be invertible). The resulting collection $(\mathbf{V}; \mathbf{a}; \mathbf{R}; \rho)_\Omega$ is a stable splitting with $B = 2\max(1, \|R_2\|_V^2)$ and $A = 1$, because

$$\|v_1 + R_2 v_2\|^2 \le 2(\|v_1\|^2 + \|R_2 v_2\|^2) \le 2\max(1, \|R_2\|_V^2)(\|v_1\|^2 + \|v_2\|^2),$$

and $v = v + R_2 0$ is an admissible decomposition in (2.17) for any $v \in V$. Taking $\Omega' = \{2\}$ and $U_2 = V_2$, we have an unstable subsplitting. This is because the compact operator R_2 is not surjective.

The second example is a family of stable space splittings for $\mathbb{R}^2$ (with the usual Euclidean scalar product as $a(\cdot,\cdot)$) induced by the parameter-dependent dictionary

$$D = \{(1,0)^T,\ (1,\epsilon)^T,\ (0,1)^T\}, \qquad \epsilon > 0,$$

and the weight sequence $\rho_i = 1$, $i \in \Omega = \{1, 2, 3\}$. The optimal constants $A = 1$ and $B = 2 + \epsilon^2$ in (2.17) and (2.2) are given by the smallest and largest positive singular value of the matrix

$$\begin{pmatrix} 1 & 1 & 0 \\ 0 & \epsilon & 1 \end{pmatrix} \begin{pmatrix} 1 & 0 \\ 1 & \epsilon \\ 0 & 1 \end{pmatrix} = \begin{pmatrix} 2 & \epsilon \\ \epsilon & 1 + \epsilon^2 \end{pmatrix},$$

respectively. For the subsplitting with $\Omega' = \{1, 2\}$, i.e., the smaller dictionary $D' = \{(1,0)^T,\ (1,\epsilon)^T\}$, the same calculation results in the matrix

$$\begin{pmatrix} 2 & \epsilon \\ \epsilon & \epsilon^2 \end{pmatrix},$$

and

$$A' = \frac{\epsilon^2}{2} + \mathrm{O}(\epsilon^4), \qquad B' = 2 + \mathrm{O}(\epsilon^2).$$

In other words, while the condition number $\kappa = 2 + \epsilon^2$ of the original space splitting remains uniformly bounded, we have an arbitrarily large condition number $\kappa' = 4\epsilon^{-2} + \mathrm{O}(1)$ for its subsplitting, when $\epsilon \longrightarrow 0$.

Other examples relevant to the asymptotic behavior of multilevel methods are given in Chapters 3 and 7. According to Lemma 2.6, Riesz splittings offer some obvious advantages in this respect.

2.3.4 Tensor product splittings

Stable space splittings for Hilbert spaces induce stable splittings on their tensor product. We state this simple fact for the tensor product $V \otimes U$ of two Hilbert spaces using the notation introduced in this chapter. The result is easily transferred to the tensor product of finitely many Hilbert spaces.

Lemma 2.7 *Let* $(\mathbf{V};\mathbf{a};\mathbf{R};\rho)_{\Omega_V}$ *and* $(\mathbf{U};\mathbf{b};\mathbf{T};\theta)_{\Omega_U}$ *be stable space splittings for* $(V;a)$ *and* $(U;b)$*, respectively. Then the tensor product*

$$\begin{aligned}(\mathbf{V};\mathbf{a};\mathbf{R};\rho)_{\Omega_V} &\otimes (\mathbf{U};\mathbf{b};\mathbf{T};\theta)_{\Omega_U}\\ &:= (V_i \otimes U_j; a_i \otimes b_j; R_i \otimes T_j; \rho_i\theta_j)_{(i,j)\in\Omega_V\times\Omega_U}\end{aligned}$$

of the two collections is a stable space splitting for $(V \otimes U; a \otimes b)$*.*

If the associated Schwarz operators of these space splittings for $(V;a)$, $(U;b)$, *and* $(V \otimes U; a \otimes b)$ *are temporarily denoted by* P_V, P_U, *and* $P_{V\otimes U}$, *then we have*

$$\lambda_{\min}(P_{V\otimes U}) = \lambda_{\min}(P_V)\lambda_{\min}(P_U), \qquad \lambda_{\max}(P_{V\otimes U}) = \lambda_{\max}(P_V)\lambda_{\max}(P_U),$$

and, consequently, their condition numbers satisfy

$$\kappa(P_{V\otimes U}) = \kappa(P_V)\kappa(P_U).$$

Proof All statements follow from the facts collected in Section A.4.3 when we establish the formula

$$P_{V\otimes U} = P_V \otimes P_U. \tag{2.39}$$

It is sufficient to verify (2.39) by testing it on the elementary tensors $v \otimes u$ with $v \in V$ and $u \in U$. First, we have

$$(R_i \otimes T_j)^* = R_i^* \otimes T_j^*,$$

since by definition of the tensor product constructions involved

$$\begin{aligned}a \otimes b(v \otimes u, (R_i \otimes T_j)(v_i \otimes u_j)) &= a(v, R_i v_i)b(u, T_j u_j) = a_i(R_i^* v, v_i)b_j(T_j^* u, u_j)\\ &= a_i \otimes b_j((R_i^* \otimes T_j^*)(v \otimes u), v_i \otimes u_j).\end{aligned}$$

So,

$$\begin{aligned} P_{V\otimes U}(v\otimes u) &= \sum_{(i,j)\in\Omega_V\times\Omega_U} \rho_i\theta_j(R_i\otimes T_j)(R_i\otimes T_j)^*(v\otimes u) \\ &= \sum_{(i,j)\in\Omega_V\times\Omega_U} \rho_i\theta_j(R_i\otimes T_j)(R_i^*v\otimes T_j^*u) \\ &= \sum_{(i,j)\in\Omega_V\times\Omega_U} \rho_i\theta_j(R_iR_i^*v\otimes T_jT_j^*u) \\ &= \left(\sum_{i\in\Omega_V}\rho_iR_iR_i^*v\right)\otimes\left(\sum_{j\in\Omega_U}\theta_jT_jT_j^*u\right) = (P_V\otimes P_U)(v\otimes u), \end{aligned}$$

which concludes the proof. □

2.4 Scales of Hilbert spaces associated with space splittings

There are at least two types of families of Hilbert spaces related to a space splitting with infinite Ω that have important applications. The first type is related to the theory of approximation spaces and harmonic analysis in a basic Hilbert space $(H;(\cdot,\cdot)_H)$, where a family $\mathbf{H}$ of subspaces of H is fixed, and Hilbert spaces $H_{\mathbf{H},\rho}$ equipped with the appropriately defined norm $|||\cdot|||_{\mathbf{H},\rho}$ are defined. For some weight sequences ρ, such spaces can be identified as Hilbert spaces $(V;a)$ of practical interest, whose scalar products $a(\cdot,\cdot)$ are spectrally equivalent to those in $H_{\mathbf{H},\rho}$, see the next chapter for examples related to Sobolev spaces. The second type appears in the analysis of stochastic iterative schemes, where the Schwarz operator P_ρ, with ρ determined from the underlying probability distribution on Ω, is of trace class, and its powers P_ρ^s, $s\in\mathbb{R}$, generate a scale of spaces that governs the convergence theory of these methods.

2.4.1 Varying the weights ρ

In this subsection, we fix a separable Hilbert space H with scalar product and norm denoted by $(\cdot,\cdot)_H$ and $\|\cdot\|_H$, and consider a family $\mathbf{H}=(H_i)_{i\in\Omega}$ of countably many of its subspaces such that their span is dense in H. Recall that, for subspaces of H, we set $a_i(\cdot,\cdot)=(\cdot,\cdot)_H$, and simply take the natural injections as $R_i : H_i \to H$ (we omit them in the notation). Consequently, $R_i^* = Q_{H_i} : H \to H_i$ are the ortho-projections onto H_i with respect to the scalar product in H.

Given some positive weight sequence ρ, we adopt the notation introduced in connection with (2.35) and (2.36) at the end of Subsection 2.3.1 and consider the set $\tilde{X}$ of all $\tilde{x}\in\tilde{H}_{\rho^{-1}}$ such that the series

$$\bar{R}\tilde{x} := \sum_{i \in \Omega} x_i$$

converges unconditionally in H. Clearly, $\tilde{X}$ is a linear set in $\tilde{H}_{\rho^{-1}}$ that contains all H_i, and $\bar{R} : \tilde{X} \to H$ is a linear operator (we do not claim at this point that $\tilde{X}$ is a subspace of $\tilde{H}_{\rho^{-1}}$, nor that $\bar{R}$ is bounded, see Problem 2.10). Thus, the linear set $X := \mathrm{ran}(\bar{R}) \subset H$ is dense in H since it contains $\mathrm{span}(\mathbf{H})$. On X, we define the Hilbert semi-norm

$$|||x|||_{\mathbf{H},\rho} := \inf_{\tilde{x} \in \tilde{X}:\, x = \bar{R}\tilde{x}} \left(\sum_{i \in \Omega} \rho_i^{-1} \|x_i\|_H^2 \right)^{1/2}, \qquad x \in X. \tag{2.40}$$

Considered as a map from $\tilde{X}$ (with norm inherited from $\tilde{H}_{\rho^{-1}}$) onto X (equipped with the semi-norm $||| \cdot |||_{\mathbf{H},\rho}$), the linear operator $\bar{R}$ is bounded with norm ≤ 1 by the definition of the semi-norm in (2.40). To arrive at a meaningful Hilbert space, we will assume that (2.40) defines a norm on X, i.e., that

$$\{x \in X : \; |||x|||_{\mathbf{H},\rho} = 0\} = \{0\}, \tag{2.41}$$

see below for an example where this condition is not satisfied. With (2.41) at hand, we define $H_{\mathbf{H},\rho}$ as the completion of X with respect to the norm (2.40). The main motivation for this definition is evident from the following obvious statements.

Lemma 2.8 *Let (2.41) hold.*
a) $H_{\mathbf{H},\rho}$ is a separable Hilbert space.
b) If $V = H_{\mathbf{H},\rho}$, and $a : V \times V \to \mathbb{R}$ is a spectrally equivalent scalar product, namely

$$c_1 |||x|||_{\mathbf{H},\rho}^2 \leq a(x,x) \leq c_2 |||x|||_{\mathbf{H},\rho}^2, \qquad x \in H_{\mathbf{H},\rho},$$

with some constants $0 < c_1 \leq c_2 < \infty$, then $(\mathbf{H}; \rho)$ is a stable space splitting for $(V; a)$. Its condition number is bounded by $\kappa \leq c_2/c_1$.

This way, we created a whole scale of separable Hilbert spaces $H_{\mathbf{H},\rho}$ based on an arbitrary subspace family $\mathbf{H}$ of a given separable Hilbert space H. However, in this generality this may not be a very useful concept, since an identification of $H_{\mathbf{H},\rho}$ with $(V; a)$ of practical or theoretical relevance, as needed for the application of Lemma 2.8 b), is possible only in special cases. See Section 3.1 for some examples.

We proceed with the special case of an orthogonal subspace splitting $\mathbf{W}$ for H, i.e., $W_i \perp W_j$ for $i \neq j$, and

$$H = \bigoplus_{i \in \Omega} W_i.$$

Then, since each $x \in H$ has a unique (unconditionally converging in H) series decomposition with respect to $\mathbf{W}$ given by

$$x = \sum_{i \in \Omega} Q_{W_i} x,$$

the space $H_{\mathbf{W},\rho}$ is the closure of the set

$$x \in H : \ |||x|||^2_{\mathbf{W},\rho} = \sum_{i\in\Omega} \rho_i^{-1} \|Q_{W_i} x\|^2 < \infty$$

with respect to the norm $||| \cdot |||_{\mathbf{W},\rho}$ (note that (2.41) is automatically satisfied). It is not difficult to see that we have continuous embeddings

$$H_{\mathbf{W},\theta} \subset H_{\mathbf{W},\rho} \subset H$$

if and only if

$$\sup_{i\in\Omega} \frac{\theta_i}{\rho_i} < \infty, \qquad \sup_{i\in\Omega} \rho_i < \infty,$$

and that $H'_{\mathbf{W},\rho}$ can be identified by isometry with $H_{\mathbf{W},\rho^{-1}}$. We will see instances of $H_{\mathbf{W},\rho}$ spaces based on orthogonal subspace splittings for H in later chapters.

Another important case that stems from multiscale analysis and approximation theory is sequences of nested subspaces

$$V_0 \subset V_1 \subset \cdots \subset V_i \subset \cdots \subset H,$$

whose span is dense in H. Naturally, the index set of the resulting subspace collection $\mathbf{V}$ is then $\Omega = \mathbb{Z}_+$. We will discuss this case in more detail in the context of stable space splittings for Sobolev spaces, see Section 3.1. Here we just state the following result:

Lemma 2.9 *Let* $\mathbf{V}$ *be a ladder of nested subspaces as defined above. Then (2.41) holds if and only if*

$$\sum_{i\in\mathbb{Z}_+} \rho_i < \infty. \tag{2.42}$$

In this case, if the associated orthogonal subspace splitting $\mathbf{W}$ *is defined by* $W_0 = V_0$, *and* $W_i = V_i \ominus V_{i-1}$, $i \in \mathbb{N}$, *then* $H_{\mathbf{V},\rho}$ *is isomorphic to* $H_{\mathbf{W},\theta}$, *where the weight sequence* θ *is given by*

$$\theta_j = \sum_{i\geq j} \rho_i, \qquad j \in \mathbb{Z}_+.$$

Proof We begin with the observation that

$$|||u_j|||^2_{\mathbf{V},\rho} = 0$$

for any $u_j \in V_j$, $j \in \mathbb{Z}_+$, if (2.42) is violated. In fact, for any $k \geq j$, we can set

$$v_{k;i} = \begin{cases} \frac{\rho_i}{\sum_{l=j}^k \rho_l} u_j, \ i = j, \ldots, k \\ 0, \qquad\quad i < j \text{ or } i > k, \end{cases}$$

and get an element $\tilde{v}_k = (v_{k;i})_{i\in\mathbb{Z}_+} \in \tilde{X}$ such that $u_j = \bar{R}\tilde{v}_k \in X$ (here we used $V_j \subset V_i$ for $i \geq j$). The meaning of X, $\tilde{X}$, $\bar{R}$ is the same as defined at the beginning of this subsection. So,

$$|||u_j|||^2_{\mathbf{V},\rho} \leq \sum_{i=j}^{k} \rho_i^{-1} \|v_{k;i}\|^2_H = \frac{\|u_j\|^2_H}{\sum_{l=j}^{k} \rho_l}.$$

Since the denominator in the right-hand side converges to infinity as $k \to \infty$ if (2.42) does not hold, we see that the semi-norm $|||u_j|||_{\mathbf{V},\rho}$ vanishes. Thus, (2.41) is violated. Additionally, the semi-norm vanishes on the dense set span($\mathbf{V}$) in H.

Now assume that (2.42) holds. Let us compute $|||u|||_{\mathbf{V},\rho}$ for arbitrary $u \in X$. Take any $\tilde{v} \in \tilde{X}$ such that $\bar{R}\tilde{v} = u$. Set $w_{ji} := Q_{W_j} v_i \in W_j$, $j \leq i$. Then

$$\|\tilde{v}\|^2_{\tilde{V}_\rho^{-1}} = \sum_{i \in \mathbb{Z}_+} \rho_i^{-1} \|v_i\|^2_H = \sum_{j \in \mathbb{Z}_+} \underbrace{\sum_{i \geq j} \rho_i^{-1} \|w_{ji}\|^2_H}_{A_j :=}.$$

Since we have

$$Q_{W_j} u = \sum_{i \in \mathbb{Z}_+} Q_{W_j} v_i = \sum_{i \geq j} w_{ji},$$

the infimum of $\|\tilde{v}\|^2_{\tilde{V}_{\rho^{-1}}}$ is equal to the sum of the minimum values of the expressions A_j for given ortho-projection $Q_{W_j} u$. These minimum values, which can be determined separately for each $j \in \mathbb{Z}_+$ using the statements in Problem 2.11, are given by

$$\min A_j = \theta_j \|Q_{W_j} u\|^2_H, \qquad j \in \mathbb{Z}_+,$$

and are attained when we set

$$w_{ji} = \frac{\rho_i}{\sum_{l \geq j} \rho_l} Q_{W_j} u = \frac{\rho_i}{\theta_j} Q_{W_j} u, \qquad i \geq j, \quad j \in \mathbb{Z}_+.$$

Thus, for $u \in X$ we have

$$|||u|||^2_{\mathbf{V},\rho} = \inf_{\tilde{v} \in \tilde{X}:\, u = \bar{R}\tilde{v}} \|\tilde{v}\|^2_{\tilde{V}_{\rho^{-1}}} = \sum_{j \in \mathbb{Z}_+} \theta_j^{-1} \|Q_{W_j} u\|^2_H = |||u|||^2_{\mathbf{W},\theta}.$$

This shows (2.41), and the remaining statements of the lemma follow. Our considerations show that under the condition (2.42) we have $X = H_{\mathbf{W},\theta}$, so there is no need for a completion step in the definition of $H_{\mathbf{W},\theta}$. □

2.4.2 Hilbert scales induced by P_ρ

This subsection is preparation for analyzing stochastic iterative methods, where the Schwarz operator P_ρ of interest is related to the expectation of a random operator associated with a space splitting for $(V; a)$ applied in each iteration step. As a rule, P_ρ is typically a symmetric positive trace class operator with dense range but is not invertible (unless V is finite-dimensional).

From now on, let us assume in this subsection that V is infinite-dimensional, and $P_\rho \in \mathcal{B}_{pos}(V)$ is compact and injective, i.e., $\ker(P_\rho) = \{0\}$. Both assumptions are not essential but simplify technicalities. Thus, there is a CONS $\{\psi_k\}_{k\in\mathbb{N}}$ of eigenvectors of P_ρ that correspond to the positive eigenvalues $\lambda_1 \geq \lambda_2 \geq \cdots > 0$ of P_ρ, such that for each $u \in V$

$$u = \sum_{k\in\mathbb{N}} a(u,\psi_k)\psi_k, \qquad P_\rho u = \sum_{k\in\mathbb{N}} \lambda_k a(u,\psi_k)\psi_k.$$

Then, for any $s \in \mathbb{R}$, we define $V^s_{P_\rho}$ as the completion of the set

$$Y^s_{P_\rho} = \{u \in V : \ \|u\|_{V^s_{P_\rho}} := \left(\sum_{k\in\mathbb{N}} \lambda_k^{-s} a(u,\psi_k)^2\right)^{1/2} < \infty\} \subset V$$

with respect to the norm $\|u\|_{V^s_{P_\rho}}$. In other words, $V^s_{P_\rho}$ is the isometric image of the sequence spaces

$$\ell^2_{\lambda^{-s}}(\mathbb{N}) = \{\mathbf{c} : \ \|\mathbf{c}\|_{\ell^2_{\lambda^{-s}}} = \left(\sum_{k\in\mathbb{N}} \lambda_k^{-s} c_k^2\right)^{1/2} < \infty\}$$

under the mapping $\mathbf{c} \to u = \sum_{k\in\mathbb{N}} c_k\psi_k$ (extended by continuity, if $s < 0$). The following properties of the spaces $V^s_{P_\rho}$ are obvious from the corresponding properties of the spaces $\ell^2_{\lambda^{-s}}(\mathbb{N})$.

Theorem 2.7 *a) The spaces $V^s_{P_\rho}$ form a decreasing family of separable Hilbert spaces, and the embedding*

$$V^s_{P_\rho} \subset V^{s'}_{P_\rho}, \qquad -\infty < s' < s < \infty, \tag{2.43}$$

is compact. The dual space of $V^s_{P_\rho}$ can be identified with $V^{-s}_{P_\rho}$ for all $s \in \mathbb{R}$. Furthermore, the powers $P^t_\rho = \sum_{k\in\mathbb{N}} \lambda^t_k a(\cdot,\psi_k)\psi_k, t \in \mathbb{R}$, act as isometries between $V^s_{P_\rho}$ and $V^{s+2t}_{P_\rho}$ for any $s \in \mathbb{R}$. In particular,

$$\|u\|^2_{V^s_{P_\rho}} = a(P^{-s}_\rho u, u) = \|P^{-s/2}_\rho u\|^2, \qquad u \in V^s_{P_\rho}, \quad s \in \mathbb{R}. \tag{2.44}$$

b) If $s \geq 0$, then $V^s_{P_\rho} = Y^s_{P_\rho}$ is a linear, dense subset of V, we have $V^0_{P_\rho} = V$, and

$$\|u\|_{V^1_{P_\rho}} = |||u|||_\rho, \qquad u \in V^1_{P_\rho}.$$

c) If $s < 0$, then $Y^s_{P_\rho} = V$ is a linear, dense subset of $V^s_{P_\rho}$. In particular, we have

$$\|u\|^2_{V^{-1}_{P_\rho}} = a(P_\rho u, u) = \|P^{1/2}_\rho u\|^2, \qquad u \in V.$$

Note that $P_\rho^{(s-s')/2}$ realizes the embedding (2.43) and that P_ρ^t, $t > 0$, is equivalent to the compact diagonal operator

$$\mathbf{c} = (c_k)_{k\in\mathbb{N}} \longrightarrow (\lambda_k^t c_k)_{k\in\mathbb{N}}$$

acting on $\ell^2(\mathbb{N})$. The expression (2.44) for the $V_{P_\rho^s}$ norm will be used frequently in the following.

Prerequisite to using the spaces $V_{P_\rho}^s$ is the verification of the compactness property of P_ρ. Here is a criterion for P_ρ to be of trace class. For the relevant definitions, see Section A.2.3 in the appendix.

Lemma 2.10 *Let* $(\mathbf{V}; \mathbf{a}, \mathbf{R}; \rho)$ *be a space splitting associated with* $(V; a)$. *Then* P_ρ *is of trace class if and only if*

$$\|P_\rho\|_{\mathrm{Tr}} = \sum_{i\in\Omega} \rho_i \|R_i\|_{\mathrm{HS}}^2 < \infty. \tag{2.45}$$

If the scaling condition (2.37) holds, then

$$\sum_{i\in\Omega} \rho_i \dim(V_i) < \infty \tag{2.46}$$

is a sufficient condition for (2.45) to hold. If $(\mathbf{V}; \mathbf{R}; \rho)$ *is a space splitting associated with* $(V; a)$ *generated by Galerkin projection and, in particular, if* $(\mathbf{V}; \rho)$ *is a subspace splitting, then (2.45) and (2.46) are equivalent.*

Proof This follows directly from the definition of trace class and Hilbert-Schmidt norms (see Appendix A.2.3) applied to $P_\rho : V \to V$ and $R_i^* : V \to V_i$, $i \in \Omega$, respectively: For an arbitrary CONS $\{\phi_k\}$ in $(V; a)$, we have

$$\begin{aligned}\|P_\rho\|_{\mathrm{Tr}} &= \sum_k a(P_\rho\phi_k, \phi_k) = \sum_k \tilde{a}_\rho(R_\rho^*\phi_k, R_\rho^*\phi_k) \\ &= \sum_{i\in\Omega} \rho_i \sum_k a_i(R_i^*\phi_k, R_i^*\phi_k) = \sum_{i\in\Omega} \rho_i \|R_i^*\|_{\mathrm{HS}}^2.\end{aligned}$$

To get (2.45), it remains to substitute the equality

$$\|R_i^*\|_{\mathrm{HS}}^2 = \|R_i\|_{\mathrm{HS}}^2.$$

For space splittings satisfying (2.37), we have

$$\|R_i\|_{\mathrm{HS}}^2 = \sum_k a(R_i\phi_{ik}, R_i\phi_{ik}) \le \Lambda \sum_k a_i(\phi_{ik}, \phi_{ik}) = \Lambda \dim(V_i), \qquad i \in \Omega,$$

for any CONS $\{\phi_{ik}\}$ in $(V_i; a_i)$. For space splittings $(\mathbf{V}; \mathbf{R}; \rho)$ generated by Galerkin projection and subspace splittings $(\mathbf{V}; \rho)$ we have $\Lambda = 1$ and equality. This shows the remaining statements. □

Problems

2.1 Properties of space splittings generated by Galerkin projection.
Let $(\mathbf{V};\mathbf{R};\rho)$ be a space splitting for $(V;a)$ generated by Galerkin projection. For further definition and notation, see Example 3 in Subsection 2.1.2. Set $U_i := \mathrm{ran}(R_i)$, $i \in \Omega$.
a) Prove that $\|R_i\|_{V_i \to V} = 1$, $i \in \Omega$.
b) Show that P_i and $I - P_i$, $i \in \Omega$, are ortho-projections in $(V;a)$ onto the subspaces U_i and $U_i^{\perp}$, respectively.
c) Show that the subspace splitting $(\mathbf{U};\rho)$ for $(V;a)$ generated by $\mathbf{U} := (U_i)_{i\in\Omega}$ is equivalent to $(\mathbf{V};\mathbf{a};\mathbf{R};\rho)$ (i.e., they have the same Schwarz operators).

2.2 Every stable space splitting is equivalent to a frame.
Let $(\mathbf{V};\mathbf{a};\mathbf{R};\rho)$ be a stable space splitting for $(V;a)$. Show that there exists a frame D in V such that the associated frame operator (2.25) and the Schwarz operator P_ρ coincide.

2.3 Strengthened Cauchy-Schwarz inequalities and the Gramian operator $\tilde{P}_\rho$.
a) Show that (2.14) is equivalent to the condition

$$\|R_j^* R_i\|_{V_i \to V_j} \le \gamma_{ij}, \qquad i, j \in \Omega.$$

Recall that the operators $R_j^* R_i$ are the entries of the operator matrix representation of $\tilde{P}_\rho$.
b) Use a) to show that, if $(\mathbf{V};\mathbf{a};\mathbf{R};\rho)$ satisfies strengthened Cauchy-Schwarz inequalities for $(V;a)$, then the Gramian operator $\tilde{P}_\rho$ possesses the norm bound

$$\|\tilde{P}_\rho\|_{\tilde{V}_\rho} \le \|\Gamma_\rho\|_{\ell^2}.$$

2.4 Splittings with two subspaces.
For a subspace splitting $(\mathbf{V};\rho)$ associated with $(V;a)$ with two subspaces, i.e., $\mathbf{V} = (V_1, V_2)$, $\Omega = \{1, 2\}$, show that

$$\|P_\rho\|_V = \|\Gamma_\rho\|_{\ell^2},$$

if the γ_{ij} in (2.14) are best possible. Provide an example of a subspace splitting with three subspaces, where equality does not hold.

2.5 A space splitting for $V = \ell^2(\mathbb{Z})$ that does not satisfy strengthened Cauchy-Schwarz inequalities.
a) Verify that the Toeplitz operator

$$T_a : x = (x_j)_{j\in\mathbb{Z}} \to T_a x = (\sum_{j\in\mathbb{Z}} a_{k-j} x_j)_{k\in\mathbb{Z}}$$

induced by a real-valued sequence $a = (a_j)_{j\in\mathbb{Z}}$ is a bounded linear operator on $\ell^2(\mathbb{Z})$ if and only if the Fourier transform

$$\hat{a}(\theta) := \sum_{j\in\mathbb{Z}} a_j e^{ij\theta}$$

of a belongs to $L^\infty(-\pi,\pi)$ (here, i temporarily denotes the imaginary unit). Furthermore, verify that T_a is symmetric if and only if $a_j = a_{-j}$, $j \in \mathbb{Z}$. Also check that T_a is coercive on $\ell^2(\mathbb{Z})$ if $\hat{a}(\theta) \geq c > 0$ for almost every $\theta \in (-\pi,\pi)$.
b) By evaluating the Fourier coefficients, check that

$$\hat{a}(\theta) = \begin{cases} 1+\alpha, & |\theta| \leq \pi/2, \\ 1-\alpha, & \pi/2 < |\theta| < \pi, \end{cases}$$

for the sequence a given by $a_0 = 1$, $a_{2k} = a_{-2k} = 0$, and

$$a_{2k-1} = a_{1-2k} = \frac{2\alpha(-1)^{k-1}}{\pi(2k-1)}, \qquad k \in \mathbb{N}.$$

c) Conclude from a) that the Toeplitz operator T_a induced by the sequence a from Part b) generates a spectrally equivalent scalar product $a(x,y) := (T_a x, y)_{\ell^2}$ on $\ell^2(\mathbb{Z})$ satisfying

$$(1-\alpha)\|x\|_{\ell^2}^2 \leq a(x,x) \leq (1+\alpha)\|x\|_{\ell^2}^2,$$

if $0 < \alpha < 1$. The constants in this norm equivalence are best possible.
d) Consider the space splitting $\mathbf{V}_D$ generated from the CONS $D = \{e_j\}_{j\in\mathbb{Z}}$ of unit coordinate sequences in $(V;a)$ with the weight sequence $\rho_i = 1$ for $i \in \mathbb{Z}$. Here $V = \ell^2(\mathbb{Z})$ and the scalar product $a(\cdot,\cdot)$ is defined as in Part c) with some fixed $0 < \alpha < 1$. Check that $\mathbf{V}_D$ is a stable space splitting for $(V;a)$.
e) Finally, check that for (2.14) to hold, one must have $\gamma_{ij} \geq |a_{i-j}|$, and that the associated matrix operator Γ_ρ therefore cannot be bounded on $\ell^2(\mathbb{Z})$. In other words, this stable space splitting does not satisfy strengthened Cauchy-Schwarz inequalities!

2.6 Angle and gap between subspaces.
Let $(\mathbf{V};\rho)$ be a subspace splitting associated with $(V;a)$. Then the best possible constants γ_{ij} in the definition of strengthened Cauchy-Schwarz inequalities (2.14) are given by

$$\gamma_{ij} = \sup_{0\neq u_i\in V_i,\ 0\neq v_j\in V_j} \frac{a(u_i,v_j)}{\|u_i\|\|v_j\|}, \qquad i,j\in\Omega.$$

Obviously, $\gamma_{ij} \in [0,1]$, $\gamma_{ij} = 0$ is equivalent to $V_i \perp V_j$, and $\gamma_{ij} = 1$ if $V_i \cap V_j \neq \{0\}$. Furthermore, $\gamma_{ij} < 1$ if and only if $V_i \cap V_j \neq \{0\}$ and the algebraic sum $V_i + V_j$ is closed in V. The quantity $\theta_{ij} := \arccos(\gamma_{ij}) \in [0,\pi/2]$ is called the *Dixmier angle* between the subspaces V_i and V_j. A related quantity is the *Friedrichs angle*

$$\tilde{\theta}_{ij} = \arccos(\tilde{\gamma}_{ij}), \qquad \tilde{\gamma}_{ij} := \sup_{0\neq u_i\in V_i\cap V_{ij}^\perp,\ 0\neq v_j\in V_j\cap V_{ij}^\perp} \frac{a(u_i,v_j)}{\|u_i\|\|v_j\|}, \qquad i,j\in\Omega,$$

where $V_{ij} = V_i \cap V_j$ for short. Finally, another way to measure the relative position of two subspaces is by the *gap* defined by the operator norm

$$\Theta(V_i, V_j) := \|Q_{V_i} - Q_{V_j}\|_V$$

of the difference of the ortho-projections Q_{V_i} and Q_{V_j} onto them. The Friedrichs angle plays a central role in the convergence theory of the alternating direction method for two subspaces, see [65, Chapter 9], [77, Chapter 7] or [37].

Show that

$$\gamma_{ij} = \|Q_{V_i}Q_{V_j}\|_V = \|Q_{V_j}Q_{V_i}Q_{V_j}\|_V^{1/2} \geq \tilde{\gamma}_{ij} = \|Q_{V_j}Q_{V_i} - Q_{V_{ij}}\|_V.$$

2.7 Dual space splitting.

If $(\mathbf{V}; \mathbf{a}; \mathbf{R}; \rho)$ is a stable space splitting for $(V; a)$, then the collection $(\mathbf{V}; \mathbf{a}; \mathbf{S}; \rho)$ with $S_i = P_\rho^{-1} R_i$, $i \in \Omega$, is also a stable space splitting for $(V; a)$, called the *canonical dual* of the stable space splitting $(\mathbf{V}; \mathbf{a}; \mathbf{R}; \rho)$.

a) Establish this fact by proving the equalities

$$\sum_{i\in\Omega} \rho_i R_i S_i^* = \sum_{i\in\Omega} \rho_i S_i R_i^* = I, \qquad \sum_{i\in\Omega} \rho_i S_i S_i^* = P_\rho^{-1}.$$

In particular, a stable space splitting and its canonical dual have the same condition number.

b) The above equalities characterize the canonical dual: If they are satisfied for a set of bounded operators $S_i : V_i \to V$, then $S_i = P_\rho^{-1} R_i$, $i \in \Omega$.

c) Verify that $\tilde{u} = (S_i^* u)_{i\in\Omega}$ provides the "best" decomposition of u in $\tilde{V}_\rho$, in the sense that

$$\tilde{u} \in \tilde{V}_\rho, \qquad u = R_\rho \tilde{u}, \qquad |||u|||_\rho^2 = \|\tilde{u}\|_{\tilde{V}_\rho}^2.$$

d) Finally, show that for a stable subspace splitting $(\mathbf{V}; \rho)$ with a ladder of nested subspaces $V_0 \subset V_1 \subset \ldots$ (see Subsection 2.4.1, where the notation is explained), the canonical dual is given by

$$S_k = \sum_{i=0}^{k} (\sum_{j\geq i} \rho_j)^{-1} Q_{W_i}.$$

Note that the concept of the canonical dual is utilized in frame and g-frame theory (see [175]), but it is less relevant to our approach of iterative methods based on space splittings.

2.8 Subspace splittings with linearly independent V_i are not necessarily Riesz splittings.

Recall that a system $\mathbf{V} = \{V_i\}_{i\in\Omega}$ of subspaces of V is called *linearly independent* if

$$\sum_{i\in\Omega'} v_i = 0, \qquad v_i \in V_i,$$

implies $v_i = 0$, $i \in \Omega'$, for all finite $\Omega' \subset \Omega$. If $(\mathbf{V}; \rho)$ is a stable subspace splitting for $(V; a)$ and Ω is finite or V is finite-dimensional, then the linear independence of $\mathbf{V}$ implies the Riesz property of the splitting, i.e., $\ker(R_\rho) = \{0\}$. Construct an

example of a stable space splitting of $\ell^2(\mathbb{N})$ with linearly independent $\mathbf{V}$ such that $\ker(R_\rho) \neq \{0\}$.
Hint: Consider the family $\mathbf{V}$ of one-dimensional subspaces of $\ell^2(\mathbb{N})$ generated from a dictionary D consisting of the unit coordinate sequences e_k, $k \in \mathbb{N}$, and an arbitrary element $x = (x_k)_{k\in\mathbb{N}} \in \ell^2(\mathbb{N})$ with infinitely many $x_k \neq 0$ and set $\rho_i = 1$.

2.9 The span of two subspaces may not be a subspace.
Consider a separable Hilbert space H with scalar product $(\cdot,\cdot)_H$ equipped with a CONS $\{e_k\}_{k\in\mathbb{N}}$. Define dictionaries $D_f = \{f_k\}_{k\in\mathbb{N}}$ and $D_g = \{g_k\}_{k\in\mathbb{N}}$ by

$$f_k = e_{2k}; \qquad g_k = \frac{1}{k}e_{2k-1} - e_{2k}, \qquad k \in \mathbb{N},$$

and subspaces H_f and H_g as the closure in H of $\mathrm{span}(D_f)$ and $\mathrm{span}(D_g)$, respectively.
a) Show that the subspaces H_g and H_f are linearly independent in H, which is equivalent to $H_f \cap H_g = \{0\}$.
b) Show that their span

$$\mathrm{span}(\{H_f, H_g\}) = \{f + g : \ f \in H_f,\ g \in H_g\}$$

is a dense linear subset, but not a subspace of H. Conclude that $((H_f, H_g); (1, 1))$ is a subspace splitting for $(H; (\cdot,\cdot)_H)$, which is not stable.
c) A dictionary $D = \{\phi_i\}_{i\in\Omega} \subset H$, $\Omega \subset \mathbb{N}$ is called a *Riesz system* in H if an inequality analogous to (2.26) holds: For all *finite* linear combinations we have

$$A\sum_{i\in\Omega}\alpha_i^2 \le \|\sum_{i\in\Omega}\alpha_i\phi_i\|_H^2 \le B\sum_{i\in\Omega}\alpha_i^2.$$

Show that D_f, D_g are Riesz systems in H_f, $H_g \subset H$, but that their union is not a Riesz system in H.

2.10 Exercises for Subsection 2.4.1.
Consider a subspace collection $(\mathbf{V};\rho)$ for H with countable Ω, i.e., V_i are subspaces of H, $a_i(\cdot,\cdot) = (\cdot,\cdot)_H$, and $R_i : V_i \to H$ are the natural inclusions. Notation such as $\tilde{X}$, $\bar{R}$ is defined in Subsection 2.4.1.
a) Show that $\bar{R} : \tilde{X} \subset V_{\rho^{-1}} \to H$ is not bounded if $\sup_{i\in\Omega}\rho_i = \infty$ and is bounded if $\sum_{i\in\Omega}\rho_i < \infty$. What happens if ρ satisfies neither one of the two conditions, for example, what happens if $\rho_i = 1$, $i \in \Omega$?
b) Show that the linear subset $\tilde{X} \subset \tilde{V}_{\rho^{-1}}$ cannot be a subspace of $\tilde{V}_{\rho^{-1}}$ if $\sup_{i\in\Omega}\rho_i = \infty$, and show that $\tilde{X} = \tilde{V}_{\rho^{-1}}$ if $\sum_{i\in\Omega}\rho_i < \infty$. What happens for $\rho_i = 1$, $i \in \Omega$?

2.11 Details for Lemma 2.9.
Let $\rho = (\rho_i)_{i\in\Omega}$ ($\Omega \subset \mathbb{N}$) be a positive sequence. For any Hilbert space H and any $u \in H$, solve the constrained minimization problem

$$\sum_{i\in\Omega}\rho_i^{-1}\|u_i\|_H^2 \to \min, \qquad u = \sum_{i\in\Omega}u_i, \quad u_i \in H,\ i \in \Omega.$$

Hints. The minimum value of this problem equals $|||u|||^2_{\mathbf{H},\rho}$ if we set $H_i = H$ in (2.40). If $\sum_{i\in\Omega}\rho_i = \infty$, then $|||u|||_{\mathbf{H},\rho} = 0$ for all $u \in H$, see the proof of Lemma 2.9. In the opposite case, fix $u \in H$, denote H_u as the one-dimensional subspace of H generated by u, and write each $u_i = \alpha_i u + v_i$ as the orthogonal sum of a multiple of u and a $v_i \in H_u^{\perp}$. Then

$$\sum_{i\in\Omega}\rho_i^{-1}\|u_i\|_H^2 = \left(\sum_{i\in\Omega}\frac{\alpha_i^2}{\rho_i}\right)\|u\|_H^2 + \sum_{i\in\Omega}\rho_i^{-1}\|v_i\|_H^2.$$

On the other hand

$$u = \sum_{i\in\Omega} u_i = \left(\underbrace{\sum_{i\in\Omega}\alpha_i}_{=1}\right) u + \underbrace{\sum_{i\in\Omega} v_i}_{=0}.$$

This allows us to reduce the given the minimization problem in H to a minimization problem for the sequence $(\alpha_i)_{i\in\Omega}$ and to find the unique minimizing sequence $(u_i)_{i\in\Omega}$.

Chapter 3
Hilbert space splittings: Examples and extensions

Abstract We present some important examples of space splittings. They are motivated mainly by their relevance to the solution of large linear systems that arise from discretizations of PDEs.

3.1 Multiscale splittings

Multiscale decompositions have their roots in time-frequency analysis and the modelling with partial differential equations, where the "scales" are generally related to frequency bands and spectral decompositions. In this section, our approach to multiscale analysis is a nested sequence of finite-dimensional subspaces

$$V_0 \subset V_1 \subset \cdots \subset V_j \subset \cdots \subset H$$

of a separable Hilbert space H, such that $\mathrm{span}(\{V_j\}_{j\in\mathbb{Z}_+})$ is dense in $(H;(\cdot,\cdot)_H)$. This formally defines a subspace collection $\mathbf{V}$ related to $(H;(\cdot,\cdot)_H)$ with index set $\Omega = \mathbb{Z}_+$, i.e., the bilinear forms $a_j(\cdot,\cdot)$ on V_j are inherited from the scalar product on H, the operators R_i are the natural injections, and $\rho_i = 1$. Instead of working with nested subspaces V_j and redundant representations $u = \sum_j v_j$ with $v_j \in V_j$, one can also switch to a collection $\mathbf{W} := \{W_j\}_{j\in\mathbb{Z}_+}$ of complement spaces defined by $W_0 = V_0$ for $j = 0$ and $W_j := \mathrm{ran}(Q_j - Q_{j-1})$ for $j = 1, 2, \ldots$. Here and in the following, $Q_j : V \to V_j$, $j = 0, 1, \ldots$, denotes a sequence of projections onto the subspaces in $\mathbf{V}$. Under suitable conditions on the Q_j, this leads to direct sum splittings

$$H = \sum_{j\in\mathbb{Z}_+} W_j, \qquad V_j = V_{j-1} \dotplus W_j, \quad j = 1, 2, \ldots,$$

and to unique decompositions $u = \sum_j w_j$, $w_j \in W_j$. This has other advantages as well, such as tighter control of norms of u via the building blocks w_j of the decomposition with respect to $\mathbf{W}$.

M. Griebel, P. Oswald, *Hilbert Space Splittings and Iterative Methods*, Springer Series in Computational Mathematics 62, https://doi.org/10.1007/978-3-031-74370-2_3

In Subsection 3.1.1, we develop an abstract theory that establishes a connection between space splittings generated by $\mathbf{V}$ or $\mathbf{W}$ and another family of Besov-type spaces B^s_ω. This connection is based on approximation properties of $\mathbf{V}$ in H. The weight sequences used in the space splittings and in the definition of B^s_ω depend on a parameter $q \in (0,1)$, which is linked to the subspaces V_i through a geometric growth condition on their dimensions, so that $\dim(V_j) \approx \dim(W_j) \approx q^{-jd}$ for some $d > 0$. The presentation of this material is largely based on [53, 54].

In Subsections 3.1.2, 3.1.3, and 3.1.4, we review standard examples with trigonometric, linear finite element, and wavelet-type spaces V_j and W_j, respectively, where the assumptions of the abstract theory can be verified and the spaces B^s_ω are equivalent to Sobolev spaces. Further details and the discussion of the resulting algorithms can be found in Chapter 7.

3.1.1 Abstract theory

Following [53, 54], we assume for the collection $\mathbf{V}$ of nested subspaces V_j the existence of a sequence of uniformly bounded projections $Q_j : H \to H$ such that $\operatorname{ran}(Q_j) = V_j$ and

$$Q_j Q_l = Q_l Q_j = Q_j, \qquad 0 \le j \le l < \infty. \tag{3.1}$$

A standard example is provided by the sequence of ortho-projections $Q_j = Q_{V_j}$ onto the spaces V_j. This choice induces an orthogonal subspace splitting $\mathbf{W}$ of $(H; (\cdot,\cdot)_H)$ by setting

$$W_0 := V_0, \quad W_j := \operatorname{ran}(Q_j - Q_{j-1}), \qquad j = 1, 2, \ldots. \tag{3.2}$$

For general Q_j, the main goal of multiscale analysis, as developed in [53, 54], is to find criteria under which the subspace collection $\mathbf{W}$ defined in (3.2) is a Riesz splitting for $(H; (\cdot,\cdot)_H)$. For the definition of the latter, see Subsection 2.2.1.2. There is also an important result about space splittings associated with $\mathbf{V}$ which does not require knowledge of specific projections Q_j, see Part a) of Theorem 3.1 below.

Now, under the above conditions on $\mathbf{V}$ and Q_j, there is another biorthogonal subspace collection $\mathbf{W}^*$ given by

$$W_0^* = \operatorname{ran}(Q_0^*), \qquad W_j^* = \operatorname{ran}(Q_j^* - Q_{j-1}^*), \quad j = 1, 2, \ldots.$$

It is generated by the sequence of Hilbert adjoints Q_j^* defined by

$$(Q_j^* u, v)_H = (u, Q_j v)_H \qquad \forall\, u, v \in H,$$

and it induces another nested sequence of subspaces

$$V_0^* \subset V_1^* \subset \cdots \subset V_j^* \subset \cdots \subset H,$$

where $V_j^* := \mathrm{ran}(Q_j^*) = \sum_{l=0}^{j} W_l^*$. It is easy to prove (see [53, Section 2-3]) that the subspace collection $\mathbf{V}^*$ is also nested and dense in $(H; (\cdot,\cdot)_H)$. The projections Q_j^* are uniformly bounded and satisfy the counterpart of (3.1). Furthermore, $\mathbf{W}$ and $\mathbf{W}^*$ are biorthogonal. Indeed, let

$$\Delta_0 = Q_0, \qquad \Delta_j = Q_j - Q_{j-1}, \quad j = 1, 2, \ldots,$$

be the projections associated with $\mathbf{W}$. Then, for arbitrary $w_l \in W_l = \mathrm{ran}(\Delta_l)$ and $w_j^* \in W_j^* = \mathrm{ran}(\Delta_j^*)$, there are $u, v \in V$ such that $w_l = \Delta_l u$, $w_j^* = \Delta_j^* v$, and we have

$$(w_l, w_j^*)_H = (\Delta_l u, \Delta_j^* v)_H = (\Delta_j \Delta_l u, v)_H = \begin{cases} (w_j, w_j^*)_H, & l = j, \\ 0, & l \neq j, \end{cases}$$

since $\Delta_j \Delta_l = 0$, $l \neq j$, as follows from (3.1). This shows the biorthogonality of $\mathbf{W}$ and $\mathbf{W}^*$.

Next, we use the construction of the spaces $H_{\mathbf{W},\rho}$ and $H_{\mathbf{W}^*,\rho}$ from Subsection 2.4.1 with specific ρ to provide criteria for the Riesz basis property of $\mathbf{W}$ and, automatically, of $\mathbf{W}^*$. We follow [54, Section 5]: Fix $0 < q < 1$, let $s \in \mathbb{R}$, and define the inner product

$$(u, v)_{\mathbf{W},s} = \sum_{j=0}^{\infty} q^{-2sj} (\Delta_j u, \Delta_j v)_H, \qquad u, v \in \mathrm{span}(\mathbf{W}).$$

Denote by $A_{\mathbf{W}}^s$ the closure of $\mathrm{span}(\mathbf{W})$ with respect to the norm $\|.\|_{\mathbf{W},s}$ induced by this inner product. Obviously,

$$A_{\mathbf{W}}^s = H_{\mathbf{W},\mathbf{q}^s},$$

where the weight sequence $\mathbf{q}^s$ is given by $(\mathbf{q}^s)_j = q^{2sj}$, $j = 0, 1, \ldots$. If H is a separable Hilbert space, then $A_{\mathbf{W}}^s$ is a separable Hilbert space as well. Moreover, we have continuous embeddings

$$A_{\mathbf{W}}^t \subset A_{\mathbf{W}}^s \subset H, \quad t > s > 0,$$

where the first inclusion is obvious and the second inclusion follows from

$$\|u\|_H^2 = \| \sum_j q^{sj} (q^{-sj} \Delta_j u) \|_H^2 \leq (\sum_j q^{2js}) \|u\|_{\mathbf{W},s}^2 .$$

The spaces $A_{\mathbf{W}^*}^s$ are defined in a similar way and have analogous properties.

To proceed, we introduce approximation-theoretic assumptions for the nested sequence V_j: We call a function $\omega : \mathbb{R}_+ \times H \to \mathbb{R}_+$ a *modulus* on H, if it is bounded and subadditive in the second argument, i.e.,

$$\omega(t, u) \leq C \|u\|_H, \quad \omega(t, u + v) \leq \omega(t, u) + \omega(t, v), \qquad u, v \in H, \quad t > 0,$$

and satisfies

$$\omega(t', u) \geq \omega(t, u) \geq 0, \quad t' \geq t \geq 0, \qquad \lim_{t \to 0} \omega(t, u) = \omega(0, u) = 0,$$

for all $u \in H$. With constants uniform in $u \in H$, $v_n \in V_n$, and $n = 0, 1, \ldots$, let

$$\inf_{v_n \in V_n} \|u - v_n\|_H \leq C\omega(q^n, u), \qquad u \in H, \quad n = 0, 1, \ldots. \tag{3.3}$$

and, with some fixed $\gamma > 0$, let

$$\omega(t, v_n) \leq C \min(1, (tq^{-n})^\gamma)\|v_n\|_H, \quad t \geq 0, \quad v_n \in V_n, \quad n = 0, 1, \ldots. \tag{3.4}$$

Following tradition, we call (3.3) a Jackson-type inequality and (3.4) a Bernstein-type inequality in H associated with $\mathbf{V}$.

The classic example is the space $H = L^2(T)$ of periodic functions f on the univariate torus $T := (-\pi, \pi]$, where

$$\begin{aligned} w(t, f) &:= \sup_{0<\delta \leq t} \|f(x + \delta) - f(x)\|_{L^2} \\ &\approx \inf_{g \in H^1(T)} \|f - g\|_{L^2} + t\|g'\|_{L^2} =: \omega(t, f) \end{aligned}$$

are suitable choices for a modulus, if we define V_n as the set of trigonometric polynomials of degree $< 2^n$ and take $q = 1/2$ and $\gamma = 1$. This function $w(t, f)$ is the standard L^2 modulus of continuity, along with its equivalent expression $\omega(t, f)$ as the K-functional for the pair $(L^2(T); H^1(T))$.

In a similar spirit, we can also take a K-functional for the pair $(V; A^\gamma_{\mathbf{W}})$, $\gamma > 0$, as modulus, namely

$$\omega(t, u) := \inf_{v \in A^\gamma_{\mathbf{W}}} \|u - v\|_H + t^\gamma \|v\|_{\mathbf{W},\gamma}, \qquad u \in H.$$

Indeed, the Jackson-type inequality holds since, due to the uniform boundedness of the projection sequence, we have

$$\begin{aligned} \inf_{v_n \in V_n} \|u - v_n\|_H &\leq \|u - Q_n u\|_H \\ &\leq \inf_{v \in A^\gamma_{\mathbf{W}}} (\|u - v\|_H + \|v - Q_n v\|_H + \|Q_n(u - v)\|_H) \\ &\leq C \inf_{v \in A^\gamma_{\mathbf{W}}} (\|u - v\|_H + \|v - Q_n v\|_H) . \end{aligned} \tag{3.5}$$

In fact, by definition of the $A^\gamma_{\mathbf{W}}$ norm, the last term (squared) can be bounded as

$$\|v - Q_n v\|_H^2 \leq (\sum_{j>n} \|\Delta_j v\|_H)^2 \leq \left(\sum_{j>n} q^{2\gamma j}\right)\left(\sum_{j>n} q^{-2\gamma j}\|\Delta_j v\|_H^2\right) \leq Cq^{2\gamma n}\|v\|_{\mathbf{W},\gamma}^2.$$

Then substitute into (3.5) and take the infimum with respect to $v \in A^\gamma_{\mathbf{W}}$.

The Bernstein-type inequality (3.4) is obtained if, on one hand, we set $v = v_n$ in the definition of $\omega(t, v_n)$, which results in the estimate

$$\omega(t, v_n) \le t^\gamma \|v_n\|_{\mathbf{W},\gamma} = t^\gamma \left(\sum_{j\le n} q^{-2\gamma j} \|\Delta_j v_n\|_H^2 \right)^{1/2}$$
$$\le C \left(\sum_{j\le n} q^{-2\gamma j} \right)^{1/2} \|v_n\|_H \le C(tq^{-n})^\gamma \|v_n\|_H.$$

Here, the uniform boundedness of the projections Δ_j was used again. On the other hand, if we take $v = 0$, then $\omega(t, v_n) \le \|v_n\|_H$. Together, this gives the bound in (3.4).

We are now ready to formulate the first abstract result about multiscale space splittings related to $\mathbf{V}$. To this end, denote by $A^s_{\mathbf{V}} := H_{\mathbf{V},\mathbf{q}^s}$ the Hilbert space with norm

$$|||u|||^2_{\mathbf{V},s} := \inf_{v_j \in V_j : u=\sum_j v_j} \sum_{j=0}^\infty q^{-2sj} \|v_j\|_H^2,$$

as defined in Subsection 2.4.1. Moreover, define B^s_ω as the complete linear metric space obtained by completion of $\mathrm{span}(\mathbf{V})$ with respect to the metric $d(u, v) := \|u - v\|_{B^s_\omega}$, where

$$\|u\|^2_{B^s_\omega} := \|u\|^2_H + \sum_{j=0}^\infty q^{-2sj} \omega(q^j, u)^2.$$

Note that the definition of B^s_ω only involves the modulus ω and q, but not the subspace collection $\mathbf{V}$. It follows from the subadditivity of $\omega(t, \cdot)$ that $d(u, v)$ is a metric. Often, $\omega(t, u)$ is also positive homogeneous in the second argument, i.e.,

$$\omega(t, \lambda u) = |\lambda| \omega(t, u), \qquad t \in \mathbb{R}_+, \quad \lambda \in \mathbb{R}, \quad u \in H.$$

In this case, $\|\cdot\|_{B^s_\omega}$ is a norm and B^s_ω becomes a Banach space.

Theorem 3.1 *Let* $\mathbf{V}$ *be a nested dense sequence of subspaces of* H *for which the inequalities (3.3) and (3.4) hold for some modulus* ω *and some* $\gamma > 0$.
a) Then we have $A^s_{\mathbf{V}} = B^s_\omega$ *and*

$$|||u|||_{\mathbf{V},s} \approx \|u\|_{B^s_\omega}, \qquad u \in B^s_\omega, \quad 0 < s < \gamma. \tag{3.6}$$

b) If, in addition, uniformly bounded projections Q_j *with* $\mathrm{ran}(Q_j) = V_j$ *and property (3.1) are given, then we also have* $A^s_{\mathbf{W}} = B^s_\omega$ *and*

$$\|u\|_{\mathbf{W},s} \approx \|u\|_{B^s_\omega}, \qquad u \in B^s_\omega, \quad 0 < s < \gamma,$$

for the associated sequence $\mathbf{W}$ *of subspaces* $W_j = \mathrm{ran}(\Delta_j)$.

Proof For Part a), it is sufficient to establish (3.6) for $u \in \mathrm{span}(\mathbf{V})$ and to consider arbitrary finite decompositions $u = \sum_j v_j$ of such u. Then, by (3.4) and the properties of ω, we have

$$\omega(q^l, u) \le \sum_j \omega(q^l, v_j) \le C\left(\sum_{j\le l} q^{\gamma(l-j)}\|v_j\|_H + \sum_{j>l}\|v_j\|_H\right).$$

For $0 < \epsilon < \min(s, \gamma - s)$, we can continue with

$$\begin{aligned}\omega(q^l, u)^2 &\le C\left(\sum_{j\le l} q^{-\epsilon j}(q^{\gamma(l-j)+\epsilon j}\|v_j\|_H)\right)^2 + C\left(\sum_{j>l} q^{\epsilon j}(q^{-\epsilon j}\|v_j\|_H)\right)^2\\ &\le C\left(\sum_{j\le l} q^{-2\epsilon j}\right)\left(\sum_{j\le l} q^{2\gamma(l-j)+2\epsilon j}\|v_j\|_H^2\right)\\ &\qquad + C\left(\sum_{j>l} q^{2\epsilon j}\right)\left(\sum_{j>l} q^{-2\epsilon j}\|v_j\|_H^2\right)\\ &\le Cq^{-2\epsilon l}\left(\sum_{j\le l} q^{2\gamma(l-j)+2\epsilon j}\|v_j\|_H^2\right) + Cq^{2\epsilon l}\left(\sum_{j>l} q^{-2\epsilon j}\|v_j\|_H^2\right).\end{aligned}$$

Similarly,

$$\begin{aligned}\|u\|_H^2 &= \left(\sum_j q^{\epsilon j}(q^{-\epsilon j}\|v_j\|_H)\right)^2 \le C\left(\sum_j q^{2\epsilon j}\right)\left(\sum_j q^{-2\epsilon j}\|v_j\|_H^2\right)\\ &\le C\left(\sum_j q^{-2\epsilon j}\|v_j\|_H^2\right).\end{aligned}$$

Note again that the constants C are generic and can change from line to line. All that remains is to substitute the last two bounds into the definition of $\|u\|_{B^s_\omega}^2$, change the order of summation and take the infimum with respect to all representations of u. This proves $\|u\|_{B^s_\omega} \le C|||u|||_{\mathbf{V},s}$.

The opposite direction is easy: Set $v_0 = Q_{V_0}u$ and $v_j = Q_{V_j}u - Q_{V_{j-1}}u$, $j = 1, 2, \dots$. Then $u = \sum_j v_j$ is a finite decomposition of u and, by (3.3), we get

$$\|v_j\|_H^2 \le 2(\|u - Q_{V_j}u\|_H^2 + \|u - Q_{V_{j-1}}u\|_H^2) \le C(\omega(q^j, u)^2 + \omega(q^{j-1}, u)^2)$$

for $j = 1, 2 \dots$, and, similarly, $\|v_0\|_H^2 \le C(\|u\|_H^2 + \omega(1, u)^2)$ for $j = 0$. This gives us the desired estimate

$$|||u|||_{\mathbf{V},s}^2 \le \sum_j q^{2sj}\|v_j\|_H^2 \le C\left(\|u\|_H^2 + \sum_j q^{2sj}\omega(q^j, u)^2\right) = C\|u\|_{B^s_\omega}^2.$$

For Part b), it is sufficient to show that

$$\|u\|_{\mathbf{W},s} \approx |||u|||_{\mathbf{V},s}, \qquad u \in A^s_{\mathbf{V}}, \qquad s > 0, \tag{3.7}$$

holds for all finitely represented $u = \sum_j v_j \in \mathrm{span}(\mathbf{V}) = \mathrm{span}(\mathbf{W})$. The argument is similar to the proof of Lemma 2.9 in Subsection 2.4.1: Since $W_j \subset V_j$, the inequality $|||u|||_{\mathbf{V},s} \le \|u\|_{\mathbf{W},s}$ is trivially satisfied. For the other direction, recall that $w_l = \Delta_l u = \Delta_l(\sum_{j\ge l} v_j)$ and that the projections Δ_l are uniformly bounded. As in the proof for Part a), for $0 < \epsilon < s$, we therefore have

$$\|w_l\|_H^2 \le C\left(\sum_{j\ge l} \|v_j\|_H\right)^2 \le Cq^{2\epsilon l} \sum_{j\ge l} q^{-2\epsilon j}\|v_j\|_H^2.$$

After substitution, change of summation, and taking the infimum with respect to arbitrary decompositions of u, the desired opposite inequality is proven. Then (3.7) follows from (3.6). The remaining statements are obvious by the completion procedure. □

Theorem 3.1 falls short of proving that $\mathbf{W}$ is a Riesz splitting for H. It is clear that additional assumptions are needed, as demonstrated by examples of Schauder bases in H that are not Riesz bases, see [53]. However, if the biorthogonal collection $\mathbf{W}^*$ satisfies Jackson- and Bernstein-type inequalities with the same modulus ω and the same q, but possibly with another value $\gamma^* > 0$ instead of γ, then we can employ duality arguments with respect to the duality pairing $(\cdot,\cdot)_H$. The setup is as follows: Let U be a complete linear metric space with metric induced by a subadditive functional $u \longrightarrow \|u\|_U$. Assume that there exists a set $X \subset H \cap U$ which is dense in both U and H. Then we define U' as the completion of the set $\{u \in X : \|u\|_{U'} < \infty\}$ under the norm

$$\|u\|_{U'} := \sup_{v\in X:\, \|v\|_U\le 1} (v,u)_H, \qquad u \in X.$$

If $\|\cdot\|_U$ is a norm of a Banach space U continuously embedded into H, then $\|\cdot\|_{U'}$ is a norm and $U' \supset H$ is a Banach space as well. Below, U will be one of the spaces $A^s_{\mathbf{W}}$, $A^s_{\mathbf{W}^*}$ or B^s_ω and $X = \mathrm{span}(\mathbf{W}) = \mathrm{span}(\mathbf{V})$ or $X = \mathrm{span}(\mathbf{W}^*) = \mathrm{span}(\mathbf{V}^*)$. Moreover, in these and other concrete examples of Hilbert spaces U considered in this book, the definition of U' by duality with respect to $(\cdot,\cdot)_H$ is isometric to the definition of U' as a space of continuous linear functionals, which justifies the notation.

With this at hand, we can formulate the following results for the subspace collections $\mathbf{W}$ and $\mathbf{W}^*$ which were originally proven in [53]:

Theorem 3.2 *Assume that the sequence of projections Q_j is uniformly bounded in H and satisfies (3.1). Furthermore, let the collection $\mathbf{V}$ of nested subspaces $V_j = \mathrm{ran}(Q_j)$ be dense in H. If, for some modulus ω and some $0 < q < 1$, both $\mathbf{V}$ and $\mathbf{V}^*$ satisfy the Jackson-type inequalities (3.3) and the Bernstein-type inequalities (3.4) with parameters $\gamma > 0$ and $\gamma^* > 0$, respectively, then $A^0_{\mathbf{W}} = A^0_{\mathbf{W}^*} = H$ with equivalent norms*

$$\|u\|_H^2 \approx \|u\|_{\mathbf{W},0}^2 = \sum_{j=0}^{\infty} \|\Delta_j u\|_H^2 \approx \|u\|_{\mathbf{W}^*,0}^2 = \sum_{j=0}^{\infty} \|\Delta_j^* u\|_H^2, \quad u \in H. \tag{3.8}$$

Furthermore, $(A_{\mathbf{W}}^s)' = A_{\mathbf{W}^*}^{-s}$ *and* $(A_{\mathbf{W}^*}^s)' = A_{\mathbf{W}}^{-s}$ *with norm equivalences*

$$\|u\|_{(A_{\mathbf{W}}^s)'} \approx \|u\|_{\mathbf{W}^*,-s}, \quad u \in A_{\mathbf{W}^*}^{-s}, \qquad \|v\|_{(A_{\mathbf{W}^*}^s)'} \approx \|v\|_{\mathbf{W},-s}, \quad v \in A_{\mathbf{W}}^{-s}, \tag{3.9}$$

for all $s \in \mathbb{R}$. *In particular, for* $0 < s < \gamma$, *we have* $(B_\omega^s)' = (A_{\mathbf{W}}^s)' = A_{\mathbf{W}^*}^{-s}$ *and*

$$\|u\|_{(B_\omega^s)'} \approx \|u\|_{\mathbf{W}^*,-s}, \qquad u \in (B_\omega^s)', \tag{3.10}$$

while, for $0 < s < \gamma^*$, *we have* $(B_\omega^s)' = (A_{\mathbf{W}^*}^s)' = A_{\mathbf{W}}^{-s}$ *and*

$$\|u\|_{(B_\omega^s)'} \approx \|u\|_{\mathbf{W},-s}, \qquad u \in (B_\omega^s)'. \tag{3.11}$$

Before proving Theorem 3.2, we formulate a corollary which, together with the result for $A_{\mathbf{V}}^s$ in Part a) of Theorem 3.1, is at the heart of applications to multilevel solvers for elliptic PDE problems in Sobolev spaces. The case $s > 0$ in Corollary 3.1 is covered by Theorem 3.1, while the case $s \leq 0$ follows from Theorem 3.2.

Corollary 3.1 *Under the conditions of Theorem 3.2, we have*

$$A_{\mathbf{W}}^s = \begin{cases} B_\omega^s, & 0 < s < \gamma, \\ H, & s = 0, \\ (B_\omega^{-s})', & -\gamma^* < s < 0, \end{cases} \quad \text{and} \quad A_{\mathbf{W}^*}^s = \begin{cases} B_\omega^s, & 0 < s < \gamma^*, \\ H, & s = 0, \\ (B_\omega^{-s})', & -\gamma < s < 0. \end{cases}$$

In all cases, the corresponding norm equivalences hold.

Proof of Theorem 3.2. We show only the essential steps, and refer to [53] for further details. First, we consider (3.9). For $u \in \mathrm{span}(\mathbf{V})$, $v \in \mathrm{span}(\mathbf{V}^*)$ and arbitrary $s \in \mathbb{R}$, we have

$$(u, v)_H = \sum_j (\Delta_j u, \Delta_j^* v)_H \leq \sum_j (q^{js} \|\Delta_j u\|_H)(q^{-js} \|\Delta_j^* v\|_H) \leq \|u\|_{\mathbf{W},s} \|v\|_{\mathbf{W}^*,-s}$$

by using biorthogonality. According to our definition of dual norms, we see that

$$\|u\|_{(A_{\mathbf{W}^*}^{-s})'} \leq \|u\|_{\mathbf{W},s}, \qquad \|v\|_{(A_{\mathbf{W}}^s)'} \leq \|v\|_{\mathbf{W}^*,-s}, \qquad s \in \mathbb{R}.$$

By completion, these inequalities hold for all $u \in A_{\mathbf{W}}^s$, $v \in A_{\mathbf{W}^*}^{-s}$, and all $s \in \mathbb{R}$. This shows one direction in (3.9).

The converse inequalities involving constants have been established in [53] using interpolation theory. Here is the direct argument: Obviously, every $u \in \mathrm{span}(\mathbf{V})$ belongs to $(A_{\mathbf{W}^*}^{-s})'$ by the previous estimate, since the summation in

$$\|u\|_{\mathbf{W},s}^2 = \sum_j q^{-2sj} \|\Delta_j u\|_H^2 = \sum_j q^{-2sj} (u, \Delta_j^* \Delta_j u)_H = (u, \bar{v})_H$$

is finite. Here, we have set $\bar{v} = \sum_j q^{-2sj}\Delta_j^*\Delta_j u$. Then, due to the uniform boundedness of the projections Δ_j^*, we have

$$\|\bar{v}\|^2_{\mathbf{W}^*,-s} = \sum_j q^{2sj}q^{-4sj}\|\Delta_j^*\Delta_j u\|_H^2 \le C\sum_j q^{-2sj}\|\Delta_j u\|_H^2 = C\|u\|^2_{\mathbf{W},s}.$$

Consequently,

$$\|u\|^2_{(A^{-s}_{\mathbf{W}^*})'} = \sup_{\|v\|_{A^{-s}_{\mathbf{W}^*}}\le 1} (u,v)_H^2 \ge \frac{(u,\bar{v})_H^2}{\|\bar{v}\|^2_{\mathbf{W}^*,-s}} \ge \frac{\|u\|^4_{\mathbf{W},s}}{C\|u\|^2_{\mathbf{W},s}} = c\|u\|^2_{\mathbf{W},s}.$$

The converse inequality $\|u\|_{(A^{-s}_{\mathbf{W}})'} \ge C\|u\|_{\mathbf{W}^*,s}$ follows in a similar way.

With (3.9) at hand, (3.10) follows from $\|u\|_{B^s_\omega} \approx \|u\|_{\mathbf{W},s}$, which was established in Theorem 3.1 for $0 < s < \gamma$, and from the definition of dual norms. Statement (3.11) follows in a similar way.

It remains to establish the Riesz property of $\mathbf{W}$ and $\mathbf{W}^*$ expressed by (3.8). Consider an arbitrary $u \in \mathrm{span}(\mathbf{V})$ and its finite decomposition $u = \sum_j \Delta_j u$ with respect to $\mathbf{W}$. Then we have

$$\|u\|_H^2 = \sum_j\sum_l (\Delta_j u, \Delta_l u)_H,$$

where each term will be estimated by duality with $U = A^s_{\mathbf{W}}$, $0 < s < \min(\gamma,\gamma^*)$. In fact, for this range of s, the dual norms for $A^s_{\mathbf{W}}$, B^s_ω, $A^s_{\mathbf{W}^*}$ are equivalent due to (3.10) and (3.11). So (3.9) implies

$$\|u\|_{(A^s_{\mathbf{W}})'} \approx \|u\|_{\mathbf{W},-s} \approx \|u\|_{(A^s_{\mathbf{W}^*})'} \approx \|u\|_{\mathbf{W}^*,-s}, \qquad 0 < s < \min(\gamma,\gamma^*).$$

Now we are ready to proceed as follows: We have

$$\begin{aligned}(\Delta_j u, \Delta_l u)_H &\le \|\Delta_{\min(j,l)}u\|_{\mathbf{W},s}\|\Delta_{\max(j,l)}u\|_{(A^s_{\mathbf{W}})'}\\ &\le C\|\Delta_{\min(j,l)}u\|_{\mathbf{W},s}\|\Delta_{\max(j,l)}u\|_{\mathbf{W},-s}\\ &= Cq^{s(\max(j,l)-\min(j,l))}\|\Delta_{\min(j,l)}u\|_H\|\Delta_{\max(j,l)}u\|_H\\ &= Cq^{s|j-l|}\|\Delta_j u\|_H\|\Delta_l u\|_H.\end{aligned}$$

Here, we used the obvious fact that $\|w_j\|_{\mathbf{W},t} = q^{-tj}\|w_j\|_H$ for any $w_j \in W_j$ and $t \in \mathbb{R}$. After substitution we get the continuous embedding $A^0_{\mathbf{W}} \subset H$, since

$$\begin{aligned}\|u\|_H^2 &\le C\sum_j\sum_l q^{s|j-l|}\|\Delta_j u\|_H\|\Delta_l u\|_H\\ &\le C\sum_{r=0}^{\infty} q^{sr}\sum_j \|\Delta_j u\|_H(\|\Delta_{j+r}u\|_H + \|\Delta_{j-r}u\|_H)\end{aligned}$$

$$\leq C(\sum_{r=0}^{\infty} q^{sr}) \sum_j \|\Delta_j u\|_H^2 \leq C\|u\|_{\mathbf{W},0}^2.$$

Due to the symmetry of the assumptions, an analogous result holds for $\mathbf{W}^*$.

Consequently, together with (3.9) for $s = 0$ and standard duality arguments, we have established the continuous embeddings

$$A_{\mathbf{W}}^0 \subset H \subset (A_{\mathbf{W}}^0)' = A_{\mathbf{W}^*}^0, \qquad A_{\mathbf{W}^*}^0 \subset H \subset (A_{\mathbf{W}^*}^0)' = A_{\mathbf{W}}^0.$$

This finally gives $A_{\mathbf{W}}^0 = A_{\mathbf{W}^*}^0 = H$, together with the norm equivalence (3.8), and completes the proof of Theorem 3.2. □

The abstract results above need further specification before they become relevant for the computational solution of operator equations. First, the Besov-type spaces B_ω^s need to be connected with Hilbert spaces $(V; a)$ that appear in applications. For example, if $H = L^2(G)$, where G is a suitable bounded Lipschitz domain in $\mathbb{R}^d$ and the V_j are standard hierarchies of discretization spaces for elliptic PDE problems based on finite elements, wavelets, or spectral decompositions, then these spaces allow characterizations in terms of Sobolev spaces $H^s(G)$. This leads to stable space splittings for $H^s(G)$ via Part b) of Lemma 2.8. Moreover, since $\dim(V_j) \to \infty$ in the above setting, these target spaces $(V; a)$ are infinite-dimensional, while computational processes can only access finite-dimensional parts of them. The finite-dimensional subspaces V_j used as examples in the following subsections are standard spaces of trigonometric functions, finite element functions, and spline wavelets. For implementations, it is also necessary to choose appropriate bases in the finite-dimensional subspaces V_j and W_j, and to consider the complexity of the resulting matrix representations of the embeddings R_i and the operators A_i. We will be brief on these issues and refer to Chapter 7 for further discussions in the context of PDE applications.

3.1.2 Example: Multivariate Fourier series

A simple but instructive example is the multivariate trigonometric Fourier series in $H = L^2(T^d)$. For simplicity, we again identify T with the periodic interval $(-\pi, \pi]$ and use the complex-valued trigonometric system $F := \{e^{i(k,x)_2}\}_{k\in\mathbb{Z}^d}$ with argument $x = (x_1, \ldots, x_d) \in T^d$ and exponent $(k, x)_2 = k_1x_1 + \cdots + k_dx_d$ as well as complex Fourier coefficients $c_k(f) = (2\pi)^{-d} \int_{T^d} f(x)e^{-i(k,x)_2}\,dx$ to represent real-valued periodic functions $f \in L^2(T^d)$. Equivalently, this requires that the coefficients $c_k(f)$ and $c_{-k}(f)$ are complex conjugate to each other which is silently assumed in all series representations below.

There are several classical ways to define subspaces V_j of trigonometric polynomials by degree restrictions. We mention two of them related to the summation of Fourier series with respect to cubes and hyperbolic crosses, respectively. In the first case we set

$$V_j := \operatorname{span}(\{e^{i(k,x)_2} : |k|_\infty < 2^j,\ k \in \mathbb{Z}^d\}) \cap L^2(T^d) \tag{3.12}$$

for $j = 0, 1, \dots$. Since $\dim(V_j) = (2^{j+1} - 1)^d \approx 2^{jd}$, we set $q = 1/2$. Since the system F is orthogonal in $L^2(T^d)$, we can define

$$W_j := V_j \ominus V_{j-1} = \operatorname{span}(\{e^{i(k,x)_2} : 2^{j-1} \le |k|_\infty < 2^j,\ k \in \mathbb{Z}^d\}) \cap L^2(T^d)$$

for $j = 1, 2, \dots$. This corresponds to the choice

$$Q_j = Q_{V_j} = Q_j^*$$

of L^2 ortho-projections on V_j and implies

$$\mathbf{V} = \mathbf{V}^* \quad \text{and} \quad \mathbf{W} = \mathbf{W}^*.$$

Although the Riesz property in $L^2(T^d)$ for the orthogonal space splitting associated with $\mathbf{W} = \mathbf{W}^*$ and the relation of the spaces $A^s_{\mathbf{V}}$ and $A^s_{\mathbf{W}}$ to the Sobolev spaces $H^s(T^d)$ are obvious for this example, it is still instructive to see what the above abstract theory tells us.

The norm in Sobolev spaces $H^s(T^d)$ can be defined in several, equivalent ways. For our purposes, we use the definition by isometry to some weighted $\ell^2(\mathbb{Z}^d)$ space in terms of Fourier coefficients. More precisely, $H^s(T^d)$, $s \ge 0$, is the set of all $f \in L^2(T^d)$ with coefficients in their Fourier series

$$f(x) = \sum_{k \in \mathbb{Z}^d} c_k(f) e^{i(k,x)_2}$$

so that

$$\|f\|^2_{H^s} = \sum_{k \in \mathbb{Z}^d} (1 + |k|_\infty)^{2s} |c_k(f)|^2 < \infty.$$

For $s < 0$, $H^s(T^d)$ can be defined as the dual space $H^{-s}(T^d)'$ or by completion of $L^2(T^d)$ with respect to the above norm. In other common definitions, the weights $(1+|k|_\infty)^{2s}$ are replaced by $\max(1, |k|^{2s}_\infty)$ or $(1+k_1^2+\dots+k_d^2)^s$. The latter is closest to the spherical summation of Fourier series and the definition in terms of powers of the Laplace operator. Obviously, the classical definition of $H^s(T^d)$ as 2π periodic functions, first for integer $s = m \in \mathbb{N}$ using $L^2(T^d)$ norms of derivatives of order $\le m$, then by complex interpolation for all $s \ge 0$ and finally by duality for $s < 0$, leads to the same space, albeit with different equivalent norms. In all cases we have $H = L^2(T^d) = H^0(T^d)$, $H^s(T^d) \subset L^2(T^d)$ for $s \ge 0$, and $(H^s(T^d))' = H^{-s}(T^d)$ for all $s \in \mathbb{R}$.

To apply Theorems 3.1 and 3.2, we need to introduce an appropriate modulus ω. The following two families of moduli are standard: The first is the K-functional

$$\omega_m(t, f) := \inf_{g \in H^m(T^d)} \|f - g\|_{L^2} + t^m |g|_{H^m}, \qquad t \ge 0, \quad f \in L^2(T^d),$$

for the pair $(L^2(T^d); H^m(T^d))$, where $|g|_{H^m}^2 := \sum_{k\in\mathbb{Z}^d} |k|_\infty^{2m}|c_k(g)|^2$. The other is given by the so-called m-th order L^2 modulus of continuity

$$w_m(t,f)_{L^2} = \sup_{\|h\|_2\le t} \|\Delta_h^m f\|_{L^2}, \qquad t\ge 0, \quad f\in L^2(T^d).$$

Here $\Delta_h^m = (\Delta_h)^m$ denotes the multivariate finite difference operator of order m with step-size $h\in\mathbb{R}^d$, where $(\Delta_h f)(\cdot) = f(\cdot+h) - f(\cdot)$. These two families of moduli are equivalent, i.e., we have

$$\omega_m(t,f) \approx w_m(t,f)_{L^2}, \qquad t\ge 0, \quad f\in L^2(T^d), \tag{3.13}$$

with constants depending on m. The proof of (3.13) is left as an exercise in Problem 3.1. In the following we use ω_m. Since in our example $\mathbf{V} = \mathbf{V}^*$ and $\mathbf{W} = \mathbf{W}^*$, we only need to check Jackson- and Bernstein-type inequalities for $\mathbf{V}$. If we set $g = v_n \in V_n$ in the definition of ω_m, we get

$$\begin{aligned}\omega_m(t,v_n)^2 \le t^{2m}|v_n|_{H^m}^2 &= t^{2m} \sum_{k\in\mathbb{Z}^d:\,|k|_\infty<2^n} |k|_\infty^{2m}|c_k(v_n)|^2 \\ &\le (t2^n)^{2m}\|v_n\|_{L^2}^2,\end{aligned}$$

while choosing $g=0$ simply gives $\omega_m(t,v_n) \le \|v_n\|_{L^2}$. So (3.4) holds with $\gamma = m$ and $q = 1/2$. As for (3.3), we can see that

$$\inf_{v_n\in V_n} \|f - v_n\|_{L^2}^2 = \|f - Q_{V_n}f\|_{L^2}^2, \qquad f\in L^2(T^d),$$

and that

$$\|g - Q_{V_n}g\|_{L^2}^2 = \sum_{k\in\mathbb{Z}^d:\,|k|_\infty\ge 2^n} |c_k(g)|^2 \le 2^{-2mn}|g|_{H^m}^2$$

for every $g\in H^m(T^d)$. So we get

$$\begin{aligned}\|f - Q_{V_n}f\|_{L^2} &\le \|f-g\|_{L^2} + \|g - Q_{V_n}g\|_{L^2} + \|Q_{V_n}(f-g)\|_{L^2} \\ &\le 2\|f-g\|_{L^2} + 2^{-nm}|g|_{H^m}.\end{aligned}$$

If we now take the infimum with respect to $g\in H^m(T^d)$, we have shown (3.3).

It is a basic fact of interpolation theory that the spaces $B_{\omega_m}^s$ for $0 < s < m$ coincide with $H^s(T^d)$ up to equivalent norms. The elementary proof is left to the reader as Problem 3.2. With this, Theorem 3.1 and Corollary 3.1 yield the following result:

Corollary 3.2 *For the Sobolev spaces $H^s(T^d)$ and $\mathbf{V}$ defined in (3.12), we have*

$$A_{\mathbf{V}}^s = H^s(T^d), \quad 0 < s < \infty, \qquad A_{\mathbf{W}}^s = H^s(T^d), \quad -\infty < s < \infty,$$

with equivalent norms when V_j is given by (3.12), $Q_j = Q_{V_j}$, and $q = 1/2$.

Similar considerations apply to the second family of subspaces V_j, the so-called hyperbolic cross spaces

$$V_j = \mathrm{span}(\{e^{i(k,x)_2} : \prod_{l=1}^{d} \max(1,|k_l|) \le 2^j,\ k \in \mathbb{Z}^d\}) \cap L^2(T^d), \qquad (3.14)$$

$j = 0, 1, \ldots$, but lead to another family of Sobolev spaces $H^s_{\mathrm{mix}}(T^d)$. Compared to the spaces in (3.12), the dimension of V_j grows less rapidly. One can easily verify that $\dim(V_j) \approx 2^j(j+1)^{d-1}$, $j = 0, 1, \ldots$ (with constants depending on d). We again set $q = 1/2$, and take as modulus

$$\omega_{m,\mathrm{mix}}(t, f) := \inf_g \|f - g\|_{L^2} + t^m \|D^m_{\mathrm{mix}} g\|_{L^2}, \qquad m = 1, 2, \ldots.$$

Here the infimum is taken with respect to all g with $D^m_{\mathrm{mix}} g \in L^2(T^d)$, where D^m_{mix} is shorthand for the mixed derivative $\partial^m_{x_1} \cdots \partial^m_{x_d}$.

The subspaces (3.14) satisfy Jackson- and Bernstein-type inequalities with respect to the modulus $\omega_{m,\mathrm{mix}}(t, f)$ with $\gamma = m$. Repeating the same argument that led to Corollary 3.2 for the family (3.12), we obtain the following characterization of the Sobolev spaces $H^s_{\mathrm{mix}}(T^d)$ of mixed smoothness, where the norm in $H^s_{\mathrm{mix}}(T^d)$ is conveniently defined in terms of the coefficients $c_k(f)$ by

$$\|f\|^2_{H^s_{\mathrm{mix}}} = \sum_{k \in \mathbb{Z}^d} \left(\prod_{l=1}^{d} \max(1, |k_l|) \right)^{2s} |c_k(f)|^2, \qquad s \ge 0,$$

with similar modifications for $s < 0$.

Corollary 3.3 *For Sobolev spaces $H^s_{\mathrm{mix}}(T^d)$, we have*

$$A^s_{\mathbf{V}} = H^s_{\mathrm{mix}}(T^d), \quad 0 < s < \infty, \qquad A^s_{\mathbf{W}} = H^s_{\mathrm{mix}}(T^d), \quad -\infty < s < \infty,$$

with equivalent norms when the subspaces V_j are given by (3.14), $Q_j = Q_{V_j}$, and $q = 1/2$.

A quick comparison of the weighted ℓ^2 coefficient norm expressions for $H^s_{\mathrm{mix}}(T^d)$ and $H^s(T^d)$ shows that we have strictly continuous embeddings

$$H^{ds}(T^d) \subset H^s_{\mathrm{mix}}(T^d) \subset H^s(T^d), \qquad s > 0,$$

with the corresponding counterparts for $s < 0$ obtained by duality. In applications, the use of hyperbolic cross subspaces V_j and W_j instead of the subspaces defined in (3.12) is often motivated by their reduced dimension, which depends only weakly on d. However, this has to be weighed against the additional smoothness requirements expressed by the above embeddings. Analogs of hyperbolic cross spaces can also be defined for other approximation schemes, see Subsection 7.3.2 for the finite element case.

3.1.3 Example: Finite elements

Among the many solution methods for elliptic and parabolic PDEs, finite element discretizations occupy a prominent place. Their popularity among engineers is due in part to the fact that they are typically given by variational problems in finite element subspaces of Sobolev spaces on domains $G \subset \mathbb{R}^d$, which inherit energy minimization properties from the continuous formulation of the PDE. This fits right in with the theme of this book.

In this subsection, we present results on multiscale splittings for Sobolev spaces $H^s(G)$ based on sequences of nested finite element spaces V_j and adhere closely to the abstract theory of Subsection 3.1.1. To ensure clarity, we will only provide full proofs for specific cases, such as C^0 Lagrange finite elements on dyadically refined partition sequences. For information on the definition and basic properties of Sobolev spaces used throughout this chapter, see Section A.5 in the appendix.

Let G be a bounded polyhedral Lipschitz domain in $\mathbb{R}^d$. Further assume that G is equipped with a sequence of quasi-uniform and regular partitions

$$\mathcal{T}_0 \prec \mathcal{T}_1 \prec \cdots \prec \mathcal{T}_j \prec \cdots \tag{3.15}$$

with cell diameter $h_j := \mathrm{diam}(\mathcal{T}_j) \approx 2^{-j}$ and cell count $n_j := |\mathcal{T}_j| \approx 2^{jd}$. Let us briefly explain the terminology routinely used in the finite element context: A partition $\mathcal{T}$ is a finite collection of closed polyhedral cells C_i, $i = 1, \ldots, |\mathcal{T}|$, such that

(a) their union is the closure of G,
(b) their interiors are non-empty and mutually disjoint, and
(c) $C_i \cap C_{i'}$, $i \neq i'$, is either empty or a common lower-dimensional boundary complex (vertex, edge, face, etc.) of the two cells.

We define

$$\mathrm{diam}(C_i) = \max_{x,y \in C_i} \|x - y\|_2, \qquad \mathrm{diam}(\mathcal{T}) = \max_i \mathrm{diam}(C_i),$$

and call $\mathcal{T}$ quasi-uniform with constant $0 < \beta \leq 1$ if

$$\mathrm{diam}(\mathcal{T}) \geq \mathrm{diam}(C_i) \geq \beta\, \mathrm{diam}(\mathcal{T}) \qquad \forall\, C_i \in \mathcal{T}.$$

Furthermore, $\mathcal{T}$ is called regular (with constant $\beta' > 0$) if the ratio of the radii of the largest inscribed ball and the smallest circumscribed ball is not less than β' for any cell in $\mathcal{T}$. For the partitions $\mathcal{T}_j$ we assume that the quasi-uniformity and regularity constants β and β' can be chosen independently of j. Finally, $\mathcal{T} \prec \mathcal{T}'$ means that each cell $C_i \in \mathcal{T}$ is the (finite) union of cells $C'_l \in \mathcal{T}'$. Figure 3.1 illustrates these definitions.

Although there are exotic partition sequences that satisfy these conditions, the common cases are partition sequences consisting of simplices (typically for domains in $\mathbb{R}^2$ or $\mathbb{R}^3$) and partition sequences consisting of d-dimensional rectangles (typically

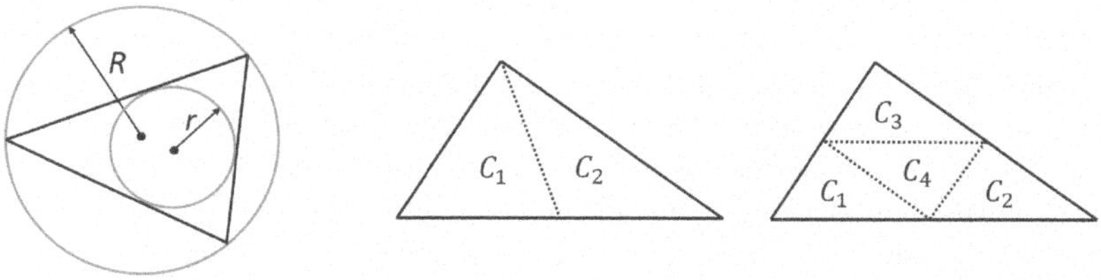

Fig. 3.1 Left: Regularity condition $R/r \le \beta'$. Right: Longest edge bisection and quadrisection of one triangle C.

for domains $G \subset \mathbb{R}^d$ composed of d-dimensional rectangles) obtained by some kind of dyadic refinement. The most common refinement schemes for simplicial partitions are newest vertex bisection (see [129, Section 4], [172]), conforming longest edge bisection [110] and standard quadrisection (subdivision of a simplex into 2^d congruent subsimplices). Figure 3.2 shows some steps of refinement for the case $d = 2$.

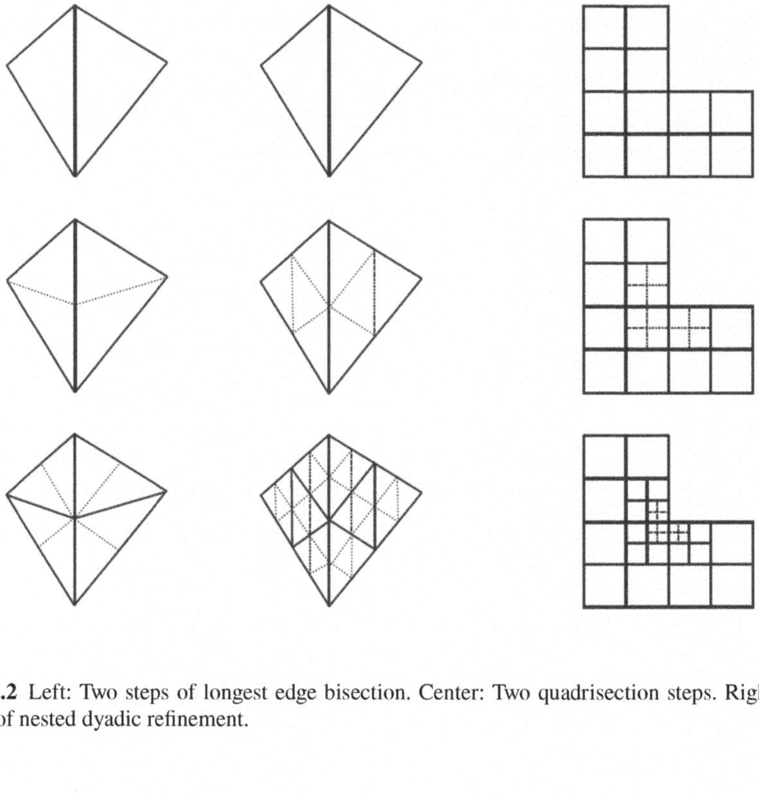

Fig. 3.2 Left: Two steps of longest edge bisection. Center: Two quadrisection steps. Right: Two steps of nested dyadic refinement.

Note that to satisfy the above assumptions, in the case of bisection, several single steps of bisecting triangles are needed to get to the next $\mathcal{T}_j$. These schemes have counterparts for simplicial partitions of domains $G \subset \mathbb{R}^d$. The last graph in Figure 3.2 shows a sequence of rectangular partitions of an L-shaped domain in $\mathbb{R}^2$ obtained by *nested dyadic refinement* based on quadrisection, which does not satisfy property (c) (the boundaries of some intersecting cells share only part of the edge of one of the cells) and also violates the quasi-uniformity assumption. However, such partition sequences arise naturally in adaptive finite element methods. We will show in Section 7.1 how to modify the following constructions to handle such cases.

Given a partition sequence (3.15), finite element subspaces $V_j := V(\mathcal{T}_j)$ are defined as piecewise polynomial functions on each cell with certain smoothness requirements over the cell boundaries. Traditionally, the definition of $V(\mathcal{T})$ is cast in terms of local polynomial interpolation problems on each cell, with interpolation conditions at the cell boundaries defined in such a way that the desired global smoothness properties of finite element functions are automatically satisfied. Many options are available. The most common case is so-called C^0 Lagrange finite elements of degree $p = 1, 2, \ldots$, where $V(\mathcal{T})$ consists of all globally continuous functions whose restrictions to each cell are polynomials of degree $\leq p$. More precisely, for simplicial partitions the total degree $\leq p$ is assumed, while for rectangular partitions polynomials of coordinate-wise degree $\leq p$ are allowed. With p fixed, the nestedness property $V(\mathcal{T}) \subset V(\mathcal{T}')$ is obvious whenever $\mathcal{T} \prec \mathcal{T}'$. The associated local interpolation problem returns a unique polynomial of the respective degree for given function values (called *nodal values*) at a set of *nodal points* uniformly distributed in a cell and on its boundary. Figure 3.3 shows the nodal points for $p = 1, 2, 3$ and $d = 2$ for triangular and rectangular cells. By construction of the local nodal interpolation problems, the global continuity of the locally defined piecewise polynomial function is automatically guaranteed.

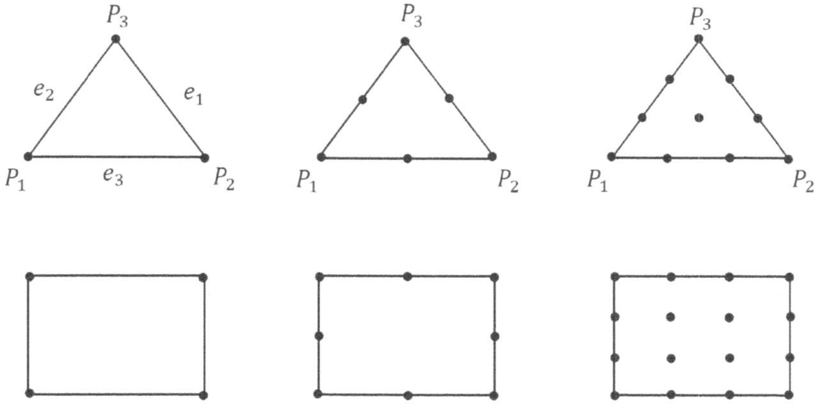

Fig. 3.3 Nodal points for $p = 1, 2, 3$ (from left to right) and $d = 2$. Top: Triangles. Bottom: Rectangles.

Also, for each nodal point P a so-called finite element *nodal basis function* $N_P(x)$ is defined by cardinal interpolation (i.e., $N_P(P') = 0$ for nodal points $P' \neq P$ and $N_P(P) = 1$), with the cells sharing P as its local support. These nodal basis functions can be assembled from a finite number of fixed shape functions associated with a standard d-dimensional simplex or unit cube by affine-linear transformations. This, together with the assumption of quasi-uniformity and regularity, ensures that the sets of nodal basis functions $N_{P,j}$ associated with $\mathcal{T}_j$ form L^2 stable bases for the C^0 Lagrange finite element spaces V_j of degree p. More precisely, one has $||N_{P,j}||_{L^2}^2 \approx h_j^d \approx 2^{-jd}$ and

$$||\sum_P c_{P,j} N_{P,j}||_{L^2}^2 \approx 2^{-jd} \sum_{P\in\mathcal{V}_j} c_{P,j}^2, \tag{3.16}$$

where $\mathcal{V}_j$ denotes the set of all nodal points P associated with $\mathcal{T}_j$. The constants in the two-sided bound (3.16) are independent of the coefficients $c_{P,j}$ and $j = 0, 1, \ldots,$ but depend on p and on the characteristics of the refinement process, and in particular on the constants in the quasi-uniformity and regularity assumption. Note that (3.16) can be interpreted as a uniform stability statement for splittings for $(V_j; (\cdot,\cdot)_{L^2})$ into one-dimensional subspaces generated by the nodal basis functions $N_{P,j}$ associated with $\mathcal{T}_j$. This has consequences that we will discuss later.

As we will see, C^0 Lagrange finite element spaces over a domain G are suitable for handling $H^s(G)$-elliptic problems with smoothness parameter $0 < s < 3/2$ and, after some modifications, also with larger range $-3/2 < s < 3/2$. This includes second-order elliptic PDE boundary value problems for which subspaces of $H^1(G)$ reflecting the type of boundary conditions are relevant. For example, the Poisson equation with homogeneous Dirichlet boundary data

$$-\Delta u = f \text{ in } G, \qquad u = 0 \text{ on } \Gamma := \partial G,$$

can be formulated as a variational problem in

$$V = H_0^1(G) := \{u \in H^1(G) : u = 0 \text{ on } \Gamma\},$$

with the bilinear form

$$a(u, v) = \int_G (\nabla u(x), \nabla v(x))_2 \, dx, \qquad u, v \in H_0^1(G),$$

and norm $||u||_{H_0^1} := |u|_{H^1} = \sqrt{a(u,u)}$. Here, the right-hand side f can formally be any element of $(H_0^1(G))' = H^{-1}(G)$. Similarly, the Neumann problem can be treated in $H^1(G)$.

Boundary integral formulations for PDEs involving the Laplace operator lead to H^s elliptic problems with values $s \in (-3/2, 3/2)$, see [103, 169]. Higher order elliptic problems and elliptic systems of PDEs appearing in applications to elasticity, fluid dynamics and electromagnetics require more complicated finite element constructions and are not considered in this book, we refer to [4, 27, 41, 196].

Non-conforming finite element methods, where the condition $V_j \subset V$ is violated for the target Hilbert space V, will be mentioned later in Subsection 3.2.2.2 and Subsection 7.1.3.

The following partial result for linear finite elements ($p = 1$) can be obtained as a corollary to the more general Theorem 3.3 formulated below. However, since it admits a direct proof in the spirit of standard finite element analysis [25, 41], we present it separately. The restriction to linear finite elements on triangulations and to convex domains in $\mathbb{R}^2$ is for convenience and simplifies the technical details, see the appendices in [190] for more information.

Proposition 3.1 *Consider a sequence (3.15) of dyadically refined quasi-uniform and regular triangulations $\mathcal{T}_j$ of a bounded convex polygonal domain $G \subset \mathbb{R}^2$, let $p = 1$ in the construction of C^0 Lagrange finite elements, and consider the associated nested sequence of linear finite element spaces $V_{j,0} = \{v_j \in V(\mathcal{T}_j) : v_j = 0 \text{ on } \Gamma\}$ in the Sobolev space $H_0^1(G)$. Then we have*

$$\|u\|_{H_0^1}^2 \approx |||u|||_{1,0}^2 := \inf_{v_j \in V_{j,0}:\, u=\sum_j v_j} \sum_{j=0}^{\infty} 2^{2j}\|v_j\|_{L^2}^2, \qquad u \in H_0^1(G).$$

Proof For the upper inequality we use Theorem 2.3. Taking into account the rescaling result of Subsection 2.3.1, the strengthened Cauchy-Schwarz inequalities to be established are of the form

$$(v_j, v_l)_{H_0^1} \le \gamma_{jl}(2^j\|v_j\|_{L^2})(2^l\|v_l\|_{L^2}), \qquad v_j \in V_{j,0},\ v_l \in V_{l,0}, \quad l, j \in \mathbb{Z}_+.$$

We will prove them with parameters $\gamma_{jl} \le C2^{-|j-l|/2}$. In fact, by symmetry it is sufficient to consider the case $l \ge j$. Green's formula can be applied to any triangle in $\mathcal{T}_j$. Since, for linear finite element functions, we have $\Delta v_j = 0$ in the interior of each triangle of $\mathcal{T}_j$, we get

$$\begin{aligned}(v_j, v_l)_{H_0^1} &= \sum_{C_i \in \mathcal{T}_j} \int_{C_i} \nabla v_j \cdot \nabla v_l \, dx = \sum_e \int_e \left[\frac{\partial v_j}{\partial n_e}\right]_e v_l \, ds \\ &\le \left(\sum_e \int_e \left[\frac{\partial v_j}{\partial n_e}\right]_e^2 ds\right)^{1/2} \left(\sum_e \int_e v_l^2 \, ds\right)^{1/2},\end{aligned}$$

where the sums are over all interior edges in $\mathcal{T}_j$, and $\left[\frac{\partial v_j}{\partial n_e}\right]_e$ denotes the appropriately signed constant jump of the normal derivative of v_j with respect to the edge e. Obviously, since v_j is linear on every triangle attached to e, we have

$$\int_e \left[\frac{\partial v_j}{\partial n_e}\right]_e^2 ds \le C2^{-j}2^{2j} \sum_P c_{P,j}^2,$$

where $c_{P,j} = v_j(P)$ and the sum is over the interior nodal points P belonging to the two triangles attached to the edge e. It follows from the quasi-uniformity and regularity assumptions that all edges in $\mathcal{T}_j$ have length $\approx 2^{-j}$. Since the number of triangles in $\mathcal{T}_j$ to which a given nodal point belongs is uniformly bounded by the regularity of $\mathcal{T}_j$, summing over all edges in conjunction with (3.16) gives

$$\sum_e \int_e \left[\frac{\partial v_j}{\partial n_e}\right]_e^2 ds \le C2^j \sum_{P \in \mathcal{V}_j} c_{P,j}^2 \le C2^{3j}\|v_j\|_{L^2}^2.$$

Similarly,

$$\sum_e \int_e v_l^2\, ds \le C \sum_e 2^{-l} \sum_{P' \in e} c_{P',l}^2 \le 2^{-l} \sum_{P' \in \mathcal{V}_l} c_{P',l}^2 \le C2^l \|v_l\|_{L^2}^2,$$

where now $c_{P',l} = v_l(P')$ and P' denotes nodal points associated with $\mathcal{T}_l$. Substituting the last two estimates we get

$$(v_j, v_l)_{H_0^1} \le C2^{3j/2}\|v_j\|_{L^2} \cdot 2^{l/2}\|v_l\|_{L^2} = C2^{-(l-j)/2}(2^j\|v_j\|_{L^2})(2^l\|v_l\|_{L^2}),$$

which proves the desired statement. Thus, from Theorem 2.3 we conclude that

$$\|u\|_{H_0^1}^2 \le C|||u|||_{1,0}^2, \qquad u \in H_0^1(G).$$

The lower inequality follows from the full regularity of the Dirichlet problem for convex domains G and from the approximation properties of the linear finite element spaces $V_{j,0}$. The full regularity implies that for any $w \in L^2(G)$ there exists a $v \in H^2(G) \cap H_0^1(G)$ such that

$$-\Delta v = w \quad \text{and} \quad \|v\|_{H^2} \le C\|w\|_{L^2}. \tag{3.17}$$

The approximation result we need for $v \in H^2(G) \cap H_0^1(G)$ is

$$\inf_{v_j \in V_{j,0}} \|v - v_j\|_{H_0^1} \le C2^{-j}\|v\|_{H^2}. \tag{3.18}$$

The estimate (3.18) for linear finite elements is covered as a special case by [25, Theorem 4.4.20]. In Problem 3.5, a proof based on Lemma A.3 is suggested for the nodal interpolation operator. In the proof of Theorem 3.3 below, we define quasi-interpolant operators Q_j with domain $L^2(G)$, which provide approximation results in $H^s(G)$ spaces.

With (3.17) and (3.18) at hand, we can argue as follows: Denote by $\tilde{Q}_{V_{j,0}}$ the $H_0^1(G)$ ortho-projection onto $V_{j,0}$. Then, with $v \in X := H^2(G) \cap H_0^1(G)$ and $w = -\Delta v$, we have

$$\|u - \tilde{Q}_{V_{j,0}}u\|_{L^2} = \sup_{0 \ne w \in L^2(G)} \frac{(u - \tilde{Q}_{V_{j,0}}u, w)_{L^2}}{\|w\|_{L^2}} \le C \sup_{0 \ne v \in X} \frac{(u - \tilde{Q}_{V_{j,0}}u, \Delta v)_{L^2}}{\|v\|_{H^2}}.$$

But

$$(u - \tilde{Q}_{V_{j,0}}u, \Delta v)_{L^2} = (\nabla(u - \tilde{Q}_{V_{j,0}}u), \nabla(v - v_j))_{L^2} \le \|u - \tilde{Q}_{V_{j,0}}u\|_{H_0^1}\|v - v_j\|_{H_0^1},$$

where it follows from the definition of the $H_0^1(G)$ ortho-projection $\tilde{Q}_{V_{j,0}}$ that we can subtract an arbitrary $v_j \in V_{j,0}$. Now we take the infimum with respect to $v_j \in V_{j,0}$, consider the estimate (3.18), and substitute the result in the above bound for $\|u-\tilde{Q}_{V_{j,0}}u\|_{L^2}$. Since $\|u-\tilde{Q}_{V_{j,0}}u\|_{H_0^1} \le \|u\|_{H_0^1}$, this gives us the L^2 approximation estimate

$$\|u - \tilde{Q}_{V_{j,0}}u\|_{L^2} \le C2^{-j}\|u\|_{H_0^1}, \qquad u \in H_0^1(G), \quad j \in \mathbb{Z}_+. \tag{3.19}$$

For the final step in the proof of Proposition 3.1 we use Lemma 2.1. There we identify $(V;a)$ with $(H_0^1(G);(\cdot,\cdot)_{H_0^1})$, set $\mathbf{V}_0 = (V_{j,0})_{j\in\mathbb{Z}_+}$, $\rho_j = 1$, and take $a_j(\cdot,\cdot) := 2^{2j}(\cdot,\cdot)_{L^2}$ for the bilinear forms on $V_{j,0}$, $j = 0, 1, \ldots$. It was established in the first part of the proof that $(\mathbf{V}_0;\mathbf{a};\rho)$ is a space splitting for $(H_0^1(G);(\cdot,\cdot)_{H_0^1})$. Furthermore, let $W_0 = V_{0,0}$ and $W_j = \operatorname{ran}(\tilde{Q}_{V_{j,0}} - \tilde{Q}_{V_{j-1,0}}) \subset V_{j,0}$, $j = 1, 2, \ldots$, be the auxiliary orthogonal splitting. To verify the condition (2.27) of Lemma 2.1, note that for any $w_j \in W_j$ with $j \ge 1$ we have $\tilde{Q}_{V_{j-1,0}}w_j = 0$. Thus, by (3.19) we get

$$\|w_j\|_j^2 = 2^{2j}\|w_j\|_{L^2}^2 = 2^{2j}\|w_j - \tilde{Q}_{V_{j-1,0}}w_j\|_{L^2}^2 \le C\|w_j\|_{H_0^1}^2, \quad j = 1, 2, \ldots.$$

For $j = 0$ we also have such a bound by the equivalence of norms on finite-dimensional spaces, with a constant depending on $\mathcal{T}_0$. This completes the verification of the conditions of Lemma 2.1 and establishes the stability of the space splitting $(\mathbf{V}_0;\mathbf{a};\rho)$. □

Our main goal in this subsection is to prove a result for the sequence of C^0 Lagrange finite element spaces based on Part a) of Theorem 3.1.

Theorem 3.3 *Let $G \subset \mathbb{R}^d$ be a bounded polyhedral Lipschitz domain and let the partition sequence $\{\mathcal{T}_j\}$ in (3.15) satisfy the above mentioned quasi-uniformity and regularity conditions. Then, for the nested sequence of C^0 Lagrange finite element spaces $V_j = V(\mathcal{T}_j)$ of fixed degree $p = 1, 2, \ldots$ and the parameter range $0 < s < 3/2$, we have the norm equivalence*

$$\|u\|_{H^s}^2 \approx |||u|||_s^2 := \inf_{v_j\in V_j:\, u=\sum_j v_j} \sum_{j=0}^{\infty} 2^{2js}\|v_j\|_{L^2}^2, \qquad u \in H^s(G). \tag{3.20}$$

Proof The starting point is the identification of Sobolev spaces $H^s(G)$ and Besov spaces $B^s_{\omega_m}(G)$. In fact, for $0 < s < m$, $m \ge 2$, we have $H^s(G) = B^s_{\omega_m}(G)$ together with the norm equivalence

$$\|u\|_{H^s}^2 \approx \|u\|_{B^s_{\omega_m}}^2 := \|u\|_{L^2}^2 + \sum_{j=0}^{\infty} 2^{2sj}\omega_m(2^{-j}, u)_{L^2}^2,$$

where

$$\omega_m(t,u) := \inf_{v\in H^m(G)} (\|u-v\|_{L^2} + t^m|v|_{H^m}), \qquad t \geq 0, \quad u \in L^2(G),$$

is the K-functional for the pair $(L^2(G); H^m(G))$. In the definition of the $B^s_{\omega_m}$ norm, the modulus $\omega_m(t,u)$ can be replaced by the m-th order L^2 modulus of continuity $w_m(t,u)$. For details, see Section A.5 in the appendix.

As can be seen from the material in Subsection 3.1.1, the spaces $B^s_{\omega_m}(G)$ connect more naturally to various approximation processes via the proof of certain Jackson- and Bernstein-type inequalities. This is what we now need to do in the special case of a collection $\mathbf{V}$ of C^0 Lagrange finite element spaces V_j. To be consistent with the notation in Subsection 3.1.2, we set $H = L^2(G)$. We will relate m to the polynomial degree p associated with $\mathbf{V}$ by setting $m = p+1$, which allows us to obtain an optimal Jackson-type inequality. However, if we are only interested in proving the norm equivalence given in Theorem 3.3, taking $m = 2$ would suffice. The reason for this is the low L^2 smoothness of C^0 Lagrange finite element functions, which is reflected in the best possible value of $\gamma = 3/2$ in the Bernstein-type inequality for C^0 finite element spaces.

We start with the Jackson-type estimate, for which we construct suitable quasi-interpolant operators $Q_j : L^2(G) \to V_j$. The properties we need are *local uniform L^2 boundedness* and *reproduction of polynomials of total degree $p = m-1$*. For our finite element setting, this can be formalized as follows: For each cell C_i in $\mathcal{T}_j$ there exists a domain $\hat{C}_i \supset C_i$ of comparable diameter $\approx 2^{-j}$ such that

(a) $Q_j u = 0$ on C_i whenever u vanishes outside $\hat{C}_i$ (*locality*),

(b) we have

$$\int_{C_i} |Q_j u|^2\,dx \leq C \int_{\hat{C}_i} u^2\,dx$$

with a constant independent of u, j, and C_i (*uniform L^2 boundedness*), and

(c) $Q_j q = q$ on C_i for any polynomial q of total degree $\leq m-1$ (*polynomial reproduction*).

For the C^0 Lagrange elements considered here and for the Q_j considered below, the sets

$$\hat{C}_i = \bigcup_{i' : C_{i'} \cap C_i \neq \emptyset} C_{i'}$$

formed by the union of all cells intersecting with C_i will suffice. Figure 3.4 illustrates this choice. Depending on the actual construction of Q_j, even smaller $\hat{C}_j$ may be sufficient. The quasi-uniformity and regularity assumptions on the partition sequence $(\mathcal{T}_j)_{j\geq 0}$, together with $\operatorname{diam}(\hat{C}_i) \approx 2^{-j}$, ensure that, for any j, any point in G is contained in a uniformly bounded number of $\hat{C}_i$. This is called the *finite overlap property* of the cover $\{\hat{C}_i\}$ of G, and is silently used below. Thus, the local L^2 boundedness assumption implies the uniform boundedness of Q_j as operators on $L^2(G)$, i.e., we have

$$\|Q_j u\|_{L^2}^2 = \sum_i \int_{C_i} |Q_j u|^2\,dx \le C \sum_i \int_{\hat{C}_i} u^2\,dx \le C\|u\|_{L^2}^2, \qquad u \in L^2(G),$$

for all $j \in \mathbb{Z}_+$.

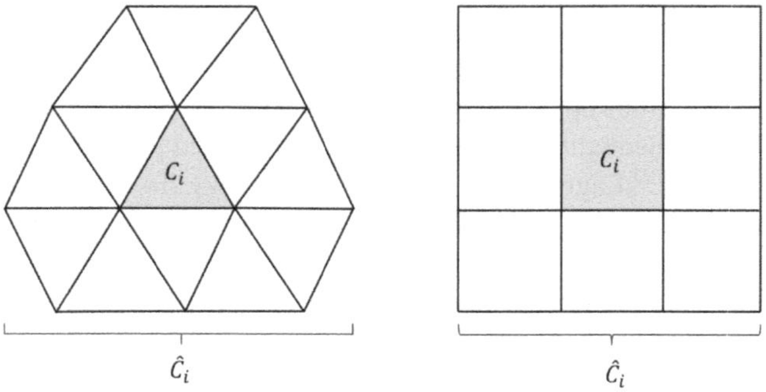

Fig. 3.4 The neighborhood $\hat{C}_i$ for triangular (left) and rectangular C^0 Lagrange finite elements.

With these properties at hand, using the linearity and uniform boundedness of Q_j, we have

$$\begin{aligned}\|u - Q_j u\|_{L^2} &\le \|u - v\|_{L^2} + \|Q_j v - Q_j u\|_{L^2} + \|v - Q_j v\|_{L^2}\\ &\le C\|u - v\|_{L^2} + \|v - Q_j v\|_{L^2}, \qquad u \in L^2(G), \quad v \in H^m(G).\end{aligned}$$

For smooth functions $v \in H^m$, the last term can be estimated using the local polynomial reproduction and L^2 boundedness properties. In fact,

$$\|v - Q_j v\|_{L^2}^2 \le \sum_i \int_{C_i} 2(|v - q_i|^2 + |Q_j(v - q_i)|^2)\,dx \le C \sum_i \int_{\hat{C}_i} |v - q_i|^2\,dx,$$

where each q_i is a polynomial of total degree $m-1$. We will now choose them to optimally bound the sum in the right-hand side. This can be done by Lemma A.3 applied on each $\hat{C}_i$, which gives the existence of polynomials q_i such that

$$\|v - q_i\|_{L^2(\hat{C}_i)}^2 \le C2^{-2mj} \sum_{|\alpha|=m} \|v^{(\alpha)}\|_{L^2(\hat{C}_i)}^2.$$

So, with the finite overlap property in mind, we get

$$\begin{aligned}\|v - Q_j v\|_{L^2}^2 &\le C2^{-2mj} \sum_{|\alpha|=m} \sum_i \|v^{(\alpha)}\|_{L^2(\hat{C}_i)}^2\\ &\le C2^{-2mj} \sum_{|\alpha|=m} \|v^{(\alpha)}\|_{L^2}^2 \le C2^{-2mj}|v|_{H^m}^2,\end{aligned}$$

and finally, since $v \in H^m(G)$ was arbitrary, we get

$$\|u - Q_j u\|_{L^2} \le C \inf_{v \in H^m(G)} (\|u - v\|_{L^2} + 2^{-mj} |v|_{H^m}) = C\omega_m(2^{-j}, u).$$

This shows the Jackson-type estimate for V_j with $H = L^2(G)$ and the K-functional ω_m.

We now provide a family of examples of Q_j which satisfy the assumptions (a)-(c) for C^0 Lagrange finite elements of degree m under the conditions of Theorem 3.3. It has the additional, very useful property that the Q_j are not just linear maps into V_j, but projections onto V_j. We give the details for the case of simplicial partitions. The rectangular case is completely analogous, the only difference being that polynomials of coordinate-wise degree $\le m$ replace polynomials of total degree $\le m$, and rectangular cells appear instead of simplices. Thus, consider the standard d-dimensional simplex $\Sigma = \{x : 0 \le x_i \le \sum_{l=1}^d x_l \le 1, i = 1, \dots, d\}$ and the family $\mathcal{B}_{r_0}$ of all closed d-dimensional balls $B(x_0, r) = \{x \in \mathbb{R}^d : \|x - x_0\|_2 \le r\}$ of radius $r \ge r_0$ contained in Σ, where $r_0 > 0$ is fixed such that $\mathcal{B}_{r_0} \ne \emptyset$. For any $B(x_0, r) \in \mathcal{B}_{r_0}$, there is a unique set of polynomials $q_{P;x_0,r}$ of total degree m such that

$$\int_{B(x_0,r)} N_{P'}(x) q_{P;x_0,r}(x)\, dx = \delta_{P'P} = \begin{cases} 1, & P' = P, \\ 0, & P' \ne P. \end{cases} \tag{3.21}$$

Here, P and P' pass through the nodal points of the standard simplex Σ, and $N_{P'}$ denotes the Lagrange nodal basis function corresponding to the nodal point P'. Outside of $B(x_0, r)$ we extend the polynomials $q_{P;x_0,r}$ by zero. Note that choosing a family of balls in this construction is for convenience, one could equally well replace $\mathcal{B}_{r_0}$ by the family $\mathcal{S}_{t_0}$ of all simplices contained in Σ which are congruent to $t\Sigma$ for some $t \ge t_0 > 0$. An example, where $B(x_0, r)$ in (3.21) is replaced by certain simplices $t\Sigma$, is treated in Problem 3.4.

Next, for each simplex C_i in $\mathcal{T}_j$, we fix an affine-linear bijection $T_i : C_i \to \Sigma$, choose a ball $B(x_0, r) \in \mathcal{B}_{r_0}$ (with x_0, r possibly depending on C_i) and define functions

$$q_{P,C_i}(x) \equiv q_{P;x_0,r,C_i}(x) := \alpha_i q_{T_i P;x_0,r}(T_i x)$$

such that

$$\int_{C_i} N_{P',j}(x) q_{P,C_i}(x)\, dx = \delta_{P'P} = \begin{cases} 1, & P' = P, \\ 0, & P' \ne P, \end{cases} \tag{3.22}$$

where P and P' now pass through the nodal points in C_i and $\mathcal{V}_j$, respectively, and $\alpha_i > 0$ are scaling factors determined by T_i. We have decided to suppress the dependence of q_{P,C_i} on $B(x_0, r)$ since the results below hold for any choice of $B(x_0, r) \in \mathcal{B}_{r_0}$. The freedom of choice for $B(x_0, r)$ will be used in Subsection 7.1.1. By the usual scaling argument, and as long as the quasi-uniformity and regularity assumptions hold, we have

$$\|q_{P,C_i}\|_{L^\infty(C_i)} \le C 2^{jd}, \qquad P \in C_i, \tag{3.23}$$

uniformly in $P \in C_i$, $C_i \in \mathcal{T}_j$ and j. The constant C in (3.23) also depends on r_0, it will grow if $r_0 \to 0$. The elementary proof of (3.23) is sketched in Problem 3.3.

Now define

$$Q_j u(x) := \sum_{P \in \mathcal{V}_j} \lambda_{P,j}(u) N_{P,j}(x), \tag{3.24}$$

where

$$\lambda_{P,j}(u) := \sum_{i:\, P \in C_i} a_{P,i} \int_{C_i} u(x) q_{P,C_i}(x)\, dx, \quad a_{P,i} \geq 0, \quad \sum_{i:\, P \in C_i} a_{P,i} = 1. \tag{3.25}$$

The linear functionals $\lambda_{P,j} : L^1(G) \to \mathbb{R}$ are called dual functionals for the nodal basis in V_j, since (3.22) implies

$$\lambda_{P,j}(N_{P',j}) = \sum_{i:\, P \in C_i} a_{P,i} \int_{C_i} N_{P',j}(x) q_{P,C_i}(x)\, dx = \sum_{i:\, P \in C_i} a_{P,i} \delta_{P'P} = \delta_{P'P}$$

for all $P, P' \in \mathcal{V}_j$. This shows that Q_j is a projection onto V_j. The locality property (a) is also obvious by construction, one can take as $\hat{C}_i$ the above defined union of the cells $C_{i'}$ intersecting with C_i. Together with the projection property, this implies (c), since polynomials of total degree $p = m - 1$ belong to V_j and are exactly reproduced by Q_j. Finally, (b) follows from

$$\begin{aligned}
\int_{C_i} |Q_j u|^2\, dx &\leq C \sum_{P \in C_i} |\lambda_{P,j}(u)|^2 2^{-dj} \\
&\leq C 2^{-dj} \sum_{P \in C_i} \left| \sum_{i':\, P \in C_{i'}} a_{P,i'} \int_{C_{i'}} u(x) q_{P,C_{i'}}(x)\, dx \right|^2 \\
&\leq C 2^{-dj} \sum_{P \in C_i} \sum_{i':\, P \in C_{i'}} \|q_{P,i'}\|^2_{L^\infty(C_{i'})} \cdot 2^{-dj} \|u\|^2_{L^2(C_{i'})} \\
&\leq C 2^{-2dj} 2^{2dj} \sum_{i':\, C_{i'} \cap C_i \neq \emptyset} \|u\|^2_{L^2(C_{i'})} \leq C \int_{\hat{C}_i} |u(x)|^2\, dx,
\end{aligned}$$

where we used $0 \leq a_{P,i} \leq 1$, the estimate (3.23), and the finite overlap property. Thus, Q_j satisfies all the conditions needed for the above proof of the Jackson-type inequality.

We next turn to the proof of the Bernstein-type inequality. Since $\omega_m(t,u) \leq \|u\|^2_{L^2}$ is obvious by definition, we can concentrate on the range $0 < t \leq c2^{-j}$, where c will be fixed below. Instead of ω_m, we consider the equivalent m-th order L^2 modulus of continuity w_m and assume $m \geq 2$. Then, using the properties of w_m, we have

$$\omega_m(t,u) \leq C w_m(t,u)_{L^2} \leq C w_2(t,u)_{L^2} \leq C t w_1(t, \nabla u)_{L^2},$$

if $u \in H^1(G)$. Next, to estimate the first-order modulus of continuity $w_1(t, \nabla v_j)_{L^2}$ for C^0 Lagrange finite element functions $v_j \in V_j \subset H^1(G)$, we denote by $q_i := v_j|_{C_i}$

the polynomials of total degree m representing v_j on the cells C_i of $\mathcal{T}_j$. Then we have

$$\begin{aligned}\|\Delta_h^1(\nabla v_j)\|^2_{L^2(G_{h,1})} &= \sum_i \int_{x\in C_i:\, x+h\in G} \|\nabla v_j(x+h) - \nabla v_j(x)\|_2^2\, dx \\ &= \sum_i \int_{x\in C_i:\, x+h\in C_i} \|\nabla q_i(x+h) - \nabla q_i(x)\|_2^2\, dx \\ &\quad + \sum_i \int_{x\in C_i:\, x+h\in G\setminus C_i} \|\nabla q_{i'}(x+h) - \nabla q_i(x)\|_2^2\, dx,\end{aligned}$$

where the summation is with respect to all cells C_i in $\mathcal{T}_j$ and the indices $i' \neq i$ appearing in the last integral can change according to $x + h \in C_{i'}$ by the geometry of $\mathcal{T}_j$. It follows from the regularity and the quasi-uniformity assumption that for $0 < |h| \leq t \leq c2^{-j}$ with appropriately fixed $c > 0$, the corresponding cells $C_{i'}$ must be among those directly attached to C_i. Thus since the set $\{x \in C_i : x + h \in G\setminus C_i\}$ is in a boundary layer of width t of ∂C_j, it has measure $\leq Ct2^{-j(d-1)}$ and we get

$$\begin{aligned}&\sum_i \int_{x\in C_i:\, x+h\in G\setminus C_i} \|\nabla q_{i'}(x+h) - \nabla q_i(x)\|_2^2\, dx \\ &\qquad \leq C \sum_i t2^{-j(d-1)} \sum_{i'} \left(\|\nabla q_i\|^2_{L^\infty(C_i)} + \|\nabla q_{i'}\|^2_{L^\infty(C_{i'})}\right) \\ &\qquad \leq Ct2^{-j(d-1)} \sum_i \|\nabla q_i\|^2_{L^\infty(C_i)},\end{aligned}$$

where the finite overlap property was used in the last step. Similarly, for the other sum of integrals, we have

$$\sum_i \int_{x,x+h\in C_i} \|\nabla q_i(x+h) - \nabla q_i(x)\|_2^2\, dx \leq C|h|^2 \sum_i \int_{C_i} \|\nabla^2 q_i(x)\|_2^2\, dx.$$

It remains to replace the various norms of the polynomials q_i locally on C_i by their local L^2 norms, using the equivalence of norms on spaces of polynomials of fixed degree for a fixed standard cell, together with the usual scaling arguments. Under the assumptions of quasi-uniformity and regularity, this gives

$$\|\nabla q_i\|^2_{L^\infty(C_i)} \leq C2^{2j}2^{jd}\|q_i\|^2_{L^2(C_i)}, \qquad \int_{C_i} \|\nabla^2 q_i(x)\|_2^2\, dx \leq C2^{2j}\|q_i\|^2_{L^2(C_i)},$$

for the cells C_i that belong to $\mathcal{T}_j$. In total, for $0 \leq |h| \leq t \leq c2^{-j}$, we have

$$\|\Delta_h^1(\nabla v_j)\|^2_{L^2(G_{h,1})} \leq C \sum_i \left(t^2 2^{2j} + t2^{-j(d-1)}2^{j(d+2)}\right) \|q_i\|^2_{L^2(C_i)} \leq Ct2^{3j}\|v_j\|^2_{L^2},$$

and finally

$$\omega_m(t, v_j)^2 \leq Ct^2 \sup_{|h|\leq t} \|\Delta_h^1(\nabla v_j)\|_{L^2(G_{h,1})}^2 \leq C(t2^j)^3\|v_j\|_{L^2}^2, \qquad 0 < t \leq c2^{-j}.$$

This proves the Bernstein-type inequality with $\gamma = 3/2$ for the modulus ω_m. All the necessary ingredients for applying Theorem 3.1 are now available, and its Part a) implies the statement of Theorem 3.3. □

We add a few remarks. The projection property of the quasi-interpolant Q_j defined in (3.24-3.25), together with the additional freedom to choose the balls $B(x_0, r) \subset \Sigma$ and the coefficients $a_{P,i}$, will be used in Subsection 7.1.1 to derive stability results for splittings of finite element spaces that occur in adaptive finite element methods. Unfortunately, one drawback is that Q_j, when restricted to functions from $H_0^1(G)$, does not map to $V_{j,0} = V_j \cap H_0^1(G)$. Therefore, it does not automatically imply a Jackson-type inequality for the family $\mathbf{V}_0 = \{V_{j,0}\}_{j\geq 0}$. How to derive from Theorem 3.3 a result about stable space splittings involving $\mathbf{V}_0$ for the spaces $H_0^s(G)$ with $1/2 < s < 3/2$ is shown in Subsection 7.1.2.

The reader may ask about extensions of Theorem 3.3 to the range $s \leq 0$ according to the abstract theory presented in Subsection 3.1.1. This is not automatic since the projections Q_j used in its proof do not satisfy the commutator property (3.1). The nodal interpolation projections

$$Q_j u = \sum_{P\in\mathcal{V}_j} u(P)N_{P,j}(x),$$

satisfy (3.1) but are neither well-defined and bounded on $L^2(G)$ nor well-defined on $H^s(G)$ if $s \leq d/2$. This implies that the resulting subspace family $\mathbf{W}$ is not a stable Riesz splitting for $H^s(G)$ for $s \leq d/2$ and fails in particular for the most interesting case $H^1(G)$, $d \geq 2$. Nevertheless, the study of this subspace family $\mathbf{W}$ has attracted some attention in connection with the popular *hierarchical basis finite element method*, see Subsection 7.1.1.3.

Next, we formulate an extension to the parameter range $-3/2 < s < 3/2$ based on the use of the L^2 ortho-projections $Q_{V_j} : L^2(G) \to V_j$. The main difficulty in applying this result is to find a stable basis in the orthogonal complement space $W_j = \mathrm{ran}(Q_{V_j} - Q_{V_{j-1}})$ which is suitable for computational purposes. In many cases such bases have been successfully constructed and are known as *finite element prewavelet bases*, see the next subsection for more details.

Theorem 3.4 *Under the conditions of Theorem 3.3, for $u \in H^s(G)$, $0 \leq s < 3/2$, we have the norm equivalence*

$$\|u\|_{H^s}^2 \approx |||u|||_{s,\mathbf{Q}}^2 := \|Q_{V_0}u\|_{L^2}^2 + \sum_{j=1}^{\infty} 2^{2js}\|(Q_{V_j} - Q_{V_{j-1}})u\|_{L^2}^2, \tag{3.26}$$

where $\mathbf{Q} := (Q_{V_j})_{j\in\mathbb{Z}_+}$ is the sequence of L^2 ortho-projections onto the subpaces V_j. For $-3/2 < s < 0$, a similar result holds for $u \in L^2(G)$ if the $H^s(G)$ norm in the left-hand side of (3.26) is replaced by the $H^{-s}(G)'$ norm.

Proof For $0 < s < 3/2$ the norm equivalence follows from (3.20), whereas only the uniform L^2 boundedness of the projections Q_j is needed. In fact, setting $v_0 = Q_{V_0}u$ and $v_j = (Q_{V_j} - Q_{V_{j-1}})u$ gives an admissible decomposition for u in the definition of the $|||\cdot|||_s$ norm. Thus we obviously have

$$c\|u\|_{H^s}^2 \leq |||u|||_s^2 \leq |||u|||_{s,\mathbf{Q}}^2.$$

To establish the opposite direction, let $u = \sum_l v_l$ be an arbitrary admissible decomposition of $u \in H^s(G) \subset L^2(G)$ with $v_l \in V_l$, $l = 0, 1, \ldots$. Then, repeating the argument used earlier in Subsection 3.1.1, we have

$$\|Q_{V_0}u\|_{L^2}^2 \leq \|u\|_{L^2}^2 \leq (\sum_l \|v_l\|_{L^2})^2 \leq C\sum_l 2^{2\epsilon l}\|v_l\|_{L^2}^2$$

for $0 < \epsilon < s$, and since $Q_{V_j}v_l = v_l$ for $l \leq j$, we get

$$\begin{aligned}\|(Q_{V_j} - Q_{V_{j-1}})u\|_{L^2}^2 &= \|(Q_{V_j} - Q_{V_{j-1}})(\sum_{l\geq j} v_l)\|_{L^2}^2 \leq 2\left(\sum_{l\geq j}\|v_l\|_{L^2}\right)^2 \\ &\leq C2^{-2\epsilon j}\sum_{l\geq j} 2^{2\epsilon l}\|v_l\|_{L^2}^2\end{aligned}$$

for $j = 1, 2, \ldots$. Substituting into the definition of $|||\cdot|||_{s,\mathbf{Q}}$, changing the order of summation, and taking the infimum with respect to all admissible representations of u yields

$$|||u|||_{s,\mathbf{Q}}^2 \leq C|||u|||_s^2 \leq C\|u\|_{H^s}^2,$$

where the last step is again based on (3.20).

For $s = 0$, (3.26) is obvious, since the Q_{V_j} in Theorem 3.4 are L^2 ortho-projections, while for $-3/2 < s < 0$ duality with respect to the duality pairing induced by $H = L^2(G)$ gives the result. Note that $|||\cdot|||_{s,\mathbf{Q}}$ equals the norm $\|\cdot\|_{\mathbf{W},s}$ in $A_{\mathbf{W}}^s$, where the L^2 orthogonal subspace collection $\mathbf{W} = \mathbf{W}^*$ is generated by $\mathbf{Q}$. Since, for $s < 0$, we have $A_{\mathbf{W}}^{-s} = H^{-s}(G)$ with equivalent norms by the already available result for positive smoothness exponents, using (3.9) implies $A_{\mathbf{W}}^s = (A_{\mathbf{W}}^{-s})' = (H^{-s}(G))'$ with equivalent norms. This concludes the argument. □

Although computing the action of the L^2 ortho-projections Q_j leads to uniformly well-conditioned linear systems (this follows from (3.16)), using the operator

$$C_s := Q_{V_j} + \sum_{j\geq 1} 2^{-2sj}(Q_{V_j} - Q_{V_{j-1}})$$

to construct preconditioners for $H^s(G)$ elliptic problems is not very practical yet. In the literature, more explicit quasi-interpolants have been proposed by suitable substitutions of the L^2 ortho-projections in the definition of the preconditioner C_s, leading to more economical implementations, see [190, Appendix C].

3.1.4 Example: Wavelets and their relatives

In the 1980s, a new type of universal decomposition system for functions on $\mathbb{R}^d$ gained attention under the name *wavelets*. Roughly speaking, wavelet systems represent localized versions of the trigonometric system and are organized in a multiscale fashion, with building blocks computed similarly to the windowed Fourier transform. In the univariate case, they are constructed from a single locally supported function ψ by shift, dilation, and scaling, leading to a rich theory and fast algorithms. Initially, the main applications were in time-frequency analysis, especially in signal and image processing, and in renormalization in quantum field theory [59, 122]. Meanwhile, wavelets and their relatives (framelets, quarks) have also been used with good success for the discretization of operator equations and variational problems.

The main feature we are interested in in this subsection is the fact that wavelet systems form Riesz bases (i.e., yield Riesz splittings into one-dimensional subspaces) for Sobolev spaces H^s with parameter range $s \in (-\tilde{\gamma}, \gamma)$, where, under some technical conditions, one can have $\gamma, \tilde{\gamma} > 0$ arbitrarily large. In particular, there exist L^2 orthogonal wavelet systems on $\mathbb{R}^d$ with the Riesz basis property in $H^s(\mathbb{R}^d)$ for $s \in (-\gamma, \gamma)$ for any $\gamma > 0$. However, the adaptation of wavelets to more general domains $G \subset \mathbb{R}^d$ and boundary conditions is a subtle issue, making their use for real-world applications a challenging task. Nevertheless, together with the multiscale structure of wavelet systems, this has led to the development of a mathematical theory of fast adaptive wavelet methods for many types of elliptic and parabolic equations, see [42] for an introduction. Moreover, the ideas behind the analysis of adaptive wavelet methods have also been fruitful for the analysis of adaptive finite element methods used in engineering applications. We refer to [129, 173] and to Subsection 7.1.4 for further information.

Here we limit ourselves to some results on wavelet systems in Sobolev spaces, mostly without proof, and give certain examples where the wavelet systems are built from piecewise polynomial functions. We avoid technical details as much as possible and have adapted our notation to the rest of the chapter. The breakthrough result of [58] (see [59, 102] for an early account of wavelet theory) was the construction of complete $L^2(\mathbb{R})$ orthogonal systems consisting essentially of dyadic shifts and dilates of a single locally supported function ψ which, after appropriate scaling, forms an unconditional basis for a wide range of Besov-Sobolev spaces. This is in contrast to classical orthonormal systems consisting of polynomials, trigonometric functions or splines, which are nonlocal or fail to be unconditional in many spaces of interest. The only exception is the classical Haar system, which was originally defined with respect to the interval $[0, 1]$ (see below for its definition). However, since it consists of discontinuous piecewise constant functions, it serves only a rather limited number of applications.

A wavelet function ψ is chosen so that it is represented by a finite linear combination of the integer shifts of another, so-called *scaling function* ϕ, i.e.,

$$\psi(x) = \sum_{l \in \mathbb{Z}} b_l \phi(2x - l),$$

where the sequence $\mathbf{b} = (b_l)_{l\in\mathbb{Z}}$ is finitely supported. In particular, this ϕ must satisfy the *refinement equation*

$$\phi(x) = \sum_{l\in\mathbb{Z}} a_l \phi(2x - l),$$

where the sequence $\mathbf{a} = (a_l)_{l\in\mathbb{Z}}$, called the *mask*, is again finitely supported. Under mild conditions, the mask $\mathbf{a}$ uniquely determines ϕ, and in many applications an explicit pointwise representation of ϕ and ψ is not needed. A direct consequence of this assumption is that the spaces

$$V_j := \left\{ v_j = \sum_{l\in\mathbb{Z}} c_{j,l}\phi(2^j \cdot -l),\ (c_{j,l})_{l\in\mathbb{Z}} \in \ell^2(\mathbb{Z}) \right\}, \qquad j \ge 0,$$

form a nested sequence of closed subspaces of $L^2(\mathbb{R})$.

With an appropriate choice of the mask $\mathbf{a}$, the scaling function can be made arbitrarily smooth, and the spaces V_j can be forced to locally contain polynomials of higher order. For the construction of orthogonal wavelets, it is additionally assumed that the integer shifts $\phi(\cdot - l)$, $l \in \mathbb{Z}$, form an $L^2(\mathbb{R})$ orthogonal basis in V_0 (this assumption is restrictive and will be relaxed later). Then, by properly choosing $\mathbf{b}$, one can show that

$$V_j = V_{j-1} \oplus W_j, \qquad W_j := \{w_j = \sum_{l\in\mathbb{Z}} d_{j,l}\psi(2^{j-1} \cdot -l),\ (d_{j,l})_{l\in\mathbb{Z}} \in \ell^2(\mathbb{Z})\}$$

is an $L^2(\mathbb{R})$ orthogonal decomposition of V_j, and that the $\psi_{j,l} := \psi(2^{j-1} \cdot -l)$ are mutually orthogonal (and orthogonal to functions in V_{j-1}). Thus, iterating we have

$$V_j = V_0 \oplus W_1 \oplus \cdots \oplus W_j, \qquad j \ge 1,$$

and a function $v_j \in V_j$ can be written both in terms of the orthogonal system

$$\Phi_j := \{\phi_{j,l} := \phi(2^j \cdot -l)\}_{l\in\mathbb{Z}}$$

and with respect to

$$\Psi_{j,0} := \Phi_0 \cup \Psi_1 \cup \cdots \cup \Psi_j, \qquad \Psi_k := \{\psi_{k,l}\}_{l\in\mathbb{Z}}.$$

Under the density condition

$$\mathrm{clos}_{L^2(\mathbb{R})}(\cup_j V_j) = L^2(\mathbb{R}),$$

the system

$$\tilde{\Psi} := \Phi_0 \cup \Psi_1 \cup \Psi_2 \cup \cdots,$$

forms a complete orthogonal system in $L^2(\mathbb{R})$. For short, we call $\tilde{\Psi}$ a *wavelet system* associated with the pair (ϕ, ψ). Note that, contrary to the cited literature, we have not normalized the functions $\phi_{j,l}$ and $\psi_{j,l}$ in $L^2(\mathbb{R})$, which leads to different scaling factors in some of the formulas below.

An important side effect of the construction is that the transformation between the two representations is extremely efficient due to the locality assumption and refinement properties. Representing approximations in V_j of smooth functions f with respect to $\Psi_{j,0}$ instead of Φ_j has the additional advantage that the coefficients $(f, \psi_{k,l})_{L^2)}$ associated with Ψ_k typically decay rapidly as k grows, which can be used for compression purposes or singularity detection. This is due to the *vanishing moments property*, since $\psi_{k,l} \perp V_{k-1}$ and V_{k-1} locally contain polynomials up to a certain degree by construction. Similarly, discretization matrices of standard differential and integral operators have better compressibility when computed with respect to $\Psi_{j,0}$.

For the Haar system mentioned above, we have

$$\phi(x) = \chi_{[0,1)}, \qquad \psi(x) = \chi_{[0,1/2)} - \chi_{[1/2,1)},$$

where χ_A is the characteristic function of a set A, and the masks **a** and **b** can be read from

$$\phi(x) = \phi(2x) + \phi(2x-1), \qquad \psi(x) = \phi(2x) - \phi(2x-1).$$

It can be proved using Theorem 3.2 that the Haar system $\tilde{\Psi}$ with appropriate normalization is a Riesz basis in $H^s(\mathbb{R})$ for $-1/2 < s < 1/2$. For this, one sets $H = L^2(\mathbb{R})$, $q = 1/2$, and considers as V_j and $W_j = V_j \ominus V_{j-1}$ the closures of the spans of the systems Φ_j and Ψ_j, respectively. By restriction to $[0, 1]$, this implies that the classical Haar system is automatically a Riesz basis in $H^s((0, 1))$ for the same range of smoothness parameters s. In Subsection 3.1.4.1, we consider the more general case of similar systems generated by splines, where ϕ is a B-spline on a uniform grid. Unfortunately, if this B-spline ϕ is continuous, achieving L^2 orthogonality for Ψ is only possible by sacrificing the local support property of ψ and its mask **b**. Since the latter property is crucial in many applications, the orthogonality property is replaced by semi-orthogonality or biorthogonality properties. Alternatively, there is the possibility to consider multiwavelet systems with several ψ spanning W_1 or framelets, see [99] for a recent exposition. Furthermore, the restriction of the constructions to bounded intervals and boundary conditions requires more care. For higher dimensions $d > 1$, wavelet systems can be obtained either by using tensor products of univariate wavelets or by a direct generalization of the construction principles for the one-dimensional case. The adaptation to bounded domains in $\mathbb{R}^d$ and boundary conditions is theoretically possible. However, its implementation for solving complex PDE problems faces difficulties. We will discuss some aspects of the higher-dimensional case in Subsection 3.1.4.2.

3.1.4.1 Univariate spline wavelets

For the classical aspects of univariate spline theory, see [61, 159]. We recall some basic facts about smooth splines of fixed order $m = 1, 2, \ldots$ on uniform partitions belonging to $C^{m-2}(\mathbb{R})$. The notation is adapted to our needs.

Throughout this subsection, we denote by $\mathcal{T}_j$ the partition of $\mathbb{R}$ into dyadic intervals $C_{j,l} := [l2^{-j}, (l+1)2^{-j})$ of length $h_j = 2^{-j}$. Let V_j be the subspace of smooth splines of order m with respect to $\mathcal{T}_j$, i.e., $v_j \in V_j$ if $v_j \in C^{m-2}(\mathbb{R}) \cap L^2(\mathbb{R})$ is a polynomial of degree $\le m-1$ on each $C_{j,l}$. The nestedness property is automatic by definition.

Let ϕ be the m-th order B-spline with respect to $\mathcal{T}_0$ with support $[0,m]$, which we define as the m-fold convolution of the characteristic function of the unit interval $C_{0,0} = [0,1)$, i.e., we have

$$\phi = \underbrace{\chi_{[0,1)} * \cdots * \chi_{[0,1)}}_{m \text{ times}} = \mathcal{F}^{-1}(\hat{\phi}), \tag{3.27}$$

where the Fourier transform $\hat{\phi} := \mathcal{F}\phi$ of the B-spline ϕ is given by

$$\hat{\phi}(\theta) = \left(\frac{1-e^{-i\theta}}{i\theta}\right)^m = e^{-im\theta/2}\left(\frac{\sin(\theta/2)}{\theta/2}\right)^m, \qquad \theta \in \mathbb{R}.$$

From the definition (3.27) by convolution it is obvious that ϕ has support $[0,m]$, is in $C^{m-2}(\mathbb{R})$, and is a polynomial of degree $m-1$ on the intervals of the partition $\mathcal{T}_0$. There are many remarkable properties of ϕ, in particular

$$\int_{\mathbb{R}} \phi(x)\,dx = 1, \qquad \sum_{l\in\mathbb{Z}} \phi(x-l) = 1, \qquad 0 \le \phi(x) \le 1, \quad x \in \mathbb{R},$$

which follow from (3.27) by induction in m. The refinement property, with the non-zero mask coefficients $a_l = 2^{-m+1}\binom{m}{l}$, $l = 0,\ldots,m$, follows from the factorization

$$\hat{\phi}(\theta) = 2^{-m}(1+e^{-i\theta/2})^m\hat{\phi}(\theta/2) = \sum_{l=0}^{m} 2^{-m}\binom{m}{l} e^{-il\theta/2}\hat{\phi}(\theta/2)$$

together with $\mathcal{F}(\phi(2\cdot - l)) = \frac{1}{2}e^{-il\theta/2}\hat{\phi}(\theta/2)$. A more explicit definition of ϕ in terms of divided differences of truncated power functions is often preferred [61, 159], but we do not need it here.

The system of shifts $\Phi_j = \{\phi_{j,l} := \phi(2^j \cdot - l)\}_{l\in\mathbb{Z}}$ is a Riesz basis in $V_j \subset L^2(\mathbb{R})$. More precisely, we have

$$B_m^{-1}\|v_j\|_{L^2}^2 \le 2^{-2j}\sum_{l\in\mathbb{Z}} c_l^2 \le A_m^{-1}\|v_j\|_{L^2}^2, \qquad v_j = \sum_{l\in\mathbb{Z}} c_l\phi_{j,l}, \tag{3.28}$$

with constants $0 < A_m \le B_m < \infty$ independent of $v_j \in V_j$ and j. It follows from the local linear independence property (3.29) below that the span of Φ_0 is indeed dense in V_0. The relation (3.28) can be shown by Fourier transform techniques, first for $j = 0$ and then using the dilation argument for arbitrary j (see Problem 3.6). Alternatively, one can use the local support property of ϕ for the lower bound, and establish the upper bound by proving the existence of suitable locally supported dual functionals $\lambda_l : V_0 \to \mathbb{R}$ such that they define a quasi-interpolant projection

$$v_0(x) = (Q_0 v_0)(x) := \sum_{l\in\mathbb{Z}} \lambda_l(v_0)\phi(x-l), \qquad v_0 \in V_0,$$

which is bounded in $L^2(\mathbb{R})$, i.e.,

$$\sum_{l\in\mathbb{Z}} |\lambda_l(v_0)|^2 \le C\|v_0\|_{L^2}^2.$$

Quasi-interpolant constructions for spline spaces on arbitrary partitions satisfying these conditions can be found in [61].

We note without proof that, as a consequence of the local linear independence property of Φ_0, see [61], there exists a biorthogonal system of polynomials q_k, $k = 0, \ldots, m-1$, such that

$$\int_0^1 \phi(x+l)q_k(x)\,dx = \delta_{lk}, \qquad l \in \mathbb{Z}. \tag{3.29}$$

So, using these polynomials, the coefficients c_{-k}, $k = 0, \ldots, m-1$, of a spline function $v_0(x) = \sum_l c_l\phi(x-l) \in V_0$ can be recovered by restricting v_0 to the unit interval $[0, 1)$. Now denote by $\tilde{q}_k(x)$ the extension by zero of $q_k(x)$ to $\mathbb{R}$. Then any function of the form

$$\lambda(x) = \sum_{k=0}^{m-1} \alpha_k \tilde{q}_k(x-k), \qquad \sum_{k=0}^{m-1} \alpha_k = 1,$$

has support in $[0, m]$ and generates dual functionals

$$\lambda_l(v_0) = \int_{\mathbb{R}} \lambda(x-l)v_0(x)\,dx, \quad l \in \mathbb{Z},$$

which gives the above properties with a constant C depending on ϕ and the choice of coefficients α_k in the definition of $\lambda(x)$.

As in the case of finite elements, the quasi-interpolant projection

$$Q_0 : f \in L^2(\mathbb{R}) \to Q_0 f := \sum_{l\in\mathbb{Z}} \lambda_l(f)\phi(x-l) \in V_0$$

and its relatives $Q_j f(x) := (Q_0(f(2^{-j}\cdot))(2^{-j}x))$ projected onto V_j can be used to establish the Jackson-type estimate

$$\inf_{v_j\in V_j} \|f - v_j\|_{L^2} \le C\omega_m(2^{-j}, f), \qquad j \in \mathbb{Z},$$

where the K-functional for the pair $(L^2(\mathbb{R}), H^m(\mathbb{R}))$ serves as modulus, i.e.,

$$\omega_m(t,f) := \inf_{g\in H^m(\mathbb{R})} \|f-g\|_{L^2} + t^m|g|_{H^m}, \qquad t>0.$$

A corresponding Bernstein-type inequality is again easier to establish in terms of the m-th order modulus of smoothness $w_m(t,f)_{L^2}$, which is equivalent to $\omega_m(t,f)$. Note that $v_0 \in V_0$ can be differentiated $m-1$ times, where $v^{(m-1)}$ is piecewise constant on $\mathcal{T}_0$. In fact, differentiating the convolution definition (3.27) of ϕ yields

$$\phi^{(m-1)}(x) = \sum_{k=0}^{m-1}(-1)^{m-1-k}\binom{m-1}{k}\chi_{[0,1)}(x-k).$$

So,

$$\begin{aligned} v_0^{(m-1)}(x) &= \sum_{l\in\mathbb{Z}} c_l\phi^{(m-1)}(x-l) = \sum_{l\in\mathbb{Z}} c_l \sum_{k=0}^{m-1}(-1)^{m-1-k}\binom{m-1}{k}\chi_{[0,1)}(x-k-l) \\ &= \sum_{r\in\mathbb{Z}}\chi_{[0,1)}(x-r)\underbrace{\sum_{k=0}^{m-1}(-1)^{m-1-k}\binom{m-1}{k}c_{r-k}}_{d_r:=}. \end{aligned}$$

Obviously, with the expressions for d_r we have

$$\|v_0^{(m-1)}\|_{L^2}^2 = \sum_{r\in\mathbb{Z}} d_r^2 \le C\sum_{l\in\mathbb{Z}} c_l^2 \le C\|v_0\|_{L^2}^2,$$

where the Riesz basis property of Φ_0 in V_0 was used in the last step. Thus, for $0<h\le 1$, we can estimate

$$\|\Delta_h^m v_0\|_{L^2}^2 \le h^{2(m-1)}\|\Delta_h v_0^{(m-1)}\|_{L^2}^2 = h^{2m-1}\sum_{r\in\mathbb{Z}} d_r^2 \le Ch^{2m-1}\|v_0\|_{L^2}^2.$$

For $h\ge 1$ we simply use $\|\Delta_h^m v_0\|_{L^2} \le 2^m\|v_0\|_{L^2}$. Together this gives

$$\|\Delta_h^m v_0\|_{L^2}^2 \le C\min(1,h^{2m-1})\|v_0\|_{L^2}, \qquad h>0, \quad v_0\in V_0.$$

Since $v_j\in V_j$ is equivalent to $v_0(x) := v_j(2^{-j}x)\in V_0$ and $\Delta_h^m v_j(x) = (\Delta_{2^jh}^m v_0)(2^jx)$ by the dilation property of the construction, this transfers to

$$\|\Delta_h^m v_j\|_{L^2}^2 \le C\min(1,(2^jh)^{2m-1})\|v_j\|_{L^2}^2, \qquad h>0, \quad v_j\in V_j, \quad j\in\mathbb{Z}.$$

Taking the supremum with respect to $0<h\le t$ in the definition of $w_m(t,v_j)_{L^2}$, we get the desired Bernstein-type inequality

$$\omega_m(t,v_j) \le Cw_m(t,v_j) \le C\min(1,(2^jt)^{m-1/2})\|v_j\|_{L^2}, \quad t>0,$$

with parameters $\gamma = m-1/2$ and $q = 1/2$.

We are now in a position to formulate the following consequence of Theorems 3.1 and 3.2, taking into account that the spaces $B^s_{\omega_m}$ appearing in their formulation can be identified with $H^s(\mathbb{R})$, $0 < s < m$.

Corollary 3.4 *Let V_j, $j \geq 0$, denote the sequence of spline spaces of order m with respect to the uniform dyadic partitions $\mathcal{T}_j$ of $\mathbb{R}$.*
a) If V_j is equipped with the scalar product $a_j(u_j, v_j) := 2^{2sj}(u_j, v_j)_{L^2}$, $u_j, v_j \in V_j$, then $(\mathbf{V}; \mathbf{a})$ is a stable splitting for $H^s(\mathbb{R})$, $0 < s < m - 1/2$. In other words, we have

$$\|f\|^2_{H^s} \approx \inf_{v_j \in V_j : u = \sum_{j \geq 0} v_j} \sum_{j=0}^{\infty} 2^{2sj} \|v_j\|^2_{L^2}, \quad f \in H^s(\mathbb{R}), \quad 0 < s < m - 1/2. \quad (3.30)$$

b) Let $W_j = V_j \ominus V_{j-1}$, $j \geq 1$, be the L^2 orthogonal complement spaces. Then $(\tilde{\mathbf{W}}; \mathbf{a})$ is a Riesz splitting for $H^s(\mathbb{R})$, $|s| < m - 1/2$, where $\tilde{\mathbf{W}} = (V_0, W_1, W_2, \ldots)$. For $f \in H^s(\mathbb{R}) \cap L^2(\mathbb{R})$, we have

$$\|f\|^2_{H^s} \approx \|Q_0 f\|^2_{L^2} + \sum_{j=1}^{\infty} 2^{2sj} \|(Q_j - Q_{j-1}) f\|^2_{L^2}, \qquad |s| < m - 1/2, \qquad (3.31)$$

where $Q_j = Q_{V_j}$ are the L^2 ortho-projections onto V_j.

A couple of comments are in order. First, for $m = 1$, the above-mentioned properties of the Haar system $\tilde{\Psi}$ are covered. In particular, the scaled Haar system

$$\tilde{\Psi}_s = \{\phi_{0,l}, 2^{j(1/2-s)}\psi_{j,l} : l \in \mathbb{Z}, j \geq 1\}$$

forms an L^2 orthogonal Riesz basis in $H^s(\mathbb{R})$, $|s| < 1/2$. This is because, for $m = 1$, the systems Φ_j and Ψ_j form complete L^2 orthogonal bases for V_j and W_j, respectively, which allows us to replace the L^2 norms in Part b) of Corollary (3.31) by the corresponding coefficient expressions.

The statement of Corollary 3.4 remains valid if the construction is properly restricted to the unit interval $[0, 1]$ and the nested partitions $\mathcal{T}_j$ are only quasi-uniform with step-size $h_j \approx 2^{-j}$ as in the finite element case considered above. This is relevant for applications to functions on bounded intervals and, through tensor product techniques, to functions on domains composed of d-dimensional rectangles. Modifications are necessary for small j and near the endpoints $x = 0$ and $x = 1$, and Fourier techniques typically used in the shift-invariant setting require appropriate substitutions, see below for some examples for $m = 2$.

Corollary 3.4 allows a reinterpretation in terms of Riesz bases and frames for Sobolev spaces for arbitrary $m > 1$. For Part a), this follows from the Riesz basis property (3.28) of Φ_j in V_j. In fact, consider the scaled union

$$\tilde{\Phi}_s = \{2^{j(1/2-s)}\phi_{j,l} : l \in \mathbb{Z}, j \geq 0\}$$

of the systems Φ_j with $j \geq 0$. Then, by rewriting $v_j \in V_j$ with respect to the basis $2^{j(1/2-s)}\Phi_j$, we have

$$\begin{aligned}\|f\|_{H^s}^2 &\approx \inf_{f=\sum_{j\ge0}\sum_l c_{j,l}2^{j(1/2-s)}\phi_{j,l}} \sum_{j=0}^{\infty} 2^{2sj}\|\sum_l c_{j,l}2^{j(1/2-s)}\phi_{j,l}\|_{L^2}^2\\ &\approx \inf_{f=\sum_{j\ge0}\sum_l c_{j,l}2^{j(1/2-s)}\phi_{j,l}} \sum_{j=0}^{\infty} 2^{2sj}2^{-j}\sum_l c_{j,l}^2 2^{j(1-2s)}\\ &= \inf_{f=\sum_{j\ge0}\sum_l c_{j,l}2^{j(1/2-s)}\phi_{j,l}} \sum_{j=0}^{\infty}\sum_l c_{j,l}^2\end{aligned}$$

from (3.30) and (3.28). Thus, $\tilde{\Phi}_s$ is a frame in $H^s(\mathbb{R})$ for the parameter range $0 < s < m - 1/2$. Since $\|\phi_{j,l}\|_{H^s}^2 \approx 2^{j(2s-1)}$, an equivalent statement is that $\tilde{\Phi} := \tilde{\Phi}_0$ is a frame in $H^s(\mathbb{R})$, $0 < s < m - 1/2$, after scaling its elements to unit norm in $H^s(\mathbb{R})$.

Regarding Part b) of Corollary 3.4, the difficulty is to find a locally supported $\psi \in V_1$ such that it is L^2 orthogonal to V_0, and the system Ψ_1 of its integer shifts forms a Riesz basis in W_1, considered as subspace of $L^2(\mathbb{R})$. If one additionally requires L^2 orthogonality of Φ_0 or Ψ_1 itself, this is not possible for $m > 1$. Following this path (see [59, 102] or [99, 131]) leads to L^2 orthogonal wavelet systems $\tilde{\Psi}$ consisting of appropriately scaled shifts and dilates of globally supported but exponentially fast decaying smooth spline functions ϕ and ψ. They are known as Battle-Lemarie and Strömberg wavelet systems and form Riesz bases in $H^s(\mathbb{R})$ for $|s| < m - 1/2$ after similar scaling. The global support makes it difficult to adapt these constructions to finite intervals and to design fast algorithms, which slow down as the number of significant non-zero mask coefficients in **a** and **b** increases.

However, if one gives up orthogonality, then the local support of ψ and the Riesz basis property of Ψ_1 in W_1 can be obtained. This leads to so-called *semi-orthogonal spline wavelets* or simply *spline prewavelets*, which were first constructed in an explicit way in [40]. The associated prewavelet $\psi \in V_1 \ominus V_0$ is supported in $[0, 2m - 1]$ (these properties already uniquely determine ψ up to a scaling factor) and is symmetric about the midpoint of its support. The mask **b** has an explicit expression, see formula (2.4) in [40], i.e., we have

$$b_l = (-1)^l 2^{-(m-1)} \sum_{k=0}^{m} \binom{m}{k} \tilde{\phi}(l-k+1), \qquad l = 0, 1, \ldots, 3m-2,$$

where $\tilde{\phi}$ denotes the B-spline of order $2m$. The Riesz basis property of the associated system Ψ_1 in W_1 follows from the properties of another basis in W_1 that is biorthogonal to Ψ_1. Its construction is given in [40, Section 3]. In the case of linear splines with $m = 2$, the above formula gives

$$\psi(x) = \frac{4}{5}\phi(2x-2) - \frac{1}{2}(\phi(2x-1) + \phi(2x-3)) + \frac{1}{12}(\phi(2x) + \phi(2x-4)).$$

Another construction has been proposed in [57]. Its advantage is that it also works for non-uniform partitions and higher-dimensional situations. For C^0 Lagrange finite element spaces, this extension is briefly described in the next subsection. Its disadvantage is that when $m > 2$, the resulting masks $\mathbf{b}$ and support intervals of the associated ψ become larger. In the shift and dilation invariant setting, the core idea is to start with two locally supported functions θ and ξ in V_1, such that the systems

$$\Theta_1 = \{\theta(\cdot - l)\}_{l\in\mathbb{Z}} \quad \text{and} \quad \Xi_1 = \{\xi(\cdot - l)\}_{l\in\mathbb{Z}}$$

of their integer shifts satisfy two properties: First, $\Theta_1 \cup \Xi_1$ forms a Riesz basis in V_1 and second, Θ_1 is biorthogonal to Φ_0, i.e.,

$$(\theta, \phi(\cdot - l))_{L^2} = \delta_{0l}, \qquad l \in \mathbb{Z}.$$

If this is satisfied, then the system Ψ_1 generated by the integer shifts from

$$\psi = \xi - \sum_{k\in\mathbb{Z}} (\xi, \phi(\cdot - k))_{L^2}\theta(\cdot - k)$$

is a Riesz basis in $W_1 = V_1 \ominus V_0$, see [57, Theorem 3.1] for a more general statement. To give an example, let $m = 2$ in the above setting, where V_0 and V_1 are spaces of linear splines with respect to $\mathcal{T}_0$ and $\mathcal{T}_1$, respectively. For reasons of symmetry and to stay close to the construction of [40], set $\xi(x) = \phi(2x - 2)$ and determine $\theta(x) = \alpha_0\phi(2x) + \alpha_1\phi(2x - 1) + \alpha_2\phi(2x - 2)$, so that the required biorthogonality condition holds. This gives $\alpha_0 = \alpha_2 = -1/2$ and $\alpha_1 = 3$. The Riesz basis property of $\Theta_1 \cup \Xi_1$ can be derived from the Riesz basis property of Φ_1 in a straightforward way. Since $(\xi, \phi(\cdot - k))_{L^2} = 1/4$ for $k = 0, 1$ are the only non-zero scalar products in the above formula, we get

$$\psi(x) = \frac{5}{4}\phi(2x - 2) - \frac{3}{4}(\phi(2x - 1) + \phi(2x - 3)) + \frac{1}{8}(\phi(2x) + \phi(2x - 4)),$$

which, not surprisingly, is a multiple of the corresponding result from [40].

Another example is the construction [44] of *biorthogonal wavelet systems* with compactly supported functions ϕ, ψ and ϕ^*, ψ^* for generating the subspace collections $\mathbf{V}$, $\mathbf{W}$ and $\mathbf{V}^*$, $\mathbf{W}^*$ to which Theorem 3.2 with $H = L^2(\mathbb{R})$ applies. The result is that the appropriately scaled systems $\tilde{\Psi}$ and $\tilde{\Psi}^*$ are Riesz bases for $H^s(\mathbb{R})$ in the range $-\tilde{\gamma} < s < \gamma$ and $-\gamma < s < \tilde{\gamma}$, respectively, where

$$\gamma = \sup(\{s : \ \phi \in H^s(\mathbb{R})\}), \qquad \tilde{\gamma} = \sup(\{s : \ \tilde{\phi} \in H^s(\mathbb{R})\}).$$

One also has wavelet decompositions

$$f = \sum_{l\in\mathbb{Z}} (f, \tilde{\phi}_{0,l})_{L^2}\phi_{0,l} + \sum_{j=1}^{\infty}\sum_{l\in\mathbb{Z}} (f, \tilde{\psi}_{j,l})_{L^2}\psi_{j,l}$$

with respect to $\tilde{\Psi}$, which converge unconditionally in the respective $H^s(\mathbb{R})$ norms, similarly for $\tilde{\Psi}^*$. The operators Q_j and Q_j^* needed for the application of Theorem 3.2 are given by the respective partial sums of these wavelet decompositions.

The wavelet coefficients with respect to $\tilde{\Psi}$ depend on the properties of the biorthogonal system $\tilde{\Psi}^*$. More importantly, for better compression of functions and operators it is desirable to enforce *moment conditions*

$$\int_{\mathbb{R}} \psi_{j,l}(x)x^k\,dx = 0, \qquad k = 0, 1, \ldots, m^*,$$

up to high order m^* for the wavelets $\psi_{j,l}$ in $\tilde{\Psi}$. Since by biorthogonality all such $\psi_{j,l}$ are L^2 orthogonal to V_0^*, this condition is satisfied if V_0^* contains polynomials of order m^* (i.e., of degree $m^* - 1$). In the orthogonal case, however, we have $V_0^* = V_0$ and there is no freedom to ensure arbitrarily high order of the moment conditions.

The algebraic part of the construction is relatively simple due to the shift and dilation invariance assumptions and boils down to properties of the polynomials associated with the masks. We give some details for the most popular example of biorthogonal spline wavelets [44], where ϕ is again the m-th order B-spline, while the order m^* of the moment conditions for ψ is arbitrary. However, we do not provide formal proofs and neglect technical details such as the subtle issues of verifying the existence of $\tilde{\phi}^*$ and determining the parameter $\tilde{\gamma}^*$ for a given finitely supported mask $\tilde{\mathbf{a}}$, see [131] for a more recent account on wavelet theory. Now, introduce the *symbol* associated with a mask $\mathbf{a}$ by the formula

$$m_{\mathbf{a}}(\theta) = \frac{1}{2}\sum_{l\in\mathbb{Z}} a_l e^{-il\theta},$$

which, in the finitely supported case, is a complex-valued 2π-periodic trigonometric polynomial. Alternatively, the z-transform of $\mathbf{a}$, a Laurent polynomial on the unit circle $|z| = 1$ in $\mathbb{C}$, is used, i.e.,

$$p_{\mathbf{a}}(z) = \frac{1}{2}\sum_{l\in\mathbb{Z}} a_l z^l \quad \text{and} \quad m_{\mathbf{a}}(\theta) = p_{\mathbf{a}}(e^{-i\theta}).$$

It is easy to verify [131, Chapter 1] that the refinability of two functions ϕ and ϕ^* from $L^2(\mathbb{R})$ and the biorthogonality of the associated systems Φ_0 and Φ_0^* of their integer shifts is equivalent to verifying

$$\hat{\phi}(\theta) = m_{\mathbf{a}}(\theta/2)\hat{\phi}(\theta/2), \qquad \hat{\phi}^*(\theta) = m_{\mathbf{a}^*}(\theta/2)\hat{\phi}^*(\theta/2)$$

for a.e. $\theta \in \mathbb{R}$ and $m_{\mathbf{a}}(\theta)\overline{m_{\mathbf{a}^*}(\theta)} + m_{\mathbf{a}}(\theta+\pi)\overline{m_{\mathbf{a}^*}(\theta+\pi)} = 1, \theta \in [0, \pi)$, respectively. Note the slight differences in the scaling constants and in the definition of Fourier transforms compared to [44, 131]. In terms of z-transforms, assuming real-valued masks $\mathbf{a}$ and $\mathbf{a}^*$, the biorthogonality condition can be rewritten as

$$p_{\mathbf{a}}(z)p_{\mathbf{a}^*}(1/z) + p_{\mathbf{a}}(-z)p_{\mathbf{a}^*}(-1/z) = 1 \tag{3.32}$$

for $|z| = 1$. See Problem 3.7 for details.

Moreover, in order for ψ to have moment conditions of order m^* or, equivalently, for V_0^* to contain polynomials of degree m^*-1 locally, the Laurent polynomial $p_{\mathbf{a}^*}(z)$ must factorize as

$$p_{\mathbf{a}^*}(z) = \left(\frac{1+1/z}{2}\right)^{m^*} q_{\mathbf{a}^*}(1/z),$$

where $q_{\mathbf{a}^*}$ is another Laurent polynomial. This is a necessary and sufficient condition if Φ_0^* is a Riesz basis in V_0^*, see for instance [131, Theorem 3.4.16]. Finally, if (3.32) holds and Φ_0 and Φ_0^* are Riesz bases in V_0 and V_0^*, respectively, then

$$\psi(x) := \sum_{n\in\mathbb{Z}} (-1)^n a^*_{-(n+1)} \phi(2x-n)$$

and

$$\psi^*(x) := \sum_{n\in\mathbb{Z}} (-1)^n a_{-(n+1)} \phi^*(2x-n)$$

automatically generate Riesz bases in W_1 and W_1^*, respectively, and are biorthogonal to each other, see [131, Theorem 1.3.7]. The corresponding symbols take the form

$$p_{\mathbf{b}}(z) = -zp_{\mathbf{a}^*}(-1/z) = -z\left(\frac{1-z}{2}\right)^{m^*} q_{\mathbf{a}^*}(-z), \qquad p_{\mathbf{b}^*}(z) = -zp_{\mathbf{a}}(-1/z).$$

In other words, the wavelet masks $\mathbf{b}$ and $\mathbf{b}^*$ can be derived from the masks $\mathbf{a}^*$ and $\mathbf{a}$ in a straightforward manner.

We now apply this to derive biorthogonal spline wavelet systems, starting from the m-th order B-spline ϕ with symbol

$$p_{\mathbf{a}}(z) = \left(\frac{1+z}{2}\right)^m.$$

We know that Φ_0 has the required Riesz basis property for V_0 and that $\gamma = m-1/2$. To construct an appropriate ϕ^*, we must satisfy (3.32). By substitution, this is equivalent to the condition

$$q_{\mathbf{a}^*}(z)\left(\frac{1+z}{2}\right)^{m+m^*} + q_{\mathbf{a}^*}(-z)\left(\frac{1-z}{2}\right)^{m+m^*} = 1$$

for $q_{\mathbf{a}^*}(-z)$ or to the condition $q^*(z)+q^*(-z)=1$ for the Laurent polynomial

$$q^*(z) := q_{\mathbf{a}^*}(z)\left(\frac{1+z}{2}\right)^{m+m^*}.$$

So we need to look for $q^*(z)$ of the form

$$q^*(z) = \frac{1}{2} + \sum_{k\in\mathbb{Z}} \beta_k z^{2k-1},$$

with finitely many real coefficients $\beta_k \neq 0$ which has an algebraic root of multiplicity $m + m^*$ at $z = -1$. Thus, determining q^* and, by factorization, $q_{\mathbf{a}^*}$ and masks $\mathbf{a}^*$ associated with candidate dual scaling functions becomes a linear algebra problem with infinitely many solutions. Masks $\mathbf{a}^*$ of smallest possible length are obtained by the postulate

$$q^*(z) = \frac{1}{2} + \sum_{k=-\lfloor (m+m^*)/2 \rfloor+1}^{m+m^*-\lfloor (m+m^*)/2 \rfloor} \beta_k z^{2k-1}$$

and are (up to integer shifts and reflections) the masks for the (m, m^*) biorthogonal spline wavelets constructed in [44, Section 6A]. The crux is that for each $\mathbf{a}^*$ found in this way, the existence of a solution ϕ^* in $L^2(\mathbb{R})$ of the refinement equation and the Riesz basis property of Φ^* in V_0 must be checked. In essence, this amounts to finding estimates for the parameter γ^*. We refer to [131] for a detailed treatment of this non-trivial problem. In [44], under additional symmetry conditions on ϕ^*, detailed derivations and explicit formulas are given for even values of $m + m^*$, including lower bounds on γ^*.

We provide some examples for $m = 2$ (linear splines). For $m^* = 0$ (no moment conditions), the construction leads directly to

$$q^*(z) = \frac{1}{z}\left(\frac{1+z}{2}\right)^2, \quad q_{a^*}(z) = \frac{1}{z}, \quad p_{a^*}(z) = z \quad p_{\mathbf{b}}(z) = 1.$$

So we get $\psi(x) = 2\phi(2x)$. The Riesz basis property of the system Ψ_1 in its L^2 closure W_1 is obvious from the Riesz basis property of Φ_1 in V_1. The associated system $\tilde{\Psi}$ is the well-known hierarchical basis of hat functions on $\mathbb{R}$, which is known to be a Riesz basis in $H^s(\mathbb{R})$ if and only if $1/2 < s < 3/2$. Its restriction to $[0, 1]$ is essentially the classical Faber-Schauder system. The associated refinement equation $\phi^*(x) = 2\phi^*(2x - 1)$ has no solution in $L^2(\mathbb{R})$, its distributional solution is the Dirac function $\phi^*(x) = \delta(x - 1)$ with Fourier transform $\hat{\phi}^*(\theta) = e^{-i\theta}$. This results in $\gamma^* = -1/2$, which is consistent with the above statement.

Next, if $m^* = 1$, we have $m + m^* = 3$ and the above ansatz gives $\beta_0 = 3/16$, $\beta_1 = 6/16$ and $\beta_2 = -1/16$. From this we calculate $q_{\mathbf{a}^*}(z) = (3/z - 1)/2$ and

$$p_{\mathbf{a}^*}(z) = \frac{3}{4}z + \frac{1}{2} - \frac{1}{4z}.$$

This gives the non-zero mask coefficients $a^*_{-1} = -\frac{1}{2}$, $a^*_0 = 1$ and $a^*_1 = \frac{3}{2}$, and the formula for the candidate wavelet is

$$\psi(x) = \frac{3}{2}\phi(2x - 2) - \phi(2x - 1) - \frac{1}{2}\phi(2x) = 2\phi(2x - 2) - \phi(x).$$

This $\psi(x)$ is not symmetric and has not appeared in the literature, probably because the associated $\phi^*(x)$ does not belong to $L^2(\mathbb{R})$ and $\gamma^* = 1 - \log_2(\sqrt{5}) = -0.1609\ldots$ is negative.

Now let $m^* = 2$. Then one checks $\beta_{-1} = \beta_2 = -\frac{1}{32}$, $\beta_0 = \beta_1 = \frac{9}{32}$. After some elementary calculations, one recovers the symmetric biorthogonal $(2, 2)$ spline wavelet from [44] with

$$\psi(x) = \frac{3}{2}\phi(2x) - \frac{1}{2}(\phi(2x-1) + \phi(2x+1)) - \frac{1}{4}(\phi(2x-2) + \phi(2x+2)),$$
$$= 2\phi(2x) - \frac{1}{2}(\phi(x) + \phi(x+1))$$

for which the Riesz basis property of Ψ_1 in W_1 follows with some $\gamma^* > 0.158$. Differences in constant scaling factors are due to different normalizations. Note that this ψ has the same mask support as the semi-orthogonal prewavelet for $m = 2$. The latter has a larger value $\gamma^* = \gamma = 3/2$ and is L^2 orthogonal to all functions of V_0 instead of just linear polynomials, but its dual ψ^* has global support. Figure 3.5 shows graphs of some of the semi-orthogonal and biorthogonal linear spline wavelets $(m = 2)$ discussed above.

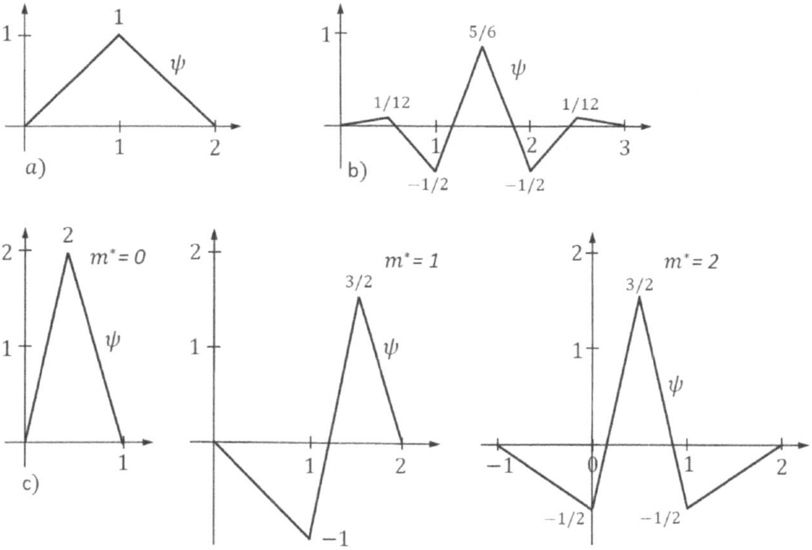

Fig. 3.5 Linear spline wavelets ($m = 2$). a) B-spline scaling function ϕ. b) Semi-orthogonal ψ from [40]. c) Biorthogonal ψ with vanishing moments of order $m^* = 0, 1, 2$ from [44].

It is no coincidence that the biorthogonal wavelets ψ in the above examples all have the form of a linear combination of a B-spline basis function $\phi_{1,2k}$ from V_1 and a few basis functions $\phi_{0,l}$ from V_0. There are general methods (the lifting scheme [176] and stable completions [33]) to construct compactly supported biorthogonal wavelets with this typical property.

There are other approaches to the construction of wavelet systems. If one is only interested in $\tilde{\Psi}$ as a tool for dicretizing operator equations in Sobolev spaces, then explicit knowledge of the system $\tilde{\Psi}^*$ is not necessary. In this case, the construction of $\tilde{\Psi}$ often starts with a suitable pair of compactly supported functions $(\phi, \psi) \in V_0 \times V_1$ with the property that the system $\Phi_0 \cup \Psi_1$ forms a Riesz basis in V_1. Its symbols $m_{\mathbf{a}}$ and $m_{\mathbf{b}}$ determine a biorthogonal pair (ϕ^*, ψ^*) in a canonical way (see Problem 3.7). It can be shown that the resulting $m_{\mathbf{a}^*}$ is a rational trigonometric function or, equivalently, $p_{\mathbf{a}^*}$ is a rational function in z. This means that in general $\mathbf{a}^*$ is not finitely supported, but that the mask coefficients a_k^* decay exponentially as $|k| \to \infty$. Under these circumstances, the existence and smoothness properties of the dual scaling function ϕ^* can still be investigated in a similar way, and (if the assumptions of Theorem 3.2 hold) the Riesz basis property of $\tilde{\Psi}_s$ in $H^s(\mathbb{R})$ for the range $-\gamma^* < s < \gamma$ can again be shown. We refer to [43, 116] for details, generalizations to $d > 1$, and references.

We will only briefly mention a family of such pairs (ϕ, ψ) related to Butterworth filters in signal processing, denoted by the symbols

$$m_{\mathbf{a}}(\theta) = \left(\frac{1 + e^{-i\theta}}{2}\right)^m, \qquad m_{\mathbf{b}}(\theta) = e^{ia\theta}\left(\frac{1 - e^{-i\theta}}{2}\right)^{m^*},$$

where $m, m^* \in \mathbb{Z}_+$ and the choice of $a = 0$ or $a = 1$ depends on $m + m^*$. Obviously, the associated scaling function ϕ is again the B-spline of order m (the case $m = 0$ formally corresponds to the distribution $\phi(x) = \delta(x)$). The complementing $\psi \in V_1$ automatically satisfies moment conditions of order m^*. Compared to the biorthogonal spline wavelets with the same parameters m, m^* from [44], which have similar properties, the new masks $\mathbf{b}$ have smaller support, but the dual scaling function ϕ^* has global support. It turns out [116, Section 3.3] that for $m = 2$ the associated linear spline system $\tilde{\Psi}$ forms a Riesz basis in $H^s(\mathbb{R})$ for $-\gamma_{m^*}^* < s < 3/2$, $m^* \geq 1$. Figure 3.6 shows the graphs of these linear spline wavelets ($m = 2$) for $1 \leq m^* \leq 3$ vanishing moments, where

$$\gamma_1^* = 0.044\ldots, \quad \gamma_2^* = 0.990\ldots, \quad \gamma_3^* = 1.648\ldots.$$

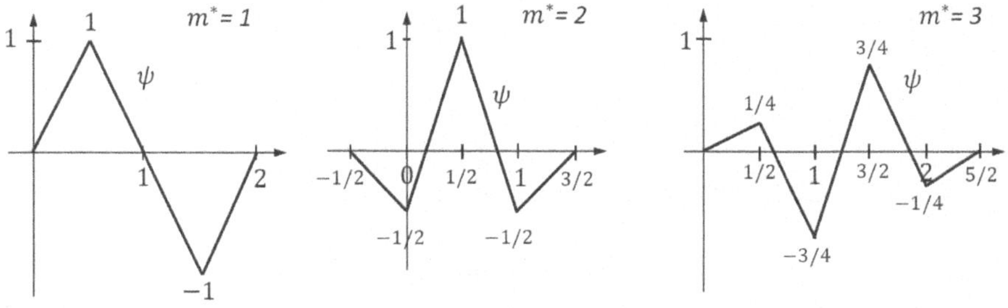

Fig. 3.6 Linear spline wavelets ($m = 2$) with vanishing moments of order $m^* = 1, 2, 3$ and small support from [116].

3.1.4.2 Wavelet systems on domains

In this subsection we briefly discuss the adaptation of the wavelet concept to bounded domains $G \subset \mathbb{R}^d$, which means to modify (or give up) the shift and dilation invariance of the construction. The latter is at the core of wavelet analysis and allows the use of Fourier transform techniques. To avoid problems with our notation, let us assume that G contains at least one cube of side-length ≈ 1, which makes our above fixation of V_0 as the coarsest scale of resolution plausible.

First, in accordance with the definition of function spaces on domains, for $d = 1$ wavelet systems can be constructed by restriction to intervals. For the Haar system and $[0, 1]$ this is a trivial fact, for spline systems with $m > 1$ there is obviously a problem near the endpoints of the interval. The remedy is boundary modifications that preserve L^2 orthogonality or biorthogonality and moment conditions, and can also account for essential boundary conditions, see [56] and the recent paper [98]. In Figure 3.7 we show the boundary modification for the semi-orthogonal linear spline wavelet ($m = 2$) on uniform partitions, compare with Figure 3.6 b) where the corresponding "inner" semi-orthogonal spline wavelets are given.

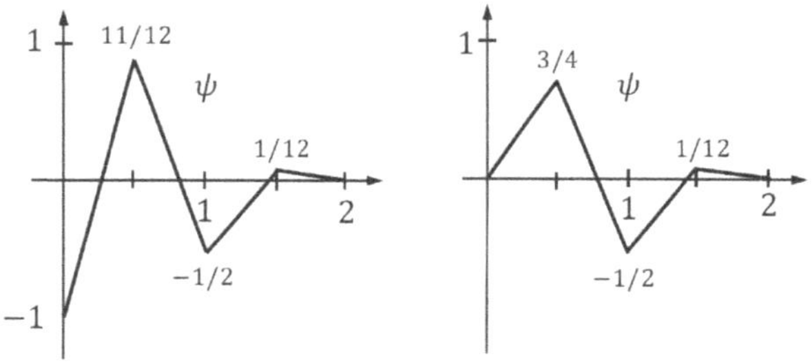

Fig. 3.7 Boundary modification for semi-orthogonal linear spline wavelets ($m = 2$) at the left interval endpoint $x = 0$ without (left) and with zero boundary conditions (right) serving H^1 and H_0^1 elliptic problems, respectively.

For $d > 1$, one possibility is to use tensor product constructions. We describe the idea for $d = 2$. Given two univariate wavelet systems $\tilde{\Psi}^i$ on $\mathbb{R}$ or on two intervals $G_i = [a_i, b_i]$, $i = 1, 2$, we can either define

$$\tilde{\Psi}_S := \tilde{\Psi}^1 \otimes \tilde{\Psi}^2 = \cup_{(l,k)\in\mathbb{Z}^2} \Psi_l^1 \otimes \Psi_k^2,$$

where we have set $\Psi_0^i = \Phi_0^i$ for notational convenience, or

$$\tilde{\Psi}_N := \cup_{j\geq 0} \Psi_{N,j}$$

as a wavelet system on $\mathbb{R}^2$ or on the rectangle $G = G_1 \times G_2$, respectively. Here,

$$\Psi_{N,0} = \Phi_0^1 \otimes \Phi_0^2, \quad \Psi_{N,j} = (\Phi_{j-1}^1 \otimes \Psi_j^2) \cup (\Psi_j^1 \otimes \Phi_{j-1}^2) \cup (\Psi_j^1 \otimes \Psi_j^2), \quad j \geq 1.$$

These two versions, which lead to the so-called standard and non-standard wavelet representations of operators [12], differ in the two-dimensional supports of the resulting wavelet functions and in the order of the moment conditions. On one hand, while wavelet functions in $\Psi_{N,j}$ are supported on rectangles of size $\approx 2^{-j}$ in both directions (isotropic case), the functions in $\Psi_l^1 \otimes \Psi_k^2$ are supported on rectangles of length $\approx 2^{-l}$ in x_1-direction and $\approx 2^{-k}$ in x_2-direction (anisotropic case) which is not always desirable in some applications. Note that

$$\operatorname{span}(\cup_{j=0}^{n}\Psi_{N,j}) = \operatorname{span}(\cup_{0\le k,l\le n}\Psi_l^1 \otimes \Psi_k^2)$$

and

$$\operatorname{span}(\Psi_{N,j}) = \operatorname{span}(\cup_{k=j \text{ or } l=j}\Psi_l^1 \otimes \Psi_k^2),$$

which can be used to transform representations with respect to $\tilde{\Psi}_N$ into representations with respect to $\tilde{\Psi}_S$ and vice versa.

On the other hand, if the univariate wavelets ψ^i satisfy moment conditions of order m^*, then moment conditions

$$\int_{\mathbb{R}^2} v_{k,l}\, p\, dx = 0, \qquad v_{k,l} \in \operatorname{span}(\Psi_l^1 \otimes \Psi_k^2), \quad k,l \ge 1,$$

hold for all polynomials p of coordinatewise degree $m^* - 1$, while the wavelets in $\Psi_{N,j}$, $j \ge 0$, satisfy only moment conditions of order m^*, i.e., they are only orthogonal to polynomials of total degree $m^* - 1$. This is relevant for the compression of functions and operators.

Now that we have

$$H^s(\mathbb{R}^2) \approx H^s(\mathbb{R}) \otimes L^2(\mathbb{R}) \cap L^2(\mathbb{R}) \otimes H^s(\mathbb{R}), \qquad H^s_{mix}(\mathbb{R}^2) \approx H^s(\mathbb{R}) \otimes H^s(\mathbb{R}),$$

for $s \ge 0$, Riesz basis properties in Sobolev spaces of the univariate systems carry over to the two-dimensional setting. For $s < 0$ duality arguments must be used. A similar reasoning applies to tensor product wavelet systems in Sobolev spaces on $\mathbb{R}^d$ and on rectangular domains $G \subset \mathbb{R}^d$. Continuing in this direction, attempts have been made to generalize the construction of wavelet systems to a certain class of domains $G \subset \mathbb{R}^d$ and d-dimensional manifolds, suitably composed of a finite number of d-dimensional rectangles. We refer to [101] for an introduction, with references and applications to the solution of boundary integral equations.

Alternatively, multivariate wavelet constructions without tensor product structure (called *non-separable*) have attracted attention. Again, one starts with a scaling function $\phi(x) \in L^2(\mathbb{R}^d)$, this time satisfying a refinement equation of the form

$$\phi(x) = \sum_{\alpha\in\mathbf{Z}^d} a_\alpha \phi(Mx - \alpha), \qquad x \in \mathbb{R}^d,$$

where M is a general *dilation matrix* of size $d \times d$ with integer entries, determinant $\det(M) > 1$, and spectral radius $r(M) > 1$. There is also the possibility of starting from several scaling functions (the so-called multiwavelet setting), which we will not go into, but see [99]. For $d = 2$, typical examples of dilation matrices are

$$M = \begin{pmatrix} 1 & -1 \\ 1 & 1 \end{pmatrix}, \quad M = \begin{pmatrix} 2 & -1 \\ 1 & 1 \end{pmatrix}, \quad M = \begin{pmatrix} 2 & 0 \\ 0 & 2 \end{pmatrix},$$

with $\det(M) = 2, 3, 4$, which are related to so-called quincunx (or $\sqrt{2}$-), $\sqrt{3}$- and dyadic refinement, respectively. Box splines are often used as scaling functions. The spaces V_j and systems Φ_j are constructed by analogy (in particular, $v_j \in V_j$ if and only if $v_j(M^{-j}\cdot) \in V_0$), and the Riesz basis property of Φ_0 in V_0 is assumed.

Given M and ϕ, one then tries to find $\det(M) - 1$ wavelets $\psi^k \in V_1$ that satisfy

$$\psi^k(x) = \sum_{\alpha \in \mathbf{Z}^d} b^k_\alpha \phi(Mx - \alpha), \qquad x \in \mathbb{R}^d, \quad k = 1, \ldots, \det(M) - 1,$$

so that the union of Φ_0 and the system

$$\Psi_1 = \{\psi^k(\cdot - l) : l \in \mathbb{Z}^d, \, k = 1, \ldots, \det(M) - 1\}$$

is a Riesz basis in V_1. For further analysis, Fourier transform techniques can be used as for $d = 1$. Again, properties such as biorthogonality, moment conditions, and smoothness of the scaling functions, which lead to statements about Riesz splittings for Sobolev spaces $H^s(\mathbb{R}^d)$ based on Theorem 3.2, can be formulated in terms of symbols. Of course, there are technical difficulties, especially when determining parameters such as γ^*, which increase with the dimension d.

Concerning the characterization of Sobolev spaces on domains $G \subset \mathbb{R}^d$ by wavelets, there are results [183, 184, 185] for quite general domains G, including bounded Lipschitz domains. They are formulated in terms of compactly supported wavelets, but involve a rather complicated boundary modification, which makes them unattractive for numerical applications. Moreover, they do not always characterize $H^s(G)$ as defined in Section A.5 of the appendix, but only certain proper subspaces of $H^s(G)$.

This remark also applies to the characterization of function spaces on domains by so-called *quarkonial decompositions*. We mention them briefly here because of their connection with hp-finite element methods [160]. The results in [183, Chapter 2 and 6] on the frame property of suitably normalized quarkonial systems in Besov-Lizorkin-Triebel spaces are formulated for bounded C^∞ domains $G \subset \mathbb{R}^d$ and can be generalized to Lipschitz domains. Roughly speaking, a quarkonial system in G is a collection of functions (called quarks)

$$\phi_{j,l,p}(x) = (x - x_{j,l})^p \phi_{j,l}(x), \quad l = 1, \ldots, n_j, \quad j \geq 0, \quad p \geq 0,$$

where the $\phi_{j,l}$, $l = 1, \ldots, n_j$, are supported in balls (or cubes) inside G of diameter $\approx 2^{-j}$ with center $x_{j,l}$ and distance $\geq c2^{-j}$ from Γ, and form (up to a boundary strip of thickness $\approx 2^{-j}$) a smooth approximate partition of unity in G. A detailed description is given in [183, Chapter 6]. For each $j \geq 0$, the centers $x_{j,l}$ form an approximate lattice of step-size $\approx 2^{-j}$ that replaces the uniform lattices $2^{-j}\mathbb{Z}^d$ in $\mathbb{R}^d$

to which the wavelet functions in $\Psi_{N,j}$ can be attached. The new feature compared to the wavelet setting is the additional degree of freedom p, which allows us to locally create polynomials of arbitrarily high degree by linear combination of quarks via the prefactors $(x-x_{j,l})^p$. This is similar to hp-finite element methods, which use spaces of continuous piecewise polynomial functions defined over a cell partition $\mathcal{T}$ of a polyhedral domain G, where, in contrast to C^0 Lagrange finite element spaces of fixed degree p, the polynomial degree can vary from cell to cell. We refer to [52] for a specific construction of quarks suitable for numerical applications. It starts with quarkonial systems on intervals in $\mathbb{R}$ including the appropriate boundary modification and the incorporation of moment conditions, and discusses the extension to certain classes of domains $G \subset \mathbb{R}^d$.

As a further example of wavelet systems on domains $G \subset \mathbb{R}^d$, we conclude this subsection with some details on the construction of semi-orthogonal linear finite element wavelets on partition sequences associated with the collection $\mathbf{V}$ of C^0 Lagrange finite element subspaces considered in Theorem 3.3 (case $p = 1$). We follow [57], which also treats the general case $p > 1$. We restrict ourselves to simplicial partitions $\mathcal{T}_j$ of a polyhedral domain G obtained by uniform quadrisection refinement from $\mathcal{T}_0$. This allows us to obtain the Riesz basis property of the resulting scaled system $\tilde{\Psi}_s$ in Sobolev spaces for all $|s| < 3/2$ via Theorem 3.2. In essence, this means to construct the systems Ψ_j, $j \geq 0$, consisting of locally supported $\psi_{j,l} \in W_j = V_j \ominus V_{j-1}$ and to prove that they form a Riesz basis for W_j (as subspace of $L^2(G)$) with Riesz constants independent of $j \geq 1$.

As in the univariate shift-invariant case discussed in the previous subsection, we first construct appropriate systems $\Theta_j := \{\theta_{j,k}\} \subset V_j$ and $\Xi_j := \{\xi_{j,l}\} \subset V_j$ such that Θ_j (after suitable scaling of the $\theta_{j,i}$) is biorthogonal with the nodal basis Φ_{j-1} in V_{j-1}. Moreover, the combined system $\Theta_j \cup \Xi_j$ must form a Riesz basis in V_j with uniform control on the Riesz constants. A dimension check suggests to associate the $\theta_{j,k}$ with the suitably enumerated nodal points P_k in V_{j-1} (i.e., the vertices P_k of $\mathcal{T}_{j-1}$) and to associate the $\xi_{j,l}$ with the "new" nodal points in $\mathcal{T}_j$ (i.e., the midpoints M_{e_l} of the correspondingly enumerated edges e_l in $\mathcal{T}_{j-1}$). Then, the candidates for the semi-orthogonal finite element wavelets are determined by

$$\psi_{j,l} := \xi_{j,l} - \sum_k \frac{(\xi_{j,l}, N_{j-1,P_k})_{L^2}}{(\theta_{j,k}, N_{j-1,P_k})_{L^2}} \theta_{j,k}, \tag{3.33}$$

where N_{j-1,P_k} denotes the nodal basis functions in V_{j-1}.

It is easy to check that the functions

$$\theta_{j,k} = N_{j,P_k} - a_d N_{j-1,P_k}, \qquad a_d := 2^{-(d+1)} \in (0, \frac{1}{4}], \tag{3.34}$$

satisfy the orthogonality condition with respect to the remaining nodal basis functions $N_{j-1,P_{k'}}$ in V_{j-1}, i.e.,

$$(N_{j-1,P_{k'}}, \theta_{j,k})_{L^2} = 0, \quad k' \neq k,$$

for all nodal points $P_k, P_{k'}$ in $\mathcal{T}_{j-1}$, while

$$(N_{j-1,P_k}, \theta_{j,k})_{L^2} = (1, \theta_{j,k})_{L^2} = \frac{1}{2}(1, N_{j,P_k}) > 0$$

for all nodal points P_k in $\mathcal{T}_{j-1}$. This is sufficient for (3.33) to be well-defined and to ensure that the $\psi_{j,l}$ are orthogonal to V_{j-1}. We refer to Problem 3.9 for details.

It is customary to now set $\xi_{j,l} = N_{j,M_{e_l}}$. To check the Riesz basis property of $\Theta_j \cup \Xi_j$ in V_j, consider an arbitrary linear combination with respect to this system and rewrite it with respect to Φ_j, i.e.,

$$\begin{aligned} v_j &= \sum_k c_k \theta_{j,k} + \sum_l d_l \xi_{j,l} \\ &= (1 - a_d) \sum_k c_k N_{j,P_k} + \sum_l (d_l - \frac{a_d}{2}(c_l^+ + c_l^-)) N_{j,M_{e_l}}, \end{aligned}$$

where $c_l^{\pm}$ is a shorthand notation for the coefficients c_k in front of the two $\theta_{j,k}$ associated with the endpoints of the edge e_l. Note that except of the constant a_d, this coefficient transformation is independent of the dimension d. By the Riesz basis property of the nodal basis Φ_j in V_j, we therefore have

$$\begin{aligned} \|v_j\|_{L^2}^2 &\approx 2^{-2dj} \left((1 - a_d)^2 \textstyle\sum_k c_k^2 + \sum_l (d_l - \frac{a_d}{2}(c_l^+ + c_l^-))^2\right) \\ &\approx 2^{-2dj} \left(\textstyle\sum_k c_k^2 + \sum_l d_l^2\right). \end{aligned} \tag{3.35}$$

Here the last step in (3.35) follows from the elementary estimates

$$(d_l - \frac{a_d}{2}(c_l^+ + c_l^-))^2 \le C(d_l^2 + (c_l^+)^2 + (c_l^-)^2)$$

and

$$d_l^2 \le C((d_l - \frac{a_d}{2}(c_l^+ + c_l^-))^2 + (c_l^+)^2 + (c_l^-)^2),$$

and the fact that the number of edges e_l emanating from any given vertex P_k in $\mathcal{T}_{j-1}$ is uniformly bounded. Obviously, (3.35) establishes the Riesz basis property of $\Theta_j \cup \Xi_j$ in V_j, with lower and upper Riesz bounds $\approx 2^{-2j}$.

It follows almost automatically that the system $\Psi_j = \{\psi_{j,l}\}$ defined by (3.33) is a Riesz basis in W_1. Consider an arbitrary linear combination from Ψ_j and rewrite it in the basis $\Theta_j \cup \Xi_j$ using (3.33)

$$w_j := \sum_l d_l \psi_{j,l} = \sum_l d_l N_{j,M_{e_l}} - \sum_k \underbrace{\left(\sum_l d_l \frac{(\xi_{j,l}, N_{j-1,P_k})_{L^2}}{(\theta_{j,k}, N_{j-1,P_k})_{L^2}} \right)}_{c_k :=} \theta_{j,k}.$$

The sum defining c_k has a uniformly bounded number of non-zero contributions, since by construction $(\xi_{j,l}, N_{j-1,P_k})_{L^2} \neq 0$ only for the edges e_l emanating from P_k. Moreover, the quotients $(\xi_{j,l}, N_{j-1,P_k})_{L^2} / (\theta_{j,k}, N_{j-1,P_k})_{L^2}$ are also uniformly bounded due to our construction of Θ_j and Ξ_j and the regularity and quasi-uniformity properties of the triangulations. Thus we have

$$\sum_k c_k^2 \leq C \sum_k \sum_{P_k \in e_l} d_l^2 \leq C \sum_l d_l^2,$$

and, together with the already established Riesz basis property (3.35) of $\Theta_j \cup \Xi_j$ in $V_j \supset W_j$, we arrive at

$$\|w_j\|_{L^2}^2 \approx 2^{-2dj} \left(\sum_k c_k^2 + \sum_l d_l^2 \right) \approx 2^{-2dj} \sum_l d_l^2.$$

This proves the claim.

These considerations, in conjunction with Theorems 3.3 and 3.2, lead to the following result:

Theorem 3.5 *Assume in Theorem 3.3 that $G \subset \mathbb{R}^d$ is a bounded polyhedral Lipschitz domain equipped with a finite simplicial partition $\mathcal{T}_0$ into simplices of diameter ≈ 1 and that the $\mathcal{T}_j$, $j \geq 1$, are obtained by uniform dyadic refinement from $\mathcal{T}_0$. Set $p = 1$ and consider the linear finite element spaces $V_j = V(\mathcal{T}_j)$ equipped with the nodal bases $\Phi_j = \{N_{j,P}\}$, where P passes through the set of nodal points associated with V_j. Let $\Psi_j = \{\psi_{j,l}\}$ denote the semi-orthogonal finite element wavelet systems in $W_j = V_j \ominus V_{j-1}$, $j \geq 1$, defined by (3.34) and (3.33). Then the scaled system*

$$\tilde{\Psi}_s = \Phi_0 \cup 2^{d/2-s}\Psi_1 \cup \cdots \cup 2^{j(d/2-s)}\Psi_j \cup \cdots$$

is a Riesz basis in $H^s(G)$ if $0 \leq s < 3/2$ and is a Riesz basis in $H^{-s}(G)'$ if $-3/2 < s < 0$.

Note that as a consequence of Theorems 3.1 and 3.3, the similarly defined system

$$\tilde{\Phi}_s = \Phi_0 \cup 2^{d/2-s}\Phi_1 \cup \cdots \cup 2^{j(d/2-s)}\Phi_j \cup \cdots$$

consisting of the properly scaled hat functions of all spaces V_j, $j \geq 0$, is a frame in $H^s(G), 0 < s < 3/2$.

Since both V_{j-1} and W_j are subspaces of V_j, with basis functions of local support, the transformation of the representations with respect to the bases Φ_{j-1} and Ψ_j of finite element functions in V_{j-1} and W_j, respectively, to their representations with respect to Φ_j is governed by sparse matrices. The cost complexity of these transformations, which enters into the execution time of iterative methods based on multiscale splittings, can be roughly estimated by the support size of the basis functions from Φ_{j-1} and Ψ_j in $\mathcal{T}_j$. For Φ_{j-1}, the support of N_{j-1,P_k} is given by the union of all simplices in $\mathcal{T}_{j-1}$ that share the vertex P_k. For Ψ_j, the support of $\psi_{j,l}$ is the union of all simplices in $\mathcal{T}_{j-1}$ that share one of the endpoints of the edge e_l, which is about twice as large as the support of a nodal basis function from Φ_{j-1}. This leads to some overhead when computing with semi-orthogonal finite element wavelet systems.

3.2 Auxiliary space splittings

Auxiliary space splittings are the formalization of a simple strategy often used in applications. Suppose we want to solve a variational problem

$$\hat{a}(\hat{u}, \hat{v}) = \langle \hat{f}, \hat{v} \rangle_{\hat{V}' \times \hat{V}} \qquad \forall\, \hat{v} \in \hat{V}, \tag{3.36}$$

which is given by a symmetric positive definite operator $\hat{A} : \hat{V} \to \hat{V}'$, and we are looking for an efficient iterative solver or just an approximate inverse for $\hat{A}$. If there is another similar, smaller or even larger, problem

$$a(u, v) = \langle f, v \rangle_{V' \times V} \qquad \forall\, v \in V, \tag{3.37}$$

with a cheap, not necessarily iterative solution method or preconditioner, one is tempted to use it for (3.36) as well. The *auxiliary space method* provides rules for connecting these two problems by introducing another Hilbert space $(W; b(\cdot, \cdot))$ in which both $(\hat{V}; \hat{a})$ and $(V; a)$ are embedded by certain linear operators $\hat{R}$ and R. Finally, bounded linear maps $Q : \hat{V} \to V$ and $\hat{Q} : V \to \hat{V}$ with nice approximation properties in W enter the construction of the switch from (3.36) to (3.37). Whether one ends up with a computationally efficient preconditioning scheme depends on the application.

The name auxiliary space method was coined in [195], but the idea is old and has been used in many papers. Early work was done in the context of the fictitious domain method, see [125]. Other references can be found in [190, Subsection 7.11], [24] and [137], from which the material of the next subsection is adapted.

3.2.1 Theory

To formulate the precise conditions for a theoretical framework of the auxiliary space method, we use the following notation: With

$$\|v\|_V^2 := a(v, v), \qquad \|\hat{v}\|_{\hat{V}}^2 := \hat{a}(\hat{v}, \hat{v}), \qquad \|w\|_W^2 := b(w, w),$$

we denote the norms for $v \in V$, $\hat{v} \in \hat{V}$, $w \in W$, respectively. Let $R : V \to W$ and $\hat{R} : \hat{V} \to W$ be bounded linear operators realizing the embeddings. In the following proof we assume that we have

$$\|\hat{v}\|_{\hat{V}}^2 \le C_{\hat{R}} \|\hat{R}\hat{v}\|_W^2, \qquad \hat{v} \in \hat{V}, \tag{3.38}$$

with a positive constant $C_{\hat{R}}$. In other words, $\hat{R}$ is coercive. Furthermore, we need bounded linear approximation operators $Q : \hat{V} \to V$ and $\hat{Q} : V \to \hat{V}$ which satisfy

$$\|Q\hat{v}\|_V^2 \le C_Q \|\hat{v}\|_{\hat{V}}^2, \qquad \|\hat{Q}v\|_{\hat{V}}^2 \le C_{\hat{Q}} \|v\|_V^2, \tag{3.39}$$

and

$$\|Rv - \hat{R}\hat{Q}v\|_W^2 \le C_V \|v\|_V^2, \qquad \|\hat{R}\hat{v} - RQ\hat{v}\|_W^2 \le C_{\hat{V}} \|\hat{v}\|_{\hat{V}}^2, \tag{3.40}$$

for arbitrary $v \in V$ and $\hat{v} \in \hat{V}$. The constants in (3.38), (3.39), and (3.40) will enter the constants in the norm equivalence that we will establish in the next theorem. The obvious relation to a stable space splitting for $(\hat{V}; \hat{a})$ involving the two spaces $(V; a)$ and $(W; b)$ will be discussed later.

Theorem 3.6 *Let the Hilbert spaces $(V; a)$, $(\hat{V}; \hat{a})$ and $(W; b)$ be given and assume that the operators $R, \hat{R}, Q, \hat{Q}$ introduced above satisfy the conditions (3.38), (3.39) and (3.40). Then the norm equivalence*

$$\hat{a}(\hat{u}, \hat{u}) \approx |||\hat{u}|||^2 := \inf_{v \in V,\ \hat{v} \in \hat{V}:\ \hat{u} = \hat{v} + \hat{Q}v} b(\hat{R}\hat{v}, \hat{R}\hat{v}) + \lambda a(v, v) \tag{3.41}$$

holds for $\hat{u} \in \hat{V}$ with constants depending on the constants in (3.38), (3.39) and (3.40) and the scaling factor $\lambda > 0$.

Proof The upper bound for $\hat{a}(\hat{u}, \hat{u}) = \|\hat{u}\|_{\hat{V}}^2$ can be obtained if, for an arbitrary representation $\hat{u} = \hat{v} + \hat{Q}v$, we first use the triangle inequality and then continue with (3.38) and the boundedness of $\hat{Q}$ from (3.39). This gives

$$\begin{aligned}\|\hat{u}\|_{\hat{V}}^2 &\le 2(\|\hat{v}\|_{\hat{V}}^2 + \|\hat{Q}v\|_{\hat{V}}^2) \le 2C_{\hat{R}} \|\hat{R}\hat{v}\|_W^2 + 2C_{\hat{Q}} \|v\|_V^2 \\ &\le 2 \max(C_{\hat{R}}, \lambda^{-1} C_{\hat{Q}})(\|\hat{R}\hat{v}\|_W^2 + \lambda \|v\|_V^2).\end{aligned}$$

It remains to take the infimum with respect to all representations of $\hat{u}$.

For the lower bound in (3.41), we define an appropriate representation of $\hat{u}$ as follows: Set $v = Q\hat{u}$ and $\hat{v} = \hat{u} - \hat{Q}v$. Consequently,

$$\hat{R}\hat{v} = (\hat{R}\hat{u} - RQ\hat{u}) + (Rv - \hat{R}\hat{Q}v).$$

So we see from (3.40) that the first term on the right-hand side of (3.41) is bounded by

$$\begin{aligned}b(\hat{R}\hat{v}, \hat{R}\hat{v}) = \|\hat{R}\hat{v}\|_W^2 &\le 2(\|\hat{R}\hat{u} - RQ\hat{u}\|_W^2 + \|Rv - \hat{R}\hat{Q}v\|_W^2) \\ &\le 2C_{\hat{V}} \|\hat{u}\|_{\hat{V}}^2 + 2C_V \|v\|_V^2.\end{aligned}$$

All that remains is to use the boundedness condition of Q from (3.39), i.e.,

$$\|v\|_V^2 = \|Q\hat{u}\|_V^2 \le C_Q \|\hat{u}\|_{\hat{V}}^2,$$

and we arrive at

$$|||\hat{u}|||^2 \le \|\hat{R}\hat{v}\|_W^2 + \lambda \|v\|_V^2 \le (2C_{\hat{V}} + (2C_V + \lambda)C_Q) \|\hat{u}\|_{\hat{V}}^2.$$

This shows the desired lower bound in (3.41) and concludes the proof. □

If one chooses $\lambda = C_{\hat{Q}}/C_{\hat{R}}$ as scaling parameter, then (3.41) holds with the concrete constants

$$(2C_{\hat{R}})^{-1}\|\hat{u}\|_{\hat{V}}^2 \le |||\hat{u}|||^2 \le (2(C_{\hat{V}} + C_V C_Q) + C_{\hat{Q}} C_Q / C_{\hat{R}})\|\hat{u}\|_{\hat{V}}^2.$$

This formula can be useful if one wants to optimize the condition number

$$\kappa \le 4C_{\hat{R}}(C_{\hat{V}} + C_V C_Q) + 2C_{\hat{Q}} C_Q$$

of the additive Schwarz operator associated with the stable space splitting behind the norm equivalence (3.41). To specify this space splitting, let $n = 2$, $\Omega = \{1, 2\}$, $V_1 = \hat{V}$, $a_1(\hat{u}, \hat{v}) = b(\hat{R}\hat{u}, \hat{R}\hat{v})$, $V_2 = V$, and $a_2(u, v) = \lambda a(u, v)$. As operators R_i, take the identity on $\hat{V}$ for R_1, and set $R_2 = \hat{Q}$. This defines a space splitting $(\mathbf{V}; \mathbf{a}; \mathbf{R})$ for $(\hat{V}; \hat{a})$ with weights $\rho_i = 1$ which is stable according to (3.41). The corresponding Schwarz operator is

$$P = ((\hat{R}'B\hat{R})^{-1} + \lambda^{-1}\hat{Q}A^{-1}\hat{Q}')\hat{A} : \hat{V} \to \hat{V},$$

where $B : W \to W'$ is the operator that produces the bilinear form b on W. This is because the Hilbert adjoint $R_1^* : \hat{V} \to \hat{V}$ is given by

$$\langle B\hat{R}R_1^*\hat{u}, \hat{R}\hat{v}\rangle_{W' \times W} = \langle \hat{R}'B\hat{R}R_1^*\hat{u}, \hat{v}\rangle_{\hat{V}' \times \hat{V}} = \langle \hat{A}\hat{u}, \hat{v}\rangle_{\hat{V}' \times \hat{V}},$$

i.e., it satisfies $R_1^* = (\hat{R}'B\hat{R})^{-1}\hat{A}$, where the existence of the inverse follows from the coercivity of $\hat{R}$, which is a consequence of (3.39). Similarly, $R_2^* : \hat{V} \to V$ is given as $R_2^* = A^{-1}\hat{Q}'\hat{A}$, since by definition

$$\langle AR_2^*\hat{u}, v\rangle_{V' \times V} = \langle \hat{A}\hat{u}, \hat{Q}v\rangle_{\hat{V}' \times \hat{V}} = \langle \hat{Q}'\hat{A}\hat{u}, v\rangle_{V' \times V}.$$

In other words, if A and $\hat{R}'B\hat{R}$ are easy to invert (or can be replaced by cheap approximate inverses), then

$$C = (\hat{R}'B\hat{R})^{-1} + \lambda^{-1}\hat{Q}A^{-1}\hat{Q}' : \hat{V}' \to \hat{V}$$

can be used as an approximate inverse for $\hat{A}$. Equivalently, one can use additive or multiplicative Schwarz methods with subproblems governed by variational problems in W and V, respectively, to solve the original variational problem (3.36). Their convergence properties can be judged by the constants in the norm equivalence (3.41). Note that these constants also implicitly depend on Q and R, even though the latter do not directly enter the algorithms.

The framework of Theorem 3.6 looks rather general and complicated, but it allows us to deal with situations where the relationship between the problems is not straightforward. It emphasizes the role of the auxiliary bilinear form b on W, which is only implicitly present in [24, 190, 195], and is related to a so-called smoother on $\hat{V}$. Furthermore, if we assume (3.39) and (3.40) but replace (3.38) by

$$\|v\|_V^2 \le C_R\|Rv\|_W^2, \qquad v \in V,$$

then by symmetry of the argument we also have the norm equivalence

$$a(u,u) \approx \inf_{\hat v \in \hat V,\, v\in V:\, u=v+Q\hat v} b(Rv,Rv) + \lambda \hat a(\hat v,\hat v), \qquad u \in V,$$

which can be used to switch from a problem on $(V;a)$ back to a problem on $(\hat V;\hat a)$. This is the form in which Theorem 3.6 was presented in [137].

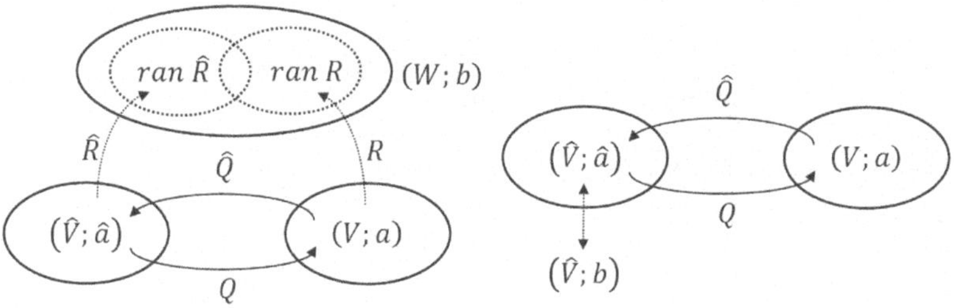

Fig. 3.8 Mapping schemes in Theorem 3.6 (left) and Corollary 3.5 (right).

However, a closer look shows that W and the mappings $R : V \to W$ and $Q : \hat V \to V$ are only used in the proof, and that the bilinear form b must essentially be defined only on $\mathrm{ran}(\hat R)$. This leads to the following simplified version of Theorem 3.6 with $W = \hat V$, which is closer to the formulations in [195, Theorem 2.1] and [190, Theorem 7.18]. The mapping schemes in Theorem 3.6 and Corollary 3.5 are shown schematically in Figure 3.8.

Corollary 3.5 *Let $(\hat V;\hat a)$ and $(V;a)$ be given as before and assume that the linear operators $Q : \hat V \to V$ and $\hat Q : V \to \hat V$ satisfy (3.39). If the auxiliary scalar product b defined on $W = \hat V$ has the properties*

$$\|\hat v\|_{\hat V}^2 \le \hat C_b \|\hat v\|_W^2, \qquad \hat v \in \hat V, \tag{3.42}$$

and

$$\|\hat v - \hat Q Q \hat v\|_W^2 \le \hat C_{\hat V} \|\hat v\|_{\hat V}^2, \qquad \hat v \in \hat V, \tag{3.43}$$

then the norm equivalence

$$\hat a(\hat u,\hat u) \approx |||\hat u|||^2 := \inf_{v\in V,\, \hat v \in \hat V:\, \hat u = \hat v + \hat Q v} b(\hat v,\hat v) + \lambda a(v,v), \qquad u \in V, \tag{3.44}$$

holds with constants that depend on λ and the constants in (3.39), (3.42), and (3.43).

The proof follows the argument of Theorem 3.6 and is left to the reader as Problem 3.4. The constants in the norm equivalence (3.44) are given by

$$\frac{1}{2\max(\hat C_b, C_{\hat Q}/\lambda)} \le \frac{|||\hat u|||^2}{\|\hat u\|_{\hat V}^2} \le 2(\hat C_{\hat V} + \lambda C_Q). \tag{3.45}$$

The formulation of Theorem 3.6 is further simplified if V is a subspace of $\hat{V}$, where it is natural to set $W = \hat{V}$ as in Corollary 3.5, to take for $\hat{Q}$ the natural embedding of V into $\hat{V}$, and to assume that $a(u, v) = \hat{a}(u, v)$ is generated by Galerkin projection. Thus $\hat{C}_{\hat{Q}} = 1$ in (3.39), while (3.43) reduces to an explicit condition on Q, namely

$$\|\hat{v} - Q\hat{v}\|_W^2 \le \hat{C}_Q \|\hat{v}\|_{\hat{V}}, \qquad \hat{v} \in \hat{V}. \tag{3.46}$$

Consequently, (3.45) is further simplified to

$$\frac{1}{2\max(\hat{C}_b, \lambda^{-1})} \le \frac{|||\hat{u}|||^2}{\|\hat{u}\|_{\hat{V}}^2} \le 2(\hat{C}_Q + \lambda C_Q). \tag{3.47}$$

The mapping scheme for this special case is shown in Figure 3.9.

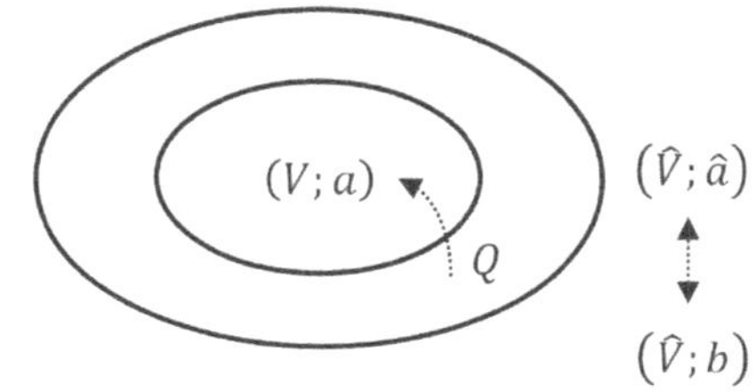

Fig. 3.9 Mapping scheme if $V \subset \hat{V} = W$.

Note that the map $Q : \hat{V} \to V$ must satisfy two conditions, boundedness and approximation properties in the auxiliary scalar product $b(\cdot, \cdot)$ on $\hat{V}$, and it only affects the upper bound in (3.47). The same applies to the estimates in (3.41) and (3.44). Their assumed linearity is not used. It is clear that Q is just needed as a vehicle to prove the associated norm equivalence, which in this particular case compares the norm of $\hat{u}$ in $(\hat{V}; \hat{a})$ with the K-functional

$$|||\hat{u}||| \approx \inf_{\hat{u}=\hat{v}+v} \|\hat{v}\|_W + \sqrt{\lambda}\|v\|_V$$

for the pair $((W; b), (V; a))$. This is also confirmed by the fact that Q does not enter the representation of the Schwarz operator $\hat{P} = \hat{C}\hat{A}$ associated with the norm equivalence nor the resulting approximate inverse $\hat{C}$ for $\hat{A}$, which now has the form

$$\hat{C} = \hat{B}^{-1} + \lambda^{-1}\hat{Q}(\hat{A}|_V)^{-1}\hat{Q}'.$$

Here, $\hat{Q} : V \to \hat{V}$ is the natural embedding operator for $V \subset \hat{V}$, and $\hat{B}$ is the operator representing b on $\hat{V}$. The choice of λ is usually determined by approximately equilibrating the upper spectral bounds of the two parts $\hat{B}^{-1}\hat{A}$ and $\lambda^{-1}\hat{Q}(\hat{A}|_V)^{-1}\hat{Q}'\hat{A}$ of the Schwarz operator $\hat{P}$. For simplicity, we set $\lambda = 1$ below.

3.2.2 Applications

We discuss some simple examples to demonstrate the use of Theorem 3.6 and its corollaries. Their relevance will become clear in Chapter 7, where we give more details on iterative solvers for PDE discretizations. As our reference Hilbert space, we take the subspace $V = V_{J,0}$ of $H_0^1(G)$ for some fixed $J > 0$ from the sequence of linear finite element spaces introduced in Proposition 3.1. Recall that the underlying triangulation $\mathcal{T} = \mathcal{T}_J$ of the polygonal domain $G \subset \mathbb{R}^2$ is quasi-uniform and regular, consisting of triangles with diameter $\approx 2^{-J}$. The fact that it is embedded in the sequence $\{\mathcal{T}_j\}_{j\in\mathbb{Z}_+}$ obtained by a special refinement procedure and the result of Proposition 3.1 will lead to efficient multigrid methods and multilevel preconditioners for the solution of linear systems arising from discretizations of second-order elliptic PDE boundary value problems. Note that the following considerations hold for any quasi-uniform and regular triangulation $\mathcal{T}$ of triangle diameters $\approx 2^{-J}$, and can be generalized to higher-dimensional situations.

We will consider some standard examples of other finite element spaces $(\hat{V}; \hat{a})$ that can be mapped in a spectrally equivalent way to the reference space $(V; a)$ by using Theorem 3.6 or Corollary 3.5. In the first example, we take $\hat{V}$ to be a linear finite element subspace of $H_0^1(G)$ with respect to another quasi-uniform and regular triangulation $\hat{\mathcal{T}}$ of triangles of similar diameter $\approx 2^{-J}$. We will also consider non-conforming linear finite elements and Lagrange elements of higher degree.

3.2.2.1 Non-matching triangulations

Let $V(\hat{\mathcal{T}})_0 := \{\hat{v} \in V(\hat{\mathcal{T}}) : \hat{v} = 0 \text{ on } \Gamma\}$ be another linear finite element subspace of $H_0^1(G)$, whose underlying triangulation $\hat{\mathcal{T}}$ is again quasi-uniform and regular, with triangles of diameter $\approx 2^J$, but not obtained by any refinement process. This often happens in engineering applications, where $\hat{\mathcal{T}}$ is the result of a standard mesh generator or is obtained by some optimization procedure. Then Proposition 3.1 is not readily available. However, it is intuitively clear that discretizations based on $\hat{V} := V(\hat{\mathcal{T}})_0$ and $V := V_{J,0}$ should have similar properties, since the underlying triangulations are similar and the spaces have similar approximation capabilities. It is therefore natural to ask whether one can reuse an existing solver for the V discretization and its theory to deal with the $\hat{V}$ discretization as well. Theorem 3.6 provides the tools to answer this question. Let $W = V + \hat{V}$ be the algebraic sum of V and $\hat{V}$, which is another, larger finite-dimensional subspace of $H_0^1(G)$. For R and $\hat{R}$ we can then take the natural embedding operators. Without loss of generality, we assume that the norms $\|\cdot\|_V$ and $\|\cdot\|_{\hat{V}}$ are induced by the norm in $H_0^1(G)$. The approximation properties of linear finite element spaces with respect to $H_0^1(G)$ suggest

$$\|w\|_W^2 = b(w, w) := 2^{2J}\|w\|_{L^2}^2$$

as the definition of the norm and the auxiliary bilinear form b on W. Then (3.38) follows from the inverse inequality

$$\|\hat{v}\|_{H_0^1} \le C2^J \|\hat{v}\|_{L^2}, \qquad \hat{v} \in \hat{V},$$

for the finite element space $\hat{V}$, where the constant C depends only on the quasi-uniformity and regularity constants of $\hat{\mathcal{T}}$. This estimate can be proved in an elementary way by estimating the norms locally on each triangle in $\hat{\mathcal{T}}$.

It turns out that we have great freedom in the choice of Q and $\hat{Q}$. Since in practical applications finite element functions are represented by their coefficients in their respective nodal bases, we will describe some choices in these terms. We do this for $\hat{Q} : V \to \hat{V}$, the construction for Q follows by symmetry of the argument. Let N_P denote the nodal basis functions for V and $\hat{N}_{\hat{P}}$ the nodal basis functions for $\hat{V}$ associated with the respective sets of nodal points including the nodal points on Γ. Then each $v \in V$ has a unique representation

$$v = \sum_P c_P N_P, \qquad c_P = v(P),$$

where $c_P = 0$ if $P \in \Gamma$. We define

$$\hat{Q}v = \sum_{\hat{P}} \lambda_{\hat{P}}(v) \hat{N}_{\hat{P}}, \tag{3.48}$$

where the linear functionals $\lambda_{\hat{P}}$ are defined as convex combinations of nodal values c_P of v at a few nodal points P close to $\hat{P}$. A canonical choice is the nodal interpolation operator for $\hat{V}$, for which we set

$$\lambda_{\hat{P}}(v) = v(\hat{P}) = \sum_{P_i \in \Delta_{\hat{P}}} \hat{\lambda}_i v(P_i),$$

where P_i are the vertices of the triangle $\Delta_{\hat{P}}$ in $\mathcal{T}$, to which the nodal point $\hat{P}$ belongs, and $\hat{\lambda}_i$ are the barycentric coordinates of $\hat{P}$ with respect to $\Delta_{\hat{P}}$, $i = 1, 2, 3$. Alternatively, take one of the vertices P_i of $\Delta_{\hat{P}}$ and set

$$\lambda_{\hat{P}}(v) = v(P_i), \tag{3.49}$$

where, in the case of $\hat{P} \in \Gamma$, we also choose $P_i \in \Gamma$. Figure 3.10 illustrates the latter construction of $\hat{Q}$. With a proper assignment $\hat{P} \to P_i$, this $\hat{Q}$ admits an extremely sparse matrix representation.

The claim is that such choices for $\hat{Q}$ (and Q) satisfy (3.39) and (3.40) with constants depending only on the quasi-uniformity and regularity constants of $\mathcal{T}$ and $\hat{\mathcal{T}}$. We start with the boundedness of $\hat{Q}$, given in (3.39). We use the fact that

$$\|v\|_V^2 = \|v\|_{H_0^1}^2 \approx \sum_e d_e(v)^2, \quad d_e(v) = v(P^+) - v(P^-), \tag{3.50}$$

where the summation is with respect to all (interior) edges in $\mathcal{T}$ and $P^\pm$ denotes the endpoints of e (the dependence of $P^\pm$ on e is not made explicit). The constants in the norm equivalence (3.50) depend on the regularity and quasi-uniformity constants

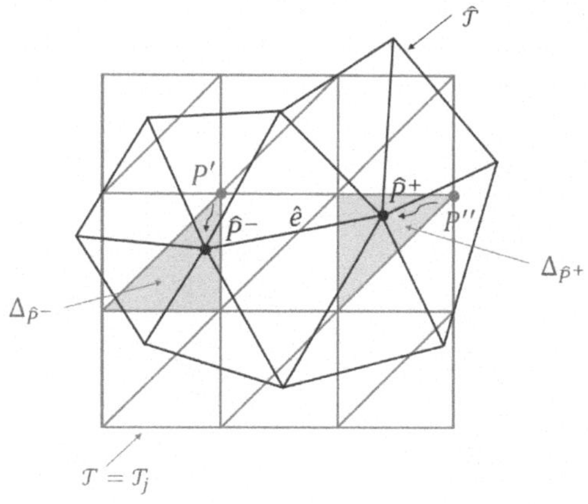

Fig. 3.10 Notation used in the construction of $\hat{Q}$ in (3.48) using (3.49). Here, $\lambda_{\hat{P}}(v) = v(P_i)$ and $\lambda_{\hat{P}'}(v) = v(P_i')$.

of $\mathcal{T}$. A similar notation is used for $\hat{\mathcal{T}}$, and a norm equivalence analogous to (3.50) holds for $\hat{v} \in \hat{V}$. Applied to $\hat{Q}v$, we get

$$\|\hat{Q}v\|_{\hat{V}}^2 \leq C \sum_{\hat{e}} d_{\hat{e}}(\hat{Q}v)^2 = C \sum_{\hat{e}} (\lambda_{\hat{P}^+}(v) - \lambda_{\hat{P}^-}(v))^2,$$

where the summation is with respect to the (interior) edges $\hat{e}$ of $\hat{\mathcal{T}}$. Now, if $\lambda_{\hat{P}^+}(v)$ and $\lambda_{\hat{P}^-}(v)$ are convex combinations of values $v(P)$ at a few nodal nodal points P near the two endpoints of the edge $\hat{e}$ whose length is comparable to the diameters of the triangles in $\mathcal{T}$, then the difference $d_{\hat{e}}(\hat{Q}v)$ can be represented as a convex combination of differences $v(P')-v(P'')$ for nodal points P', P'' located at distance $\leq C_0 2^{-J}$ from $\hat{e}$. The constant C_0 depends only on the quasi-uniformity and regularity constants of $\mathcal{T}$ and $\hat{\mathcal{T}}$ and our choice of $\lambda_{\hat{P}}$. Let the set of all such points P be temporarily denoted by $E_{\hat{e}}$. Note that in each $E_{\hat{e}}$ the number of nodal points and edges in $\mathcal{T}$ connecting them is uniformly bounded, and that the family of all such nodal point sets $E_{\hat{e}}$ has finite overlap. This will be used implicitly in the following. Thus

$$d_{\hat{e}}(\hat{Q}v)^2 \leq \sum_{P', P'' \in E_{\hat{e}}} |v(P') - v(P'')|^2.$$

If P', P'' are endpoints of an edge in $\mathcal{T}$, then $d_e(v) = v(P') - v(P'')$. For other pairs $P', P'' \in E_{\hat{e}}$, we can connect them by a chain of edges e_i with endpoints P_{i-1} and P_i, $i = 1, \ldots, k$, so that $P_i \in E_{\hat{e}}$, $P_0 = P'$ and $P_k = P''$. Obviously, $k \leq C_1$ with a constant C_1 depending only on the regularity and quasi-uniformity constants of $\mathcal{T}$. Then

$$|v(P') - v(P''|^2 \leq C_1 \sum_{i=1}^{k} d_{e_i}(v)^2 \leq C_1 \sum_{e:\, P^\pm \in E_{\hat{e}}} d_e(v)^2.$$

Figure 3.11 shows an example for the construction of $\hat{Q}$ depicted in Figure 3.10, where $k = 2$ is sufficient. It also demonstrates that the minimal $E_{\hat{e}}$ is often much smaller.

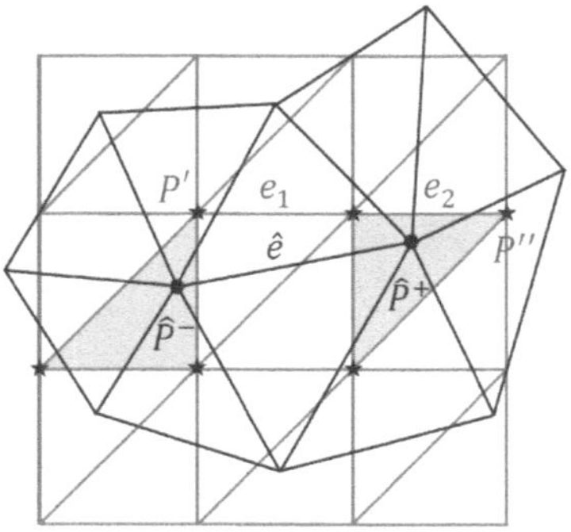

Fig. 3.11 The minimal set $E_{\hat{e}}$ and the chain $\{e_1, e_2\}$ connecting P' and P'' corresponding to the situation shown in Figure 3.10. Nodal points in $E_{\hat{e}}$ are indicated by ⋆.

Together with the mentioned properties of the sets $E_{\hat{e}}$, after substitution we arrive at

$$\begin{aligned}\|\hat{Q}v\|^2_{\hat{V}} &\le C\sum_{\hat{e}}\sum_{P',P''\in E_{\hat{e}}} |v(P')-v(P'')|^2\\ &\le C\sum_{\hat{e}}\sum_{e:\,P^{\pm}\in E_{\hat{e}}} d_e(v)^2 \le C\sum_{e} d_e(v)^2 \le C\|v\|^2_V,\end{aligned}$$

where in the last step we again used (3.50). This proves (3.39).

The above argument is completely elementary and does not use any finite element approximation theory. Similar considerations apply in the proof of (3.40). Take an arbitrary triangle Δ in $\mathcal{T}$ with vertices P_i, $i = 1,2,3$. Then, as before,

$$v(x) = \sum_{i=1}^{3} \lambda_i(x)v(P_i), \qquad x \in \Delta,$$

is a convex combination of the nodal values $v(P_i)$. The same x also belongs to a triangle $\hat{\Delta}$ of $\hat{\mathcal{T}}$ with vertices $\hat{P}_j$, $j = 1,2,3$. Now

$$\hat{Q}v(x) = \sum_{j=1}^{3} \hat{\lambda}_j(x)\hat{Q}v(\hat{P}_i) = \sum_{j=1}^{3} \hat{\lambda}_j(x)\lambda_{\hat{P}_j}(v),$$

is a convex combination of the functionals $\lambda_{\hat{P}_j}(v)$, which in turn are convex combinations of the nodal values $v(P_i)$ in the vicinity of $\hat{P}_j$, and thus in the vicinity of Δ. Consequently, $v(x) - \hat{Q}v(x)$ can be represented by a convex combination of differences $v(P') - v(P'')$ with nodal points P', P'' at distance $\leq C_0' 2^{-J}$ from Δ. Let the set of all these points be denoted by E_Δ. The family of all such E_Δ now plays the role of the family $E_{\hat{e}}$ and has similar properties. For $x \in \Delta$ this results in

$$|v(x) - \hat{Q}v(x)|^2 \leq \sum_{P',P'' \in E_\Delta} |v(P') - v(P'')|^2 \leq C \sum_{e:\, P^\pm \in E_\Delta} d_e(v)^2,$$

and after integrating over Δ, followed by summing with respect to the triangles in $\mathcal{T}$, we get

$$\begin{aligned} \|v - \hat{Q}v\|_W^2 &= 2^{2J} \sum_\Delta \int_\Delta |v(x) - \hat{Q}v(x)|^2\, dx \\ &\leq C 2^{2J} \sum_\Delta 2^{-2J} \sum_{e:\, P^\pm \in E_\Delta} d_e(v)^2 \\ &\leq C \sum_e d_e(v)^2 \leq C \|v\|_V^2. \end{aligned}$$

This establishes (3.40), and completes the verification of the assumptions of Theorem 3.6.

As a result, if all operators are represented with respect to the nodal bases in $\hat{V}$ and V, respectively, we get the formulas

$$\hat{P} = (2^{-2J} \hat{B}^{-1} + \lambda^{-1} \hat{Q} A^{-1} \hat{Q}^T) \hat{A}$$

for the matrix representation $\hat{P} = \hat{C}\hat{A}$ of the Schwarz operator and

$$\hat{C} = 2^{-2J} \hat{B}^{-1} + \lambda^{-1} \hat{Q} A^{-1} \hat{Q}^T$$

for the preconditioner for $\hat{A}$ resulting from the norm equivalence (3.41). Here, the matrix representation of $\hat{Q}$ is sparse (in the case of the second choice for the functionals $\lambda_{\hat{P}}$ it is just an incidence matrix), and $\hat{B}$ denotes the Gram matrix of the nodal basis in $\hat{V}$ with respect to the L^2 scalar product. It is easy to see that $2^{2J} B$ is spectrally equivalent to the identity matrix due to the L^2 stability of nodal bases (3.16), and if C denotes now a spectrally equivalent preconditioner for A, then $\hat{C}$ can be replaced by

$$\hat{C} = I + \lambda^{-1} \hat{Q} C \hat{Q}^T.$$

We note that an application of Corollary 3.5 is also possible, since the crucial estimate (3.43) follows from a similar argument. In fact, we have by (3.16) and with some constant C the bound

$$\|\hat{v} - \hat{Q}Q\hat{v}\|_W^2 = 2^{2J} \|\hat{v} - \hat{Q}Q\hat{v}\|_{L^2}^2 \leq C \sum_{\hat{P}} |\hat{v}(\hat{P}) - (\hat{Q}Q\hat{v})(\hat{P})|^2.$$

Now, by definition of $\hat{Q}$ and Q, the nodal values $(\hat{Q}Q\hat{v})(\hat{P})$ are convex combinations of nodal values of $Q\hat{v}$ at nearby nodal points P in $\mathcal{T}$, which are also convex combinations of nodal values $\hat{v}(\hat{P}')$ at nearby $\hat{P}'$ in $\hat{\mathcal{T}}$. So the difference $\hat{v}(\hat{P}) - (\hat{Q}Q\hat{v})(\hat{P})$ can be represented as a convex combination of differences $v(P') - v(P'')$ for nodal P', P'' in $\hat{\mathcal{T}}$ at distance $\leq C_0''2^{-J}$ from $\hat{P}$. The rest of the argument is the same as before.

The above arguments also work if $\mathcal{T}$ and $\hat{\mathcal{T}}$ are quasi-uniform regular triangulations of approximately the same triangle diameter h for two different polygonal domains G and $\hat{G}$, whose boundaries Γ and $\hat{\Gamma}$ are at distance $O(h)$. Such a situation was treated in [195] in the case of $\hat{G} \supset G$. The definition of Q and $\hat{Q}$ needs to be modified only near the boundary. To be precise, for $\hat{Q}$ we set $\lambda_{\hat{P}}(v) = v(P_i) = 0$ with $P_i \in \Gamma$ for all $\hat{P} \in \hat{\Gamma} \cup \hat{G}\backslash G$, similarly for Q. With this and the discrete H_0^1 and L^2 norm representations for $\hat{V}$ and V at hand, the conditions in Theorem 3.6 or Corollary 3.5 can be easily verified. We leave this to the reader as Problem 3.8.

3.2.2.2 Non-conforming linear elements

Let $\hat{V}_{J,0}$ denote the space of all piecewise linear functions with respect to $\mathcal{T} = \mathcal{T}_J$ that are continuous across edge midpoints M_e of the interior edges e and vanish at the midpoints of the boundary edges $e \subset \Gamma$. Figure 3.12 shows the nodal points and nodal basis functions for the conforming and non-conforming cases. Since $\hat{V}_{J,0} \not\subset H_0^1(G)$, the scalar product on $\hat{V}_{J,0}$ cannot be directly inherited from $H_0^1(G)$. So it is common to take the broken H_0^1 scalar product

$$\hat{a}_J(\hat{u}, \hat{v}) := \sum_{\Delta \in \mathcal{T}_J} \int_\Delta (\nabla \hat{u}, \nabla \hat{v})_2 \, dx.$$

One might be tempted to define these spaces not only on $\mathcal{T}_J$ but on all $\mathcal{T}_j$, denote them by $\hat{V}_{j,0}$, and ask for a multilevel space splitting for $(\hat{V}_{J,0}; \hat{a}_J)$ using the family $\{\hat{V}_{j,0}\}_{j \leq J}$. However, since the sequence is not nested, i.e., $\hat{V}_{j,0} \not\subset \hat{V}_{j+1,0}$, this presents many difficulties. We will return to this option in Subsection 7.1.3.

Here we demonstrate the use of the auxiliary space method. Note that $V := V_{J,0} \subset \hat{V} := \hat{V}_{J,0}$ and that restricting $\hat{a}_J$ to V reproduces the $H_0^1(G)$ scalar product on V (from now on we drop the subscript J to simplify the notation). This fits the framework mentioned as a special case for Corollary 3.5 in Subsection 3.2.1. For example, we can take $W = \hat{V}$ and identify $\hat{Q}$ with the natural embedding of V in $\hat{V}$, so that $C_{\hat{Q}} = 1$ in (3.38). For the scalar product b on $\hat{V}$, we again use the scaled L^2 scalar product

$$b(\hat{u}, \hat{v}) := 2^{2J}(\hat{u}, \hat{v})_{L^2}, \qquad \hat{u}, \hat{v} \in \hat{V}.$$

The condition (3.42) follows directly from the L^2 stability of the nodal basis $\{\hat{N}_{M_e}\}$ for $\hat{V}$, expressed by

$$\|\hat{v}\|_{L^2}^2 \approx 2^{-2J} \sum_e \hat{v}(M_e)^2, \qquad \hat{v} \in \hat{V}, \tag{3.51}$$

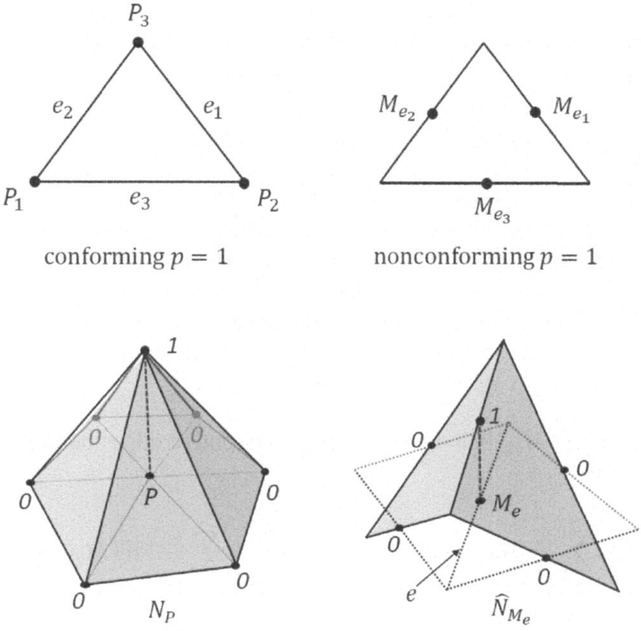

Fig. 3.12 Nodal points and nodal basis functions for conforming (left) and non-conforming (right) linear finite elements on triangular meshes.

and the corresponding norm equivalence

$$\|\hat{v}\|_{\hat{V}}^2 = \hat{a}_J(\hat{v}, \hat{v}) \approx \sum_{e,e'} |\hat{v}(M_e) - \hat{v}(M_{e'}|^2, \qquad \hat{v} \in \hat{V} \tag{3.52}$$

for the broken $H_0^1(G)$ norm on $\hat{V}$, where the latter summation is over all edge pairs belonging to one of the triangles in $\mathcal{T}$. The verification of (3.51) and (3.52) can be done locally on each triangle, with constants depending only on the quasi-uniformity and regularity constants of $\mathcal{T}$. For example, (3.51) follows directly from the midpoint quadrature rule on triangles, which is exact for quadratic polynomials.

It remains to construct an operator $Q : \hat{V} \to V$ for which the bounds in (3.39) and (3.43) hold. The procedure is similar to what we did in the previous subsection. A standard choice is to define the nodal values $Q\hat{v}(P)$ at the interior vertices of $\mathcal{T}$ by convex combinations of the limit values of $\hat{v}$ when approaching P from inside of each of the triangles attached to P (this was done in [134]). The attractive feature of this choice is that it produces a projection onto V_J. Unfortunately, the analysis is a bit more complicated, since $Q\hat{v}(P)$ is no longer a convex combination, but only an affine-linear combination of the values $\hat{v}(M_e)$. It is simpler to use

$$Q\hat{v} = \sum_P \lambda_P(\hat{v}) N_P, \qquad \lambda_P(\hat{v}) := \hat{v}(M_{e_P}),$$

where e_P is one of the edges attached to P, with the restriction that $e_P \subset \Gamma$ whenever $P \in \Gamma$. Obviously, the boundedness condition (3.39) for Q follows from (3.50) and (3.52), since we have

$$d_e(Q\hat{v}) = \hat{v}(M_{e_{P^+}}) - \hat{v}(M_{e_{P^-}}) = v(M_{e'}) - v(M_{e''})$$

for any edge e with endpoints $P^{\pm}$ in $\mathcal{T}$, where e' and e'' share an endpoint with e. Let the set of midpoints of all such edges be denoted by E_e. These sets are uniformly bounded in size and have finite overlap, with constants depending only on the regularity and quasi-uniformity constants of $\mathcal{T}$. The edges e' and e'' must not belong to the same triangle, but can be connected by a chain

$$e_0 = e', e_1, \ldots, e_{m-1}, e_m = e'',$$

where the midpoints M_{e_k} belong to E_e, and consecutive edges belong to the same triangle. Therefore,

$$d_e(Q\hat{v})^2 \leq C \sum_{k=1}^{m} |\hat{v}(M_{e_k}) - \hat{v}(M_{e_{k-1}})|^2 \leq C \sum_{M_{e'}, M_{e''} \in E_e} |\hat{v}(M_{e'}) - \hat{v}(M_{e''})|^2,$$

where the summation is only over all edge pairs that belong to a triangle in the neighborhood C_e of the edge e. In other words, the right-hand side represents the part of the discrete H_0^1 norm of $\hat{v}$ associated with E_e. The notation is illustrated in Figure 3.13. It remains to substitute into (3.50), to take the finite overlap property of the edge neighborhoods C_e into account, and to compare with (3.52).

The verification of (3.43) follows in a similar way. With (3.51) we have

$$\|\hat{v} - Q\hat{v}\|_W^2 \leq C \sum_e |\hat{v}(M_e) - (\hat{Q}\hat{v})(M_e)|^2 = C \sum_e |\hat{v}(M_e) - \frac{1}{2}(\hat{v}(M_{e_{P^+}}) + \hat{v}(M_{e_{P^-}}))|^2.$$

Now the estimate

$$|\hat{v}(M_e) - \frac{1}{2}(\hat{v}(M_{e_{P^+}}) + \hat{v}(M_{e_{P^-}}))|^2 \leq C \sum_{e', e'' \in E_e} |\hat{v}(M_{e'}) - \hat{v}(M_{e''})|^2$$

follows by the same argument as above, and we can proceed as in the boundedness proof for Q to finally arrive at (3.43).

Thus the norm equivalence (3.47) holds with the constants $\hat{C}_b$, C_Q and $\hat{C}_Q$, which depend only on the quasi-uniformity and regularity constants of $\mathcal{T} = \mathcal{T}_J$ and are independent of J. The matrix representation of the resulting preconditioner for the discretization matrix $\hat{A}_J$ associated with the non-conforming linear finite element space $(\hat{V}_{J,0}; \hat{a}_J)$ is of the form

$$\hat{C}_J = 2^{-2J} \hat{B}_J^{-1} + \lambda^{-1} \hat{Q}_J A_J^{-1} \hat{Q}_J^T.$$

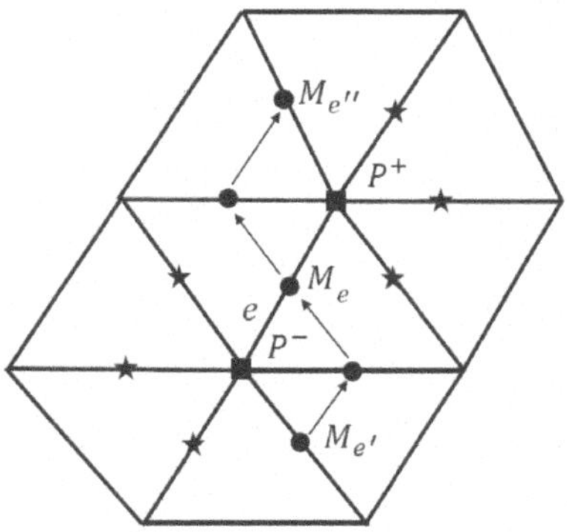

Fig. 3.13 Notation for the estimation of $d_e(Q\hat{v})$: Chain of edges connecting two midpoints $M_{e'}$ and $M_{e''}$ in E_e.

All matrices are the representations of the corresponding operators with respect to the nodal bases in V and $\hat{V}$. The part $2^{-2J}\hat{B}_J^{-1}$, where $\hat{B}_J$ is the L^2 Gramian matrix of the nodal basis in $\hat{V}$, is spectrally equivalent to the identity on $\hat{V}$, which follows from (3.51). Since

$$\hat{Q}v(M_e) = \frac{1}{2}(v(P_e^+) + v(P_e^-))$$

for all interior edges, the matrix representation $\hat{Q}_J$ of the quasi-interpolant map $\hat{Q}$ contains two entries $1/2$ per row, its application is almost cost and memory free. As before, A_J^{-1} can be replaced by suitable preconditioners C_J for A_J. Similar preconditioners were proposed in [134] and [195].

3.2.2.3 Preconditioning higher-order C^0 finite elements

In Theorem 3.3, we treated nested sequences of C^0 Lagrange finite element spaces of arbitrary degree. In principle, this leads to efficient multilevel preconditioners for the associated PDE discretizations. However, the complexity of their implementation increases with p. An alternative is to switch from a finite element discretization of degree $p > 1$ to a linear finite element discretization by the auxiliary space method and use the less complex multilevel preconditioner for the latter.

We show this for $p = 2$ and finite element spaces on the same triangulation as an example. Let V be the linear finite element space on $\mathcal{T} = \mathcal{T}_J$ as before, and take as $\hat{V}$ the subspace of quadratic C^0 Lagrange finite element functions with respect to the same $\mathcal{T}$ with vanishing boundary values on Γ. Figure 3.14 shows the nodal points and the nodal basis functions N_P and N_{M_e} for the quadratic case. The bilinear forms a and $\hat{a}$ are again inherited from the scalar product in $H_0^1(G)$. Since $V \subset \hat{V}$, it is

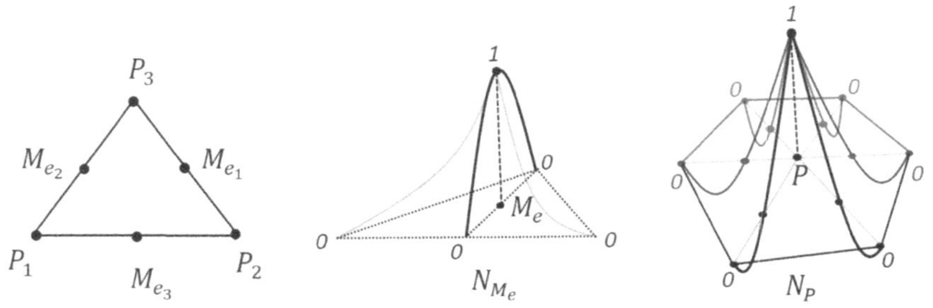

Fig. 3.14 Nodal points and nodal basis functions for quadratic C^0 Lagrange finite elements ($p = 2$) on triangulations.

possible to use the simplified theory of Corollary 3.5, where we set $W = \hat{V}$, equip W with the L^2 scalar product scaled by 2^{2J} as before, and use the natural embedding of V in $\hat{V}$ as $\hat{Q}$. Since (3.42) is obvious from the inverse inequality for $\hat{V}$, we only need to worry about defining a bounded Q that satisfies (3.46). Again, any operator

$$Q\hat{v} = \sum_{P} \lambda_P(\hat{v}) N_P$$

with functionals λ_P representing convex combinations (or even affine-linear combinations with uniformly bounded coefficients) of the nodal values $\hat{v}(\hat{P})$ will suffice. The simplest choice is $\lambda_P(\hat{v}) = \hat{v}(P)$, which we will use in the sequel. This Q is the usual nodal interpolation operator and represents a projection onto V. Recall that the nodal points for $\hat{V}$ are the vertices P and the edge midpoints M_e of $\mathcal{T}$, which, incidentally, are the nodal points for the linear finite element space V' on the dyadically refined partition $\mathcal{T}'$, i.e., $\mathcal{T}'$ is obtained by connecting the midpoints of each triangle in $\mathcal{T}$ with new edges. If $\hat{v}' \in V'$ denotes the nodal interpolant of $\hat{v} \in \hat{V}$, then we have

$$\|\hat{v}\|_{\hat{V}}^2 = \int_G |\nabla \hat{v}|^2\, dx \approx \int_G |\nabla \hat{v}'|^2\, dx = \|\hat{v}'\|_{V'}^2, \quad \hat{v} \in \hat{V}, \tag{3.53}$$

with constants depending only on the regularity and quasi-uniformity constants of $\mathcal{T}$. In fact, this is true for the fixed reference triangle

$$\Delta_0 = \{(x_1, x_2) : \ 0 \le x_1, x_2 \le x_1 + x_2 \le 1\},$$

because

$$\left(\int_{\Delta_0} |\nabla \hat{v}|^2\, dx\right)^{1/2} \quad \text{and} \quad \left(\int_{\Delta_0} |\nabla \hat{v}'|^2\right)^{1/2}$$

are two norms on the 5-dimensional space of quadratic polynomials modulo constants and are thus equivalent. This also holds by an affine-linear mapping for any Δ, with constants depending on the regularity of $\mathcal{T}$, and leads to (3.53). Together with (3.50), this implies

$$\|\hat{v}\|_{\hat{V}}^2 \approx \sum_{e'} d_{e'}(\hat{v})^2, \qquad \hat{v} \in \hat{V}, \tag{3.54}$$

where the sum is over all edges in $\mathcal{T}'$.

Now we are ready to examine the necessary mapping properties of Q. By (3.50) and (3.54) we have

$$\|Q\hat{v}\|_V^2 \leq C\sum_e d_e(\hat{v})^2 \leq 2C\sum_{e'} d_{e'}(\hat{v})^2 \leq C\|\hat{v}\|_{\hat{V}},$$

since

$$d_e(\hat{v}) = \hat{v}(P^+) - \hat{v}(M_e) + \hat{v}(M_e) - \hat{v}(P^-) = d_{e^+}(\hat{v}) + d_{e^-}(\hat{v})$$

and the two half-edges $e^\pm$ of e are edges in $\mathcal{T}'$. This gives the boundedness of Q. The approximation property (3.46) holds in a similar way using the L^2 stability of the nodal basis in $\hat{V}$, i.e., we have

$$\begin{aligned}\|\hat{v} - Q\hat{v}\|_W^2 &\leq C\sum_e |\hat{v}(M_e) - \frac{1}{2}(\hat{v}(P^+) + \hat{v}(P^-))|^2 \\ &\leq \frac{C}{2}\sum_e (d_{e^+}(\hat{v})^2 + d_{e^+}(\hat{v})^2) \leq C\|\hat{v}\|_{\hat{V}}^2.\end{aligned}$$

So,

$$\hat{C} = 2^{-2J}\hat{B}^{-1} + \hat{Q}A^{-1}\hat{Q}^T \approx I + \hat{Q}C\hat{Q}^T$$

can be used as a preconditioner for $\hat{A}$, while the condition number of $\hat{C}\hat{A}$ depends only on the regularity and quasi-uniformity constants of $\mathcal{T}$. As before, $\hat{B}$ denotes the well-conditioned Gram matrix of the nodal basis in $\hat{V}$, C is a preconditioner for the matrix A associated with the nodal basis discretization in $(V; a)$, and $\hat{Q}$, which implements the natural embedding $V \to \hat{V}$, has rows with one or two non-zero elements.

The switch from a more complicated or higher-degree finite element space to a linear finite element space is specific to H^s elliptic problems with $|s| < 3/2$, and can be done in several ways. In the special case of quadratic finite elements, it is also possible to use the spectral equivalence of the spaces $(\hat{V}; a)$ and $(V'; a')$, which is guaranteed by (3.53). This means that any preconditioner for A' (the discretization matrix of an H_0^1-elliptic problem on the linear finite element space V' on the refined triangulation $\mathcal{T}'$) can be used as a preconditioner for $\hat{A}$. The above splitting was also proposed in [137] for the construction of cheaply computable a posteriori error estimators for linear finite element discretizations.

Finally, note that a more sophisticated approach is needed to make this work for arbitrary $p > 1$ and for the hp-finite element method with constants in the norm equivalences that do *not* depend on p. We refer to [73, 106] for recent work in this direction and links to the literature.

3.2.3 Algebraic two-grid methods

We will briefly touch on the connection between the auxiliary space method and the algebraic two-grid method. The latter is a prerequisite to the algebraic multigrid method (AMG), which is one of the few efficient solution methods for certain classes of large linear systems with a sparse, but not necessarily structured, symmetric positive definite matrix $\hat{A}$. To explain this relationship, we simplify a bit and refer to [190, Chapter 6] and [199] for more details and references. In contrast to the examples above, where $(\hat{V}, \hat{a})$ and $(V; a)$ were given and a space splitting was constructed by introducing $(W; b)$, the goal now is to *construct* a suitable smaller space $(V; a)$ (called the coarse-grid space) from $(\hat{V}; \hat{a})$ (called the fine-grid space) and a given auxiliary scalar product b on $\hat{V} = W$. This b is chosen in advance and associated with a so-called smoother, usually a few steps of a simple iterative solver for $\hat{A}\hat{x} = \hat{b}$, such as the Jacobi-Richardson or the (symmetrized) Gauss-Seidel method.

The standing assumptions for AMG are that $\hat{V} = \mathbb{R}^N$, $\hat{a}(x, y) = (\hat{A}\hat{x}, \hat{y})_2 = \hat{y}^T \hat{A}\hat{x}$, and $b(\hat{x}, \hat{y}) = (B\hat{x}, \hat{y})_2 = \hat{y}^T B\hat{x}$ is given by another spd matrix B such that (3.42) is satisfied with a constant $\hat{C}_b \le 1$. For instance, $B = \lambda_{\max}(\hat{A})I$ corresponds to the damped Richardson iteration and yields $\hat{C}_b = 1$. Next, we assume that the coarse-grid space V can be identified with R^n for some $n < N$. Consequently, V can be regarded as a subspace of $\hat{V}$ via the extension-by-zero map

$$y \in V \quad \longrightarrow \quad \hat{y} = \begin{pmatrix} 0 \\ y \end{pmatrix} \in \hat{V}.$$

Similarly, any $\hat{x} \in \hat{V}$ can be decomposed as

$$\hat{x} = \begin{pmatrix} z \\ y \end{pmatrix}, \qquad y \in V, \ z \in \mathbb{R}^{N-n},$$

where $\hat{x}_c := y$ contains the coarse-grid variables, while $\hat{x}_f := z$ contains the fine-grid variables that are not included in the coarse grid. All operators can then be identified with their matrix representations in a corresponding block fashion.

This means that we are left with finding a suitable $n \times N$ matrix $Q : \hat{V} \longrightarrow V$ (called a restriction), an $N \times n$ matrix $\hat{Q} : V \longrightarrow \hat{V}$ (called a prolongation) and a coarse-grid $n \times n$ matrix A which defines the bilinear form a on V. Since the main application is to linear systems with sparse $\hat{A}$, there are questions about the sparsity of the matrix representations of $\hat{Q}$ and the resulting A, which we will not address here. Normally, AMG uses prolongations $\hat{Q}$, which are interpolatory, i.e., they preserve the coarse grid variables via

$$\hat{Q} = \begin{pmatrix} \hat{Q}_{fc} \\ I \end{pmatrix}.$$

Furthermore, the scalar product on V and its matrix representation A are generated by the Galerkin projection, i.e., we have

$$a(x, y) = \hat{a}(\hat{Q}x, \hat{Q}y), \qquad A = \hat{Q}^T \hat{A} \hat{Q} = A_{cc} + \hat{Q}_{fc}^T A_{fc} + A_{cf}\hat{Q}_{fc} + \hat{Q}_{fc}^T A_{ff}\hat{Q}_{fc}.$$

This automatically gives the boundedness of $\hat{Q}$ in (3.39) with constant $C_{\hat{Q}} = 1$.

To finally find a reasonable choice for $(V; a)$, in the light of Corollary 3.5, besides the decision on the set of coarse variables and on $\hat{Q}_{fc}$, one needs an appropriate matrix Q such that the conditions on Q in (3.39) and (3.43) required in Corollary 3.5 are satisfied with small constants C_Q and $\hat{C}_{\hat{V}}$, respectively. Then

$$\hat{C} = B^{-1} + \lambda^{-1}\hat{Q}A^{-1}\hat{Q}^T$$

may be a reasonable preconditioner for $\hat{A}$. Candidates for Q that come to mind are $Q = \hat{Q}^T$ or simple restriction $Q = \begin{pmatrix} 0 & I \end{pmatrix}$. The first choice leads to checking the norm of $\hat{Q}\hat{Q}^T$ as an operator from $\hat{V}$ to $\hat{V}$ and of $I - \hat{Q}\hat{Q}^T$ as an operator from $\hat{V}$ to W. For the second choice, the required properties of Q are equivalent to checking the inequalities

$$\|\hat{x}_c\|_V^2 \le C_Q \|\hat{x}\|_{\hat{V}}^2, \qquad \|\hat{x}_f\|_W^2 \le \hat{C}_{\hat{V}} \|\hat{x}\|_{\hat{V}}^2, \qquad \hat{v} \in \hat{V}.$$

We will not go into further detail. Our goal was just to point out the connection to the auxiliary space method, and we believe that this view can be helpful when looking at the AMG literature, see e.g. [190, 199].

Note that AMG theory, which deals with iterative solvers for general sparse $\hat{A}$ built by recursively applying the two-grid construction, does not focus on two-grid preconditioning. Rather, it tries to find optimal bounds on the spectrum of the error propagation operator

$$M = (I - B^{-1}A)(I - \hat{Q}A^{-1}\hat{Q}^T)$$

of another iterative process related to the auxiliary space method. We refer to the next chapter for the definition of additive and multiplicative Schwarz methods based on space splittings which provides a link to the preconditioner construction via the auxiliary space theory. Needless to say, given the many unknowns in AMG design, there is still no definitive, fully satisfying convergence result for general classes of sparse matrices $\hat{A}$.

3.3 Paving and sketching

In this section, we consider two more recent approaches to the generation of space splittings for a given Hilbert space, which offer advantages such as improved conditioning or the construction of iterative solvers with desirable properties. In its traditional form, *paving* aims to produce a stable direct decomposition $\mathbf{V}$ of $(V; a)$ into finitely many subspaces V_i, $i = 1, \ldots, n$, with small n, such that the subspace operators A_i generated by Galerkin projection are uniformly well-conditioned with respect to the designated scalar product $(\cdot, \cdot)_V$ on V. Although the Schwarz operator associated with the resulting splitting inherits its condition number from the representation of the original problem with respect to $(\cdot, \cdot)_V$, i.e., there is no preconditioning effect, one benefits from $\kappa(A_i) \approx 1$ and a small number of subspaces V_i when solving the subproblems iteratively. The existence of such pavings for matrix operators on ℓ^2 spaces is guaranteed by the recently confirmed Kadison-Singer conjecture, but efficient methods for creating them are not available. There are, however, randomized and deterministic constructions that give close to optimal results.

For structured applications, such as PDE discretizations, pavings can often be guessed. For example, the red-black ordering of variables in the finite-difference discretization $Ax = b$ with $2d+1$-stencils of a diffusion equation on a tensor product grid over a d-dimensional product domain leads to a paving where the subproblem matrices are diagonal. The paving condition $A_i \approx I_i$ can be enforced by appropriate scaling with the diagonal elements of A. This example also shows that paving does not necessarily lead to significant gains: Other than rewriting each Jacobi step as two parallel Jacobi substeps associated with red and black variables, the algorithm and its performance guarantees have not changed. The workaround for variational PDE discretizations is to consider overlapping pavings and multigrid approaches. These alternatives are discussed in Chapter 7.

Sketching is the popular name for projecting a given high-dimensional problem into a low-dimensional subspace while still preserving enough structure. Typically the projection onto a subspace of dimension $\geq k$ is chosen randomly and is supposed to approximately capture the best rank-k approximation of the representation of A with respect to $(\cdot,\cdot)_V$.

Here is a toy example: Consider a linear system $Ax = b$ with $n\times n$ spd matrix A. Set $V = \mathbb{R}^n$ and $a(x,y) = y^T Ax = (Ax,y)_2$, where $(x,y)_V := (x,y)_2$ is the designated scalar product mentioned above for representing operators and vectors in V. Let $\psi_1,\ldots,\psi_n$ be the normalized eigenvectors of A, corresponding to the eigenvalues $\lambda_1 \geq \cdots \geq \lambda_n > 0$. Assume that the first k eigenpairs $(\psi_i;\lambda_i)$, $i = 1,\ldots,k$, are available and set $V_k = \mathrm{span}(\{\psi_i\}_{i=1,\ldots,k}) = \mathrm{ran}(\Psi_k)$, where Ψ_k is the $n \times k$ matrix with columns given by ψ_i, $i = 1,\ldots,k$. The best rank-k approximation of A is given by the matrix $A_k = \Psi_k D_k \Psi_k^T$, where D_k is the diagonal $k \times k$ matrix with λ_i, $i = 1,\ldots,k$, on the diagonal. Furthermore, $Q_k = \Psi_k \Psi_k^T$ is the ortho-projection onto V_k.

Formally, if we consider $(V;(\cdot,\cdot)_2)$ as space splitting for $(V;a)$, then the associated Schwarz iteration schemes are variants of the standard Richardson iteration for $Ax = b$, whose performance is governed by the condition number $\kappa = \kappa(A) = \lambda_1/\lambda_n$ of A. However, if we now switch to a direct splitting $(\mathbf{V};\mathbf{a})$ of $(V;a)$ into the two spaces $(V_k;(A_k\cdot,\cdot)_2)$ and $(V_k^\perp;\lambda_{k+1}(\cdot,\cdot)_2)$, then the condition of the splitting improves to $\kappa = \lambda_{k+1}/\lambda_n$. In fact, by construction, the representation $x = x_k + x_k^\perp$ with $x_k = Q_k x \in V_k$ and $x_k^\perp = (I - Q_k)x \in V_k^\perp$ is unique. So we have

$$|||x|||^2 = (A_k x_k, x_k)_2 + \lambda_{k+1}(x_k^\perp, x_k^\perp)_2,$$

and

$$a(x,x) = (Ax_k,x_k)_2 + (Ax_k^\perp,x_k^\perp)_2 \begin{cases} \leq (A_k x_k, x_k)_2 + \lambda_{k+1}(x_k^\perp,x_k^\perp)_2, \\ \geq (A_k x_k, x_k)_2 + \lambda_n(x_k^\perp,x_k^\perp)_2. \end{cases}$$

Comparing the two expressions yields the stability constants $\lambda_{\min} = 1$ and $\lambda_{\max} = \lambda_{k+1}/\lambda_n$ for this splitting, which proves the above claim. The preconditioner associated with this splitting is

$$C = \Psi_k D_k^{-1} \Psi_k^T + \lambda_{k+1}^{-1}(I - Q_k) = \lambda_{k+1}^{-1}(I - \Psi_k(I_k - \lambda_{k+1} D_k^{-1})\Psi_k^T).$$

This is not a practical algorithm, since it requires expensive computation of parts of the eigensystem of A. However, the above analysis suggests possible improvements if reliable approximations of A_k can be computed.

In the following two subsections, we provide more details on these two approaches. Since research in this area is still ongoing, and the necessary theoretical background in random matrix theory is beyond the scope of this book, we will be brief and limit the exposition to some standard results directly related to iterative Schwarz methods. More can be found in the surveys [117, 119, 193] which we rely on.

3.3.1 Paving

The selection of a large, well-conditioned part of a bounded operator on a separable Hilbert space is a challenging problem with a long history. Even more challenging is the problem of decomposing a bounded operator into a direct sum of a small number of well-structured and well-conditioned subproblems. The reader is cautioned that finding a decomposition into well-conditioned subproblems is not the same as finding a well-conditioned space splitting. Of course, spectral methods based on knowledge of the associated eigenvalue problem are a classical, albeit expensive, way to go.

For $n \times n$ matrices B, the following result from [118] solved a long-standing problem known as Anderson paving conjecture:

Proposition 3.2 *For any $n \times n$ matrix B with zero diagonal and any $0 < \epsilon < 1$, there exists a partition*

$$\{1, \ldots, n\} = I_1 \cup \cdots \cup I_k, \qquad I_r \cap I_s = \emptyset, \quad r \neq s,$$

of the index set $\{1, \ldots, n\}$ into $k \leq k(\epsilon)$ pairwise disjoint index subsets I_r, $r = 1, \ldots, k$, such that the diagonal submatrices $B_r := B_{I_r, I_r}$ have spectral norm

$$\|B_r\|_2 \leq \epsilon \|B\|_2, \qquad r = 1, \ldots, k. \tag{3.55}$$

We call this a (k, ϵ)-paving for B. The important fact is that the guaranteed upper bound $k(\epsilon)$ for the size k of the paving does not depend on n, but only on ϵ. This can be used to establish the same result for any bounded operator $B \in \mathcal{B}(H)$ on a separable Hilbert space H, for which $(Be_i, e_i)_H = 0$ in a certain CONS $\{e_i\}_{i \in \mathbb{N}}$. For the discussion in this subsection, we will restrict ourselves to the case of $n \times n$ matrices with arbitrarily large finite n.

The consequence of this paving result is that any $n \times n$ matrix A with unit diagonal can be similarly paved using $k \leq k(\epsilon/(\|A\|_2 + 1))$ index subsets, such that the diagonal submatrices $A_r := A_{I_r,I_r}$ are invertible and uniformly well-conditioned, i.e.,

$$\kappa(A_r) := \|A_r\|_2\|A_r^{-1}\|_2 \le \frac{1+\epsilon}{1-\epsilon}, \qquad r = 1,\dots,k.$$

More precisely, we have $\|A_r\|_2 \le 1+\epsilon$ and $\|A_r^{-1}\|_2 \le (1-\epsilon)^{-1}$ for all r. In fact, set $B = A - I$, note that now $b_{ii} = 0$ and that $\|B\|_2 \le \|A\|_2 + 1$, apply (3.55) with B and $\epsilon' = \epsilon/(\|A\|_2 + 1)$, and check that

$$\|A_r\|_2 = \|I + B_r\|_2 \le 1 + \|B_r\|_2 \le 1 + \epsilon'\|B\|_2 \le 1 + \epsilon.$$

Due to $\|B_r\|_2 \le \epsilon < 1$, the existence of $A_r^{-1} = (I+B_r)^{-1} = I + \sum_{l=1}^{\infty}(-1)^l B_r^l$ follows from the convergence of the Neumann series, which also implies

$$\|A_r^{-1}\|_2 \le 1 + \sum_{l=1}^{\infty} \epsilon^l = \frac{1}{1-\epsilon}.$$

For symmetric positive definite $n \times n$ matrices A with condition number $\kappa(A)$ this has the following consequence: Consider the symmetrically scaled matrix $\tilde{A} = D^{-1/2}AD^{-1/2}$, which now has unit diagonal and condition number

$$\kappa(\tilde{A}) = \kappa(D^{-1/2}AD^{-1/2}) \le \kappa(A)\kappa(D) \le \kappa(A)^2.$$

Note that this is a very rough worst-case estimate, it may well be that diagonal scaling actually improves the condition number. Now apply the paving result above to $\tilde{A}$, and denote the respective diagonal submatrices of $\tilde{A}$, A, and D by $\tilde{A}_r$, A_r, and D_r, respectively, and the size of the index sets I_r by n_r. Here $r = 1,\dots,k$, where $k \le k(\epsilon/(\|\tilde{A}\|_2+1)$ is the size of the paving. This generates a space splitting $(\mathbf{V};\mathbf{a};\mathbf{R})$ for the Hilbert space $V = \mathbb{R}^n$ equipped with the scalar product $a(x,y) = (Ax,y)_2$ if we set $V_r = \mathbb{R}^{n_r}$, $a_r(x_r,y_r) = (A_r x_r, y_r)_2$, and take for R_r the natural extension-by-zero embedding of $\mathbb{R}^{n_r}$ in $\mathbb{R}^n$, $r = 1,\dots,k$. Obviously, this is a Riesz splitting for $(V;a)$ with the forms inherited by the Galerkin projection. By interpreting $\mathbb{R}^{n_r}$ as a subspace of $\mathbb{R}^n$, it can be turned into a subspace splitting $\mathbf{V}$ for $(V;a)$. The condition number κ of this splitting is bounded by

$$\frac{1-\epsilon}{1+\epsilon}\kappa(D^{-1/2}AD^{-1/2}) \le \kappa \le \frac{1+\epsilon}{1-\epsilon}\kappa(D^{-1/2}AD^{-1/2}), \tag{3.56}$$

since the norm associated with the splitting satisfies

$$\begin{aligned}|||x|||^2 &:= \sum_{r=1}^{k}(A_r x_r, x_r)_2 = \sum_{r=1}^{k}(\tilde{A}_r D_r^{1/2}x_r, D_r^{1/2}x_r)_2 \\ &\le (1+\epsilon)\sum_{r=1}^{k}(D_r x_r, x_r)_2 = (1+\epsilon)(Dx,x)_2.\end{aligned}$$

Here the decomposition $x = \sum_{r=1}^{k} R_r x_r$ into $x_r \in \mathbb{R}^{n_r}$ is unique. Similarly, we get $|||x|||^2 \geq (1+\epsilon)(Dx,x)_2$ and thus

$$\frac{\lambda_{\min}(\tilde{A})}{1+\epsilon} \leq \frac{(Ax,x)_2}{(1+\epsilon)(Dx,x)_2} \leq \frac{a(x,x)}{|||x|||^2} \leq \frac{(Ax,x)_2}{(1-\epsilon)(Dx,x)_2} \leq \frac{\lambda_{\max}(\tilde{A})}{1-\epsilon}.$$

This gives the upper bound in (3.56). The lower bound follows by taking the supremum of the above lower bound for $a(x,x)/|||x|||^2$ and the infimum of the corresponding upper bound. In other words, for $\epsilon \ll 1$ we have $\kappa(P) \approx \kappa(\tilde{A})$, where

$$P = \sum_{r=1}^{k} R_r R_r^* = (\sum_{r=1}^{k} R_r A_r^{-1} R_r')A = (\sum_{r=1}^{k} R_r D_r^{1/2} \tilde{A}_r^{-1} D_r^{1/2} R_r')A$$

is the associated Schwarz operator. This shows that the paving procedure does not improve the condition of the problem. However, if we switch from iteratively solving $Ax = b$ to iteratively solving the Schwarz equation $Px = CAx = Cb$ with the preconditioner

$$C := \sum_{r=1}^{k} R_r D_r^{1/2} \tilde{A}_r^{-1} D_r^{1/2} R_r',$$

the possible advantage is the restructuring of the iteration process by using subproblems governed by extremely well-conditioned matrices $\tilde{A}_r$.

Unfortunately, the proof of the Anderson paving conjecture given in [118] is not constructive. Regarding the size k of the paving, the best asymptotically optimal estimate known to us for symmetric B with zero diagonal follows from results in [149] and guarantees the existence of a (k,ϵ)-paving with some k satisfying

$$k \leq \frac{20.8}{\epsilon^2}. \tag{3.57}$$

To see this bound, one must carefully follow the proof of Theorem 1.1 in [149, Section 5.2], specialized to the case of the pair $[B,-B]$. In particular, one must replace the final estimate in the statement of [149, Theorem 5.6] with its slight improvement, which appears in the last step of its proof. This gives a (k,ϵ)-paving for B if

$$2(\frac{2}{\sqrt{k}} + \frac{1}{k}) = 2((\frac{1}{\sqrt{k}} + 1)^2 - 1) \leq \epsilon.$$

After rearranging, this is equivalent to

$$k \geq \frac{1}{(\sqrt{\epsilon/2+1} - 1)^2} = \frac{4(\sqrt{\epsilon/2+1} + 1)^2}{\epsilon^2}.$$

Thus, the existence of a (k,ϵ)-paving is guaranteed for some k with

$$k < \frac{4(\sqrt{\epsilon/2+1} + 1)^2}{\epsilon^2} + 1 \leq \frac{4(\sqrt{3/2} + 1)^2 + 1}{\epsilon^2} < \frac{20.8}{\epsilon^2}.$$

The best constant in (3.57) for the existence of (k,ϵ)-pavings with respect to the class of all symmetric B with zero diagonal is currently unknown. It cannot be less than 1, as shown in [34] on examples of symmetric conference matrices.

To get (k,ϵ)-pavings in practice, one must either use structural properties of a matrix, or give up the independence of k from the size n of the matrix. For example, if the spd matrix A is sparse with $\leq k$ non-zero entries per row, then $\tilde{A}$ admits a $(k,0)$ paving, i.e., $\tilde{A}_r = I$ and $A_r = D_r$ for all $r = 1,\ldots,k$. The elementary argument is outlined in Problem 3.10. For general matrices A, the usual approach is to iteratively construct submatrices of large size using a strong form of the Bourgain-Tzafriri restricted invertibility principle. For matrices, it is common to formulate it for column selection of an $m\times n$ matrix with non-zero columns $B_i \in \mathbb{R}^m, i = 1,\ldots,n$. The following result can be found in [201]. To formulate it, let D be the $n\times n$ diagonal matrix with entries $d_{ii} = \|B_i\|_2 > 0$ on the diagonal. Consequently, $\tilde{B} = BD^{-1}$ has unit norm columns and $\mathrm{Tr}(D^2) = \mathrm{Tr}(B^T B) = \|B\|_F^2$ is the Frobenius norm of B.

Proposition 3.3 *For every such B, there exists a subset $J \subset \{1,\ldots,n\}]$ of column indices of size*

$$|J| \geq \frac{\epsilon^2}{(1+\epsilon)^2}\frac{\|B\|_F^2}{\|B\|_2^2}$$

such that the $m \times |J|$ submatrix $\tilde{B}_J := [\tilde{B}_i]_{i\in J}$ of $\tilde{B}$ satifies

$$1-\epsilon \leq \lambda_{\min}(\tilde{B}_J^T\tilde{B}_J) \leq \lambda_{\max}(\tilde{B}_J^T\tilde{B}_J) \leq 1+\epsilon.$$

Note that $\tilde{B}_J = B_J D_J^{-1}$, where B_J and D_J are the similarly defined submatrices of B and D, respectively. If B already has normalized columns, i.e., if D is the identity, then $\|B\|_F^2 = n$ and Proposition 3.3 yields a B_J with $\sigma(B_J^T B_J) \subset [1-\epsilon, 1+\epsilon]$, where $|J| \geq (\epsilon/((1+\epsilon)\|B\|_2))^2 n$. So the number of columns in the remainder matrix $B_{J^c} = [\tilde{B}_i]_{i\notin J}$ is $\leq cn$ with $c = 1-(\epsilon/((1+\epsilon)\|B\|_2))^2$. Now apply Proposition 3.3 to B_{J^c}. This gives another index subset $J' \subset J^c$ disjoint from J, so again $\sigma(B_{J'}^T B_{J'}) \subset [1-\epsilon, 1+\epsilon]$. Since $\|B_{J^c}\|_2 \leq \|B\|_2$, the new remainder matrix $B_{(J\cup J')^c}$ has $\leq c^2 n$ columns. Continuing this way, after k steps, the number of columns in the remainder matrix is $\leq c^k n$. Thus if k satisfies $c^k n < 1$, then we have exhausted the entire column space of B. Moreover, by setting $I_1 = J$, $I_2 = J'$ and so on, we have obtained a partition $I_1 \cup \cdots \cup I_k$ such that $B_r = B_{I_r}$ satisfies

$$1-\epsilon \leq \lambda_{\min}(B_r^T B_r) \leq \lambda_{\max}(B_r^T B_r) \leq 1+\epsilon, \quad r = 1,\ldots,k.$$

Obviously, the smallest k for which $c^k n < 1$ satisfies

$$k \leq \frac{\log n}{\log(1/c)} + 1 \leq \frac{(1+\epsilon)^2\|B\|_2^2}{\epsilon^2}\log n + 1.$$

As shown in [201], this result can be used to obtain a paving for any $n \times n$ spd matrix A. Take $B = \tilde{A}^{1/2}$, where $\tilde{A} = \|T\|_2^{-1} T + I$ and $T = D_A^{-1/2} A D_A^{-1/2} - I$ with D_A being the $n \times n$ diagonal matrix with entries $A_{ii}, i = 1,\ldots,n$, on the diagonal. Note that T has zero diagonal elements, with $T \neq 0$ (unless A is diagonal, in which case the statement below holds with $k = 1$). Moreover, $\|T\|_2 \leq \max(1, \|D_A^{-1/2} A D_A^{-1/2}\|_2 - 1)$. Obviously, B is spd and column normalized, since $B_i^T B_i = (\tilde{A})_{ii} = 1$. Also,

$$\|B\|_2^2 = \|\tilde{A}\|_2 = \sup_{\|x\|_2=1} \|T\|_2^{-1}(Tx, x)_2 + \|x\|_2^2 = 2.$$

With this at hand, we can determine the partition $\{I_r\}_{r=1,\ldots,k}$ with some

$$k \leq \frac{2(1+\epsilon)^2}{\epsilon^2} \log n + 1$$

by applying the above result to B, and checking that

$$B_r^T B_r = \tilde{A}_r = \|T\|_2^{-1}(T_r - I) + I, \qquad T_r = D_r^{-1/2} A_r D_r^{-1/2} - I,$$

where $\tilde{A}_r$, A_r and D_r are the diagonal submatrices of $\tilde{A}$, A and D_A associated with I_r, respectively. Rearranging, we have $T_r = \|T\|_2(B_r^T B_r - I) + I$ and the spectral properties of the B_r yield

$$1 - \epsilon\|T\|_2 \leq \lambda_{\min}(T_r) \leq \lambda_{\max}(T_r) \leq 1 + \epsilon\|T\|_2$$

for $r = 1, \ldots, k$. If we set $\epsilon = \epsilon'/\|T\|_2$, we get essentially the same result as in Proposition 3.2, but with a weaker guarantee on the admissible size k of the partition, which may depend logarithmically on the size of A. On the positive side, there are several probabilistic and deterministic algorithms that realize the statement of Proposition 3.3. An algorithm that performs a kind of rank-revealing factorization on a submatrix $B_{J'}$ for a randomly chosen subset J' of column indices, leading with high probability to an appropriate $J \subset J'$ in Proposition 3.3, was proposed in [186]. Later, deterministic algorithms were developed in [168, 201]. In [124], the authors show that for certain classes of matrices B satisfying an additional incoherence condition, a random permutation σ of the column set $\{B_i\}$, followed by subdividing $\{\sigma(1), \ldots, \sigma(n)\}$ into k subsets of approximately equal size $\approx n/k$, leads with high probability to a paving of $A = B^T B$ with control of the condition numbers of the submatrices A_r.

3.3.2 Sketching

As explained above, although sketching is not directly related to a space splitting, it can be a first step in constructing a space splitting with improved condition number. To explain the basic ideas, we follow [97, 193] and [119, Section 10]. We restrict ourselves to the matrix case and focus first on finding the least-squares solution of an overdetermined linear system $Bx = b$ with $m \times n$ matrix B and $m \gg n$ as our model application. In other words, we are looking for a minimizer $x_B \in \mathbb{R}^n$ of the quadratic functional $J_B(x) := \|Bx - b\|_2^2$. Note that this scenario is different from the introductory example of preconditioning a linear system $Ax = b$ with an spd matrix based on a low-rank approximation of A. We will return to that situation later.

A *sketch* is provided by a $k \times m$ matrix S, where typically $n < k \ll m$, and $Bx = b$ is replaced by $SBx = Sb$. The new matrix SB is $k \times n$ in size, which hopefully makes the new least-squares problem

$$J_{SB}(x) = \|S(Bx - b)\|_2^2 \longrightarrow \min$$

faster to solve. This depends on k, the cost of creating and storing S, and the complexity of the matrix-vector multiplication with S, issues we will not address here. The basic requirement on S that we focus on is that it provides an $(1 + \epsilon)$-isometric embedding of the subspace $\mathrm{ran}(B)$ to $\mathbb{R}^k$, i.e., we have

$$(1 - \epsilon)\|x\|_2 \le \|Sx\|_2 \le (1 + \epsilon)\|x\|_2, \qquad x \in \mathrm{ran}(B).$$

With this at hand, if $b \in \mathrm{ran}(B)$ (the consistent case), the minimizers x_{SB} of the new least-squares problem satisfy

$$\begin{aligned} 0 = \|Bx_B - b\|_2 &\le \|Bx_{SB} - b\|_2 \le (1 - \epsilon)^{-1}\|S(Bx_{SB} - b)\|_2 \\ &\le (1 - \epsilon)^{-1}\|S(Bx_B - b)\|_2 \le \frac{1 + \epsilon}{1 - \epsilon}\|Bx_B - b\|_2. \end{aligned}$$

In other words, x_{SB} is another minimizer of the original problem. In the more interesting inconsistent case, the same argument yields

$$\|b^\perp\|_2 = \|Bx_B - b\|_2 \le \|Bx_{SB} - b\|_2 \le \frac{1 + \epsilon}{1 - \epsilon}\|b^\perp\|_2,$$

assuming that S is an $(1 + \epsilon)$-isometric embedding of the subspace spanned by b and $\mathrm{ran}(B)$. Here $b^\perp$ denotes the orthogonal projection of b onto $\mathrm{ran}(B)^\perp = \ker(B^T)$.

It remains to find S such that it provides a $(1 + \epsilon)$-isometry for $\mathrm{ran}(B)$ with sufficiently small ϵ and k as small as possible. A generic result is that random Gaussian $k \times m$ matrices S, whose entries are i.i.d. centered Gaussian variables with variance $1/k$, produce $(1 + \epsilon)$-isometric embeddings of $\mathrm{ran}(B)$ for a given $m \times n$ matrix B of rank $r \le n$ with probability

$$\mathbb{P}(\forall\, 0 \ne x \in \mathrm{ran}(B) : \; 1 - \epsilon \le \frac{\|Sx\|_2}{\|x\|_2} \le 1 + \epsilon) > 1 - 2\delta, \quad 0 < \delta < 1/2,$$

if $k \ge (1 + \sqrt{2\log(1/\delta)} + \sqrt{r})^2/\epsilon^2$. This follows from [119, Theorem 8.4] if one sets $t = \sqrt{2\log(1/\delta)/k}$ there and uses the property $w(E) \le \sqrt{r}$ for the Gaussian width w of the set $E = \{x \in \mathrm{ran}(B) : \; \|x\|_2 = 1\}$ as explained in [119]. So, for fixed $0 < \epsilon < 1$, sketch sizes $k = \mathrm{O}(\log(1/\delta) + r)$ will suffice. The ϵ^{-2} dependence of k is prohibitive when $\epsilon \longrightarrow 0$, since Gaussian matrices are dense and expensive to multiply. Cheaper and more structured replacements for Gaussian sketch matrices S are discussed in [119, Section 9]. Recall that forming SB by collecting k randomly selected rows of B into SB is a very simple sketching strategy, which already appeared in the context of

Proposition 3.3 for selecting columns instead of selecting rows. Its analysis is related to coherence assumptions on B or requires importance sampling using so-called leverage scores, see [117]. The latter are the appropriately scaled diagonal entries of the $m \times m$ ortho-projection matrix $Q_{\mathrm{ran}(B)} = UU^T$, where the columns of the $m \times r$ matrix U are given by any orthogonal basis in $\mathrm{ran}(B)$, e.g. obtained from the SVD or QR factorization of B. Sampling $k \geq 2r\log(r)/\epsilon^2$ rows using the leverage score distribution yields a $(1+\epsilon)$-isometric embedding of $\mathrm{ran}(B)$ in $\mathbb{R}^k$ with high probability. Note that the simplicity of S is compensated by an additional logarithm $\log(r)$ in the lower bound for k. For more details on structured sketching matrices S and their properties, see [119, Section 9].

How to actually compute the solution of the least-squares problem is described in [119]. We will only mention the following application of the sketching idea, originally proposed in [151]: Suppose for simplicity that B has full rank $r = n$, i.e., x_B is the unique solution of the $n \times n$ system of normal equations $B^T Bx = B^T b$. Let S be any of the $k \times m$ sketch matrices with $k = \mathrm{O}(n)$ or $k = \mathrm{O}(n\log(n))$ (depending on their construction), which provides a $(1+\epsilon)$-isometric embedding of $\mathrm{ran}(B)$ for some $\epsilon < 1$. Then run the QR factorization on the matrix SB, i.e., $SB = QR$, where the $k \times n$ matrix Q has orthogonal columns and R is an invertible upper triangular $n \times n$ matrix. Note that the rank-revealing QR factorization can be used in the rank-deficient case $r < n$. In addition, other factorizations have been proposed in the literature instead of the QR factorization. The $(1+\epsilon)$-isometry property of S yields

$$(B^T Bx, x)_2 = \|Bx\|_2^2 \approx \|SBx\|_2^2 = \|QRx\|_2^2 = \|Rx\|_2^2 = (R^T Rx, x)_2.$$

This means that $R^T R$ is spectrally equivalent to $B^T B$, and $(R^T R)^{-1}$ can be used as a preconditioner for the system of normal equations. The condition number of the preconditioned system

$$R^{-1}R^{-T}B^T Bx = R^{-1}R^{-T}B^T b$$

is bounded by $\kappa(R^{-1}R^{-T}B^T B) \leq \left(\frac{1+\epsilon}{1-\epsilon}\right)^2 \leq 4$ with a moderate choice of $\epsilon \leq 1/3$. In other words, sketching can lead to preconditioning. Here the application of the preconditioner $R^{-1}R^{-T}$ requires the solution of two linear systems with triangular matrices and involves $\mathrm{O}(n^2)$ operations, which is negligible compared to the $\mathrm{O}(mn)$ cost of multiplying by B.

Let us briefly touch on another application of sketching. It is inspired by our introductory example and is based on computing an approximate low-rank approximation of an $n \times n$ spd matrix A at a cost that beats the use of standard SVD or QR algorithms. We refer to Section 14 and Subsection 17.3 in [119] for details on the actual algorithm and further references. We form $Y = AS$, where S is a $n \times l$ sketching matrix. For analysis purposes, we assume that S is standard Gaussian, i.e., its entries are i.i.d. normally distributed random variables. We now compute a symmetric positive semidefinite Nyström approximation N in factorized form, namely

$$N := Y(S^T Y)^\dagger Y^T = \tilde{U}\tilde{D}\tilde{U}^T \approx U_k D_k U_k^T =: A_k,$$

where $\tilde{U}$ is an $n \times l$ orthogonal matrix and $\tilde{D}$ is an $l \times l$ diagonal matrix with non-negative diagonal elements, while U_k and D_k are the $n \times k$ and $k \times k$ submatrices of $\tilde{U}$ and $\tilde{D}$, respectively, corresponding to the largest k diagonal entries of $\tilde{D}$. Theoretically, this factorization can be obtained by first performing an SVD on the symmetric semidefinite $l \times l$ matrix

$$S^T Y = S^T A S = V \Sigma^2 V^T,$$

followed by a second SVD on the $n \times l$ matrix

$$Y((S^T Y)^\dagger)^{1/2} = Y V \Sigma^\dagger = \tilde{U} \tilde{\Sigma} \tilde{V}^T,$$

which gives

$$Y(S^T Y)^\dagger Y^T = \tilde{U} \tilde{D} \tilde{U}^T, \qquad \tilde{D} = \tilde{\Sigma}^2.$$

The actual computation reported in [119, Algorithm 16] uses a shift for reasons of stability, i.e., A is replaced by $A + \nu I$ with an appropriate value of ν, and the first SVD is replaced by a Cholesky factorization of $S^T(A + \nu I)S$.

According to [119, Theorem 14.1], this procedure satisfies the error bound

$$\mathbb{E}(\|A - A_k\|_2) \le \epsilon_k := \lambda_{k+1} + \frac{k}{l-k-1} \sum_{j=k+1}^{n} \lambda_j, \quad 1 \le k < l-1 < n, \tag{3.58}$$

where λ_j are the eigenvalues of A in decreasing order. Recall that the optimal rank-k approximation obtained from the SVD of A gives the lower bound $\epsilon_k \ge \lambda_{k+1}$. Thus, choosing a sufficiently large l reduces the influence of the tail term and can bring ϵ_k closer to this optimal lower bound. The approximation is especially efficient when the spectrum decays rapidly, i.e., when

$$\sum_{j>k} \lambda_j = \mathrm{O}(\lambda_{k+1})$$

holds with reasonable constants. A refined analysis of a similar algorithm for approximate rank-k SVD of general $m \times n$ matrices can be found in [94].

As in the least-squares problem and in our motivating example, the low-rank approximation A_k can be used to provide a preconditioner and an associated space splitting with reduced condition number. In fact, consider the space splitting of $(V; a)$ given by $V = \mathbb{R}^n$ and $a(x, y) = (Ax, y)_2$ into the two orthogonal subspaces $(V_k := \mathrm{ran}(U_k); a_k)$ and $(V_k^\perp; \epsilon_k(\cdot,\cdot)_2)$, where the scalar product a_k is defined by $a_k(x_k, y_k) = (A_k x_k, y_k)_2 + \epsilon_k (x_k, y_k)_2, x_k, y_k \in \mathrm{ran}(U_k)$. Then, according to (3.58), its condition number is bounded in expectation by

$$\mathbb{E}(\kappa) \le 1 + \frac{2\epsilon_k(A)}{\lambda_{\min}(A)} \le 1 + \frac{\lambda_{k+1} + \frac{k}{l-k-1} \sum_{j=k+1}^{n} \lambda_j}{\lambda_n}$$

and often compares favorably to the condition number $\kappa(A) = \lambda_1/\lambda_n$ of A. To see this condition number estimate, let $x = x_k + x_k^\perp$ denote the unique decomposition of $x \in V$ into components $x_k = UU^T x \in V_k$ and $x_k^\perp = (I - UU^T)x \in V_k^\perp$. Note that by definition of A_k we have $(A_k x, x)_2 = (A_k x_k, x_k)$ for all $x \in V$, while $\|x\|_2^2 = \|x_k\|_2^2 + \|x_k^\perp\|_2^2$ holds due to the orthogonality of V_k and V_k^T. The norm $|||\cdot|||$ associated with the subspace splitting satisfies

$$\begin{aligned}|||x|||^2 &= (A_k x_k, x_k)_2 + \epsilon_k \|x_k\|^2 + \epsilon_k \|x_k^\perp\|^2 = (A_k x, x) + \epsilon_k \|x\|_2^2 \\ &\le (Ax, x)_2 + 2\epsilon_k \|x\|^2 \le (1 + \frac{2\epsilon_k}{\lambda_{\min}(A)}) a(x, x)\end{aligned}$$

on the one hand, while

$$a(x,x) = (Ax,x)_2 \le (A_k x, x) + \epsilon_k \|x\|_2^2 = (A_k x_k, x_k)_2 + \epsilon_k \|x\|_2^2 = |||x|||^2$$

on the other hand. This proves the claim. The associated Schwarz operator P and preconditioner C_k are therefore

$$P = C_k A, \qquad C_k := U_k (D_k + \epsilon_k I)^{-1} U_k^T + \epsilon_k^{-1} (I - U_k U_k^T).$$

Except for the choice of the scaling parameter ϵ_k and minor variations, this is the preconditioner mentioned in [119, Section 17.3]. Note that the cost of computing a matrix-vector product with this preconditioner is $O(kn)$. This is negligible compared to the cost of computing Ax for a dense matrix A. For a sparse matrix A, other approaches are needed that avoid densely populated factorizations of low-rank approximations to A.

Problems

3.1 Equivalence of K-functional and $L^2(T^d)$ moduli of continuity.
Prove the two-sided inequality (3.13).
Hints. Since

$$\Delta_h e^{i(k,x)_2} = (e^{i(k,h)_2} - 1) e^{i(k,x)_2},$$

the Fourier series of $\Delta_h^m f(x)$ and its L^2 norm can be computed explicitly in terms of the Fourier coefficients $c_k(f)$ of f. Convince yourself that

$$w_m(t,f)_{L^2}^2 \approx t^{-d} \int_{\|h\|_2 \le t} \|\Delta_h^m f\|_{L^2}^2 \, dh, \qquad t > 0,$$

replace the resulting expression for $\|\Delta_h^m f\|_{L^2}^2$, and compare with the definition of $\omega_m(t,f)$. Note that the right-hand side in

$$\omega_m(t,f)^2 \approx \inf_{g \in H^m(T^d)} \|f - g\|_{L^2}^2 + t^{2m} |g|_{H^m}^2$$

can also be expressed only in terms of the Fourier coefficients, see Problem 3.2.

3.2 Proof of $B^s_{\omega_m} = H^s(T^d), 0 < s < m$.
Show that for $f \in L^2(T^d)$ we have

$$\|f\|_{B^s_{\omega_m}} \approx \|f\|_{H^s}, \qquad 0 < s < m,$$

where the constants depend on s and m. For notation and definitions, see Subsection 3.1.2.
Hints. First, prove that

$$\omega_m(t, f)^2 \approx \sum_{k \in \mathbb{Z}^d} \frac{(|k|_2 t)^{2m}}{1 + (|k|_2 t)^{2m}} |c_k(f)|^2, \quad t > 0,$$

for $f \in L^2(T^d)$. Substitute this into the definition of the $B^s_{\omega_m}(T^d)$ norm and compare with the definition of the $H^s(T^d)$ norm.

3.3 Property (3.23) used in the construction of quasi-interpolants Q_j for C^0 Lagrange finite element spaces V_j.
The notation is explained in the proof of Theorem 3.3 in Subsection 3.1.3.
a) First establish the bound

$$\|q_{P;x_0,r}\|_{L^\infty(\Sigma)} \le C, \tag{3.59}$$

where C is independent of the nodal point P of the standard simplex and of the choice of the ball $B(x_0, r) \in \mathcal{B}_{r_0}$ that appears in the construction of the family $\{q_{P;x_0,r}\}$.
b) Derive (3.23) from the result in Part a).
c) Argue why similar constructions and bounds hold for rectangular finite element spaces (with the standard cell $[0, 1]^d$), or if $\mathcal{B}_{r_0}$ in the definition of $\{q_{P;x_0,r}\}$ is replaced by other appropriate families of subdomains of the standard cell.
Hints. For any fixed $B(x_0, r) \in \mathcal{B}_{r_0}$, the existence of polynomials $q_{P;x_0,r}$ satisfying (3.21) follows from the linear independence of the set $\{N_P\}$ of nodal basis functions associated with Σ. That the constant C in (3.59) can be chosen independently of P and $B(x_0, r) \in \mathcal{B}_{r_0}$ follows from the continuous dependence of $q_{P;x_0,r}$ on the parameters x_0 and r associated with $B(x_0, r) \in \mathcal{B}_{r_0}$. For Part b), use the change of variable formula for integrals over C_i and Σ associated with the affine linear map T_i. In particular, find the scaling factor α_i and show that $\alpha_i \approx 2^{jd}$. For Part c), note that the argument for Part a) holds for any family of subsets of the standard cell that can be parameterized over a compact set such that the associated family of polynomials biorthogonal to $\{N_P\}$ depends continuously on the parameters.

3.4 Another quasi-interpolant for linear finite element spaces.
Consider the case of linear finite elements ($p = 1$) on a quasi-uniform and regular triangulation $\mathcal{T}_j$ of a bounded polygonal domain in the plane ($d = 2$) and fix the subset $\Sigma_t, 0 < t < 1$, of the standard simplex Σ in $\mathbb{R}^2$ as shown in Figure 3.15. In the construction of quasi-interpolants via (3.24-3.25), as suggested in the proof of Theorem 3.3, replace the circles $B(x_0, r)$ by Σ_t.

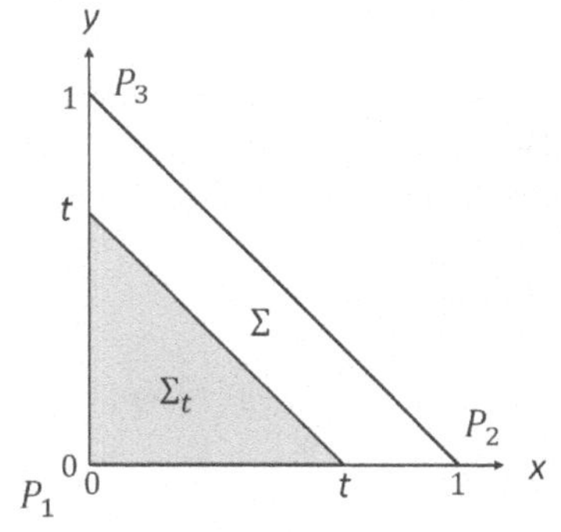

Fig. 3.15 The subset Σ_t, $0 < t < 1$, for the standard simplex Σ.

a) Compute the linear polynomials $q_{P_i,t}$, $i = 1, 2, 3$, defined by (3.21) with Σ_t instead of $B(x_0, r)$, and check how the $L^\infty(\Sigma)$ norm of $q_{P_i,t}$ depends on the parameter t.
b) Give estimates for the L^2 norm of the resulting quasi-interpolant projection Q_j onto $V(\mathcal{T}_j)$ and its dependence on t when in (3.25) all but one coefficient $a_{P,i}$ vanishes, i.e., when

$$\lambda_P(u) = \int_{C_i} u(x) q_{P,C_i}(x)\,dx$$

for some triangle C_i attached to P, and the map T_i used for the construction of q_{P,C_i} transforms $P \in C_i$ into $P_1 \in \Sigma$ for all vertices P of $\mathcal{T}_j$.
Hints. For Part b), use the midpoint rule for triangles (which is exact for polynomials of total degree ≤ 2) to show that

$$\|Q_j u\|_{L^2}^2 = \frac{1}{3} \sum_e |Q_j u(M_e)|^2 |C_e| \leq \frac{1}{3} \sum_P \lambda_P(u)^2 |\text{supp}(N_P)|,$$

where $|C_e|$ is the area of the union of triangles that share the edge e and the summation is with respect to all edges in $\mathcal{T}_j$. The estimate may deteriorate if the triangle C_i associated with P is "small" compared to the support of N_P.

3.5 Nodal interpolation and other quasi-interpolant operators for simplicial C^0 Lagrange finite element spaces.
Suppose $V(\mathcal{T})$ is a C^0 Lagrange finite element space of degree $p \geq 1$ over a quasi-uniform and regular simplicial partition $\mathcal{T}$ of a bounded polyhedral Lipschitz domain $G \subset \mathbb{R}^d$, see Subsection 3.1.3 for the definitions and notation used below.
a) Consider the nodal interpolation operator given by

$$Qf = \sum_{P \in \mathcal{V}} f(P) N_P, \qquad f \in C(\bar{G}).$$

By the embedding (A.25) it is well-defined and bounded on $H^s(G)$, $s > d/2$. Using Lemma A.3, show the bounds

$$|f - Qf|_{H^k} \leq C\mathrm{diam}(\mathcal{T})^{m-k}|f|_{H^m}, \qquad f \in H^m(G),$$

where $k = 0, 1,\ p+1 \geq m \geq k, m > d/2$, and the constant C depends only on p, d, and the quasi-uniformity and regularity constants of $\mathcal{T}$. Note that this bound implies the inequality (3.18) used in the proof of Proposition 3.1, if one sets $k = 1$, $m = 2$, $p = 1$ and $d = 2$.
b) Consider quasi-interpolants of the form

$$Qf = \sum_{P \in \mathcal{V}} \lambda_P(f) N_P,$$

where the dual functionals $\lambda_P(f)$ are given by integrals over the cells in $\mathcal{T}$ or over $(d-1)$-dimensional boundary simplices of cells C_P (as in the construction of the Scott-Zhang quasi-interpolant [162]), such that

$$\lambda_P(N_{P'}) = \delta_{PP'}, \qquad P, P' \in \mathcal{V},$$

and Q is a projection on $V(\mathcal{T})$. More precisely, if P is in the interior of a cell in $\mathcal{T}$, denote that cell by C_P, and set

$$\lambda_P(f) = \int_{C_P} f(x) q_{P,C_P}(x)\, dx,$$

where $q_{P,C_P}(x)$ is constructed as in the proof of Theorem 3.3, see the explanations following (3.21), but now with $B(x_0, r)$ replaced by Σ. If P is not in the interior of any cell, then it must be on a $(d-1)$-dimensional boundary simplex Γ_P of some cell C_P (if it belongs to several cells, choose one arbitrarily). Then we have two options: Either we define $\lambda_P(f)$ again by the above formula as an integral over C_P, or alternatively we set

$$\lambda_P(f) = \int_{\Gamma_P} f(x) q_{P,\Gamma_P}(x)\, dx.$$

Here $q_{P,\Gamma_P}(x)$ is supported on Γ_P and constructed similarly, but now with respect to Γ_P and the nodal basis functions restricted to Γ_P. Obviously, for the latter functionals to be well-defined, we need the trace $T_{\Gamma_P} f$ to be well-defined in the sense of $L^1(\Gamma_P)$. This is guaranteed if

$$f \in H^s(G), \quad s > 1/2,$$

see Subsection A.5.2 in the appendix. In the Scott-Zhang quasi-interpolant construction [162], the alternative rule is used when $P \in \Gamma = \partial G$.
Show that

$$|f - Qf|_{H^k} \leq C\mathrm{diam}(\mathcal{T})^{m-k}|f|_{H^m}, \qquad f \in H^m(G),$$

where $k = 0, 1$, $p + 1 \geq m \geq 1$, and the constant C depends only on p, d, and the quasi-uniformity and regularity constants of $\mathcal{T}$.

c) A simple quasi-interpolant operator is given by

$$Qf = \sum_{P\in\mathcal{V}} \Big(\int_G f(x)\hat{N}_P(x)\,dx\Big) N_P, \qquad f \in L^1(G),$$

where $\hat{N}_P$ denotes the nodal basis function N_P scaled so that $\int_G \hat{N}_p(x)\,dx = 1$. For $m = 1$, verify the properties (a)-(c) of a quasi-interpolant operator formulated in the proof of Theorem 3.3. Furthermore show the Jackson-type inequality

$$\|f - Qf\|_{L^2} \le C\mathrm{diam}(\mathcal{T})|f|_{H^1}, \qquad f \in H^1(G).$$

Hints. For a), it is sufficient to consider as G the standard simplex Σ equipped with the partition $\mathcal{T} = \{\Sigma\}$ consisting of a single cell. The result is then obtained by affine-linear transformation to the cells of $\mathcal{T}$ and summation of all local estimates. The proof for the standard simplex starts with taking the polynomial q of degree $\le m-1$ from Lemma A.3. Then we have

$$\begin{aligned}|f - Qf|_{H^k} &\le |f-q|_{H^k} + |Q(f-q)|_{H^k} \le |f-q|_{H^k} + C \max_{P\in\mathcal{V}_\Sigma} |f(p) - q(P)| \\ &\le |f-q|_{H^k} + C\|f-q\|_{H^m} \le C|f|_{H^m}\end{aligned}$$

under the restrictions on k and m. In the last step, the estimate of Lemma A.3 was used (note that $\mathrm{diam}(\Sigma) \approx 1$). The bound for $|Q(f-q)|_{H^k}$ uses the fact that $Q(f-q)$ belongs to a fixed finite-dimensional linear space $V_{\mathcal{T}_\Sigma}$, on which every semi-norm is bounded by a constant multiple of every other norm. We also have $\max_{P\in\mathcal{V}_\Sigma} |f(p) - q(P)| \le \|f-q\|_{C(\bar{\Sigma})} \le C\|f-q\|_{H^m}$ from (A.25).
For b) and c), follow the proof of the Jackson-type estimate for Theorem 3.3. Since Q is not as local as the quasi-interpolants considered in a), you must again work with cell neighborhoods $\hat{C}_i \supset C_i$. For b), the dual functionals associated with Γ_P can first be estimated using the embedding

$$\|T_{\partial\Sigma} f\|_{L^2(\partial\Sigma)} \le C\|T_{\partial\Sigma} f\|_{H^{1/2}(\partial\Sigma)} \le C\|f\|_{H^1(\Sigma)},$$

for the standard simplex Σ, followed by affine-linear transformation. Note that the Q in c) is not a projection onto $V(\mathcal{T})$. However, it reproduces constants.

3.6 Riesz basis property of the B-spline basis.
a) Let $\phi \in L^2(\mathbb{R})$. Show that the system $\Phi_0 = \{\phi(\cdot - l)\}_{l\in\mathbb{Z}}$ of integer shifts of ϕ is a Riesz basis with constants $0 < A \le B < \infty$ in the subspace V_0 of $L^2(\mathbb{R})$ generated by Φ_0 if and only if

$$A \le \sum_{k\in\mathbb{Z}} |\hat{\phi}(\theta + 2\pi k)|^2 \le B$$

for a.e. $\theta \in (-\pi, \pi]$. Furthermore, show that Φ_0 is an orthonormal basis in V_0 if and only if $A = B = 1$ in this condition.
b) Use this criterion to show the Riesz basis property of the B-spline systems Φ_j in the spline spaces V_j from Subsection 3.1.4.1.

3.7 Properties of integer shifts.
Let ϕ, ϕ^* be in $L^2(\mathbb{R})$.
a) Show that the associated systems Φ_0, Φ_0^* of integer shifts are biorthogonal, i.e.,

$$(\phi_{0,l}, \phi_{0,k}^*)_{L^2} = \delta_{lk}, \quad l, k \in \mathbb{Z},$$

if and only if

$$\sum_{l \in \mathbb{Z}} \hat{\phi}(\theta + 2\pi k)\overline{\hat{\phi}^*(\theta + 2\pi k)} = 1$$

for a.e. $\theta \in (-\pi, \pi]$.
b) Show that $f \in L^2(\mathbb{R})$ is orthogonal to Φ_0, i.e.,

$$(\phi_{0,l}, f)_{L^2} = 0, \quad l \in \mathbb{Z},$$

if and only if

$$\sum_{l \in \mathbb{Z}} \hat{\phi}(\theta + 2\pi k)\overline{\hat{f}(\theta + 2\pi k)} = 0$$

for a.e. $\theta \in (-\pi, \pi]$.
c) If in addition ϕ, ϕ^*, ψ satisfy refinement equations with masks $\mathbf{a}, \mathbf{a}^*, \mathbf{b}$, respectively, then the biorthogonality of the systems $\tilde{\Psi}$ and $\tilde{\Psi}^*$ implies

$$m_{\mathbf{a}}(\theta)\overline{m_{\mathbf{a}^*}(\theta)} + m_{\mathbf{a}}(\theta + \pi)\overline{m_{\mathbf{a}^*}(\theta + \pi)} = 1,$$
$$m_{\mathbf{b}}(\theta)\overline{m_{\mathbf{a}^*}(\theta)} + m_{\mathbf{b}}(\theta + \pi)\overline{m_{\mathbf{a}^*}(\theta + \pi)} = 0.$$

d) Under the assumptions in c), derive formulas for $m_{\mathbf{b}}$ with $m_{\mathbf{a}}$ and $m_{\mathbf{a}^*}$ given and for $m_{\mathbf{a}^*}$ with $m_{\mathbf{a}}$ and $m_{\mathbf{b}}$ given, respectively.

3.8 Auxiliary space method for finite element discretizations on different domains $G \neq \hat{G}$.
Let $G \neq \hat{G}$ be two bounded polygonal domains in $\mathbb{R}^2$, equipped with quasi-uniform and regular triangulations $\mathcal{T}$ and $\hat{\mathcal{T}}$ of comparable diameter $h \approx \hat{h}$, respectively, whose boundaries satisfy $\mathrm{dist}(\Gamma, \hat{\Gamma}) \leq Ch$ (see Figure 3.16 for an example). Construct a preconditioner for a $H_0^1(\hat{G})$-elliptic problem on the linear finite element space $V(\hat{\mathcal{T}})_0$ by reducing it to an already available preconditioner for a $H_0^1(G)$-elliptic problem on $V(\mathcal{T})_0$.
Hint. Follow the instructions at the end of Subsection 3.2.2.1.

3.9 Semi-orthogonal linear finite element wavelets.
Prove the properties of the functions $\theta_{j,k}$ and $\psi_{j,l}$ defined in (3.34) and (3.33), respectively.
Hints. Using the affine-linear transformation, it is sufficient to solve the problem for the standard d-dimensional simplex Σ (identified with $\mathcal{T}_0$), set $j = 1$, and take the origin as P_k. Then, the associated function $\theta_{1,k}$ is given by

$$\theta(x) = \phi(2x) - a_d\phi(x), \qquad \phi(x) = \begin{Bmatrix} 1 - \sum_{i=1}^d x_i. & x \in \Sigma, \\ 0, & x \notin \Sigma. \end{Bmatrix}$$

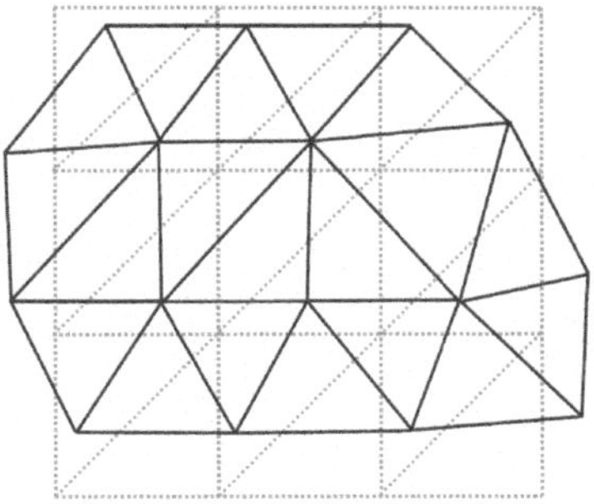

Fig. 3.16 Two domains with comparable triangulations.

By symmetry, the orthogonality of θ and the remaining nodal basis functions $N_{0,P}$, $P \neq 0$, is equivalent to checking

$$0 = (\theta, 1-\phi)_{L^2(\Sigma)} = (\phi(2\cdot), \frac{1}{2}(1-\phi(2\cdot)))_{L^2(\frac{1}{2}\Sigma)} - a_d(\phi, 1-\phi)_{L^2(\Sigma)}.$$

From this, $a_d = 2^{-(d+1)}$ can be concluded by a dilation argument, without computing any integrals explicitly. The same trick applies to the verification of the formula for $(\theta_{j,k}, N_{j-1,P_k})_{L^2}$ since

$$(\theta, \phi)_{L^2(\Sigma)} = (\phi(2\cdot), \frac{1}{2}(1+\phi(2\cdot)))_{L^2(\frac{1}{2}\Sigma)} - a_d(\phi, \phi)_{L^2(\Sigma)}.$$

Also check the uniform boundedness of the coefficients in (3.33).

3.10 Paving of sparse matrices.
Let A be an $n \times n$ spd matrix with $\leq k$ non-zero entries per row (and column).
a) Consider the following algorithm to find an index subset $J \subset \{1, \ldots, n\}$. Start with $J = \emptyset$, $J' = \{1, \ldots, n\}$. As long as $J' \neq \emptyset$, pick any $i \in J'$, add i to J, remove all indices $j \in J'$ with $a_{ij} \neq 0$ from J', and repeat. Obviously, the recursion stops after $\leq n$ steps. Prove that the remainder matrix A_{J^c,J^c} has $\leq k-1$ non-zero elements per row and that $A_{J,J}$ is diagonal.
b) Denote the output J and $A_{J,J}$ of the recursion in a) by I_1 and A_1, respectively. If $I_1^c \neq \emptyset$, apply it again to the matrix $A_{I_1^c,I_1^c}$. This gives I_2 and A_2 and (if $(I_1 \cup I_2)^c \neq \emptyset$) continue with the new remainder matrix $A_{(I_1\cup I_2)^c,(I_1\cup I_2)^c}$ and so on. Show that this produces a partition $I_1 \cup \cdots \cup I_{k'} = \{1, \ldots, n\}$ such that all $A_r = A_{I_r,I_r}$ are diagonal and $k' \leq k$.

Chapter 4
Schwarz iterative methods: Finite Ω

Abstract This chapter reviews the convergence theory of subspace correction methods for finitely many subproblems. Deterministic, greedy, randomized, and accelerated versions are discussed.

4.1 Schwarz iterative methods: Generalities

In this chapter, we develop the convergence theory of Schwarz iterative methods for solving linear operator equations $Au = f$ in V, based on a fixed stable space splitting $(\mathbf{V}; \mathbf{a}; \mathbf{R}; \rho)$ for $(V; a)$ into *finitely many* spaces V_i, $i \in \Omega := \{1, \dots, n\}$, with bilinear forms $a_i(\cdot, \cdot)$, bounded linear operators $R_i : V_i \to V$, and a weight sequence $\rho = \{\rho_i\}_{i=1,\dots,n}$. We follow the notational conventions of Chapter 2. In particular, recall the Schwarz operator P_ρ and the Gramian operator $\tilde{P}_\rho$ of a space splitting $(\mathbf{V}; \mathbf{a}; \mathbf{R}; \rho)$ from Subsection 2.1.1, and recall the Schwarz equation $P_\rho u = g$ and the Gramian equation $\tilde{P}_\rho \tilde{u} = \tilde{g}$ from Subsection 2.2.3, which correspond to $Au = f$. The common feature of the Schwarz iterative methods considered in this chapter is that, on the one hand, their convergence properties essentially depend on metric properties of the underlying space splitting and, in particular, on the spectral bounds for the Schwarz operator P_ρ. On the other hand, they are obtained by applying the operators

$$P_i := R_i R_i^* = R_i A_i^{-1} R_i' A, \qquad i = 1, \dots, n,$$

in a certain order. In this way, solving $Au = f$ is reduced to repeatedly solving operator equations with the operators A_i on the spaces V_i. Splitting the original problem on V into many simpler problems on the spaces V_i is also attractive for parallel and distributed computing. In other words, if the spaces $(V_i; a_i)$ and the operators R_i are properly chosen so that the application of the P_i is cheap and the space splitting is well-conditioned, then attractive iterative schemes can result.

We first focus on a generic one-step Schwarz iteration scheme, which is described as follows: Given the current iterate $u^{(m)}$, one computes the next iterate $u^{(m+1)}$ by

M. Griebel, P. Oswald, *Hilbert Space Splittings and Iterative Methods*, Springer Series in Computational Mathematics 62, https://doi.org/10.1007/978-3-031-74370-2_4

the formula

$$u^{(m+1)} = u^{(m)} + \omega_m d^{(m)}, \quad d^{(m)} := \sum_{i \in I_m} \rho_i P_i e^{(m)}, \quad e^{(m)} := u - u^{(m)}, \tag{4.1}$$

where I_m is a subset of Ω, $m = 0, 1, \ldots$, and u denotes the solution of the operator equation $Au = f$ or, equivalently, of the variational problem

$$a(u, v) = \langle f, v \rangle_{V' \times V} \qquad \forall v \in V.$$

Using terminology from the optimization literature, see for example [107], we call $d^{(m)}$ the *search direction* and ω_m the *step-size*. Note that

$$d^{(m)} = \sum_{i \in I_m} \rho_i R_i A_i^{-1} R_i' A e^{(m)} = \sum_{i \in I_m} \rho_i R_i A_i^{-1} R_i' (f - Au^{(m)})$$

is computable without knowing the error $e^{(m)}$ (and thus the exact solution u), but can be derived from residual information available at the m-th iteration step.

Different strategies for choosing I_m and ω_m lead to different classical and new algorithms. For example, in the extreme case $I_m = \Omega$, the iteration (4.1) can be reformulated as

$$u^{(m+1)} = u^{(m)} + \omega_m P_\rho e^{(m)}, \qquad m = 0, 1, \ldots, \tag{4.2}$$

since

$$d^{(m)} = \sum_{i \in \Omega} \rho_i P_i e^{(m)} = P_\rho e^{(m)}$$

in (4.1). Setting $\omega_m = \omega$ with fixed step-size $\omega > 0$ leads to the standard Richardson iteration for solving the associated Schwarz equation $P_\rho u = g$. Alternatively, the variable step-size

$$\omega_m = \frac{a(e^{(m)}, d^{(m)})}{a(d^{(m)}, d^{(m)})} \tag{4.3}$$

minimizes the error $\|e^{(m+1)}\|$ of the next iterate $u^{(m+1)}$ for the given search direction $d^{(m)}$. We will call (4.3) the *steepest descent* rule.

In the literature, iterations (4.1) with $I_m = \Omega$ appear under different names, i.e., *additive* or *parallel* or *asynchronous* or *simultaneous* (Schwarz) iteration. Since all P_i are executed independently in each iteration step, there is no ordering aspect and thus clear potential for parallelization. The classical convergence result for (4.2) is as follows.

Theorem 4.1 *Let* $\Omega = \{1, \ldots, n\}$ *and let* $(\mathbf{V}; \mathbf{a}; \mathbf{R}; \rho)$ *be a finite stable space splitting for* $(V; a)$ *with condition number* $\kappa = \kappa(P_\rho)$ *and spectral bounds* $\lambda_{\max} = \lambda_{\max}(P_\rho)$, $\lambda_{\min} = \lambda_{\min}(P_\rho)$. *Then the additive Schwarz iteration (4.2) satisfies*

$$\frac{\|e^{(m+1)}\|}{\|e^{(m)}\|} \leq \max(|1 - \omega_m \lambda_{\max}|, |1 - \omega_m \lambda_{\min}|) < 1, \qquad 0 < \omega_m < 2/\lambda_{\max}, \tag{4.4}$$

$m = 0, 1, \ldots$. *This bound is minimized for the fixed step-size*

$$\omega_m = \omega := 2/(\lambda_{\max} + \lambda_{\min}),$$

and results in

$$\|e^{(m)}\| \le (1 - 2/(1+\kappa))^m \|e^{(0)}\|, \qquad m = 1, 2, \ldots. \tag{4.5}$$

The bound (4.5) also holds if ω_m is given by the steepest descent rule (4.3).

Proof By (4.2) we have

$$e^{(m+1)} = u - u^{(m)} - \omega_m P_\rho e^{(m)} = (I - \omega_m P_\rho) e^{(m)},$$

and hence $\|e^{(m+1)}\| \le \|I - \omega_m P_\rho\| \|e^{(m)}\|$. The spectrum of the symmetric iteration operator $I - \omega_m P_\rho$ is tightly contained in the interval $[1 - \omega_m \lambda_{\max}, 1 - \omega_m \lambda_{\min}]$, and thus

$$\|I - \omega_m P_\rho\| = \max(|1 - \omega_m \lambda_{\max}|, |1 - \omega_m \lambda_{\min}|).$$

Substitution then gives (4.4). Obviously, the minimum in this formula is attained for $\omega_m = 2/(\lambda_{\max} + \lambda_{\min})$, which yields (4.5). Since the quotient $\|e^{(m+1)}\|/\|e^{(m)}\|$ is minimized when ω_m is given by (4.3), the bound (4.5) also holds in this case. □

The other extreme for choosing the index set I_m in (4.1) is to pick exactly one index i_m from Ω in each step and set $I_m = \{i_m\}$. This results in the *multiplicative* or *sequential* or *synchronous* (Schwarz) iteration. Now the order of execution of the P_i becomes relevant. Classically, the order in which Ω is traversed was deterministic. For example, in the standard cyclic ordering one chooses $i_m = i$ if $m = kn + i - 1$ for some $k = 0, 1, \ldots$. Then (4.1) leads to executing the inner loop

$$u^{(kn+i)} = u^{(kn+i-1)} + \omega_k \rho_i P_i e^{(kn+i-1)}, \qquad i = 1, \ldots, n, \tag{4.6}$$

repeatedly for $k = 0, 1, \ldots$. The step-size is kept constant in each inner loop for convenience. The convergence theory of (4.6) and some of its variants is discussed in Section 4.2.

Randomized and greedy choices for I_m in (4.1), which have received a lot of attention in recent years, are studied in Sections 4.3 and 4.4, respectively. Here, the tightest convergence bounds are obtained for the steepest descent version of (4.1) in the case of a singleton $I_m = \{i_m\}$, namely

$$u^{(m+1)} = u^{(m)} + \frac{a(P_{i_m} e^{(m)}, e^{(m)})}{a(P_{i_m} e^{(m)}, P_{i_m} e^{(m)})} P_{i_m} e^{(m)}, \qquad m = 0, 1, \ldots. \tag{4.7}$$

This guarantees maximal error reduction given by the formula

$$\|e^{(m+1)}\|^2 = \|e^{(m)}\|^2 - \Delta_{i_m}^{(m)}, \quad \Delta_i^{(m)} := \frac{a(P_i e^{(m)}, e^{(m)})^2}{a(P_i e^{(m)}, P_i e^{(m)})}, \ i = 1, \ldots, n, \tag{4.8}$$

in each iteration step. In addition, the quantities $\Delta_i^{(m)}$ will serve as error indicators guiding the selection of i_m in the greedy case.

Unlike (4.1), this algorithm does not depend on an a priori choice of the weight sequence ρ nor does it require the choice of certain step-size parameters ω_m. One may wonder if the weight sequence ρ, which influences the relative scaling of the subproblems, is needed at all. Since the proofs in this chapter will show a dependence of the convergence bounds on spectral properties of the Schwarz operator P_ρ, and in particular on condition numbers such as $\kappa = \kappa(P_\rho)$, it is in some cases instructive to keep ρ as free design variable in the derivations. If an algorithm like (4.7) does not depend on the choice of ρ, then the convergence bounds can be optimized with respect to ρ. On the other hand, to simplify the formulas and make the exposition more readable, we will make a compromise in many places and set $\rho_i = 1$. One can think of this as combining ω_m and ρ_{i_m} into a single step-size parameter. In Sections 4.3 and 4.4 of this chapter, and in Chapter 5, a variety of other randomized and greedy versions of (4.1) and variably sized I_m are considered.

In practice, more iterations are needed to achieve the desired error reduction for larger κ. When the preconditioning effect of the space splitting is not optimal, one looks for *acceleration techniques*, similar to the switch from gradient methods to conjugate gradient methods in numerical linear algebra. What one can hope for is to reduce the dependence of the convergence estimates on κ to a dependence on $\sqrt{\kappa}$. We examine two-step Schwarz methods with accelerated convergence in Section 4.5.

When comparing the convergence of various Schwarz iterative methods in this and the next chapter, we implicitly assume that the essential work per iteration is associated with solving the subproblems, i.e., applying P_i, $i \in I_m$, and that solving a single subproblem is much less costly than solving many or all subproblems. Thus, it would be misleading to compare the error reduction per single step for, say, additive Schwarz iterations with $|I_m| = n$ to multiplicative Schwarz iterations with $|I_m| = 1$. In the concrete examples above, it makes more sense to compare the relative error reduction $\|e^{(m+1)}\|/\|e^{(m)}\|$ for (4.2) with the relative error reduction $\|e^{((k+1)n)}\|/\|e^{(kn)}\|$ per loop for (4.6). More generally, we will compare the error reduction for different Schwarz iterative methods in terms of the total number of subproblem solves performed. Of course, concrete applications may require better models for estimating the work per iteration.

4.2 Deterministic orderings

Throughout this section, we consider the multiplicative Schwarz method (4.6) with *cyclic ordering*, where $I_m := \{i_m\} \subset \Omega$, $m = 0, 1, \ldots$ is a singleton, and the sequence

$$\{i_m\}_{m\in\mathbb{Z}_+} := \{1, 2, \ldots, n, 1, 2, \ldots, n, 1, 2, \ldots, n, 1, 2, \ldots, n, \ldots\}$$

is fixed. Furthermore, we restrict ourselves to constant step-sizes $\omega_k = \omega$, $k = 0, 1, \ldots$ in (4.6). For this case we give different representations and norm estimates for the error iteration operator associated with the subsequence $\{e^{(kn)}\}_{k\in\mathbb{Z}_+}$ generated by (4.6), i.e., we lump the n steps of the inner loop in (4.6) into one iteration step for the whole system of subproblems. This makes the iteration stationary and more directly comparable to the additive Schwarz iteration (4.2). Similar results hold for the symmetrized multiplicative Schwarz iteration with the inner loop

$$\begin{aligned} u^{(2kn+i)} &= u^{(2kn+i-1)} + \omega\rho_i P_i e^{(2kn+i-1)}, & i &= 1,\dots,n,\\ u^{(2kn+n+i)} &= u^{(2kn+n+i-1)} + \omega\rho_{n+1-i} P_{n+1-i} e^{(2kn+n+i-1)}, & i &= 1,\dots,n, \end{aligned} \tag{4.9}$$

$k = 0, 1, \dots$, and the error propagation operator associated with the subsequence $\{e^{(2kn)}\}_{m\in\mathbb{Z}_+}$. A generalization for arbitrary step-sizes is mentioned in Problem 4.2.

4.2.1 Representations of the error propagation operator

First we consider the plain iteration (4.6). Let $M_{\rho,\omega} : V \to V$ denote the error propagation operator

$$e^{((k+1)n)} = M_{\rho,\omega} e^{(kn)}, \qquad k = 0, 1, \dots,$$

associated with (4.6). Since

$$e^{(kn+i)} = u - \left(u^{(kn+i-1)} + \omega\rho_i P_i e^{(kn+i-1)}\right) = (I - \omega\rho_i P_i) e^{(kn+i-1)},$$

$i = 1, \dots, n$, we get the product representation

$$M_{\rho,\omega} = (I - \omega\rho_n P_n) \cdots (I - \omega\rho_2 P_2)(I - \omega\rho_1 P_1). \tag{4.10}$$

Similarly, the error propagation operator

$$e^{(2(k+1)n)} = \bar{M}_{\rho,\omega} e^{(2kn)}, \qquad k = 0, 1, \dots,$$

for the symmetrized multiplicative Schwarz iteration (4.9) takes the form

$$\bar{M}_{\rho,\omega} = (I - \omega\rho_1 P_1) \cdots (I - \omega\rho_n P_n)(I - \omega\rho_n P_n) \cdots (I - \omega\rho_1 P_1) = M^*_{\rho,\omega} M_{\rho,\omega}.$$

Note that, unlike $M_{\rho,\omega}$, the operator $\bar{M}_{\rho,\omega}$ is symmetric in $(V; a)$ and thus

$$r(M_{\rho,\omega})^2 \le \|M_{\rho,\omega}\|_V^2 = \|\bar{M}_{\rho,\omega}\|_V = r(\bar{M}_{\rho,\omega}). \tag{4.11}$$

Norm estimates for product operators such as (4.10), especially when the P_i are projections, have a long history in connection with the alternating directions method [37, 65, 77]. In the next subsection, we will prove a special identity for $\|M_{\rho,\omega}\|_V^2$ that goes back to [198, 199]. It is often used in the analysis of multiplicative Schwarz algorithms.

But first we will look at the error propagation operator $M_{\rho,\omega}$ in more detail. The key step in the analysis of (4.10) is to rewrite the above Schwarz iterations for solving the Schwarz equation $P_\rho u = g$ as iterations for solving the associated Gramian equation $\tilde{P}_\rho \tilde{u} = \tilde{g}$ in V_ρ. Interpreting the latter as a linear system with the operator matrix $\tilde{P}_\rho$ reveals their analogy to classical iterative methods in numerical linear algebra. To see this, consider the additive Schwarz iteration (4.2). Let $u^{(m)} = R_\rho \tilde{u}^{(m)}$, $u = R_\rho \tilde{u}$, where $\tilde{u}$ is an arbitrary solution of the Gramian equation, and set $\tilde{e}^{(m)} = \tilde{u} - \tilde{u}^{(m)}$. Then, by multiplying the iteration

$$\tilde{u}^{(m+1)} = \tilde{u}^{(m)} + \omega_m \tilde{P}_\rho \tilde{e}^{(m)}, \qquad m = 0, 1, \ldots, \tag{4.12}$$

by R_ρ, we get (4.2), because

$$R_\rho \tilde{P}_\rho \tilde{e}^{(m)} = P_\rho R_\rho \tilde{e}^{(m)} = P_\rho e^{(m)}.$$

Similar representations hold for multiplicative Schwarz iterations with cyclic ordering. To see this, decompose the Gramian as

$$\tilde{P}_\rho = \tilde{L}_\rho + \tilde{D}_\rho + \tilde{L}_\rho^*,$$

where $\tilde{D}_\rho$ is the diagonal part

$$(\tilde{D}_\rho \tilde{u})_i = \rho_i R_i^* R_i u_i, \qquad i = 1, \ldots, n,$$

and $\tilde{L}_\rho$ represents the strictly lower diagonal part of $\tilde{P}_\rho$ viewed as an operator matrix, i.e.,

$$(\tilde{L}_\rho \tilde{u})_i = \sum_{j=1}^{i-1} \rho_j R_i^* R_j u_j, \qquad i = 1, \ldots, n.$$

Finally, denote by $\tilde{I}$ the identity operator acting on $\tilde{V}_\rho$. To make the proofs more readable, we will keep the subscript ρ only in the formulation of the statements and the surrounding text. This should cause no confusion here, and we will follow this convention also in later sections.

Lemma 4.1 *For the error propagation operators associated with the multiplicative Schwarz iterations (4.6) and (4.9), we have the representations*

$$M_{\rho,\omega} = I - R_\rho (\omega^{-1} \tilde{I} + \tilde{L}_\rho)^{-1} R_\rho^* \tag{4.13}$$

and

$$\bar{M}_{\rho,\omega} = I - R_\rho (\omega^{-1} \tilde{I} + \tilde{L}_\rho^*)^{-1} (2\omega^{-1} \tilde{I} - \tilde{D}_\rho)(\omega^{-1} \tilde{I} + \tilde{L}_\rho)^{-1} R_\rho^*, \tag{4.14}$$

respectively.

Proof First, since $\tilde{L}$ is strictly lower triangular, the inverse of $\omega^{-1}\tilde{I} + \tilde{L}$ exists for any $\omega > 0$. Then we have

$$(\omega^{-1}\tilde{I}+\tilde{L})^{-1} = \omega(\tilde{I}+\omega\tilde{L})^{-1} = \omega\sum_{k=0}^{n-1}(-\omega)^k\tilde{L}^k \qquad (\tilde{L}^0 := \tilde{I}),$$

since $\tilde{L}^n = 0$.

By induction in $k = 0, 1, \ldots, n-1$ we show

$$(\tilde{L}^k R^*)_j = \begin{cases} 0, & j = 1, \ldots, k, \\ \sum_{0<j_1<\cdots<j_k<j} R_j^* \hat{P}_{j_k} \cdots \hat{P}_{j_1}, & j = k+1, \ldots, n, \end{cases} \tag{4.15}$$

where we temporarily set $\hat{P}_i = \rho_i P_i$, $i = 1, \ldots, n$. For $k = 0$ this holds by the definition of R^*. Now suppose that (4.15) has been established for some $0 \le k < n-1$. Since $\tilde{L}$ is strictly lower triangular, we have $(\tilde{L}^{k+1})_{l,j} = 0$ for all $l = 1, \ldots, k+1$ and $j = 1, \ldots, n$, and thus also $(\tilde{L}^{k+1}R^*)_l = 0$. By definition of $\tilde{L}$ and (4.15), for $l = k+2, \ldots, n$ we have

$$\begin{aligned}(\tilde{L}^{k+1}R^*)_l &= \sum_{j=k+1}^{l-1} \rho_j R_l^* R_j (\tilde{L}^k R^*)_j \\ &= \sum_{j=k+1}^{l-1} R_l^*(\rho_j R_j R_j^*) \sum_{0<j_1<\cdots<j_k<j} \hat{P}_{j_k}\cdots\hat{P}_{j_1} \\ &= \sum_{0<j_1<\cdots<j_k<j<l} R_l^* \hat{P}_j \hat{P}_{j_k}\cdots\hat{P}_{j_1}.\end{aligned}$$

Renaming j as j_{k+1} and then renaming l as j yields (4.15) with k replaced by $k+1$. This completes the induction step.

Now, the substitution of (4.15) into the formula for $(\tilde{I}+\omega\tilde{L})^{-1}$ results in

$$\begin{aligned}I - \omega R(\tilde{I}+\omega\tilde{L})^{-1}R^* &= I + \sum_{k=1}^{n}(-\omega)^k R_\rho \tilde{L}^{k-1}R^* \\ &= I + \sum_{k=1}^{n}(-\omega)^k \sum_{j=k}^{n} \sum_{0<j_1<\cdots<j_{k-1}<j} \rho_j R_j R_j^* \hat{P}_{j_{k-1}}\cdots\hat{P}_{j_1} \\ &= I + \sum_{k=1}^{n}(-\omega)^k \sum_{0<j_1<\cdots<j_k\le n} \hat{P}_{j_k}\cdots\hat{P}_{j_1}.\end{aligned}$$

Exactly the same expression is obtained if the product formula (4.10) for M_ω is multiplied out to a finite sum. This proves (4.13).

It remains to derive (4.14). Simple algebraic manipulations give

$$\begin{aligned}\bar{M}_\omega = M_\omega^* M_\omega &= (I - R(\omega^{-1}\tilde{I}+\tilde{L}^*)^{-1}R^*)(I - R(\omega^{-1}\tilde{I}+\tilde{L})^{-1}R^*) \\ &= I - R((\omega^{-1}\tilde{I}+\tilde{L}^*)^{-1} + (\omega^{-1}\tilde{I}+\tilde{L})^{-1})R^* \\ &\quad + R((\omega^{-1}\tilde{I}+\tilde{L}^*)^{-1}\tilde{P}(\omega^{-1}\tilde{I}+\tilde{L})^{-1})R^*.\end{aligned}$$

Substituting

$$\tilde{P} = (\omega^{-1}\tilde{I} + \tilde{L}) + (\omega^{-1}\tilde{I} + \tilde{L}^*) - (2\omega^{-1}\tilde{I} - \tilde{D})$$

into the last expression and cancelling equal terms then gives the desired result. □

The formula (4.13) corresponds to an iteration for solving $\tilde{P}_\rho \tilde{u} = \tilde{g}$ in $\tilde{V}_\rho$ with inner loop

$$\tilde{u}_j^{(kn+i)} = \tilde{u}_j^{(kn+i-1)} + \begin{cases} \omega R_i^* R_\rho \tilde{e}^{(kn+i-1)}, & j = i, \\ 0, & j \neq i, \end{cases} \qquad i = 1, \ldots, n, \tag{4.16}$$

where $k = 0, 1, \ldots$ is the index of the outer loop. In operator notation, (4.16) can also be written as

$$\tilde{u}^{((k+1)n)} = \tilde{u}^{(kn)} + \omega_k (\tilde{I} + \omega_k \tilde{L}_\rho)^{-1} \tilde{P}_\rho \tilde{e}^{(kn)}, \qquad k = 0, 1, \ldots. \tag{4.17}$$

In fact, using the same notational conventions as in the above discussion of (4.12), we see that applying R_ρ to (4.16) gives (4.6). Similarly, applying R_ρ to (4.17) gives (4.13). The formula shows the analogy of (4.6) to the classical Gauss-Seidel and SOR methods.

4.2.2 Convergence estimates

The following considerations mimic the convergence proofs for the Gauss-Seidel and SOR methods in numerical linear algebra. We again resort to (4.11). Together with (4.14) we have

$$\begin{aligned} \|M_{\rho,\omega}\|_V^2 = \|\bar{M}_{\rho,\omega}\|_V &= \sup_{u \in V:\, \|u\|=1} a(\bar{M}_{\rho,\omega} u, u) \\ &= 1 - \inf_{u \in V:\, \|u\|=1} a(R_\rho(\omega^{-1}\tilde{I} + \tilde{L}_\rho^*)^{-1}(2\omega^{-1}\tilde{I} - \tilde{D}_\rho)(\omega^{-1}\tilde{I} + \tilde{L}_\rho)^{-1} R_\rho^* u, u) \\ &= 1 - \inf_{u \in V:\, \|u\|=1} \tilde{a}_\rho((2\omega^{-1}\tilde{I} - \tilde{D}_\rho)(\omega^{-1}\tilde{I} + \tilde{L}_\rho)^{-1} R_\rho^* u, (\omega^{-1}\tilde{I} + \tilde{L}_\rho)^{-1} R_\rho^* u). \end{aligned}$$

So to prove $\|M_{\rho,\omega}\|_V^2 < 1$, we must require that $2\tilde{I} - \omega\tilde{D}_\rho$ is coercive on $\tilde{V}_\rho$, which is equivalent to the condition

$$0 < \omega < \omega_{\max} := \frac{2}{\|\tilde{D}_\rho\|_{\tilde{V}_\rho}}, \tag{4.18}$$

where

$$\|\tilde{D}_\rho\|_{\tilde{V}_\rho} = \max_{i=1,\ldots,n} \rho_i \|R_i\|_{V_i \to V}^2 \leq \lambda_{\max}. \tag{4.19}$$

In fact,

$$\begin{aligned}\tilde a_\rho(\tilde D_\rho \tilde u, \tilde u) &= \sum_{i=1}^n \rho_i^2 a_i(R_i^* R_i u_i, u_i) = \sum_{i=1}^n \rho_i^2 a(R_i u_i, R_i u_i)\\ &\le \sum_{i=1}^n \rho_i^2 \|R_i\|^2_{V_i\to V} a_i(u_i,u_i) \le (\max_{i=1,\dots,n} \rho_i \|R_i\|^2_{V_i\to V})\tilde a_\rho(\tilde u, \tilde u).\end{aligned}$$

Since both inequality signs cannot be improved on the set of all $\tilde u \in \tilde V_\rho$ with unit norm $\|\tilde u\|_{\tilde V_\rho} = 1$, one obtains the equality for the norm of $\tilde D_\rho$ stated in (4.19). Furthermore, since $\tilde D_\rho \in \mathcal{B}_{pos}(\tilde V_\rho)$ is the diagonal part of $\tilde P_\rho$, we have

$$\|\tilde D_\rho\|_{\tilde V_\rho} = \lambda_{\max}(\tilde D_\rho) \le \lambda_{\max}(\tilde P_\rho) = \lambda_{\max}.$$

After these preparations, we are now in a position to prove a formula for $\|M_{\rho,\omega}\|_V^2$, which is known as the XZ-identity and was first established in [198], see also [38].

Theorem 4.2 *Let* $(\mathbf{V};\mathbf{a};\mathbf{R};\rho)$ *be a finite stable splitting for* $(V;a)$ *with* $\Omega = \{1,\dots,n\}$ *and assume that* ω *satisfies (4.18). Then we have the identity*

$$\|M_{\rho,\omega}\|_V^2 = 1 - \frac{1}{c_1(\omega)}, \tag{4.20}$$

where

$$\begin{aligned} c_1(\omega) &= \sup_{u\in V:\,\|u\|=1}\ \inf_{\tilde v\in \tilde V_\rho: u=R_\rho \tilde v} \tilde a_\rho((2\omega^{-1}\tilde I - \tilde D_\rho)^{-1}(\omega^{-1}\tilde I + \tilde L_\rho^*)\tilde v, (\omega^{-1}\tilde I + \tilde L_\rho^*)\tilde v)\\ &= \sup_{u\in V:\,\|u\|=1}\ \inf_{\tilde v\in \tilde V_\rho: u=R_\rho \tilde v} \sum_{i=1}^n \rho_i a_i((2\omega^{-1} I - \rho_i R_i^* R_i)^{-1} w_i, w_i),\end{aligned}$$

with $w_i = v_i + R_i^*(\sum_{j=i+1}^n \rho_j R_j v_j) \in V_i$, $i = 1,\dots,n$.

Proof Throughout the proof we will again drop the subscript ρ. Starting with the above expression for $\|M_\omega\|_V^2$, we get

$$\begin{aligned} c_1(\omega) &= \left(\inf_{u\in V:\,\|u\|=1} \tilde a((2\omega^{-1}\tilde I - \tilde D)(\omega^{-1}\tilde I + \tilde L)^{-1}R^*u, (\omega^{-1}\tilde I + \tilde L)^{-1}R^*u)\right)^{-1}\\ &= \sup_{0\ne u\in V} \frac{a(u,u)}{\tilde a((\omega^{-1}\tilde I + \tilde L^*)^{-1}(2\omega^{-1}\tilde I - \tilde D)(\omega^{-1}\tilde I + \tilde L)^{-1}R^*u, R^*u)}\\ &= \sup_{\tilde v\in \tilde V: 0\ne R\tilde v} \frac{a(R\tilde v, R\tilde v)}{\tilde a((\omega^{-1}\tilde I + \tilde L^*)^{-1}(2\omega^{-1}\tilde I - \tilde D)(\omega^{-1}\tilde I + \tilde L)^{-1}\tilde P\tilde v, \tilde P\tilde v)}\\ &= \sup_{\tilde v\in \tilde V: 0\ne R\tilde v} \frac{\tilde a(\tilde P\tilde v, \tilde v)}{\tilde a(\tilde B_\omega \tilde P\tilde v, \tilde P\tilde v)},\end{aligned}$$

where $\tilde B_\omega$ stands for

$$\tilde B_\omega := (\omega^{-1}\tilde I + \tilde L^*)^{-1}(2\omega^{-1}\tilde I - \tilde D)(\omega^{-1}\tilde I + \tilde L)^{-1}.$$

By the assumption (4.18), we know that the operators $2\omega^{-1}\tilde{I} - \tilde{D}$ and $\tilde{B}_\omega$ are invertible and positive definite on $(\tilde{V}; \tilde{a})$.

It remains to show that the above expression for $c_1(\omega)$ can be rewritten as

$$c_1(\omega) = \sup_{0 \neq \tilde{u} \in \mathrm{ran}(R^*)} \frac{\inf_{\tilde{w} \in \ker(R)} \tilde{a}(\tilde{B}_\omega^{-1}(\tilde{u} + \tilde{w}), \tilde{u} + \tilde{w})}{\tilde{a}(\tilde{P}\tilde{u}, \tilde{u})}. \tag{4.21}$$

Since for each $u \in V$ there exists a unique $\tilde{u} = R^* P^{-1} u \in \mathrm{ran}(R^*)$ with $u = R\tilde{u}$, we have for $\tilde{v} = \tilde{u} + \tilde{w}$ with $\tilde{w} \in \ker(R)$ on the one hand

$$a(u, u) = a(R\tilde{u}, R\tilde{u}) = a(R\tilde{v}, R\tilde{v}) = \tilde{a}(\tilde{P}\tilde{v}, \tilde{v}) = \tilde{a}(\tilde{P}\tilde{u}, \tilde{u}), \quad u = R\tilde{v}.$$

On the other hand, we obtain

$$\begin{aligned} \tilde{a}(\tilde{B}_\omega^{-1}\tilde{v}, \tilde{v}) &= \tilde{a}((\omega^{-1}\tilde{I} + \tilde{L})(2\omega^{-1}\tilde{I} - \tilde{D})^{-1}(\omega^{-1}\tilde{I} + \tilde{L}^*)\tilde{v}, \tilde{v}) \\ &= \tilde{a}((2\omega^{-1}\tilde{I} - \tilde{D})^{-1}(\omega^{-1}\tilde{I} + \tilde{L}^*)\tilde{v}, (\omega^{-1}\tilde{I} + \tilde{L}^*)\tilde{v}) \end{aligned}$$

by the definition of $\tilde{B}_\omega$. Substituting these formulas into the right-hand side of (4.21) immediately implies the expression for $c_1(\omega)$ in Theorem 4.2.

The identity (4.21) is straightforward if $\ker(R) = \{0\}$, since $R\tilde{v} \neq 0$ is then equivalent to $\tilde{v} \neq 0$, and $\tilde{P}$ is invertible and positive definite on $\tilde{V}$. In fact, in this case, the square root $\tilde{P}^{1/2}$ and its inverse are well-defined positive definite operators on $\tilde{V}$, and we have

$$\begin{aligned} c_1(\omega) &= \sup_{\tilde{v} \in \tilde{V}: 0 \neq R\tilde{v}} \frac{\tilde{a}(\tilde{P}\tilde{v}, \tilde{v})}{\tilde{a}(\tilde{B}_\omega \tilde{P}\tilde{v}, \tilde{P}\tilde{v})} = \sup_{0 \neq \tilde{v} \in \tilde{V}} \frac{\tilde{a}(\tilde{v}, \tilde{v})}{\tilde{a}(\tilde{P}^{1/2}\tilde{B}_\omega \tilde{P}^{1/2}\tilde{v}, \tilde{v})} \\ &= \frac{1}{\lambda_{\min}(\tilde{P}^{1/2}\tilde{B}_\omega \tilde{P}^{1/2})} = \lambda_{\max}(\tilde{P}^{-1/2}\tilde{B}_\omega^{-1}\tilde{P}^{-1/2}) \\ &= \sup_{0 \neq \tilde{v} \in \tilde{V}} \frac{\tilde{a}(\tilde{P}^{-1/2}\tilde{B}_\omega^{-1}\tilde{P}^{-1/2}\tilde{v}, \tilde{v})}{\tilde{a}(\tilde{v}, \tilde{v})} = \sup_{0 \neq \tilde{v} \in \tilde{V}} \frac{\tilde{a}(\tilde{B}_\omega^{-1}\tilde{v}, \tilde{v})}{\tilde{a}(\tilde{P}\tilde{v}, \tilde{v})}. \end{aligned}$$

The last supremum coincides with the right-hand side in (4.21), since $\ker(R) = \{0\}$ and $\mathrm{ran}(R^*) = \tilde{V}$.

In the general case, using again the notation $\tilde{X} = \mathrm{ran}(R^*) = \ker(R)^\perp$, the same consideration yields the equality

$$\begin{aligned} \sup_{\tilde{v} \in \tilde{V}: 0 \neq R\tilde{v}} \frac{\tilde{a}(\tilde{P}\tilde{v}, \tilde{v})}{\tilde{a}(\tilde{B}_\omega \tilde{P}\tilde{v}, \tilde{P}\tilde{v})} &= \sup_{0 \neq \tilde{v} \in \tilde{X}} \frac{\tilde{a}(\tilde{P}\tilde{v}, \tilde{v})}{\tilde{a}(Q_{\tilde{X}}\tilde{B}_\omega \tilde{P}\tilde{v}, \tilde{P}\tilde{v})} \\ &= \sup_{0 \neq \tilde{v} \in \tilde{X}} \tilde{a}(\frac{(\tilde{P}|_{\tilde{X}})^{-1/2}(Q_{\tilde{X}}\tilde{B}_\omega|_{\tilde{X}})^{-1}(\tilde{P}|_{\tilde{X}})^{-1/2}\tilde{v}, \tilde{v})}{\tilde{a}(\tilde{v}, \tilde{v})} \\ &= \sup_{0 \neq \tilde{v} \in \tilde{X}} \frac{\tilde{a}((Q_{\tilde{X}}\tilde{B}_\omega|_{\tilde{X}})^{-1}\tilde{v}, \tilde{v})}{\tilde{a}(\tilde{P}\tilde{v}, \tilde{v})}. \end{aligned}$$

Here, $Q_{\tilde{X}} : \tilde{V} \to \tilde{X}$ denotes the ortho-projection operator in $(\tilde{V}; \tilde{a})$. Indeed, both $Q_{\tilde{X}} \tilde{B}_\omega|_{\tilde{X}}$ and $\tilde{P}|_{\tilde{X}}$ are invertible and positive definite operators on $(\tilde{X}; \tilde{a})$. To get from here to (4.21), it is sufficient to check that

$$\tilde{a}((Q_{\tilde{X}} \tilde{B}_\omega|_{\tilde{X}})^{-1}\tilde{v}, \tilde{v}) = \min_{\tilde{w} \in \ker(R)} \tilde{a}(\tilde{B}_\omega^{-1}(\tilde{v}+\tilde{w}), (\tilde{v}+\tilde{w})), \qquad \tilde{v} \in \tilde{X}. \tag{4.22}$$

Let $\tilde{v}' = (Q_{\tilde{X}} \tilde{B}_\omega|_{\tilde{X}})^{-1}\tilde{v} \in \tilde{X}$. Then $\tilde{B}_\omega \tilde{v}' = \tilde{v} - \tilde{w}'$ and $\tilde{v}' = \tilde{B}_\omega^{-1}\tilde{v} - \tilde{B}_\omega^{-1}\tilde{w}'$ for some unique $\tilde{w}' \in \ker(R)$. Note that

$$\tilde{a}(\tilde{B}_\omega^{-1}\tilde{v}, \tilde{w}) = \tilde{a}(\tilde{B}_\omega^{-1}\tilde{w}', \tilde{w}) + \tilde{a}(\tilde{v}', \tilde{w}) = \tilde{a}(\tilde{B}_\omega^{-1}\tilde{w}', \tilde{w}) \quad \forall\, \tilde{w} \in \ker(R), \tag{4.23}$$

since $\tilde{v}' \in \tilde{X}$ and $\ker(R) = \tilde{X}^\perp$. So we have

$$\begin{aligned}\tilde{a}((Q_{\tilde{X}} \tilde{B}_\omega|_{\tilde{X}})^{-1}\tilde{v}, \tilde{v}) &= \tilde{a}(\tilde{v}', \tilde{v}) = \tilde{a}(\tilde{B}_\omega^{-1}\tilde{v}, \tilde{v}) - \tilde{a}(\tilde{B}_\omega^{-1}\tilde{w}', \tilde{v}) \\ &= \tilde{a}(\tilde{B}_\omega^{-1}\tilde{v}, \tilde{v}) - \tilde{a}(\tilde{B}_\omega^{-1}\tilde{w}', \tilde{w}').\end{aligned}$$

A straightforward calculation shows that we have

$$\min_{\tilde{w} \in \ker(R)} \tilde{a}(\tilde{B}_\omega^{-1}(\tilde{v}+\tilde{w}), (\tilde{v}+\tilde{w})) = \tilde{a}(\tilde{B}_\omega^{-1}\tilde{v}, \tilde{v}) - \tilde{a}(\tilde{B}_\omega^{-1}\tilde{w}', \tilde{w}'),$$

since the unique minimizer of this minimization problem is characterized by (4.23). This shows (4.22) and completes the proof of Theorem 4.2. □

The formula for $c_1(\omega)$ in Theorem 4.2 is slightly simplified if the auxiliary forms $a_i(\cdot,\cdot)$ on V_i are generated by Galerkin projection from $(V; a)$, i.e., if

$$a_i(u_i, v_i) = a(R_i v_i, R_i u_i), \qquad u_i, v_i \in V_i, \quad i = 1, \ldots, n. \tag{4.24}$$

Since in this case $R_i^* R_i = I_i$ are the identity operators on V_i, the operator $\tilde{D}_\rho$ has diagonal entries $\rho_i I_i$, $i = 1, \ldots$. So $\|\tilde{D}_\rho\|_{\tilde{V}_\rho} = \max(\rho_i)$, and the formula for $c_1(\omega)$ in Theorem 4.2 is reduced to

$$\|M_{\rho,\omega}\|_V^2 = 1 - \frac{1}{\sup_{u \in V:\, \|u\|=1} \inf_{\tilde{v} \in \tilde{V}_\rho : u = R_\rho \tilde{v}} \sum_{i=1}^n \frac{\rho_i}{2/\omega - \rho_i} a_i(w_i, w_i)} < 1,$$

if $0 < \omega < 2/\max(\rho_i)$. Here the $w_i \in V_i$ have the same meaning as before. For easier comparison with the statements in [198, 199], set $\omega = \rho_i = 1$. Note that these papers also consider product-type operators (4.10) with non-symmetric P_i, which result, for example, from replacing A_i^{-1} by approximate inverses of A_i.

Although identities such as (4.20) have merit on their own and can be used as a guide for optimizing Schwarz iterative methods, in most cases they are not effectively applicable due to their complicated structure. An exception is the case $n = 2$, see Problem 4.1. For $n \geq 3$, more explicit estimates of the convergence speed are desirable, which are easier to verify. For (4.6), this can be achieved by first deriving bounds in terms of $\|\tilde{D}_\rho\|_{\tilde{V}_\rho}$ and $\|\tilde{L}_\rho\|_{\tilde{V}_\rho}$, and then replacing these operator norms by upper bounds in terms of properties of the underlying space splitting. Continuing with the formula for $c_1(\omega)$ given in Theorem 4.2, we then have

$$\begin{aligned} c_1(\omega) &= \sup_{\tilde v\in\tilde V_\rho:\, \|R_\rho \tilde v\|=1} \tilde a_\rho((2\omega^{-1}\tilde I - \tilde D_\rho)^{-1}(\omega^{-1}\tilde I + \tilde L_\rho^*)\tilde v, (\omega^{-1}\tilde I + \tilde L_\rho^*)\tilde v) \\ &\le (2/\omega - \|\tilde D_\rho\|_{\tilde V_\rho})^{-1} \sup_{0\neq \tilde v\in \operatorname{ran}(R_\rho^*)} \frac{\tilde a_\rho((\omega^{-1}\tilde I + \tilde L_\rho^*)\tilde v, (\omega^{-1}\tilde I + \tilde L_\rho^*)\tilde v)}{\tilde a(\tilde P_\rho \tilde v, \tilde v)} \\ &\le \omega^{-1}(2-\omega\|\tilde D_\rho\|_{\tilde V_\rho})^{-1} \frac{\|\tilde I + \omega \tilde L_\rho\|^2_{\tilde V_\rho}}{\lambda_{\min}} \le \frac{(1+\omega\|\tilde L_\rho\|_{\tilde V_\rho})^2}{\omega(2-\omega\|\tilde D_\rho\|_{\tilde V_\rho})\lambda_{\min}}. \end{aligned}$$

This gives the following convergence result for the multiplicative Schwarz method (4.6) with a constant step-size parameter.

Theorem 4.3 *Let* $(\mathbf{V};\mathbf{a};\mathbf{R};\rho)$ *be a finite stable splitting for* $(V;a)$ *with* $\Omega = \{1,\dots,n\}$. *Then the multiplicative Schwarz iteration (4.6) with constant step-size* $\omega_k = \omega$ *converges if (4.18) is satisfied. The iteration error can be bounded as*

$$\|e^{(k+1)n}\|^2 \le \left(1 - \frac{\omega(2-\omega\|\tilde D_\rho\|_{\tilde V_\rho})\lambda_{\min}}{(1+\omega\|\tilde L_\rho\|_{\tilde V_\rho})^2}\right)\|e^{(kn)}\|^2, \quad k=0,1,\dots. \tag{4.25}$$

The upper bound (4.25) is optimized when we set $\omega = \omega_{opt} := (\|\tilde L_\rho\|_{\tilde V_\rho} + \|\tilde D_\rho\|_{\tilde V_\rho})^{-1}$. *It then takes the form*

$$\|e^{(k+1)n}\|^2 \le \left(1 - \frac{\lambda_{\min}}{2\|\tilde L_\rho\|_{\tilde V_\rho} + \|\tilde D_\rho\|_{\tilde V_\rho}}\right)\|e^{(kn)}\|^2, \quad k=0,1,\dots. \tag{4.26}$$

The bounds (4.25) and (4.26) are a direct consequence of the above norm estimates for the error propagation operator $M_{\rho,\omega}$. Analogous results hold for (4.9), we leave their formulation to the reader. For the case of iterations with cyclic orderings with arbitrary step-size sequences, we refer to the Problem 4.2.

To get more explicit bounds, we need information about the norm $\|\tilde L_\rho\|_{\tilde V_\rho}$ of the lower triangle operator $\tilde L_\rho$, while we can rely on (4.19) for $\|\tilde D_\rho\|_{\tilde V_\rho}$. We present two estimates for $\|\tilde L_\rho\|_{\tilde V_\rho}$. The first uses strengthened Cauchy-Schwarz inequalities (2.14), as discussed in Subsection 2.1.3. The second uses properties of the lower triangular projection operator for operator matrices.

Theorem 4.4 *a) We have*

$$\|\tilde L_\rho\|_{\tilde V_\rho} \le \frac{1}{2}\|\Gamma_\rho\|_{\ell^2}, \tag{4.27}$$

where Γ_ρ *is the operator on* $\ell^2(\Omega)$ *defined in (2.15).*
b) We have

$$\|\tilde L_\rho\|_{\tilde V_\rho} \le \frac{1}{2}\lceil \log_2(n)\rceil \|\tilde P_\rho\|_{\tilde V_\rho} = \frac{1}{2}\lceil \log_2(n)\rceil \lambda_{\max}. \tag{4.28}$$

Proof For Part a), consider any $\tilde{u}, \tilde{v} \in \tilde{V}$ and use the definition of $\tilde{L}$ and (2.14). This gives

$$
\begin{aligned}
\tilde{a}(\tilde{L}\tilde{u}, \tilde{v}) &= \sum_{i=2}^{n} \rho_i a_i(\sum_{j=1}^{i-1} \rho_j R_i^* R_j u_j, v_i) \\
&= \sum_{i=2}^{n} \sum_{j=1}^{i-1} \rho_i \rho_j a(R_j u_j, R_i v_i) \le \sum_{i=2}^{n} \sum_{j=1}^{i-1} \rho_i \rho_j \gamma_{ij} \|u_j\|_j \|v_i\|_i \\
&\le \frac{1}{2} \sum_{i=1}^{n} \sum_{j=1}^{n} \sqrt{\rho_i}\sqrt{\rho_j}\gamma_{ij}(\sqrt{\rho_j}\|u_j\|_j)(\sqrt{\rho_i}\|v_i\|_i) \le \frac{1}{2}\|\Gamma_\rho\|_{\ell^2} \|\tilde{u}\|_{\tilde{V}_\rho} \|\tilde{v}\|_{\tilde{V}_\rho}.
\end{aligned}
$$

Now set $\tilde{v} = \tilde{L}\tilde{u}$ and use the definition of the operator norm $\|\tilde{L}\|_{\tilde{V}}$ to get the estimate (4.27).

The inequality in (4.28) holds for any operator

$$
\tilde{P} := (P_{ij})_{i,j=1,\dots,n} \in \mathcal{B}_{pos}(\tilde{V})
$$

with the entries $P_{ij} : V_j \longrightarrow V_i$ and its strictly lower triangular part $\tilde{L}$. To prove this by induction, we need the auxiliary estimate

$$
\|\tilde{P}_{I,J}\|_{\tilde{V}} \le \frac{1}{2}\|\tilde{P}\|_{\tilde{V}}, \qquad I \cap J = \emptyset, \quad I, J \subset \{1, \dots, n\}, \tag{4.29}
$$

where $\tilde{P}_{I,J} : \tilde{V} \longrightarrow \tilde{V}$ denotes the restriction of the operator matrix representation of $\tilde{P}$ to the index set $I \times J$. To see this, we use $\tilde{u}_I \in \tilde{V}$ to denote the restriction of an element $\tilde{u} \in \tilde{V}$ to the index set $I \subset \{1, \dots, n\}$. Then we have

$$
\begin{aligned}
\tilde{a}(\tilde{P}_{I,J}\tilde{u}, \tilde{v}) &= \tilde{a}(\tilde{P}\tilde{u}_J, \tilde{v}_I) = \frac{1}{4}(\tilde{a}(\tilde{P}(\tilde{u}_J + \tilde{v}_I), \tilde{u}_J + \tilde{v}_I) - \tilde{a}(\tilde{P}(\tilde{u}_J - \tilde{v}_I), \tilde{u}_J - \tilde{v}_I)) \\
&\le \frac{1}{4}\tilde{a}(\tilde{P}(\tilde{u}_J + \tilde{v}_I), \tilde{u}_J + \tilde{v}_I) \le \frac{1}{4}\|\tilde{P}\|_{\tilde{V}} \|\tilde{u}_J + \tilde{v}_I\|_{\tilde{V}}^2 \\
&= \frac{1}{4}\|\tilde{P}\|_{\tilde{V}}(\|\tilde{u}_J\|_{\tilde{V}}^2 + \|\tilde{v}_I\|_{\tilde{V}}^2) \le \frac{1}{4}\|\tilde{P}\|_{\tilde{V}}(\|\tilde{u}\|_{\tilde{V}}^2 + \|\tilde{v}\|_{\tilde{V}}^2).
\end{aligned}
$$

The last equality holds because the index sets I and J are disjoint. Since

$$
\|\tilde{P}_{I,J}\|_{\tilde{V}} = \sup_{\|\tilde{u}\|_{\tilde{V}} = \|\tilde{v}\|_{\tilde{V}} = 1} \tilde{a}(\tilde{P}_{I,J}\tilde{u}, \tilde{v}),
$$

we get (4.29) by taking the supremum of all $\tilde{u}, \tilde{v} \in \tilde{V}$ with unit norm in the above estimate.

With (4.29) at hand, we use induction with respect to the size $n = 2^k$, $k = 1, 2, \dots,$ of the operator matrix $\tilde{P} \in \mathcal{B}_{pos}(\tilde{V})$ to establish

$$
\|\tilde{L}\|_{\tilde{V}} \le \frac{k}{2}\|\tilde{P}\|_{\tilde{V}}, \qquad \tilde{P} \in \mathcal{B}_{pos}(\tilde{V}). \tag{4.30}
$$

In fact, for $k = 1$, the inequality (4.30) follows from (4.29), since $\tilde{L} = \tilde{P}_{\{2\},\{1\}}$. Now suppose that (4.30) holds for $n = 2^k$. Take any $\tilde{P} \in \mathcal{B}_{pos}(\tilde{V})$ with $n = 2^{k+1}$ and consider the disjoint index sets $I = \{1, \ldots, 2^k\}$, $J = \{2^k + 1, \ldots, 2^{k+1}\}$. Then $\tilde{L}$ is decomposed as

$$\tilde{L} = \tilde{L}_{I,I} + \tilde{P}_{J,I} + \tilde{L}_{J,J}.$$

By (4.29), the second term satisfies $\|\tilde{P}_{J,I}\|_{\tilde{V}} \le \frac{1}{2}\|\tilde{P}\|_{\tilde{V}}$. For the remaining terms we have

$$\|\tilde{L}_{I,I} + \tilde{L}_{J,J}\|_{\tilde{V}} = \max(\|\tilde{L}_{I,I}\|_{\tilde{V}}, \|\tilde{L}_{J,J}\|_{\tilde{V}}),$$

since I and J are disjoint. But $\tilde{L}_{I,I}$ is the strictly lower triangular part of $\tilde{P}_{I,I}$, which is a positive symmetric operator of size $n = 2^k$ when considered on the corresponding $\tilde{V}$. Thus, by the induction hypothesis, we obtain

$$\|\tilde{L}_{I,I}\|_{\tilde{V}} \le \frac{k}{2}\|\tilde{P}_{I,I}\|_{\tilde{V}} \le \frac{k}{2}\|\tilde{P}\|_{\tilde{V}}.$$

The same estimate holds for $\tilde{L}_{J,J}$. Altogether this gives

$$\|\tilde{L}\|_{\tilde{V}} \le \|\tilde{P}_{J,I}\|_{\tilde{V}} + \max(\|\tilde{L}_{I,I}\|_{\tilde{V}}, \|\tilde{L}_{J,J}\|_{\tilde{V}}) \le (\frac{1}{2} + \frac{k}{2})\|\tilde{P}\|_{\tilde{V}},$$

which completes the induction step.

Since for $n = 2^k + 1, \ldots, 2^{k+1} - 1$ the estimate for $n = 2^{k+1}$ still holds (just pad $\tilde{P}$ with zero operators to this larger size), (4.28) follows from (4.30). This completes the proof of Part b). □

Substituting (4.19) and the results from Theorem 4.4 into (4.26), we obtain the following convergence estimates for the multiplicative Schwarz method (4.6) with fixed step-sizes $\omega_k = \omega$. They depend on $\|\Gamma_\rho\|_{\ell^2}$ and on the spectral properties of the underlying space splitting, respectively.

Corollary 4.1 *Let the conditions of Theorem 4.2 hold.*
a) If $\omega_k = (\lambda_{\max} + \|\Gamma_\rho\|_{\ell^2}/2)^{-1}$, *then (4.6) satisfies*

$$\|e^{(k+1)n}\|^2 \le \left(1 - \frac{\lambda_{\min}}{\|\Gamma_\rho\|_{\ell^2} + \lambda_{\max}}\right)\|e^{kn}\|^2, \quad k = 0, 1, \ldots. \tag{4.31}$$

b) If $\omega_k = ((1 + \lceil\log_2(n)\rceil/2)\lambda_{\max})^{-1}$, *then (4.6) satisfies*

$$\|e^{(k+1)n}\|^2 \le \left(1 - \frac{1}{(1 + \lceil\log_2(n)\rceil)\kappa}\right)\|e^{kn}\|^2, \quad k = 0, 1, \ldots. \tag{4.32}$$

Proof Part a) follows if we substitute the bound

$$\|\tilde{D}_\rho\|_{\tilde{V}_\rho} \le \lambda_{\max}$$

from (4.19) and (4.27) into (4.25), and optimize the resulting error reduction factor with respect to ω. Similarly, Part b) follows if (4.28) is used instead of (4.27). □

Part a) of Theorem 4.4 and of Corollary 4.1 goes back to [20, 194], see also [72, 202] for the role of strengthened Cauchy-Schwarz inequalities in this context. The proof for Part b) is taken from [86].

A comparison between the optimized upper bounds for the error reduction factor for additive and multiplicative Schwarz iterative methods suggests that (4.5) in Theorem 4.1 is better than the bounds from Corollary 4.1. In fact, according to Theorem 2.3, we have $\|\Gamma_\rho\|_{\ell^2} \geq \lambda_{\max}$, while $\log_2(n) \geq 1$ for $n \geq 2$. Therefore

$$\begin{aligned}\left(1 - \frac{2}{1+\kappa}\right)^2 &= 1 - \frac{4\kappa}{(1+\kappa)^2} \leq 1 - \frac{1}{\kappa} < 1 - \frac{1}{2\kappa} \\ &\leq 1 - \max\left(\frac{\lambda_{\min}}{\|\Gamma_\rho\|_{\ell^2} + \lambda_{\max}}, 1 - \frac{1}{(1 + \lceil \log_2(n) \rceil)\kappa}\right).\end{aligned}$$

As a matter of fact, the examples from [136, 141] reproduced in Subsection 5.4.2 below demonstrate the necessity of the logarithmic dependence on n in the bound (4.32). In many numerical applications, however, multiplicative Schwarz iterations slightly outperform their additive counterpart.

The actual convergence speed depends critically on the choice of the step-size parameters. The optimized bounds in Theorem 4.3 and Corollary 4.1 assume a fixed $\omega_k = \omega$ with a value ω that depends on the knowledge of the spectral properties of the underlying space splitting, in particular $\lambda_{\max}$. This information is not always available. For the additive Schwarz iteration, however, there is the alternative of using the steepest descent rule (4.3), which, at some additional computational cost, does not require a priori knowledge. For multiplicative Schwarz iterations with cyclic orderings, this option is not available. Furthermore, the best performance is not necessarily obtained with values of ω that result from optimizing the worst-case upper bound for the error reduction per iteration step. Note also that the bounds in Corollary 4.1 do not depend on the order in which the subproblem solvers are applied within the inner loop in (4.6), while changing the enumeration of the spaces V_i may indeed affect the convergence behavior in applications.

A final remark, which also applies to the content of the next sections, concerns quantities such as $\|R_i\|^2_{V_i \to V}$, $\lambda_{\max}$, $\lambda_{\min}$, and κ that appear in the convergence results: Often the exact values of these constants are not available, and especially $\lambda_{\min}$ is difficult to access. However, all statements remain valid if we substitute upper bounds for $\|R_i\|^2_{V_i \to V}$, $\lambda_{\max}$, κ, and lower bounds for $\lambda_{\min}$. This also applies to the formulas for step-size parameters, which lead to optimized estimates.

4.3 Randomized orderings

An alternative to deterministic orderings is to choose I_m randomly, according to some probability distribution defined on a suitably fixed family $\mathcal{I}_m$ of admissible index sets $I_m \subset \Omega$. We always assume that the choice of I_m is independent of I_k, $k < m$. The cost of sampling I_m is often negligible compared to the cost of solving the subproblems. Compared to deterministic orderings, where there is no cost associated with the choice of I_m, randomized orderings may have advantages since, as shown below, the randomized Schwarz iteration may have better convergence guarantees in expectation.

Furthermore, the consideration of randomized Schwarz iterations has another, more direct justification than just as an option to save costs or improve the convergence speed of Schwarz iterations. There are situations in which the choice of I_m during the iteration process (4.1) is not in our hands, but is dictated by an independent random process. Examples include machine learning, uncertainty quantification, and hardware and software fault mitigation on distributed and parallel platforms. For proximal coordinate and stochastic gradient descent methods, see for example [75] and [79].

Randomized Schwarz iterations do not have deterministic convergence guarantees, and their analysis must be based on probabilistic error measures. Note that the error $e^{(m)}$ is now a *V-valued random variable* and the error sequence $\{e^{(m)}\}_{m\in\mathbb{Z}_+}$ represents a *random process in V*. In this chapter, we focus on estimates for the expectation of the squared error norm of the Schwarz iteration (4.1) in $(V;a)$. Because

$$\mathbb{E}(\|e^{(m)}\|^2) = \|\mathbb{E}(e^{(m)})\|^2 + \mathbb{E}(\|e^{(m)} - \mathbb{E}(e^{(m)})\|^2),$$

monitoring the decay of the expectation of the squared error $\mathbb{E}(\|e^{(m)}\|^2)$ guarantees control of both the expectation and the variance of the error $e^{(m)}$. Other measures of probabilistic convergence and a link to the literature on stochastic approximation are briefly discussed in Section 5.6 of the next chapter.

The following subsections deal with different types of randomization. We start in Subsection 4.3.1 with the singleton case $I_m = \{i_m\}$ in (4.1), and derive convergence results for uniform and non-uniform probability distributions on $\Omega = \{1, \ldots, n\}$ governing the random choice of i_m. Subsection 4.3.2 discusses the case of index subsets $I_m \subset \Omega_m$ of size $1 \leq k_m \leq n$ chosen uniformly at random. The influence of random permutations of Ω within the inner loops of a multiplicative Schwarz method with cyclic ordering, as considered in the previous section, is studied in Subsection 4.3.3. Then, various greedy strategies for the index selection are discussed in Section 4.4. Finally, more sophisticated randomized Schwarz iterations with improved convergence are analyzed in the later Section 4.5 on acceleration techniques.

Throughout this section, if $X^{(m)}$ is a random variable depending on the result of the first m iteration steps of a randomized Schwarz iteration, then $\mathbb{E}(X^{(m)})$ stands for its expectation with respect to the joint distribution of the random variables I_k, $k = 0, 1, \ldots, m-1$. Since we assume independence of I_k and $I_{k'}$, $k \neq k'$, this joint distribution is a product measure and estimating $\mathbb{E}(X^{(m+1)})$ can be reduced to estimating the conditional expectation $\mathbb{E}(X^{(m+1)}|I_{m-1}, \ldots, I_0)$. The latter is often a function of the state $X^{(m)}$ itself, which leads to recursive estimates for the sequence $\{\mathbb{E}(X^{(m)})\}$. This strategy is the basis of the proofs in this section. To reduce the notational overhead, we will denote such conditional expectations by

$$\mathbb{E}'(X^{(m+1)}) := \mathbb{E}(X^{(m+1)}|I_{m-1}, \ldots, I_0).$$

4.3.1 Singleton I_m

We first consider the randomized version of (4.7) with singleton $I_m = \{i_m\}$, where the i_m are independent and identically distributed (i.i.d.) random samples with a fixed underlying probability distribution on $\Omega = \{1, \dots, n\}$, i.e.,

$$\mathbb{P}(i_m = i) = p_i, \qquad i = 1, \dots, n, \qquad \sum_{i=1}^{n} p_i = 1. \tag{4.33}$$

Thus, the choice of i_m is independent of the current error $e^{(m)}$ and of the indices i_k with $k < m$. The first bound for $\mathbb{E}(\|e^{(m)}\|^2)$ is obtained in terms of geometric quantities and does not depend on ρ. It will allow us later to derive more familiar bounds for certain choices in (4.33) that are related to spectral properties of a given P_ρ. To formulate this result, we introduce the cosine of the angle $\angle(v, w)$ between two non-zero elements $v, w \in V$ by the formula

$$\cos(\angle(v, w)) = \frac{a(v, w)}{\|v\|\|w\|}.$$

If at least one of the two elements vanishes, we set $\cos(\angle(v, w)) = 0$, i.e., $\angle(v, w) = \pi/2$, which corresponds to the definition of orthogonality in V. Note that $\cos(\angle(v, w))$ refers to the distance between the two unit elements $v' = v/\|v\|$ and $w' = w/\|w\|$ by the formula

$$1 - \cos(\angle(v, w)) = \frac{1}{2}\|v' - w'\|^2.$$

Similarly, the cosine of the angle between a linear subset $W \subset V$ and a non-zero element v is defined by

$$\cos(\angle(v, W)) = \sup_{w \in W:\, \|w\|=1} \cos(\angle(w, v)) = 1 - \frac{1}{2} \inf_{w \in W:\, \|w\|=1} \|w - v'\|^2.$$

The reader may find more information about angles and their role in the theory of alternating projection methods in Hilbert spaces in [37, 65, 77].

We are now in a position to formulate the following result:

Theorem 4.5 *Let* $(\mathbf{V}; \mathbf{a}; \mathbf{R}; \rho)$ *be a finite stable splitting for* $(V; a)$ *with* $\Omega = \{1, \dots, n\}$. *Then the steepest descent Schwarz iteration (4.7) with i.i.d.* $i_m \in \Omega$ *and underlying probability distribution (4.33) possesses the error estimate*

$$\mathbb{E}(\|e^{(m)}\|^2) \le (1 - \mu_R)^m \|e^{(0)}\|^2, \quad m = 1, 2, \dots, \tag{4.34}$$

where

$$\mu_R := \inf_{v \in V:\|v\|=1} \sum_{i=1}^{n} p_i \cos^2(\angle(P_i v, v)). \tag{4.35}$$

Proof The result follows directly from the formula (4.8) for the error reduction in one step, namely

$$\mathbb{E}'(\|e^{(m+1)}\|^2) = \|e^{(m)}\|^2 - \sum_{i=1}^{n} p_i \Delta_i^{(m)}$$
$$= \left(1 - \sum_{i=1}^{n} p_i \frac{a(P_i e^{(m)}, e^{(m)})^2}{\|P_i e^{(m)}\|^2 \|e^{(m)}\|^2}\right) \|e^{(m)}\|^2$$
$$= \left(1 - \sum_{i=1}^{n} p_i \cos^2(\angle(P_i e^{(m)}, e^{(m)}))\right) \|e^{(m)}\|^2.$$

Now, replacing the forefactor with its largest possible value $1 - \mu_R$ and using the independence assumption gives the result (4.34). □

Admittedly, the constant μ_R that appears in the bound (4.34) is not very explicit. However, if the V_i are one-dimensional, or if the forms $a_i(\cdot,\cdot)$ are generated from $a(\cdot,\cdot)$ by Galerkin projection, then

$$|\cos(\angle(P_i v, v))| = \cos(\angle(\mathrm{ran}(R_i), v)) = 1 - \frac{1}{2} \inf_{v_i \in V_i:\, \|R_i v_i\|=1} \|R_i v_i - v'\|^2. \quad (4.36)$$

Thus, μ_R becomes a measure for the smallest average angle of a non-zero v with the n linear sets $\mathrm{ran}(R_i) \subset V$ or, equivalently, for the largest average distance of a point on the unit sphere $\Sigma_V = \{v \in V : \|v\| = 1\}$ from V to the union of the intersections of $\mathrm{ran}(R_i)$ with Σ_V. The averaging depends on the chosen probability distribution (4.33). The proof of (4.36) is left as an exercise, see Part a) of Problem 4.3.

Bounds in terms of properties of the Schwarz operator P_ρ can now be easily derived. In fact, since

$$\|P_i v\|^2 = \|R_i R_i^* v\|^2 \le \|R_i\|^2_{V_i \to V} \|R_i^* v\|_i^2 = \|R_i\|^2_{V_i \to V} a(P_i v, v), \quad (4.37)$$

we get

$$\mu_R = \inf_{v \in V:\|v\|=1} \sum_{i=1}^{n} p_i \frac{a(P_i v, v)^2}{\|P_i v\|^2} \ge \inf_{v \in V:\|v\|=1} \sum_{i=1}^{n} \frac{p_i}{\|R_i\|^2_{V_i \to V}} a(P_i v, v)$$
$$\ge \min_{i=1,\dots,n} \frac{p_i}{\rho_i \|R_i\|^2_{V_i \to V}} \inf_{v \in V:\|v\|=1} \frac{a(P_\rho v, v)}{\|v\|^2} = \lambda_{\min} \min_{i=1,\dots,n} \frac{p_i}{\rho_i \|R_i\|^2_{V_i \to V}}.$$

This lower bound for μ_R is maximized when we use

$$p_i = \frac{\rho_i \|R_i\|^2_{V_i \to V}}{\sum_{j=1}^{n} \rho_j \|R_j\|^2_{V_j \to V}}, \qquad i = 1, \dots, n, \quad (4.38)$$

which yields

$$\mu_R \geq \mu'_R := \frac{\lambda_{\min}}{\sum_{j=1}^n \rho_j \|R_j\|_{V_j\to V}^2}. \tag{4.39}$$

The analog of (4.34) with μ_R replaced by μ'_R also holds when we replace the steepest descent update formula (4.7) with

$$u^{(m+1)} = u^{(m)} + \frac{1}{\|R_{i_m}\|_{V_{i_m}\to V}^2} P_{i_m} e^{(m)}, \qquad m = 0, 1, \ldots, \tag{4.40}$$

where the i_m are randomly and independently chosen according to (4.38). Update formulas with precomputable step-sizes, such as (4.40), are often preferred in practice, since their computational cost is slightly lower compared to (4.7). In fact, using (4.37) with $i = i_m$, we have

$$\begin{aligned} \|e^{(m+1)}\|_2^2 &= \|e^{(m)}\|_2^2 - 2\frac{a(P_{i_m}e^{(m)}, e^{(m)})}{\|R_{i_m}\|_{V_{i_m}\to V}^2} + \frac{\|P_{i_m}e^{(m)}\|^2}{\|R_{i_m}\|_{V_{i_m}\to V}^4} \\ &\leq \|e^{(m)}\|_2^2 - \frac{a(P_{i_m}e^{(m)}, e^{(m)})}{\|R_{i_m}\|_{V_{i_m}\to V}^2}, \end{aligned}$$

where it remains to compute the expectations with respect to (4.38). This gives the following canonical result for randomized Schwarz methods with singleton I_m.

Theorem 4.6 *Let* $(\mathbf{V}; \mathbf{a}; \mathbf{R}; \rho)$ *be a finite stable splitting for* $(V; a)$ *with* $\Omega = \{1, \ldots, n\}$. *Then the Schwarz iteration (4.40) with i.i.d.* $i_m \in \Omega$ *and the underlying probability distribution (4.38) possesses the error estimate*

$$\mathbb{E}(\|e^{(m)}\|^2) \leq (1 - \mu'_R)^m \|e^{(0)}\|^2, \quad m = 1, 2, \ldots, \tag{4.41}$$

where $\mu'_R \leq \mu_R$ *is defined in (4.39).*

The statements for random Schwarz iterations remain valid if the exact values $\|R_i\|_{V_i\to V}$ are replaced by upper bounds. For example, by (4.19) we have

$$\|R_i\|_{V_i\to V}^2 \leq \frac{\lambda_{\max}}{\rho_i}, \qquad i = 1, \ldots, n.$$

Then, by substituting these bounds into the appropriate definitions and estimates, the probability distribution in (4.38) becomes even uniform, i.e.,

$$\mathbb{P}(i_m = i) = \frac{1}{n}, \qquad i = 1, \ldots, n. \tag{4.42}$$

The corresponding iteration then has the form

$$u^{(m+1)} = u^{(m)} + \frac{\rho_{i_m}}{\lambda_{\max}} P_{i_m} e^{(m)}, \qquad m = 0, 1, \ldots, \tag{4.43}$$

the quantity μ'_R in (4.39) is replaced by

$$\bar{\mu} := \frac{\lambda_{\min}}{n\lambda_{\max}} = \frac{1}{n\kappa} \leq \mu'_R,$$

and (4.41) changes to

$$\mathbb{E}(\|e^{(m)}\|^2) \leq \left(1 - \frac{1}{n\kappa}\right)^m \|e^{(0)}\|^2, \quad m = 1, 2, \ldots. \tag{4.44}$$

Obviously, this new bound is weaker than (4.41), but it holds for the uniform distribution (4.42), which is often used in applications.

4.3.2 Non-singleton I_m

We now give a result for a Schwarz iteration method where in each iteration step (4.1) a subset $I_m \subset \Omega = \{1, \ldots, n\}$ of size $1 \leq k_m := |I_m| \leq n$ is chosen independently and uniformly at random. Here, uniformly means that $\mathbb{P}(i \in I_m) = k_m/n$ for all $i = 1, \ldots, n$. If $k_m = 1$, this is just the uniform distribution (4.42) for the singleton case. If $k_m = n$, there is no randomness in the choice of I_m, and we are back to the additive Schwarz iteration (4.2). In applications, k_m is usually fixed in advance, or at least known at each iteration step. For general k_m we get the following result:

Theorem 4.7 *Let* $(\mathbf{V}; \mathbf{a}; \mathbf{R}; \rho)$ *be a finite stable splitting for* $(V; a)$ *with* $\Omega = \{1, \ldots, n\}$. *Then the Schwarz iteration (4.1) with constant step-size* $\omega_m = 1/\lambda_{\max}$ *and with index set* I_m *of size* k_m, *drawn independently and uniformly at random from* Ω, *possesses the error estimate*

$$\mathbb{E}'(\|e^{(m+1)}\|^2) \leq \left(1 - \frac{k_m}{n\kappa}\right) \|e^{(m)}\|^2, \qquad m = 0, 1, \ldots. \tag{4.45}$$

Because of the independence assumption, this results in

$$\mathbb{E}(\|e^{(m)}\|^2) \leq \prod_{k=0}^{m-1} \left(1 - \frac{r_k}{\kappa}\right) \|e^{(0)}\|^2, \quad r_m = \frac{k_m}{n}, \quad m = 0, 1, \ldots. \tag{4.46}$$

Proof In the error equation

$$\|e^{(m+1)}\|^2 = \|e^{(m)}\|^2 - 2\omega_m a(d^{(m)}, e^{(m)}) + \omega_m^2 a(d^{(m)}, d^{(m)}),$$

where $d^{(m)} = \sum_{i \in I_m} \rho_i P_i e^{(m)}$, we estimate the last term using

$$\|R_\rho\|^2_{V_\rho \to V} = \|P_\rho\|_V = \lambda_{\max}.$$

In fact, define $\tilde{v} \in \tilde{V}_\rho$ by setting $v_i = R_i^* e^{(m)}$ if $i \in I_m$ and $v_i = 0$ otherwise. Then

$$a(d^{(m)}, d^{(m)}) = \|R_\rho \tilde{v}\|^2 \le \lambda_{\max} \|\tilde{v}\|_{V_\rho}^2 = \lambda_{\max} \sum_{i \in I_m} \rho_i \|R_i^* e^{(m)}\|_i^2$$
$$= \lambda_{\max} a(\sum_{i \in I_m} \rho_i P_i e^{(m)}, e^{(m)}) = \lambda_{\max} a(d^{(m)}, e^{(m)}).$$

Note here that for small k_m this may be a crude bound, which could be further improved by using norm bounds for the operators

$$R_{\rho, I_m} := \sum_{i \in I_m} \rho_i R_i : V_\rho \to V,$$

if available.

After substituting into the above error equation for $\|e^{(m+1)}\|^2$, we take conditional expectations on both sides of the resulting inequality, and also use the step-size formula $\omega_m = 1/\lambda_{\max}$. We then obtain

$$\mathbb{E}'(\|e^{(m+1)}\|^2) \le \|e^{(m)}\|^2 - \omega_m(2 - \omega_m \lambda_{\max}) a(\mathbb{E}'(d^{(m)}), e^{(m)})$$
$$= \|e^{(m)}\|^2 - \frac{k_m}{n\lambda_{\max}} a(P_\rho e^{(m)}, e^{(m)}) \le \left(1 - \frac{r_m}{\kappa}\right) \|e^{(m)}\|^2.$$

In the last step we again relied on $\lambda_{\min} = \lambda_{\min}(P_\rho)$. This leads to (4.45) and, using the independence assumption, finally gives the bound in (4.46). □

Compared to the special case (4.44), the result says that lumping several random subproblem solves with the same residual into one step does not impose a penalty, since the guaranteed progress grows proportionally with k_m, i.e.,

$$\left(1 - \frac{1}{n\kappa}\right)^{k_m} \approx \left(1 - \frac{r_m}{\kappa}\right).$$

Note that this comment refers to the quality of the upper bound for convergence in expectation, not necessarily the actual convergence speed. However, depending on the compute platform, the actual cpu-time to perform one step of (4.1) may be significantly less than the cpu-time for k_m sequential steps in (4.43).

4.3.3 Random permutations

The independence assumption for the rules of choosing i_m in random Schwarz iterations with singleton $I_m = \{i_m\}$ does not prevent us from repeating the same index in subsequent iterations. Intuitively, this random choice with repetition may not be the best way to quickly reduce the error, and practice shows that random algorithms can perform better when repetition is reduced. However, rigorously showing this improvement is a delicate matter, and general results are still missing.

We mention a special case where randomization is enforced without repetition. It complements the results on deterministic orderings in Section 4.2. Let $\{\sigma_k\}_{k \in \mathbb{Z}_+}$

be a sequence of permutations of the index set $\Omega = \{1, \ldots, n\}$, drawn independently and uniformly at random from the set Σ_n of all $n!$ such permutations. Note that a permutation $\sigma = (\sigma(1), \ldots, \sigma(n)) \in \Sigma_n$ is a bijection defined on Ω and σ^{-1} denotes its inverse. Then, for $k = 0, 1, \ldots,$ execute the inner loop

$$u^{(kn+i)} = u^{(kn+i-1)} + \omega\rho_{\sigma_k(i)} P_{\sigma_k(i)} e^{(kn+i-1)}, \qquad i = 1, \ldots, n. \tag{4.47}$$

Compared to the deterministic ordering in (4.6), we now have a different random order in each inner loop, but each subproblem is solved exactly once, without repetition. The following convergence result for (4.47) should be compared with the estimate (4.32) for the deterministic Schwarz iteration (4.6) from Corollary 4.1.

Theorem 4.8 *Let* $(\mathbf{V}; \mathbf{a}; \mathbf{R}; \rho)$ *be a finite stable splitting for* $(V; a)$ *with* $\Omega = \{1, \ldots, n\}$. *If* $0 < \omega < 2/\|\tilde{D}_\rho\|_{\tilde{V}_\rho}$, *the expected squared error of the random Schwarz iteration (4.47) with the permutations* σ_k *drawn independently and uniformly at random from* Σ_n *is reduced according to*

$$\mathbb{E}'(\|e^{((k+1)n)}\|^2) \le \left(1 - \frac{\omega(2 - \omega\|\tilde{D}_\rho\|_{\tilde{V}_\rho})\lambda_{\min}}{1 + \omega\lambda_{\max} + \frac{1}{2}\omega^2\lambda_{\max}^2}\right) \|e^{(kn)}\|^2, \quad k = 0, 1, \ldots. \tag{4.48}$$

This bound yields

$$\mathbb{E}(\|e^{(kn)}\|^2) \le \left(1 - \frac{2}{5\kappa}\right)^k \|e^{(0)}\|^2, \qquad k = 1, 2, \ldots, \tag{4.49}$$

for the step-size value $\omega = 1/\lambda_{\max}$.

Proof Adapting the operator notation from Section 4.2 and dropping the subscript ρ in the notation as before, we can now write the error propagation for (4.47) in the form

$$e^{((k+1)n)} = (I - \omega R \Pi_{\sigma_k}^* (\tilde{I} + \omega \tilde{L}_{\sigma_k})^{-1} \Pi_{\sigma_k} R^*) e^{(kn)}, \quad k = 0, 1, \ldots.$$

Here Π_σ denotes the permutation operator associated with $\sigma \in \Sigma_n$, which maps from $(\tilde{V}; \tilde{a})$ to its "permutation" by the formula $\Pi_\sigma \tilde{v} = (v_{\sigma(1)}, \ldots, v_{\sigma(n)})$, $\Pi_\sigma^* = \Pi_\sigma^{-1} = \Pi_{\sigma^{-1}}$ is its inverse and

$$\tilde{P}_\sigma := \Pi_\sigma \tilde{P} \Pi_\sigma^* = \tilde{L}_\sigma + \tilde{D}_\sigma + \tilde{L}_\sigma^*$$

is the decomposition of the symmetric permutation $\tilde{P}_\sigma$ of the Gramian $\tilde{P}$ into strictly lower triangular, diagonal, and strictly upper triangular parts. Note that, independently of σ, we have $\tilde{D} = \Pi_\sigma^* \tilde{D}_\sigma \Pi_\sigma$ for the diagonal part, and

$$\tilde{H} := \tilde{L} + \tilde{L}^* = \Pi_\sigma^* (\tilde{L}_\sigma + \tilde{L}_\sigma^*) \Pi_\sigma \tag{4.50}$$

for the non-diagonal part $\tilde{H}$ of $\tilde{P} = \Pi_\sigma^* \tilde{P}_\sigma \Pi_\sigma$. However, for the triangular parts $\tilde{L}_\sigma$ and $\tilde{L}_\sigma^*$ separately, this is generally not true.

For the squared error we have

$$\|e^{((k+1)n)}\|^2 \le \|e^{(kn)}\|^2 - \omega\tilde{a}(\Pi^*_{\sigma_k}\tilde{N}_{\sigma_k,\omega}\Pi_{\sigma_k}R^*e^{(kn)}, R^*e^{(kn)}), \tag{4.51}$$

$k = 0, 1, \dots$, where we have written

$$\tilde{N}_{\sigma,\omega} = (\tilde{I} + \omega\tilde{L}^*_\sigma)^{-1}(2\tilde{I} - \omega\tilde{D}_\sigma)(\tilde{I}_\sigma + \omega\tilde{L}_\sigma)^{-1}.$$

Now set $\tilde{v} = R^*e^{(kn)} \in \tilde{X} = \operatorname{ran}(R^*)$ and consider an arbitrary $\sigma \in \Sigma_n$. As for Theorem 4.4, for $0 < \omega < 2/\|\tilde{D}\|_{\tilde{V}}$ we get

$$\begin{aligned}
a(\Pi^*_\sigma\tilde{N}_{\sigma,\omega}\Pi_\sigma\tilde{v}, \tilde{v}) &= \tilde{a}((2\tilde{I} - \omega\tilde{D})\Pi^*_\sigma(\tilde{I} + \omega\tilde{L}_\sigma)^{-1}\Pi_\sigma\tilde{v}, \Pi^*_\sigma(\tilde{I} + \omega\tilde{L}_\sigma)^{-1}\Pi_\sigma\tilde{v}) \\
&\ge (2 - \omega\|\tilde{D}\|_{\tilde{V}})\tilde{a}(\Pi^*_\sigma(\tilde{I} + \omega\tilde{L}^*_\sigma)^{-1}(\tilde{I} + \omega\tilde{L}_\sigma)^{-1}\Pi_\sigma\tilde{v}, \tilde{v}) \\
&\ge (2 - \omega\|\tilde{D}\|_{\tilde{V}})\frac{\tilde{a}(\tilde{v}, \tilde{v})^2}{\tilde{a}(\Pi^*_\sigma(\tilde{I} + \omega\tilde{L}_\sigma)(\tilde{I} + \omega\tilde{L}^*_\sigma)\Pi_\sigma\tilde{v}, \tilde{v})}.
\end{aligned}$$

In the last step, we used the inequality

$$\tilde{a}(\tilde{v}, \tilde{v})^2 = \tilde{a}(\tilde{S}^{-1/2}\tilde{v}, \tilde{S}^{1/2}\tilde{v})^2 \le \tilde{a}(\tilde{S}^{-1}\tilde{v}, \tilde{v})\tilde{a}(\tilde{S}\tilde{v}, \tilde{v})$$

with the invertible and positive definite operator $\tilde{S} = \Pi^*_\sigma(\tilde{I} + \omega\tilde{L}_\sigma)(\tilde{I} + \omega\tilde{L}^*_\sigma)\Pi_\sigma$ and its inverse $\tilde{S}^{-1} = \Pi^*_\sigma(\tilde{I} + \omega\tilde{L}^*_\sigma)^{-1}(\tilde{I} + \omega\tilde{L}_\sigma)^{-1}\Pi_\sigma$.

Obviously $\tilde{v} = R^*e^{(kn)}$ as well as $\tilde{a}(\tilde{v}, \tilde{v})$ are independent of σ, and we have

$$\tilde{a}(\tilde{v}, \tilde{v}) = a(Pe^{(kn)}, e^{(kn)}) \ge \lambda_{\min}\|e^{(kn)}\|^2.$$

Thus, the previous lower bound for $a(\Pi^*_\sigma\tilde{N}_{\sigma,\omega}\Pi_\sigma\tilde{v}, \tilde{v})$ takes the form

$$a(\Pi^*_\sigma\tilde{N}_{\sigma,\omega}\Pi_\sigma\tilde{v}, \tilde{v}) \ge \frac{(2 - \omega\|\tilde{D}\|_{\tilde{V}})\lambda_{\min}\|e^{(kn)}\|^2\tilde{a}(\tilde{v}, \tilde{v})}{\tilde{a}(\Pi^*_\sigma(\tilde{I} + \omega\tilde{L}_\sigma)(\tilde{I} + \omega\tilde{L}^*_\sigma)\Pi_\sigma\tilde{v}, \tilde{v})}.$$

If we substitute this for $\sigma = \sigma_k$ in (4.51) and take the conditional expectation with respect to the choice of σ_k, then

$$\mathbb{E}'(\|e^{((k+1)n)}\|^2) \le \left(1 - \omega(2 - \omega\|\tilde{D}\|_{\tilde{V}})\lambda_{\min}\mu_{\omega,\tilde{v}}\right)\|e^{(kn)}\|^2 \tag{4.52}$$

for $k = 0, 1, \dots$, where

$$\begin{aligned}
\mu_{\omega,\tilde{v}} &:= \frac{1}{n!}\sum_{\sigma\in\Sigma_n}\frac{\tilde{a}(\tilde{v}, \tilde{v})}{\tilde{a}(\Pi^*_\sigma(\tilde{I} + \omega\tilde{L}_\sigma)(\tilde{I} + \omega\tilde{L}^*_\sigma)\Pi_\sigma\tilde{v}, \tilde{v})} \\
&\ge \inf_{\tilde{u}\in\tilde{X}:\,\|\tilde{u}\|_{\tilde{V}}=1}\frac{1}{n!}\sum_{\sigma\in\Sigma_n}\tilde{a}(\Pi^*_\sigma(\tilde{I} + \omega\tilde{L}_\sigma)(\tilde{I} + \omega\tilde{L}^*_\sigma)\Pi_\sigma\tilde{u}, \tilde{u})^{-1} \\
&\ge \inf_{\tilde{u}\in\tilde{X}:\,\|\tilde{u}\|_{\tilde{V}}=1}\frac{n!}{\sum_{\sigma\in\Sigma_n}\tilde{a}(\Pi^*_\sigma(\tilde{I} + \omega\tilde{L}_\sigma)(\tilde{I} + \omega\tilde{L}^*_\sigma)\Pi_\sigma\tilde{u}, \tilde{u})}.
\end{aligned}$$

In the last step we used the inequality between harmonic and arithmetic mean.

To continue, we need an upper bound for

$$\begin{aligned}\frac{1}{n!}\sum_{\sigma\in\Sigma_n} &\tilde{a}(\Pi_\sigma^*(\tilde{I}+\omega\tilde{L}_\sigma)(\tilde{I}+\omega\tilde{L}_\sigma^*)\Pi_\sigma\tilde{u},\tilde{u})\\ &= \tilde{a}((\tilde{I}+\omega\tilde{H}+\frac{\omega^2}{n!}\sum_{\sigma\in\Sigma_n}\Pi_\sigma^*\tilde{L}_\sigma\tilde{L}_\sigma^*\Pi_\sigma)\tilde{u},\tilde{u})\\ &= \tilde{a}((\tilde{I}+\omega\tilde{H}+\omega^2\tilde{B})\tilde{u},\tilde{u}) \le \|\tilde{I}+\omega\tilde{H}+\omega^2\tilde{B}\|_{\tilde{V}}\|\tilde{u}\|^2,\end{aligned}$$

where (4.50) has been taken into account and

$$\tilde{B} := \frac{1}{n!}\sum_{\sigma\in\Sigma_n}\Pi_\sigma^*\tilde{L}_\sigma\tilde{L}_\sigma^*\Pi_\sigma$$

is the expectation of the operator family $\{\Pi_\sigma^*\tilde{L}_\sigma\tilde{L}_\sigma^*\Pi_\sigma\}_{\sigma\in\Sigma_n}$ with respect to the uniform distribution on Σ_n. Since the entries of the operator matrix representation of $\tilde{P}_\sigma$ with respect to the "permuted" $\tilde{V}$ are given by

$$(\tilde{P}_\sigma)_{i,j} = \rho_{\sigma(j)}R_{\sigma(i)}^*R_{\sigma(j)}, \quad i,j=1,\ldots,n,$$

one easily sees that the entries of the matrix operator representation of $\Pi_\sigma^*\tilde{L}_\sigma\tilde{L}_\sigma^*\Pi_\sigma$, now with respect to $\tilde{V}$, satisfy

$$(\Pi_\sigma^*\tilde{L}_\sigma\tilde{L}_\sigma^*\Pi_\sigma)_{i,j} = R_i^*\left(\sum_{l=1}^{\min(\sigma^{-1}(i),\sigma^{-1}(j))-1}\rho_{\sigma(l)}P_{\sigma(l)}\right)\rho_jR_j,$$

$i,j=1,\ldots,n$. Consequently, the entries of the operator matrix $\tilde{B}$ are given by

$$\begin{aligned}(\tilde{B})_{i,j} &= R_i^*\left(\frac{1}{n!}\sum_{\sigma\in\Sigma_n}\sum_{l=1}^{\min(\sigma^{-1}(i),\sigma^{-1}(j))-1}\rho_{\sigma(l)}P_{\sigma(l)}\right)\rho_jR_j\\ &= R_i^*\left(\sum_{m=1}^{n}N_{m;i,j}\rho_mP_m\right)\rho_jR_j,\end{aligned}$$

where $N_{m;i,j}$ is the cardinality of the set of all $\sigma\in\Sigma_n$ such that

$$l=\sigma^{-1}(m) < \min(\sigma^{-1}(i),\sigma^{-1}(j)).$$

It turns out that

$$N_{m;i,i} = \frac{1}{2}n!,\quad m\ne i=j, \qquad N_{m;i,j} = \frac{1}{3}n!,\quad m\ne i\ne j\ne m,$$

while for all remaining cases with $m=i$ or $m=j$ we have $N_{m;i,j}=0$. In fact, if $m\ne i=j$, then any σ in the associated set is obtained by first choosing an arbitrary pair $l < s$ from the index set Ω and setting $\sigma^{-1}(m) = l$, $\sigma^{-1}(i) = s$, followed by arbitrarily assigning the remaining $n-2$ positions of σ. This results in exactly

$$\frac{n(n-1)}{2} \cdot (n-2)! = \frac{1}{2}n!$$

different σ and gives the formula for $N_{m;i,i}$. In the case of $m \neq i \neq j \neq m$, to get all associated σ, first choose an arbitrary triple $l < s < t$ from Ω and set $\sigma^{-1}(m) = l$ and then either $\sigma^{-1}(i) = s$, $\sigma^{-1}(j) = t$ or $\sigma^{-1}(j) = s$, $\sigma^{-1}(i) = t$, followed by arbitrarily assigning the remaining $n-3$ positions of σ. This gives

$$\frac{n(n-1)(n-2)}{6} \cdot 2 \cdot (n-3)! = \frac{1}{3}n!$$

different σ and gives the formula for $N_{m;i,j}$. For $m = i$ or $m = j$ the statement is obvious. This gives us a compact formula for $\tilde{B}$, namely

$$\tilde{B} = \frac{1}{2}\tilde{H}^2 + \frac{1}{6}\tilde{D}_{\tilde{H}^2}, \tag{4.53}$$

where $\tilde{D}_{\tilde{H}^2}$ denotes the diagonal part of $\tilde{H}^2$. In fact,

$$(\tilde{H}^2)_{i,j} = \sum_{m \neq i,j} (R_i^* \rho_m R_m)(R_m^* \rho_j R_j) = R_i^* \left(\sum_{m \neq i,j} \rho_m P_m \right) \rho_j R_j,$$

and comparing with the formula for $\tilde{B}_{i,j}$ we see (4.53).

Since $\tilde{H} = \tilde{P} - \tilde{D}$ is the difference of two positive operators acting on $\tilde{V}$, and $\lambda_{\max} = \|\tilde{P}\|_{\tilde{V}} \geq \|\tilde{D}\|_{\tilde{V}}$, we have

$$-\lambda_{\max}\|\tilde{u}\|_{\tilde{V}}^2 \leq -\tilde{a}(\tilde{D}\tilde{u}, \tilde{u}) \leq \tilde{a}(\tilde{H}\tilde{u}, \tilde{u}) \leq \tilde{a}(\tilde{P}\tilde{u}, \tilde{u}) \leq \lambda_{\max}\|\tilde{u}\|_{\tilde{V}}^2.$$

This gives us $\|\tilde{H}\|_{\tilde{V}} \leq \lambda_{\max}$. Also,

$$\|\tilde{D}_{\tilde{H}^2}\|_{\tilde{V}} \leq \|\tilde{H}^2\|_{\tilde{V}} \leq \|\tilde{H}\|_{\tilde{V}}^2 \leq \lambda_{\max}^2.$$

So,

$$\|\tilde{I} + \omega\tilde{H} + \omega^2\tilde{B}\|_{\tilde{V}} \leq 1 + \omega\lambda_{\max} + \frac{1}{2}\omega^2\lambda_{\max}^2,$$

and we get the lower bound

$$\mu_{\omega,\tilde{V}} \geq \frac{1}{1 + \omega\lambda_{\max} + \frac{1}{2}\omega^2\lambda_{\max}^2}.$$

Substitution in (4.52) then results in (4.48). Also (4.49) is immediately obtained by using the bound $\|\tilde{D}\|_{\tilde{V}} \leq \lambda_{\max}$ and setting $\omega = 1/\lambda_{\max}$. □

Theorem 4.8 shows that (4.47) has a better worst-case convergence rate in expectation than the deterministic Schwarz iteration (4.6). It also suggests that the ordering of the subspaces in the space splitting may affect the convergence properties of (4.6). In practice, applying a random permutation to the space splitting beforehand, followed by running (4.6) in the new ordering, often leads to improvements. We will mention a result supporting this observation later in Section 5.1.

4.4 Greedy strategies

To improve convergence in an iterative method, it is intuitive to choose a search direction that potentially reduces the error the most. In Schwarz methods such as (4.1), the search directions $d^{(m)}$ are linear combinations of $d_i^{(m)} := \rho_i P_i e^{(m)}$, $i \in I_m \subset \Omega$. The maximal error reduction in the direction of a single $d_i^{(m)}$ is equal to

$$\Delta_i^{(m)} := \|e^{(m)}\|^2 - \min_{\xi\in\mathbb{R}} \|e^{(m)} - \xi d_i^{(m)}\|^2 = \frac{a(P_i e^{(m)}, e^{(m)})^2}{\|P_i e^{(m)}\|^2}, \qquad i = 1, \dots, n.$$

It is achieved by the steepest descent rule and is independent of ρ_i. A natural greedy strategy is then to choose I_m so that it contains indices with large $\Delta_i^{(m)}$. Computing $\Delta_i^{(m)}$ requires computing the partial residual $r_i^{(m)} := R_i^* e^{(m)} \in V_i$ by solving the i-th subproblem

$$a_i(r_i^{(m)}, v_i) = a(e^{(m)}, R_i v_i) = \langle f, R_i v_i\rangle_{V'\times V} - a(u^{(m)}, R_i v_i), \qquad i = 1, \dots, n.$$

Using this notation, we have

$$\Delta_i^{(m)} = \frac{a_i(r_i^{(m)}, r_i^{(m)})^2}{a(R_i r_i^{(m)}, R_i r_i^{(m)})} = \frac{\|r_i^{(m)}\|_i^4}{\|R_i r_i^{(m)}\|^2} \geq \frac{\|r_i^{(m)}\|_i^2}{\|R_i\|_{V_i\to V}^2} =: \bar{\Delta}_i^{(m)}, \tag{4.54}$$

with equality if (4.24) holds, i.e., if the subproblems are generated by Galerkin projection. In practice, I_m is often determined by choosing i with large $\bar{\Delta}_i^{(m)}$ instead of large $\Delta_i^{(m)}$. Finally, since

$$\sum_{i=1}^n \rho_i \|R_i\|_{V_i\to V}^2 \bar{\Delta}_i^{(m)} = \sum_{i=1}^n \rho_i \|r_i^{(m)}\|_i^2 = \|R_\rho^* e^{(m)}\|_{\tilde{V}_\rho}^2 = a(P_\rho e^{(m)}, e^{(m)}), \tag{4.55}$$

and since $\lambda_{\min}\|e^{(m)}\|^2 \leq a(P_\rho e^{(m)}, e^{(m)}) \leq \lambda_{\max}\|e^{(m)}\|^2$ relates the sum in (4.55) to the error term $\|e^{(m)}\|^2$, one could also compose I_m of indices with large values of

$$\tilde{\Delta}_i^{(m)} := \rho_i \|R_i\|_{V_i\to V}^2 \bar{\Delta}_i^{(m)} = \rho_i \|r_i^{(m)}\|_i^2, \qquad i = 1, 2, \dots. \tag{4.56}$$

Note that all three quantities $\Delta_i^{(m)}$, $\bar{\Delta}_i^{(m)}$, and $\tilde{\Delta}_i^{(m)}$ represent just different scalings of the squared V_i norm of the partial residuals $r_i^{(m)}$. Roughly speaking, they can be interpreted by (4.55) as *residual error indicators* associated with the i-th subproblem.

For the singleton case, covered in Subsection 4.4.1, we first derive a geometric bound on the error decay of the so-called weak greedy version of the steepest descent Schwarz iteration (4.7) based on the residual error indicators $\Delta_i^{(m)}$. It then turns out that the bound for the deterministic error sequence $\|e^{(m)}\|^2$ in the pure greedy case is generally better than the bound for $\mathbb{E}(\|e^{(m)}\|^2)$ obtained for the same Schwarz iteration with randomized choice, which is covered in Theorem 4.5. We also discuss greedy schemes using the other two error indicators. In Subsection 4.4.2 we briefly discuss bulk chasing strategies, where index sets I_m with respect to the largest $k_m > 1$ residual error indicators are used in the Schwarz iteration (4.1).

4.4.1 Singleton I_m

We first consider the greedy version for the steepest descent Schwarz iteration (4.7) with singleton $I_m = \{i_m\}$, where, for some parameters $0 < \beta_m \leq 1$, the index i_m is chosen so that

$$\Delta_{i_m}^{(m)} \geq \beta_m \max_{i=1,\dots,m} \Delta_i^{(m)}. \tag{4.57}$$

Following the literature on abstract greedy algorithms [179], we call the resulting method (4.7) *weak greedy Schwarz iteration*, and we call β_m the *weakness parameters*. If $\beta_m = 1$ in (4.57), then the method (4.7) is called *pure greedy Schwarz iteration*. Similar to Theorem 4.5, a geometric bound on the convergence speed of this method is readily available:

Theorem 4.9 *Let* $(\mathbf{V}; \mathbf{a}; \mathbf{R}; \rho)$ *be a finite stable splitting for* $(V; a)$ *with* $\Omega = \{1, \dots, n\}$. *Then, the weak greedy Schwarz iteration (4.7) with weakness parameters* β_m *possesses the error estimate*

$$\|e^{(m)}\|^2 \leq \left(\prod_{k=0}^{m-1} (1 - \beta_k \mu_G)\right) \|e^{(0)}\|^2, \quad m = 1, 2, \dots, \tag{4.58}$$

where

$$\mu_G := \inf_{v \in V: \|v\|=1} \max_{i=1,\dots,n} \cos^2(\angle(P_i v, v)). \tag{4.59}$$

The convergence $u^{(m)} \to u$, $m \to \infty$, *in* V *is guaranteed if* $\sum_{m=0}^{\infty} \beta_m = \infty$.

Proof If $e^{(m)} \neq 0$, using the error decay formula for the steepest descent update rule (4.7) together with (4.57), we get

$$\begin{aligned}\|e^{(m+1)}\|^2 &= \|e^{(m)}\|^2 - \Delta_{i_m}^{(m)} \leq \|e^{(m)}\|^2 - \beta_m \max_{i=1,\dots,n} \Delta_i^{(m)} \\ &= \left(1 - \beta_m \max_{i=1,\dots,n} \frac{a(P_i e^{(m)}, e^{(m)})^2}{\|P_i e^{(m)}\|^2 \|e^{(m)}\|^2}\right) \|e^{(m)}\|^2.\end{aligned}$$

It remains to recall the definition of $\cos(\angle(P_i e^{(m)}, e^{(m)}))$ to see that

$$\max_{i=1,\ldots,n} \frac{a(P_i e^{(m)}, e^{(m)})^2}{\|P_i e^{(m)}\|^2 \|e^{(m)}\|^2} = \max_{i=1,\ldots,n} \cos^2(\angle(P_i e^{(m)}, e^{(m)})) \geq \mu_G.$$

So $\|e^{(m+1)}\|^2 \leq (1-\beta_m\mu_G)\|e^{(m)}\|^2$ if $e^{(m)} \neq 0$. Since this trivially holds if $e^{(m)} = 0$ and $m = 0, 1, \ldots$, is arbitrary, we arrive at (4.58). The convergence statement follows directly from (4.58), since

$$\prod_{k=0}^{m-1}(1 - \beta_m\mu_G) \to 0 \quad \Leftrightarrow \quad \sum_{k=0}^{m-1} \beta_m \to \infty.$$

This completes the proof. □

A comparison with the randomized case covered by Theorem 4.5 shows that $\mu_G \geq \mu_R$ for any probability distribution (4.33). This is consistent with numerical observations: In terms of convergence speed, greedy Schwarz methods with β_m uniformly away from 0 beat Schwarz iterations with deterministic and random orderings. However, enforcing the weak greedy rule (4.57) at each iteration step is computationally expensive, since it may require solving many subproblems just to find an index with relatively large $\Delta_i^{(m)}$.

Next, we formulate results for weak greedy versions of the iteration (4.40), which rely on properties of P_ρ and involve the residual error indicators (4.54) and (4.56). They should be compared with Theorem 4.6.

Theorem 4.10 *Let* $(\mathbf{V}; \mathbf{a}; \mathbf{R}; \rho)$ *be a finite stable splitting for* $(V; a)$ *with* $\Omega = \{1, \ldots, n\}$.
a) If i_m *is chosen by the weak greedy rule*

$$\bar{\Delta}_{i_m}^{(m)} \geq \beta_m \max_{i=1,\ldots,n} \bar{\Delta}_i^{(m)} \tag{4.60}$$

with weakness parameters β_m, *then the Schwarz iteration (4.40) possesses the error estimate*

$$\|e^{(m)}\|^2 \leq \prod_{k=0}^{m-1}(1 - \beta_k\mu_G')\|e^{(0)}\|^2, \quad m = 1, 2, \ldots, \tag{4.61}$$

where

$$\mu_G \geq \mu_G' := \inf_{v \in V} \frac{a(P_\rho v, v)^2}{\|v\|^2 \left(\sum_{i=1}^n \rho_i \|R_i\|_{V_i \to V} \sqrt{a(P_i v, v)}\right)^2} \geq \mu_R'. \tag{4.62}$$

b) If i_m *satisfies*

$$\tilde{\Delta}_{i_m}^{(m)} \geq \beta_m \max_{i=1,\ldots,n} \tilde{\Delta}_i^{(m)}, \tag{4.63}$$

then for $m = 1, 2, \ldots$ *the Schwarz iteration (4.40) possesses the error estimate*

$$\|e^{(m)}\|^2 \leq \prod_{k=0}^{m-1}(1 - \beta_k\mu_G'')\|e^{(0)}\|^2 \leq \prod_{k=0}^{m-1}\left(1 - \frac{\beta_k}{n\kappa}\right)\|e^{(0)}\|^2. \tag{4.64}$$

Here,

$$\mu''_G := \frac{1}{\max_{i=1,\dots,n} \rho_i \|R_i\|^2_{V_i \to V}} \inf_{v \in V} \frac{a(P_\rho v, v)^2}{\|v\|^2 \left(\sum_{i=1}^n \sqrt{\rho_i}\sqrt{a(P_i v, v)}\right)^2} \le \mu'_G.$$

In both cases the convergence $u^{(m)} \to u$*,* $m \to \infty$*, in* V *is guaranteed if* $\sum_{m=0}^{\infty} \beta_m = \infty$.

Proof From (4.40), taking into account that $P_i = R_i R_i^*$, $P_i e^{(m)} = R_i r_i^{(m)}$ and $r_i^{(m)} = R_i^* e^{(m)}$, we have

$$\begin{aligned}\|e^{(m+1)}\|^2 &= \|e^{(m)}\|^2 - 2\|R_{i_m}\|^{-2}_{V_{i_m}\to V} a(e^{(m)}, P_{i_m} e^{(m)}) + \|R_{i_m}\|^{-4}_{V_{i_m}\to V} \|P_{i_m} e^{(m)}\|^2 \\ &\le \|e^{(m)}\|^2 - \|R_{i_m}\|^{-2}_{V_{i_m}\to V} \|r_{i_m}^{(m)}\|^2_{i_m} = \|e^{(m)}\|^2 - \bar{\Delta}_{i_m}^{(m)},\end{aligned}$$

compare with the definition of $\bar{\Delta}_i^{(m)}$ in (4.54). Thus, if i_m satisfies (4.60), then

$$\|e^{(m+1)}\|^2 \le \|e^{(m)}\|^2 - \beta_m \max_{i=1,\dots,n} \bar{\Delta}_{i_m}^{(m)} = \|e^{(m)}\|^2 - \beta_m \max_{i=1,\dots,n} \frac{\|r_i^{(m)}\|_i^2}{\|R_i\|^2_{V_i \to V}}.$$

The maximum is estimated as follows: Since

$$\sum_{i=1,\dots,n} \theta_i x_i^2 \le \max_{i=1,\dots,n} x_i \sum_{i=1,\dots,n} \theta_i x_i$$

holds for arbitrary $x_i, \theta_i \ge 0$, $i = 1, \dots, n$, we have the inequality

$$\max_{i=1,\dots,n} x_i^2 \ge \frac{(\sum_{i=1,\dots,n} \theta_i x_i^2)^2}{(\sum_{i=1,\dots,n} \theta_i x_i)^2}, \tag{4.65}$$

if the denominator in (4.65) does not vanish. If $r^{(m)} = 0$, then $e^{(m)} = 0$ and there is nothing to prove. If $r^{(m)} \neq 0$, we set

$$x_i = \frac{\|r_i^{(m)}\|_i}{\|R_i\|_{V_i \to V}}, \qquad \theta_i = \rho_i \|R_i\|^2_{V_i \to V}, \qquad i = 1, \dots, n,$$

and get

$$\begin{aligned}\sum_{i=1,\dots,n} \theta_i x_i^2 &= \sum_{i=1}^n \rho_i \|r_i^{(m)}\|_i^2 = a(P_\rho e^{(m)}, e^{(m)}), \\ \sum_{i=1,\dots,n} \theta_i x_i &= \sum_{i=1}^n \rho_i \|R_i\|_{V_i \to V} \|r_i^{(m)}\| = \sum_{i=1}^n \rho_i \sqrt{a(P_i e^{(m)}, e^{(m)})} > 0.\end{aligned}$$

With (4.65) we arrive at

$$\max_{i=1,\ldots,n} \frac{\|r_i^{(m)}\|_i^2}{\|R_i\|_{V_i\to V}^2} \geq \frac{a(P_\rho e^{(m)}, e^{(m)})^2}{\left(\sum_{i=1}^n \rho_i \|R_i\|_{V_i\to V}\sqrt{a(P_i e^{(m)}, e^{(m)})}\right)^2}.$$

Substituting this into the above estimate for $\|e^{(m+1)}\|^2$ and considering the definition of μ'_G in (4.62), we end up with

$$\|e^{(m+1)}\|^2 \leq (1 - \beta_m \mu'_G)\|e^{(m)}\|^2, \qquad m = 0, 1, \ldots.$$

This gives (4.61).

Since $\Delta_i^{(m)} \geq \bar{\Delta}_i^{(m)}$ for all i, it is obvious from the proofs of (4.58) and (4.61) that $\mu_G \geq \mu'_G$. Moreover, the relation $\mu'_G \geq \mu'_R$ is obtained using

$$\left(\sum_{i=1}^n \rho_i \|R_i\|_{V_i\to V}\sqrt{a(P_i v, v)}\right)^2 \leq \left(\sum_{i=1}^n \rho_i \|R_i\|_{V_i\to V}^2\right)\left(\sum_{i=1}^n \rho_i a(P_i v, v)\right)$$
$$= \left(\sum_{i=1}^n \rho_i \|R_i\|_{V_i\to V}^2\right) a(P_\rho v, v) = \lambda_{\min}\left(\sum_{i=1}^n \rho_i \|R_i\|_{V_i\to V}^2\right)\|v\|^2$$

in the definition of μ'_G, and then comparing with the definition of μ'_R in (4.39). This completes the proof of the statement in Part a).

For Part b), we first observe that

$$\|e^{(m+1)}\|^2 \leq \|e^{(m)}\|^2 - \frac{\|r_{i_m}^{(m)}\|_{i_m}^2}{\|R_{i_m}\|_{V_{i_m}\to V}^2} = \|e^{(m)}\|^2 - \frac{\tilde{\Delta}_{i_m}^{(m)}}{\rho_{i_m}\|R_{i_m}\|_{V_{i_m}\to V}^2}$$
$$\leq \|e^{(m)}\|^2 - \beta_m \frac{\max_{i=1,\ldots,n} \tilde{\Delta}_i^{(m)}}{\max_{i=1,\ldots,n} \rho_i \|R_i\|_{V_i\to V}^2}$$

if the weak greedy rule (4.63) is used in (4.40). Now apply (4.65) with

$$x_i = \sqrt{\tilde{\Delta}_i^{(m)}} = \sqrt{\rho_i}\|r_i\|_i, \qquad \theta_i = 1,$$

to get

$$\max_{i=1,\ldots,n} \tilde{\Delta}_i^{(m)} \leq \frac{\left(\sum_{i=1}^n \rho_i \|r_i^{(m)}\|_i^2\right)^2}{\left(\sum_{i=1}^n \sqrt{\rho_i}\|r_i^{(m)}\|_i\right)^2} = \frac{a(P_\rho e^{(m)}, e^{(m)})^2}{\left(\sum_{i=1}^n \sqrt{\rho_i}\sqrt{a(P_i e^{(m)}, e^{(m)})}\right)^2}.$$

Proceeding as in Part a), we obtain

$$\|e^{(m+1)}\|^2 \leq (1 - \beta_m \mu''_G)\|e^{(m)}\|^2, \qquad m = 0, 1, \ldots,$$

which, used recursively, leads to the first bound in (4.64). Since

$$\left(\sum_{i=1}^{n} \rho_i \|R_i\|_{V_i\to V}\sqrt{a(P_i v, v)}\right)^2 \le \max_{i=1,\dots,n} \rho_i \|R_i\|^2_{V_i\to V}\left(\sum_{i=1}^{n}\sqrt{\rho_i}\sqrt{a(P_i v, v)}\right)^2$$
$$\le \lambda_{\max} a(P_\rho v, v),$$

we have on the one hand $\mu'_G \ge \mu''_G$, and on the other hand

$$\mu''_G \ge \bar{\mu} = \frac{1}{n\kappa}.$$

This proves the statement in Part b). □

4.4.2 Non-singleton I_m

Applied to the setting of Schwarz iterative methods, another greedy strategy, similar to bulk chasing in adaptive numerical methods, is as follows: Instead of using only one of the search directions with large residual error indicator in the iteration step, as was the case in the above weak greedy rules with singleton I_m, we now use several of them at once. For example, since

$$\sum_{i=1}^{n} \tilde{\Delta}_i^{(m)} = a(P_\rho e^{(m)}, e^{(m)})$$

is a measure of the current error $\|e^{(m)}\|^2$, one could take I_m so that

$$\sum_{i\in I_m} \tilde{\Delta}_i^{(m)} \ge \beta_m \sum_{i=1}^{n} \tilde{\Delta}_i^{(m)}, \tag{4.66}$$

with some $0 < \beta_m \le 1$. With this choice for I_m we have

$$\|e^{(m+1)}\|^2 \le (1 - \frac{\beta_m}{\kappa})\|e^{(m)}\|^2, \qquad m = 0, 1, \dots. \tag{4.67}$$

Moreover, a small modification in the proof of Part b) in Theorem 4.10 yields the estimate

$$\|e^{(m+1)}\|^2 \le (1 - \frac{k_m}{n\kappa})\|e^{(m)}\|^2, \qquad m = 0, 1, \dots, \tag{4.68}$$

for (4.1), if I_m is given by the k_m *largest* $\tilde{\Delta}_i^{(m)}$. The proof of the estimates (4.67) and (4.68) is left to the reader, see Problem 4.4 at the end of this chapter.

Comparing (4.68) with the estimate (4.61) corresponding to the case $k_m = 1$, it becomes clear that taking non-singleton I_m in a weak greedy fashion does not entail a serious penalty. On the contrary, in many cases the work for executing a single iteration step does not grow significantly with k_m, and is less than the work required for k_m weak greedy iteration steps with singleton I_m.

4.5 Acceleration

The convergence results for one-step Schwarz iterations (4.1) obtained so far all depend, in one way or another, on lower and upper bounds for the spectrum of the Schwarz operator P_ρ. In most cases, this dependence is unavoidable. It also manifests itself in the practical performance of these methods, which deteriorates for ill-conditioned P_ρ. Here, multi-step extensions promise some improvement, called *acceleration*. Roughly speaking, while the number of iterations of one-step methods like (4.2) or (4.6) scales linearly with the condition number κ of the space splitting, the number of iterations of accelerated Schwarz iterations scales only linearly with $\sqrt{\kappa}$.

For linear operator equations $Au = f$ with symmetric positive definite $A : V \to V'$, the classical recipes for acceleration are the use of conjugate gradient (cg) and semi-iterative methods. We discuss them briefly in Subsection 4.5.1. Acceleration can also be achieved by a special family of two-step iterations originally introduced in [126] for convex optimization problems, see Section 5.3 for more details. In Subsection 4.5.2, we present their adaptation to the case of Schwarz iterations based on a finite stable splitting. In contrast to the conjugate gradient and classical semi-iterative methods, the analysis of such Nesterov-type schemes is not based on Krylov subspace arguments, which makes them also candidates for speeding up greedy and randomized Schwarz iterations.

4.5.1 Conjugate gradient and semi-iterative Schwarz methods

Since the Schwarz operator $P_\rho = C_\rho A$ associated with a stable splitting is symmetric positive and invertible on $(V; a)$, the cg-method for the Schwarz equation $P_\rho u = g$ is an option to achieve acceleration. Obviously, other reformulations of the equation $Au = f$ leading to an operator equation in a Hilbert space with a symmetric positive and invertible operator offer the same opportunity. For example, the Gramian equation $\tilde{P}_\rho \tilde{u} = \tilde{g}$ complemented by $u = R_\rho \tilde{u}$ is a candidate if we consider the subspace $\tilde{X}_\rho = \mathrm{ran}(R_\rho^*)$ of $(\tilde{V}_\rho; \tilde{a}_\rho)$. The resulting cg-iteration is equivalent to the cg-iteration for the Schwarz equation $P_\rho u = g$. Moreover, the symmetric multiplicative Schwarz iteration (4.9) with constant $\omega_k = \omega$ satisfying (4.18) yields such a reformulation, if one factors $\bar{M}_{\rho,\omega} = \bar{C}_{\rho,\omega} A$ and considers the operator equation

$$\bar{M}_{\rho,\omega} u = \bar{g} := \bar{C}_{\rho,\omega} f$$

in $(V; a)$. Note that the cg-iteration for $P_\rho u = g$ and $\bar{M}_{\rho,\omega} u = \bar{g}$ can also be viewed as a preconditioned conjugate gradient (pcg) iteration for the original equation $Au = f$ with preconditioners C_ρ and $\bar{C}_{\rho,\omega}$, respectively. For the latter, we refer to Problem 4.5.

We formulate the convergence result for the cg-iteration in the case of the Schwarz equation $P_\rho u = g$. The algorithm is as follows: For a given initial guess $u^{(0)} \in V$, set $r^{(0)} = p^{(0)} = g - P_\rho u^{(0)}$, and repeat

$$u^{(k+1)} = u^{(k)} + \alpha_k p^{(k)}, \quad \alpha_k = \frac{a(r^{(k)}, r^{(k)})}{a(p^{(k)}, p^{(k)})},$$
$$r^{(k+1)} = r^{(k)} - \alpha_k P_\rho p^{(k)},$$
$$p^{(k+1)} = r^{(k+1)} + \beta_k p^{(k)}, \quad \beta_k = \frac{a(r^{(k+1)}, r^{(k+1)})}{a(r^{(k)}, r^{(k)})}$$

for $k = 0, 1, \ldots$, until $\|r^{(k+1)}\|^2 = a(r^{(k+1)}, r^{(k+1)})$ is sufficiently small.

The analysis of this cg-iteration is based on two main properties, namely the orthogonality of the residual sequence $\{r^{(k)}\}$ in $(V; a)$ and the orthogonality of the sequence $\{p^{(k)}\}$ of search directions with respect to the scalar product $(\cdot, \cdot)_{P_\rho} := a(P_\rho \cdot, \cdot)$ induced by the Schwarz operator P_ρ. As in [181, Lecture 38], one proves that the error $e^{(k)} = u - u^{(k)}$ satisfies the bound

$$\|e^{(k)}\|_{P_\rho} = \inf_{p_k} \|p_k P_\rho e^{(0)}\|_{P_\rho} \le (\inf_{p_k} \sup_{\lambda \in \sigma(P_\rho)} |p_k(\lambda)|)\|e^{(0)}\|_{P_\rho},$$

where the infimum is taken with respect to all polynomials p_k of degree $\le k$ with the normalization condition $p_k(0) = 1$. This follows since it can be shown by induction that $e^{(k)}$ coincides with the element of minimal distance from the origin with respect to the $\|\cdot\|_{P_\rho}$ norm in the affine Krylov subspace

$$\mathcal{K}_k := e^{(0)} + \text{span}(\{P_\rho e^{(0)}, \ldots, P_\rho^k e^{(0)}\})$$

generated by $e^{(0)}$ and P_ρ, i.e.,

$$e^{(k)} = \text{argmin}_{v \in \mathcal{K}_k} \|v\|_{P_\rho}.$$

In terms of the condition number $\kappa = \kappa(P_\rho)$, this yields the bound

$$\|e^{(k)}\|_{P_\rho} \le 2\left(1 - \frac{2}{\sqrt{\kappa} + 1}\right)^k \|e^{(0)}\|_{P_\rho}, \quad k = 1, 2, \ldots. \tag{4.69}$$

The proof for linear equations in $\mathbb{R}^n$, given in textbooks on numerical linear algebra such as [181], is easily transferred to linear equations with positive definite and invertible operators in Hilbert spaces.

Note at this point that the cg-iteration is a special instance of the two-step iteration

$$u^{(k+1)} = u^{(k)} + \eta_k(u^{(k)} - u^{(k-1)}) + \xi_k P_\rho e^{(k)}, \qquad k = 0, 1, \ldots, \tag{4.70}$$

where η_k and ξ_k are determined such that $\|e^{(k+1)}\|_{P_\rho}$ is minimized and $u^{(-1)} = u^{(0)}$. This follows by induction from the minimization property of the error norms with respect to the affine Krylov subspaces $\mathcal{K}_k$, see Problem 4.6. There are whole families of two- and multi-step methods that differ in the choice of parameters (such as ξ_k and η_k) and in the number of additional search directions $u^{(l)} - u^{(l-1)}$, $l \le k$, in the update formula for $u^{(k+1)}$. In particular, one can choose constant values $\xi_k = \xi > 0$ and $0 < \eta_k = \eta < 1$ in (4.70) such that the resulting stationary iteration admits an error bound that is qualitatively comparable to the estimate for the cg-iteration. This can be seen by rewriting (4.70) in vector form as an iteration for $U^{(k)} := (u^{(k)}, u^{(k-1)})^T$, $k \geq 1$, in $V \times V$ (with elements written as column vectors). The result is a 2×2 block iteration for the associated error vector

$$E^{(k+1)} := \begin{pmatrix} e^{(k+1)} \\ e^{(k)} \end{pmatrix} = \begin{pmatrix} (1+\eta)I - \xi P_\rho & -\eta I \\ I & 0 \end{pmatrix} E^{(k)}, \quad k \geq 1, \tag{4.71}$$

where the starting vector $E^{(1)}$ is determined from $e^{(0)}$ by setting $e^{(1)} = (I - \xi P_\rho)e^{(0)}$. Denoting the operator matrix in (4.71) by $M_{\eta,\xi}$ and the identity operator on $V \times V$ by $I_{V\times V}$, we see that $\lambda \in \sigma(M_{\eta,\xi})$ if and only if

$$-\frac{\eta - \lambda(1+\eta-\lambda)}{\lambda\xi} \in \sigma(P_\rho), \tag{4.72}$$

since formally

$$(M_{\eta,\xi} - \lambda I_{V\times V})^{-1} = ((\eta - \lambda(1+\eta-\lambda))I + \lambda\xi P_\rho)^{-1} \begin{pmatrix} -\lambda I & \eta I \\ -I & (1+\eta-\lambda)I - \xi P_\rho \end{pmatrix}.$$

The elementary analysis of the condition (4.72) can be found in many papers, we leave it as an exercise to the reader, see Problem 4.7 at the end of this chapter. One particular outcome is that, for the values

$$\eta = \left(1 - \frac{2}{1+\sqrt{\kappa}}\right)^2, \qquad \xi = \frac{(1+\sqrt{\eta})^2}{\lambda_{\max}}, \tag{4.73}$$

the spectrum of $M_{\eta,\xi}$ is contained in a circle with radius $\sqrt{\eta} = 1 - 2/(1+\sqrt{\kappa})$ around the origin in $\mathbb{C}$. Furthermore, the spectral radius of $M_{\eta,\xi}$ is minimized for these parameters. This gives the following result:

Theorem 4.11 *Let* $(\mathbf{V}; \mathbf{a}; \mathbf{R}; \rho)$ *be a finite stable splitting for* $(V; a)$ *with spectral bounds* $\lambda_{\min}$, $\lambda_{\max}$, *and condition* κ. *The error propagation operator* $M_{\eta,\xi}$ *associated with the vector form of the two-step iteration (4.70) with the constant parameters* $\eta_k = \eta \in (0,1)$ *and* $\xi_k = \xi > 0$ *has minimal spectral radius*

$$r(M_{\eta,\xi}) = 1 - \frac{2}{1+\sqrt{\kappa}}, \tag{4.74}$$

if the parameters η *and* ξ *are given by (4.73).*

Since the spectral radius of the error propagation operator governs the asymptotic convergence rate of a stationary iteration process, this shows that the two-step iteration (4.70) with the indicated parameters is qualitatively as good as the cg-method, at least if we compare with the bound (4.69). However, since the error $e^{(k)}$ of the iteration (4.70) is an element of the affine Krylov subspace $\mathcal{K}_k$, the two-step method (4.70) cannot yield a better error decay than the cg-method with respect to the $\|\cdot\|_{P_\rho}$ norm.

The literature is full of special cases where weaker results hold. For example, a popular choice is $\xi = 1/\lambda_{\max}$, which, complemented by $\eta = (1 - 1/\sqrt{\kappa})^2$, yields $r(M_{\eta,\xi}) = 1 - 1/\sqrt{\kappa}$. Achieving such rates in practice requires knowledge of the spectral bounds $\lambda_{\min}$, $\lambda_{\max}$ of the splitting, which is a drawback compared to the cg-method. As commented on in the previous sections, if instead of exact values only estimates

$$0 < \underline{\lambda} \le \lambda_{\min} \le \lambda_{\max} \le \bar{\lambda}$$

are available for the spectral bounds, then the corresponding upper bounds for $r(M_{\eta,\xi})$ still hold if one replaces $\lambda_{\max}$ and κ in (4.73) and (4.74) by $\bar{\lambda}$ and $\bar{\kappa} := \bar{\lambda}/\underline{\lambda}$, respectively.

4.5.2 Nesterov-type acceleration

Conjugate gradient and semi-iterative methods are Krylov subspace methods that are sensitive to changes of the underlying operator during the course of the iteration. This is the case when we consider randomized and greedy Schwarz methods, where only parts of P_ρ appear in the update formulas in each step. To the best of our knowledge, it is currently unknown whether modifications of the two-step method (4.70) can be used for acceleration in these situations.

A slightly more sophisticated two-step method [126, 127, 128] is more robust in this respect. We write it in vector form as follows: Starting with $u^{(0)} = v^{(0)}$, let

$$\begin{aligned} u^{(m+1)} &= v^{(m)} + \zeta_m d_v^{(m)}, \\ v^{(m+1)} &= u^{(m+1)} + \eta_m (u^{(m+1)} - u^{(m)}) + \xi_m d_v^{(m)}, \end{aligned} \qquad d_v^{(m)} := \sum_{i \in I_m} \rho_i P_i e_v^{(m)}, \tag{4.75}$$

for $m = 0, 1, \ldots$. Again, $I_m \subset \Omega = \{1, \ldots, n\}$ is the index set used and ζ_m, η_m, ξ_m are parameters at our disposal. They may depend on properties of the space splitting and the sequence $\{I_m\}$. The parameters in the following error bounds may also depend on $\{I_m\}$, but some estimation steps are universal.

Note that setting $I_m = \Omega$ and $\zeta_m = 0$ in (4.75) essentially reduces this iteration to (4.70). If $I_m = \Omega$ for all $m = 0, 1, \ldots$, then the error propagation operator on $V \times V$ corresponding to (4.75), written in matrix form, is

$$M'_{\eta_m,\zeta_m,\xi_m} := \begin{pmatrix} 0 & I - \zeta_m P_\rho \\ -\eta_m I & (1+\eta_m)(I - \zeta_m P_\rho) - \xi_m P_\rho \end{pmatrix},$$

which for $\zeta_m = 0$ is a permutation of the error propagation operator M_{η_m,ξ_m} of (4.70). The additional parameter ζ_m is now crucial for the following analysis of (4.75). The main "trick" is to show a geometric decay for a certain linear combination of the error norms $\|e_u^{(m)}\|^2$ and $|||e_w^{(m)}|||_\rho^2 = a(P_\rho^{-1} e_w^{(m)}, e_w^{(m)})$, where

$$w^{(m)} := (1 + \gamma_m) v^{(m)} - \gamma_m u^{(m)}, \quad \gamma_m > 0, \quad m = 0, 1, \ldots, \tag{4.76}$$

is an auxiliary sequence with an additional parameter γ_m, which will be determined later. Here and in the following let

$$e_u^{(m)} := u - u^{(m)}, \quad e_v^{(m)} := u - v^{(m)}, \quad \text{and} \quad e_w^{(m)} := u - w^{(m)}$$

denote the terms in the respective error sequences.

For $d_v^{(m)}$ with arbitrary I_m, we have

$$|||d_v^{(m)}|||_\rho^2 = |||\sum_{i\in I_m} \rho_i R_i R_i^* e_v^{(m)}|||_\rho^2 \le \sum_{i\in I_m} \rho_i \|R_i^* e_v^{(m)}\|_i^2 = a(d_v^{(m)}, e_v^{(m)})$$

by the definition of $|||\cdot|||_\rho$. We also have

$$\|d_v^{(m)}\|^2 \le \lambda_{\max} |||d_v^{(m)}|||_\rho^2 \le \lambda_{\max} a(d_v^{(m)}, e_v^{(m)})$$

by the upper stability bound of the space splitting. Together with the first formula in (4.75), this gives

$$\begin{aligned}\|e_u^{(m+1)}\|^2 &= \|e_v^{(m)}\|^2 - 2\zeta_m a(d_v^{(m)}, e_v^{(m)}) + \zeta_m^2 \|d_v^{(m)}\|^2 \\ &\le \|e_v^{(m)}\|^2 - \zeta_m(2 - \zeta_m\lambda_{\max}) a(d_v^{(m)}, e_v^{(m)}).\end{aligned}$$

So we have

$$|||d_v^{(m)}|||_\rho^2 \le a(d_v^{(m)}, e_v^{(m)}) \le \sigma_m^{-1}(\|e_v^{(m)}\|^2 - \|e_u^{(m+1)}\|^2), \tag{4.77}$$

where

$$\sigma_m := \zeta_m(2 - \zeta_m\lambda_{\max}) \in (0, 1/\lambda_{\max}] \qquad \text{if} \quad 0 < \zeta_m < 2/\lambda_{\max}. \tag{4.78}$$

Note that if $I_m = \{i_m\}$ is a singleton then (4.78) holds with $\lambda_{\max}$ replaced by the smaller quantity $\rho_{i_m}\|R_{i_m}\|_{V_i\to V}^2$, since in this case

$$\begin{aligned}\|d_v^{(m)}\|^2 &= \|\rho_{i_m} R_{i_m} R_{i_m}^* e_v^{(m)}\|^2 \le \|R_{i_m}\|_{V_i\to V}^2 \|\rho_{i_m} R_{i_m}^* e_v^{(m)}\|_i^2 \\ &= \rho_{i_m}\|R_{i_m}\|_{V_i\to V}^2 a(d_v^{(m)}, e_v^{(m)}).\end{aligned}$$

To estimate $|||e_w^{(m+1)}|||^2$, we first use (4.75-4.76) and rewrite $e_w^{(m+1)}$ as a linear combination of $e_w^{(m)}$, $e_v^{(m)}$ and $d_v^{(m)}$ as follows:

$$\begin{aligned}e_w^{(m+1)} &= (1+\gamma_{m+1})e_v^{(m+1)} - \gamma_{m+1}e_u^{(m+1)} \\ &= ((1+\gamma_{m+1})(1+\eta_m) - \gamma_{m+1})e_u^{(m+1)} - (1+\gamma_{m+1})(\eta_m e_u^{(m)} + \xi_m d_v^{(m)}) \\ &= (1 + (1+\gamma_{m+1})\eta_m)e_v^{(m)} - (1+\gamma_{m+1})\eta_m e_u^{(m)} \\ &\qquad -((1+\gamma_{m+1})\xi_m + (1+(1+\gamma_{m+1})\eta_m)\zeta_m)d_v^{(m)} \\ &= \delta_m e_w^{(m+1)} + (1-\delta_m)e_v^{(m)} - \tau_m d_v^{(m)},\end{aligned}$$

where

$$\delta_m := \frac{(1+\gamma_{m+1})\eta_m}{\gamma_m}, \qquad \tau_m := (1+\gamma_{m+1})\xi_m + (1+(1+\gamma_{m+1})\eta_m)\zeta_m \tag{4.79}$$

are new positive parameters. In the last step we substituted

$$-e_u^{(m)} = \gamma_m^{-1}(e_w^{(m)} - (1+\gamma_m)e_v^{(m)})$$

and rearranged the terms.

With this at hand, we have

$$\begin{aligned} |||e_w^{(m+1)}|||_\rho^2 &= |||\delta_m e_w^{(m)} + (1-\delta_m)e_v^{(m)}|||_\rho^2 + \tau_m^2 |||d_v^{(m)}|||_\rho^2 \\ &\quad - 2\tau_m a(P_\rho^{-1}\delta_m e_w^{(m)} + (1-\delta_m)e_v^{(m)}, d_v^{(m)}). \end{aligned} \tag{4.80}$$

The first term on the right-hand side in (4.80) is estimated using the convexity of the squared norm $|||\cdot|||_\rho^2$ and the lower stability bound

$$|||e_v^{(m)}|||_\rho^2 \le \lambda_{\min}^{-1} \|e_v^{(m)}\|^2,$$

while we use (4.77) for the second term. This way we get

$$\begin{aligned} |||e_w^{(m+1)}|||_\rho^2 &\le \delta_m |||e_w^{(m)}|||_\rho^2 + \tfrac{1-\delta_m}{\lambda_{\min}} \|e_v^{(m)}\|^2 + \tfrac{\tau_m^2}{\sigma_m}(\|e_v^{(m)}\|^2 - \|e_u^{(m+1)}\|^2) \\ &\quad - 2\tau_m a(\delta_m e_w^{(m)} + (1-\delta_m)e_v^{(m)}, P_\rho^{-1} d_v^{(m)}). \end{aligned} \tag{4.81}$$

To estimate the term

$$a_m := a(\delta_m e_w^{(m)} + (1-\delta_m)e_v^{(m)}, P_\rho^{-1} d_v^{(m)}), \tag{4.82}$$

we have to use knowledge about I_m. First we consider the case $I_m = \Omega$. Then $d_v^{(m)} = P_\rho e_v^{(m)}$, and a_m simplifies to

$$\begin{aligned} a_m &= a(e_v^{(m)} - \delta_m\gamma_m(e_v^{(m)} - e_u^{(m)}), e_v^{(m)}) \\ &= a(e_v^{(m)} - \delta_m\gamma_m(e_v^{(m)} - e_u^{(m)}), e_v^{(m)}) \\ &\ge \|e_v^{(m)}\|^2 + \frac{\delta_m\gamma_m}{2}(\|e_v^{(m)}\|^2 - \|e_u^{(m)}\|^2) \\ &= (1 + \frac{\delta_m\gamma_m}{2})\|e_v^{(m)}\|^2 - \frac{\delta_m\gamma_m}{2}\|e_u^{(m)}\|^2. \end{aligned}$$

Substituting this into the above estimate (4.81) and moving the term involving $\|e_u^{(m+1)}\|^2$ to the left-hand side, we arrive at

$$\begin{aligned} |||e_w^{(m+1)}|||_\rho^2 + \frac{\tau_m^2}{\sigma_m}\|e_u^{(m+1)}\|^2 &\le \delta_m \left(|||e_w^{(m)}|||_\rho^2 + \gamma_m\tau_m \|e_u^{(m)}\|^2\right) \\ &\quad + \left(\frac{1-\delta_m}{\lambda_{\min}} + \frac{\tau_m^2}{\sigma_m} - \tau_m(2+\delta_m\gamma_m)\right)\|e_v^{(m)}\|^2. \end{aligned}$$

In the special case $I_m = \Omega$, we can assume that the parameters appearing in (4.75), (4.76), (4.78), and (4.79) are independent of m, so we drop the subscript m. If we enforce that

$$\alpha := \frac{\tau^2}{\sigma} = \gamma\tau, \tag{4.83}$$

and if

$$\frac{1-\delta}{\lambda_{\min}} + \frac{\tau^2}{\sigma} - \tau(2+\delta\gamma) = 0 \tag{4.84}$$

holds, then the iteration of the combined error norms becomes stationary. We obtain

$$|||e_w^{(m+1)}|||_\rho^2 + \alpha\|e_u^{(m+1)}\|^2 \le \delta(|||e_w^{(m)}|||_\rho^2 + \alpha\|e_u^{(m)}\|^2), \quad m = 0, 1, \ldots.$$

All that remains is to choose the parameters so that the conditions (4.83-4.84) are satisfied and δ is minimized. Using (4.83) we see that (4.84) is equivalent to

$$\frac{1}{\lambda_{\min}} - \frac{2}{1-\delta}\tau + \frac{\tau^2}{\sigma} = 0.$$

For this quadratic equation to have real solutions τ, the discriminant must be non-negative. This gives

$$(1-\delta)^2 \le \sigma\lambda_{\min}.$$

To minimize δ, we need to take the largest possible $\sigma = 1/\lambda_{\max}$ in (4.78) by setting $\zeta = 1/\lambda_{\max}$. This results in

$$\delta = 1 - 1/\sqrt{\kappa},$$

and by solving the above quadratic equation we obtain $\tau = (\lambda_{\min}\lambda_{\max})^{-1/2}$. Next, from (4.83), we get

$$\gamma := \sqrt{\kappa}, \qquad \alpha = \frac{1}{\lambda_{\min}}.$$

Finally, substitution in (4.79) yields the remaining constants

$$\eta = \frac{\sqrt{\kappa}-1}{\sqrt{\kappa}+1}, \qquad \xi = 0.$$

Thus we have proved the following result, which complements Theorem 4.11:

Theorem 4.12 *Let* $(\mathbf{V}; \mathbf{a}; \mathbf{R}; \rho)$ *be a finite stable splitting for* $(V; a)$ *with spectral bounds* $\lambda_{\min}$, $\lambda_{\max}$ *and condition* κ. *In the case*

$$I_m = \Omega,$$

the iteration (4.75) with constant parameters

$$\eta_m = \frac{\sqrt{\kappa}-1}{\sqrt{\kappa}+1}, \quad \zeta_m = \frac{1}{\lambda_{\max}}, \quad \xi_m = 0, \tag{4.85}$$

possesses the error estimate

$$\|e_u^{(m)}\|^2 + \lambda_{\min}|||e_w^{(m)}|||_\rho^2 \le \left(1 - \tfrac{1}{\sqrt{\kappa}}\right)^m (\|e_u^{(0)}\|^2 + \lambda_{\min}|||e_w^{(0)}|||_\rho^2)$$
$$\le 2\left(1 - \tfrac{1}{\sqrt{\kappa}}\right)^m \|e_u^{(0)}\|^2, \quad m = 0, 1, \ldots,$$

where $w^{(m)} = v^{(m)} + \sqrt{\kappa}(v^{(m)} - u^{(m)})$.

The last inequality follows from the lower stability bound of the space splitting: Since $u^{(0)} = v^{(0)} = w^{(0)}$, we have

$$\lambda_{\min}|||e_w^{(0)}|||_\rho^2 \le \|e_w^{(0)}\|^2 = \|e_u^{(0)}\|^2.$$

Compared to Theorem 4.11, we have a slightly worse decay factor $1 - 1/\sqrt{\kappa} > 1-2/(1+\sqrt{\kappa})$, but the estimate concerns a specific norm instead of the spectral radius of the error propagation matrix $M_{\eta,\xi}$. However, it can be shown (see Problem 4.8) that the spectral radius of the error propagation operator $M'_{\eta,\zeta,0}$ associated with the stationary iteration (4.75) with the parameters (4.85) described in Theorem 4.12 is also equal to $1 - 1/\sqrt{\kappa}$ and thus larger than the optimal value of $r(M_{\eta,\xi})$ stated in Theorem 4.11.

It turns out that under certain assumptions on the index selection process, the argument leading to Theorem 4.12 holds with little change. The crucial element is how to estimate the term a_m in (4.82). Note that the scalar product in the definition of a_m is not directly accessible, since its calculation involves P_ρ^{-1}. However, if I_m is chosen randomly such that for some $0 < r_m \le 1$ we have

$$\mathbb{E}'(\sum_{i\in I_m} \rho_i P_i u) = r_m P_\rho u, \qquad u \in V, \tag{4.86}$$

then, with $u = e_v^{(m)}$, $d_v^{(m)} = \sum_{i\in I_m} \rho_i P_i e_v^{(m)}$ and because of the linearity, we obtain

$$\begin{aligned}\mathbb{E}'(a_m) &= a(\delta_m e_w^{(m)} + (1-\delta_m)e_v^{(m)}, P_\rho^{-1}\mathbb{E}'(d_v^{(m)}))\\ &= r_m a(e_v^{(m)} - \delta_m\gamma_m(e_v^{(m)} - e_u^{(m)}), e_v^{(m)})\\ &\ge r_m((1 + \frac{\delta_m\gamma_m}{2})\|e_v^{(m)}\|^2 - \frac{\delta_m\gamma_m}{2}\|e_u^{(m)}\|^2).\end{aligned}$$

Recall that $\mathbb{E}'$ denotes the conditional expectation with respect to fixed I_k, $k < m$, as before. Typically, the value of r_m is related to the size of I_m relative to n.

Thus, taking into account the additional factor r_m, the proof strategy of Theorem 4.12 can be repeated. In fact, if one takes conditional expectations in (4.80) and (4.77) and repeats the estimation steps leading to (4.81), then one obtains

$$\begin{aligned}\mathbb{E}'(|||e_w^{(m+1)}|||_\rho^2) &= |||\delta_m e_w^{(m)} + (1-\delta_m)e_v^{(m)}|||_\rho^2 + \tau_m^2\mathbb{E}'(a_m)\\ &\le \delta_m|||e_w^{(m)}|||_\rho^2 + \frac{1-\delta_m}{\lambda_{\min}}\|e_v^{(m)}\|^2 + \frac{\tau_m^2}{\sigma_m}(\|e_v^{(m)}\|^2 - \mathbb{E}'(\|e_u^{(m+1)}\|^2))\\ &\quad - r_m\tau_m((2+\delta_m\gamma_m)\|e_v^{(m)}\|^2 - \delta_m\gamma_m\|e_u^{(m)}\|^2).\end{aligned}$$

After rearranging the terms, we have

$$\mathbb{E}'(|||e_w^{(m+1)}|||_\rho^2 + \frac{\tau_m^2}{\sigma_m}\|e_u^{(m+1)}\|^2) \le \delta_m(|||e_w^{(m)}|||_\rho^2 + r_m\tau_m\gamma_m\|e_u^{(m)}\|^2)$$
$$+ \left(\frac{1-\delta_m}{\lambda_{\min}} + \frac{\tau_m^2}{\sigma_m} - r_m\tau_m(2+\delta_m\gamma_m)\right)\|e_v^{(m)}\|^2.$$

For simplicity, we replace (4.83) with the new condition

$$\alpha' := \frac{\tau_m^2}{\sigma_m} = r_m\gamma_m\tau_m, \qquad m = 0, 1, \ldots, \tag{4.87}$$

where α' is independent of m, and assume that $0 < \delta_m < 1$ satisfies

$$\frac{1-\delta_m}{\lambda_{\min}} + \frac{\tau_m^2}{\sigma_m} - \tau_m(2 + r_m\delta_m\gamma_m) = 0. \tag{4.88}$$

This gives

$$\mathbb{E}'(|||e_w^{(m+1)}|||_\rho^2 + \alpha'\|e_u^{(m+1)}\|^2) \le \delta_m(|||e_w^{(m)}|||_\rho^2 + \alpha'\|e_u^{(m)}\|^2), \tag{4.89}$$

for $m = 0, 1, \ldots$.

Following the same strategy as in the proof of Theorem 4.12, we now determine the parameters so that the conditions (4.87-4.88) hold and δ_m is minimized. The result of these calculations is that

$$\delta_m = 1 - \frac{r_m}{\sqrt{\kappa}}, \qquad \alpha' = \alpha = \frac{1}{\lambda_{\min}}$$

holds if we use in (4.75)

$$\eta_m = \frac{\sqrt{\kappa} - r_m}{\sqrt{\kappa} + r_{m+1}}, \quad \zeta_m = \frac{1}{\lambda_{\max}}, \quad \xi_m = -\frac{(1 - r_{m+1})\sqrt{\kappa} - r_m + r_{m+1}}{\sqrt{\kappa} + r_{m+1}}, \tag{4.90}$$

$m = 0, 1, \ldots$. The remaining constants are

$$\sigma_m = \frac{1}{\lambda_{\max}}, \quad \gamma_m = \frac{\sqrt{\kappa}}{r_m}, \quad \tau_m = \frac{1}{\sqrt{\lambda_{\min}\lambda_{\max}}}, \quad m = 0, 1, \ldots.$$

The result of these derivations is summarized in the following theorem.

Theorem 4.13 *Let* $(\mathbf{V}; \mathbf{a}; \mathbf{R}; \rho)$ *be a finite stable splitting for* $(V; a)$ *with spectral bounds* $\lambda_{\min}$, $\lambda_{\max}$ *and condition* κ. *Consider the randomized Nesterov-type iteration (4.75), where the random selection of the index sets* I_m *satisfies (4.86) and is independent for different* $m = 0, 1, \ldots$. *If the parameters in (4.75) are chosen according to (4.90), then*

$$\mathbb{E}'(\|e_u^{(m+1)}\|^2 + \lambda_{\min}|||e_w^{(m+1)}|||_\rho^2) \le (1 - \frac{r_m}{\sqrt{\kappa}})(\|e_u^{(m)}\|^2 + \lambda_{\min}|||e_w^{(m)}|||_\rho^2), \tag{4.91}$$

where

$$w^{(m)} = v^{(m)} + \frac{\sqrt{\kappa}}{r_m}(v^{(m)} - u^{(m)}) \quad and \quad m = 0, 1, \ldots.$$

Also,

$$\begin{aligned}\mathbb{E}(\|e_u^{(m)}\|^2 + \lambda_{\min}|||e_w^{(m)}|||_\rho^2) &\le \textstyle\prod_{k=0}^{m-1}(1 - \frac{r_k}{\sqrt{\kappa}})(\|e_u^{(0)}\|^2 + \lambda_{\min}|||e_w^{(0)}|||_\rho^2) \\ &\le 2\textstyle\prod_{k=0}^{m-1}(1 - \frac{r_k}{\sqrt{\kappa}})\|e_u^{(0)}\|^2, \quad m = 1, 2, \ldots.\end{aligned} \tag{4.92}$$

The estimate (4.91) is a reformulation of (4.89) and, together with the independence assumption for the choice of I_m, implies (4.92). The additional factor 2 appears since $w^{(0)} = u^{(0)}$ and

$$\lambda_{\min}|||e_w^{(0)}|||_\rho^2 \le \|e_w^{(0)}\|^2 = \|e_u^{(0)}\|^2$$

by the lower stability bound of the splitting.

In the special case $I_m = \Omega$ with probability one, we have $r_m = 1$ in (4.86), and recover Theorem 4.12. A typical scenario where the assumption (4.86) is satisfied and r_m is known was already discussed in Section 4.4 in connection with Theorem 4.7, namely when I_m is a uniformly at random drawn subset of Ω of size $1 \le k_m \le n$. Then $r_m = k_m/n$, and comparing the estimate (4.92) with the result of Theorem 4.7 demonstrates the acceleration effect.

In practice, we may not have access to exact values for the spectral bounds $\lambda_{\min}$, $\lambda_{\max}$ and the factors r_m in (4.86), which enter the parameters in (4.90). The proof shows that, at the expense of slightly worse convergence guarantees, the statement of Theorem 4.13 still holds if we replace $\lambda_{\min}$ and r_m by suitable lower bounds $\underline{\lambda} \le \lambda_{\min}$ and $\underline{r}_m \le r_m$, and if we replace $\lambda_{\max}$ by its upper bound $\bar{\lambda} \ge \lambda_{\max}$. Attempts to design accelerated two-step methods that do not require knowledge of these constants while providing the same theoretical convergence guarantees have not yet been realized. Recall that for one-step methods, the steepest descent rule for choosing the step-size solves this problem.

4.5.3 Improvements for singleton I_m

In the rest of this section we consider the case of the singleton $I_m = \{i_m\}$. If the i_m are chosen as i.i.d. with respect to the uniform distribution, then $r_m = 1/n$ and the bound on the error reduction factor in each iteration step implied by Theorem 4.13 is $1 - (n\sqrt{\kappa})^{-1}$. In analogy to Theorem 4.6, one may ask what happens if other probability distributions are used that better fit the relative scaling of the subproblems. Similarly, one may ask what happens if greedy selection rules are used?

A review of the proof of Theorem 4.13 reveals the difficulty one then faces: On the one hand, estimates for $a(P_\rho e^{(m)}, e^{(m)})$ replacing (4.77) are easy to obtain for both random and greedy cases. For example, if we use the counterpart for the u-update in (4.40), namely

$$u^{(m+1)} = v^{(m)} + \frac{1}{\|R_{i_m}\|^2_{V_{i_m}\to V}} P_{i_m} e_v^{(m)}, \qquad m = 0, 1, \ldots, \tag{4.93}$$

where i_m is chosen according to the probability distribution (4.38), then

$$\begin{aligned}\mathbb{E}'(\|e_u^{(m+1)}\|^2) &\le \|e_v^{(m)}\|^2 - \mathbb{E}'\left(\frac{a(P_{i_m} e_v^{(m)}, e_v^{(m)})}{\|R_{i_m}\|^2_{V_{i_m}\to V}}\right) \\ &= \|e_v^{(m)}\|^2 - \frac{a(P_\rho e_v^{(m)}, e_v^{(m)})}{\sum_{i=1}^n \rho_i \|R_i\|^2_{V_i\to V}}.\end{aligned}$$

The same upper bound for $\mathbb{E}'(\|e_u^{(m+1)}\|^2)$ is obtained if the steepest descent Schwarz iteration (4.7) is used instead of (4.93). For details, see the proofs of Theorems 4.5 and 4.6. So for $m = 0, 1, \ldots,$ we have

$$a(P_\rho e_v^{(m)}, e_v^{(m)}) \le \frac{1}{\nu_R'}(\|e_v^{(m)}\|^2 - \mathbb{E}'(\|e_u^{(m+1)}\|^2), \tag{4.94}$$

where

$$\nu_R' := \left(\sum_{i=1}^n \rho_i \|R_i\|^2_{V_i\to V}\right)^{-1}.$$

Similarly, if a weak greedy rule (4.60) based on the residual error indicators

$$\bar{\Delta}_{i,v}^{(m)} := \frac{\|R_i^* e_v^{(m)}\|_i^2}{\|R_i\|^2_{V_i\to V}} = \frac{a(P_i e_v^{(m)}, e_v^{(m)})}{\rho_i \|R_i\|^2_{V_i\to V}}, \qquad i = 1, \ldots, n,$$

is used to select i_m in (4.93), then one obtains the bound

$$\|e_u^{(m+1)}\|^2 \le \|e_v^{(m)}\|^2 - \frac{\beta_m a(P_\rho e_v^{(m)}, e_v^{(m)})^2}{\left(\sum_{i=1}^n \rho_i \|R_i\|_{V_i\to V} \sqrt{a(P_i e_v^{(m)}, e_v^{(m)})}\right)^2}$$

as in the proof of Theorem 4.10. This time the result is

$$a(P_\rho e_v^{(m)}, e_v^{(m)}) \le \frac{1}{\beta_m \nu_G'}(\|e_v^{(m)}\|^2 - \|e_u^{(m+1)}\|^2), \qquad m = 0, 1, \ldots, \tag{4.95}$$

where

$$\nu_G' := \inf_{v\in V} \frac{a(P_\rho v, v)}{\left(\sum_{i=1}^n \rho_i \|R_i\|_{V_i\to V} \sqrt{a(P_i v, v)}\right)^2}. \tag{4.96}$$

Again, one can easily check $\nu_G' \ge \nu_R' \ge \nu_R''$ and sees that (4.95) improves on (4.94) for $\beta_m = 1$.

Finally, if one uses the counterpart

$$u^{(m+1)} = v^{(m)} + \frac{a(P_{i_m}e_v^{(m)}, e_v^{(m)})}{a(P_{i_m}e_v^{(m)}, P_{i_m}e_v^{(m)})} P_{i_m}e_v^{(m)}, \qquad m = 0, 1, \ldots,$$

of the steepest descent update formula (4.7) for the u-update, then a weak greedy rule based on the quantities

$$\Delta_{i,v}^{(m)} = \frac{a(P_i e_v^{(m)}, e_v^{(m)})^2}{a(P_i e_v^{(m)}, P_i e_v^{(m)})}, \qquad i = 1, \ldots, n,$$

is a natural choice. Again, as in the proof of Theorem 4.9, in this case one derives the bound

$$a(P_\rho e_v^{(m)}, e_v^{(m)}) \le \frac{1}{\beta_m \nu_G}(\|e_v^{(m)}\|^2 - \|e_u^{(m+1)}\|^2), \qquad m = 0, 1, \ldots, \tag{4.97}$$

where

$$\nu_G := \inf_{v \in V} \frac{a(P_\rho v, v)}{\left(\sum_{i=1}^n \rho_i \|P_i v\|\right)^2} \ge \nu_G'. \tag{4.98}$$

On the other hand, handling the estimation of $|||e_w^{(m+1)}|||_\rho^2$ without the assumption (4.86) faces difficulties: It is not clear how to then bound the term $|||d_{i_m,v}^{(m)}|||_\rho^2$ (or its conditional expectation) appearing in the above proofs by a multiple of $a(P_\rho e_v^{(m)}, e_v^{(m)})$. Recall that, according to our notation, the search directions in the singleton case $I_m = \{i_m\}$ are given by

$$d_{i,v}^{(m)} = \rho_i P_i e_v^{(m)}.$$

Moreover, (4.86) is crucial in the estimation of the term a_m, where it cancels the action of P_ρ^{-1}. Unfortunately, replacements for (4.86) are hard to come up with if we use non-uniform probability distributions or weak greedy rules to generate the index i_m.

An alternative suggested in the optimization literature for the randomized case is to replace the $|||\cdot|||_\rho$ norm with more explicit auxiliary norms tied to the probability distribution (4.38), which avoid the appearance of P_ρ^{-1}. We defer a discussion of this approach to Section 5.3. Another remedy that serves both randomized and greedy selection rules is to use two different search directions $d_{i_m,v}^{(m)}$ and $d_{i_m',v}^{(m)}$ for the u- and w-updates, respectively, where the second index i_m' is uniformly at random and independently chosen from Ω. This ensures that we have

$$\mathbb{E}'(d_{i_m',v}^{(m)}) = \frac{1}{n} P_\rho e_v^{(m)}, \tag{4.99}$$

i.e., the analog of (4.86) with $r_m = 1/n$ holds for all m. Note that it is not necessary to insist on the singleton $\{i_m'\}$ for the w-update. It is important, however, that an analog of (4.86) holds with known values of r_m. The changes in the estimates below are minimal and are left to the reader, see Problem 4.9.

The first index i_m can be chosen randomly or in a weak greedy way, so that an estimate holds for $a(P_\rho e_v^{(m)}, P_\rho e_v^{(m)})$ in terms of $\|e_v^{(m)}\|^2 - \|e_u^{(m+1)}\|^2$. To make this precise, we rewrite (4.75) as an iteration for u and w, with an auxiliary variable v: Starting with $u^{(0)} = w^{(0)} = v^{(0)}$, repeat for $m = 0, 1, \ldots,$

$$\begin{aligned} u^{(m+1)} &= v^{(m)} + \zeta_m d_{i_m,v}^{(m)}, \\ w^{(m+1)} &= \delta_m w^{(m)} + (1-\delta_m)v^{(m)} + \tau_m d_{i'_m,v}^{(m)}, \\ v^{(m+1)} &= \alpha_{m+1}u^{(m+1)} + (1-\alpha_{m+1})w^{(m+1)}. \end{aligned} \tag{4.100}$$

Here we have kept the notation already used in Subsection 4.5.2 and introduced the new parameter $\alpha_m := \gamma_m/(1+\gamma_m) \in (0,1)$, $m = 1, 2, \ldots$. Depending on the choice of i_m, we set

$$\zeta_m = \frac{1}{\rho_{i_m}\|R_{i_m}\|^2_{V_{i_m}\to V}} \tag{4.101}$$

or use the steepest descent rule

$$\zeta_m = \frac{a(e_v^{(m)}, d_{i_m,v}^{(m)})}{a(d_{i_m,v}^{(m)}, d_{i_m,v}^{(m)})}. \tag{4.102}$$

With this, the u-update in (4.100) does not depend on the weight ρ_{i_m}, although the w-update still does.

The next result states convergence rates for the accelerated iteration (4.100), where i'_m is chosen randomly and satisfies (4.99), while i_m is chosen in a random or greedy way. In both cases, we can only give estimates for the expected squared errors due to the random choice of i'_m.

Theorem 4.14 *Let* $(\mathbf{V};\mathbf{a};\mathbf{R};\rho)$ *be a finite stable splitting for* $(V;a)$ *with spectral bounds* $\lambda_{\min}$, $\lambda_{\max}$ *and condition* κ. *Furthermore let* $\Omega = \{1,\ldots,n\}$. *Consider the Nesterov-type iteration (4.100) with i.i.d.* $i'_m \in \Omega$ *drawn from the uniform distribution.*

a) If ζ_m *is defined by (4.101), then, for i.i.d.* $i_m \in \Omega$ *with underlying probability distribution (4.38), we have*

$$\begin{aligned} \mathbb{E}(\|e_u^{(m)}\|^2 + \lambda_{\min}|||e_w^{(m)}|||_\rho^2) &\le \left(1 - \sqrt{\frac{\lambda_{\min}\nu'_R}{n}}\right)^m (\|e_u^{(0)}\|^2 + \lambda_{\min}|||e_w^{(0)}|||_\rho^2) \\ &\le 2\left(1 - \sqrt{\frac{\lambda_{\min}\nu'_R}{n}}\right)^m \|e_u^{(0)}\|^2, \quad m = 1, 2, \ldots, \end{aligned} \tag{4.103}$$

where ν'_R *is defined as in (4.94) and the parameters* $\alpha_m, \delta_m, \tau_m$ *are chosen as in (4.105) and (4.106) below.*

b) If ζ_m *is defined by (4.101) and* i_m *is chosen according to a weak greedy rule with weakness parameters* β_m *for the residual error indicators* $\bar{\Delta}_{i,v}^{(m)}$, *then we have*

$$\mathbb{E}(\|e_u^{(m)}\|^2 + \lambda_{\min}|||e_w^{(m)}|||_\rho^2) \le \prod_{k=0}^{m-1}\left(1 - \sqrt{\frac{\lambda_{\min}\beta_k \nu'_G}{n}}\right)(\|e_u^{(0)}\|^2 + \lambda_{\min}|||e_w^{(0)}|||_\rho^2)$$
$$\le 2\prod_{k=0}^{m-1}\left(1 - \sqrt{\frac{\lambda_{\min}\beta_k \nu'_G}{n}}\right)\|e_u^{(0)}\|^2, \quad m = 1, 2, \ldots, \tag{4.104}$$

with ν'_G defined in (4.96) and with the parameters $\alpha_m, \delta_m, \tau_m$ determined in (4.105) and (4.106) below.
c) Finally, if ζ_m is defined by (4.102) and i_m is selected according to a weak greedy rule with weakness parameters β_m for the residual error indicators $\Delta_{i,v}^{(m)}$, then (4.103) holds with ν'_G replaced by ν_G defined in (4.98) and the same parameters $\alpha_m, \beta_m, \tau_m$ as in Part b).

Proof Up to different constants in the estimates (4.94), (4.95), and (4.97) for $a(P_\rho e_v^{(m)}, e_v^{(m)})$ used to prove Parts a), b), and c), respectively, the reasoning is the same as in the proof of Theorem 4.13. We only show the crucial steps.

As in the proof of Theorem 4.13, we have the bound

$$\begin{aligned}
\mathbb{E}'(|||e_w^{(m+1)}|||_\rho^2) &= |||\delta_m e_w^{(m)} + (1-\delta_m)e_v^{(m)}|||_\rho^2 + \tau_m^2 \mathbb{E}'(|||d_{i'_m,v}^{(m)}|||_\rho^2) \\
&\quad -2\tau_m a(P_\rho^{-1}\mathbb{E}'(d_{i'_m,v}^{(m)}), \delta_m e_w^{(m)} + (1-\delta_m)e_v^{(m)}) \\
&\le \delta_m |||e_w^{(m)}|||_\rho^2 + \frac{(1-\delta_m)}{\lambda_{\min}}\|e_v^{(m)}\|^2 + \frac{\tau_m^2}{n} a(P_\rho e_v^{(m)}, e_v^{(m)}) \\
&\quad - \frac{2\tau_m}{n} a(e_v^{(m)}, \delta_m e_w^{(m)} + (1-\delta_m)e_v^{(m)}) \\
&\le \delta_m |||e_w^{(m)}|||_\rho^2 + \frac{(1-\delta_m)}{\lambda_{\min}}\|e_v^{(m)}\|^2 + \frac{A\tau_m^2}{n}(\|e_v^{(m)}\|^2 - \mathbb{E}'(\|e_u^{(m+1)}\|^2)) \\
&\quad - \frac{2\tau_m}{n}((1 + \frac{\delta_m\gamma_m}{2})\|e_v^{(m)}\|^2 - \frac{\delta_m\gamma_m}{2}\|e_u^{(m)}\|^2)
\end{aligned}$$

due to the assumptions about i'_m. In the first estimation step, in addition to (4.99) and the convexity of $|||\cdot|||_\rho^2$, we used that

$$\mathbb{E}'(|||d_{i'_m,v}^{(m)}|||_\rho^2) \le \mathbb{E}'(a(P_{i'_m} e_v^{(m)}, e_v^{(m)})) = \frac{1}{n} a(P_\rho e_v^{(m)}, e_v^{(m)}).$$

In the last step, we used the lower bound

$$a(e_v^{(m)}, \delta_m e_w^{(m)} + (1-\delta_m)e_v^{(m)}) \ge (1 + \frac{\delta_m\gamma_m}{2})\|e_v^{(m)}\|^2 - \frac{\delta_m\gamma_m}{2}\|e_u^{(m)}\|^2,$$

which was already used in the estimation of the term $\mathbb{E}'(a_m)$ in the proof leading to Theorem 4.13, and we substituted a generic version of the estimates (4.94), (4.95), (4.97). More specifically, for Parts a), b), and c), we have set $A = 1/\nu'_R$, $A = 1/(\beta_m\nu'_G)$ and $A = 1/(\beta_m\nu_G)$, respectively. In the case of Parts b) and c), the conditional expectation $\mathbb{E}'(\|e_u^{(m+1)}\|^2)$ can be silently replaced by $\|e_u^{(m+1)}\|^2$.

Rearranging the terms, we get

$$\mathbb{E}'(|||e_w^{(m+1)}|||_\rho^2 + \frac{A\tau_m^2}{n}\|e_u^{(m+1)}\|^2) \le \delta_m(|||e_w^{(m)}|||_\rho^2 + \frac{\tau_m\gamma_m}{n}\|e_u^{(m)}\|^2) + \left(\frac{(1-\delta_m)}{\lambda_{\min}} - \frac{\tau_m}{n}(2+\delta_m\gamma_m) + \frac{A\tau_m^2}{n}\right)\|e_v^{(m)}\|^2.$$

As before, we require $A\tau_m^2/n = \tau_m\gamma_m/n$, i.e., $\gamma_m = A\tau_m$, and

$$\begin{aligned} 0 &= \frac{(1-\delta_m)}{\lambda_{\min}} - \frac{\tau_m}{n}(2+\delta_m\gamma_m) + \frac{A\tau_m^2}{n} \\ &= \frac{(1-\delta_m)}{n}\left(\frac{n}{\lambda_{\min}} - \frac{2\tau_m}{(1-\delta_m)} + A\tau_m^2\right), \end{aligned}$$

which has a real solution $\tau_m > 0$ if and only if

$$\frac{1}{(1-\delta_m)^2} \ge \frac{An}{\lambda_{\min}} \quad \Leftrightarrow \quad \delta_m \ge 1 - \sqrt{\frac{\lambda_{\min}}{An}}.$$

So the smallest possible δ_m is

$$\delta_m = 1 - \sqrt{\frac{\lambda_{\min}}{An}}, \tag{4.105}$$

while the remaining constants are then given by

$$\tau_m = \sqrt{\frac{n}{A\lambda_{\min}}}, \quad \gamma_m = \sqrt{\frac{nA}{\lambda_{\min}}}, \quad \alpha_m = \sqrt{\frac{nA}{nA+\lambda_{\min}}}. \tag{4.106}$$

Since $A\tau_m^2/n = 1/\lambda_{\min}$, we arrive at

$$\mathbb{E}'(\|e_u^{(m+1)}\|^2 + \lambda_{\min}|||e_w^{(m+1)}|||_\rho^2) \le \left(1 - \sqrt{\frac{\lambda_{\min}}{An}}\right)(\|e_u^{(m)}\|^2 + \lambda_{\min}|||e_w^{(m)}|||_\rho^2), \tag{4.107}$$

with the corresponding constants A for Parts a), b), and c).

To complete the proof of the error estimates (4.103) and (4.104) stated in Theorem 4.14, all that remains is to use the independence assumption and substitute the appropriate values for A. □

A few comments are in order. For the case of the singleton I_m with $r_m = 1/n$ in (4.86) and a uniform probability distribution, Theorem 4.13 implies (4.103) with ν_R' replaced by the value $(n\lambda_{\max})^{-1}$. Thus, Theorem 4.14 improves the bound in Theorem 4.13, since

$$\lambda_{\max} \le (\nu_R')^{-1} = \sum_{i=1}^n \rho_i \|R_i\|_{V_i\to V}^2 \le n\lambda_{\max}.$$

Although in general $\nu_G \geq \nu'_G \geq \nu'_R$, these constants coincide under mild conditions. In fact, if the auxiliary forms of the splitting are generated by the Galerkin projection, then $\|P_i\|_V = \|R_i\|^2_{V_i\to V} = 1$ and

$$\|P_i v\|^2 = a(P_i v, P_i v) = a_i(R_i^* v, R_i^* v) = a(P_i v, v),$$

which implies $\nu_G = \nu'_G$. Moreover, if there is a $v \in V$ such that $a(P_i v, v) = \|R_i\|^2_{V_i\to V}$ for all $i \in \{1, \ldots, n\}$, then

$$\nu'_G = \nu'_R.$$

Thus, in many cases of interest, the estimates of Theorem 4.14 are the same for the randomized and greedy versions of the accelerated Schwarz iterations.

To compare the results for the accelerated methods with the results for the one-step Schwarz methods from Section 4.3, we focus on the methods with singleton I_m and check the error reduction factor in n single iteration steps. For the accelerated randomized Schwarz iteration based on (4.38), which was treated in Part a) of Theorem 4.14, we have

$$\lambda_{\min}\nu'_R = \mu'_R.$$

So we get

$$\left(1 - \sqrt{\frac{\lambda_{\min}\nu'_R}{n}}\right)^n \approx 1 - \sqrt{n\mu'_R}.$$

This improves the bound obtained for the corresponding one-step method from Theorem 4.6, since

$$(1 - \mu'_R)^n \approx 1 - n\mu'_R.$$

For the weak greedy Schwarz methods covered by Part b) in Theorem 4.14 and Theorem 4.10, we can only show that

$$\lambda_{\min}\nu'_G \geq \mu'_G,$$

but we expect the same effect. Note here that the work per iteration for the accelerated Schwarz methods of Theorem 4.14 is doubled, since two subproblem solves are needed to compute the search directions $d^{(m)}_{i_m,v}$ and $d^{(m)}_{i'_m,v}$ in each step. However, this is compensated by the improved convergence speed.

To implement (4.100) with the parameters specified in (4.105) and (4.106), one needs to know the values of A and $\lambda_{\min}$. As before, good upper bounds for A and lower bounds for $\lambda_{\min}$ would suffice to ensure (4.103) with slightly worse constants. Research is underway to theoretically ensure the same performance guarantees for accelerated Schwarz iterations without relying on a priori knowledge of $\lambda_{\min}$. It is also not known whether the Nesterov-type iteration (4.75) with a single search direction $d^{(m)}_{i_m,v}$, which is obtained by a weak greedy rule, can satisfy deterministic error bounds similar to those given in Parts b) or c) of Theorem 4.14. Finally, it is not clear whether there are variants of the simpler scheme (4.70) that provide acceleration in the randomized or greedy case.

Problems

4.1 Subspace splittings with $n = 2$.
Let $(\mathbf{V}; \rho)$ be a Riesz splitting for $(V; a)$ with two subspaces V_1 and V_2 ($n = 2$).
a) Show that the condition number of the splitting is minimized if $\rho_1 = \rho_2$. In this case,

$$\kappa = \frac{1 + \cos(\alpha)}{1 - \cos(\alpha)}, \qquad \alpha := \angle(V_1, V_2),$$

where the (Dixmier) angle $\angle(X_1, X_2) \in [0, \pi/2]$ between two linear sets X_1, X_2 in $(V; a)$ is defined by

$$\cos(\angle(X_1, X_2)) := \sup_{0 \neq v_i \in X_i,\, i=1,2} \frac{a(v_1, v_2)}{\|v_1\| \|v_2\|}.$$

b) Show that if $\rho_1 = \rho_2 = 1$, then the optimal step-size for (4.2) is $\omega_m = 1$ and (4.5) becomes

$$\|e^{(m)}\| \leq \cos(\alpha)^m \|e^{(0)}\|, \qquad m = 1, 2, \ldots.$$

Provide an example that this estimate is sharp, i.e., becomes an equality.
c) Verify the XZ-identity for (4.6), again assuming $\rho_1 = \rho_2 = 1$ and $\omega_k = 1$.
d) Using the same parameters as in c), show that for (4.6)

$$\|e^{(2k)}\| \leq \cos(\alpha)^{2k-1} \|e^{(0)}\|, \qquad k = 1, 2, \ldots.$$

Give an example that this estimate is sharp. The result is the same if the order of the subproblems is reversed.
e) What changes if the subspace splitting $(\mathbf{V}; \rho)$ is only stable, i.e., if the condition $V_1 \cap V_2 = \{0\}$ is dropped?
Hints: For a), use the definition of $|||u|||_\rho^2$ in which the infimum can be dropped since the decomposition $u = u_1 + u_2$, $u_i \in V_i$, is unique due to the Riesz property. Then solve the optimization problems that define the spectral bounds for the splitting. For b)-d) it is helpful to visualize the situation by drawing a picture for $V = \mathbb{R}^2$ and two one-dimensional subspaces. For d), note that the iteration is equivalent to alternately projecting the initial $e^{(0)}$ onto the orthogonal complements $V_1^\perp$ and $V_2^\perp$. For e), consider $W = V_1 \cap V_2$, $W_i = V_i \ominus W$, $i = 1, 2$. Except for the situation when one or both V_i coincide with V, the W_i are non-trivial and form a Riesz splitting of $W^\perp$. The statements a)-d) remain valid if the angle between V_1 and V_2 is replaced by the angle between W_1 and W_2. Note that in each single step of the multiplicative Schwarz iteration with $\rho_1 = \rho_2 = 1$, the error component in W is reduced by the factor $|1 - \omega|$.

4.2 Counterpart of Theorem 4.3 for arbitrary step-size parameters.
Prove a convergence result, similar to Theorem 4.3, for the iteration

$$u^{(kn+i)} = u^{(kn+i-1)} + \omega_{kn+i} \rho_i P_i e^{(kn+i-1)}, \quad i = 1, \ldots, n, \quad k = 0, 1, \ldots.$$

Compared to (4.6), the step-size parameters are now allowed to vary within each inner loop.
Hint: Note that for the norm of the error propagation operator associated with the k-th inner loop, an analog of the XZ-identity holds if in the proof of Theorem 4.2 the operator $\omega^{-1}\tilde{I}$ is replaced by the diagonal operator $\mathrm{diag}((\omega_{kn+i}^{-1} I_i)_{i\in\Omega})$, where I_i denotes the identity operator on V_i.

4.3 Angles between subspaces and cyclic orderings.
Consider the special case of a finite stable space splitting $(\mathbf{V}, \mathbf{R}; \rho)$ with forms $a_i(\cdot,\cdot)$ generated from $a(\cdot,\cdot)$ by Galerkin projection.
a) Prove (4.36).
b) Show that $P_i^\perp := I - P_i$ is the ortho-projection in V onto the subspace $W_i := (\mathrm{ran}(R_i))^\perp \subset V$.
c) For $\rho_i = 1$, express the matrix Γ_ρ appearing in the definition of the strengthened Cauchy-Schwarz inequality in terms of angles between the subspaces W_i, and formulate the resulting error estimate for the multiplicative Schwarz iteration (4.6) with cyclic ordering and step-size parameter $\omega = 1$ using Theorem 4.3 and Part a) of Theorem 4.4.
d) Prove the following result from [65, Theorem 9.33]: If $\rho_i = 1$ and $\omega = 1$, then the error propagation matrix $M_{\rho,\omega}$ for (4.6) satisfies

$$\|M_{\rho,\omega}\|^2 \le 1 - \prod_{i=1}^{n-1} (1 - \cos^2(\angle(W_i, \cap_{j=i+1}^n W_j)_F), \tag{4.108}$$

where the Friedrichs angle $\angle(X_1, X_2)_F \in [0, \pi/2]$ between two *subspaces* X_1, X_2 of $(V; a)$ is defined by

$$\cos(\angle(X_1, X_2)_F) := \sup_{0\neq v_i \in X_i \cap (X_1\cap X_2)^\perp,\, i=1,2} \frac{a(v_1, v_2)}{\|v_1\| \|v_2\|}.$$

Obviously, $\cos(\angle(X_1, X_2)_F) \le \cos(\angle(X_1, X_2))$, with equality if $X_1 \cap X_2 = \{0\}$. For more about angles between subspaces and their properties see [65, Chapter 9], improvements of the bound (4.108) are discussed in [66].
Hints For a), express

$$\|v\|^2 \cos^2(\angle(v, R_i u_i)) = \frac{a(v, R_i u_i)^2}{a(R_i u_i, R_i u_i)}, \qquad 0 \neq u_i \in V_i, \quad 0 \neq v \in V,$$

with respect to $a_i(\cdot,\cdot)$, and check that the maximizer of the resulting expression is $u_i = R_i^* v$. For b), use the definition of R_i^* and (4.24) to show the required properties $(P_i^\perp)^2 = P_i^\perp$, $(P_i^\perp)^* = P_i^\perp$, and $P_i^\perp w_i = w_i$, $w_i \in W_i$, of a projection onto W_i. For c), note that the optimal γ_{ij} in the definition of the strengthened Cauchy-Schwarz inequalities is $\cos(\angle(\mathrm{ran}(R_i), \mathrm{ran}(R_j)))$. Finally, for d), set $U_i = W_n \cap \cdots \cap W_i$, $T_i = P_n^\perp \cdots P_i^\perp$, and denote

$$d_i := \inf_{u\in U_i^\perp:\, \|u\|=1} \|u\|^2 - \|T_i u\|^2, \qquad i = n, \ldots, 1.$$

Note that $d_n = 1$, $d_1 = 1 - \|M_{\rho,\omega}\|^2$, and establish the inequality

$$d_i \geq d_{i+1}(1 - \cos^2(\angle(W_i, U_{i+1})_F)), \qquad i = 1, \ldots, n-1.$$

To prove the latter, decompose $u \in U_i^\perp$ orthogonally into three parts as follows:

$$u_1 = Q_{U_i^\perp \cap W_i^\perp} u, \quad u_2 = Q_{U_i^\perp \cap W_i \cap U_{i+1}} u, \quad u_3 = Q_{U_{i+1}^\perp \cap W_i} u.$$

Then observe that $T_i u_1 = 0$, $T_i u_2 = u_2 \in U_{i+1}$, and $T_i u_3 = T_{i+1} u_3 \in U_{i+1}^\perp$, and verify that

$$\begin{aligned}\|u_2\|^2 &= a(u_2 + u_3, u_2) \leq \cos(\angle(W_i, U_{i+1})_F)\|u_2 + u_3\|\|u_2\| \\ &\leq \cos^2(\angle(W_i, U_{i+1})_F)\|u\|^2.\end{aligned}$$

4.4 Estimates for bulk chasing.
Prove the estimates (4.68) and (4.67) for the bulk chasing Schwarz methods outlined in Subsection 4.4.2.

4.5 Symmetric multiplicative Schwarz preconditioner.
Verify that the symmetric multiplicative Schwarz iteration (4.9) gives rise to a preconditioned conjugate gradient iteration.
a) Show that the operator

$$\bar{P}_{\rho,\omega} = \bar{C}_{\rho,\omega} A := I - \bar{M}_{\rho,\omega}$$

is symmetric positive definite in $(V; a)$ if ω satisfies (4.18). Also estimate its condition number.
b) Using the product representation of $\bar{M}_{\rho,\omega}$, derive the recursion for computing the action of the approximate inverse $\bar{C}_{\rho,\omega} : V' \to V$.
Hint: For a), note that lower bounds for the spectrum of

$$\bar{P}_{\rho,\omega} = \omega R_\rho (\tilde{I} + \omega \tilde{L}_\rho^*)^{-1}(2\tilde{I} - \omega \tilde{D}_\rho)(\tilde{I} + \omega \tilde{L}_\rho)^{-1} R_\rho^*$$

were obtained in Section 4.2. Find a complementing upper bound by estimating the operator norm of $(\tilde{I} + \omega \tilde{L}_\rho)^{-1}$. For b), set $C_j := \rho_j R_j A_j^{-1} R_j'$ and use the representation of $\bar{M}_{\rho,\omega} = I - \bar{C}_{\rho,\omega} A$ as the product of the operators $I - C_j A$ to obtain $\bar{C}_{\rho,\omega}$ by recursive application of C_j and A.

4.6 cg-method as two-step method.
Show that the iteration (4.70) with parameters

$$(\eta_k, \xi_k) = \operatorname{argmin}_{(\eta,\xi)} \|u - u^{(k)} - \eta(u^{(k)} - u^{(k-1)}) - \xi P_\rho e^{(k)}\|^2$$

is identical to the cg-iteration.
Hint: Use induction in k and the minimization properties of the cg-iterates with respect to $\mathcal{K}_k$ to show that the error sequences $\{e^{(k)}\}$ of the two iterations are identical.

4.7 Analysis of (4.72).
a) Let $\eta \in (0, 1)$ and $t > 0$. Show that the quadratic equation

$$\lambda^2 - (1 + \eta - t)\lambda + \eta = 0$$

has two complex conjugate solutions of absolute value

$$|\lambda_1| = |\lambda_2| = \sqrt{\eta}$$

if and only if $(1 - \sqrt{\eta})^2 \le t \le (1 + \sqrt{\eta})^2$. If t is outside this range, then the roots are real, and

$$\sqrt{\eta} < \max(|\lambda_1|, |\lambda_2|) = \begin{cases} p + \sqrt{p^2 - \eta}, & 0 < t < (1 - \sqrt{\eta})^2, \\ -p + \sqrt{p^2 - \eta}, & t > (1 + \sqrt{\eta})^2, \end{cases}$$

where $p := (1 + \eta - t)/2$ for short.
b) Using the result from Part a), show that for fixed $\eta \in (0, 1)$ and $\xi > 0$ we have

$$r(M_{\eta,\xi}) = \sqrt{\eta}$$

if and only if

$$(1 - \sqrt{\eta})^2 \le \xi\lambda_{\min} \le \xi\lambda_{\max} \le (1 + \sqrt{\eta})^2. \tag{4.109}$$

If this condition is violated, then we have

$$\sqrt{\eta} < r(M_{\eta,\xi}) = \max(p_{\max} + \sqrt{p_{\max}^2 - \eta}, p_{\min} + \sqrt{p_{\min}^2 - \eta}),$$

where $p_{\max} = (1 + \eta - \xi\lambda_{\min})/2$, $p_{\min} = (1 + \eta - \xi\lambda_{\max})/2$.
c) Using the result from Part b), find a formula for

$$r(\eta) := \min_{\xi > 0} r(M_{\eta,\xi}), \qquad 0 < \eta < 1.$$

d) By minimizing $r(\eta)$, show that the spectral radius $r(M_{\eta,\xi})$ is minimized for the parameters in (4.73).
Hints. For Part b), note that, according to (4.72), the spectrum of $M_{\eta,\xi}$ consists of all λ that solve the above quadratic equation with some $t = \xi\lambda'$, where $\lambda' \in \sigma(P_\rho)$ is arbitrary. For Part c), first consider the case

$$\kappa = \frac{\lambda_{\max}}{\lambda_{\min}} \le \frac{(1 + \sqrt{\eta})^2}{(1 - \sqrt{\eta})^2},$$

or, equivalently,

$$\sqrt{\eta} \ge \frac{\sqrt{\kappa} - 1}{\sqrt{\kappa} + 1} = 1 - \frac{2}{\sqrt{\kappa} + 1},$$

and check that ξ can be chosen such that (4.109) holds, and consequently $r(\eta) = \sqrt{\eta}$. For larger κ (or smaller η), the minimum of $r(M_{\eta,\xi})$ is attained when $p_{\min} = p_{\max}$. This gives a formula for $r(\eta)$ that involves η and κ.

4.8 Spectral radius of the error propagation operator for the stationary Nesterov-type iteration of Theorem 4.12.
For the error propagation operator $M'_{\eta,\zeta,0}$ associated to the iteration (4.75) with constant parameters $\eta_m = \eta = (\sqrt{\kappa}-1)/(\sqrt{\kappa}+1)$, $\zeta_m = \zeta = \lambda_{\max}^{-1}$ and $\xi_m = 0$, show that

$$r(M'_{\eta,\zeta,0}) = 1 - \frac{1}{\sqrt{\kappa}}.$$

4.9 Extension of Theorem 4.14.
Consider the modification of the iteration (4.100), where the search direction $d^{(m)}_{i'_m,\nu}$ is replaced by any random direction

$$d_\nu^{(m)} = \sum_{i\in I'_m} \rho_i P_i e_\nu^{(m)}$$

such that the analog of (4.86) holds with known $0 < r'_m \leq 1$. State and prove an analog of Theorem 4.14 by determining the appropriate parameters α_m, β_m, and τ_m. *Hint*: (4.86) replaces (4.99) and enters the proof of Theorem 4.14 in only two places, namely for the estimation of $\mathbb{E}'(|||d_\nu^{(m)}|||_\rho^2)$ and the simplification of $\mathbb{E}'(a(P_\rho^{-1}d_\nu^{(m)}, \beta_m e_w^{(m)} + (1-\beta_m)e_\nu^{(m)}))$.

Chapter 5
Special topics and extensions

Abstract We discuss the application of the theory developed in Chapter 4 to linear systems, and establish links to block-coordinate descent methods in convex optimization and stochastic approximation algorithms. Finally, lower bounds and the sensitivity of Schwarz iterations to errors are briefly touched upon.

5.1 Linear systems with symmetric positive definite matrices

Although iterative methods for solving systems of linear algebraic equations have a long history, they became an important branch of numerical linear algebra only with the availability of computers and the need to deal with large-scale discretizations of partial differential and integral equations. There are many excellent textbooks and monographs on the classical algorithms, e.g., [5, 80, 95, 153, 181], to name a few.

In this section, we use the space splittings from Example 1 in Subsection 2.1.2 and formulate corollaries to the general theorems on Schwarz iterative methods applicable to linear systems $Ax = b$ with spd matrix A. Compared to the standard textbook literature on SOR-type methods, the more recent results on randomized and greedy orderings are also covered.

In the notation of the space splitting of Example 1 in Subsection 2.1.2, the method (4.1) for solving a linear system $Ax = b$ with $n\times n$ spd matrix A for its unique solution $x \in \mathbb{R}^n$ is as follows: With starting vector $x^{(0)} \in \mathbb{R}^n$, for $m = 0, 1, \dots,$ compute

$$x^{(m+1)} = x^{(m)} + \omega_m \sum_{i \in I_m} \rho_i (b - Ax)_i e_i, \qquad (5.1)$$

until a stopping criterion is met. At each step, only the coordinates in I_m are updated. Obviously, for different choices of index subsets $I_m \subset \Omega = \{1, \dots, n\}$, step-size parameters $\omega_m > 0$, and scaling parameters ρ_i, this includes classical methods such as Jacobi-Richardson, Gauss-Seidel, and SOR methods. For example, if $\rho_i = a_{ii}^{-1}$ and $\omega_m = 1$, then (5.1) with $I_m = \Omega$ becomes the classical Jacobi update, while

M. Griebel, P. Oswald, *Hilbert Space Splittings and Iterative Methods*, Springer Series in Computational Mathematics 62, https://doi.org/10.1007/978-3-031-74370-2_5

for singleton $I_m = \{i_m\}$ with cyclic ordering we get the Gauss-Seidel method. If constant step-size parameters $\omega_m = \omega \neq 1$ are allowed, then we have the so-called weighted or damped Jacobi and SOR methods, respectively. If the scaling parameters are set to $\rho_i = 1$ in the case $I_m = \Omega$, we have the Richardson method, and so on.

Greedy versions of the Gauss-Seidel method had already been proposed by Gauss and Seidel [164], who used it to solve small least-squares problems by hand. These approaches were popularized for engineering applications by Southwell [167] in the 1940s. For a long time, randomization was not systematically considered as an option to improve the performance of classical iterative schemes. Interest was sparked with the paper [174], which is devoted to the analysis of a randomized Kaczmarz method for overdetermined linear systems, see also [114]. Research on randomized matrix computations has intensified over the last two decades, we refer to the surveys [117, 119, 193].

Furthermore, if $1 < |I_m| < n$, then (5.1) is a special instance of a *block-iterative method* for solving $Ax = b$, which in its simplest form can be written as

$$x^{(m+1)} = x^{(m)} + \omega_m R_{I_m} C_{I_m} (b - Ax)_{I_m}, \qquad m = 0, 1, \ldots .$$

Here x_I denotes the restriction of the vector x with respect to the index subset $I \subset \Omega$, while R_I denotes the associated extension-by-zero operator. The choice of the $|I| \times |I|$ matrix C_I is up to us, classically one chooses $C_I = A_{I,I}^{-1}$ as the inverse of the block-diagonal submatrix of A associated with I. In the update formula (5.1), it is just a diagonal scaling matrix associated with the index subset I_m.

From now on we set $\rho_i = 1$ and drop the subscript ρ to simplify the notation where appropriate. Then we have $P = \tilde{P} = A$ for the Schwarz and Gramian operators, respectively, and their spectral properties are inherited from A. The reader can easily recover the results for the case of general weights ρ by replacing A with $I_\rho A$. In places where the choice of ρ is important, we will comment on it. Recall that

$$\|x\| := (x^T Ax)^{1/2} = (Ax, x)_2^{1/2} = \|A^{1/2}x\|_2$$

is the energy norm induced by A, while $\|x\|_p$, $1 \leq p \leq \infty$, denotes the p-norm for $x \in \mathbb{R}^n$. Furthermore, $A = L + D + L^T$ denotes the decomposition of A into strictly lower triangular, diagonal, and strictly upper triangular parts, respectively.

Deterministic Jacobi-Richardson, Gauss-Seidel and SOR methods (Theorems 4.1, 4.3 and Corollary 4.1): If $I_m = \Omega = \{1, \ldots, n\}$ and $\omega_m = \omega$, then (5.1) has the form

$$x^{(m+1)} = x^{(m)} + \omega I_\rho^{-1}(b - Ax), \qquad m = 0, 1, \ldots .$$

Setting $\rho_i = 1$ makes it the stationary Richardson iteration, while $\rho_i = a_{ii}^{-1}$ makes it the weighted Jacobi method. According to Theorem 4.1, the latter converges for $0 < \omega < 2/\lambda_{\max}(D^{-1}A)$ and arbitrary $x^{(0)}$ with the error bound

$$\|e^{(m)}\| \leq (\max(|1 - \omega\lambda_{\max}(D^{-1}A)|, |1 - \omega\lambda_{\min}(D^{-1}A)|))^m \|e^{(0)}\|,$$

$m = 1,2,\ldots$, where $e^{(m)} = x - x^{(m)}$. This bound is minimized for $\omega = 2/(\lambda_{\min}(D^{-1}A) + \lambda_{\max}(D^{-1}A))$, and is then equal to $1 - 2/(1 + \kappa(D^{-1}A))$. For the Richardson method, the same estimate holds with $D^{-1}A$ replaced by A.

Similarly, for the singleton $I_m = \{i_m\}$ obtained by cyclic ordering, the iteration (5.1) with constant $\omega_m = \omega$ becomes the execution of the loops

$$x^{(kn+i)} = x^{(kn+i-1)} + \omega\rho_i(b - Ax)_i e_i, \qquad i = 1,2,\ldots,n, \tag{5.2}$$

of n single-coordinate updates for $k = 0,1,\ldots$. Thus, in each step of the k-th loop, only one coordinate is overwritten. For $\rho_i = a_{ii}^{-1}$, this is the classical SOR method, of which the Gauss-Seidel method is a special case ($\omega = 1$). As a consequence of Corollary 4.1 b), the iteration (5.2) converges for $0 < \omega < 2$. Furthermore, for $\omega = ((1 + \lceil\log_2(n)\rceil/2)\lambda_{\max}(D^{-1}A))^{-1}$ it possesses the error bound

$$\|e^{(kn)}\|^2 \le \left(\frac{1}{1 + \lceil\log_2(n)\rceil)\kappa(D^{-1}A)}\right)^k \|e^{(0)}\|^2, \qquad k = 1,2,\ldots.$$

Analogous estimates hold for the counterpart of the symmetric multiplicative Schwarz iteration (4.9). The appearance of the logarithmic factor in front of the condition number is inevitable, see Subsection 5.4.2 below. In some situations, Part a) of Corollary 4.1 can be used to remove it from the estimate, see Problem 5.1.

Random orderings (Theorems 4.5-4.8): Let us first compute the constant μ_R which, according to Theorem 4.5, governs the convergence speed of the following iteration: Starting with $x^{(0)}$, for $m = 0,1,\ldots$, randomly and independently choose the index i_m according to the general probability distribution (4.33), set $x^{(m+1)} = x^{(m)}$, and overwrite the i_m-th coordinate with

$$x_{i_m}^{(m+1)} = x_{i_m}^{(m)} + \frac{b - (Ax^{(m)})_{i_m}}{a_{i_m,i_m}}. \tag{5.3}$$

This update formula corresponds to the steepest descent Schwarz iteration (4.7), which coincides with (4.40), since $\|R_i\|_{V_i\to V}^2 = a_{ii}$. Now let $D_{\mathbf{p}} = I_{\mathbf{p}}^{-1}D$, where $I_{\mathbf{p}}$ stands for the diagonal matrix with the probabilities p_i, $i = 1,\ldots,n$, as diagonal entries. By definition of μ_R in (4.35), we have

$$\begin{aligned}\mu_R &= \inf_{\|x\|=1} \sum_{i=1}^n p_i \frac{(Ax)_i^2}{a_{ii}} = \inf_{0\neq x\in\mathbb{R}^n} \frac{\|D_{\mathbf{p}}^{-1/2}Ax\|_2^2}{\|A^{1/2}x\|_2^2} = \inf_{0\neq y\in\mathbb{R}^n} \frac{\|D_{\mathbf{p}}^{-1/2}A^{1/2}y\|_2^2}{\|y\|_2^2} \\ &= \|A^{-1/2}D_{\mathbf{p}}^{1/2}\|_{2\to2}^{-2} = \lambda_{\min}(D_{\mathbf{p}}^{-1/2}A^{1/2})^2 = \lambda_{\min}(I_{\mathbf{p}}D^{-1}A),\end{aligned}$$

where the (p,q) norm $\|B\|_{p\to q}$ of a matrix B denotes the norm of the operator from $\ell^p(\Omega)$ to $\ell^q(\Omega)$ induced by B. In particular, $\|\cdot\|_{2\to2}$ is the spectral matrix norm induced by the $\ell^2(\Omega)$ norm in $\mathbb{R}^n$. Thus, as long as the iteration error is measured in the energy norm, the choice of the probability distribution in (4.33) affects the error bound via the row scaling in A. Consider the special case of Theorem 4.6, where the probability distribution (4.38) is used. Because of $\|R_i\|_{V_i\to V}^2 = a_{ii}$, we have

$$p_i = \frac{a_{ii}}{\sum_{j=1}^n a_{jj}} = \frac{a_{ii}}{\mathrm{Tr}(A)} \quad \Leftrightarrow \quad I_{\mathrm{p}} = \mathrm{Tr}(A)^{-1}D,$$

and get

$$\mu_R = \mu_R' = \frac{\lambda_{\min}(A)}{\mathrm{Tr}(A)} \geq \frac{1}{n\kappa(A)}.$$

This is essentially the result of [174].

For the non-singleton case, Theorem 4.7 gives the estimate

$$\mathbb{E}(\|e^{(m)}\|^2) \leq \prod_{k=0}^{m-1}(1 - \frac{k_m}{n\kappa(A)})\|e^{(0)}\|^2, \quad m = 1, 2, \ldots,$$

for the iteration (5.1) with $\rho_i = 1$ and $\omega_m = 1/\lambda_{\max}(A)$, if $I_m \subset \Omega$ is an independently and uniformly at random chosen index set of size k_m. The update step has the form

$$x_i^{(m+1)} = \begin{cases} x_i^{(m)} + \frac{(b-Ax^{(m)})_i}{\lambda_{\max}(A)}, & i \in I_m, \\ x_i^{(m)}, & i \notin I_m. \end{cases}$$

The reformulation of Theorem 4.8 for linear systems with spd matrix A basically says that by randomly and independently permuting rows and columns before executing the inner loop (5.2), the logarithmic dependence on n in the convergence bound for SOR-type methods with cyclic ordering can be removed. For further details, see [142]. Theorem 4 b) in [142] states a complementary result about the role permutations can play for the SOR method, which we quote without proof.

Theorem 5.1 *There is an absolute constant C_1 such that, for any spd matrix A with unit diagonal $D = I$, there exists a permutation σ of the index set $\Omega = \{1, \ldots, n\}$ for which the iteration (5.2) with $\rho_i = 1$ applied to the symmetrically permuted system*

$$A_\sigma x = b_\sigma, \qquad A_\sigma = (a_{\sigma(i),\sigma(j)})_{i,j\in\Omega}, \qquad b_\sigma = (b_{\sigma_i})_{i\in\Omega},$$

converges to the permuted solution vector x_σ for $0 < \omega < 2$. Furthermore, it has the error bound

$$\|e^{(kn)}\|^2 \leq \left(1 - \frac{\omega(2-\omega)\lambda_{\max}(A)}{(1 + C_1\omega\lambda_{\max}(A))^2\kappa(A)}\right)^k \|e^{(0)}\|^2, \qquad k = 1, 2, \ldots.$$

If one sets $\omega = 1/C_1\lambda_{\max}(A)$, then this bound is optimized and implies that a suitable one-time permutation of the rows and columns in A is sufficient to achieve the same qualitative improvement as guaranteed by Theorem 4.8 in the bound for SOR-type methods. The constant C_1 corresponds to the best constant in the inequality

$$\min_\sigma \|L_\sigma\|_{2\to 2} \leq C_1\|A\|_{2\to 2},$$

where L_σ denotes the strictly lower triangular part of A_σ. The proof that C_1 is uniformly bounded with respect to n is a corollary to the recently confirmed Anderson paving conjecture, see Proposition 3.2 in Subsection 3.3.1. Unfortunately, the current estimates of C_1 are too pessimistic, and the permutation σ is determined in a non-constructive way.

Greedy orderings (Theorems 4.9 and 4.10): First, in the special situation of the space splitting in Example 1, we have the identity

$$\Delta_i^{(m)} = \bar{\Delta}_i^{(m)} = \frac{(Ae^{(m)})_i^2}{a_{ii}}$$

for the residual error indicators introduced in Section 4.4, since $\|R_i\|^2_{V_i\to V} = a_{ii}$. The weak greedy algorithm covered in Theorem 4.9 and Theorem 4.10 is as follows: For given weakness parameters $0 < \beta_m \leq 1$ and a starting vector $x^{(0)} \in \mathbb{R}^n$, for $m = 0, 1, \ldots$, choose the index i_m such that

$$\Delta_{i_m}^{(m)} \geq \beta_m \max_{i=1,\ldots,n} \Delta_i^{(m)},$$

set $x^{(m+1)} = x^{(m)}$, and overwrite the i_m-th coordinate of $x^{(m+1)}$ according to (5.3) as in the randomized case.

Next, we give a formula for the quantity μ_G that governs the error decay of this method. Recall that D is the diagonal of A. Since

$$\max_{i=1,\ldots,n} \Delta_i^{(m)} = \max_i \frac{(Ae^{(m)})_i^2}{a_{ii}} = \|D^{-1/2}Ae^{(m)}\|_\infty^2 = \left(\frac{\|D^{-1/2}Ae^{(m)}\|_\infty}{\|A^{1/2}e^{(m)}\|_2}\right)^2 \|e^{(m)}\|^2,$$

by the definition of μ_G in (4.59) we get the identity

$$\mu_G = \inf_y \frac{\|D^{-1/2}Ay\|_\infty^2}{\|A^{1/2}y\|_2^2} = \inf_z \frac{\|z\|_\infty^2}{\|A^{-1/2}D^{1/2}z\|_2^2} = (\|A^{-1/2}D^{1/2}\|_{\infty\to2})^{-2}.$$

So the error of the weak greedy iteration satisfies the recursion

$$\|e^{(m+1)}\|^2 \leq \|e^{(m)}\|^2 - \beta_m \max_{i=1,\ldots,n} \Delta_i^{(m)} \leq \left(1 - \frac{\beta_m}{\|A^{-1/2}D^{1/2}\|_{\infty\to2}^2}\right)\|e^{(m)}\|^2.$$

Formally, this estimate compares favorably with the bounds obtained for deterministic Jacobi and Gauss-Seidel-SOR iterations and for randomized methods. In fact, for any probability distribution $\mathbf{p}$ we have

$$\begin{aligned}\mu_G^{-1} &= \|A^{-1/2}D^{1/2}\|_{\infty\to2}^2 \leq \|A^{-1/2}D^{1/2}I_{\mathbf{p}}^{-1/2}\|_{2\to2}^2\|I_{\mathbf{p}}^{1/2}\|_{\infty\to2}^2 \\ &= \lambda_{\min}(I_{\mathbf{p}}D^{-1}A)^{-1}\sum_{i=1}^n p_i = \mu_R^{-1} \leq n\kappa(D^{-1}A).\end{aligned} \tag{5.4}$$

This implies

$$(1-\mu_G)^n \le \inf_{\mathbf{p}}(1-\mu_R)^n \le \sup_{\mathbf{p}}(1-\mu_R)^n \le (1-\frac{1}{n\kappa(D^{-1}A)})^n \approx 1-\frac{1}{\kappa(D^{-1}A)},$$

and we see that the convergence bounds for the greedy iteration with $\beta_m = 1$ are at least as good as the convergence bounds for Jacobi-Richardson and SOR-type methods and their randomized versions. Recall that n steps of the weak greedy iteration should be compared to one step of the Jacobi-Richardson and one sweep through the inner loop of an SOR-type method.

In practice, the observed convergence speed for greedy iterations is often superior to their deterministic and randomized counterparts, even for well-conditioned $D^{-1}A$. Possible explanations for this could be its independence from $\mathbf{p}$ or significant differences in the size of the various estimates for $\|A^{-1/2}D_A^{1/2}\|_{\infty\to 2}$ in (5.4). On the one hand, the quotient μ_G/μ_R does not have a finite upper bound that holds independently of A and $\mathbf{p}$. To see this, consider diagonal matrices A. On the other hand, it follows from the little Grothendieck inequality (see [146, Theorem 5.2, (i)] for an equivalent formulation) that for any non-singular $n \times n$ matrix B there exists a diagonal matrix Λ with diagonal elements $\lambda_i > 0$ and $\sum_{i=1}^n \lambda_i = 1$ such that

$$\|Bx\|_2^2 \le \frac{\pi}{2}\|B\|_{\infty\to 2}^2 \sum_{i=1}^n \lambda_i x_i^2, \qquad x \in \mathbb{R}^n, \tag{5.5}$$

or, in terms of matrix factorization,

$$\|B\Lambda^{-1/2}\|_{2\to 2}^2 \le \frac{\pi}{2}\|B\|_{\infty\to 2}^2.$$

The existence of this so-called Pietsch factorization is a non-constructive result based on a probabilistic argument. The inequality (5.5), applied to $B = A^{-1/2}D^{1/2}$ with $\mathbf{p}$ defined by $p_i = \lambda_i$, $i = 1, \ldots, n$, or, equivalently, $I_{\mathbf{p}} = \Lambda$, implies that

$$\frac{2}{\pi}\mu_G \le \mu_R (\le \mu_G). \tag{5.6}$$

In fact, setting $x = I_{\mathbf{p}}^{-1/2}y$ yields (5.6), if one takes $\mu_G = \|B\|_{\infty\to 2}^{-2}$ and $\mu_R = \|BI_{\mathbf{p}}^{-1/2}\|_{2\to 2}^{-2}$ into account.

In other words, for any A there exists a probability distribution $\mathbf{p}$ such that our bounds for greedy and randomized SOR iterations are close in the sense that the gap $\mu_G/\mu_R \le \pi/2$ is bounded by a small constant, independently of n. Unfortunately, this $\mathbf{p}$ is not available in practice. It can also be shown that the gap between μ_R and μ_G is bounded by $\kappa(D^{-1}A)$ if the uniform distribution is taken as $\mathbf{p}$, see Problem 5.2. Finally, the toy example of 2×2 matrices A considered in Problem 5.3 shows that the constants μ_R and μ_G in our estimates are far from explaining the actual behavior of the underlying iteration. The reason is that we relied solely on estimates of $\|e^{(m+1)}\|^2$ for *arbitrarily given* $e^{(m)}$ and ignored the iteration history. This is a weakness of the approach in this and the previous chapter, especially for nonlinear iterations such as Schwarz iterations with greedy and randomized orderings.

Acceleration (Theorems 4.13 and 4.14). For linear systems $Ax = b$ with spd matrix A, acceleration is typically achieved by the cg-method. Alternatively, one can use the randomized Nesterov-type iteration (4.75). In the setting of the space splitting of Example 1 and with $\rho_i = 1$, (4.75) takes the form

$$\begin{aligned} x^{(m+1)} &= y^{(m)} + \zeta_m d_y^{(m)}, \\ y^{(m+1)} &= x^{(m+1)} + \eta_m (x^{(m+1)} - x^{(m)}) + \xi_m d_y^{(m)}, \end{aligned} \qquad d_y^{(m)} := \sum_{i \in I_m} (b - (Ay^{(m)})_i, \tag{5.7}$$

$m = 0, 1, \ldots,$ where $x^{(0)} = y^{(0)}$ at the start. Theorem 4.13 guarantees an error decay in expectation as

$$\mathbb{E}(\|e_x^{(m)})\|^2) \leq 2 \prod_{k=0}^{m-1} (1 - \frac{k_m}{\sqrt{n\kappa(A)}}) \|e_x^{(0)})\|^2,$$

if the index sets I_m are independently and uniformly at random chosen subsets of $\Omega = \{1, \ldots, n\}$ of size k_m, and if the parameters in (5.7) are given by (4.90), where $r_m = k_m/n$. These parameters depend on k_{m+1} and the spectral properties of A. In practice, exact values for $\kappa(A)$, $\lambda_{\max}(A)$, and k_{m+1} may not be available. However, a similar estimate holds if these constants are replaced in the above formulas by their respective upper and lower bounds.

As for the application of Theorem 4.14, we first note that

$$\nu_G = \nu'_G = \nu'_R = \left(\sum_{i=1}^{n} a_{ii} \right)^{-1} = \mathrm{Tr}(A)^{-1} \geq \frac{1}{n\lambda_{\max}(A)}.$$

The first two equality signs follow from the arguments at the end of Subsection 4.5.3, since $\|R_i\|^2_{V_i \to V} = a_{ii}$ and the condition $a(P_i x, x) = a_{ii}$ is satisfied for the vector $x = \mathbf{1}^T$. In particular, the two versions of the weak greedy algorithm covered by Part b) and c) of Theorem 4.14 are identical. For $\beta_m = 1$ this leads to the same estimate for $\mathbb{E}(\|e_x^{(m)}\|^2)$ as for the fully random algorithm specified in Part a). We only formulate the result for the latter algorithm, which according to (4.100) is as follows: Starting with $x^{(0)} = y^{(0)} = z^{(0)} \in \mathbb{R}^n$, for $m = 0, 1, \ldots,$ choose i_m independently and randomly according to the probability distribution

$$\mathbb{P}(i_m = i) = \frac{a_{ii}}{\sum_{j=1}^{n} a_{jj}} = \frac{a_{ii}}{\mathrm{Tr}(A)}, \qquad i = 1, \ldots, n,$$

choose i'_m independently and uniformly at random, and update

$$\begin{aligned} x^{(m+1)} &= y^{(m)} + \frac{(b - Ay)_{i_m}}{a_{i_m, i_m}}, \\ z^{(m+1)} &= \delta z^{(m)} + (1 - \delta) y^{(m)} + \tau (b - Ay)_{i'_m}, \\ y^{(m+1)} &= \alpha x^{(m+1)} + (1 - \alpha) z^{(m+1)}. \end{aligned} \tag{5.8}$$

The parameters are given by

$$\delta = 1 - \sqrt{\tilde{\mu}'_R}, \qquad \tau = \frac{1}{\mathrm{Tr}(A)\sqrt{\tilde{\mu}'_R}}, \qquad \alpha = \frac{1}{\sqrt{1+\tilde{\mu}'_R}}, \tag{5.9}$$

where

$$\tilde{\mu}'_R = \frac{\lambda_{\min}(A)}{n\,\mathrm{Tr}(A)}.$$

This algorithm possesses the error estimate

$$\mathbb{E}(\|e_x^{(m)}\|^2) \le 2\left(1 - \sqrt{\frac{\lambda_{\min}(A)}{n\,\mathrm{Tr}(A)}}\right)^m \|e_x^{(0)}\|^2, \qquad m = 1, 2, \ldots.$$

Similar bounds hold for the weak greedy version of (5.8), where i_m is chosen so that

$$\frac{(b_{i_m} - (Ay)_{i_m})^2}{a_{i_m i_m}} \ge \beta_m \max_{i=1,\ldots,n} \frac{(b_i - (Ax)_i)^2}{a_{ii}}$$

with weakness parameters $\beta_m \in (0, 1]$.

To run the algorithm in practice, the dependence of the parameters in (5.7) on $\lambda_{\min}(A)$ is inconvenient. However, it is sufficient to replace it by a lower bound $0 < \underline{\lambda} \le \lambda_{\min}(A)$, in which case the convergence estimate still holds with $\lambda_{\min}(A)$ replaced by $\underline{\lambda}$.

5.2 Kaczmarz methods and least-squares problems

There are several slightly different ways to treat general linear systems $Ax = b$ with $n' \times n$ matrix A by least-squares methods within the space splitting approach. Recall that the least-squares solution x^{LS} of such a system is defined as the minimizer of the quadratic functional $J(x) := \|Ax - b\|_2^2$ of least Euclidean norm, or equivalently, the minimizer belonging to $\mathrm{ran}(A^T)$. One distinguishes between the consistent case $b \in \mathrm{ran}(A)$ and the inconsistent case $b \notin \mathrm{ran}(A)$.

In the consistent case, the space splitting of Example 2 in Subsection 2.1.2 is the appropriate framework to cover the convergence theory of the block-versions of the Kazcmarz method. We again set $\rho_i = 1$ to simplify the formulas, keeping in mind that an additional diagonal scaling matrix I_ρ appears in the general case. The update formula (4.1) is then

$$x^{(m+1)} = x^{(m)} + \omega_m \sum_{i \in I_m} (b_i - A_i x^{(m)}) A_i^T, \qquad m = 0, 1, \ldots, \tag{5.10}$$

where A_i denotes the row vectors of A, $i = 1, \ldots, n'$. Recall that in this example the index set of the splitting is $\Omega' := \{1, \ldots, n'\}$, and that now $I_m \subset \Omega'$. This is because the underlying linear equation $x = x^{LS}$ is considered on $V := \mathrm{ran}(A^T)$, with the bilinear form on V given by $a(x,x') = (x')^T x$. The operators $P_i x = (A_i x)A_i^T, x \in V$, induced by the space splitting of Example 2, are associated with the row vectors A_i, $i \in \Omega'$. Thus, if $x^{(0)} \in V$, then all iterates $x^{(m)}$ according to (5.10) belong to V, and the convergence theory for Schwarz methods developed in Chapter 4 can be used to prove error estimates for $\|e_x^{(m)}\| = \|e_x^{(m)}\|_2$, where $e_x^{(m)} := x^{LS} - x^{(m)}$. Remember that the Schwarz and Gramian operators are given by

$$Px = \sum_{i=1}^{n'} P_i x = A^T A x, \quad \tilde{P} y = AA^T y, \qquad x \in V, \quad y \in \tilde{V} = \mathbb{R}^{n'} = \ell^2(\Omega').$$

If $x^{(0)}$ is not contained in $V = \mathrm{ran}(A^T)$, then obviously $x^{(m)} \longrightarrow x^{LS} + Q_{\ker(A)} x^{(0)}$, where $Q_{\ker(A)}$ denotes the ortho-projection onto the subspace $\ker(A)$ in $\ell^2(\Omega)$.

Alternatively, one can consider the system

$$AA^T y = b, \qquad y \in V := \mathbb{R}^{n'},$$

with the bilinear form $a(y,y') := (y')^T AA^T y = (A^T y, A^T y')_2$. As a rule, this form is only positive semi-definite and $\|y\| = \|A^T y\|_2$ is thus only a semi-norm. Formally, the developed convergence theory does not apply if we use the setting of Example 1 of Subsection 2.1.2. Note, however, that (4.1) would have the form

$$y^{(m+1)} = y^{(m)} + \omega_m \sum_{i \in I_m} \rho_i (b - AA^T y^{(m)})_i e_i, \qquad m = 0, 1, \ldots,$$

where e_i are the unit coordinate vectors in $\mathbb{R}^{n'}$. If we now apply A^T to this update formula and set

$$x^{(m)} = A^T y^{(m)},$$

we get the Kaczmarz iteration (5.10). Thus, to obtain convergence results, one would have to extend the theory of the previous section to the semi-definite case. This is not difficult (see Problem 5.4) and we leave it to the reader to work out the details.

In the inconsistent case $b \notin \mathrm{ran}(A)$, Kaczmarz methods (if they converge at all) do not recover x^{LS}. Then one possibility is to consider the extended system

$$Ax + y = b, \quad A^T y = 0, \qquad x \in V_x := \mathrm{ran}(A^T) \subset \mathbb{R}^n, \quad y \in V_y := \mathbb{R}^{n'}. \tag{5.11}$$

The unique solution of (5.11) is $x = x^{LS}$ and $y = Q_{\ker(A^T)} b$. The extended Kaczmarz algorithm introduced in [148] starts with arbitrary $x^{(0)} \in V_x$ and $y^{(0)} = b$. At the m-th iteration step, it first updates $y^{(m)}$ to $y^{(m+1)}$ by a Kaczmarz iteration step using the second equation in (5.11), and then updates $x^{(m)}$ to $x^{(m+1)}$ by a Kaczmarz iteration step applied to the equation $Ax = b - y^{(m+1)}$. Random versions of this extended Kaczmarz algorithm are described in [204], convergence results for deterministic and greedy versions are given in [144]. Of course, instead of alternating between the two equations, one could just as well apply the Kaczmarz method directly to the extended system (5.11), i.e., one could iterate in $V := V_x \oplus V_y \subset \mathbb{R}^{n+n'}$ using the setup from Example 2.

Next, we list some consequences of the general convergence results of Chapter 4 for Kaczmarz iterations, focusing on random and greedy versions. Because of the obvious analogy to the results for linear systems with spd matrix A, we keep the presentation brief.

Randomized Kaczmarz iterations (consistent case, Theorems 4.5-4.8): Using the setup of Example 2 in Subsection 2.1.2, we first compute the constant μ_R, which, according to Theorem 4.5, governs the convergence bound for the steepest descent version

$$x^{(m+1)} = x^{(m)} + \sum_{i \in I_m} \frac{(b_{i_m} - A_{i_m} x^{(m)}) A_{i_m}^T}{\|A_{i_m}\|_2}, \qquad m = 0, 1, \ldots . \tag{5.12}$$

of (5.10) with singleton $I_m = \{i_m\}$. This is the counterpart of (4.7), where the indices i_m are chosen independently at random with respect to a general probability distribution (4.33), this time on the index set Ω'. With the formula for $P_i x$ we have

$$\begin{aligned}
\mu_R &= \inf_{0 \neq x \in V} \sum_{i=1}^{n'} p_i \frac{a(P_i x, x)^2}{\|P_i x\|^2 \|x\|_2^2} = \inf_{0 \neq x \in V} \sum_{i=1}^{n'} p_i \frac{|A_i x|^4}{|A_i x|^2 \|A_i\|_2^2 \|x\|_2^2} \\
&= \inf_{0 \neq x \in \mathrm{ran}(A^T)} \frac{\|I_{\mathbf{p}}^{1/2} D^{-1/2} A x\|_2^2}{\|x\|_2^2} = \inf_{0 \neq y \in \mathrm{ran}(A)} \frac{\|y\|_2^2}{\|I_{\mathbf{p}}^{-1/2} D^{1/2} A^+ y\|_2^2} \\
&= \frac{1}{\|A^+ D^{1/2} I_{\mathbf{p}}^{-1/2}\|_{2 \to 2}^2} = \lambda_{\min}^+ (A^T D^{-1} I_{\mathbf{p}} A),
\end{aligned}$$

where A^+ denotes the pseudo-inverse of A, D is now the diagonal of AA^T, and $\lambda_{\min}^+(B)$ is the smallest *non-zero* eigenvalue of a matrix B.

In particular, if the uniform probability distribution $p_i = 1/n'$, $i = 1, \ldots, n'$, is chosen, then (5.12) possesses the bound

$$\mathbb{E}(\|e_x^{(m)}\|_2^2) \le (1 - \mu_R)^m \|e_x^{(0)}\|_2^2, \qquad m = 1, 2, \ldots,$$

where

$$\mu_R = \frac{\lambda_{\min}^+(A^T D^{-1} A)}{n'} = \frac{\sigma_{\min}^+(D^{-1/2} A)^2}{n'}$$

and $\sigma_{\min}^+(B)$ denotes the smallest *non-zero* singular value of a matrix B.

Thus, μ_R depends on the number n' of rows in A and the smallest non-zero singular value of A after scaling its rows A_i to unit Euclidean norm. If the counterpart of the probability distribution (4.38) is considered, that is, if

$$p_i = \frac{\|A_i\|_2^2}{\sum_{j=1}^{n'} \|A_j\|_2^2} = \frac{\|A_i\|_2^2}{\|A\|_F^2} = \frac{\|A_i\|_2^2}{\mathrm{Tr}(A^T A)}, \tag{5.13}$$

then the same result holds with

$$\mu_R = \mu_R' = \frac{\lambda^+_{\min}(A^T A)}{\mathrm{Tr}(A^T A)},$$

compare Theorem 4.6.

Now, if the index sets $I_m \subset \{1, \ldots, n'\}$ satisfy the conditions of Theorem 4.7, i.e., if they are independently and uniformly at random chosen subsets of size $1 \le k_m \le n'$, then the iteration

$$x^{(m+1)} = x^{(m)} + \frac{1}{\lambda_{\max}(A^T A)} \sum_{i \in I_m} (b_i - A_i x^{(m)}) A_i^T, \qquad m = 0, 1, \ldots, \tag{5.14}$$

starting with $x^{(0)} \in \mathrm{ran}(A^T)$, possesses the error bound

$$\mathbb{E}(\|e_x^{(m)}\|_2^2) \le \prod_{k=0}^{m-1} (1 - \frac{k_m}{n' \kappa^+(A^T I_\rho A)}) \|e_x^{(0)}\|_2^2, \qquad m = 1, 2, \ldots,$$

where $\kappa^+(B) := \lambda_{\max}(B)/\lambda^+_{\min}(B)$.

Finally, Theorem 4.8 asserts that randomly and independently permuting the rows of A after each round of n' singleton Kaczmarz steps executed in cyclic order improves the guaranteed convergence bound compared to the deterministic Kaczmarz iteration with cyclic ordering. For further discussion, see [141].

Greedy Kaczmarz iterations (consistent case, Theorems 4.9 and 4.10): The residual error indicators have the form

$$\Delta_i^{(m)} = \bar{\Delta}_i^{(m)} = \frac{|b_i - A_i x^{(m)}|^2}{\|A_i\|_2^2}, \qquad \tilde{\Delta}_i^{(m)} = |b_i - A_i x^{(m)}|^2.$$

The constant μ_G defined in (4.59) is equal to

$$\mu_G = \inf_{x \in V} \max_{i=1,\ldots,n'} \frac{|A_i x|^2}{\|A_i\|_2^2 \|x\|_2^2} = \inf_{x \in \mathrm{ran}(A^T)} \frac{\|D^{-1/2} A x\|_\infty^2}{\|x\|_2^2} = \frac{1}{\|A^+ D^{1/2}\|_{\infty \to 2}^2}.$$

Consequently, Theorem 4.9 implies that the Kaczmarz iteration (5.12) with $x^0 \in \mathrm{ran}(A^T)$ and i_m satisfying the weak greedy rule

$$\Delta_{i_m}^{(m)} \ge \beta_m \max_{i=1,\ldots,n'} \Delta_i^{(m)}$$

with weakness parameters $\beta_m \in (0, 1]$ converges with the bound

$$\|e_x^{(m)}\|_2^2 \le \prod_{k=0}^{m-1} (1 - \beta_k \mu_G) \|e_x^{(0)}\|_2^2, \qquad m = 1, 2, \ldots.$$

We do not give the corresponding corollary to Part a) of Theorem 4.10 for the weak greedy Kaczmarz iteration based on the residual error indicators $\bar{\Delta}_i^{(m)}$, because the iteration itself is the same but the computation of $\mu_G' \le \mu_G$ leads to a less explicit expression. Regarding Part b) of Theorem 4.10 and the statements in Subsection 4.4.2, we only give the following result: If I_m is chosen according to the bulk chasing rule (4.66) with the residual error indicators $\tilde{\Delta}_i^{(m)}$, then (5.14) converges with the bound

$$\|e_x^{(m)}\|_2^2 \le \prod_{k=0}^{m-1}(1 - \frac{\beta_k}{\kappa^+(A^TA)})\|e_x^{(0)}\|_2^2, \qquad m = 1,2,\ldots.$$

Acceleration (consistent case, Theorem 4.13 and 4.14): With the space splitting of Example 2 in Subsection 2.1.2 in mind, the accelerated method covered by Theorem 4.13 takes the form

$$x^{(m+1)} = y^{(m)} + \lambda_{\max}^{-1} d_y^{(m)}, \qquad d_y^{(m)} := \sum_{i\in I_m}(b_i - A_i y^{(m)})A_i^T,$$
$$y^{(m+1)} = x^{(m)} + \eta_m(x^{(m+1)} - x^{(m)}) + \xi_m d_y^{(m)},$$

$m = 0,1,\ldots$, with $x^{(0)} = y^{(0)} \in \mathrm{ran}(A^T)$ at the start. Here the parameters are $\lambda_{\max} = \lambda_{\max}(A^TA)$,

$$\eta_m = \frac{\sqrt{\kappa^+} - r_m}{\sqrt{\kappa^+} - r_{m+1}}, \quad \xi_m = -\frac{(1-r_m)\sqrt{\kappa^+} - r_m + r_{m+1}}{\sqrt{\kappa^+} + r_{m+1}}, \quad \kappa^+ = \kappa^+(A^TA),$$

and $r_m = |I_m|/n'$ denotes the relative size of the index subset $I_m \subset \Omega'$. The latter is chosen independently and uniformly at random for $m = 0,1,\ldots$. The convergence bound guaranteed by Theorem 4.13 for this iteration is

$$\mathbb{E}(\|e_x^{(m)}\|_2^2) \le 2\prod_{k=0}^{m-1}(1 - \frac{r_k}{\sqrt{\kappa^+}})\|e_x^{(0)}\|_2^2, \qquad m = 1,2,\ldots.$$

Similar statements follow from Theorem 4.14. Since the convergence bounds for both random and greedy versions are identical, we give only the corollary to Part a) of Theorem 4.14. The algorithm reads as follows: Starting with $x^{(0)} = y^{(0)} = z^{(0)} \in \mathrm{ran}(A^T)$, for $m = 0,1,\ldots$, choose the index i_m independently and randomly according to the probability distribution (5.13), choose another index i'_m independently and uniformly at random from $\Omega = \{1,\ldots,n'\}$ and perform the update

$$x^{(m+1)} = y^{(m)} + \|A_{i_m}\|_2^{-2}(b_{i_m} - A_{i_m}y^{(m)})A_{i_m}^T,$$
$$z^{(m+1)} = \delta z^{(m)} + (1-\delta)y^{(m)} + \tau(b_{i'_m} - A_{i'_m}y^{(m)})A_{i'_m}^T,$$
$$y^{(m+1)} = \alpha x^{(m+1)} + (1-\alpha)z^{(m+1)}.$$

The parameters involved are given by (5.9), where this time

$$\tilde{\mu}'_R := \frac{\lambda_{\min}^+(A^TA)}{n'\,\mathrm{Tr}(AA^T)}.$$

With these specifications we have

$$\mathbb{E}(\|e_x^{(m)}\|_2^2) \le 2(1-\sqrt{\tilde{\mu}_R'})^m \|e_x^{(0)}\|_2^2, \qquad m = 1,2,\ldots.$$

This bound compares favorably with the bound for the corresponding one-step method covered by Theorem 4.6.

Extended Kaczmarz iterations (inconsistent case): As outlined at the beginning of this section, we have two ways to develop extended Kaczmarz iterations based on the system (5.11). First, since (5.11) is a consistent system, any of the above Kaczmarz iteration schemes can be applied, at the expense of working in a space $V := \mathrm{ran}(A^T) \oplus \mathbb{R}^{n'} \subset \mathbb{R}^{n+n'}$ of larger dimension and a corresponding larger space splitting in Example 2. This possibility has not been in the focus of attention in the literature. Therefore, we will not go into details and refer to Problem 5.5.

Instead, we explore the second possibility, where the Kaczmarz idea is applied to the two systems $A^T y = 0$ and $Ax = b - y$ in some alternating fashion. Note that the consistent system $A^T y = 0$ does not involve x. Thus, according to the above considerations, the Kaczmarz iterates $y^{(m)}$ for the system $A^T y = 0$ with starting vector $y^{(0)} = b$ would converge to $y^{LS} = Q_{\ker(A^T)} b$. If this limit is substituted, then the other system $Ax = b - y^{LS} = Q_{\mathrm{ran}(A)}$ also becomes consistent, and the Kaczmarz iteration applied to it would yield x^{LS} in the limit. This is the rationale behind the success of the extended Kaczmarz methods.

We focus on the following one-step version: Starting with $x^{(0)} \in V_x = \mathrm{ran}(A^T)$ and $y^{(0)} = b$, at each step we perform an update of the form

$$\begin{aligned} x^{(m+1)} &= x^{(m)} + \omega_m \sum_{i \in I_m} (b_i - \omega y_i^{(m)} - A_i x^{(m)}) A_i^T, \\ y^{(m+1)} &= y^{(m)} - \theta_m \sum_{j \in J_m} (A^j)^T y^{(m)} A^j, \end{aligned} \tag{5.15}$$

where A_i, A^j denote the row and column vectors of A, respectively, $J_m \subset \{1,\ldots,n\}$, $I_m \in \{1,\ldots,n'\}$ are index sets, and ω_m, θ_m are step-size parameters at our disposal. In practice, it is also possible to do the y-update first, followed by the x-update with $y^{(m)}$ replaced by $y^{(m+1)}$, or repeat several y-updates before switching to x-updates.

To demonstrate the corresponding small modifications in the convergence analysis, we consider two concrete instances of (5.15) with singleton index sets $J_m = \{j_m\}$ and $I_m = \{i_m\}$. First, let j_m and i_m be chosen independently at random according to the probability distributions

$$q_j := \mathbb{P}(j_m = j) = \frac{\|A^j\|_2^2}{\mathrm{Tr}(A^T A)}, \quad j = 1,\ldots,n,$$

and

$$p_i := \mathbb{P}(i_m = i) = \frac{\|A_i\|_2^2}{\mathrm{Tr}(A A^T)}, \quad j = 1,\ldots,n',$$

respectively. Recall that $\mathrm{Tr}(A^TA) = \mathrm{Tr}(AA^T) = \|A\|_F^2$. We also use the steepest descent rule in (5.15), i.e., we set $\theta_m = \|A^{j_m}\|_2^{-2}$ and $\omega_m = \|A_{i_m}\|_2^{-2}$. With these specifications, this randomized extended Kaczmarz algorithm was considered in [204].

The application of Theorem 4.6 to the Kaczmarz iteration for the system $A^Ty = 0$, starting with $y^{(0)} = b$, yields the following estimate for the associated error sequence $e_y^{(m)} := y^{LS} - y^{(m)}$: For $m = 1, 2, \ldots$, we have

$$\mathbb{E}(\|e_y^{(m)})\|_2^2) \le (1 - \mu_R')^m \|Q_{\mathrm{ran}(A)}b\|_2^2, \qquad \mu_R' = \frac{\lambda_{\min}^+(A^TA)}{\mathrm{Tr}(A^TA)}.$$

Estimates for the error sequence $e_x^{(m)} := x^{LS} - x^{(m)}$ are more delicate, because the sequence $\{x^{(m)}\}$ is not the result of a one-step random Kaczmarz iteration for the system $Ax = b - y^{LS} = Q_{\mathrm{ran}(A)}b$ with fixed right-hand side. Instead, the x-update is a randomized Kaczmarz iteration step applied to the randomly perturbed system $Ax = b - y^{(m)} = Q_{\mathrm{ran}(A)}b + e_y^{(m)}$, where the norm of the perturbation $e_y^{(m)}$ decays exponentially according to the previous result. In [204], this difficulty was overcome by studying Kaczmarz iterations with perturbed right-hand sides.

We follow a different strategy that leads to almost the same convergence bound. Since the x-update uses the steepest descent rule, we have

$$\|e_x^{(m+1)}\|_2^2 = \|e_x^{(m)}\|_2^2 - \frac{(A_{i_m}e_x^{(m)} + (e_y^{(m)})_{i_m})^2}{\|A_{i_m}\|_2^2}.$$

Using the corresponding formula for $\|e_y^{(m+1)}\|_{\ell^2}^2$ and taking conditional expectations, we get

$$\begin{aligned}&\mathbb{E}'(\|e_x^{(m+1)}\|_2^2 + \gamma\|e_y^{(m+1)}\|_2^2)\\ &\quad = \|e_x^{(m)}\|_2^2 + \gamma\|e_y^{(m)}\|_2^2 - \|A\|_F^{-2}(\|Ae_x^{(m)} + e_y^{(m)}\|_2^2 + \gamma\|A^Te_y^{(m)}\|_2^2),\end{aligned}$$

where $\gamma > 0$ is an auxiliary parameter that will be used later to optimize the final bound for $\mathbb{E}(\|e_x^{(m)}\|_2)$. Since $e_x^{(m)} \in V_x := \mathrm{ran}(A^T)$ and $e_y^{(m)} \in V_y := \mathrm{ran}(A)$, together with the independence assumption, this yields

$$\mathbb{E}(\|e_x^{(m)}\|_2^2 + \gamma\|e_y^{(m)}\|_2^2) \le \left(1 - \frac{\nu_{R,\gamma}}{\|A\|_F^2}\right)^m \left(\|Q_{V_x}b - x^{(0)}\|_2^2 + \gamma\|Q_{V_y}b\|_2^2\right), \quad (5.16)$$

$m = 1, 2, \ldots$, where

$$\nu_{R,\gamma} := \inf_{x\in V_x,\, y\in V_y} \frac{\|Ax + y\|_2^2 + \gamma\|A^Ty\|_{\ell^2}^2}{\|x\|_2^2 + \gamma\|y\|_2^2}. \quad (5.17)$$

Using the reduced singular value decomposition of A, the constant $\nu_{R,\gamma}$ in (5.17) can be calculated explicitly as a function of γ and $\lambda_{\min}^+(A^TA) = \sigma_{\min}^+(A)^2$. The result is

$$\nu_{R,\gamma} = \lambda^+_{\min}\left(\begin{pmatrix} A^TA & \gamma^{-1}A^T \\ \gamma^{-1}A & \gamma^{-2}I + AA^T \end{pmatrix}\right) = \frac{\sigma^4}{\sigma^2 + \frac{1}{2\gamma} + \sqrt{\frac{\sigma^2}{\gamma} + \frac{1}{4\gamma^2}}}, \tag{5.18}$$

where $\sigma = \sigma^+_{\min}(A)$ for short. The elementary calculations are left to the reader, see Problem 5.6. The choice $\gamma = 2/\sigma^2 = 2/\lambda^+_{\min}(A^TA)$ results in the final estimate

$$\begin{aligned}\mathbb{E}(\|e_x^{(m)}\|_2^2) &\le \left(1 - \frac{\lambda^+_{\min}(A^TA)}{2\|A\|_F^2}\right)^m \left(\|Q_{V_x}b - x^{(0)}\|_2^2 + \frac{2}{\lambda^+_{\min}(A^TA)}\|Q_{V_y}b\|_2^2\right) \\ &\le \left(1 - \frac{\lambda^+_{\min}(A^TA)}{2\|A\|_F^2}\right)^m \left(\|x^{LS} - x^{(0)}\|_2^2 + 2\kappa^+(A^TA)\|x^{LS}\|_2^2\right),\end{aligned}$$

since $\|Q_{V_y}b\|_2^2 = \|Ax^{LS}\|_2^2 \le \lambda_{\max}(A^TA)\|x^{LS}\|_2^2$. This is essentially the estimate of [204, Theorem 8].

Similar estimates can be obtained for the greedy counterpart of this algorithm, where the indices i_m and j_m maximize the residual error estimators

$$\Delta_{i,x}^{(m)} := \frac{|b_i - A_i x^{(m)} - y_i^{(m)}|^2}{\|A_i^T\|_2^2}, \quad i = 1, \dots, n',$$

and

$$\Delta_{j,y}^{(m)} := \frac{|(A^j)^T y^{(m)}|^2}{\|A^j\|_2^2}, \quad j = 1, \dots, n,$$

respectively. Except for differences in the scaling of the residual error indicators, this greedy version of the extended Kaczmarz iteration was introduced in [148]. The error recursion formulas are now

$$\begin{aligned}\|e_x^{(m+1)}\|_2^2 &= \|e_x^{(m)}\|_2^2 - \|D^{-1/2}Ae_x^{(m)} + e_y^{(m)}\|_\infty^2, \\ \|e_y^{(m+1)}\|_2^2 &= \|e_y^{(m)}\|_2^2 - \|D^{-1/2}A^Ty\|_\infty^2.\end{aligned}$$

Arguing as in the randomized case, we get

$$\|e_x^{(m)}\|_2^2 + \gamma\|e_y^{(m)}\|_2 \le (1 - \mu_{G,\gamma})^m(\|Q_{V_x}b - x^{(0)}\|_2^2 + \gamma\|Q_{V_y}b\|_2^2), \tag{5.19}$$

$m = 1, 2, \dots,$ where

$$\mu_{G,\gamma} := \inf_{x\in V_x,\, y\in V_y} \frac{\|D^{-1/2}(Ax+y)\|_\infty^2 + \gamma\|D^{-1/2}A^Ty\|_\infty^2}{\|x\|_2^2 + \gamma\|y\|_2^2}.$$

A more explicit expression for $\mu_{G,\gamma}$ is not known. Since

$$\|D^{-1/2}(Ax+y)\|_\infty^2 \ge \|A\|_F^{-2}\|Ax+y\|_2^2 \quad \text{and} \quad \|D^{-1/2}A^Ty\|_\infty^2 \ge \|A\|_F^{-2}\|A^Ty\|_2^2,$$

we have $\mu_{G,\gamma} \ge \|A\|_F^{-2}\nu_{R,\gamma}$, which means that the bound (5.19) for the greedy version is at least as good as the bound (5.16) for its randomized counterpart.

5.3 Block-coordinate descent methods in convex optimization

Many results and proof techniques presented in this and the previous chapter have their origin in convex optimization. In recent years, with large-scale problems and machine learning applications in mind, compare for example [112], there has been a particular focus on the convergence theory of first-order block-coordinate descent methods. The simplest setting of block-coordinate descent is to consider an unconstrained convex minimization problem

$$J(x) \longrightarrow \min$$

on $\mathbb{R}^N$, and consider the coordinate vectors x as composed of n subvectors or coordinate blocks x^i, which is

$$\tilde{x} := (x^1, \ldots, x^n), \qquad x^i \in \mathbb{R}^{N_i}, \quad i = 1, \ldots, n, \quad \sum_{i=1}^{n} N_i = N.$$

In each step of a block-coordinate descent method, a single coordinate block x^i is updated by approximately minimizing $J(x)$ only with respect to the variables in the i-th subvector. Obviously, this corresponds to a Schwarz iterative scheme with singleton I_m, $V = \mathbb{R}^N$ and $V_i = \mathbb{R}^{N_i}$, the respective Euclidean scalar products as $a(\cdot, \cdot)$ and $a_i(\cdot, \cdot)$, and R_i given by the natural embeddings

$$R_i x^i = (0, \ldots, x^i, \ldots, 0) \in \mathbb{R}^N, \qquad x^i \in \mathbb{R}^{N_i}, \quad i = 1, \ldots, n.$$

If one sets $\rho_i = 1$ and removes the subscript ρ from the notation, then P and $\tilde{P}$ are identity maps on the respective spaces, since $\|R\tilde{x}\|^2 = \sum_{i=1}^n \|x^i\|_i^2 = \|\tilde{x}\|^2$. According to the results in Subsection 5.3.1, the minimization of the functional

$$\tilde{J}(\tilde{x}) := J(R\tilde{x})$$

over $\tilde{V}$ is equivalent to the original minimization problem, if one recovers the minimizers x^* from the minimizers $\tilde{x}^*$ of the new problem by the formula $x^* = R\tilde{x}^*$.

We give some generalizations of the results for Schwarz iterative methods applied to linear equations $Au = f$ for minimizing convex functionals $J : V \longrightarrow \mathbb{R}$ by block-coordinate descent methods. For the framework and notation used, see the next subsection. Throughout this section we fix a finite stable splitting $(\mathbf{V}; \mathbf{a}; \mathbf{R})$ for $(V; a)$ with index set $\Omega = \{1, \ldots, n\}$. The reason for setting $\rho_i = 1$ is that the weights ρ_i were previously fixed in order to have a Schwarz operator P_ρ with preferred spectral properties. In particular, it was implicitly assumed that the weights were chosen so that the condition number of the space splitting is as small as possible. Now, considering the convex minimization problem

$$J(u) \longrightarrow \min, \qquad u \in V, \tag{5.20}$$

and its equivalent form

$$\tilde{J}(\tilde{u}) := J(R\tilde{u}) \to \min, \qquad \tilde{u} \in \tilde{V}, \tag{5.21}$$

the role of the Schwarz operator is played by the gradient

$$\nabla \tilde{J}(\tilde{u}^{(m)}) = R^* \nabla J(R\tilde{u}^{(m)}),$$

which depends on the current iterate, and we generally do not expect to have a priori knowledge that could lead to good choices for ρ_i. However, as we will see, certain weights related to the Lipschitz constants of the partial gradients of the functional $\tilde{J}$ will prove useful. They will be used as parameters in the iterative methods and in some auxiliary norms.

5.3.1 Space splittings and convex optimization

We noted in A.3 that linear equations in Hilbert spaces can often be cast as quadratic minimization problems. The space splitting theory developed in Chapter 2 deals with a linear equation

$$Au = f,$$

where $f \in V'$ and $A \in \mathcal{B}(V, V')$ is symmetric and coercive. It provides a strategy for reformulating this problem into the Schwarz equation $Pu = g$ in $(V; a)$ and the Gramian equation $\tilde{P}\tilde{u} = \tilde{g}$ in $(\tilde{V}; \tilde{a})$, with control of the spectral properties of P and $\tilde{P}$, see Theorem 2.5. We have dropped the subscript ρ since $\rho_i = 1$ was assumed. It is easy to see that this transition corresponds to replacing the quadratic minimization problem

$$J(u) := \frac{1}{2} a(u, u) - \langle f, u \rangle_{V' \times V} \to \min, \qquad u \in V, \tag{5.22}$$

by an equivalent one, namely

$$\tilde{J}(\tilde{u}) := \frac{1}{2} \tilde{a}(\tilde{P}\tilde{u}, \tilde{u}) - \tilde{a}(\tilde{g}, \tilde{u}) \to \min, \qquad \tilde{u} \in \tilde{V}, \quad u = R\tilde{u}. \tag{5.23}$$

Now note that

$$\tilde{J}(\tilde{u}) = J(R\tilde{u}) \qquad \forall \tilde{u} \in \tilde{V}.$$

In fact, using the notation of Theorem 2.5, we have

$$\begin{aligned} J(R\tilde{u}) &= \frac{1}{2} a(R\tilde{u}, R\tilde{u}) - \langle f, R\tilde{u} \rangle_{V' \times V}) \\ &= \sum_{i \in \Omega} \left(\frac{1}{2} a(R\tilde{u}, R_i u_i) - \langle f, R_i u_i \rangle_{V' \times V} \right) \\ &= \sum_{i \in \Omega} \left(\frac{1}{2} a_i(R_i^* R\tilde{u}, u_i) - a_i(g_i, u_i) \right) = \frac{1}{2} \tilde{a}(\tilde{P}\tilde{u}, \tilde{u}) - \tilde{a}(\tilde{g}, \tilde{u}) = \tilde{J}(\tilde{u}). \end{aligned}$$

This leads us to the following extension of the stable splitting concept to the classes $\mathcal{F}_L^{1,1}(V;a)$ and $\mathcal{S}_{\mu,L}^{1,1}(V;a)$ of C^1-smooth convex optimization problems on the Hilbert space $(V;a)$. Recall the following definitions given in [128, Section 2.1]: A function $J:\ V\longrightarrow\mathbb{R}$ *belongs to the class* $\mathcal{F}_L^{1,1}(V;a)$ if it possesses a Lipschitz-continuous derivative $\nabla J:\ V\longrightarrow V$ such that

$$0\le J(v)-J(u)-a(\nabla J(u),v-u)\le\frac{L}{2}\|u-v\|^2,\qquad u,v\in V.\tag{5.24}$$

The lower estimate states the convexity of J, the upper estimate is equivalent to the Lipschitz continuity of ∇J with constant L. A function $J:\ V\longrightarrow\mathbb{R}$ *belongs to the class* $\mathcal{S}_{\mu,L}^{1,1}(V;a)$ if $J\in\mathcal{F}_L^{1,1}(V;a)$ is strongly convex with constant $\mu>0$, i.e., if

$$\frac{\mu}{2}\|u-v\|^2\le J(v)-J(u)-a(\nabla J(u),v-u)\le\frac{L}{2}\|u-v\|^2,\quad u,v\in V.$$

These classes play a central role in the convergence theory of first-order methods for smooth convex optimization, see [128, Chapter 2]. When $J\in\mathcal{S}_{\mu,L}^{1,1}(V;a)$, the condition number $\kappa_J:=L/\mu$ is crucial for their performance. Note that the constants L and μ, the explicit form of $\nabla J(u)$, the norm in V', and the duality map R_V depend on the choice of the scalar product $a(\cdot,\cdot)$ in V, but the linear functional $w\longrightarrow a(\nabla J(u),w)$ is independent of a. Thus, examining the linear approximation to $J(u+w)-J(u)$ can lead to a choice of the scalar product with favorable values for μ and L. Finally, note that $J\in\mathcal{S}_{\mu,L}^{1,1}(V;a)$ is equivalent to

$$0<\mu\le\frac{a(\nabla J(u+w)-\nabla J(u),w)}{a(w,w)}\le L<\infty,\qquad u,w\in V.$$

In the quadratic minimization example, the function J in (5.22) belongs to $\mathcal{S}_{1,1}^{1,1}(V;a)$ and is optimally conditioned ($\kappa_J=1$), while $\tilde J$ in (5.23) belongs to $\mathcal{F}_{\lambda_{\max}}^{1,1}(\tilde V;\tilde a)$ and to $\mathcal{S}_{\lambda_{\min},\lambda_{\max}}^{1,1}(\tilde X;\tilde a)$ if we restrict to the subspace $\tilde X=\operatorname{ran}(R^*)$ of $\tilde V$. Here, $\lambda_{\min}$ and $\lambda_{\max}$ are the lower and upper stability constants of the underlying stable space splitting $(\mathbf{V};\mathbf{a};\mathbf{R})$ for $(V;a)$. Consequently, $\kappa_{\tilde J}=\kappa$ coincides with the condition number of the stable space splitting, and it looks like we have made the problem worse. The crux is that although (5.22) is optimally conditioned by the choice of the scalar product, computing the gradient $\nabla J(u)=u-R_V^{-1}f$ involves computing $R_V^{-1}f$, which is the same as solving $Au=f$ since $R_V=A$. In other words, with this choice of scalar product on V, the space splitting theory transforms a well-conditioned but hard to solve problem into a moderately conditioned problem, which, due to the structure of the derivative

$$\nabla\tilde J(\tilde u)=\tilde P\tilde u-\tilde g=(R_i^*R\tilde u-g_i)_{i\in\Omega}=(A_i^{-1}R_i'(AR\tilde u-f))_{i\in\Omega}$$

of the functional $\tilde J$, is accessible in a distributed way.

Now, returning to convex optimization problems, we have the following result with a similar interpretation:

Theorem 5.2 *Let* $(\mathbf{V}; \mathbf{a}; \mathbf{R})$ *be a stable space splitting for* $(V; a)$ *with stability constants* $\lambda_{\min}$ *and* $\lambda_{\max}$.
a) If

$$J(u) \longrightarrow \min, \qquad u \in V$$

is a minimization problem with strongly convex C^1*-smooth* $J \in \mathcal{S}^{1,1}_{\mu,L}(V; a)$*, then the new problem*

$$\tilde{J}(\tilde{u}) := J(R\tilde{u}) \longrightarrow \min, \quad \tilde{u} \in \tilde{X} = \mathrm{ran}(R^*), \qquad u = R\tilde{u},$$

has a strongly convex C^1*-smooth* $\tilde{J} \in \mathcal{S}^{1,1}_{\mu',L'}(\tilde{X}; \tilde{a})$ *with* $\mu' = \mu\lambda_{\min}$ *and* $L' = L\lambda_{\max}$*. Consequently,*

$$\kappa_{\tilde{J}} \leq \kappa \cdot \kappa_J.$$

b) Suppose $J_i \in \mathcal{S}^{1,1}_{\mu_i,L_i}(V_i; a_i)$ *with* $\underline{\mu} \leq \mu_i \leq L_i \leq \bar{L}$ *for all* $i \in \Omega$*. Then the composite minimization problem*

$$J(u) := \sum_{i \in \Omega} J_i(R_i^* u) \longrightarrow \min, \qquad u \in V,$$

satisfies $J \in \mathcal{S}^{1,1}_{\mu,L}(V; a)$*, where* $\mu = \underline{\mu}\lambda_{\min}$ *and* $L = \bar{L}\lambda_{\max}$*. Consequently,*

$$\kappa_J \leq \kappa \cdot \bar{L}/\underline{\mu}.$$

Proof For Part a), we first compute the derivative of $\tilde{J}$ in $(\tilde{V}; \tilde{a})$ from the derivative of J in $(V; a)$: We have

$$\begin{aligned}\tilde{J}(\tilde{v}) - \tilde{J}(\tilde{u}) &= J(R\tilde{v}) - J(R\tilde{u}) = a(\nabla J(R\tilde{u}), R(\tilde{v} - \tilde{u})) + \mathrm{o}(\|R(\tilde{v} - \tilde{u})\|) \\ &= \sum_{i \in \Omega} a_i(R_i^* \nabla J(R\tilde{u}), v_i - u_i) + \mathrm{o}(\|\tilde{v} - \tilde{u})\|_{\tilde{V}}).\end{aligned}$$

Here we have only used that $R \in \mathcal{B}(\tilde{V}, V)$, and have not relied on the lower stability bound of the space splitting. This gives $\nabla \tilde{J}(\tilde{u}) = (R_i^* \nabla J(Ru))_{i \in \Omega}$, and with $J \in \mathcal{F}^{1,1}_L(V; a)$ we get

$$\begin{aligned}\tilde{J}(\tilde{v}) - \tilde{J}(\tilde{u}) - \tilde{a}(\nabla\tilde{J}(\tilde{u}), \tilde{v} - \tilde{u}) &= J(R\tilde{v}) - J(R\tilde{u}) - a(\nabla J(R\tilde{u}), R(\tilde{v} - \tilde{u})) \\ &\leq \frac{L}{2}\|R(\tilde{v} - \tilde{u})\|^2 \leq \frac{L}{2}\lambda_{\max}\|\tilde{v} - \tilde{u}\|^2_{\tilde{V}}, \end{aligned} \tag{5.25}$$

again using only the upper stability bound. This shows $\tilde{J} \in \mathcal{F}^{1,1}_{L\lambda_{\max}}(\tilde{V}; \tilde{a})$.

If $J \in \mathcal{S}^{1,1}_{\mu,L}(V; a)$, then we also have

$$\tilde{J}(\tilde{v}) - \tilde{J}(\tilde{u}) - \tilde{a}(\nabla\tilde{J}(\tilde{u}), \tilde{v} - \tilde{u}) \geq \frac{\mu}{2}\|R(\tilde{v} - \tilde{u})\|^2. \tag{5.26}$$

Here we have to restrict ourselves to the subspace $\tilde{X} = \mathrm{ran}(R^*)$ of $\tilde{V}$. In fact, if we set $\tilde{w} = R^* w$, $w = v - u$, and use the formula $R^* = R^{-1}P$ (see the proof of Theorem 2.6), then

$$\|R\tilde{w}\|^2 = \|Pw\|^2 \geq \lambda_{\min} a(Pw, w) = \lambda_{\min}\tilde{a}(R^* w, R^* w) = \lambda_{\min}\|\tilde{w}\|_{\tilde{V}}^2.$$

Substituting in (5.26), we have

$$\tilde{J}(\tilde{v}) - \tilde{J}(\tilde{u}) - \tilde{a}(\nabla\tilde{J}(\tilde{u}), \tilde{v} - \tilde{u}) \geq \frac{\mu\lambda_{\min}}{2}\|\tilde{v} - \tilde{u}\|_{\tilde{V}}^2, \qquad \tilde{u}, \tilde{v} \in \tilde{X}.$$

Together with the upper estimate (5.25), this shows $\tilde{J} \in \mathcal{S}_{\mu',L'}^{1,1}(\tilde{X}; \tilde{a})$.

For Part b), if $J_i \in \mathcal{F}_{L_i}^{1,1}(V_i; a_i)$ with $L_i \leq \bar{L}$ for all $i \in \Omega$, then by the upper bound in (5.24) we have

$$\begin{aligned} J(v) - J(u) &= \sum_{i\in\Omega}(J_i(R_i^* v) - J_i(R_i^* u)) \\ &\leq \sum_{i\in\Omega}\left(a_i(\nabla J_i(R_i^* u), R_i^*(v-u)) + \frac{L_i}{2}\|R_i^*(v-u)\|_i^2\right) \\ &\leq a(\sum_{i\in\Omega} R_i\nabla J_i(R_i^* u), v - u) + \frac{\bar{L}}{2}\|R^*(v-u)\|_{\tilde{V}}^2 \\ &\leq a(\sum_{i\in\Omega} R_i\nabla J_i(R_i^* u), v - u) + \frac{\bar{L}\lambda_{\max}}{2}\|v-u\|^2, \end{aligned}$$

where only the upper stability estimate of the space splitting has been used, namely

$$\sum_{i\in\Omega}\|R_i^*(v-u)\|_i^2 = \|R^*(v-u)\|_{\tilde{V}}^2 = a(P(v-u), v-u) \leq \lambda_{\max}\|v-u\|^2.$$

From the lower bound in (5.24) we have by similar reasoning

$$J(v) - J(u) \geq \sum_{i\in\Omega} a_i(\nabla J_i(R_i^* u), R_i^*(v-u)) = a(\sum_{i\in\Omega} R_i\nabla J_i(R_i^* u), v-u).$$

So, $\nabla J(u) = \sum_{i\in\Omega} R_i\nabla J_i(R_i^* u)$ and

$$0 \leq J(v) - J(u) - a(\nabla J(u), v-u) \leq \frac{\bar{L}\lambda_{\max}}{2}\|v-u\|^2, \qquad u, v \in V. \qquad (5.27)$$

This shows $J \in \mathcal{F}_L^{1,1}(V; a)$ with $L = \bar{L}\lambda_{\max}$ under weaker conditions on the J_i and the space splitting than stated in the formulation of Part b).

Similarly, if $J_i \in \mathcal{S}_{\mu_i,L_i}^{1,1}(V_i; a_i)$ with $\mu_i \geq \underline{\mu}$ for all $i \in \Omega$, then

$$J(v) - J(u) - a(\nabla J(u), v-u) \geq \frac{\underline{\mu}}{2}a(P(v-u), v-u) \geq \frac{\underline{\mu}\lambda_{\min}}{2}\|v-u\|^2.$$

Together with (5.27), this implies $J \in \mathcal{S}_{\mu,L}^{1,1}$, which completes the proof. □

The following corollary to the proof of Theorem 5.2 is worth mentioning.

Corollary 5.1 *Let* $(\mathbf{V}; \mathbf{a}; \mathbf{R})$ *be a space splitting for* $(V; a)$ *with associated constant* $\lambda_{\max}$.
a) Suppose we have a convex minimization problem $J(u) \to \min$ *for* $u \in V$ *with* $J \in \mathcal{F}_L^{1,1}(V; a)$. *Then the problem*

$$\tilde{J}(\tilde{u}) := J(R\tilde{u}) \to \min, \qquad \tilde{u} \in \tilde{X} = \mathrm{ran}(R^*),$$

is convex C^1*-smooth and* $\tilde{J} \in \mathcal{F}_{L'}^{1,1}(\tilde{V}; \tilde{a})$, *where* $L' = L\lambda_{\max}$.
b) Suppose, $J_i \in \mathcal{F}_{L_i}^{1,1}(V_i; a_i)$ *with* $L_i \le \bar{L}$ *for all* $i \in \Omega$. *Then the composite convex minimization problem*

$$J(u) := \sum_{i \in \Omega} J_i(R_i^* u) \to \min, \qquad u \in V,$$

satisfies $J \in \mathcal{F}_L^{1,1}(V; a)$, *where* $L = \bar{L}\lambda_{\max}$.

Let us add some comments. Part a) has been used in the optimization literature in connection with block-coordinate descent methods as follows: Reuse the notation introduced at the beginning of this section, set $V = \mathbb{R}^N$ and $V_i = \mathbb{R}^{N_i}$, and take the usual Euclidean scalar products as $a(\cdot,\cdot)$ and $a_i(\cdot,\cdot)$, $i \in \Omega = \{1, \ldots, N\}$. Denote by $R_i : \mathbb{R}^{N_i} \to \mathbb{R}^N$ and $R_i^* : \mathbb{R}^N \to \mathbb{R}^{N_i}$ the associated extension-by-zero and restriction operators, respectively. Then the resulting family $(\mathbf{V}; \mathbf{a}; \mathbf{R})$ is a Riesz splitting for $(V; a)$ with $\lambda_{\min} = \lambda_{\max} = 1$. Furthermore, $\tilde{X} = \mathrm{ran}(R^*)$ coincides with $\tilde{V}$. So Part a) just says that the new function $\tilde{J}$ belongs to the same $\mathcal{S}_{\mu,L}^{1,1}$ class as J. The advantage is that the derivative calculation $\nabla\tilde{J}(\tilde{u}) = (R_i^*\nabla J(R\tilde{u}))_{i \in \Omega}$ can be decomposed into the computation of partial derivatives $R_i^*\nabla J(R\tilde{u})$ with respect to subsets of variables, and allows first-order methods to be executed in a distributed way. Obviously, Part a) provides supporting theory also for other choices of V_i and R_i, e.g., it easily generalizes to splittings of $\mathbb{R}^N$ using partitions into overlapping index sets.

Part b) meets another trend in distributed optimization, where the structure of an optimization problem is taken into account: Suppose $J(u)$ is a composite functional,

$$J(u) = \sum_{i \in \Omega} J_i(T_i u),$$

where the functionals J_i are convex on certain spaces V_i (usually of smaller dimension), and $T_i : V \to V_i$ are bounded linear restriction operators for simplicity. Assume that one can fix scalar products $a(\cdot,\cdot)$ and $a_i(\cdot,\cdot)$ on the spaces V and V_i, respectively, such that the space splitting $(\mathbf{V}; \mathbf{a}; \mathbf{R})$ with $R_i = T_i^*$ is well conditioned. Further assume that the J_i belong to $\mathcal{S}_{\underline{\mu},\bar{L}}^{1,1}(V_i; a_i)$ for all $i \in \Omega$ with moderate constants $0 < \underline{\mu} \le \bar{L} < \infty$ (and allow for a cheap application of operators like T_i, R_i, and A_i^{-1}). Then we can provide performance guarantees for first-order methods that work on the derivatives ∇J_i in a distributed way. Note that the requirement $\mu_i \ge \underline{\mu} > 0$

in Part b) of Theorem 5.2 is not necessary for the strong convexity of $J(u)$ with $\mu > 0$. In fact, a situation often encountered in practice concerns functionals of the form $J(u) = J_1(u) + J_2(u)$, where $J_1 \in \mathcal{S}^{1,1}_{\mu_1,L_1}(V;a)$ is strongly convex, but $J_2 \in \mathcal{F}^{1,1}_{L_2}(V;a)$ is not (i.e., $\mu_2 = 0$, thus $\underline{\mu} = 0$). By setting $(V_i;a_i) = (V;a)$, $R_i = I$, $i = 1,2$, one can easily check that $J(u) \in \mathcal{S}^{1,1}_{\mu_1,L_1+L_2}(V;a)$ is nevertheless strongly convex.

5.3.2 One-step block-coordinate descent methods

Throughout this subsection we fix a stable space splitting $(\mathbf{V};\mathbf{a};\mathbf{R})$ for $(V;a)$ with index set $\Omega = \{1,\dots,n\}$ and a functional $J \in \mathcal{S}^{1,1}_{\mu,L}(V;a)$. According to Theorem 5.2, the transformed functional $\tilde{J}$ belongs to $\mathcal{S}^{1,1}_{\tilde{\mu},\tilde{L}}(\tilde{X};\tilde{a})$, where $\tilde{\mu},\tilde{L}$ are some new parameters and $\tilde{X} = \mathrm{ran}(R^*) \subset \tilde{V}$. Recall that the smallest $\tilde{L}$ is equal to the Lipschitz constant of $\nabla\tilde{J}(\tilde{u}) = R^*\nabla J(R\tilde{u})$, given by

$$\tilde{L}_{\min} := \sup_{\tilde{u},\tilde{w}\neq 0} \frac{\tilde{a}(R^*(\nabla J(R(\tilde{u}+\tilde{w})) - \nabla J(R\tilde{u})),\tilde{w})}{\|\tilde{w}\|^2_{\tilde{V}}}.$$

Furthermore, the partial gradients with respect to V_i and their smallest Lipschitz constants are given by

$$\nabla_i\tilde{J}(\tilde{u}) = R_i^*\nabla J(R\tilde{u}), \qquad i = 1,\dots,n, \tag{5.28}$$

and

$$\tilde{L}_{i,\min} := \sup_{\tilde{u},w_i\neq 0} \frac{a_i(R_i^*(\nabla J(R\tilde{u}+R_iw_i) - \nabla J(R\tilde{u})),w_i)}{\|w_i\|_i^2}, \qquad i = 1,\dots,n,$$

respectively.

We assume that good upper bounds $\tilde{L}_i \geq \tilde{L}_{i,\min}$ are known and that the computation of partial gradients is cheap compared to the computation of the full gradient $\nabla J(u)$. The standard one-step iteration process of a coordinate descent method in $\tilde{V}$ to solve (5.21) is then of the form

$$\tilde{u}_i^{(m+1)} = \tilde{u}_i^{(m)} + \begin{cases} -\dfrac{1}{\tilde{L}_{i_m}}\nabla_{i_m}\tilde{J}(\tilde{u}^{(m)}), & i = i_m, \\ 0, & i \neq i_m, \end{cases} \qquad m = 0,1,\dots, \tag{5.29}$$

where the index i_m is chosen according to some rule (cyclic, greedy, random). Obviously, other step-size rules (e.g., steepest descent) and group updates (replacing i_m by an index set I_m, and forming the search direction as a linear combination of partial gradients $\nabla_i\tilde{J}(\tilde{u}^{(m)})$ with $i \in I_m$) are also possible. However, we will not consider them here. The resulting sequence of iterates $(\tilde{u}^{(m)})_{m\in\mathbb{Z}_+}$ belongs to $\tilde{V}$

and not necessarily to $\tilde{X}$, but the value of the objective $\tilde{J}(\tilde{u}^{(m)})$ does not depend on the component of $\tilde{u}^{(m)}$ in $\tilde{X}^{\perp} = \ker(R)$. If we apply R to (5.29) and denote $u^{(m)} := R\tilde{u}^{(m)}$, then, due to (5.28), we get the iteration

$$u^{(m+1)} = u^{(m)} + \frac{1}{\tilde{L}_{i_m}} P_{i_m} \nabla J(u^{(m)}), \qquad m = 0, 1, \ldots, \tag{5.30}$$

in V. Except for components in $\tilde{u}^{(m)}$ that belong to $\tilde{X}^{\perp} = \ker(R)$ and are annihilated with each application of R, the two sequences of iterates are equivalent. Obviously, one-step Schwarz iterations for the linear case, such as (4.40), are a special case of (5.30). This follows from the example at the beginning of Subsection 5.3.1, where $\nabla J(u) = I$ and $\tilde{L}_{i,\min} = \|R_i\|^2_{V_i \to V}$.

The analysis of these iterations was initiated in [113, 127]. It uses auxiliary norms of the form

$$\|\tilde{u}\|_{\tilde{V},s} := \left(\sum_{i=1}^{n} \tilde{L}_i^s \|u_i\|_i^2 \right)^{1/2}, \qquad s \in \mathbb{R}. \tag{5.31}$$

We have the duality with respect to the scalar product in $\tilde{V}$ as

$$\tilde{a}(\tilde{u}, \tilde{v}) = \sum_{i=1}^{n} a_i(u_i, v_i) \le \|\tilde{u}\|_{\tilde{V},s} \|\tilde{v}\|_{\tilde{V},-s}.$$

Obviously, this family of equivalent Hilbert space norms on $\tilde{V}$ is only relevant if the partial Lipschitz bounds $\tilde{L}_i$ have significantly different sizes. The most interesting special cases are $s = 0, \pm 1/2, \pm 1$. For $s = 0$ this is the norm in $\tilde{V}$.

Error decay is now measured in terms of the error in the objective to be minimized, i.e.,

$$\Delta\tilde{J}^{(m)} := \tilde{J}(\tilde{u}^{(m)}) - \tilde{J}(\tilde{u}) = J(u^{(m)}) - J(u) =: \Delta J^{(m)}, \tag{5.32}$$

where $u = R\tilde{u} \in V$ and $\tilde{u} \in \tilde{X}$ denote the unique minimizers for (5.20) and (5.21), respectively. Keeping the equality in (5.32) in mind, from now on we will only give estimates for $\Delta\tilde{J}^{(m)}$. Since $\tilde{L}_i$ is a bound on the Lipschitz constant of $\nabla_i \tilde{J}$, we have from (5.24)

$$\tilde{J}(\tilde{u}^{(m)} + \tilde{w}_i) \le \tilde{J}(\tilde{u}^{(m)}) + a_i(\nabla_i \tilde{J}(\tilde{u}^{(m)}), w_i) + \frac{\tilde{L}_i}{2} a_i(w_i, w_i), \tag{5.33}$$

where we have written for short $\tilde{w}_i := (0, \ldots, w_i, \ldots, 0)$, $i \in \Omega$. Now substitute $w_i = -\frac{1}{\tilde{L}_i} \nabla_i \tilde{J}(\tilde{u}^{(m)})$, set $i = i_m$, subtract $\tilde{J}(\tilde{u})$ from both sides, and consider the update formula (5.29). This results in

$$\Delta\tilde{J}^{(m+1)} \le \Delta\tilde{J}^{(m)} - \frac{1}{2\tilde{L}_{i_m}} \|\nabla_{i_m} \tilde{J}(\tilde{u}^{(m)})\|^2_{i_m}. \tag{5.34}$$

The inequality (5.34) establishes descent in the objective error, no matter what the choice of i_m is.

To obtain decay rates, one quantifies strong convexity with respect to various norms. For random choices of i_m governed by the probability distribution

$$\mathbb{P}(i_m = i) = \frac{\tilde{L}_i^s}{\tilde{S}_s}, \qquad \tilde{S}_s := \sum_{i=1}^{n} \tilde{L}_i^s, \tag{5.35}$$

one quantifies the strong convexity of $\tilde{J}$ in its dependence on s as

$$\tilde{J}(\tilde{v} + \tilde{w}) \geq \tilde{J}(\tilde{v}) + \tilde{a}(\nabla\tilde{J}(\tilde{v}), \tilde{w}) + \frac{\tilde{\sigma}_{1-s}}{2}\|\tilde{w}\|^2_{\tilde{V},1-s}, \quad \tilde{v} \in \tilde{V},\ \tilde{w} \in \tilde{X}, \tag{5.36}$$

with some $\tilde{\sigma}_{1-s} > 0$. The restriction $\tilde{w} \in \tilde{X}$ is justified because $\tilde{J}(\tilde{v} + \tilde{w}) = \tilde{J}(\tilde{v})$ for $\tilde{w} \in \tilde{X}^\perp = \ker(R)$, and $\nabla\tilde{J}(\tilde{v}) = R^*\nabla J(R\tilde{u}) \in \tilde{X}$. Since $\tilde{J} \in \mathcal{S}^{1,1}_{\tilde{\mu},\tilde{L}}(\tilde{X}; \tilde{a})$, we have $\tilde{\sigma}_0 = \tilde{\mu}$ and (5.36) holds for any $s \in \mathbb{R}$ due to the mutual equivalence of the $\|\cdot\|_{\tilde{V},s}$ norms, but the constants $\tilde{\sigma}_{1-s}$ may depend on s.

To proceed, we consider the case of Riesz splittings, where $\ker(R) = \{0\}$ implies $\tilde{X} = \tilde{V}$. Minimizing on both sides of (5.36) with respect to $\tilde{w}$ for fixed $\tilde{v}$ gives

$$\tilde{J}(\tilde{u}) \geq \tilde{J}(\tilde{v}) - \frac{1}{2\tilde{\sigma}_{1-s}}\|\nabla\tilde{J}(\tilde{v})\|^2_{\tilde{V},s-1}.$$

This is because the minimum of the left-hand side of the inequality is equal to the minimum of $\tilde{J}$ attained at $\tilde{u}$, while the minimizer $\tilde{w}$ for the quadratic problem on the right-hand side is given componentwise by $w_i = -\dfrac{1}{\tilde{\sigma}_{1-s}\tilde{L}_i^{1-s}} R_i^*\nabla J(\tilde{v})$. At this point, the condition $\tilde{X} = \tilde{V}$ is used here, because otherwise it is not clear that the minimizer $\tilde{w}$ belongs to $\tilde{X} = \mathrm{ran}(R^*)$. An exception is the parameter $s = 1$, where the space splitting can be arbitrary. If we set $\tilde{v} = \tilde{u}^{(m)}$, we get

$$2\tilde{\sigma}_{1-s}\Delta\tilde{J}^{(m)} \leq \|\nabla\tilde{J}(\tilde{u}^{(m)})\|^2_{\tilde{V},s-1}. \tag{5.37}$$

After these preparations, we estimate the conditional expectation $\mathbb{E}'(\Delta J^{(m+1)})$ with i_m randomly chosen according to (5.35) using (5.34), and get

$$\begin{aligned}
\mathbb{E}'(\Delta J^{(m+1)}) &\leq \Delta J^{(m)} - \frac{1}{2\tilde{S}_s}\sum_{i=1}^{n}\frac{\tilde{L}_i^s}{\tilde{L}_i}\|\nabla_i\tilde{J}(\tilde{u}^{(m)})\|_i^2 \\
&= \Delta J^{(m)} - \frac{1}{2\tilde{S}_s}\|\nabla\tilde{J}(\tilde{u}^{(m)})\|_{\tilde{V},s-1} \leq \left(1 - \frac{\tilde{\sigma}_{1-s}}{\tilde{S}_s}\right)\Delta J^{(m)}.
\end{aligned}$$

Similarly, let i_m be determined by the weak greedy rule

$$\bar{\Delta}_{i_m}^{(m)} \geq \beta_m \max_{i=1,\ldots,n} \bar{\Delta}_i^{(m)}, \qquad \bar{\Delta}_i^{(m)} := \tilde{L}_i^{-1}\|\nabla_i\tilde{J}(\tilde{u}^{(m)})\|_i^2, \quad i = 1,\ldots,n. \tag{5.38}$$

The error indicators $\bar{\Delta}_i^{(m)}$ are suggested by (5.34). If we now use the inequality (4.65) with $x_i = (\bar{\Delta}_i^{(m)})^{1/2}$ and $\theta_i = \tilde{L}_i^s$, then (5.34) combined with (5.37) yields

$$
\begin{aligned}
\Delta J^{(m+1)} &\le \Delta J^{(m)} - \frac{\beta_m}{2} \max_{i=1,\dots,n} \bar{\Delta}_i^{(m)} \\
&\le \Delta J^{(m)} - \frac{\beta_m}{2} \frac{\left(\sum_{i=1}^n \tilde{L}_i^{s-1} \|\nabla_i \tilde{J}(\tilde{u}^{(m)})\|_i^2\right)^2}{\left(\sum_{i=1}^n \tilde{L}_i^{s-1/2} \|\nabla_i \tilde{J}(\tilde{u}^{(m)})\|_i\right)^2} \\
&\le \left(1 - \beta_m \tilde{\sigma}_{1-s} \inf_{\tilde{w} \in \tilde{X}} \frac{\|\nabla \tilde{J}(\tilde{w})\|^2_{\tilde{V},s-1}}{\left(\sum_{i=1}^n \tilde{L}_i^{s-1/2} \|\nabla_i \tilde{J}(\tilde{w})\|_i\right)^2} \right) \Delta J^{(m)}.
\end{aligned}
$$

This estimate holds for all $s \in \mathbb{R}$ if $\tilde{X} = \tilde{V}$, and for arbitrary space splittings if $s = 1$.

Thus, we have generalized the results on one-step Schwarz iterations for linear equations in Subsection 4.5.2 to block-coordinate descent methods for convex minimization problems as follows.

Theorem 5.3 *Let* $(\mathbf{V}; \mathbf{a}; \mathbf{R})$ *be a finite Riesz splitting for* $(V; a)$ *with index set* $\Omega = \{1, \dots, n\}$. *For the convex functional* $\tilde{J}(\tilde{v}) = J(R\tilde{v})$ *we assume (5.33) and strong convexity with constant* $\tilde{\sigma}_{1-s}$ *as defined in (5.36). Consider the iterations (5.29) in* $\tilde{V}$ *and (5.30) in* V *with starting point* $\tilde{u}^{(0)}$ *and* $u^{(0)} = R\tilde{u}^{(0)}$, *respectively.*
a) If the index i_m *is drawn randomly and independently according to the probability distribution (5.35) with fixed* $s \in \mathbb{R}$, *then we have*

$$
\mathbb{E}(J(u^{(m)}) - J(u)) = \mathbb{E}(\tilde{J}(\tilde{u}^{(m)}) - \tilde{J}(\tilde{u})) \le \left(1 - \frac{\tilde{\sigma}_{1-s}}{\tilde{S}_s}\right)^m (J(u^{(0)}) - J(u)),
$$

$m = 1, 2, \dots,$ *where* $\tilde{S}_s$ *is defined in (5.35).*
b) If the index i_m *is chosen by the weak greedy rule (5.38), then*

$$
J(u^{(m)}) - J(u) = \tilde{J}(\tilde{u}^{(m)}) - \tilde{J}(\tilde{u}) \le \prod_{k=0}^{m-1} \left(1 - \frac{\beta_k \tilde{\sigma}_{1-s}}{\tilde{A}_s}\right) (J(u^{(0)}) - J(u)),
$$

$m = 1, 2, \dots,$ *where*

$$
\tilde{A}_s := \sup_{\tilde{w} \in \tilde{X}} \frac{\left(\sum_{i=1}^n \tilde{L}_i^{s-1/2} \|\nabla_i \tilde{J}(\tilde{w})\|_i\right)^2}{\|\nabla \tilde{J}(\tilde{w})\|^2_{\tilde{V},s-1}} \le \tilde{S}_s, \qquad s \in \mathbb{R}. \tag{5.39}
$$

c) For arbitrary stable space splittings $(\mathbf{V}; \mathbf{a}; \mathbf{R})$, *i.e., without assuming* $\ker(R) = \{0\}$, *the statements in Part a) and b) remain valid for* $s = 1$.

The inequality $\tilde{A}_s \le \tilde{S}_s$ in (5.39) follows from the application of the Cauchy-Schwarz inequality, namely

$$
\left(\sum_{i=1}^n \tilde{L}_i^{s-1/2} \|\nabla_i \tilde{J}(\tilde{w})\|_i\right)^2 \le \left(\sum_{i=1}^n \tilde{L}_i^s\right) \left(\sum_{i=1}^n \tilde{L}_i^{s-1} \|\nabla_i \tilde{J}(\tilde{w})\|_i^2\right) = \tilde{S}_s \|\nabla \tilde{J}(\tilde{w})\|^2_{\tilde{V},s-1}.
$$

We also see that the greedy rule (5.38) with $\beta_m = 1$ has a better per-step decay estimate than randomized choices for i_m. This follows directly from the estimate (5.34) and the definition of $\bar{\Delta}_i^{(m)}$, since $\mathbb{E}'(\bar{\Delta}_i^{(m)}) \le \max_{i\in\Omega} \bar{\Delta}_i^{(m)}$ for any probability distribution on Ω.

For $s = 1$, Theorem 5.3 directly generalizes the statement of Theorem 4.6 in the randomized case and the statement of Theorem 4.10 in the weak greedy case. For Riesz splittings, additional estimates are obtained for $s \neq 1$. In fact, consider

$$J(u) = \frac{1}{2}a(u,u) - \langle f, u\rangle_{V'\times V}.$$

Then we have $\nabla J(v) = v - A^{-1}f$, $\nabla_i \tilde{J}(\tilde{v}) = R_i^*(R\tilde{v} - A^{-1}f)$, and

$$\tilde{L}_i = \sup_{w_i\neq 0} \frac{a_i(\nabla_i\tilde{J}(\tilde{v}+\tilde{w}_i) - \nabla_i\tilde{J}(\tilde{v}), w_i)}{\|w_i\|_i^2} = \sup_{w_i\neq 0} \frac{a_i(R_i^*R_iw_i, w_i)}{\|w_i\|_i^2} = \|R_i\|_{V_i\to V}^2.$$

So the iteration (5.30) reduces to (4.40), and

$$\Delta\tilde{J}^{(m)} = J(u^{(m)}) - J(u) = \frac{1}{2}\|e^{(m)}\|^2.$$

For $s = 1$ we have $\tilde{S}_1 = \sum_{i=1} \|R_i\|_{V_i\to V}^2$ and the best constant $\tilde{\sigma}_0$ in (5.37) is equal to

$$\tilde{\sigma}_0 = \inf_{\tilde{v}\in\tilde{X}} \frac{\|\nabla\tilde{J}(\tilde{u}^{(m)})\|_{\tilde{V}}^2}{2(J(u^{(m)}) - J(u))} = \sup_{\tilde{v}\in\tilde{X}} \frac{a(Pe^{(m)}, e^{(m)})}{\|e^{(m)}\|^2} = \lambda_{\min}.$$

Here we used that $\nabla\tilde{J}(\tilde{u}^{(m)}) = R^*(R\tilde{u}^{(m)} - A^{-1}f) = -R^*e^{(m)}$ and

$$\|\nabla\tilde{J}(\tilde{u}^{(m)})\|_{\tilde{V}}^2 = \tilde{a}(R^*e^{(m)}, R^*e^{(m)}) = a(Pe^{(m)}, e^{(m)}).$$

As for the estimate for the greedy version, we similarly have

$$\tilde{A}_1 = \sup_{\tilde{w}\in\tilde{X}} \frac{\left(\sum_{i=1}^n \tilde{L}_i^{1/2}\|\nabla_i\tilde{J}(\tilde{w})\|_i\right)^2}{\|\nabla\tilde{J}(\tilde{w})\|_{\tilde{V}}^2} = \sup_{w=R\tilde{w}\in V} \frac{\left(\sum_{i=1}^n \|R_i\|_{V_i\to V}\sqrt{a(P_iw,w)}\right)^2}{a(Pw,w)}.$$

Compared to the estimate (4.61) in Theorem 4.10 a), we have a slightly worse bound, since

$$\begin{aligned}\mu_G' &= \inf_{v\in V} \frac{a(Pv,v)^2}{\|v\|^2\left(\sum_{i=1}^n \|R_i\|_{V_i\to V}\sqrt{a(P_iv,v)}\right)^2}\\ &\ge \lambda_{\min} \inf_{v\in V} \frac{a(Pv,v)}{\left(\sum_{i=1}^n \|R_i\|_{V_i\to V}\sqrt{a(P_iv,v)}\right)^2} = \frac{\tilde{\sigma}_0}{\tilde{A}_1}.\end{aligned}$$

The constant $\tilde{A}_1^{-1}$ already appeared in the analysis of accelerated weak greedy iterations in Section 4.5, see (4.96) for $\rho_i = 1$.

When $\tilde{X} = \tilde{V}$, we can use other values of s. For example, for $s = 0$ we compute $\tilde{S}_0 = n$,

$$\tilde{\sigma}_1 = \inf_{\tilde{v}\in\tilde{X}} \frac{\|\nabla\tilde{J}(\tilde{u}^{(m)})\|^2_{\tilde{V},1}}{2(J(u^{(m)}) - J(u))} = \inf_{\tilde{v}\in\tilde{X}} \frac{\sum_{i=1}^n \|R_i\|^2_{V_i\to V} a(P_i e^{(m)}, e^{(m)})}{\|e^{(m)}\|^2},$$

and

$$\tilde{A}_0 = \sup_{\tilde{w}\in\tilde{X}} \frac{\left(\sum_{i=1}^n \tilde{L}_i^{-1/2}\|\nabla_i\tilde{J}(\tilde{w})\|_i\right)^2}{\|\nabla\tilde{J}(\tilde{w})\|^2_{\tilde{V},-1}} = \sup_{w(=R\tilde{w})\in V} \frac{\left(\sum_{i=1}^n \|R_i\|^{-1}_{V_i\to V}\sqrt{a(P_i w, w)}\right)^2}{\sum_{i=1}^n \|R_i\|^{-2}_{V_i\to V} a(P_i w, w)}.$$

5.3.3 Acceleration

The material of this subsection is an adaptation of the original results in [127] (case $s = 0$) and [113] (arbitrary s, see the proof of Theorem 5 in [113] for $\epsilon = 0$). For simplicity, we assume $s \geq 0$ and the Riesz property of the underlying space splitting, i.e., $\tilde{V} = \tilde{X}$.

We keep the notation of Subsection 5.3.2, with one technical change: We replace the values $\tilde{L}_i$ with their upper bounds

$$\tilde{L}_i \leq \tilde{L}_i' := \max(\tilde{L}_i, (\tilde{S}_s/n)^{1/s}), \qquad i = 1, \dots, n,$$

and use these new values to define the s-dependent probabilities

$$\mathbb{P}(i_m = i) = \frac{(\tilde{L}_i')^s}{\tilde{S}_s'}, \qquad \tilde{S}_s' := \sum_{i=1}^n (\tilde{L}_i')^s, \tag{5.40}$$

where

$$\tilde{S}_s \leq \tilde{S}_s' < \sum_{i=1}^n \tilde{L}_i^s + n\frac{\tilde{S}_s}{n} = 2\tilde{S}_s.$$

Instead of (5.31), the auxiliary norms are now given by

$$\|\tilde{v}\|_{\tilde{V},s} := \left(\sum_{i=1}^n (\tilde{L}_i')^s a_i(v_i, v_i)\right)^{1/2}, \qquad \tilde{v} \in \tilde{V}. \tag{5.41}$$

Concerning the convex functional $\tilde{J}$, we again need (5.33) and (5.36), the latter with respect to the modified norm (5.41), which majorizes the one defined in (5.31). Thus, (5.36) holds with a new constant $\tilde{\sigma}_{1-s}' \leq \tilde{\sigma}_{1-s}$ instead of $\tilde{\sigma}_{s-1}$. Under these conditions, we now analyze the following counterpart of the Nesterov-type iteration (4.100) to solve the minimization problem (5.21):

Starting with $\tilde{u}^{(0)} = \tilde{w}^{(0)} = \tilde{v}^{(0)} \in \tilde{V}$, compute

$$\begin{aligned} \tilde{u}^{(m+1)} &= \tilde{v}^{(m)} + \zeta_m \nabla_{i_m} \tilde{J}(\tilde{v}^{(m)}), \\ \tilde{w}^{(m+1)} &= \delta_m \tilde{w}^{(m)} + (1-\delta_m)\tilde{v}^{(m)} + \tau_m \nabla_{i_m} \tilde{J}(\tilde{v}^{(m)}), \\ \tilde{v}^{(m+1)} &= \alpha_{m+1}\tilde{u}^{(m+1)} + (1-\alpha_{m+1})\tilde{w}^{(m+1)}, \end{aligned} \tag{5.42}$$

where the index i_m is chosen randomly and independently according to (5.40) for some fixed $s \in \mathbb{R}$, $m = 0, 1, \ldots$. By setting $u^{(m)} = R\tilde{u}^{(m)}$, this becomes an accelerated coordinate descent method for solving (5.20). Note that, unlike the analysis in Subsection 4.5.2, only one partial gradient is needed per iteration step. The parameters ζ_m and τ_m are conveniently set to the values

$$\zeta_m = \frac{1}{\tilde{L}'_{i_m}}, \qquad \tau_m = \frac{\tau'_m}{\tilde{L}'_{i_m}}, \qquad m = 0, 1, \ldots, \tag{5.43}$$

where τ'_m is determined below.

Theorem 5.4 *Let* $(\mathbf{V}; \mathbf{a}; \mathbf{R})$ *be a finite Riesz splitting for* $(V; a)$ *with index set* $\Omega = \{1, \ldots, n\}$. *For the convex functional* $\tilde{J}(\tilde{v}) = J(R\tilde{v})$ *we assume (5.33) and strong convexity defined by*

$$\tilde{J}(\tilde{v} + \tilde{w}) \geq \tilde{J}(\tilde{v}) + \tilde{a}(\nabla \tilde{J}(\tilde{v}), \tilde{w}) + \frac{\tilde{\sigma}'_{1-s}}{2} \|\tilde{w}\|^2_{\tilde{V},1-s}, \qquad \tilde{v} \in \tilde{V}, \ \tilde{w} \in \tilde{X}, \tag{5.44}$$

for some fixed $s \in \mathbf{R}$ *with respect to the modified norm (5.41). Then the iteration (5.42) with* i_m *chosen i.i.d. according to (5.40) satisfies the estimate*

$$\begin{aligned} \mathbb{E}(\Delta\tilde{J}_u^{(m)} + \tfrac{\tilde{\sigma}'_{1-s}}{2} \|\tilde{e}_w^{(m)}\|^2_{1-s}) &\leq \left(1 - \sqrt{\frac{\tilde{S}_s \tilde{\sigma}'_{1-s}}{n(\tilde{S}'_s)^2}}\right)^m \left(\Delta\tilde{J}_u^{(0)} + \tfrac{\tilde{\sigma}'_{1-s}}{2} \|\tilde{e}_w^{(0)}\|^2_{1-s}\right) \\ &\leq 2\left(1 - \sqrt{\frac{\tilde{S}_s \tilde{\sigma}'_{1-s}}{n(\tilde{S}'_s)^2}}\right)^m \Delta\tilde{J}_u^{(0)}, \quad m = 1, 2, \ldots, \end{aligned} \tag{5.45}$$

if the parameters in (5.42) are given by (5.43) and

$$\delta_m = 1 - \sqrt{\frac{\tilde{S}_s \tilde{\sigma}'_{1-s}}{n(\tilde{S}'_s)^2}}, \qquad \tau'_m = \sqrt{\frac{\tilde{S}_s}{n\tilde{\sigma}'_{1-s}}}, \qquad \alpha_m = \frac{\tilde{S}'_s \sqrt{n}}{\tilde{S}'_s \sqrt{n} + \sqrt{\tilde{S}_s \tilde{\sigma}'_{1-s}}}. \tag{5.46}$$

For $s = 1$, *the estimates hold for arbitrary space splittings.*

Proof The structure of the proof is the same as in Theorem 4.14, with a few changes: Namely, $\|e_u^{(m)}\|^2$ is replaced by $\Delta\tilde{J}_u^{(m)} := \tilde{J}(\tilde{u}^{(m)}) - \tilde{J}(\tilde{u}) = J(u^{(m)}) - J(u)$, while the role of $|||e_w^{(m)}|||^2$ is played by $\|\tilde{e}_w^{(m)}\|^2_{1-s}$ as defined in (5.41), where $\tilde{e}_w^{(m)} := \tilde{u} - \tilde{w}^{(m)}$. We also set $\tilde{e}_v^{(m)} := \tilde{u} - \tilde{v}^{(m)}$ and $\Delta\tilde{J}_v^{(m)} := \tilde{J}(\tilde{v}^{(m)}) - \tilde{J}(\tilde{u})$. Other changes result from replacing the lower spectral bound for P by the new strong convexity assumption (5.44). With this, the basic strategy for proving the decay estimate is the same as for Theorem 4.14, and we will therefore skip some intermediate steps.

First, the update formula for the $\tilde{u}$-iterates yields the estimate

$$\tilde{J}(\tilde{u}^{(m+1)}) \leq \tilde{J}(\tilde{v}^{(m)}) - \frac{1}{2\tilde{L}'_{i_m}} \|\nabla_{i_m} \tilde{J}(\tilde{v}^{(m)})\|^2_{i_m},$$

or, after subtracting $\tilde{J}(\tilde{u})$ and rearranging,

$$\|\nabla_{i_m} \tilde{J}(\tilde{v}^{(m)})\|^2_{i_m} \leq 2\tilde{L}'_{i_m} (\Delta\tilde{J}_v^{(m)} - \Delta\tilde{J}_u^{(m+1)}). \tag{5.47}$$

This is the counterpart of (5.34).

Next, we estimate the conditional expectation on both sides of the inequality

$$\|\tilde{e}_w^{(m+1)}\|^2_{1-s} \leq \delta_m \|\tilde{e}_w^{(m)}\|^2_{1-s} + (1-\delta_m)\|\tilde{e}_v^{(m)}\|^2_{1-s} + (\tau'_m)^2 (\tilde{L}'_{i_m})^{-1-s} \|\nabla_{i_m} \tilde{J}(\tilde{v}^{(m)})\|^2_{i_m}$$
$$+2\tau'_m (\tilde{L}'_{i_m})^{-s} a_{i_m}((\delta_m \tilde{e}_w^{(m)} + (1-\delta_m)\tilde{e}_v^{(m)})_{i_m}, \nabla_{i_m} \tilde{J}(\tilde{v}^{(m)})), \tag{5.48}$$

where the formula for τ_m in (5.43) has already been taken into account. We first deal with the last term on the right-hand side of (5.48) by replacing $\tilde{w}^{(m)}$ with its expression

$$\tilde{w}^{(m)} = \tilde{v}^{(m)} - \gamma_m(\tilde{u}^{(m)} - \tilde{v}^{(m)}),$$

where $\gamma_m = \alpha_m/(1-\alpha_m)$. This gives

$$\begin{aligned}
&\mathbb{E}'((\tilde{L}'_{i_m})^{-s} a_{i_m}(\delta_m \tilde{e}_w^{(m)} + (1-\delta_m)\tilde{e}_v^{(m)}, \nabla_{i_m} \tilde{J}(\tilde{v}^{(m)}))) \\
&\quad = \frac{1}{\tilde{S}'_s} \sum_{i=1}^n a_i(\tilde{u}_i - \delta_m \tilde{w}_i^{(m)} - (1-\delta_m)\tilde{v}_i^{(m)}, \nabla_i \tilde{J}(\tilde{v}^{(m)})) \\
&\quad = \frac{1}{\tilde{S}'_s} \tilde{a}(\tilde{u} - \tilde{v}^{(m)} + \delta_m \gamma_m (\tilde{u}^{(m)} - \tilde{v}^{(m)}), \nabla \tilde{J}(\tilde{v}^{(m)})) \\
&\quad = \frac{1}{\tilde{S}'_s} \left(\tilde{a}(\tilde{u} - \tilde{v}^{(m)}, \nabla \tilde{J}(\tilde{v}^{(m)})) + \delta_m \gamma_m \tilde{a}(\tilde{u}^{(m)} - \tilde{v}^{(m)}, \nabla \tilde{J}(\tilde{v}^{(m)})) \right).
\end{aligned}$$

The last two terms are estimated by convexity, i.e.,

$$\tilde{a}(\tilde{u}^{(m)} - \tilde{v}^{(m)}, \nabla \tilde{J}(\tilde{v}^{(m)})) \leq \tilde{J}(\tilde{u}^{(m)}) - \tilde{J}(\tilde{v}^{(m)}) = \Delta\tilde{J}_u^{(m)} - \Delta\tilde{J}_v^{(m)},$$

and by strong convexity (5.36) with constant $\tilde{\sigma}'_{1-s}$ instead of $\tilde{\sigma}_{1-s}$, i.e.,

$$\begin{aligned}
\tilde{a}(\tilde{u} - \tilde{v}^{(m)}, \nabla \tilde{J}(\tilde{v}^{(m)})) &\leq \tilde{J}(\tilde{u}) - \tilde{J}(\tilde{v}^{(m)}) - \frac{\tilde{\sigma}'_{1-s}}{2} \|\tilde{e}_v^{(m)}\|^2_{1-s} \\
&= -\Delta\tilde{J}_v^{(m)} - \frac{\tilde{\sigma}'_{1-s}}{2} \|\tilde{e}_v^{(m)}\|^2_{1-s},
\end{aligned}$$

respectively. Together this gives

$$\mathbb{E}'((\tilde{L}'_{i_m})^{-s}a_{i_m}((\delta_m\tilde{e}_w^{(m)}+(1-\delta_m)\tilde{e}_v^{(m)})_{i_m},\nabla_{i_m}\tilde{J}(\tilde{v}^{(m)})))$$
$$\leq\frac{1}{\tilde{S}'_s}\left(\delta_m\gamma_m\Delta\tilde{J}_u^{(m)}-(1+\delta_m\gamma_m)\Delta\tilde{J}_v^{(m)}-\frac{\tilde{\sigma}'_{1-s}}{2}\|\tilde{e}_v^{(m)}\|_{1-s}^2\right).\quad(5.49)$$

For the third term in the right-hand side of (5.48), we use (5.47) and the bound

$$(\tilde{L}'_{i_m})^{-s}\leq n/\tilde{S}_s,$$

which holds by definition of $\tilde{L}'_i$. Then we get

$$(\tilde{L}'_{i_m})^{-1-s}\|\nabla_{i_m}\tilde{J}(\tilde{v}^{(m)})\|_{i_m}^2\leq 2(\tilde{L}'_{i_m})^{-s}(\Delta\tilde{J}_v^{(m)}-\Delta\tilde{J}_u^{(m+1)})\leq\frac{2n}{\tilde{S}_s}(\Delta\tilde{J}_v^{(m)}-\Delta\tilde{J}_u^{(m+1)}).$$

Thus,

$$\mathbb{E}'((\tilde{L}'_{i_m})^{-1-s}\|\nabla_{i_m}\tilde{J}(\tilde{v}^{(m)})\|_{i_m}^2)\leq\frac{2n}{\tilde{S}_s}(\Delta\tilde{J}_v^{(m)}-\mathbb{E}'(\Delta\tilde{J}_u^{(m+1)})).\quad(5.50)$$

Putting together (5.48), (5.49), and (5.50), we get

$$\begin{aligned}\mathbb{E}'(\|\tilde{e}_w^{(m+1)}\|_{1-s}^2)&\leq\delta_m\|\tilde{e}_w^{(m)}\|_{1-s}^2+(1-\delta_m)\|\tilde{e}_v^{(m)}\|_{1-s}^2\\&\quad+\frac{2n(\tau'_m)^2}{\tilde{S}_s}\left(\Delta\tilde{J}_v^{(m)}-\mathbb{E}'(\Delta\tilde{J}_u^{(m+1)})\right)\\&\quad+\frac{2\tau'_m}{\tilde{S}'_s}(\delta_m\gamma_m\Delta\tilde{J}_u^{(m)}-(1+\delta_m\gamma_m)\Delta\tilde{J}_v^{(m)}-\frac{\tilde{\sigma}'_{1-s}}{2}\|\tilde{e}_v^{(m)}\|_{1-s}^2)\\&=\delta_m(\|\tilde{e}_w^{(m)}\|_{1-s}^2+\frac{2\tau'_m\gamma_m}{\tilde{S}'_s}\Delta\tilde{J}_u^{(m)})-\frac{2n(\tau'_m)^2}{\tilde{S}_s}\mathbb{E}'(\Delta\tilde{J}_u^{(m+1)}),\end{aligned}$$

where we have fixed the constants so that the terms with $\|\tilde{e}_v^{(m)}\|_{1-s}^2$ and $\Delta\tilde{J}_v^{(m)}$ in the above estimate cancel. For this, we need

$$(1-\delta_m)=\frac{\tau'_m\tilde{\sigma}'_{1-s}}{\tilde{S}'_s},\qquad\frac{n(\tau'_m)^2}{\tilde{S}_s}=\frac{2\tau'_m}{\tilde{S}'_s}(1+\delta_m\gamma_m).$$

If additionally

$$\frac{n(\tau'_m)^2}{\tilde{S}_s}=\frac{\tau'_m\gamma_m}{\tilde{S}'_s},$$

then the constants in front of $\mathbb{E}'(\Delta\tilde{J}_u^{(m+1)})$ and $\Delta\tilde{J}_u^{(m)}$ are the same. The last two equations are equivalent to

$$n\frac{\tilde{S}'_s}{\tilde{S}_s}\tau'_m=(1+\delta_m\gamma_m)=\gamma_m,$$

and can be used to determine τ'_m and α_m via γ_m. We then have

$$\gamma_m = \frac{1}{1-\delta_m}, \qquad \alpha_m = \frac{1}{2-\delta_m}, \qquad \tau'_m = \frac{\tilde{S}_s}{n\tilde{S}'_s(1-\delta_m)}. \tag{5.51}$$

Substitution into the remaining equation yields

$$(1-\delta_m) = \frac{\tilde{S}_s\tilde{\sigma}'_{1-s}}{n(\tilde{S}'_s)^2(1-\delta_m)},$$

from which we get

$$\delta_m = 1 - \sqrt{\frac{\tilde{S}_s\tilde{\sigma}'_{1-s}}{n(\tilde{S}'_s)^2}}. \tag{5.52}$$

This simplifies the factor in front of $\mathbb{E}'(\Delta\tilde{J}_u^{(m+1)})$ and $\Delta\tilde{J}_u^{(m)}$ to $2/\tilde{\sigma}'_{1-s}$, and we obtain a stationary recursive estimate for the conditional expectation of a fixed linear combination of $\Delta\tilde{J}_u^{(m)}$ and $\|\tilde{e}_w^{(m)}\|_{1-s}^2$, namely

$$\mathbb{E}'(\Delta\tilde{J}_u^{(m+1)} + \frac{\tilde{\sigma}'_{1-s}}{2}\|\tilde{e}_w^{(m+1)}\|_{1-s}^2) \le \delta_m\left(\Delta\tilde{J}_u^{(m)} + \frac{\tilde{\sigma}'_{1-s}}{2}\|\tilde{e}_w^{(m)}\|_{1-s}^2\right),$$

$m = 0, 1, \ldots$.

Iterating this inequality under the independence assumption proves the statements of Theorem 5.4. In fact, substituting the value for δ_m from (5.52) in (5.51) implies the formulas for τ'_m and α_m in (5.46). The final bound in terms of the initial objective error $\Delta\tilde{J}_u^{(0)}$ in (5.45) follows since, with (5.36), we have

$$\frac{\tilde{\sigma}'_{1-s}}{2}\|\tilde{e}_w^{(0)}\|_{1-s}^2 = \frac{\tilde{\sigma}'_{1-s}}{2}\|\tilde{e}_u^{(0)}\|_{1-s}^2 \le \tilde{J}(\tilde{u}^{(0)}) - \tilde{J}(\tilde{u}) = \Delta J_u^{(0)},$$

where we have used $w^{(0)} = u^{(0)}$ and $\nabla\tilde{J}(\tilde{u}) = 0$ for the minimizer $\tilde{u}$. □

The reader may wonder why we formulated (5.42) with the non-stationary parameters δ_m, α_m, and τ_m, even though they can be kept constant throughout the iteration according to (5.46). There are several reasons for this. First, with properly adjusted non-stationary values for these parameters, accelerated convergence rates can be obtained even without strong convexity conditions. This was the seminal result in [126] for first-order methods for minimizing convex functionals $\tilde{J}$ of the class $\mathcal{F}_L^{1,1}$. The case of randomized coordinate descent was considered in [127]. We refer to Problem 5.7 for the formulation of such results. Second, even in the case of strongly convex J, the non-stationary parameter settings proposed in [113, 127] seem to lead to slightly better practical performance than the stationary values from (5.46). Finally, a practical drawback of the above theory is that the algorithms analyzed involve parameters that require precise information about the partial Lipschitz constants $\tilde{L}_i$ and the strong convexity constants $\tilde{\sigma}'_{1-s}$. Especially the latter are difficult to evaluate or estimate. The Nesterov-type algorithm for $\mathcal{F}_L^{1,1}$ comes with weaker convergence rates, but does not require knowledge of the strong convexity constants. A different approach to this problem using restart techniques is discussed in [74, 132, 152].

5.4 Lower bounds

In this section, we provide lower bounds on the convergence speed of Schwarz iterative methods with finite Ω, and collect some examples that help to judge how sharp the upper convergence estimates obtained in Chapter 4 are.

5.4.1 Lower bounds for randomized Schwarz iterative methods

We begin with a lower bound for the randomized Schwarz iteration (4.40), which complements Theorem 4.6 and improves a lower bound given in [174] for the Kaczmarz iteration. We refer to [81] for similar results.

Theorem 5.5 *Let the conditions of Theorem 4.6 hold. Then the expected squared error of the iteration (4.40) satisfies*

$$\sup_{0\neq e^{(0)}\in V} \frac{\mathbb{E}(\|e^{(m)}\|^2)}{\|e^{(0)}\|^2} \geq (1-\mu_R')^{2m}, \qquad m=1,2,\ldots, \tag{5.53}$$

where μ_R' is defined in (4.39).

Proof We use the fact that

$$\mathbb{E}(\|e^{(m)}\|^2) \geq \|\mathbb{E}(e^{(m)})\|^2, \qquad m=1,2,\ldots,$$

and that the expected error trajectory $(E^{(m)} := \mathbb{E}(e^{(m)}))_{m=0,1,\ldots}$ satisfies a stationary linear iteration in V, namely

$$E^{(m+1)} = E^{(m)} - \frac{1}{\sum_{i\in\Omega}\rho_i\|R_i\|^2_{V_i\to V}} P_\rho E^{(m)}.$$

So,

$$E^{(m)} = (I-\omega P_\rho)^m E^{(0)}, \qquad \omega := (\sum_{i\in\Omega}\rho_i\|R_i\|^2_{V_i\to V})^{-1}.$$

Since $I-\omega P_\rho$ is symmetric positive definite, we get

$$\sup_{0\neq e^{(0)}\in V}\frac{\mathbb{E}(\|e^{(m)}\|^2)}{\|e^{(0)}\|^2} \geq \sup_{0\neq e^{(0)}\in V}\frac{\|\mathbb{E}(e^{(m)})\|^2}{\|e^{(0)}\|^2} = \|I-\omega P_\rho\|_V^{2m} = (1-\mu_R')^{2m},$$

because

$$\lambda_{\max}(I-\omega P_\rho) = 1-\omega\lambda_{\min}(P_\rho) = 1-\mu_R'.$$

This completes the proof. □

Since

$$(1-\mu_R')^{2m} \geq (1-2\mu_R')^m \geq (1-2m\mu_R'),$$

we get the weaker lower bound from [174, Theorem 3]. Compared to the upper bound (4.41) in Theorem 4.5, there is a gap. Another example from [174] shows that there are problem instances where (4.41) turns into equality for some $e^{(0)}$, and thus cannot be improved in general. A trivial example is given by the linear system $Ix = 0$ in $\mathbb{R}^n$ with identity matrix I and the space splitting of Example 1 in Subsection 2.1.1, where we set $\rho_i = 1$. In fact, in this case $\mu'_R = 1/n$, and if $x^{(0)}$ coincides with one of the coordinate vectors $e_j \in \mathbb{R}^n$, then $e_x^{(m)} = -e_j$ if $i_k \neq i$ for all $k = 0, \dots, m-1$ (by independence, this happens with probability $(1 - 1/n)^m$), otherwise we have $e^{(m)} = 0$. Consequently,

$$\mathbb{E}(\|e^{(m)}\|^2) = (1 - \mu'_R)^m \|e^{(0)}\|^2.$$

This still leaves open the question of whether the lower bound (5.53) can be improved.

Theorems 4.6 and 5.5 imply that the quantity μ'_R is indeed closely related to the expected convergence behavior of the randomized Schwarz iteration. This is useful when comparing different Schwarz methods that use the same space splitting. For example, if μ'_G defined in (4.62) satisfies the inequality

$$\mu'_G > 2\mu'_R,$$

then the pure greedy counterpart of the randomized Schwarz iteration (4.40) covered by Theorem 4.9 converges faster, i.e., the squared error of the pure greedy method decays faster than the expectation of the squared error of the randomized method. Since each finite section $(e^{(m)})_{m=0,1,\dots,M}$ of a pure greedy error trajectory is also realized by the randomized version with positive, albeit small probability, it can still happen by chance that an individual error trajectory of the randomized iteration progresses faster. This is rarely observed in practice.

Statements similar to Theorem 5.5 can be obtained for other randomized iterations whenever the associated iteration for the expected trajectory $(E^{(m)})_{m=0,1,\dots}$ is governed by a linear iteration in V. For a general class of randomized iterative methods for linear systems $Ax = b$, lower bounds follow this way from [81, Theorem 4.4]. As another example, for the algorithm covered by Theorem 4.7, we have

$$E^{(m+1)} = (I - \frac{r_m}{\lambda_{\max}} P_\rho) E^{(m)}, \qquad m = 0, 1, \dots,$$

and get the lower bound

$$\mathbb{E}(\|e^{(m)}\|^2) \geq \|E^{(m)}\|^2 = \|\prod_{k=0}^{m-1} (I - \frac{r_k}{\lambda_{\max}} P_\rho) e^{(0)}\|^2.$$

So,

$$\sup_{0 \neq e^{(0)} \in V} \frac{\mathbb{E}(\|e^{(m)}\|^2)}{\|e^{(0)}\|^2} \geq \|\prod_{k=0}^{m-1} (I - \frac{r_m}{\lambda_{\max}} P_\rho)\|_V^2 = \prod_{k=0}^{m-1} (1 - \frac{r_m}{\kappa})^2.$$

We leave it to the reader to formulate further results along this line of reasoning.

5.4.2 Sharpness of bounds for cyclic orderings

The upper convergence estimate (4.32) for multiplicative Schwarz iterations with cyclic ordering depends logarithmically on the number of subproblems, which is not the case for the additive Schwarz iteration and the methods with random and greedy orderings. It turns out that this is not an artifact of the proof methods. The following theorem provides a worst-case lower bound on the norm of the iteration matrix $M_\omega := M_{\rho,\omega}$ for a family of space splittings with $\rho_i = 1$.

Theorem 5.6 *There exists a sequence of $n \times n$ spd matrices A_n with unit diagonal such that, for the standard space splitting of Example 1 from Subsection 2.1.2 with the weight sequence $\rho_i = 1$, we have*

$$\kappa(A_n) \le C_0, \qquad \inf_{0<\omega<2} \|M_\omega\|_{V_n} \ge 1 - \frac{c_0}{\log n}, \qquad n = 2, 3, \ldots,$$

with some absolute constants $C_0, c_0 > 0$. Here, $V_n = \mathbb{R}^n$ is equipped with the scalar product induced by A_n.

Proof For A_n we choose the finite sections of the specific Toeplitz operator considered in Problem 2.5 in Chapter 2, where we arbitrarily fix the parameter $\alpha \in (0, 1)$. Then the entries of A_n are given by $a_{kk} = 1$ and

$$a_{kl} = \begin{cases} \frac{2\alpha(-1)^{r-1}}{\pi(2r-1)}, & |k-l| = 2r-1,\ r = 1, 2, \ldots, \\ 0, & |k-l| = 2, 4, \ldots, \end{cases}$$

independently of n. By the result in Part c) of Problem 2.5, we have

$$1 - \alpha \le \lambda_{\min}(A_n) < \lambda_{\max}(A_n) \le 1 + \alpha, \qquad n = 1, 2, \ldots,$$

and

$$\kappa(A_n) \nearrow \frac{1+\alpha}{1-\alpha} =: C_0, \qquad n \to \infty.$$

To obtain a lower bound for $\|M_\omega\|_{V_n}$, note that, for the space splitting of Example 1 with $\rho_i = 1$ from Subsection 2.1.2, the constant $c_1(\omega)$ in the XZ-identity $\|M_\omega\|_{V_n}^2 = 1 - 1/c_1(\omega)$ of Theorem 4.2 is equal to

$$\begin{aligned} c_1(\omega) &= \frac{\omega}{2-\omega} \sup_{z \in \mathbb{R}^n:\, \|z\|_2 = 1} (A_n(\omega^{-1}I + L_n^T)z, (\omega^{-1}I + L_n^T)z)_2 \\ &\ge \frac{\omega\lambda_{\min}(A_n)}{2-\omega} \|\omega^{-1}I + L_n^T\|_{2\to 2}^2, \end{aligned}$$

where L_n is the strictly lower triangular part of A_n. Consider the specific vector $z \in \mathbb{R}^n$ with the entries $z_{2r-1} = (-1)^r$ and $z_{2r} = 0$, $r = 1, 2, \ldots$. Then

$$\|z\|_2^2 = \lfloor \frac{n+1}{2} \rfloor.$$

A quick calculation of the entries of $y = (\omega^{-1}I + L_n^T)z$ shows

$$y_{2r-1}^2 = \omega^{-2}, \quad y_{2r}^2 = \left(\sum_{l=0}^{\lfloor (n-1)/2\rfloor - r} \frac{1}{2l+1}\right)^2 \geq c\log^2(n-2r), \qquad r = 1,2,\ldots.$$

This results in

$$\|\omega^{-1}I + L_n^T\|_{2\to 2}^2 \geq \frac{\|y\|_2^2}{\|z\|_2^2} \geq c(\omega^{-2} + \log^2 n), \qquad n = 2,3,\ldots,$$

with another constant $c > 0$. Since $\lambda_{\min}(A_n) \geq 1 - \alpha$ with fixed $\alpha \in (0,1)$, this altogether gives

$$c_1(\omega) \geq c\frac{\omega^{-1} + \omega\log^2 n}{2-\omega} \geq c_0 \log n, \quad 0 < \omega < 2, \quad n = 2,3,\ldots,$$

and we arrive at the desired lower bound for $\|M_\omega\|_{V_n}$. □

According to (4.11), this shows that the symmetrized multiplicative Schwarz method (i.e., in our special case, the symmetrized SOR iteration for linear systems) can exhibit worse convergence behavior than its additive counterpart. Since $r(M_\omega) \leq \|M_\omega\|_{V_n}$, the above result does not formally establish the same result for the standard Schwarz iteration (4.6) with cyclic ordering, for numerical evidence see [136].

5.4.3 Limits for acceleration

It is well-known that the bound obtained in Theorem 4.12 is best possible in the sense that for any fixed $C_0 > 4$, each $m = 1,2,\ldots,$ and $\kappa > C_0^2$, there are problem instances (i.e., variational problems and space splittings generating the operator P_ρ with condition number κ, solution u and initial approximation $u^{(0)}$) such that

$$\|e_u^{(m)}\|^2 \geq \left(1 - C_0\sqrt{\kappa}\right)^m \|e_u^{(0)}\|^2 \tag{5.54}$$

for some u and $u^{(0)}$. Moreover, such a lower bound holds for *any* method where the update formula is of the form

$$u^{(m+1)} = y^{(m)} + \theta_m P_\rho e_{z^{(m)}}, \qquad m = 0,1,\ldots, \tag{5.55}$$

where $\theta_m \in \mathbb{R}$ and both $y^{(m)}$ and $z^{(m)}$ are linear combinations of the previous iterates $u^{(k)}$, $k \leq m$, with coefficients independent of u and $u^{(0)}$. These requirements are satisfied for all standard non-adaptive linear multi-step methods. In other words, according to (5.54), except for the value of the constants in front of $\sqrt{\kappa}$, there is no hope of further reducing the dependence of the convergence estimates for additive Schwarz iterations on the condition number of the underlying space splitting.

The standard argument is as follows: Under the assumptions made for (5.55), it is easy to show by induction that the error after m iteration steps satisfies

$$e^{(m)} = (1-\gamma_m)u + p_m(P_\rho)e^{(0)}, \qquad m = 1, 2, \ldots,$$

for some $\gamma_m \in \mathbb{R}$ and some fixed sequence of algebraic polynomials p_m of degree $\leq m$ with $p_m(0) = \gamma_m$. Assuming further that $u^{(0)} = 0$ and thus $e^{(0)} = u$, we see that

$$e^{(m)} = q_m(P_\rho)e^{(0)}, \qquad m = 1, 2, \ldots,$$

for the polynomials $q_m(t) = 1 - \gamma_m + p_m(t)$ of degree $\leq m$ with $q_m(0) = 1$. According to a classical result of Bernstein, given any interval $I = [\lambda_{\min}, \lambda_{\max}]$ with $\lambda_{\min} > 0$, there exists for any integer M and $\epsilon > 0$ an increasing sequence

$$\lambda_{\min} = \lambda_1 < \lambda_2 < \cdots < \lambda_N = \lambda_{\max},$$

of sufficiently many points in I so that

$$\max_{i=1,\ldots,N} |q(\lambda_i)| \geq (1-\epsilon)\|q\|_{C(I)}$$

holds for all algebraic polynomials q of degree $\leq M$. Here it is possible to choose N of the order $M/\sqrt{\epsilon}$. Now select a problem instance such that P_ρ is compact and has eigenvalues λ_i, $i = 1, \ldots, N$. This can be realized by a linear algebra problem in $\mathbb{R}^N$ using the space splitting of Example 1. Then, by choosing as $u = e^{(0)}$ the eigenvector corresponding to the maximal value $|q_m(\lambda_i)|$, we see for given $m \leq M$ that

$$\frac{\|e_u^{(m)}\|^2}{\|e_u^{(0)}\|^2} = \max_{i=1,\ldots,N} |q_m(\lambda_i)|^2 \geq (1-2\epsilon)\|q_m\|^2_{C(I)}.$$

By properties of the Chebyshev polynomials T_m we have

$$\min_{q_m:\, q_m(0)=1} \|q_m\|_{C(I)} = \left(\max_{q_m:\, \|q_m\|_{C(I)}\leq 1} q_m(0)\right)^{-1} = T_m\left(\frac{\kappa+1}{\kappa-1}\right)^{-1} \geq \left(\frac{\sqrt{\kappa}-1}{\sqrt{\kappa}+1}\right)^{2m},$$

and arrive at

$$\frac{\|e_u^{(m)}\|^2}{\|e_u^{(0)}\|^2} \geq (1-2\epsilon)\left(1-\frac{2}{\sqrt{\kappa}+1}\right)^{2m} \geq (1-2\epsilon)\left(1-\frac{4}{\sqrt{\kappa}}\right)^{m}.$$

All that remains is to choose now ϵ sufficiently small to establish (5.54) for given constant $C_0 > 4$.

However, the Krylov space argument behind this reasoning cannot be used to prove the sharpness of the upper bounds for accelerated randomized Schwarz iterations given in Theorem 4.13 or Part a) of Theorem 4.14. The examples from [128, Chapter 2], which were used to show lower complexity estimates for first-order methods for convex minimization problems, concern the preasymptotic range $m < n$, in which the upper bounds in our theorems are not effective. They only provide evidence that much better error decay factors than those given in (4.89) and (4.107), respectively, are not to be expected. The following example is worked out in full detail in Problem 5.8 below. For arbitrarily fixed odd $n = 2N - 1$, denote by A the tridiagonal $n \times n$ spd Toeplitz matrix with entries $a_{ii} = 1$ on the main diagonal and $a_{ij} = -1/2$ for $|i - j| = 1$ on the other two diagonals. Consider the linear system $Ax = b$, where the entries of the right-hand side b are given by $b_i = 0$ if $i = 1, \ldots, n$, $i \neq N$, and $b_N = 1$. Its solution x is given by

$$x_{n\pm i} = N - i, \qquad i = 0, \ldots, N - 1.$$

If we use the standard space splitting of Example 1 from Subsection 2.1.2 with $\rho_i = 1$, then

$$P_\rho = A$$

satisfies

$$\lambda_{\min} = 1 - \cos(\frac{\pi}{2N}) \asymp \frac{\pi^2}{2(n+1)^2}, \qquad \lambda_{\max} = 1 + \cos(\frac{\pi}{2N}) \asymp 2 - \frac{\pi^2}{2(n+1)^2},$$

and thus $\sqrt{\kappa} \asymp 2(n+1)/\pi$, $n \to \infty$.

Starting with $x^{(0)} = 0$, we have

$$\|e_x^{(0)}\|^2 = (Ax, x)_2 = (b, x)_2 = N = \frac{n+1}{2}.$$

Some elementary but tedious calculations, which are postponed to Problem 5.8, show that

$$\frac{\mathbb{E}(\|e_x^{(m)}\|^2)}{\|e_x^{(0)}\|^2} \geq 1 - \frac{m}{n(n+1)}, \qquad m = 1, \ldots, N. \tag{5.56}$$

This lower bound holds for *any* algorithm of the form

$$x^{(m+1)} = y^{(m)} + \theta_m P_{i_m} e_{z^{(m)}}, \qquad m = 0, 1, \ldots, \tag{5.57}$$

where the index i_m is chosen uniformly at random and independently, and $y^{(m)}$ and $z^{(m)}$ are arbitrary linear combinations of the previous iterates $x^{(k)}$, $k \leq m$. For any fixed m, (5.56) compares well with the error reduction factors

$$\left(1 - \frac{1}{n\sqrt{\kappa}}\right)^m \asymp 1 - \frac{\pi m}{2n(n+1)}, \quad \left(1 - \frac{\sqrt{\lambda_{\min}}}{\sqrt{n\nu_R'}}\right)^m \asymp 1 - \frac{\pi m}{\sqrt{2}n(n+1)}, \qquad n \to \infty,$$

after m steps, which for this example result from the one-step estimates (4.89) and (4.107) for Nesterov-type methods in Theorem 4.13 and 4.14, respectively. Note, however, that we tacitly neglected the additional term in these estimates, which is not justified in the preasymptotic range $m \ll n$.

We do not currently know if a similar example exists for the range $m \gg n$ or for greedy Schwarz methods.

5.5 Robustness

The convergence estimates provided so far for Schwarz iterative methods do not take into account perturbations (e.g. due to rounding) and other errors. The question arises whether the described algorithms are sensitive or insensitive to errors. The short answer is that in general they are, especially for systematic errors in the execution of the iteration steps. However, for certain reasonable models of error behavior, robustness results can be obtained, sometimes at the cost of changes in the algorithms and somewhat worse guarantees for error decay. We now formulate some results along these lines. Throughout this section we consider stable space spittings with finite index set $\Omega = \{1, \dots, n\}$ and w.l.o.g. set $\rho_i = 1$.

5.5.1 Bounded additive errors

First, consider the perturbed multiplicative Schwarz iteration

$$u^{(m+1)} = u^{(m)} + \omega_m P_{i_m} e^{(m)} + \xi_m, \qquad \|\xi_m\| \le \sigma_m, \qquad m = 0, 1, \dots, \tag{5.58}$$

with singleton $I_m = \{i_m\}$, where the additive error ξ_m is uniformly bounded but otherwise arbitrary. Depending on the source of the error, different a priori information about its size may be available. For simplicity, we assume that

$$\sigma_m \le \sigma' + \sigma'' \|e^{(m)}\|, \qquad \sigma', \sigma'' \ge 0. \tag{5.59}$$

The first term could be motivated by rounding errors proportional to the size of $u^{(m)} \approx u$, while the second term refers to possible errors in the computation of $P_{i_m} e^{(m)}$, which are generally small when $e^{(m)}$ is small. If only perturbations from the evaluation of $P_{i_m} e^{(m)}$ are relevant, then the perturbed iteration can also be written as

$$u^{(m+1)} = u^{(m)} + \omega_m (P_{i_m} e^{(m)} + \xi_m), \qquad m = 0, 1, \dots. \tag{5.60}$$

This makes a difference when using decreasing step-sizes $\omega_m \to 0$, see the next subsection.

If i_m is chosen by a weak greedy rule with weakness parameter β_m uniformly away from zero, i.e., $0 < \beta \le \beta_m \le 1$, then according to the results in Section 4.4 we have the bound

$$\|e^{(m+1)}\| \le \|e^{(m)} - \omega_m P_{i_m} e^{(m)}\| + \|\xi_m\| \le q_m \|e^{(m)}\| + \sigma_m,$$

where $q_m \le q < 1$ is the error reduction factor of the unperturbed iteration. For simplicity, we assume an m-independent upper bound $q < 1$ for it, which implies that the step-sizes must satisfy $\omega_m \ge \omega > 0$. Thus, if $\tilde{q} := q + \sigma'' < 1$ and considering (5.59), we conclude by iterating that

$$\|e^{(m)}\| \le \tilde{q}^m \|e^{(0)}\| + \frac{\sigma'}{1-\tilde{q}}, \qquad m = 1, 2, \ldots.$$

In other words, we can no longer guarantee convergence, but we can still get close to u within a distance proportional to σ'. The proportionality factor depends on both σ'' and the error reduction factor q of the unperturbed iteration.

Similar statements apply to multiplicative Schwarz methods with deterministic or random ordering, as long as an error reduction estimate is available for some $q < 1$. For example, for the deterministic version with cyclic ordering, we obtain the formula

$$e^{(n)} = M_\omega e^{(0)} + \sum_{k=0}^{n-1} \prod_{l=k+2}^{n} (I - \omega P_l)\xi_k,$$

which, under mild assumptions such as $\|I - \omega P_l\|_V \le 1$, $l \in \Omega$, and $\|e^{(k)}\| \le C_0 \|e^{(0)}\|$, $k = 1, \ldots, n-1$, gives the bound

$$\|e^{(n)}\| = \|M_\omega\|_V \|e^{(0)}\| + \sum_{k=0}^{n-1} \|\xi_k\| \le (\|M_\omega\|_V + C_0\sigma'' n)\|e^{(0)}\| + n\sigma'.$$

Now assume $\tilde{q} := \|M_\omega\|_V + C_0\sigma'' n < 1$ and iterate. This gives the estimate

$$\|e^{(kn)}\| \le \tilde{q}^k \|e^{(0)}\| + \frac{\sigma' n}{1-\tilde{q}}, \qquad k = 1, 2, \ldots.$$

for the error after k inner loops.

For the randomized case, use the inequality

$$2a(v, w) \le 2\|v\|\|w\| \le \epsilon\|v\|^2 + \epsilon^{-1}\|w\|^2$$

with $v = e^{(m)} - \omega_m P_{i_m} e^{(m)}$ and $w = \xi_m$ to estimate

$$\begin{aligned} \|e^{(m+1)}\|^2 &= \|e^{(m)} - \omega_m P_{i_m} e^{(m)}\|^2 + 2a(e^{(m)} - \omega_m P_{i_m} e^{(m)}, \xi_m) + \|\xi_m\|^2 \\ &\le (1+\epsilon)\|e^{(m)} - \omega_m P_{i_m} e^{(m)}\|^2 + (1+\epsilon^{-1})\|\xi_m\|^2. \end{aligned}$$

After substituting the bound $\|\xi_m\|^2 \le 2(\sigma')^2 + 2(\sigma'')^2\|e^{(m)}\|^2$, one can take the conditional expectation on both sides and use the error estimates for the unperturbed case to the first term. Then, for sufficiently small ϵ and σ'', one can again proceed as for the greedy case and obtain a similar bound for $\|e^{(m)}\|^2$.

5.5.2 Independent random errors

We again consider the perturbed iteration (5.58), but this time we assume that ξ_m is a V-valued random noise independent of $u^{(k)}$, $k \le m$, and ξ_k, $k < m$, with zero mean and bounded variation. Thus $\tilde{\mathbb{E}}(\xi_m) = 0$, $\tilde{\mathbb{E}}(\|\xi_m\|^2) = \sigma_m^2$, where the expectation $\tilde{\mathbb{E}}$

is taken with respect to the probability distribution underlying ξ_m. Then

$$\tilde{\mathbb{E}}(a(e^{(m)} - \omega_m P_{i_m} e^{(m)}, \xi_m)) = a(e^{(m)} - \omega_m P_{i_m} e^{(m)}, \tilde{\mathbb{E}}(\xi_m)) = 0,$$

and consequently

$$\mathbb{E}'(\|e^{(m+1)}\|^2) = \mathbb{E}'(\|e^{(m)} - \omega_m P_{i_m} e^{(m)}\|^2) + \tilde{\mathbb{E}}(\|\xi_m\|^2) \le q\|e^{(m)}\|^2 + \sigma_m^2.$$

Here, $q < 1$ denotes the error reduction factor of the unperturbed iteration. Now, together with the independence assumption, the arguments of the previous subsection apply and lead to bounds of the form

$$\mathbb{E}(\|e^{(m)}\|^2) \le q^m \|e^{(0)}\|^2 + \sum_{k=0}^{m-1} q^{m-1-k}\sigma_k^2 \le q^m\|e^{(0)}\|^2 + \frac{\max_{k\in\mathbb{Z}_+}\sigma_k^2}{1-q},$$

$m = 1, 2, \ldots$. The difference to the results of the previous subsection is that σ_m now represents a probabilistic average instead of a uniform bound on the errors, while the error decay to a neighborhood of the solution is only guaranteed in expectation.

If the random noise enters the iteration as in (5.60), one can even achieve convergence in expectation. There are two easy ways to do this: Decreasing the step-sizes $\omega_m \to 0$ and averaging the iterates $u^{(m)}$. We will demonstrate this for an academic example, namely the additive Schwarz iteration with independent additive zero-mean noise and uniformly bounded noise variance $\sigma_m^2 \le \sigma^2$, which we write in the form

$$u^{(m+1)} = u^{(m)} + \omega_m(Pe^{(m)} + \xi_m), \qquad m = 0, 1, \ldots, \tag{5.61}$$

where P is the Schwarz operator associated with the underlying space splitting. Iterating this formula and introducing the product operators

$$M_k := \prod_{l=k}^{m-1}(I - \omega_l P), \quad k = 0, \ldots, m-1, \qquad M_m = I,$$

we get the error representation

$$e^{(m)} = M_0 e^{(0)} - \sum_{k=0}^{m-1} \omega_k \xi_k', \qquad \xi_k' := M_{k+1}\xi_k.$$

The random variables ξ_k', $k = 0, \ldots, m-1$, are mutually independent, independent of $M_0 e^{(0)}$, and satisfy

$$\tilde{\mathbb{E}}(\xi_k') = M_{k+1}\tilde{\mathbb{E}}(\xi_k) = 0.$$

So by our assumptions we have

$$\mathbb{E}(\|e^{(m)}\|^2) = \|M_0 e^{(0)}\|^2 + \sum_{k=0}^{m-1} \omega_k^2 \tilde{\mathbb{E}}(\|\xi_k'\|^2) \le \|M_0\|_V^2 \|e^{(0)}\|^2 + \sigma^2 \sum_{k=0}^{m-1} \omega_k^2 \|M_{k+1}\|_V^2.$$

To proceed, we estimate the norms of the above product operators under the natural restriction $0 < \omega_m \leq 1/\lambda_{\max}$ for $k = 0, \ldots, m-1$ as

$$\|M_k\|_V^2 = \prod_{l=k}^{m-1} (1 - \omega_l \lambda_{\min})^2 \leq e^{-2\lambda_{\min} \sum_{l=k}^{m-1} \omega_l},$$

where the last step follows from taking logarithms and applying the inequality $\log(1-t) \leq -t$, $0 < t < 1$. This gives

$$\mathbb{E}(\|e^{(m)}\|^2) \leq e^{-2\lambda_{\min} s_{m-1}} \|e^{(0)}\|^2 + \sigma^2 \sum_{k=0}^{m-1} \omega_k^2 e^{-2\lambda_{\min}(s_{m-1} - s_k)}, \quad s_k := \sum_{l=0}^{k} \omega_l.$$

To prove convergence for $m \to \infty$ from this estimate, we need $s_k \to \infty$, which together with $\sum_{k=0}^{\infty} \omega_k^2 < \infty$ is also sufficient. In addition, convergence rates can be obtained by considering specific step-size sequences. Typical step-size choices are $\omega_m = A(m+1)^{-t}$ with $1/2 < t \leq 1$, where $0 < A \leq 1/\lambda_{\max}$. For more detailed results see [70].

A further improvement is obtained by averaging, even with constant step-sizes $\omega_m = \omega \in (0, 1/\lambda_{\max}]$. For this purpose, given the sequence from (5.61), we define the corresponding sequence of averages

$$\bar{u}^{(m)} = \frac{1}{m+1} \sum_{k=0}^{m} u^{(k)} = \frac{m}{m+1} \bar{u}^{(m-1)} + \frac{1}{m+1} u^{(m)}, \quad m = 1, 2, \ldots,$$

$\bar{u}^{(0)} = u^{(0)}$. By iterating with (5.61) we get the representation

$$\begin{aligned}
(m+1)\bar{e}^{(m)} &= e^{(m)} + \cdots + e^{(0)} \\
&= \sum_{k=0}^{m} (I - \omega P)^k e^{(0)} - \sum_{k=0}^{m-1} \sum_{l=0}^{m-1-k} (I - \omega P)^l \xi_k \\
&= \frac{1}{\omega} P^{-1} \left((I - (I - \omega P)^{m+1}) e^{(0)} - \sum_{k=0}^{m-1} (I - (I - \omega P)^{m-k}) \xi_k \right)
\end{aligned}$$

for the error $\bar{e}^{(m)} := u - \bar{u}^{(m)}$ of the averaged sequence. From the independence assumption and using $\|P^{-1}\|_V = \lambda_{\min}^{-1}$ and $\|I - (I - \omega P)^l\|_V \leq 1$, $l = 1, 2, \ldots$, we get

$$(m+1)^2 \mathbb{E}(\|\bar{e}^{(m)}\|^2) \leq \frac{1}{(\omega \lambda_{\min})^2} \left(\|e^{(0)}\|^2 + \sum_{k=0}^{m-1} \tilde{\mathbb{E}}(\|\xi_k\|^2) \right).$$

So if we choose the largest possible value $\omega = \lambda_{\max}^{-1}$ for the step-size, then we have

$$\mathbb{E}(\|\bar{e}^{(m)}\|^2) \leq \kappa^2 \left(\frac{\|e^{(0)}\|^2}{(m+1)^2} + \frac{\sigma^2}{m+1} \right).$$

5.6 Stochastic approximation

The statements about the convergence of randomized Schwarz methods in Chapter 4 and 5 were formulated in terms of the expectations of the squared error norms. In this section, we give some details on stochastic approximation methods, of which randomized Schwarz methods are a special case, and mention some other types of probabilistic convergence. For more theoretical background and formal definitions, the reader may consult [165] or standard works on stochastic approximation [10, 111], see also [143, Chapter 6].

To stay within the framework of this book, we restrict ourselves to discrete-time stochastic approximation processes of the form

$$u^{(m+1)} = u^{(m)} + \omega_m G(u^{(m)}, \Theta_m), \qquad m = 0, 1, \ldots, \tag{5.62}$$

where $u^{(0)}$ is given, $G : V \times X \to V$ is a certain fixed map, V is a separable Hilbert space, and Θ_m are X-valued random variables defined on probability spaces $(Y, \mathcal{F}_m, \mathbb{P}_m)$. Here, $\mathcal{F}_0 \subset \mathcal{F}_1 \subset \cdots \subset \mathcal{F}$ is an increasing sequence of σ-subalgebras of a σ-algebra $\mathcal{F}$ on Y and $\mathbb{P}_m$ is the restriction of a fixed probability measure $\mathbb{P}$ defined on $(Y, \mathcal{F})$ to the σ-subalgebra $\mathcal{F}_m$. The sequence $(\mathcal{F}_m)_{m\in\mathbb{Z}_+}$ is called a filtration for $(Y, \mathcal{F}, \mathbb{P})$, with the stochastic process $(\Theta_m)_{m\in\mathbb{Z}_+}$ adapted to it. Such notions are standard in martingale theory. Obviously, the iteration (5.62) induces another, now V-valued process $(u^{(m)})_{m\in\mathbb{Z}_+}$. The goal is to study the behavior of this process as $m \to \infty$. Standard references on stochastic approximation such as [10, 111] study the finite-dimensional case $V = \mathbb{R}^d$ in a much more general way. For example, more terms can be included in (5.62), G can be replaced by a function G_m, and $u^{(0)}$ can be assumed to be a random variable itself. The form (5.62) is usually called the Robbins-Monro algorithm. It was originally designed to find zeros of univariate functions using noisy function evaluations. The generalization to the V-valued stochastic processes created by (5.62) is not too difficult. For basic facts on martingale theory see [165], information on V-valued martingales can be found in [51, 104].

Let us first clarify that the randomized Schwarz iterations considered in Chapter 4 are special instances of (5.62). For simplicity, consider the one-step Schwarz iteration

$$u^{(m+1)} = u^{(m)} + \omega_m P_{i_m} e^{(m)}, \qquad m = 0, 1, \ldots, \tag{5.63}$$

with i_m i.i.d. random variables taking values in $X := \Omega = \{1, \ldots, n\}$ according to a probability distribution (4.33). Since the underlying space splitting $(\mathbf{V}; \mathbf{a}; \mathbf{R})$ is finite, we can assume that the operators R_i are uniformly bounded, i.e., $\|P_i\|_V = \|R_i\|^2_{V_i \to V} \le \Lambda$ for some $\Lambda < \infty$. One can take as Y the set $\Omega^{\mathbb{Z}_+}$ of all infinite sequences $\mathbf{i} = (i_0, i_1, \ldots)$ with entries in Ω and define $\Theta_m(\mathbf{i}) = i_m$. Now the probability space $(Y, \mathcal{F}, \mathbb{P})$ and its filtration can be constructed as follows: Let $\mathcal{F}_m$ be the minimal σ-algebra generated by the n^{m+1} mutually disjoint sets

$$A_{j_0,\ldots,j_m} = \{\mathbf{i} \in Y : \ i_k = j_k, \ k = 0, \ldots, m\}$$

and induce the probability measure $\mathbb{P}_m$ on $\mathcal{F}_m$ by

$$\mathbb{P}_m(A_{j_0,\dots,j_m}) = p_{j_0}\cdots p_{j_m}, \quad j_k \in \Omega, \quad k = 0,\dots,m.$$

Obviously $\mathcal{F}_m$ is finite, $\mathcal{F}_m \subset \mathcal{F}_{m+1}$, and $\mathbb{P}_{m+1}|_{\mathcal{F}_m} = \mathbb{P}_m$ for every $m \in \mathbb{Z}_+$. This naturally defines a σ-finite premeasure on $\cup_{m\in\mathbb{Z}_+}\mathcal{F}_m$ whose restriction to $\mathcal{F}_m$ coincides with $\mathbb{P}_m$, and Caratheodory's unique extension theorem can be used to define the probability space $(Y,\mathcal{F},\mathbb{P})$ so that the above properties hold, where $\mathcal{F}$ is the smallest σ-algebra containing all $\mathcal{F}_m$.

The function $G(v,i) := P_i(u-v)$ has the remarkable property that

$$\mathbb{E}_{\mathcal{F}_m}(G(u^{(m)},\Theta_m)) = \sum_{i=1}^{n} p_i P_i e^{(m)} =: P_{\mathbf{p}}e^{(m)}$$

depends only on $u^{(m)}$. Here, $\mathbb{E}_{\mathcal{F}_m}$ denotes the conditional expectation with respect to $\mathcal{F}_m$. Loosely speaking, this means the expectation with respect to given Θ_k, $k < m$, and thus also to fixed $u^{(k)}$, $k \le m$. This is a nice property which, with minor modifications, is one of the standard assumptions in stochastic approximation. We refer to [111, Section 5.1], where the conditional expectations $\bar{g}(\cdot) := \mathbb{E}_{\mathcal{F}_m}(G(\cdot,\Theta_m))$ may also depend on m and include an additional additive term $\beta_m \to 0$, or to [10, Part II, Section 1.1]. Thus from (5.63) we conclude

$$\mathbb{E}_{\mathcal{F}_m}(u^{(m+1)}) = u^{(m)} + \omega_m P_{\mathbf{p}}(u-u^{(m)}), \qquad m = 0,1,\dots.$$

This shows that the sequence $(z^{(m)})_{m\in\mathbb{N}}$ given by

$$z^{(m)} = u^{(m)} - \sum_{k=0}^{m-1}\omega_k P_{\mathbf{p}}(e^{(k)}), \qquad m = 0,1,\dots,$$

forms a V-valued martingale adapted to the filtration $(\mathcal{F}_m)_{m\in\mathbb{Z}_+}$. For the definition of V-valued martingales, as well as sub- and supermartingales and martingale difference sequences, see [51, Section 3.4] or [104, Section 3]. Indeed, $z^{(m)}$ is $\mathcal{F}_m$-measurable (even $\mathcal{F}_{m-1}$-measurable, where $\mathcal{F}_{-1} = \{\emptyset, Y\}$), $\mathbb{E}(\|z^m\|) < \infty$, which is trivially satisfied since $z^{(m)}$ can take only finitely many values in V, and

$$\begin{aligned}\mathbb{E}_{\mathcal{F}_m}(z^{(m+1)}) &= \mathbb{E}_{\mathcal{F}_m}(u^{(m+1)} - \sum_{k=0}^{m}\omega_k P_{\mathbf{p}}(u-u^{(k)}))\\ &= u^{(m)} - \sum_{k=0}^{m-1}\omega_k P_{\mathbf{p}}(u-u^{(k)}) = z^{(m)},\end{aligned}$$

$m = 0,1,\dots$. Recall that $z^{(0)} = u^{(0)}$ is deterministically fixed.

Thus, martingale inequalities and convergence theorems based on them can be used to show convergence almost surely (a.s.) with respect to $(Y,\mathcal{F},\mathbb{P})$. Note that a.s. convergence does not necessarily follow from the so-called mean square convergence $\mathbb{E}(\|e^{(m)}\|^2) \to 0$, which was already established in Section 4.3 under certain conditions on the step-sizes ω_m and on the Schwarz operator $P_{\mathbf{p}}$. Traditionally, the common strategy to prove convergence of stochastic approximation algorithms is to rewrite (5.62) as

$$u^{(m+1)} = u^{(m)} + \omega_m \bar{g}(u^{(m)}) + \omega_m (G(u^{(m)}, \Theta_m) - \bar{g}(u^{(m)})).$$

Here, the last term is a martingale difference adapted to $(\mathcal{F}_m)_{m\in\mathbb{Z}_+}$, while the term $u^{(m)} + \omega_m \bar{g}(u^{(m)})$ corresponds to the Euler discretization of the continuous-time initial value problem

$$\frac{d}{dt} u(t) = \bar{g}(u(t)), \quad t > 0, \qquad u(0) = u^{(0)}, \tag{5.64}$$

in V with step-size sequence $(\omega_m)_{m\in\mathbb{Z}_+}$. Then, compactness arguments and stability theory based on Lyapunov functions for the initial value problem (5.64) can be exploited to establish convergence of the stochastic approximation method in various settings, such as convergence in distribution or in the a.s. or mean square sense. In addition to stochastic noise, which enters in particular via the martingale differences

$$\omega_m (G(u^{(m)}, \Theta_m) - \bar{g}(u^{(m)}))$$

mentioned above, the switch from the discrete to the continuous dynamical system introduces new perturbations that have to be dealt with. Note that assumptions like

$$\mathbb{E}_{\mathcal{F}_{m-1}}(\|G(v, \Theta_m)\|^2) \le C(1 + \|v\|^2), \tag{5.65}$$

for some constant C, and

$$a(u - v, \bar{g}(v)) \ge \mu \|u - v\|^2, \qquad v \in V, \tag{5.66}$$

with some $\mu > 0$ appear in the stochastic approximation literature to obtain quantitative estimates of the convergence rate in the mean square or the weak sense. Obviously, (5.66) implies absolute stability of the stationary point u of the problem (5.64). Both (5.65) and (5.66) are trivially satisfied in the case of the randomized Schwarz iteration (5.63). Moreover, the growth condition (5.65) holds in a much stronger form, namely

$$\mathbb{E}_{\mathcal{F}_m}(\|G(v, \Theta_m)\|^2) = \mathbb{E}_{\mathcal{F}_m}(\|P_i e^{(m)}\|^2) \le \Lambda a(P_{\mathbf{p}} e^{(m)}, e^{(m)})$$

for some $\Lambda < \infty$. This leads to better convergence bounds than can be expected from condition (5.65) in the general case, where only $O(\omega_m)$ rates are derived for $\mathbb{E}(\|e^{(m)}\|^2)$ if the step-size sequence satisfies

$$\sum_{m=0} \omega_m = \infty, \qquad \sum_{m=0} \omega_m^2 < \infty,$$

compare [10, Part II, 1.10] or [143, Proposition 6.2].

In our restricted setting of one-step Schwarz methods based on space splittings, one can circumvent these more sophisticated arguments and directly apply Doob's martingale convergence theorem (see, e.g., Theorem 1 and Corollary 1 in [165, Chapter VII.4.]) formulated for supermartingales: Let $(Z_m)_{m\in\mathbb{Z}_+}$ be a real-valued, non-negative supermartingale adapted to the filtration $(\mathcal{F}_m)_{m\in\mathbb{Z}_+}$, i.e., $Z_m : Y \to \mathbb{R}_+$ is $\mathcal{F}_m$-measurable with $\mathbb{E}_{\mathcal{F}_m}(Z_{m+1}) \le Z_m < \infty$. Then $Z_m \to Z$ a.s. on Y, where Z is some a.s. finite random variable on $(Y, \mathcal{F}, \mathbb{P})$.

An appropriate supermartingale for our purposes is defined by

$$Z_m = \|e^{(m)}\|^2 + \sum_{k=0}^{m-1} \omega_k(2 - \Lambda\omega_k)a(e^{(k)}, P_{\mathbf{p}}e^{(k)}), \qquad m = 0, 1, \dots,$$

where the constant $\Lambda \ge \|R_i\|^2_{V_i \to V}$ is fixed and $0 < \omega_k < 2/\Lambda$. In fact, the bounds $0 \le Z_m < \infty$ and the $\mathcal{F}_m$-measurability of Z_m are obvious, and the supermartingale property $\mathbb{E}_{\mathcal{F}_m}(Z_{m+1}) \le Z_m < \infty$ follows from the standard error bound

$$\|e^{(m+1)}\|^2 \le \|e^{(m)}\|^2 - \omega_m(2 - \Lambda\omega_m)a(e^{(k)}, P_{i_m}e^{(k)})$$

for the one-step Schwarz method (5.63) when we take conditional expectations

$$\begin{aligned}\mathbb{E}_{\mathcal{F}_m}(\|e^{(m+1)}\|^2) &\le \mathbb{E}_{\mathcal{F}_m}(\|e^{(m)}\|^2 - \omega_m(2 - \Lambda\omega_m)a(e^{(k)}, P_{i_m}e^{(k)})) \\ &= \|e^{(m)}\|^2 - \omega_m(2 - \Lambda\omega_m)a(e^{(m)}, P_{\mathbf{p}}e^{(m)}),\end{aligned}$$

and substitute this into the formula for $\mathbb{E}_{\mathcal{F}_m}(Z_{m+1})$.

Thus, $(Z_m)_{m\in\mathbb{Z}_+}$ is adapted to $(\mathcal{F}_m)_{m\in\mathbb{Z}_+}$ and satisfies the assumptions of the above version of Doob's convergence theorem. This gives $Z_m \to Z < \infty$ a.s. on $(Y, \mathcal{F}, \mathbb{P})$. Consequently, by definition of Z_m we have

$$\sum_{k=0}^{\infty} \omega_k(2 - \Lambda\omega_k)a(e^{(k)}, P_{\mathbf{p}}e^{(k)}) \le \lim_{m\to\infty} Z_m = Z < \infty.$$

on a set of full measure in Y. Thus, $a(e^{(k)}, P_{\mathbf{p}}e^{(k)}) \to 0$ a.s. on Y, if the step-size sequence satisfies

$$0 < \liminf_{m\to\infty} \omega_m \le \limsup_{m\to\infty} \omega_m < 2/\Lambda.$$

This is satisfied for the typical step-size choices discussed in Section 4.3. Now, since $\lambda_{\min}(P_{\mathbf{p}}) > 0$, also $\|e^{(m)}\| \to 0$ a.s..

Other types of convergence, such as convergence in distribution and convergence in probability, are automatic consequences of our mean square convergence results and the Chebyshev inequality. We have

$$\mathbb{P}(\|e^{(m)}\|^2 > \lambda) \le \frac{\mathbb{E}(\|e^{(m)}\|^2)}{\lambda}, \qquad \lambda > 0.$$

In fact, whenever $\mathbb{E}(\|e^{(m)}\|^2) \to 0$, we have $\mathbb{P}(\|e^{(m)}\|^2 > \lambda) \to 0$ for every $\lambda > 0$, which by definition is equivalent to $\|e^{(m)}\| \to 0$ in probability. For formal definitions and other properties of probabilistic convergence types, see [165, II.10,III.1].

Finally, note that some authors consider another, even weaker type of convergence, namely $\|\mathbb{E}(e^{(m)})\| \to 0$, i.e., the convergence to 0 of the mean error trajectory $E^{(m)} := \mathbb{E}(e^{(m)})$, $m = 1, 2, \ldots$, in V. We have already used the sequence $(\|E^{(m)}\|)_{m\in\mathbb{Z}_+}$ to prove lower bounds for mean square convergence. In fact, we always have $\|\mathbb{E}(e^{(m)})\|^2 \le \mathbb{E}(\|e^{(m)}\|^2)$, which means in particular that mean square convergence implies $E^{(m)} \to 0$ in V, but not vice versa. Bounding $\|\mathbb{E}(e^{(m)})\|$ is often much easier, but can be misleading. There are instances of randomized Schwarz methods where $E^{(m)}$ converges to 0 in V with optimal speed, while $\mathbb{E}(\|e^{(m)}\|^2)$ does not converge to 0 at all. An example is discussed in Problem 5.9.

Problems

5.1 Strengthened Cauchy-Schwarz inequality for linear systems.
Let $Ax = b$ be a linear system with spd matrix A. Consider the space splitting from Example 1 in Subsection 2.1.2.
a) Show that the best possible constants γ_{ij} for the strengthened Cauchy-Schwarz inequality to hold are given by

$$\gamma_{ij} = |a_{ij}|.$$

This yields

$$\lambda_{\max}(I_\rho(A)) \le \|\Gamma_\rho\|_{\ell^2} = \lambda_{\max}(I_\rho \mathrm{abs}(A)),$$

where $\mathrm{abs}(A)$ is the matrix with entries $|a_{ij}|$. Note that $\mathrm{abs}(A)$ is not $|A|(= \sqrt{AA^T})$.
b) Give examples of matrices A for which the estimate from Part a) is not tight, i.e., where $\lambda_{\max}(I_\rho(A))$ is much smaller than $\lambda_{\max}(I_\rho \mathrm{abs}(A))$.
Hint. Consult Problem 2.5.

5.2 Gap between μ_G and μ_R.
Under the conditions given in Section 5.1 for randomized and greedy SOR iterations, prove that the gap μ_G/μ_R satisfies

$$1 \le \mu_G/\mu_R \le \max_{i=1,\ldots,n} p_i^{-1} = (\min_{i=1,\ldots,n} p_i)^{-1}$$

for all $n \times n$ spd matrices A and all probability distributions $\mathbf{p}$, and show that

$$\sup_{A,\mathbf{p}} \mu_G/\mu_R = \infty.$$

Furthermore, show that

$$\mu_G/\mu_R \le \kappa(D^{-1}A)$$

if $\mathbf{p}$ is the uniform distribution.

5.3 Comparison of deterministic, randomized and greedy Schwarz iterations for a small linear system.
Consider a linear system $Ax = b$ with the 2×2 matrix

$$A = \begin{pmatrix} 1 & c \\ c & 1 \end{pmatrix}, \qquad 0 < c < 1,$$

and use the splitting of Example 1 in Subsection 2.1.2 to set up the multiplicative Schwarz iteration with cyclic, random, and greedy orderings. Assume that $e^{(0)} \neq 0$ and that the components of the residual vector $r^{(0)} = b - Ax^{(0)}$ satisfy $|r_1^{(0)}| \geq |r_2^{(0)}|$.
a) Show that both cyclic and greedy ordering give the same error formula

$$\|e^{(m+1)}\|^2 = \frac{c^{2m}}{1 + (1 - c^2)(r_2^{(0)}/r_1^{(0)} - c)^2}\|e^{(0)}\|^2 \leq \frac{1+c}{2}c^{2m}\|e^{(0)}\|^2$$

in the energy norm induced by A for $m = 0, 1, \ldots$. If $r_2^{(0)}/r_1^{(0)} = c$, the iteration terminates after the first step.
b) For the randomized version with $p_1 = p_2 = 1/2$ (uniform probability distribution), show that

$$\mathbb{E}(\|e^{(m+1)}\|^2) = \left(\frac{1+c^2}{2}\right)^{m+1} \|e^{(0)}\|^2, \qquad m = 0, 1, \ldots.$$

c) Check that $\mu_G = \mu_R = (1 - c)/2$. For $c \longrightarrow 0$, the comparison with the results in Part a) and b) shows the weakness of the a priori convergence estimates reported in this chapter, especially for the greedy case.
Hint. Note that after the first iteration step, the residual vectors are always multiples of either $(0\ 1)^T$ or $(1\ 0)^T$, a case that is easy to analyze.

5.4 Linear systems with semi-definite A.
Let $Ax = b$ be a consistent linear system with a symmetric positive (but not positive definite!) $n \times n$ matrix A, and consider the splitting of Example 1 in Subsection 2.1.2 with weights $\rho_i = 1$. Note that the form $a(x, y) = y^T Ax$, $x, y \in \mathbb{R}^n$, yields only a semi-norm.
a) Modify $(\mathbf{V}; \mathbf{a}; \mathbf{R})$ by replacing the definition of R_i with

$$\tilde{R}_i x_i = x_i Q_V e_i, \qquad i = 1, \ldots, n, \qquad V := \operatorname{ran}(A),$$

where $Q_V : \mathbb{R}^n \longrightarrow V$ is the ortho-projection to V with respect to the Euclidean scalar product and e_i are the unit coordinate vectors in $\mathbb{R}^n$. Check that $(\mathbf{V}; \mathbf{a}; \tilde{\mathbf{R}})$ is a stable space splitting for $(V; a)$, with stability constants

$$\lambda_{\min} = \lambda^+_{\min}(A), \quad \lambda_{\max} = \lambda_{\max}(A),$$

and condition number

$$\kappa = \kappa^+(A) = \frac{\lambda_{\max}(A)}{\lambda^+_{\min}(A)}.$$

b) Consider the update formula

$$x^{(m+1)} = x^{(m)} + \omega_m \sum_{i \in I_m} (b - Ax^{(m)})_i e_i, \qquad m = 0, 1, \ldots,$$

which (in the consistent case) is formally equivalent to (4.1) using the original space splitting $(\mathbf{V}; \mathbf{a}; \mathbf{R})$. Show that the V-component $z^{(m)} = Q_V x^{(m)}$ of $x^{(m)}$ satisfies the iteration

$$z^{(m+1)} = z^{(m)} + \omega_m \sum_{i \in I_m} (b - Az^{(m)})_i Q_V e_i, \qquad m = 0, 1, \ldots,$$

which corresponds to (4.1) for the space splitting $(\mathbf{V}; \mathbf{a}; \tilde{\mathbf{R}})$ of Part a).
c) Let x denote any solution of the consistent system $Ax = b$. Show that the iterates $z^{(m)}$ (i.e., the V-components of $x^{(m)}$) converge to the unique V-component $z := Q_V x$ of x, under conditions and with rates similar to those formulated for the spd case in Section 5.1. Also show that the components $w^{(m)} := x^{(m)} - z^{(m)}$ in $V^\perp = \ker(A)$ converge to *some* $w \in \ker(A)$.
d) Use Part c) to establish the results of Section 5.2 on the Kaczmarz method for consistent linear systems.
Hints: For a) and b), check that $\tilde{R}_i x_i = (Ax)_i$. This implies $Px = Q_V Ax$ and $P = (A)|_V$. For c), only the convergence of $w^{(m)}$ needs an additional argument. To see this, check that the limit w (if it exists) is equal to

$$w = \sum_{m=0}^{\infty} \omega_m \sum_{i \in I_m} (A(z - z^{(m)}))_i Q_{V^\perp} e_i.$$

The convergence of this series follows from the exponential convergence of $\|z - z^{(m)}\|$ and the uniform boundedness of $(\omega_m)_{m \in \mathbb{Z}_+}$. Special care is needed in the case of randomly chosen I_m, where w is a $V^\perp$-valued random variable.

5.5 Kaczmarz iteration for the extended system (5.11).
Derive the formulas and convergence estimates of Kaczmarz iterations in $V := \mathrm{ran}(A^T) \times \mathbb{R}^{n'} \subset \mathbb{R}^{n+n'}$ for solving the system (5.11).
Hints. Find formulas for μ_R, μ_G, and consider the randomized version of the corresponding Nesterov iteration.

5.6 Formulas for $\nu_{R,\gamma}$.
Prove the formula (5.18) for $\nu_{R,\gamma}$.

5.7 Nesterov acceleration for minimization of convex functionals $J \in \mathcal{F}_L^{1,1}$.
Prove that, for $J \in \mathcal{F}_L^{1,1}$, the algorithm

$$\begin{aligned}
\tilde{u}^{(m+1)} &= \tilde{v}^{(m)} + (\tilde{L}'_{i_m})^{-1} \nabla_{i_m} \tilde{J}(\tilde{v}^{(m)}), \\
\tilde{w}^{(m+1)} &= \tilde{w}^{(m)} + \tau'_m (\tilde{L}'_{i_m})^{-1} \nabla_{i_m} \tilde{J}(\tilde{v}^{(m)}), \qquad (5.67) \\
\tilde{v}^{(m+1)} &= (1 - \alpha'_{m+1}) \tilde{u}^{(m+1)} + \alpha'_{m+1} \tilde{w}^{(m+1)},
\end{aligned}$$

satisfies the convergence estimate

$$\mathbb{E}(\Delta\tilde{J}_u^{(m)}) \le \frac{C\Delta\tilde{J}_u^{(0)}}{(m+1)^2}, \qquad m = 1, 2, \ldots,$$

with some positive constant C. More precisely, as in Theorem 5.4, i_m is determined according to the probability distribution (5.40) and the parameters τ'_m, α'_{m+1} must be properly chosen.
Hints. Note that (5.67) coincides with (5.42) by setting $\delta_m = 1$ and $\alpha_{m+1} = 1 - \alpha'_{m+1}$. With this specification, follow the proof of Theorem 5.4 up to the estimate

$$\mathbb{E}'(\|\tilde{e}_w^{(m+1)}\|_{1-s}^2) \le \|\tilde{e}_w^{(m)}\|_{1-s}^2 + \frac{2n(\tau'_m)^2}{\tilde{S}_s}\left(\Delta\tilde{J}_v^{(m)} - \mathbb{E}'(\Delta\tilde{J}_u^{(m+1)})\right) + \frac{2\tau'_m}{\tilde{S}'_s}\left(\gamma_m\Delta\tilde{J}_u^{(m)} - (1+\gamma_m)\Delta\tilde{J}_v^{(m)}\right),$$

where $\gamma_m = (1-\alpha'_m)/\alpha'_m$ (have in mind that for $J \in \mathcal{F}_L^{1,1}$ the lower bound (5.44) holds with $\tilde{\sigma}'_{1-s} = 0$). If one sets $\tau'_m = (1+\gamma_m)\tilde{S}_s/(n\tilde{S}'_s)$ to cancel the terms involving $\Delta\tilde{J}_v^{(m)}$, then one obtains

$$\mathbb{E}'(\|\tilde{e}_w^{(m+1)}\|_{1-s}^2) + \frac{2\tilde{S}_s(1+\gamma_m)^2}{n(\tilde{S}'_s)^2}\mathbb{E}'(\Delta\tilde{J}_u^{(m+1)}) \le \|\tilde{e}_w^{(m)}\|_{1-s}^2 + \frac{2\tilde{S}_s(1+\gamma_m)\gamma_m}{n(\tilde{S}'_s)^2}\Delta\tilde{J}_u^{(m)}.$$

So to get a recursion, determine the sequence γ_m such that

$$(1+\gamma_{m+1})\gamma_{m+1} = (1+\gamma_m)^2,$$

or, equivalently,

$$(\alpha'_{m+1})^{-1} = 1 + \gamma_{m+1} = \frac{1}{2} + \sqrt{(\alpha'_m)^{-2} + 1/4} \ge \sqrt{(\alpha'_0)^{-2} + 1/4} + \frac{m+1}{2}.$$

This gives the result if one considers that $\|\tilde{e}_w^{(0)}\|_{1-s}^2$ can be bounded by $\Delta\tilde{J}^{(0)}$. The constant C depends on the quotient $n(\tilde{S}'_s)^2/\tilde{S}'_s$ and the choice of $\alpha'_0 \in (0, 1]$.

5.8 Proof of (5.56).
Let A be the tridiagonal $n \times n$ matrix with the entries

$$a_{ii} = 1, \quad a_{ij} = -1/2, \qquad i, j = 1, \ldots, n, \ |i-j| = 1,$$

where $n = 2N - 1$ is odd, and consider the vector x given with $x_{N\pm i} = N - i$, $i = 0, \ldots, N-1$. Set $\|z\| := \sqrt{(Az, z)_{\ell^2}}$.
a) Check that $(Ax)_i = 0$ for $i \ne N$ and $(Ax)_N = 1$.
b) For $s, r = 1, \ldots, N-1$, let $X_{s,r}$ denote the set of all vectors z such that $z_i = x_i$ for all $i \le N - s$ and $i \ge N + r$, and let

$$X_j := \cup_{s,r:\, s+r=j+1} X_{s,r}, \qquad j = 1, \ldots, N.$$

Show that

$$\min_{z \in X_j} \|z\|^2 \geq N - \frac{j+1}{4}, \qquad j = 1, \ldots, N.$$

c) Let $x^{(0)} = 0$ and assume that $x^{(m)}$ is obtained by an algorithm of the form (5.57). There are n^m different index paths $\tilde{I}_m := (i_0, \ldots, i_{m-1})$, each associated with a particular realization of the random variable $e^{(m)}$ and occurring with probability n^{-m}. We say that an index path $\tilde{I}_m$ belongs to J_0 if $i_k \neq N$ for all $k = 0, \ldots, m-1$. For $j = 1, \ldots, m$, we say that $\tilde{I}_m$ belongs to J_j if it contains a subpath $(i_{k_1}, \ldots, i_{k_j})$, $0 \leq k_1 < \cdots < k_j \leq m-1$, of j *different* indices such that, first, $i_{k_1} = N$ and second, for $l = 2, \ldots, j$ the index i_{k_l} is a left or right "neighbor" of the set $\{i_{k_1}, \ldots, i_{k_{l-1}}\}$. Formally, this means that $i_{k_l} \neq i_{k_r}$ for $r = 1, \ldots, l-1$, and

$$\min_{r=1,\ldots,l-1} |i_{k_l} - i_{k_r}| = 1.$$

Show that $e^{(m)} \in X_j$ whenever the associated $\tilde{I}_m$ does not belong to J_{j+1}. Furthermore, check that $\mathbb{P}(\tilde{I}_m \in J_0) = (1 - 1/n)^m$ and

$$\mathbb{P}(\tilde{I}_m \in J_j \backslash J_{j+1}) = \frac{2^{j-1}}{n^j} \sum_{k=0}^{m-j} \binom{m-k-1}{j-1} \left(1 - \frac{1}{n}\right)^k \left(1 - \frac{2}{n}\right)^{m-k-j},$$

if $j = 1, \ldots, m \leq N$. Here we have set $J_{m+1} = \emptyset$.
d) Using b) and c), check (5.56).
Hints: For b), check that

$$\min_{z \in X_{s,r}} \|z\|^2 = N - \frac{sr}{s+r} \geq N - \frac{s+r}{4}, \qquad s, r = 1, \ldots, N-1,$$

by examining the minimizing vector z. It turns out that the missing entries z_i, $i = N-s+1, \ldots, N+r-1$, of the latter are obtained by linearly "interpolating" between x_{N-s} and x_{N+r}. For c), use the tridiagonal structure of A to prove the first statement. The rest is to count the number of index paths $\tilde{I}_m$ in $J_j \backslash J_{j+1}$. It is instructive to do this first for $j = 0, 1, 2$.

5.9 Convergence in expectation may be misleading.
Consider the $n \times n$ spd matrix

$$A := (1-\alpha)I + \alpha \mathbf{1}\mathbf{1}^T = \begin{pmatrix} 1 & \alpha & \alpha & \cdots & \alpha & \alpha \\ \alpha & 1 & \alpha & \cdots & \alpha & \alpha \\ \alpha & \alpha & 1 & \cdots & \alpha & \alpha \\ \vdots & \vdots & \vdots & & \vdots & \vdots \\ \alpha & \alpha & \alpha & \cdots & 1 & \alpha \\ \alpha & \alpha & \alpha & \cdots & \alpha & 1 \end{pmatrix}$$

with unit diagonal, where the parameter α is fixed and satisfies $-1/n < \alpha < 1$, $\alpha \neq 0$. Consider the randomized SOR iteration

$$x^{(m+1)} = x^{(m)} + \omega(b - Ax^{(m)})_{i_m} e_{i_m}, \qquad m = 0, 1, \dots,$$

for solving $Ax = b$, where the index i_m is chosen independently and uniformly at random from $\Omega = \{1, \dots, n\}$.
a) Show that the expected error trajectory $(\mathbb{E}(e^{(m)}))_{m\in\mathbb{Z}_+}$ converges to 0 in $\mathbb{R}^n$ for all $0 < \omega < 2n/(1 + (n-1)\alpha)$ and all starting vectors $x^{(0)}$.
b) Show that

$$\mathbb{E}(\|e^{(m)}\|^2) \to \infty, \qquad m \to \infty,$$

for some $x^{(0)}$ if $2 \leq \omega < 2n/(1 + (n-1)\alpha)$. Recall that $\|\cdot\|$ is the vector norm induced by A.
Hints. For a), use the stationary iteration for $\mathbb{E}(e^{(m)})$. For b), derive a recursion for the matrices

$$B^{(m)} = a_m I + b_m \mathbf{1}\mathbf{1}^T$$

so that

$$\mathbb{E}(\|e^{(m)}\|^2) = (e^{(0)})^T B^{(m)} e^{(0)}, \qquad m = 0, 1, \dots.$$

Then show that $a_m + b_m \to \infty$ if $\omega \geq 2$, and choose $x^{(0)}$ appropriately.

Chapter 6
Schwarz approximation methods: Infinite Ω

Abstract This chapter reviews the theory of approximation methods using space splittings with countably many spaces V_i. Greedy and randomized versions are included.

6.1 New aspects

Throughout this chapter, let $(\mathbf{V}; \mathbf{a}; \mathbf{R})$ be a collection with a countable index set

$$\Omega := \mathbb{N}$$

that satisfies the conditions listed at the beginning of Section 2. To avoid artificial difficulties, we assume that the separable Hilbert space V is infinite-dimensional and that the span of the linear sets $\{\mathrm{ran}(R_i)\}_{i\in\Omega}$ is dense in $(V; a)$, i.e., we assume that every $v \in V$ can be approximated by finite linear combinations of terms $R_i v_i$, $v_i \in V_i$, with arbitrary accuracy. This is necessary for Schwarz methods to potentially converge for any $u \in V$. Furthermore, we require that the operator family $\mathbf{R}$ is uniformly bounded, i.e., there exists a positive constant Λ such that

$$\|P_i\|_V = \|R_i\|^2_{V_i \to V} \leq \Lambda, \qquad i \in \Omega. \tag{6.1}$$

This condition can be achieved by appropriate scaling. We do not assume a priori the existence of a bounded Schwarz operator, i.e., a weight sequence ρ is not involved. Weights will come into play later.

First, consider a simple example. Take a separable infinite-dimensional Hilbert space V with scalar product $a(u, v) := (u, v)_V$ and norm $\|v\| = \|v\|_V$ and fix an arbitrary CONS $D := \{\phi_i\}_{i\in\Omega}$ for it. Then the dictionary splitting $\mathbf{V}_D$ is an orthogonal Riesz splitting for $(V; (\cdot; \cdot)_V)$, see Example 4 in Subsection 2.1.2 and Subsection 2.2.1. The counterpart of the steepest descent Schwarz method (4.7) with $\mathbf{V}_D$ as underlying space splitting for solving the trivial operator equation $u = u$ in V is

M. Griebel, P. Oswald, *Hilbert Space Splittings and Iterative Methods*, Springer Series in Computational Mathematics 62, https://doi.org/10.1007/978-3-031-74370-2_6

$$u^{(m+1)} = u^{(m)} + \begin{cases} (u, \phi_{i_m})_V \phi_{i_m}, & i_m \neq i_k,\ k < m, \\ 0, & \text{otherwise,} \end{cases} \qquad m = 0, 1, \ldots. \tag{6.2}$$

Starting with $u^{(0)} = 0$, this method is equivalent to recovering u from its sequence of coefficients $\{c_i := (u, \phi_i)_V)\}_{i \in \Omega}$ with respect to D by an incremental process. We call this an *approximation method* rather than an iterative method, since we never touch all subproblems, but only finitely many in the updating process. Note at this point that the update step of every Schwarz approximation method considered in this chapter is based on the computation of one or more search directions $P_i e^{(m)} = P_i u - P_i u^{(m)}$. This again requires the computation of $R_i^* u$ via a variational problem in V_i, for which the evaluation of scalar products $a(u, R_i v_i)$ is necessary. If, as in the previous chapters, u solves an operator equation $Au = f$ and $a(u, v) = \langle Au, v \rangle_{V' \times V}$, then this is indeed possible without direct access to u, since $f \in V'$ is given. In other situations, as in our example above, we simply assume that the scalar products $a(u, R_i v_i)$ are available for each $v_i \in V_i$, $i \in \Omega$.

There is also a fundamental difference with the case of finite Ω concerning the nature of the convergence results. Let us review the three main versions of executing (6.2) based on deterministic, greedy, and randomized orderings. The natural deterministic ordering $i_m = m + 1$, $m = 0, 1, \ldots$, corresponds to the classical partial sum approximation of the orthogonal decomposition $u = \sum_{i=1}^{\infty} c_i \phi_i$ with respect to D, in which case the error

$$\|e^{(m)}\|^2 = \sum_{i=m+1}^{\infty} c_i^2, \qquad m = 1, 2, \ldots,$$

converges to zero for all $u \in V$, but with no guaranteed rate. The same is true for the pure greedy ordering, where $u^{(m)}$ is the sum of those terms from the series $u = \sum_{i \in \Omega} c_i \phi_i$ with the m largest in absolute value coefficients c_i. Then we have

$$\|e^{(m)}\|^2 = \sum_{i=m+1}^{\infty} |c_i^*|^2, \qquad m = 1, 2, \ldots,$$

where $(c_i^*)_{i \in \Omega}$ denotes the non-increasing rearrangement of $(c_i)_{i \in \Omega}$, i.e., $c_i^* = c_{\sigma(i)}$, $i \in \Omega$, for some fixed bijection $\sigma : \Omega \to \Omega$ such that $|c_1^*| \geq |c_2^*| \geq \cdots$. In other words, in this case the above method realizes the rate of the best possible *nonlinear m-term approximation* with respect to the dictionary $D = \{\phi_i\}_{i \in \Omega}$, i.e.,

$$\|e^{(m)}\| = \sigma_m(u; D)_V := \inf_{I \subset \mathbb{N}:\, |I| \leq m} \inf_{d_i \in \mathbb{R}} \|u - \sum_{i \in I} d_i \phi_i\|, \qquad m = 1, 2, \ldots.$$

Finally, if i_m is independently drawn at random according to a fixed probability distribution $\mathbf{p} = (p_i)_{i \in \Omega}$, where $p_i = \mathbb{P}(i_m = i) > 0$ and $\sum_{i \in \Omega} p_i = 1$, then

$$\mathbb{E}(\|e^{(m)}\|^2) = \sum_{i \in \Omega} (1 - p_i)^m c_i^2, \qquad m = 1, 2, \ldots.$$

This holds because the probability of not choosing a certain index i in any of the first m steps of the iteration is equal to $(1 - p_i)^m$. As long as $p_i > 0$ for all $i \in \Omega$, this guarantees $\mathbb{E}(\|e^{(m)}\|^2) \to 0$, again with no chance of obtaining a convergence rate that is valid for arbitrary $u \in V$.

To obtain convergence rates, additional structural assumptions must be made on u. For example, in the case of deterministic orderings, the assumption

$$\|\mathbf{c}\|^2_{\ell^2_{\mathbf{w}}} := \sum_{i\in\Omega} w_i c_i^2 < \infty, \qquad \mathbf{c} = (c_i)_{i\in\Omega},$$

where the weight sequence $\mathbf{w} = (w_i)_{i\in\Omega}$ is monotonically increasing to ∞, implies

$$\|e^{(m)}\|^2 \le w_{m+1}^{-1}\|\mathbf{c}\|^2_{\ell^2_{\mathbf{w}}}, \qquad m = 1, 2, \ldots.$$

In other words, we have the guaranteed convergence rate $\|e^{(m)}\|^2 = \mathrm{O}(w_{m+1}^{-1})$ for all $u \in V$ whose coefficient sequence $\mathbf{c}$ belongs to $\ell^2_{\mathbf{w}}(\Omega)$. Note that this condition is also necessary under mild growth assumptions on $\mathbf{w}$.

Similarly, for the above randomized version of (6.2), one obtains the estimate

$$\mathbb{E}(\|e^{(m)}\|^2) \le (m+r)^{-r}\|\mathbf{c}\|^2_{\ell^2_{\mathbf{p}^{-r}}}, \qquad m = 1, 2, \ldots,$$

where $r > 0$ is arbitrary and $\mathbf{p}^{-r} = (p_i^{-r})_{i\in\Omega}$. This is because

$$\sum_{i\in\Omega}^{\infty}(1-p_i)^m c_i^2 \le \left(\sup_{i\in\Omega} p_i^r(1-p_i)^m\right)\sum_{i\in\Omega}^{\infty} p_i^{-r}c_i^2 \le \frac{1}{(m+r)^r}\|\mathbf{c}\|^2_{\ell^2_{\mathbf{p}^{-r}}}.$$

Here we have used that for $r > 0$ and $m = 1, 2, \ldots$ the maximum of the function $\phi(t) = t^r(1-t)^m$ on $[0, 1]$ is reached at $t_0 = (m + r)^{-1}$.

Thus, in both deterministic and randomized cases, convergence rates can be obtained if the coefficient sequence $\mathbf{c}$ associated with $u \in V$ belongs to $\ell^2_{\mathbf{w}}(\Omega)$ for certain weight sequences $\mathbf{w}$, where in the random version this $\mathbf{w}$ is induced by the probability distribution $\mathbf{p}$. Since in our example $P_i u = (u, \phi_i)_V \phi_i$, we have

$$\|P_{\mathbf{w}}^{1/2}u\|^2 = (P_{\mathbf{w}}u, u)_V^2 = \sum_{i\in\Omega} w_i(u, \phi_i)_V^2 = \|\mathbf{c}\|^2_{\ell^2_{\mathbf{w}}}.$$

This means that the convergence conditions can be formulated using spaces defined in terms of the Schwarz operator $P_{\mathbf{w}}$ associated with the dictionary splitting $(\mathbf{V}_D; \mathbf{w})$. In particular, for the randomized version, the above bound holds if $u \in V^r_{P_{\mathbf{p}}}$, where the spaces $V^r_{P_{\mathbf{p}}}$ are defined in Subsection 2.4.2.

The convergence of the greedy version of (6.2) is determined by the properties of the non-increasing rearrangement $\mathbf{c}^*$ of $\mathbf{c}$. Thus, it is not easily captured by $\ell^2_{\mathbf{w}}$ norms of $\mathbf{c}$ or properties of some $P_{\mathbf{w}}$. One possibility, explored in the theory of greedy methods [179], is to require $\mathbf{c} \in \ell^p(\Omega)$ with $p < 2$. This allows us to obtain algebraic convergence rates of the form

$$\|e^{(m)}\|^2 = C(m+1)^{-(2/p-1)}, \qquad m = 1, 2, \ldots. \tag{6.3}$$

In fact, from the above error representation for the greedy version of (6.2) we first deduce that $\|e^{(m)}\|^2 \le C(m+1)^{-r}, m = 1, 2, \ldots$, is equivalent to

$$|c_m^*|^2 \le C' m^{-r-1} \qquad m = 2, 3, \ldots,$$

where C/C' is bounded away from 0 and ∞ with constants only depending on $r > 0$. In one direction, this is obvious from the properties of the series, since

$$\|e^{(m)}\|^2 = \sum_{i=m+1}^{\infty} |c_i^*|^2 \le C' \sum_{i=m+1}^{\infty} i^{-r-1} \le C' C_r (m+1)^{-r}.$$

For the opposite direction, we use the monotonicity of $\mathbf{c}^*$ and get for $m = 2, 3, \ldots$

$$|c_m^*|^2 \le \frac{2}{m} \sum_{i=\lfloor m/2 \rfloor+1}^{m} |c_i^*|^2 \le \frac{2}{m} \|e^{(\lfloor m/2 \rfloor)}\|^2 \le \frac{2C}{m(\lfloor m/2 \rfloor + 1)^r} \le \frac{CC_r'}{m^{r+1}}.$$

With this at hand, we arrive at (6.3), since by the properties of the non-increasing rearrangement $\mathbf{c}^*$ we have

$$|c_m^*|^2 \le \left(\frac{1}{m} \sum_{i=1}^{m} |c_i^*|^p \right)^{2/p} \le m^{-2/p} \|\mathbf{c}^*\|_{\ell^p}^2 = m^{-2/p} \|\mathbf{c}\|_{\ell^p}^2$$

for $\mathbf{c} \in \ell^p(\Omega)$, $p < 2$.

Although the setting of the above example is very special, we will show in this chapter how the above observations can be applied, at least in part, to Schwarz approximation methods for general countable collections $(\mathbf{V}; \mathbf{a}; \mathbf{R})$. In the following section, we give some abstract convergence estimates for iterations to solve an operator equation $Pu = g$ in V, where $P \in \mathcal{B}_{pos}(V)$ is compact. These estimates involve the V_p^s norms introduced in Subsection 2.4.2. The results should be compared with Theorems 4.1 and 4.12 in Chapter 4, where similar iterations with invertible operators $P = P_\rho$ were considered for solving the Schwarz equation $P_\rho u = g$ in the case of finite Ω. Of course, even if P has the structure of a Schwarz operator for some weight sequence ρ, the iterations considered in Section 6.2 are not a practical Schwarz method, since $\Omega = \mathbb{N}$ is infinite.

In the later sections on randomized and greedy versions of Schwarz approximation schemes, we will concentrate on update formulas of the form

$$u^{(m+1)} = \alpha_m u^{(m)} + \mu_m P_{i_m} e^{(m)}, \quad i_m \in \Omega, \qquad m = 0, 1, \ldots,$$

which differ from (4.1) by the restriction to singletons I_m and by an additional sequence of regularization parameters α_m. For the latter, a standard choice is $\alpha_m = (m+1)/(m+2)$. Without such a modification, optimal rates cannot be obtained in some situations.

6.2 Iterations with $P \in \mathcal{B}_{pos}(V)$

To motivate the considerations in the following sections and to provide guidance for the comparison with the convergence results for additive Schwarz methods in Chapter 4, assume for the moment that $(\mathbf{V}; \mathbf{a}; \mathbf{R})$ is a space splitting for $(V; a)$ with $\Omega = \mathbb{N}$ and with the Schwarz operator

$$P := \sum_{i \in \Omega} P_i \in \mathcal{B}_{pos}(V).$$

If this space splitting is stable, then the conclusions of Theorems 4.1 and 4.11 concerning additive Schwarz iterations remain valid without change. More generally speaking, the analysis of Krylov subspace methods for iteratively solving operator equations $Pu = g$ with invertible $P \in \mathcal{B}_{pos}(V)$ in separable Hilbert spaces is a straightforward extension of the analysis for linear systems in finite-dimensional spaces, for the latter we refer to [115]. For non-invertible and, in particular, for compact operators $P \in \mathcal{B}_{pos}(V)$ (this corresponds to the semi-definite case in numerical linear algebra or the non-strongly convex case in convex optimization), Krylov subspace methods have received some attention as iterative regularization methods for ill-posed linear operator equations governed by non-invertible linear operators $T \in \mathcal{B}(V, U)$ between Hilbert spaces. The reader is encouraged to consult [71, 100], where some semi-iterative and cg-methods with the operator $P = T^*T$ are analyzed using a similar methodology but slightly different notation.

It turns out that quantitative convergence rates for solving operator equations $Pu = g$ with non-invertible $P \in \mathcal{B}_{pos}(V)$ by Krylov subspace methods can be established only under additional smoothness assumptions on u or g. These convergence rates are weaker than the geometric decay rates obtained for invertible P, but are also applicable to the case of invertible P when we have no knowledge of $\lambda_{\min}(P)$, and in particular to the Schwarz equations treated in Chapter 4. The necessary smoothness requirements are formulated in terms of the spaces V_P^s. Although the following theorem holds for any injective operator $P \in \mathcal{B}_{pos}(V)$, we formulate and prove it only under the condition that P is compact. Then the family of Hilbert spaces V_P^s has a straightforward definition in terms of the eigenvectors and eigenvalues of P. For the properties of V_P^s used below, see Subsection 2.4.2 and in particular Theorem 2.7. Similar spaces appear in [71, 100], where they are denoted $\mathcal{X}_\mu$.

Theorem 6.1 *Consider the operator equation $Pu = g$, where $P \in \mathcal{B}_{pos}(V)$ is compact and injective. For any $s \in \mathbb{R}$, if $g \in V_P^{s+2}$, then the solution u belongs to V_P^s and the Richardson iteration*

$$u^{(m+1)} = u^{(m)} + \omega P e^{(m)}, \qquad m = 0, 1, \ldots, \qquad e^{(m)} = u - u^{(m)},$$

starting with $u^{(0)} \in V_P^s$, converges to u for $0 < \omega < 2/\lambda_{\max}(P)$ in V_P^t, $-\infty < t < s < +\infty$. If $0 < \omega < (1 + m^{-(s-t)/(2m)})/\lambda_{\max}(P)$, then we have the bound

$$\|e^{(m)}\|_{V_P^t}^2 \le \frac{C}{(\omega m)^{s-t}} \|e^{(0)}\|_{V_P^s}^2, \qquad m = 1, 2, \ldots, \tag{6.4}$$

where C depends only on $s - t$.

Proof The error propagation

$$e^{(m+1)} = (I - \omega P)e^{(m)}, \qquad m = 0, 1, \ldots, \tag{6.5}$$

is stationary and is governed by the symmetric operator

$$(I - \omega P)v = \sum_{k\in\mathbb{N}} (1 - \omega\lambda_k)a(v, \psi_k)\psi_k,$$

where λ_k and ψ_k are the eigenvalues and eigenvectors of P, respectively. Thus, by the definition of the spaces V_P^t, we have

$$\|(I - \omega P)^m\|^2_{V_P^s \to V_P^t} = \sup_{k\in\mathbb{N}} |1 - \omega\lambda_k|^{2m} \lambda_k^{s-t}.$$

We refer to Subsection 2.4.2 for details.

To bound the supremum on the right, we use the fact that for $\alpha = s - t > 0$ we have

$$\phi(x) := |1 - x|^{2m} x^\alpha \le Cm^{-\alpha}, \qquad 0 < x < x_m := 1 + m^{-\alpha/2m},$$

with some constant C depending only on α. In fact, on the subinterval $[0, 1]$, the continuous function $\phi(x)$ reaches its maximum value

$$\phi(x'_m) = \frac{(2m)^{2m}\alpha^\alpha}{(2m + \alpha)^{2m+\alpha}} \le \frac{(\alpha/2)^\alpha}{m^\alpha}$$

at $x'_m = \alpha/(2m + \alpha)$, while it increases on $[1, x_m]$. So on the whole interval $[0, x_m]$ we have

$$\phi(t) \le \max(\phi(x'_m), \phi(x_m)) \le \frac{\max((\alpha/2)^\alpha, 2^\alpha)}{m^\alpha}.$$

Since $0 < \omega\lambda_k < x_m$ under the given restrictions on ω, we get (6.4) with constant $C \max((\alpha/2)^\alpha, 2^\alpha)$. Since $x_m \to 2$ for $m \to \infty$, this also establishes the convergence of the Richardson method for all $0 < \omega < 2/\lambda_{\max}(P)$. □

Theorem 6.1 provides guidance for some of the results to be proved later in this chapter. First, we should expect algebraic convergence rates rather than geometric convergence rates, and this only under additional assumptions on u and e_0 and depending on the spectral properties of P. Second, instead of devising one-step recursions for the squared errors $\|e^{(m)}\|^2$ (or their expectations), we relied on error representations $e^{(m)} = M_m e^{(0)}$ with explicit M_m for each $m = 1, 2, \ldots$, and obtained error norm estimates by using mapping properties of the operators M_m. Note that for the above iteration, one-step error recursions of the form $\|e^{(m+1)}\|^2 \le q_m\|e^{(m)}\|^2$ with $q_m < 1$ cannot hold independently of $e^{(0)}$. Third, error estimates with respect to $\|\cdot\|_{V_P^t}$ norms may be possible. Finally, since the expected error trajectory $\mathbb{E}(e^{(m)})$ for randomized Schwarz methods often satisfies the recursion (6.5) with some appropriate P, we can derive lower bounds for $\mathbb{E}(\|e^{(m)}\|^2)$ using the same trick as in Section 5.4.

Next, we present a partial improvement on the convergence rates in Theorem 6.1. Consider the two-step iteration

$$u^{(m+1)} = 2u^{(m)} - u^{(m-1)} + \omega P e^{(m)}, \qquad m = 0, 1, \ldots, \tag{6.6}$$

starting with given $u^{(-1)} = u^{(0)}$ and fixed step-size $0 < \omega < 2/\lambda_{\max}(P)$. This is the limit case of the heavy-ball method [147]. It resembles the iteration covered by Theorem 4.12, if we set $\eta_m = 1$, $\zeta_m = \omega$ and $\xi_m = 0$. We will see that as such it does not converge in any V_P^t if $e^{(0)} \neq 0$. However, compared to the result in Theorem 6.1, the averages

$$\bar{u}^{(m+1)} = \frac{1}{m+1}\sum_{l=0}^{m} u^{(l)} = \frac{m}{m+1}\bar{u}^{(m)} + \frac{1}{m+1}u^{(m)}, \qquad m = 0, 1, \ldots, \tag{6.7}$$

converge at a higher rate for $0 < s - t < 2$.

Theorem 6.2 *Let the notation and assumptions of Theorem 6.1 hold. Then for the iteration (6.6) we have $\|e^{(m)}\|_{V_P^t} \not\to 0$ for any $u^{(0)} \neq u$. The averages (6.7) converge for $s > t$ and $0 < \omega < 2/\lambda_{\max}(P)$ with the rate*

$$\|\bar{e}^{(m)}\|_{V_P^t}^2 \le \frac{C}{(\omega(m+1)^2)^{\min(s-t,1)}}\|e^{(0)}\|_{V_P^s}^2, \qquad m = 1, 2, \ldots, \tag{6.8}$$

where C can depend on $s - t$ and $\lambda_{\max}(P)$, and $\bar{e}^{(m)} := u - \bar{u}^{(m)}$.

Proof We use expansions of the error

$$e^{(m)} = \sum_{k=0}^{\infty} c_k^{(m)} \psi_k, \qquad m = 0, 1, \ldots,$$

in the CONS of the eigenvectors ψ_k of P. Then, since (6.6) is a stationary linear two-step iteration, we can write it as a vector iteration

$$E_k^{(m)} := \begin{pmatrix} c_k^{(m)} \\ c_k^{(m-1)} \end{pmatrix} = \underbrace{\begin{pmatrix} 2 - \omega\lambda_k & -1 \\ 1 & 0 \end{pmatrix}}_{M_k :=} E_k^{(m-1)} = \cdots = M_k^m E_k^{(0)},$$

separately for each coefficient sequence $(c_k^{(m)})_{m \in \mathbb{Z}_+}$. Since $u^{(-1)} = u^{(0)}$, we get

$$c_k^{(m)} = (1 \;\; 0) M_k^m \begin{pmatrix} 1 \\ 1 \end{pmatrix} c_k^{(0)}, \qquad k = 1, 2, \ldots.$$

A quick calculation shows that M_k has two complex conjugate eigenvalues $q_{\pm,k} = e^{\pm i\phi_k}$, where $0 < \phi_k < \pi/2$ is determined from $\sin(\phi_k/2) = \sqrt{\omega\lambda_k}/2 < 1/\sqrt{2}$. Remember that $0 < \omega < 2/\lambda_{\max}(P)$. Representing

$$\begin{pmatrix} 1 \\ 1 \end{pmatrix} = \frac{1}{1+q_{-,k}} \begin{pmatrix} 1 \\ q_{+,k} \end{pmatrix} + \frac{1}{1+q_{+,k}} \begin{pmatrix} 1 \\ q_{-,k} \end{pmatrix}$$

as a linear combination of the associated eigenvectors of M_k, we can easily compute

$$(1\ 0)M_k^m \begin{pmatrix} 1 \\ 1 \end{pmatrix} = \frac{q_{+,k}^m}{1+q_{-,k}} + \frac{q_{-,k}^m}{1+q_{+,k}} = \frac{\cos((m+1/2)\phi_k)}{\cos(\phi_k/2)}.$$

Thus, for any k with $c_k^{(0)} \neq 0$, we have $\limsup_{m\to\infty} |c_k^{(m)}| > 0$, which shows the failure of convergence for the heavy-ball iteration (6.6).

Next, we consider the averages. The coefficients $\bar{c}_k^{(m)}$ of the new errors $\bar{e}^{(m)}$ satisfy

$$\begin{aligned}(m+1)\bar{c}_k^{(m)} &= \sum_{l=0}^{m} c_k^{(l)} = \sum_{l=0}^{m} \frac{\cos((l+1/2)\phi_k)}{\cos(\phi_k/2)} c_k^{(0)} \\ &= \frac{\sin((m+1)\phi_k)}{\sin(\phi_k)} c_k^{(0)}, \qquad m = 0, 1, 2, \ldots,\end{aligned}$$

which results from some standard calculations involving trigonometric sums. Thus

$$|\bar{c}_k^{(m)}|^2 = \frac{\sin^2((m+1)\phi_k)}{(m+1)^2 \sin^2(\phi_k)} |c_k^{(0)}|^2 \le |c_k^{(0)}|^2.$$

So

$$\begin{aligned}\|\bar{e}^{(m)}\|_{V_\rho^t}^2 &= \sum_{k=0}^{\infty} \underbrace{\left(\lambda_k^{s-t} \frac{\sin^2((m+1)\phi_k)}{(m+1)^2 \sin^2(\phi_k)} \right)}_{a_k} \lambda_k^{-s} |c_k^{(0)}|^2 \\ &\le \frac{C}{(\omega(m+1)^2)^{\min(s-t,1)}} \|e^{(0)}\|_{V_\rho^s}^2.\end{aligned}$$

This follows because, for $\omega\lambda_k \le (m+1)^{-2}$, the term a_k is bounded by

$$a_k \le \lambda_k^{s-t} \le (1/(\omega(m+1)^2))^{s-t}.$$

For $\omega\lambda_k > (m+1)^{-2}$, since

$$\sin^2(\phi_k) = 4\sin^2(\phi_k/2)\cos^2(\phi_k/2) \ge \frac{1}{2}\omega\lambda_k$$

by definition of $\phi_k \in (0, \pi/2)$, we have

$$a_k \le \frac{\lambda_k^{s-t}}{(m+1)^2 \sin^2(\phi_k)} \le \frac{2\lambda_k^{s-t-1}}{\omega(m+1)^2} \le \begin{cases} \frac{2}{(\omega(m+1)^2)^{s-t}}, & s-t \le 1, \\ \frac{2\lambda_{\max}^{s-t-1}}{\omega(m+1)^2}, & s-t > 1. \end{cases}$$

This completes the proof of (6.8). □

In Problem 6.1, it is proved that, under the conditions on P and g in Theorem 6.1, any Krylov subspace method cannot do better than

$$\frac{\|u - u^{(m)}\|^2_{V_P^t}}{\|u - u^{(0)}\|^2_{V_P^s}} = \mathrm{O}((m+1)^{-2(s-t)}), \qquad m \longrightarrow \infty, \quad s > t.$$

Thus, in other words, Theorem 6.2 shows that the averaged heavy-ball iteration (6.6-6.7) is optimal when $0 < s - t \le 1$, while its convergence speed saturates for $s - t > 1$.

Other improvements are possible. We refer to [100] for examples of some semi-iterative methods that are optimal in the range $0 < s - t < \nu$ for arbitrarily large $\nu > 0$.

As another example, in Problem 6.2, it is shown that further averaging of the heavy-ball iteration can help: The sequence

$$\hat{u}^{(m)} = \binom{m+2}{2}^{-1} \sum_{l=0}^{m} (l+1)\bar{u}^{(l)}, \qquad m = 1, 2, \ldots, \tag{6.9}$$

gives optimal convergence rates in the range $0 < s - t \le 2$.

Clearly, the cg-method and variants of Nesterov-type iterations are also candidates for achieving optimal rates in some situations. For example, since the error of the cg-method satisfies the minimization property

$$\|e^{(m)}\|^2_{V_P^{-1}} = a(Pe^{(m)}, e^{(m)}) = \inf_{p_m(0)=1} a(Pp_m(P)e^{(0)}, p_m(P)e^{(0)}),$$

one immediately gets

$$\|e^{(m)}\|^2_{V_P^{-1}} \le \inf_{p_m(0)=1} \|\lambda^{s+1} p_m(\lambda)^2\|_{C([0,\|P\|])} \|e^{(0)}\|^2_{V_P^s}, \qquad s > -1.$$

Setting $t = -1 < s$ in the upper bound for ϵ'_m in Problem 6.1, we obtain

$$\|\lambda^{s+1} p_m(\lambda)^2\|_{C([0,\|P\|])} \le C(m+1)^{-2(s+1)} \|P\|^{s+1}.$$

In other words, the cg-method works as an unsaturated method, but we can only prove this if the error is measured in the V_P^{-1} norm. For further aspects on the cg-method, see [71, Section 7].

Similarly, the linear non-stationary Nesterov-type iteration considered in Problem 6.3 satisfies the estimate

$$\|e^{(m)}\|^2_{V_P^t} \le \frac{\lambda_{\max}(P)}{(m+1)^2} \|e^{(0)}\|^2_{V_P^{t+1}}, \qquad m = 1, 2, \ldots,$$

i.e., it provides optimal acceleration in any V_P^t norm for $s = t + 1$, a case already covered by the method analyzed in Theorem 6.2.

6.3 Randomized Schwarz approximation

In the introduction to this chapter, we discussed some results for dictionary splittings $\mathbf{V}_D$ generated by a CONS $D = \{\phi_k\}_{k\in\mathbb{N}}$ in a Hilbert space V. In particular, for this special case, it is easy to obtain convergence rates for randomized Schwarz iterations in terms of smoothness spaces induced by the Schwarz operator $P_{\mathbf{p}}u = \sum_{i\in\mathbb{N}} p_i(u,\phi_k)_V\phi_k$, which look basically the same as the results in Theorem 6.1.

In this section, we present a partial answer to the question of whether there is an extension of the results for orthogonal space splittings to general space splittings for $(V;a)$. We consider one-step Schwarz iterations of the form

$$u^{(m+1)} = \alpha_m u^{(m)} + \mu_m P_{i_m} e^{(m)}, \qquad m = 0, 1, \ldots, \tag{6.10}$$

where $\alpha_m \to 1$ and $\mu_m > 0$ are parameters at our disposal, while $\{i_m\}$ is a sequence of i.i.d. $\mathbb{N}$-valued random variables with fixed distribution $\mathbf{p} = \{p_i\}_{i\in\mathbb{N}}$. For $(\mathbf{V};\mathbf{a};\mathbf{R})$ we assume the properties formulated at the beginning of Section 6.1. Due to (6.1) and Lemma 2.10, the operator $P_{\mathbf{p}} = \mathbb{E}(P_i)$ relevant for the analysis is trace class if $\sum_{i\in\mathbb{N}} p_i \dim(V_i) < \infty$, which is implicitly assumed in the following.

We start with results from [88, 89], which concern the case of measuring the rate of convergence in V under the smoothness assumption $u \in V^s_{P_{\mathbf{p}}}$ for some $0 < s \le 1$.

Theorem 6.3 *Under the stated assumptions, if $u \in V^1_{P_{\mathbf{p}}}$, then the iteration (6.10) satisfies*

$$\mathbb{E}(\|e^{(m)}\|^2) \le \frac{\|e^{(0)}\|^2}{(m+1)^2} + \frac{\|u\|^2 + \Lambda\|u\|^2_{V^1_{P_{\mathbf{p}}}}}{m+1}, \qquad m = 1, 2, \ldots, \tag{6.11}$$

if one sets $\alpha_m = \frac{m+1}{m+2}$ and $\mu_m = \alpha_m/\Lambda$. Here Λ is the bound in the scaling condition (6.1).

Proof With the specific values for α_m and μ_m, set $\bar{\alpha}_m = 1 - \alpha_m$ and compute

$$\begin{aligned}
\|e^{(m+1)}\|^2 &= \|\alpha_m(e^{(m)} - \frac{1}{\Lambda}P_{i_m}e^{(m)}) + \bar{\alpha}_m u\|^2 \\
&= \alpha_m^2(\|(e^{(m)}\|^2 - \frac{2}{\Lambda}a(e^{(m)}, P_{i_m}e^{(m)}) + \frac{1}{\Lambda^2}\|P_{i_m}e^{(m)}\|^2) \\
&\quad +2\alpha_m\bar{\alpha}_m a(e^{(m)} - \frac{1}{\Lambda}P_{i_m}e^{(m)}, u) + \bar{\alpha}_m^2\|u\|^2.
\end{aligned}$$

By (6.1) we have

$$\|P_{i_m}e^{(m)}\|^2 = \|R_{i_m}R^*_{i_m}e^{(m)}\|^2 \le \Lambda\|R^*_{i_m}e^{(m)}\|_i^2 = \Lambda a(P_{i_m}e^{(m)}, e^{(m)}),$$

and after substitution we can take conditional expectations on both sides. This gives

$$\mathbb{E}'(\|e^{(m+1)}\|^2) \le \alpha_m^2(\|(e^{(m)}\|^2 - \frac{1}{\Lambda}a(e^{(m)}, P_{\mathbf{p}}e^{(m)})) + \bar{\alpha}_m^2\|u\|^2 \\ + 2\alpha_m\bar{\alpha}_m a(e^{(m)} - \frac{1}{\Lambda}P_{\mathbf{p}}e^{(m)}, u).$$

For the last term, we use the fact that $u \in V^1_{P_{\mathbf{p}}}$ is equivalent to the existence of a $v \in V$ such that $u = P_{\mathbf{p}}^{1/2}v$ and $\|u\|_{V^1_{P_{\mathbf{p}}}} = \|v\|$. Therefore,

$$\begin{aligned} 2\alpha_m\bar{\alpha}_m a(e^{(m)} - \frac{1}{\Lambda}P_{\mathbf{p}}e^{(m)}, u) &= 2\alpha_m\bar{\alpha}_m a(e^{(m)}, (I - \frac{1}{\Lambda}P_{\mathbf{p}})u) \\ &= 2\alpha_m\bar{\alpha}_m a(\Lambda^{-1/2}P_{\mathbf{p}}^{1/2}e^{(m)}, \Lambda^{1/2}(I - \frac{1}{\Lambda}P_{\mathbf{p}})v) \\ &\le \frac{\alpha_m^2}{\Lambda}\|P_{\mathbf{p}}^{1/2}e^{(m)}\|^2 + \bar{\alpha}_m^2\Lambda\|(I - \frac{1}{\Lambda}P_{\mathbf{p}})v\|^2 \\ &\le \frac{\alpha_m^2}{\Lambda}a(e^{(m)}, P_{\mathbf{p}}e^{(m)}) + \bar{\alpha}_m^2\Lambda\|v\|^2 \\ &= \frac{\alpha_m^2}{\Lambda}a(e^{(m)}, P_{\mathbf{p}}e^{(m)}) + \bar{\alpha}_m^2\Lambda\|u\|^2_{V^1_{P_{\mathbf{p}}}}. \end{aligned}$$

After substitution, two terms cancel, and we finally get

$$\mathbb{E}'(\|e^{(m+1)}\|^2) \le \alpha_m^2\|e^{(m)}\|^2 + \bar{\alpha}_m^2(\|u\|^2 + \Lambda\|u\|^2_{V^1_{P_{\mathbf{p}}}}), \quad m = 0, 1, \ldots. \tag{6.12}$$

All that remains is to switch to expectations using the independence assumption and set

$$d_m := (m+1)\mathbb{E}(\|e^{(m)}\|^2).$$

Then (6.12) implies

$$d_{m+1} \le \alpha_m d_m + \bar{\alpha}_m(\|u\|^2 + \Lambda\|u\|^2_{V^1_{P_{\mathbf{p}}}}), \qquad m = 0, 1, \ldots.$$

Iterating this recursion yields

$$d_m \le \frac{1}{m+1}d_0 + \frac{m}{m+1}(\|u\|^2 + \Lambda\|u\|^2_{V^1_{P_{\mathbf{p}}}}), \quad m = 1, 2, \ldots,$$

from which the estimate (6.11) follows by the definition of d_m and $d_0 = \|e^{(0)}\|^2$. □

Since the result of Theorem 6.3 is based on the bound (6.12), an estimate identical to (6.11) holds if the parameter μ_m is obtained by a steepest descent rule, i.e., if we take

$$\mu_m := \mathrm{argmin}_\mu \|u - \alpha_m u^{(m)} - \mu P_{i_m} e^{(m)}\|^2. \tag{6.13}$$

Calculating this parameter requires additional work, but does not require knowledge of the solution u. This result was shown in [89] for $u^{(0)} = 0$, while in [88] the estimate was established under more restrictive conditions for u.

Theorem 6.3 gives a similar result for the parameters $0 < s < 1$, which is obtained by an interpolation argument. For this we derive the following result:

Corollary 6.1 *For arbitrary $u, u^{(0)} \in V$ and $h \in V^1_{P_p}$, the iteration (6.10) with α_m and μ_m as specified in Theorem 6.3 satisfies*

$$\mathbb{E}(\|e^{(m)}\|^2) \leq 4\|u-h\|^2 + \frac{B}{m+1}, \qquad m = 1, 2, \ldots, \tag{6.14}$$

with any constant

$$B \geq \max\left(2(\|u\|^2 + \Lambda\|h\|^2_{V^1_{P_p}}), \frac{17}{32}\|e^{(0)}\|^2 + \frac{1}{2}(\|u\|^2 + \Lambda\|h\|^2_{V^1_{P_p}})\right).$$

Proof We follow the argument for Theorem 6.3, but estimate the last term in the upper bound for $\mathbb{E}'(\|e^{(m+1)}\|^2)$ for any given $h \in V^1_{P_p}$ as

$$\begin{aligned}
&2\alpha_m\bar{\alpha}_m a(e^{(m)} - \frac{1}{\Lambda}P_p e^{(m)}, u) \\
&= 2\alpha_m\bar{\alpha}_m\left(a(e^{(m)}, (I - \frac{1}{\Lambda}P_p)(u-h)) + a(e^{(m)} - \frac{1}{\Lambda}P_p e^{(m)}, h)\right) \\
&\leq 2\alpha_m\bar{\alpha}_m\|e^{(m)}\|\|u-h\| + \frac{\alpha_m^2}{\Lambda}a(e^{(m)}, P_p e^{(m)}) + \bar{\alpha}_m^2\Lambda\|h\|^2_{V^1_{P_p}},
\end{aligned}$$

where we implicitly used $\|I - \Lambda^{-1}P_p\|_V = 1$. After substitution, we take unconditional expectations using the independence assumption and get for $\varepsilon_m := (\mathbb{E}(\|e^{(m)}\|^2))^{1/2} \geq \mathbb{E}(\|e^{(m)}\|)$ the recursion

$$\varepsilon_{m+1}^2 = \mathbb{E}(\|e^{(m+1)}\|^2) \leq \alpha_m^2\varepsilon_m^2 + 2\alpha_m\bar{\alpha}_m\varepsilon_m\|u-h\| + \bar{\alpha}_m^2(\|u\|^2 + \Lambda\|h\|^2_{V^1_{P_p}}), \tag{6.15}$$

$m = 1, 2, \ldots$, which allows us to prove (6.14).

To do this, we can assume w.l.o.g. $\|u - h\| > 0$ (otherwise the result follows from Theorem 6.3). Instead of repeating the reasoning in [88], borrowed from [8], we will prove by induction that for any constant $D > 1$ we have

$$\varepsilon_m^2 \leq D^2\|u-h\|^2 + B'\bar{\alpha}_{m-1}, \qquad m = 1, 2, \ldots, \tag{6.16}$$

for some B' depending on D, $\|e^{(0)}\|^2$, and $D' := \|u\|^2 + \Lambda\|h\|^2_{V^1_{P_p}}$. In fact, since $\alpha_0 = \bar{\alpha}_0 = 1/2$, we have for $m = 1$

$$\varepsilon_1^2 \leq \frac{1}{4}\|e^{(0)}\|^2 + \frac{1}{2}\|e^{(0)}\|\|u-h\| + \frac{1}{4}D' \leq D^2\|u-h\|^2 + (\frac{1}{4} + \frac{1}{16D^2})\|e^{(0)}\|^2 + \frac{1}{4}D'.$$

So (6.16) holds for $m = 1$ if

$$B' \geq \frac{1}{2}((1 + \frac{1}{4D^2})\|e^{(0)}\|^2 + D').$$

For the induction step, assume that (6.16) holds for some $m = 1, 2, \ldots$. Then we also have

$$\varepsilon_m \le D\|u-h\| \left(1 + \frac{B'\bar{\alpha}_{m-1}}{D^2\|u-h\|^2}\right)^{1/2} \le D\|u-h\| + \frac{B'\bar{\alpha}_{m-1}}{2D\|u-h\|},$$

which is used with $\alpha_m \bar{\alpha}_{m-1} = \bar{\alpha}_m$ to show the required bound for ε_{m+1}^2, namely

$$\begin{aligned}\varepsilon_{m+1}^2 &\le \alpha_m^2 (D^2\|u-h\|^2 + B'\bar{\alpha}_{m-1}) + \bar{\alpha}_m^2 D' \\ &\quad + 2\alpha_m\bar{\alpha}_m D\|u-h\|^2 + \alpha_m\bar{\alpha}_m \frac{B'\bar{\alpha}_{m-1}}{D} \\ &\le D^2\|u-h\|^2 + \bar{\alpha}_m \left(\alpha_m B' + \bar{\alpha}_m (\frac{B'}{D} + D')\right).\end{aligned}$$

So the induction step holds if $B'/D + D' \le B'$, i.e., if

$$B' \ge DD'/(D-1).$$

Together with the condition on B' from the case $m = 1$, we have established (6.16) for each

$$B' \ge \max\left(\frac{DD'}{D-1}, \frac{1}{2}\left((1 + \frac{1}{4D^2})\|e^{(0)}\|^2 + D'\right)\right). \tag{6.17}$$

Now set $D = 2$ and replace $D' = \|u\|^2 + \Lambda\|h\|^2_{V^1_{P_\mathbf{p}}}$. This gives the statement in Corollary 6.1. □

Corollary 6.2 *For arbitrary $u \in V^s_{P_\mathbf{p}}$, $0 < s < 1$, and $u^{(0)} \in V$, the iteration (6.10) with α_m and μ_m as specified in Theorem 6.3 satisfies*

$$\mathbb{E}(\|e^{(m)}\|^2) \le \frac{4\Lambda^s}{(m+1)^s}\|u\|^2_{V^s_{P_\mathbf{p}}} + \frac{\|e^{(0)}\|^2 + 2\|u\|^2}{m+1}, \qquad m = 1, 2, \ldots. \tag{6.18}$$

We also have

$$\mathbb{E}(\|e^{(m)}\|^2) \to 0$$

for any $u, u^{(0)} \in V$, with no particular rate of convergence.

Proof For the last statement, use the density of $V^1_{P_\mathbf{p}}$ in V and find for any given $\epsilon > 0$, an element $h \in V^1_{P_\mathbf{p}}$ such that $\|u-h\|^2 \le \epsilon/8$. Then the constant B in (6.14) is finite and one can choose m_0 such that $B/(m+1) \le \epsilon/2$ for $m \ge m_0$. Consequently, $\mathbb{E}(\|e^{(m)}\|^2) \le \epsilon$ for $m \ge m_0$. Since $\epsilon > 0$ was arbitrary, this shows $\mathbb{E}(\|e^{(m)}\|^2) \to 0$ for arbitrary $u, u^{(0)} \in V$.

If $u \in V^s_{P_\mathbf{p}}$, then by definition of the spaces induced by the operator $P_\mathbf{p}$ we have

$$u = P_\mathbf{p}^{s/2} v = \sum_{k=1}^{\infty} c_k \lambda_k^{s/2} \psi_k, \qquad \|u\|^2_{V^s_{P_\mathbf{p}}} = \|v\|^2_V = \sum_{k=1}^{\infty} c_k^2.$$

Now take

$$h := \sum_{k:\,\lambda_k(m+1)\geq\Lambda} c_k\lambda_k^{s/2}\psi_k.$$

This is a finite sum, so h belongs to $V^1_{P_\mathrm{p}}$ and its norm is bounded by

$$\|h\|^2_{V^1_{P_\mathrm{p}}} = \sum_{k:\,\lambda_k(m+1)\geq\Lambda} c_k^2\lambda_k^{s-1} \leq \frac{(m+1)^{1-s}}{\Lambda^{1-s}} \sum_{k:\,\lambda_k(m+1)\geq\Lambda} c_k^2.$$

Furthermore,

$$\|u-h\|^2 = \sum_{k:\,\lambda_k(m+1)<\Lambda} c_k^2\lambda_k^s \leq \frac{\Lambda^s}{(m+1)^s} \sum_{k:\,\lambda_k(m+1)<\Lambda} c_k^2.$$

Now we substitute the estimates for $\|h\|^2_{V^1_{P_\mathrm{p}}}$ and $\|u-h\|^2$ into (6.14), where we choose the sufficiently large value

$$B = 4\Lambda\|h\|^2_{V^1_{P_\mathrm{p}}} + 2\|u\|^2 + \|e^{(0)}\|^2$$

for convenience. This gives

$$\begin{aligned}
\mathbb{E}(\|e^{(m)}\|^2) &\leq 4(\|u-h\|^2 + \frac{\Lambda}{m+1}\|h\|^2_{V^1_{P_\mathrm{p}}}) + \frac{2\|u\|^2 + \|e^{(0)}\|^2}{m+1} \\
&\leq \frac{4\Lambda^s}{(m+1)^s}\sum_{k=1}^{\infty} c_k^2 + \frac{2\|u\|^2 + \|e^{(0)}\|^2}{m+1} \\
&= \frac{4\Lambda^s}{(m+1)^s}\|u\|^2_{V^s_{P_\mathrm{p}}} + \frac{2\|u\|^2 + \|e^{(0)}\|^2}{m+1},
\end{aligned}$$

and establishes (6.18). □

It is open whether inequalities such as (6.11) and (6.18) hold in this generality for pairs of $V^t_{P_\mathrm{p}}$ and $V^s_{P_\mathrm{p}}$ norms other than the cases $t = 0 < s \leq 1$ established above, or with improvements in the algebraic decay rate (see the discussion in Section 6.2). With additional assumptions on the space splitting, this seems possible, see Problem 6.4. It can also be proved that under the conditions of Theorem 6.3, for the iteration (6.10), and for each $m = 1, 2, \ldots$, there exist $u \in V^s_{P_\mathrm{p}}$, $0 < s < 2$, and $u^{(0)} \in V$ such that

$$\mathbb{E}(\|e^{(m)}\|^2) \geq \frac{1}{2}\left(\frac{\|e^{(0)}\|^2}{(m+1)^2} + \frac{\Lambda^s}{(m+1)^s}\|u\|^2_{V^s_{P_\mathrm{p}}}\right). \tag{6.19}$$

The argument is outlined in Problem 6.5. In particular, for $s = 1$, the lower bound (6.19) shows that the estimate in Theorem 6.3 cannot be improved and that the $\mathrm{O}((m+1)^{-s})$ convergence rate obtained in Corollary 6.2 for $0 < s < 1$ is also the best possible.

6.4 Greedy Schwarz approximation

The theory of greedy approximation in Hilbert and Banach spaces is now well-developed. A comprehensive treatment can be found in [179], where the exposition focuses on the case of dictionary splittings $\mathbf{V}_D$. The extension to general collections $(\mathbf{V};\mathbf{a};\mathbf{R})$ is not very difficult, and we present some results here, partly following [8, 88, 89]. Although the proof methods and results are similar to the case of the randomized Schwarz approximation considered in Section 6.3, they hold for larger subsets in V than the spaces $V^s_{P_\mathbf{p}}$.

We begin by defining the classes $\mathcal{A}^p$ that generalize $\ell^p(\Omega)$, where $\Omega = \mathbb{N}$ as before. The classes $\ell^p(\Omega)$ appeared in Section 6.1 as suitable candidates for formulating convergence estimates in the case of orthogonal space splittings. Now, for $u \in V$ and $0 < p \le 2$, define the semi-quasi-norm

$$\|u\|_{\mathcal{A}^p} = \lim_{\epsilon\to 0} \inf_{v_i \in V_i \,:\, \|u - \sum_i R_i v_i\| \le \epsilon} \left(\sum_{i\in\Omega} \|v_i\|_i^p\right)^{1/p}.$$

Obviously $\|\cdot\|_{\mathcal{A}^p}$ is a semi-norm only if $1 \le p \le 2$. Then let A^p and A^p_0 denote the completion of the sets $\{u \in V : \|u\|_{\mathcal{A}^p} < \infty\}$ and $\{u \in V : \|u\|_{\mathcal{A}^p} = 0\}$ under this semi-quasi-norm. Since it can happen that $A^p_0 \ne \{0\}$, we define

$$\mathcal{A}^p := A^p/A^p_0.$$

For convenience, we keep the notation u and $\|u\|_{\mathcal{A}^p}$ for the elements and the quasi-norm of $\mathcal{A}^p$, with the understanding that u stands for any element in the associated equivalence class $u + A^p_0$.

By construction, $\mathcal{A}^p$ is a quasi-Banach space, and

$$\operatorname{span}(\{R_i V_i\}_{i\in\Omega}) \subset \mathcal{A}^p \subset \mathcal{A}^q, \qquad 0 < p < q \le 2.$$

It becomes a Banach space if $1 \le p \le 2$. Other embeddings of interest are given in the following statement.

Lemma 6.1 *a) Under the condition (6.1) we always have $\mathcal{A}^1 \subset V$ with embedding constant $\Lambda^{1/2}$, i.e.,*

$$\|u\| \le \Lambda^{1/2}\|u\|_{\mathcal{A}^1}, \qquad u \in \mathcal{A}^1.$$

We also have

$$\sup_{i\in\Omega} \|R_i^* e\|_i \ge \sup_{v\in\mathcal{A}^1 :\, \|v\|_{\mathcal{A}^1}\le 1} a(e,v) \tag{6.20}$$

which holds for every $e \in V$.
b) If $(\mathbf{V};\mathbf{a};\mathbf{R})$ is a space splitting for $(V;a)$, then we have $\mathcal{A}^2 \subset V$ with embedding constant $\|R\|^{1/2}_{\tilde V\to V}$.
c) Assume again (6.1). For any discrete probability distribution $\mathbf{p}$ on $\mathbb{N}$, we have $V^1_{P_\mathbf{p}} \subset \mathcal{A}_1$ with embedding constant 1.

Proof The embedding in Part a) follows because, by definition of the $\mathcal{A}^1$ norm, for any $u \in V \cap \mathcal{A}^1$ and any $\epsilon > 0$ there is a $h = \sum_{i\in\mathbb{N}} R_i v_i$ such that $\|u - h\| \le \epsilon$ and $\sum_{i\in\Omega} \|v_i\|_i \le \|u\|_{\mathcal{A}^1} + \epsilon$. So,

$$\|u\| \le \|u - h\| + \sum_{i\in\Omega} \|R_i v_i\| \le (1 + \Lambda^{1/2})\epsilon + \Lambda^{1/2}\|u\|_{\mathcal{A}^1}.$$

For $\epsilon \to 0$ we get the desired embedding inequality for all $u \in V \cap \mathcal{A}^1$. By completion we conclude that $\mathcal{A}^1 \subset V$.

The inequality (6.20) follows, since $r_i = R_i^* e$ is the minimizer of the quadratic optimization problem

$$J(u_i) := \frac{1}{2} a_i(u_i, u_i) - a(e, R_i u_i) \mapsto \min$$

on V_i. Thus,

$$\|r_i\|_i^2 = a(e, R_i r_i) \ge a(e, R_i w_i) + \frac{1}{2}(\|r_i\|_i^2 - \|w_i\|^2) \ge a(e, R_i w_i)$$

for any $w_i \in V_i$ with norm $\|w_i\|_i \le \|r_i\|_i$. After dividing by $\|r_i\|_i > 0$ we have

$$\|r_i\|_i \ge a(e, R_i v_i) \qquad \forall\, v_i \in V_i : \ \|v_i\|_i \le 1.$$

For arbitrary ϵ and $v \in \mathcal{A}^1 \subset V$ with norm 1, decompose

$$v = v - h + \sum_{i\in\Omega} R_i v_i, \quad \|v - h\| \le \epsilon, \quad \sum_{i\in\Omega} \|v_i\|_i \le \|v\|_{\mathcal{A}^1} + \epsilon = 1 + \epsilon.$$

This gives

$$\begin{aligned} a(e, v) &= a(e, v - h) + \sum_{i\in\Omega} a(e, R_i v_i) \le \epsilon\|e\| + \sum_{i\in\Omega} \|r_i\|_i \|v_i\|_i \\ &\le \epsilon\|e\| + \sup_{i\in\Omega} \|r_i\|_i \sum_{i\in\Omega} \|v_i\|_i \le \epsilon\|e\| + (1 + \epsilon) \sup_{i\in\Omega} \|r_i\|_i. \end{aligned}$$

After letting $\epsilon \to 0$ and taking the supremum with respect to v with $\|v\|_{\mathcal{A}^1} \le 1$, we get (6.20).

For Part b), just remember that the assumption implies the boundedness of the operator $R = \sum_{i\in\Omega} R_i : \ \tilde{V} \to V$, i.e.,

$$\|\sum_{i\in\Omega} R_i v_i\|^2 \le \|R\|_{\tilde{V}\to V}^2 \sum_{i\in\Omega} \|v_i\|_i^2,$$

which is equivalent to $\|u\| \le \|R\|_{\tilde{V}\to V} \|u\|_{\mathcal{A}^2}$ by the definition of the $\mathcal{A}^2$ norm.

Finally, to prove the embedding in Part c), note that $u \in V_{P_\mathrm{p}}^1$ is equivalent to the existence of (unique) elements $w \in V_{P_\mathrm{p}}^{-1}$ and $v \in V$ such that $u = P_\mathrm{p} w = P_\mathrm{p}^{1/2} v$ and

$v = P_{\mathbf{p}}^{1/2} w$. By using the formula (2.21) for the norm $|||\cdot|||_{\mathbf{p}}$, we get

$$|||u|||_{\mathbf{p}}^2 = |||P_{\mathbf{p}} w|||_{\mathbf{p}}^2 = a(P_{\mathbf{p}} w, w) = \|v\|^2 = \|u\|_{V^1_{P_{\mathbf{p}}}}^2 .$$

But for any representation

$$u = \sum_{i \in \mathbb{N}} R_i v_i = \sum_{i \in \Omega} p_i R_i w_i, \qquad w_i = p_i^{-1} v_i,$$

we have

$$\Big(\sum_{i \in \Omega} \|v_i\|_i\Big)^2 = \Big(\sum_{i \in \Omega} p_i^{1/2} (p_i^{-1/2} \|v_i\|_i)\Big)^2 \le \sum_{i \in \Omega} p_i^{-1} \|v_i\|_i^2 = \sum_{i \in \Omega} p_i \|w_i\|_i^2 .$$

All that remains is to take the infimum with respect to all possible representations. This gives us the desired embedding inequality

$$\|u\|_{\mathcal{A}^1} \le |||u|||_{\mathbf{p}} = \|u\|_{V^1_{P_{\mathbf{p}}}} .$$

Since the inequality results from applying the Cauchy-Schwarz inequality to sequences, it is not hard to see that $\mathcal{A}^1$ is always strictly larger than $V^1_{P_{\mathbf{p}}}$. □

The space $\mathcal{A}^1$ is central to the convergence analysis of greedy Schwarz approximation algorithms. Since we assume (6.1), $\mathcal{A}^1$ is embedded in V by Lemma 6.1 a), and we can ask about convergence rates for its elements with respect to the norm in V. We consider again (6.10) with $\alpha_m = (m+1)/(m+2)$, but this time we determine μ_m by the steepest descent rule (6.13). Note also that an explicit step-size choice that leads to the same convergence bound can be extracted from the proof below.

Theorem 6.4 *Let* $(\mathbf{V}; \mathbf{a}; \mathbf{R})$ *satisfy (6.1). Then the iteration (6.10), with parameters* α_m *and* μ_m *as specified above and* i_m *chosen according to the weak greedy rule (4.63) with constant weakness parameter* $0 < \beta = \beta_m \le 1$, *possesses the error bound*

$$\|e^{(m)}\|^2 \le \frac{\|e^{(0)}\|^2}{(m+1)^2} + \frac{2\left(\|u\|^2 + \beta^{-1}\Lambda \|u\|_{\mathcal{A}^1}^2\right)}{m+1}, \qquad m = 1, 2, \dots . \tag{6.21}$$

Proof By (4.63), the index i_m satisfies

$$\|R_{i_m}^* e^{(m)}\|_{i_m}^2 \ge \beta \sup_{i \in \mathbb{N}} \|R_i^* e^{(m)}\|_i^2, \qquad m = 0, 1, \dots .$$

So we conclude from (6.20) that

$$\|R_{i_m}^* e^{(m)}\|_{i_m} = \frac{a(e^{(m)}, P_{i_m} e^{(m)})}{\|R_{i_m}^* e^{(m)}\|_{i_m}} \ge \beta^{1/2} \frac{a(e, u)}{\|u\|_{\mathcal{A}^1}} . \tag{6.22}$$

To derive a suitable error recursion, it is convenient to set $\mu_m = \bar{\alpha}_m \xi$, with $\xi > 0$ at our disposal. Again, we set $\bar{\alpha}_m := 1 - \alpha_m$. Since μ_m in (6.10) is determined by the steepest descent rule, we have

$$\begin{aligned}\|e^{(m+1)}\|^2 &\le \|\alpha_m e^{(m)} + \bar{\alpha}_m(u - \xi P_{i_m} e^{(m)})\|^2 \\ &= \alpha_m^2 \|e^{(m)}\|^2 + 2\alpha_m\bar{\alpha}_m a(e^{(m)}, u - \xi P_{i_m} e^{(m)}) + \bar{\alpha}_m^2 \|u - \xi P_{i_m} e^{(m)}\|^2,\end{aligned} \tag{6.23}$$

for all $\xi > 0$. The last term is approximately bounded by

$$\|u - \xi P_{i_m} e^{(m)})\|^2 \le 2(\|u\|^2 + \xi^2 \|P_{i_m} e^{(m)}\|^2) \le 2(\|u\|^2 + \xi^2 \Lambda \|R_{i_m}^* e^{(m)}\|_{i_m}^2).$$

Next, we show that with an appropriate choice of ξ we can neglect the second term in (6.23). In fact, by (6.22) we have

$$a(e^{(m)}, P_{i_m} e^{(m)}) \ge \beta^{1/2} \frac{a(e, u)}{\|u\|_{\mathcal{A}^1}} \|R_{i_m}^* e^{(m)}\|_{i_m},$$

and by setting

$$\xi = \xi_m := \beta^{-1/2} \frac{\|u\|_{\mathcal{A}^1}}{\|R_{i_m}^* e^{(m)}\|_{i_m}}, \tag{6.24}$$

we have $a(e^{(m)}, u - \xi P_{i_m} e^{(m)}) \le 0$. Substituting into (6.23) we get

$$\|e^{(m+1)}\|^2 \le \alpha_m^2 \|e^{(m)}\|^2 + 2\bar{\alpha}_m^2 \left(\|u\|^2 + \beta^{-1}\Lambda \|u\|_{\mathcal{A}^1}^2\right), \quad m = 0, 1, \ldots. \tag{6.25}$$

This recursive estimate is identical, with minor changes, to the error recursion (6.12) for the randomized Schwarz approximation, and we arrive at (6.21) by repeating the last steps of the proof for Theorem 6.3. □

A few comments are in order. First, according to the proof, one could avoid calculating μ_m by the steepest descent rule (6.13) and instead set

$$\mu_m = \frac{\|u\|_{\mathcal{A}^1}}{(m+2)\beta^{1/2} \|R_{i_m}^* e^{(m)}\|_{i_m}},$$

compare (6.24). The difficulty then is to find tight a priori bounds on the $\mathcal{A}^1$ norm of the unknown solution u.

Second, convergence guarantees similar to (6.21) can also be obtained for other greedy rules, such as (4.57) or (4.60) following the proof above. If one assumes a little more about the R_i, for example, if one has a strengthening of (6.1) of the form

$$\Lambda' \|v_i\|_i^2 \le \|R_i v_i\|^2 \le \Lambda \|v_i\|_i^2, \quad v_i \in V_i, \quad i = 1, 2, \ldots, \tag{6.26}$$

with constants $0 < \Lambda' \le \Lambda < \infty$, then the necessary changes are minimal, see Problem 6.6. For space splittings $(\mathbf{V}; \mathbf{R})$ generated by the Galerkin projection, i.e., with bilinear forms a_i inherited from $(V; a)$, we automatically have (6.26) with $\Lambda' = \Lambda = 1$.

A third comment concerns the need of choosing the parameter $\alpha_m < 1$ in the iterative method (6.10). In the introductory example of an orthogonal space splitting, the same rate of order $O((m+1)^{-1})$ was obtained by setting $\alpha_m = 1$. Sophisticated examples reported in [179, Chapter 2] and [108] show that there are space splittings $\mathbf{V}_D$ generated by dictionaries D with unit norm dictionary elements, such that for some u with $\|u\|_{\mathcal{A}^1} = 1$ and $u^{(0)} = 0$, one has

$$\|e^{(m)}\|^2 \geq c_\gamma (m+1)^{-\gamma}, \qquad m = 1, 2, \ldots. \tag{6.27}$$

for some $c_\gamma > 0$ and $\gamma > \gamma_0$, where $\gamma_0 \approx 0.364$ is a fixed constant. The lower bound (6.27) holds if $u^{(m)}$ is computed by (6.10) with $\alpha_m = 1$ and with μ_m determined according to (6.13), while i_m is greedily chosen with $\beta = 1$. In this case, all three greedy rules match. In other words, to obtain the convergence rate of optimal order $O((m+1)^{-1})$ on the space $\mathcal{A}^1$ established in Theorem 6.4, it is necessary to choose a value of $\alpha_m \neq 1$. On the positive side, for the pure greedy algorithm with $\alpha_m = 1$ and $\beta = 1$, in the dictionary case a rate of $O((m+1)^{-\gamma_0})$ can be established for $u \in \mathcal{A}^1$, see [166] or [179, Theorem 2.18]. We do not know whether similar statements hold for the randomized case considered in Section 6.3.

Finally, due to Lemma 6.1 c), we see that, for the iteration (6.10) with weak greedy choice of i_m, Theorem 6.4 guarantees the $O((m+1)^{-1})$ bound for $\|e^{(m)}\|^2$ on a larger class than Theorem 6.3 does in expectation using the randomized version. In practice, the difference in convergence speed is often significant, but enforcing the weak greedy rule at every step is far from obvious for infinite Ω, and has a higher computational cost than randomly choosing i_m.

We now turn to the case $u \in \mathcal{A}^p$ with $1 < p < 2$, for which we need an extension of Theorem 6.4 that is analogous to Corollary 6.1 in the previous section.

Corollary 6.3 *Let the conditions and specifications of Theorem 6.4 hold. Then, for any $u \in V$ and any $h \in \mathcal{A}^1$, we have*

$$\|e^{(m)}\|^2 \leq 4\|u - h\|^2 + \frac{B}{(m+1)}, \qquad m = 1, 2, \ldots, \tag{6.28}$$

if the constant B satisfies

$$B \geq \max\left(4(\|u\|^2 + \beta^{-1}\Lambda\|h\|^2_{\mathcal{A}^1}), \frac{17}{32}\|e^{(0)}\|^2 + \|u\|^2 + \beta^{-1}\Lambda\|h\|^2_{\mathcal{A}^1}\right).$$

Proof We repeat the proof of Theorem 6.4 with minor changes in the estimation of $\|e^{(m+1)}\|^2$. First, (6.22) can be replaced by

$$\|R^*_{i_m} e^{(m)}\|_{i_m} = \frac{a(e^{(m)}, P_{i_m} e^{(m)})}{\|R^*_{i_m} e^{(m)}\|_{i_m}} \geq \beta^{1/2} \frac{a(e, h)}{\|h\|_{\mathcal{A}^1}},$$

and if we set this time

$$\xi = \xi_m := \beta^{-1/2} \frac{\|h\|_{\mathcal{A}^1}}{\|R^*_{i_m} e^{(m)}\|},$$

then

$$a(e^{(m)}, u - \xi P_{i_m} e^{(m)}) = a(e^{(m)}, u - h) + a(e^{(m)}, h - \xi P_{i_m} e^{(m)})$$
$$\leq a(e^{(m)}, u - h) \leq \|e^{(m)}\|\|u - h\|.$$

After substituting into the bound for $\|e^{(m+1)}\|^2$, we now get for the error $\varepsilon_m := \|e^{(m)}\|$ instead of (6.25) the modified recursive estimate

$$\varepsilon_{m+1}^2 \leq \alpha_m^2 \varepsilon_m^2 + 2\alpha_m \bar{\alpha}_m \|u - h\| \varepsilon_m + 2\bar{\alpha}_m^2 \left(\|u\|^2 + \beta^{-1} \Lambda \|h\|_{\mathcal{A}^1}^2\right), \quad m = 0, 1, \ldots.$$

The rest is the same as in the proof of Corollary 6.1. In fact, using the same notation and comparing with (6.15), we see that the bound (6.16) holds for our ε_m^2 if we replace D' with $D' = 2(\|u\|^2 + \beta^{-1}\Lambda\|h\|_{\mathcal{A}^1}^2)$. The statement of the corollary then follows if we set $D = 2$ and substitute the new value for D' into the expression for B' in (6.17). □

In analogy to the randomized case, Corollary 6.3 can be used to establish convergence for any $u \in V$ and, under an additional condition, convergence rates for $u \in \mathcal{A}^p$, $1 < p < 2$, by interpolation. We have the following result:

Corollary 6.4 *Under the conditions and specifications of Theorem 6.4, we have* $\|e^{(m)}\|^2 \to 0$ *for any* $u \in V$. *In addition, if* $(\mathbf{V}; \mathbf{a}; \mathbf{R})$ *is a space splitting for* $(V; a)$ *and* $u \in \mathcal{A}^p$ *for some* $1 < p < 2$, *then*

$$\|e^{(m)}\|^2 \leq \frac{4 \max(\|R\|_{\tilde{V} \to V}^2 \|u\|_{\mathcal{A}^p}^p, (\Lambda/\beta)\|u\|_{\mathcal{A}^p}^{2p})}{(m+1)^{2/p-1}} + \frac{\|e^{(0)}\|^2 + 4\|u\|^2}{m+1}, \tag{6.29}$$

for $m = 1, 2, \ldots$.

Proof To establish convergence for any $u \in V$, we use the density of $\mathcal{A}^1$ in V and choose $h \in \mathcal{A}^1$ such that $\|u - h\|^2 \leq \epsilon/8$. Then, with u, h, and $u^{(0)}$ specified, the constant B in (6.28) is fixed, and we can find m_0 such that $B/(m_0 + 1) \leq \epsilon/2$. Consequently, $\|e^{(m)}\|^2 \leq \epsilon$ for all $m \geq m_0$, which proves the claim.

Now let $u \in \mathcal{A}^p$ for some $1 < p < 2$. Since $(\mathbf{V}; \mathbf{a}; \mathbf{R})$ is a space splitting for $(V; a)$, we know from Lemma 6.1 b) that $u \in V$. Taking an arbitrary representation

$$u = \sum_{i \in \Omega} R_i v_i, \qquad \sum_{i \in \Omega} \|v_i\|_i^p < \infty, \tag{6.30}$$

we set

$$h = \sum_{i:\, \|v_i\|_i \geq (m+1)^{-1/p}} R_i v_i.$$

Obviously this sum is finite and $h \in \mathcal{A}^1$. Its norm can be estimated by

$$\|h\|_{\mathcal{A}^1} \leq \sum_{i:\, \|v_i\|_i \geq (m+1)^{-1/p}} \|v_i\|_i \leq (m+1)^{1-1/p} \sum_{i \in \Omega} \|v_i\|_i^p.$$

Furthermore, due to the assumed boundedness of the operator R, we have

$$\|u-h\|^2 = \|\sum_{i:\,\|v_i\|_i<(m+1)^{-1/p}} R_i v_i\|^2 \le \|R\|^2_{\tilde{V}\to V} \sum_{i:\,\|v_i\|_i<(m+1)^{-1/p}} \|v_i\|_i^2$$
$$\le \|R\|^2_{\tilde{V}\to V}(m+1)^{-(2/p-1)} \sum_{i\in\Omega} \|v_i\|_i^p.$$

The last two estimates are now substituted into (6.28), where we simply choose

$$B = \|e^{(0)}\|^2 + 4\|u\|^2 + 4(\Lambda/\beta)\|h\|^2_{\mathcal{A}^1}.$$

The result is

$$\|e^{(m)}\|^2 \le \frac{4\|R\|^2_{\tilde{V}\to V}}{(m+1)^{2/p-1}} \sum_{i\in\Omega} \|v_i\|_i^p + \frac{\|e^{(0)}\|^2 + 4\|u\|^2}{m+1}$$
$$+ \frac{4\Lambda(m+1)^{2-2/p}}{\beta(m+1)} \left(\sum_{i\in\Omega} \|v_i\|_i^p\right)^2.$$

After taking the infimum with respect to all admissible representations of $u \in \mathcal{A}^p$ in (6.30), we arrive at

$$\|e^{(m)}\|^2 \le \frac{\|e^{(0)}\|^2 + 4\|u\|^2}{m+1} + \frac{4\max(\|R\|^2_{\tilde{V}\to V}\|u\|^p_{\mathcal{A}^p}, (\Lambda/\beta)\|u\|^{2p}_{\mathcal{A}^p})}{(m+1)^{2/p-1}},$$

$m = 1, 2, \ldots$ which gives (6.29). □

The exponents $\gamma_p = 2/p - 1$ in the obtained convergence estimates for the greedy Schwarz approximation on the classes $\mathcal{A}^p$ with $1 \le p < 2$ cannot be improved, as can be seen from the example of a subspace splitting generated by a complete orthonormal system considered in the introductory Section 6.1. However, one might expect better convergence rates for $u \in \mathcal{A}^p$ if $0 < p < 1$. The results reported in [179] for the special case of space splittings generated by dictionaries suggest that this is possible for dictionaries close to orthogonal systems, but is hopeless in the general case considered in this chapter. However, there is no rigorous result on the impossibility of obtaining convergence results of the type

$$\|e^{(m)}\|^2 = \mathrm{O}((m+1)^{\gamma}), \quad m \to \infty, \quad \forall\, u \in \mathcal{A}^p,\ u^{(0)} \in V,$$

with some $\gamma > 1$ and some $p < 1$ by a reasonable greedy method for arbitrary $(\mathbf{V}; \mathbf{a}, \mathbf{R})$. By reasonable we mean linear methods of the form

$$u^{(m+1)} = v^{(m)} + P_{i_m}(u - w^{(m)}), \qquad m = 0, 1, \ldots, \tag{6.31}$$

where $v^{(m)}, w^{(m)} \in \mathrm{span}(u^{(0)}, \ldots, u^{(m)})$ are computable in an efficient way and the index i_m is chosen by an appropriate greedy rule.

A natural lower bound on the error that can be obtained by such a method (6.31) is provided by the theory of nonlinear best m-term approximation. In our language, for a given $(\mathbf{V}; \mathbf{a}; \mathbf{R})$ associated with $(V; a)$, define

$$\sigma_m(u) = \inf_{I \subset \Omega:\, |I| \le m} \inf_{v_i \in V_i} \|u - \sum_{i \in I} R_i v_i\|, \qquad u \in V, \quad m = 1, 2, \ldots. \tag{6.32}$$

This quantity describes the best approximation by linear combinations of elements from at most m linear subsets $R_i V_i$ in V. For simplicity, let $u^{(0)} = 0$. Then obviously

$$\|e^{(m)}\|^2 \ge \sigma_m(u)^2, \qquad m = 1, 2, \ldots,$$

for any method (6.31). If $(\mathbf{V}; \mathbf{a}; \mathbf{R})$ is a space splitting for $(V; a)$, then

$$\sigma_m(u)^2 \le \frac{C_p \|R\|^2_{\tilde{V} \to V} \|u\|^p_{\mathcal{A}^p}}{(m+1)^{2/p-1}}, \qquad m = 1, 2, \ldots,$$

for all $u \in \mathcal{A}^p$ and $0 < p < 2$, where C_p depends only on p. The argument is the same as in the special case of an orthogonal space splitting treated in Section 6.1. See [64, 178, 179] for more information on best m-term approximation and its relevance to the theory of greedy methods and adaptive algorithms.

6.5 Online algorithms for kernel regression

In this section, following [91], we consider a Schwarz approximation method, where the index set Ω of the underlying space splitting is an arbitrary compact metric space equipped with a Borel probability measure ρ, and the order of subproblem solves is random. The motivation for this extension comes from the regression problem in learning theory, see [49, 50, 158, 170, 189] for the relevant background on kernel methods and [31, 32, 121] for the theoretical background in the vector-valued case we deal with here. Let there be $N \le \infty$ samples $(\omega_m, y_m) \in \Omega \times Y$, $m = 0, \ldots, N-1$, of an input-output process $\omega \to y$, which are i.i.d. with respect to a (generally unknown) probability measure μ defined on $\Omega \times Y$. For simplicity, let Ω be a compact metric space, Y a separable Hilbert space, and μ a Borel measure. What we are looking for is a so-called regression function $f_\mu : \Omega \to Y$, which in some sense optimally represents the underlying input-output process. We deal with algorithms for least-squares regression, which aim at finding approximations to the solution

$$f_\mu(\omega) = \mathbb{E}(y|\omega) \in L^2_\rho(\Omega, Y)$$

of the minimization problem

$$\mathbb{E}(\|f(\omega) - y\|_Y^2) = \int_{\Omega \times Y} \|f(\omega) - y\|_Y^2 \, d\mu(\omega, y) \to \min \tag{6.33}$$

for $f \in L^2_\rho(\Omega, Y)$ from the given samples $(\omega_m, y_m), m = 0, \ldots, N-1$, where ρ is the marginal probability measure generated by μ on Ω. On a theoretical level, for the minimization problem to make sense, one needs

$$\mathbb{E}(\|y\|_Y^2) = \int_{\Omega\times Y} \|y\|_Y^2 \, d\mu = \int_\Omega \mathbb{E}(\|y\|_Y^2|\omega) \, d\rho < \infty.$$

In the literature, stronger conditions are often imposed on the random variable y, such as uniform boundedness μ-a.e., which we will not discuss here.

Recall that ρ is usually out of our hands, and the discretized problem

$$\frac{1}{N}\sum_{m=0}^{N-1} \|f(\omega_m) - y_m\|_Y^2 \to \min \tag{6.34}$$

for $f \in L^2_\rho(\Omega, Y)$ is ill-posed and therefore needs further regularization. A remedy is to add the assumption $f \in H$, called *prior*, where $H \subset L^2_\rho(\Omega, Y)$ is a set of functions $f : \Omega \to Y$ such that point evaluations $\omega \to f(\omega)$ are well-defined. Staying within the Hilbert space setting, candidates for H are vector-valued reproducing kernel Hilbert spaces (RKHS). Such vector-valued RKHS H can be introduced by a family $\mathbf{R} = \{R_\omega\}_{\omega\in\Omega}$ of bounded linear operators $R_\omega \in \mathcal{B}(Y, V)$, called a feature map, where V is another separable Hilbert space, called a feature space. The scalar product $(\cdot,\cdot)_V$ in V now plays the role of the basic bilinear form a, which should be kept in mind when comparing with the material in other sections of this chapter. More precisely, we introduce the notation

$$f_v(\omega) := R_\omega^* v, \quad \omega \in \Omega, \quad v \in V,$$

and set

$$H := \{f_v : v \in V, \quad \|f_v\|_H := \|v\|_V\}.$$

Under the simplifying conditions

$$\|R_\omega^* v\|_Y = 0 \quad \rho-\text{a.e.} \quad \Rightarrow \quad v = 0, \qquad v \in V, \tag{6.35}$$

and

$$f_v(\omega) \in C(\Omega, Y) \qquad \forall\, v \in V, \tag{6.36}$$

we then have $H \subset C(\Omega, Y) \subset L^2_\rho(\Omega, Y)$, and $f_v = 0$ in $L^2_\rho(\Omega, Y)$ implies $v = 0$. By definition, the RKHS H and the feature space V are isometric, which allows us to easily switch between H and V in the sequel.

The condition (6.36) is the continuity of the operator family $\mathbf{R}$ in the strong operator topology and ensures the Bochner integrability of functions from Ω into Y and V, respectively, which appear in the formulas below. Because of the assumed compactness of Ω, it also implies

$$\|R_\omega\|_{Y\to V}^2 \le \Lambda < \infty, \tag{6.37}$$

with some $\Lambda < \infty$. Another consequence is that that the operator kernel

$$K(\omega,\theta) := R_\omega^* R_\theta$$

associated with H is a Mercer kernel. The uniform boundedness (6.37) of $\mathbf{R}$ is equivalent to the uniform boundedness

$$\|K(\omega,\theta)\|_Y \le \Lambda, \qquad \omega,\theta \in \Omega$$

of the operator kernel K associated with H. This, in turn, is equivalent to the uniform boundedness of the operator family $P_\omega := R_\omega R_\omega^* : V \to V$ in V, i.e.,

$$\|P_\omega\|_V \le \Lambda, \qquad \omega \in \Omega.$$

For fixed V and $\mathbf{R}$ satisfying the above properties, instead of (6.33) one now seeks the global minimizer $u \in V$ of the quadratic minimization problem

$$J(v) := \mathbb{E}(\|f_v - y\|_Y^2) = \int_{\Omega\times Y} \|R_\omega^* v - y\|_Y^2 \, d\mu(\omega,y) \to \min, \qquad v \in V. \quad (6.38)$$

If such a minimizer $u \in V$ exists, it must satisfy the necessary condition

$$\mathbb{E}((R_\omega^* u - y, R_\omega^* w)_Y) = \mathbb{E}((P_\omega u - R_\omega y, w)_V) = 0 \qquad \forall\, w \in V.$$

This condition is equivalent to the linear operator equation

$$P_\rho u = \mathbb{E}(R_\omega y), \qquad P_\rho := \mathbb{E}(P_\omega) = \mathbb{E}(R_\omega R_\omega^*) \quad (6.39)$$

in V. More precisely, using Theorem A.3, the right-hand side $\mathbb{E}(R_\omega y)$ in (6.39) and the action of P_ρ on elements $v \in V$ are defined by the equalities

$$\begin{aligned}(\mathbb{E}(R_\omega y), w)_V &:= \int_{\Omega\times Y} (R_\omega y, w)_V \, d\mu(\omega,y) = \int_\Omega (f_\mu(\omega), R_\omega^* w)_Y \, d\rho(\omega) \\ &= \int_\Omega (R_\omega f_\mu(\omega), w)_V \, d\rho(\omega)\end{aligned}$$

and

$$\begin{aligned}(P_\rho v, w)_V &:= \int_{\Omega\times Y} (P_\omega v, w)_V \, d\mu(\omega,y) = \int_\Omega (P_\omega v, w)_V \, d\rho(\omega) \\ &= \int_\Omega (R_\omega^* v, R_\omega^* w)_Y \, d\rho(\omega) = (f_v, f_w)_{L^2(\Omega,Y)}\end{aligned}$$

to hold for all $w \in V$, respectively. The fact that P_ρ is the equivalent of the Schwarz operator associated with a collection $(\mathbf{V};\mathbf{a};\mathbf{R})$ in $(V;(\cdot,\cdot)_V)$ can be seen by comparing

$$(P_\rho v, w)_V = \int_\Omega (R_\omega R_\omega^* v, w)_V \, d\rho(\omega) \qquad \forall\, v,w \in V,$$

with the corresponding formula for Schwarz operators for space splittings with finite and countable Ω in $(V;a)$, where now $a(\cdot,\cdot) = (\cdot,\cdot)_V$. For this, the corresponding collection $(\mathbf{V};\mathbf{a};\mathbf{R})$ is given by $(V_\omega;a_\omega) = (Y;(\cdot,\cdot)_Y)$, $\omega \in \Omega$, and the feature map $\mathbf{R}$ introduced for the construction of the prior RKHS H. Just as well we could have identified V_ω with the closure of $\mathrm{ran}(R_\omega^*)$ in Y. It simplifies the considerations that all V_ω are equal to a single separable Hilbert space Y. The general case requires the concept of the direct integral of Hilbert spaces, see Subsection 6.5.2.3.

The boundedness of P_ρ, together with the estimate

$$\|P_\rho\|_V \le \Lambda,$$

follows from (6.37). Thus, $P_\rho \in \mathcal{B}_{pos}(V)$, the spectrum of P_ρ is contained in $[0,\Lambda]$, and $\ker(P_\rho) = \{0\}$ due to (6.35). For simplicity, from now on we will also assume that P_ρ is compact. In many applications, P_ρ is even trace class. Note that, by analogy with Lemma 2.10, P_ρ is trace class if any of the expressions in

$$\|P_\rho\|_{\mathrm{Tr}} = \int_\Omega \|R_\omega\|_{\mathrm{HS}}^2\, d\rho(\omega) = \int_\Omega \|P_\omega\|_{\mathrm{Tr}}\, d\rho(\omega) \le \Lambda \int_\Omega \dim(\mathrm{ran}(R_\omega))\, d\rho(\omega)$$

is finite. In particular, this is true if the operators R_ω, $\omega \in \Omega$, have uniformly bounded finite rank. The injectivity and compactness of P_ρ are used as technical simplifications in the proof of Theorem 6.5 below. In particular, they allow us to use the definition of spaces $V_{P_\rho}^s$ as adopted in Section 2.4.2 and to work with a CONS $\{\psi_k\}$ of eigenvectors of P_ρ and associated eigenvalues $\lambda_1 \ge \lambda_2 \ge \cdots > 0$ with limit 0. Furthermore, $\mathrm{ran}(P_\rho)$ will then be a dense linear set in V, which is different from V. It turns out that, under the above assumptions, (6.38) possesses a minimizer $u \in V$ if and only if $u \in V$ solves the equation (6.39) or, equivalently, if

$$\mathbb{E}(R_\omega y) \in \mathrm{ran}(P_\rho). \tag{6.40}$$

In general, one can only be sure that $\mathbb{E}(R_\omega y) \in \mathrm{ran}(P_\rho^{1/2})$. If (6.40) is violated, then a minimizer $u \in V$ for (6.38) does not exist, and the infimum of the quadratic functional $J(v)$ in (6.38) is approached only if $\|v\|_V \to \infty$ in a specific way. This can be seen if one determines the (unique) minimizer u_r of the constrained minimization problem

$$J(v) = \int_{\Omega\times Y} \|R_\omega^* v - y\|_Y^2\, d\mu(\omega,y) \to \min, \qquad v \in V, \quad \|v\| \le r, \tag{6.41}$$

for $r \to \infty$, see Problem 6.7 for details. Since our main result in this section concerns the convergence of approximations built from the samples to the minimizer u of (6.38) with respect to the norm in V, we will assume that (6.40) is satisfied.

Note that the RKHS H and its isometric copy V are isometrically embedded in $L_\rho^2(\Omega,Y)$ by the action of $P_\rho^{1/2}$. This follows from

$$\|f_v\|_{L_\rho^2(\Omega,Y)}^2 = \|R_\omega^* v\|_{L_\rho^2(\Omega,Y)}^2 = \int_\Omega (P_\omega v, v)_V\, d\rho(\omega) = (P_\rho v, v)_V = \|P_\rho^{1/2} v\|_V^2.$$

Depending on the choice of the feature map and the probability measure μ, the closure of H in $L^2_\rho(\Omega, Y)$, denoted here by W, does not necessarily coincide with $L^2_\rho(\Omega, Y)$, which implies that f_μ does not necessarily coincide with f_u. We have

$$\|f_\mu - f_v\|^2_{L^2(\Omega,Y)} \geq \mathrm{dist}_{L^2(\Omega,Y)}(f_\mu, W)^2 = \|f_\mu - f_u\|^2_{L^2(\Omega,Y)}$$

for any approximation f_v from within H to the regression function f_μ, whereas this lower bound can be strictly positive.

For a fixed prior RKHS H induced by V and $\mathbf{R}$ and for given samples (ω_m, y_m), $m = 0, \dots, N-1$, a standard regularization of the problem (6.34) is to find the minimizer $u_N \in V$ of the unconstrained minimization problem

$$J_N(v) := \frac{1}{N} \sum_{m=0}^{N-1} \|R^*_{\omega_m} v - y_m\|^2_Y + \kappa_N \|v\|^2_V \to \min \tag{6.42}$$

on V, where $\kappa_N > 0$ is a suitable regularization parameter. For comparison with (6.34), recall that $f_v(\omega_m) = R^*_{\omega_m} v$ are the function values of a function in the prior RKHS H. Using the representer theorem for Mercer kernels [121], this problem leads to a linear system with a typically dense and ill-conditioned $N \times N$ operator matrix. There is a large literature, especially in the scalar-valued case $Y = \mathbb{R}$, devoted to setting up, analyzing, and solving this problem for fixed N. Alternatively, one can restrict the minimization in (6.38) to bounded closed subsets in V, which, under our assumptions on P_ρ, are compact subsets of $L^2_\rho(\Omega, Y)$, see [49].

We focus here on *online learning* algorithms for finding approximations to f_u, where $u \in V$ is the minimizer of (6.38), and are interested in their asymptotic performance, i.e., we assume that N is not fixed (set formally $N = \infty$) and that the i.i.d. samples (ω_m, y_m) are successively supplied by a random process, $m = 0, 1, \dots$. The task of an online algorithm is to recover $u \in V$ from this stream of samples, avoiding the solution of minimization problems such as (6.42).

To do this, we define the noise term

$$\epsilon_{\omega,y} := y - f_u(\omega) = y - R^*_\omega u, \qquad \omega \in \Omega.$$

Since $f_\mu(\omega) = \mathbb{E}(y|\omega)$, we have $\mathbb{E}(\epsilon_{\omega,y}|\omega) = f_\mu(\omega) - f_u(\omega)$ for any $\omega \in \Omega$. Furthermore, the noise variance with respect to $f_u \in H$, namely

$$\sigma_H^2 := \mathbb{E}(\|\epsilon_{\omega,y}\|^2_Y) = \underbrace{\mathbb{E}(\|y - f_\mu\|^2_Y)}_{\sigma^2 :=} + \|f_\mu - f_u\|^2_{L^2_\rho(\Omega,Y)}, \tag{6.43}$$

is finite, since $\mathbb{E}(\|y\|^2_Y) < \infty$ was assumed. The value of σ characterizes the average size of the noise $y - f_\mu(\omega)$ on Ω measured in the Y norm. Thus, σ_H depends on both σ and $\mathrm{dist}(f_\mu, W) = \|f_\mu - f_u\|_{L^2_\rho(\Omega,Y)}$.

Now consider the randomized Schwarz approximation algorithm (6.10) associated with the above collection $(\mathbf{V}; \mathbf{a}; \mathbf{R})$ applied to the trivial equation $u = u$ in $(V; (\cdot, \cdot)_V)$, where at each step the index i_m is replaced by $\omega_m \in \Omega$, which is i.i.d. drawn according to the marginal probability measure ρ on Ω. This gives

$$u^{(m+1)} = \alpha_m(u^{(m)} + \mu_m P_{\omega_m} e^{(m)}) = \alpha_m(u^{(m)} + \mu_m R_{\omega_m}(R^*_{\omega_m} u - R^*_{\omega_m} u^{(m)})),$$

$m = 0, 1, \ldots$. Here, to avoid confusion, we have implicitly changed the notation for the step-size parameter μ_m in (6.10) to the notation $\alpha_m \mu_m$. We do not have access to $R^*_{\omega_m} u$, but only to its noisy estimate $y_m = R^*_{\omega_m} u + \epsilon_{\omega_m, y_m}$. Therefore, the online learning algorithm for determining f_u from the samples is of the form

$$u^{(m+1)} = \alpha_m(u^{(m)} + \mu_m R_{\omega_m}(y_m - R^*_{\omega_m} u^{(m)})), \qquad m = 0, 1, \ldots. \tag{6.44}$$

This is a randomized Schwarz approximation method with additional random noise. In contrast to Section 5.5, where the robustness to deterministic and random perturbations of Schwarz iterative methods for finite Ω was studied on an elementary level, we now adapt the considerations of Section 6.3 to derive convergence results for the expected squared error $\mathbb{E}(\|u - u^{(m)}\|_V^2)$. This corresponds to convergence in the RKHS H. As expected, such estimates again require additional smoothness assumptions on u in the form $u \in V^s_{P_\rho}$ with $0 < s \le 1$, but they now also include a dependence on both the initial error $\|e^{(0)}\|_V^2$ and the noise variance σ_H^2. The price for convergence is a certain decay of the step-sizes $\mu_m \to 0$, which is typical for stochastic approximation algorithms. More specifically, in the rest of this section, we set

$$\alpha_m = \frac{m+1}{m+2}, \qquad \mu_m = \frac{A}{(m+1)^t}, \qquad m = 0, 1, \ldots, \tag{6.45}$$

where the parameters $1/2 < t < 1$ and $0 < A \le (2\Lambda)^{-1}$ will be properly fixed later. In the language of learning algorithms, this is a so-called regularized algorithm, compared to unregularized algorithms with $\alpha_m = 1$.

Our main result, Theorem 6.5 proved in Subsection 6.5.2, concerns the convergence of the online learning algorithm (6.44–6.45) to u in V (or, which is the same, to f_u in H), where u is the minimizer in (6.38). If the latter does not exist, i.e., if (6.40) does not hold, then the algorithm does not converge in V, since we have $\|\mathbb{E}(u^{(m)})\|_V \to \infty$ independently of the choices for the parameters A and t in (6.45). This is established in Problem 6.8. Better estimates for

$$\mathbb{E}(\|f_u - f_{u^{(m)}}\|^2_{L^2_\rho(\Omega,Y)}) = \mathbb{E}(\|P_\rho^{1/2} e^{(m)}\|_V^2) = \mathbb{E}((P_\rho e^{(m)}, e^{(m)})_V)$$

are also of interest, but cannot be obtained within our framework. We will comment on these difficulties in Subsection 6.5.2. There is a large literature, especially for the scalar-valued case $Y = \mathbb{R}$, devoted to various versions of the algorithm (6.10) and other types of probabilistic convergence. In particular, (6.10) is often considered in the so-called finite horizon case, where $N < \infty$ is fixed and the step-sizes μ_m are chosen in dependence on N such that expectations like $\mathbb{E}(\|u - u^{(N)}\|_V^2)$ or $\mathbb{E}(\|f_u - f_{u^{(N)}}\|^2_{L^2_\rho(\Omega,Y)})$ are optimized for the final approximation $u^{(N)}$. We refer to [91] for a discussion of these issues and a literature review.

6.5.1 Convergence in V

In this subsection we will use the notation and assumptions outlined above, with the only change that the scalar product and the norm in V will simply be denoted by $(\cdot,\cdot)$ and $\|\cdot\|$, respectively. We will prove an estimate of the form

$$\mathbb{E}(\|e^{(m)}\|^2) = \mathrm{O}((m+1)^{-s/(s+2)}), \qquad m \longrightarrow \infty, \tag{6.46}$$

under the assumption $u \in V^s_{P_\rho}$, $0 < s \le 1$, if the parameters A and t in (6.45) are chosen accordingly. The precise statement and the dependence of the constant in (6.46) on the initial error, the noise variance and the smoothness assumption are given at the end of this subsection.

The argument starts from the analog of the basic error recurrence used in the proof of Corollary 6.1, which we will derive next. From (6.44) and $y_m = R^*_{\omega_m} u + \epsilon_{\omega_m, y_m}$ we derive the error representation

$$e^{(m+1)} = \underbrace{\alpha_m(e^{(m)} - \mu_m P_{\omega_m} e^{(m)}) + \bar{\alpha}_m u}_{\bar{e}^{(m+1)}:=} \;-\; \alpha_m \mu_m R_{\omega_m} \epsilon_{\omega_m, y_m},$$

where $\bar{\alpha}_m = 1 - \alpha_m = (m+2)^{-1}$ as before. The first term $\bar{e}^{(m+1)}$ has the same structure as in the noiseless case considered in Section 6.3, while the remaining term is the noise contribution. Thus

$$\|e^{(m+1)}\|^2 = \|\bar{e}^{(m+1)}\|^2 - 2\alpha_m\mu_m(R_{\omega_m}\epsilon_{\omega_m,y_m}, \bar{e}^{(m+1)}) + \alpha_m^2\mu_m^2\|R_{\omega_m}\epsilon_{\omega_m,y_m}\|^2. \tag{6.47}$$

We now estimate the conditional expectation of $\|e^{(m+1)}\|^2$ with respect to a given $u^{(m)}$ separately for the three terms in (6.47). For the third term, by (6.37) and the definition of σ_H^2, we have

$$\mathbb{E}'(\|R_{\omega_m}\epsilon_{\omega_m,y_m}\|^2) \le \Lambda \mathbb{E}(\|\epsilon_{\omega_m,y_m}\|_Y^2) = \Lambda\sigma_H^2. \tag{6.48}$$

For the second term we need

$$\mathbb{E}((R_\omega \epsilon_{\omega,y}, w)) = \mathbb{E}((y - R^*_\omega u, R^*_\omega w)_Y) = 0 \qquad \forall\, w \in V.$$

This follows directly from the fact that $u \in V$ is the minimizer of the problem (6.38). So if we set $w = \alpha_m e^{(m)} + \bar{\alpha}_m u$, we get

$$\begin{aligned}
&\mathbb{E}'(-2\alpha_m\mu_m(R_{\omega_m}\epsilon_{\omega_m,y_m}, \bar{e}^{(m+1)})) \\
&= 2\alpha_m\mu_m(\alpha_m\mu_m\mathbb{E}'((R_{\omega_m}\epsilon_{\omega_m,y_m}, P_{\omega_m}e^{(m)})) - \mathbb{E}'((R_{\omega_m}\epsilon_{\omega_m,y_m}, w))) \\
&= 2\alpha_m^2\mu_m^2\mathbb{E}'((R_{\omega_m}\epsilon_{\omega_m,y_m}, P_{\omega_m}e^{(m)})) \\
&\le \alpha_m^2\mu_m^2(\mathbb{E}'(\|R_{\omega_m}\epsilon_{\omega_m,y_m}\|^2) + \mathbb{E}'(\|P_{\omega_m}e^{(m)}\|^2)).
\end{aligned}$$

Here the first term is estimated by (6.48). For the second term we substitute the upper bound

$$\mathbb{E}'(\|P_{\omega_m}e^{(m)}\|^2) \le \Lambda(P_\rho e^{(m)}, e^{(m)}), \tag{6.49}$$

which follows from (6.37) and the definition of P_ρ. Together this gives

$$\mathbb{E}'(-2\alpha_m\mu_m(R_{\omega_m}\epsilon_{\omega_m,y_m}, \bar{e}^{(m+1)})) \le \Lambda\alpha_m^2\mu_m^2(\sigma_H^2 + (P_\rho e^{(m)}, e^{(m)})) \tag{6.50}$$

for the second term in (6.47).

The estimate for $\mathbb{E}'(\|\bar{e}^{(m+1)}\|^2)$ is essentially the same as the estimate for $\mathbb{E}'(\|e^{(m+1)}\|^2)$ in the proof of Theorem 6.3 and Corollary 6.1 in Section 6.3. In fact, we have

$$\begin{aligned}\|\bar{e}^{(m+1)}\|^2 = \bar{\alpha}_m^2\|u\|^2 + 2\alpha_m\bar{\alpha}_m(u, e^{(m)} - \mu_m P_{\omega_m}e^{(m)})\\ + \alpha_m^2(\|e^{(m)}\|^2 - 2\mu_m(e^{(m)}, P_{\omega_m}e^{(m)}) + \mu_m^2\|P_{\omega_m}e^{(m)}\|^2).\end{aligned}$$

After taking conditional expectations and with the definition of P_ρ and (6.49), we obtain

$$\begin{aligned}\mathbb{E}'(\|\bar{e}^{(m+1)}\|^2) &= \bar{\alpha}_m^2\|u\|^2 + 2\alpha_m\bar{\alpha}_m(u, e^{(m)} - \mu_m P_\rho e^{(m)})\\ &\quad + \alpha_m^2(\|e^{(m)}\|^2 - 2\mu_m(e^{(m)}, P_\rho e^{(m)}) + \mu_m^2\mathbb{E}'(\|P_{\omega_m}e^{(m)}\|^2))\\ &\le \bar{\alpha}_m^2\|u\|^2 + 2\alpha_m\bar{\alpha}_m(u, e^{(m)} - \mu_m P_\rho e^{(m)})\\ &\quad + \alpha_m^2(\|e^{(m)}\|^2 - \mu_m(2 - \Lambda\mu_m)(e^{(m)}, P_\rho e^{(m)})).\end{aligned}$$

As in the proof of Corollary 6.1, to estimate the term $(u, e^{(m)} - \mu_m P_\rho e^{(m)})$, we take an arbitrary $h = P_\rho^{1/2}v \in V_{P_\rho}^1$, where $v \in V = V_\rho^0$ and $\|h\|_{V_{P_\rho}^1} = \|v\|$. With this we have

$$\begin{aligned}&2\alpha_m\bar{\alpha}_m(u, e^{(m)} - \mu_m P_\rho e^{(m)})\\ &\quad = 2\alpha_m\bar{\alpha}_m((u-h, (I-\mu_m P_\rho)e^{(m)}) + (h, (I-\mu_m P_\rho)e^{(m)}))\\ &\quad \le 2\alpha_m\bar{\alpha}_m\|u-h\|\|(I-\mu_m P_\rho)e^{(m)}\| + 2(\bar{\alpha}_m\mu_m^{-1/2}(I-\mu_m P_\rho)v, \alpha_m\mu_m^{1/2}e^{(m)})\\ &\quad \le 2\alpha_m\bar{\alpha}_m\|u-h\|\|e^{(m)}\| + \bar{\alpha}_m^2\mu_m^{-1}\|(I-\mu_m P_\rho)v\|^2 + \alpha_m^2\mu_m\|P_\rho^{1/2}e^{(m)}\|^2\\ &\quad \le 2\alpha_m\bar{\alpha}_m\|u-h\|\|e^{(m)}\| + \bar{\alpha}_m^2\mu_m^{-1}\|h\|_{V_{P_\rho}^1}^2 + \alpha_m^2\mu_m(P_\rho e^{(m)}, e^{(m)}).\end{aligned}$$

Here we used that $\|(I-\mu_m P_\rho)e^{(m)}\| \le \|e^{(m)}\|$, and similarly

$$\|(I-\mu_m P_\rho)v\| \le \|v\| = \|h\|_{V_{P_\rho}^1},$$

which holds since $0 < \mu_m \le A \le (2\Lambda)^{-1}$ according to (6.45). Substitution into the previous inequality results in

$$\mathbb{E}'(\|\bar{e}^{(m+1)}\|^2) \le \bar{\alpha}_m^2(\|u\|^2 + \mu_m^{-1}\|h\|^2_{V^1_{P_\rho}}) + 2\alpha_m\bar{\alpha}_m\|u-h\|\|e^{(m)}\| \\ + \alpha_m^2(\|e^{(m)}\|^2 - \mu_m(1-\Lambda\mu_m)(e^{(m)}, P_\rho e^{(m)})).$$

Now, combining this estimate for the conditional expectation of the first term in (6.47) with the bounds (6.48) and (6.50) for the third and second terms, we get after simplifications

$$\mathbb{E}'(\|\bar{e}^{(m+1)}\|^2) \le \alpha_m^2(\|e^{(m)}\|^2 + 2\Lambda\sigma_H^2\mu_m^2) \\ + 2\alpha_m\bar{\alpha}_m\|u-h\|\|e^{(m)}\| + \bar{\alpha}_m^2(\|u\|^2 + \mu_m^{-1}\|h\|^2_{V^1_{P_\rho}}). \tag{6.51}$$

Here the term with $(e^{(m)}, P_\rho e^{(m)}) \ge 0$ has been omitted, since its combined forefactor $-\alpha_m^2\mu_m(1-2\Lambda\mu_m)$ is non-positive due to the restriction on A in (6.45).

For given

$$u = \sum_k c_k\psi_k \in V^s_{P_\rho}, \qquad 0 < s \le 1,$$

in (6.51), we choose

$$h = \sum_{k:\,\lambda_k(m+1)^b \ge B} c_k\psi_k$$

with some fixed constants $b, B > 0$ specified below. This gives

$$\|h\|^2_{V^1_{P_\rho}} = \sum_{k:\,\lambda_k(m+1)^b \ge B} \lambda_k^{-(1-s)}(\lambda_k^{-s}c_k^2) \le B^{-(1-s)}(m+1)^{(1-s)b}\|u\|^2_{V^s_{P_\rho}}$$

and

$$\|u-h\|^2 = \sum_{k:\,\lambda_k(m+1)^b < B} \lambda_k^s(\lambda_k^{-s}c_k^2) \le B^s(m+1)^{-bs}\|u\|^2_{V^s_{P_\rho}}.$$

After substitution in (6.51) we get

$$\mathbb{E}'(\|\bar{e}^{(m+1)}\|^2) \le \alpha_m^2(\|e^{(m)}\|^2 + 2\Lambda\sigma_H^2\mu_m^2) + 2\alpha_m\bar{\alpha}_m B^{s/2}(m+1)^{-bs/2}\|u\|_{V^s_{P_\rho}}\|e^{(m)}\| \\ + \bar{\alpha}_m^2(\|u\|^2 + \mu_m^{-1}B^{-(1-s)}(m+1)^{(1-s)b}\|u\|^2_{V^s_{P_\rho}}). \tag{6.52}$$

Obviously, if $s = 1$, we can set $h = u$, which would greatly simplify the following considerations and lead to a more accurate final estimate. We refer to Problem 6.9.

Next, we take full expectations in (6.52) using the independence assumption for the sampling process and introduce the notation

$$\varepsilon_m := \mathbb{E}(\|e^{(m)}\|^2)^{1/2} \ge \mathbb{E}(\|e^{(m)}\|).$$

Then, together with (6.45) and $\alpha_m = (m+1)\bar{\alpha}_m$, we get

$$\begin{aligned}\varepsilon_{m+1}^2 &\le \alpha_m^2(\varepsilon_m^2 + 2A^2\Lambda\sigma_H^2(m+1)^{-2t}) + 2\alpha_m\bar{\alpha}_m B^{s/2}(m+1)^{-bs/2}\|u\|_{V_{P_\rho}^s}\varepsilon_m\|e^{(m)}\| \\ &\quad + \bar{\alpha}_m^2(\|u\|^2 + A^{-1}B^{-(1-s)}(m+1)^{(1-s)b+t}\|u\|_{V_{P_\rho}^s}^2) \\ &\le \alpha_m^2\varepsilon_m^2 + \bar{\alpha}_m^2\Big(2A^2\Lambda\sigma_H^2(m+1)^{2-2t} + 2B^{s/2}(m+1)^{-bs/2+1}\|u\|_{V_{P_\rho}^s}\varepsilon_m\|e^{(m)}\| \\ &\quad + \|u\|^2 + A^{-1}B^{-(1-s)}(m+1)^{(1-s)b+t}\|u\|_{V_{P_\rho}^s}^2\Big).\end{aligned}$$

As a final step, let us assume for a moment that

$$\varepsilon_k \le C(k+1)^{-r}, \qquad k = 0, \dots, m, \tag{6.53}$$

holds for some constants $C, r > 0$. Next we set

$$a := \max(0, 2-2t, -bs/2+1-r, (1-s)b+t) > 0$$

and set

$$D := 2A^2\Lambda\sigma_H^2 + 2CB^{s/2}\|u\|_{V_{P_\rho}^s} + \|u\|^2 + A^{-1}B^{-(1-s)}\|u\|_{V_{P_\rho}^s}^2.$$

Then, for $k = 0, 1, \dots, m$, the estimate for ε_{k+1} simplifies to

$$\varepsilon_{k+1}^2 \le \alpha_k^2\varepsilon_k^2 + D\bar{\alpha}_k^2(k+1)^a.$$

Since $\alpha_k^2\bar{\alpha}_{k-1}^2 = \bar{\alpha}_k^2$ we get

$$d_{k+1} := \bar{\alpha}_k^{-2}\varepsilon_{k+1}^2 \le \bar{\alpha}_k^{-2}\alpha_k^2\varepsilon_k^2 + D(k+1)^a = d_k + D(k+1)^a$$

and by recursion we get

$$d_{m+1} \le d_0 + D\sum_{k=0}^{m}(k+1)^a = \varepsilon_0^2 + D\sum_{k=0}^{m}(k+1)^a$$

and finally

$$\varepsilon_{m+1}^2 \le (m+2)^{-2}(\|e^{(0)}\|^2 + D(m+2)^{a+1}) < (\|e^{(0)}\|^2 + D)(m+2)^{a-1}.$$

Thus (6.53) holds by induction for all m if we ensure that $\varepsilon_0 = \|e^{(0)}\| \le C$ at the induction start for $m = 0$ and $2r = 1-a$ and $\|e^{(0)}\|^2 + D \le C^2$ for the induction step $m \to m+1$.

To maximize r, it is intuitively clear that we need

$$a = 1-2r = 2-2t = -bs/2+1-r = (1-s)b+t,$$

which has the unique solution

$$t = \frac{1+s}{2+s}, \qquad b = \frac{1}{2+s}, \qquad 2r = \frac{s}{2+s}, \qquad a = \frac{2}{2+s}.$$

Furthermore, to guarantee (6.53), the value for C must satisfy $\|e^{(0)}\|^2 + D \leq C^2$, i.e.,

$$\|e^{(0)}\|^2 + \|u\|^2 + 2A^2\Lambda\sigma_H^2 + 2CB^{s/2}\|u\|_{V^s_{P_\rho}} + A^{-1}B^{-(1-s)}\|u\|^2_{V^s_{P_\rho}} \leq C^2.$$

By choosing concrete values for $0 < A \leq (2\Lambda)^{-1}$ and $B > 0$, the constant C^2 can be made more explicit. For example, substituting the upper bound

$$2CB^{s/2}\|u\|_{V^s_{P_\rho}} \leq \frac{C^2}{2} + 2B^s\|u\|^2_{V^s_{P_\rho}}$$

and rearranging shows that

$$C^2 = 2\left(\|e^{(0)}\|^2 + \|u\|^2 + B^s(2 + (AB)^{-1})\|u\|^2_{V^s_{P_\rho}} + 2A^2\Lambda\sigma_H^2\right)$$

is a possible choice for C. If we now set A to its maximal value $A = (2\Lambda)^{-1}$ and set $B = \Lambda$, then this gives a more explicit dependence of C^2 on the assumptions about the initial value $\|e^{(0)}\|^2$, the variance σ_H^2 and the smoothness of u, namely

$$C^2 = 2\|e^{(0)}\|^2 + 2\|u\|^2 + 8\Lambda^s\|u\|_{V^s_{P_\rho}} + \sigma_H^2/\Lambda. \tag{6.54}$$

This is the constant we refer to in our main result below. Clearly, varying A and B affects the trade-off between initial error, noise variance, and smoothness assumptions in the convergence estimate (6.53). Note also that B is not part of the algorithm and can be set to any value.

The following theorem summarizes the results of our considerations above.

Theorem 6.5 *Let Y, V be separable Hilbert spaces, let Ω be a compact metric space, let μ be a Borel probability measure on $\Omega \times Y$, and let ρ be the marginal Borel probability measure on Ω induced by μ. Assume that*

$$\mathbb{E}(\|y\|_Y^2) = \int_{\Omega\times Y} \|y\|_Y^2 \, d\mu < \infty.$$

For the feature map $\mathbf{R} = \{R_\omega\}_{\omega\in\Omega} : \Omega \to \mathcal{B}(Y, V)$, we need the conditions (6.35) and (6.36). We also assume that the associated operator $P_\rho = \mathbb{E}(R_\omega R_\omega^*) \in \mathcal{B}_{pos}(V)$ is compact. Finally, we assume (6.40) and that the unique solution $u \in V$ of the minimization problem (6.38) belongs to $V^s_{P_\rho}$ for some $0 < s \leq 1$. Consider the online learning algorithm (6.44), where $u^{(0)} \in V$ is arbitrary, the parameters α_m, μ_m are given by (6.45) with $t = t_s := (1+s)/(2+s)$ and $A = 1/(2\Lambda)$, and the random samples (ω_m, y_m), $m = 0, 1, \ldots$, are i.i.d. with respect to μ. Then the expected squared error $\mathbb{E}(\|u - u^{(m)}\|^2)$ in V satisfies

$$\mathbb{E}(\|u-u^{(m)}\|^2) = \mathbb{E}(\|f_u-f_{u^{(m)}}\|_H^2) \le C^2(m+1)^{-s/(2+s)}, \qquad m = 1,2,\ldots, \quad (6.55)$$

where C^2 is given by (6.54) and the noise characteristics σ_H^2 is defined in (6.43).

In this generality, except in [91], Theorem 6.5 has not yet appeared in the literature, at least to our knowledge. A brief discussion of its relation to other convergence results for incremental learning algorithms such as (6.44) is given in [91]. The asymptotic behavior of the estimate in dependence on the quantities $\|e^{(0)}\|^2$ and $\|u\|^2$ can be refined, for $s = 1$ this is shown in Problem 6.9. Estimates of the form (6.55) hold for all values $1/2 < t < 1$ and $0 < A < (2\Lambda)^{-1}$ admissible in (6.45), but with weaker exponents and different constants C^2 depending on t and A, see Problem 6.10. Note again that the assumption of the existence of a minimizer in (6.38) is also necessary for the convergence in V of the online algorithm (6.44–6.45), since $\|\mathbb{E}(u^{(m)})\|_V \to \infty$ if (6.40) is violated, see Problem 6.8.

6.5.2 Discussion and complementary results

We begin this subsection by pointing out why our analysis of the convergence of online algorithms in H does not easily generalize to convergence results in $L^2_\rho(\Omega, Y)$. We then consider the recovery of an element $u \in V$ from noisy measurements of its coefficients with respect to a fixed CONS in V, which, under some further assumptions, is a special case of the kernel learning problem considered in this section, but allows for a more detailed analysis. Theorem 6.6 and Proposition 6.1 provide upper and lower bounds on the expected squared errors of online algorithms in $V^{\bar{s}}_{P_\rho}$ norms when $u \in V^s_{P_\rho}$, $\bar{s} < s$, and illustrate to what extent one can hope to generalize our main Theorem 6.5.

6.5.2.1 Difficulties with convergence in $L^2_\rho(\Omega, Y)$

Our result for the vector-valued case concerned convergence in the feature space V, which is isometric to the RKHS H generated by the feature map $\mathbf{R}$. What we have not been able to do is to extend our methods to obtain better asymptotic rates for convergence in $L^2_\rho(\Omega, Y)$ to f_u (or better, to f_W, the $L^2_\rho(\Omega, Y)$ orthoprojection of f_μ onto the closure W of H in $L^2_\rho(\Omega, Y)$). Note that W is isometric to $V^{-1}_{P_\rho}$. More specifically, we have

$$W = \{f_{u,-1} = \sum_k c_k f_{\psi_k} \,:\, u = \sum_k \lambda_k^{1/2} c_k \psi_k \in V\},$$

with

$$\|f_{u,-1}\|^2_{L^2_\rho(\Omega,Y)} = \sum_k c_k^2 = \|P_\rho^{-1/2} u\|^2 = \|u\|^2_{V^{-1}_{P_\rho}}.$$

This follows because the functions $f_{\psi_k} = \lambda_k^{-1/2}\psi_k \in H$ form a CONS in $W \subset L^2_\rho(\Omega, Y)$, since the $\psi_k \in V$ form the CONS of the eigenfunctions of the compact operator P_ρ acting on V. Thus, if the assumption (6.40) about the existence of the minimizer u in (6.38) is kept, then it would be justified to study $\mathbb{E}(\|P_\rho^{1/2}e^{(m)}\|^2) = \mathbb{E}((P_\rho e^{(m)}, e^{(m)}))$ instead of $\mathbb{E}(\|e^{(m)}\|^2)$ to obtain $L^2_\rho(\Omega, Y)$ estimates.

However, if, in analogy to (6.47), one looks at the error decomposition

$$\|P_\rho^{1/2}e^{(m+1)}\|^2 \le \|P_\rho^{1/2}\bar{e}^{(m+1)}\|^2 - 2\alpha_m\mu_m(P_\rho R_{\omega_m}\epsilon_{\omega_m}, \bar{e}^{(m+1)}) + \alpha_m^2\mu_m^2\|P_\rho^{1/2}R_{\omega_m}\epsilon_{\omega_m}\|^2,$$

then difficulties arise mainly from the first term on the right-hand side (we drop the subscript V again in the notation of norms and scalar products). In fact, we have

$$\|P_\rho^{1/2}\bar{e}^{(m+1)}\|^2 = \bar{\alpha}_m^2\|P_\rho^{1/2}u\|^2 + 2\alpha_m\bar{\alpha}_m(P_\rho u, e^{(m)} - \mu_m P_{\omega_m}e^{(m)}) + \alpha_m^2(\|P_\rho^{1/2}e^{(m)}\|^2 - 2\mu_m(P_\rho e^{(m)}, P_{\omega_m}e^{(m)}) + \mu_m^2\|P_\rho^{1/2}P_{\omega_m}e^{(m)}\|^2).$$

After taking conditional expectations $\mathbb{E}'(\|P_\rho^{1/2}\bar{e}^{(m+1)}\|^2)$, we get a negative term

$$-2\alpha_m^2\mu_m\|P_\rho e^{(m)}\|^2$$

on the right-hand side, which must compensate for positive contributions from terms like

$$\mathbb{E}'(\|P_\rho^{1/2}P_{\omega_m}e^{(m)}\|^2)$$

and

$$\mathbb{E}'(2\alpha_m\bar{\alpha}_m(P_\rho u, e^{(m)} - \mu_m P_{\omega_m}e^{(m)})) = 2\alpha_m\bar{\alpha}_m(P_\rho u, e^{(m)} - \mu_m P_\rho e^{(m)}).$$

Since P_ρ normally does not commute with the operators P_ω, this strategy does not work without additional assumptions, see Problem 6.4.

6.5.2.2 A special case: Upper and lower bounds

Consider the special "learning" problem of recovering an unknown element $u \in V$ from noisy measurements of its coefficients with respect to a CONS $\Psi = \{\psi_i\}_{i\in\mathbb{N}}$ in V by an online method. To do this, we assume that we are given μ-distributed random samples (i_m, y_m), where $i_m \in \mathbb{N}$ and

$$y_m = (u, \psi_{i_m}) + \epsilon_m, \qquad m = 0, 1, \dots,$$

are the noisy samples of the coefficients (u, ψ_{i_m}). Starting from $u^{(0)} = 0$, we want to approximate u by the iterates $u^{(m)}$ obtained by the online algorithm

$$u^{(m+1)} = \alpha_m u^{(m)} + \alpha_m\mu_m(y_m - (u^{(m)}, \psi_{i_m}))\psi_{i_m}, \qquad m = 0, 1, \dots, \tag{6.56}$$

where the coefficients α_m and μ_m are given by (6.45) with $\Lambda = 1$. This is a special instance of (6.44) if we set $\Omega = \mathbb{N}$, $Y = \mathbb{R}$ and define $R_i : \mathbb{R} \to V$ and $R_i^* : V \to \mathbb{R}$ by $R_i y = y\psi_i$ and $R_i^* v = (v, \psi_i)$. To simplify things further, let i_m be i.i.d. samples from Ω with respect to a discrete probability measure ρ on Ω, i.e., ρ is given w.l.o.g. by $\mathbb{P}(i_m = i) = \rho_i$, where $\sum_{i\in\Omega} \rho_i = 1$ and $\rho_1 \geq \rho_2 \geq \cdots > 0$. Moreover, let ϵ_m be i.i.d. random noise with zero mean and finite variance $\sigma^2 < \infty$, which is independent of i_m. The corresponding operator P_ρ is given by

$$P_\rho v = \sum_{i\in\Omega} \rho_i (v, \psi_i)\psi_i,$$

its eigenvalues $\lambda_i = \rho_i$ are given by ρ, and it is trace class. The spaces $V^s_{P_\rho}$, $-\infty < s < \infty$, can now be identified as sets of formal orthogonal series

$$V^s_{P_\rho} := \left\{ u \sim \sum_{i\in\Omega} c_i\psi_i \;:\; \|u\|^2_{V^s_{P_\rho}} = \sum_{i\in\Omega} \rho_i^{-s} c_i^2 \right\}.$$

Obviously, $V^s_{P_\rho} \subset V = V^0_{P_\rho}$ for $s > 0$. Since functions $f : \Omega \to \mathbb{R}$ can be identified with formal series with respect to Ψ by

$$u \sim \sum_{i\in\Omega} c_i\psi_i \quad \Leftrightarrow \quad f_u : f_u(i) = c_i,$$

we have $\|f_u\|_{L^2_\rho(\Omega)} = \|u\|_{V^{-1}_{P_\rho}}$ and we can identify $L^2_\rho(\Omega)$ with $V^{-1}_{P_\rho}$. Under the assumptions made, the underlying minimization problem (6.38) on V is

$$\mathbb{E}(|f_v - y|^2) = \|f_v - f_u\|^2_{L^2_\rho(\Omega)} + \sigma^2 \quad \to \quad \min,$$

and, as expected, has u as its unique solution. This example also shows that sometimes it can be more appropriate to consider convergence in V instead of convergence in $L^2_\rho(\Omega, Y)$.

The simplicity of this example allows a rather comprehensive convergence theory with respect to the scale of $V^s_{P_\rho}$ spaces. Our main result is as follows:

Theorem 6.6 *Let $-1 \leq \bar{s} \leq 0$, $s \geq 0$, and $\bar{s} < s \leq \bar{s} + 2$. Then, for the sampling process described above, the online algorithm (6.56) converges for $u \in V^s_{P_\rho}$ in the $V^{\bar{s}}_{P_\rho}$ norm with the bound*

$$\mathbb{E}(\|e^{(m)}\|^2_{V^{\bar{s}}_{P_\rho}}) \leq C(m+1)^{-\min(\frac{s-\bar{s}}{s+2}, \frac{2}{\bar{s}+4})}(A^{\bar{s}-s}\|u\|^2_{V^s_{P_\rho}} + A^{2+\bar{s}}\sigma^2), \quad m = 1, 2, \ldots, \tag{6.57}$$

if the parameters in (6.45) satisfy

$$t = t_s := \max((s+1)/(s+2), (\bar{s}+3)/(\bar{s}+4)), \qquad 0 < A \leq 1/2.$$

Proof The proof of (6.57) reduces to the study of the expectations of the squares of the coefficients $c_i^{(m)} = (e^{(m)}, \psi_i)$ of the error $e^{(m)} = u - u^{(m)}$ with respect to Ψ. We have

$$\mathbb{E}(\|e^{(m)}\|^2_{V^{\bar{s}}_{P_\rho}}) = \sum_i \rho_i^{-\bar{s}}\, \mathbb{E}((c_i^{(m)})^2), \qquad \|u\|^2_{V^s_{P_\rho}} = \|e^{(0)}\|^2_{V^s_{P_\rho}} = \sum_i \rho_i^{-s} c_i^2, \tag{6.58}$$

where $c_i = (u, \psi_i)$ are the coefficients in the decomposition of u. From (6.56) we find that

$$\begin{aligned} c_i^{(m+1)} &= \bar{\alpha}_m c_i + \alpha_m c_i^{(m)} + \alpha_m \begin{cases} \mu_m (y_{i_m} - (u^{(m)}, \psi_{i_m})), & i_m = i \\ 0, & i_m \neq i \end{cases} \\ &= \bar{\alpha}_m c_i + \alpha_m (c_i^{(m)} - \delta_{i,i_m} \mu_m (c_i^{(m)} + \epsilon_m)) \end{aligned}$$

for $m = 0, 1, \ldots$, where $\delta_{i,i_m} = 1$ with probability ρ_i and $\delta_{i,i_m} = 0$ with probability $1 - \rho_i$. So,

$$\begin{aligned} \mathbb{E}'((c_i^{(m+1)})^2) &= \bar{\alpha}_m^2 c_i^2 + 2\alpha_m \bar{\alpha}_m c_i c_i^{(m)} + \alpha_m^2 (c_i^{(m)})^2 \\ &\quad - 2\alpha_m \mu_m \mathbb{E}'(\delta_{i,i_m}(c_i^{(m)} + \epsilon_m)(\bar{\alpha}_m c_i + \alpha_m c_i^{(m)})) \\ &\quad + \alpha_m^2 \mu_m^2 \mathbb{E}'(\delta_{i,i_m}(c_i^{(m)} + \epsilon_m)^2) \\ &= \alpha_m^2 (1 - \rho_i \mu_m (2 - \mu_m))(c_i^{(m)})^2 + 2\alpha_m \bar{\alpha}_m (1 - \rho_i \mu_m) c_i c_i^{(m)} \\ &\quad + \bar{\alpha}_m^2 c_i^2 + \rho_i \alpha_m^2 \mu_m^2 \sigma^2, \end{aligned}$$

where we used the properties and the independence of the random variables δ_{i,i_m} and ϵ_m. We also need

$$\begin{aligned} \mathbb{E}'(c_i^{(m+1)}) &= \bar{\alpha}_m c_i + \alpha_m c_i^{(m)} - \alpha_m \mu_m \mathbb{E}'(\delta_{i,i_m}(c_i^{(m)} + \epsilon_m)) \\ &= \alpha_m (1 - \rho_i \mu_m) c_i^{(m)} + \bar{\alpha}_m c_i. \end{aligned}$$

Now, if we denote $\varepsilon_m := \mathbb{E}((c_i^{(m)})^2)$ and $\bar{\varepsilon}_m := \mathbb{E}(c_i^{(m)})$ and use the independence assumption, we get the linear recursion

$$\begin{aligned} \varepsilon_{m+1} &= \alpha_m^2 (1 - \rho_i \mu_m (2 - \mu_m)) \varepsilon_m + 2\alpha_m \bar{\alpha}_m (1 - \rho_i \mu_m) c_i \bar{\varepsilon}_m \\ &\quad + \bar{\alpha}_m^2 c_i^2 + \rho_i \alpha_m^2 \mu_m^2 \sigma^2, \\ \bar{\varepsilon}_{m+1} &= \alpha_m \bar{\varepsilon}_m + \bar{\alpha}_m c_i. \end{aligned}$$

The initial values are $\varepsilon_0 = c_i^2$ and $\bar{\varepsilon}_0 = c_i$. The recursion for $\bar{\varepsilon}_m$ can even be handled explicitly. To see this, denote the necessary finite products by Π_k^{m-1}, namely

$$\Pi_k^{m-1} = (1 - \frac{a}{m^t}) \cdots (1 - \frac{a}{(k+1)^t}), \quad 0 \leq k \leq m-1, \quad \Pi_m^{m-1} = 1, \tag{6.59}$$

where $a > 0$ is a constant. To derive the upper bound (6.57), we set $a = A\rho_i$. By the properties

$$\alpha_m\alpha_{m-1}\cdots\alpha_k = (k+1)\bar{\alpha}_m, \qquad \alpha_m\alpha_{m-1}\cdots\alpha_{k+1}\bar{\alpha}_k = \bar{\alpha}_m = (m+2)^{-1},$$

we have

$$\begin{aligned}
\bar{\varepsilon}_m &= \alpha_{m-1}(1-\rho_i\mu_{m-1})(\alpha_{m-2}(1-\rho_i\mu_{m-2})\bar{\varepsilon}_{m-2} + \bar{\alpha}_{m-2}c_i) + \bar{\alpha}_{m-2}c_i \\
&= \alpha_{m-1}\alpha_{m-2}\Pi_{m-2}^{m-1}\bar{\varepsilon}_{m-2} + \bar{\alpha}_{m-1}c_i(\Pi_{m-1}^{m-1} + \Pi_m^{m-1}) \\
&\ \ \vdots \\
&= \alpha_{m-1}\alpha_{m-2}\cdots\alpha_0\Pi_0^{m-1}\bar{\varepsilon}_0 + \bar{\alpha}_{m-1}c_i\sum_{k=1}^{m}\Pi_k^{m-1} \\
&= \bar{\alpha}_{m-1}c_iS_m, \qquad S_m := \sum_{k=0}^{m}\Pi_k^{m-1}.
\end{aligned}$$

After substituting the formula for $\bar{\varepsilon}_m$ into the above recursion for ε_m and considering that $1-\rho_i\mu_m(2-\mu_m) \le 1-\rho_i\mu_m$, the latter becomes

$$\begin{aligned}
\varepsilon_{m+1} &\le \alpha_m^2(1-\rho_i\mu_m)\varepsilon_m + 2\alpha_m\bar{\alpha}_{m-1}\bar{\alpha}_m(1-\rho_i\mu_m)S_mc_i^2 \\
&\quad + \bar{\alpha}_m^2c_i^2 + \alpha_m^2\rho_i\mu_m^2\sigma^2 \\
&= \alpha_m^2(1-\rho_i\mu_m)\varepsilon_m + \bar{\alpha}_m^2(1+2(1-\rho_i\mu_m)S_m)c_i^2 + \alpha_m^2\rho_i\mu_m^2\sigma^2,
\end{aligned}$$

$m = 0, 1, \ldots$. Similar to the considerations for $\bar{\varepsilon}_m$, by recursion and again using the properties of the coefficients α_m, this gives the bound

$$\begin{aligned}
\varepsilon_m &\le \bar{\alpha}_{m-1}^2\left(\Pi_0^{m-1}c_i^2 + \sum_{k=0}^{m-1}(\Pi_{k+1}^{m-1} + 2\Pi_k^{m-1}S_k)c_i^2\right. \\
&\qquad\qquad \left. + \rho_i\sigma^2\sum_{k=0}^{m-1}(k+1)^2\mu_k^2\Pi_{k+1}^{m-1}\right) \\
&\le \frac{1}{(m+1)^2}\left(2c_i^2\sum_{k=0}^{m}\Pi_k^{m-1}S_k + \rho_i\sigma^2\sum_{k=0}^{m-1}(k+1)^2\mu_k^2\Pi_{k+1}^{m-1}\right).
\end{aligned}$$

The further considerations use the dependence of Π_k^{m-1} and S_m on m, $a = A\rho_i$ and the exponent $t \in (1/2, 1)$, which appear in the definition (6.45) of μ_m. For the substitution in the previous estimate for $\varepsilon_m = \mathbb{E}((c_i^{(m)})^2)$, we use the bounds

$$\sum_{k=0}^{m}\Pi_k^{m-1}S_k \le C\min((m+1)^2, A^{-2}\rho_i^{-2}(m+1)^{2t}), \quad m = 1, 2, \ldots,$$

and

$$\sum_{k=0}^{m-1}(k+1)^2\mu_k^2\Pi_{k+1}^{m-1} \le CA^2\min((m+1)^{3-2t}, A^{-1}\rho_i^{-1}(m+1)^{2-t}), \quad m = 1, 2, \ldots,$$

which follow from the estimates (6.78) and (6.79) stated in Problem 6.12. Since their proof is elementary, we decided to leave it to the problem section. The constant C in these inequalities is independent of i and m, but can depend on t. This gives

$$\mathbb{E}((c_i^{(m)})^2) \le C\Big(\min(1, A^{-2}\rho_i^{-2}(m+1)^{2t-2})c_i^2 \\ + A^2(m+1)^{1-2t}\min(1, A^{-1}\rho_i^{-1}(m+1)^{t-1})\rho_i\sigma^2\Big) \tag{6.60}$$

for all $i \in \Omega$ and $m = 1, 2, \ldots$.

Next, we substitute (6.60) into the expression for $\mathbb{E}(\|e^{(m)}\|^2_{V^{\bar{s}}_{P_\rho}})$ in (6.58). First consider the index set $I_1 := \{i \in \Omega : A\rho_i \le (m+1)^{t-1}\}$. Then

$$\sum_{i\in I_1}\rho_i^{-\bar{s}}\mathbb{E}((c_i^{(m)})^2) \le C\left(\sum_{i\in I_1}\rho_i^{s-\bar{s}}(\rho_i^{-s}c_i^2) + A^2\sum_{i\in I_1}\rho_i^{-\bar{s}}(m+1)^{1-2t}(\rho_i\sigma^2)\right)$$
$$\le C\left(A^{\bar{s}-s}(m+1)^{(s-\bar{s})(t-1)}\sum_{i\in I_1}\rho_i^{-s}c_i^2 + A^{2+\bar{s}}(m+1)^{-\bar{s}(t-1)+1-2t}\sum_{i\in I_1}\rho_i\sigma^2\right),$$

where we used that $s - \bar{s} > 0$, $\bar{s} \le 0$. If we set $t = t_s = (s+1)/(s+2)$, then the exponents in the powers of $(m+1)$ are equal and best possible, i.e.,

$$(s-\bar{s})(t_s - 1) = -\bar{s}(t_s - 1) + 1 - 2t_s = -\frac{s-\bar{s}}{s+2},$$

and we get

$$\sum_{i\in I_1}\rho_i^{-\bar{s}}\mathbb{E}((c_i^{(m)})^2) \le C(m+1)^{-\frac{s-\bar{s}}{s+2}}\left(A^{\bar{s}-s}\sum_{i\in I_1}\rho_i^{-s}c_i^2 + A^{2+\bar{s}}\sum_{i\in I_1}\rho_i\sigma^2\right). \tag{6.61}$$

If $i \in I_2 := \Omega\backslash I_1$, then

$$\min(1, A^{-2}\rho_i^{-2}(m+1)^{2t-2}) = \min(1, A^{-1}\rho_i^{-1}(m+1)^{t-1})^2 \\ = A^{-2}\rho_i^{-2}(m+1)^{2t-2}$$

in (6.60). This gives

$$\sum_{i\in I_2}\rho_i^{-\bar{s}}\mathbb{E}((c_i^{(m)})^2) \le C\left(A^{-2}\sum_{i\in I_2}\rho_i^{s-\bar{s}-2}(m+1)^{2t-2}(\rho_i^{-s}c_i^2) \right. \\ \left. + A\sum_{i\in I_2}\rho_i^{-\bar{s}-1}(m+1)^{-t}(\rho_i\sigma^2)\right).$$

Now we distinguish the cases $s - \bar{s} \le 2$ and $s - \bar{s} > 2$. In the first case, we substitute

$$\max_{i \in I_2} \rho_i^{s-\bar{s}-2}(m+1)^{2t-2} \le A^{2+\bar{s}-s}(m+1)^{(s-\bar{s})(t-1)}$$

and

$$\max_{i \in I_2} \rho_i^{-\bar{s}-1}(m+1)^{-t} \le A^{\bar{s}+1}(m+1)^{-\bar{s}(t-1)+1-2t},$$

which hold because we have $\rho_i > A^{-1}(m+1)^{t-1}$ for $i \in I_2$ and $s - \bar{s} - 2 \le 0$ and $-\bar{s} - 1 \le 0$ by our restrictions on the parameters. This results in

$$\sum_{i \in I_2} \rho_i^{-\bar{s}} \mathbb{E}((c_i^{(m)})^2) \le C \left(A^{\bar{s}-s}(m+1)^{(s-\bar{s})(t-1)} \sum_{i \in I_2} \rho_i^{-s} c_i^2 \right.$$
$$\left. + A^{2+\bar{s}}(m+1)^{-\bar{s}(t-1)+1-2t} \sum_{i \in I_2} \rho_i \sigma^2 \right).$$

After substituting the value $t = t_s = (s+1)/(s+2)$, the final expression is exactly the same as shown in (6.61) for the summation with respect to I_1. Combining the results for the two cases, we arrive at (6.57) when $s \le \bar{s} + 2$.

In the second case, since $s - \bar{s} - 2 > 0$ and $\rho_i < 1$, the first maximum can only be estimated by

$$\max_{i \in I_2} \rho_i^{s-\bar{s}-2}(m+1)^{2t-2} \le (m+1)^{2(t-1)}$$

and we arrive at

$$\sum_{i \in I_2} \rho_i^{-\bar{s}} \mathbb{E}((c_i^{(m)})^2) \le C \left((m+1)^{2(t-1)} \sum_{i \in I_2} \rho_i^{-s} c_i^2 \right.$$
$$\left. + A^{2+\bar{s}}(m+1)^{-\bar{s}(t-1)+1-2t} \sum_{i \in I_2} \rho_i \sigma^2 \right).$$

All that remains is to check that $(s - \bar{s})(t-1) < 2(t-1)$ and

$$2(t_s - 1) = -\bar{s}(t_s - 1) + 1 - 2t_s = 2/(\bar{s} + 4)$$

in the second case. This concludes the proof of Theorem 6.6. □

The convergence bounds of Theorem 6.6 need to be compared with the results for the randomized algorithm reported in the introductory Section 6.1 of this chapter, which corresponds to the noiseless case $\sigma = 0$. Setting $\bar{s} = 0$, one concludes from (6.62) that the convergence estimate for the online algorithm (6.44), which holds by Theorem 6.5 for $0 < s \le 1$ in the general case, is indeed matched. For $\bar{s} = -1$, which corresponds to $L_\rho^2(\Omega)$ convergence, the rate is better and agrees with known lower bounds.

We next show that the estimate (6.57) for the online algorithm (6.56) is the best possible, in the sense that, under the conditions of Theorem 6.6, the exponent in (6.57) cannot be increased without additional assumptions. In particular, for $s > \bar{s} + 2$, no further improvement is possible, i.e., the estimate actually saturates at $s = \bar{s} + 2$.

Proposition 6.1 *Let $-1 \le \bar{s} \le 0 \le s < \infty$, $\bar{s} < s$, and $\sigma > 0$. For the online algorithm (6.56) we have*

$$\sup_{\rho} \sup_{u:\, \|u\|_{V^s_{P_\rho}}=1} (m+1)^{\min((s-\bar{s})/(2+s),2/(\bar{s}+4))} \mathbb{E}(\|e^{(m)}\|^2_{V^{\bar{s}}_{P_\rho}}) \ge c > 0, \qquad (6.62)$$

$m = 1, 2, \ldots$, where c depends on $\bar{s}$, s, σ, and the parameters t and A in (6.45), but is independent of m.

Proof The statement is a consequence of the lower bound (6.80) for the expectations $\mathbb{E}((c_i^{(m)})^2)$ established in Problem 6.13, which corresponds to the upper bound (6.60). Without loss of generality, let m be sufficiently large. For the range $0 < s - \bar{s} \le 2$, we can choose the discrete probability measure ρ such that

$$\rho_i = (m+1)^{t-1}/A$$

for $i = 1, \ldots, n$, where $n \ge 1$ satisfies $n(m+1)^{t-1}/A < 1 \le (n+1)(m+1)^{t-1}/A$. The remaining $\rho_n \ge \rho_i > 0$, $i > n$, can be arbitrary. Also define

$$u = \rho_1^{s/2} \psi_1.$$

So $c_1 = \rho_1^{s/2}$, $c_i = 0$ for $i > 1$, and $u \in V^s_{P_\rho}$ has unit norm. With these choices, the index set I_2 is empty and, according to (6.80), we get

$$\begin{aligned}
\mathbb{E}(\|e^{(m)}\|^2_{V^{\bar{s}}_{P_\rho}}) &= \sum_i \rho_i^{-\bar{s}} \mathbb{E}((c_i^{(m)})^2) \\
&\ge c \left(\rho_1^{-\bar{s}} c_1^2 + A^2 \sum_i \rho_i^{-\bar{s}} (m+1)^{1-2t} \rho_i \sigma^2 \right) \\
&\ge c \left(\rho_1^{s-\bar{s}} + A^2 \sigma^2 n \rho_1 \rho_1^{-\bar{s}} (m+1)^{1-2t} \right) \\
&\ge c \left(A^{\bar{s}-s} (m+1)^{-(s-\bar{s})(1-t)} + \frac{1}{2} A^{2+\bar{s}} \sigma^2 (m+1)^{1-2t+\bar{s}(1-t)} \right) \\
&\ge c(m+1)^{-\min((s-\bar{s})(1-t),1-2t+\bar{s}(1-t))} = c(m+1)^{-(s-\bar{s})/(s+2)},
\end{aligned}$$

where we have used that $n\rho_1 \ge 1/2$ by the definition of n. This settles the case $s \le \bar{s} + 2$.

For $s > \bar{s} + 2$ we need a different choice for ρ. This time we set $\rho_1 = 1/3$, $\rho_2 = \cdots = \rho_{n+1} = (m+1)^{t-1}/A$, where $n \ge 1$ is determined such that

$$n(m+1)^{t-1}/A < 2/3 \le (n+1)(m+1)^{t-1}/A.$$

For the remaining indices $i > n+1$ we can take arbitrary values $\rho_i > 0$. Then $u = \rho_1^{s/2}\psi_1 = 3^{-s/2}\psi_1$ has unit norm in $V^s_{P_\rho}$, the index $i = 1$ belongs to I_2, all remaining indices belong to I_1, and $n\rho_2 \ge 1/3$ by the definition of n. By (6.80) we finally have

$$
\begin{aligned}
\mathbb{E}(\|e^{(m)}\|^2_{V^{\bar{s}}_{P_\rho}}) &\geq \sum_{i=1}^{n+1} \rho_i^{-\bar{s}} \mathbb{E}((c_i^{(m)})^2) \\
&\geq c(A^{-2}\rho_1^{-\bar{s}-2}(m+1)^{2t-2}c_1^2 + nA^2\rho_2^{-\bar{s}+1}\sigma^2(m+1)^{1-2t}) \\
&\geq c(A^{-2}(m+1)^{2t-2} + A^{2+\bar{s}}\sigma^2(m+1)^{(\bar{s}+2)(1-t)-1}) \\
&\geq c(m+1)^{-\min(2(1-t),1-(\bar{s}+2)(1-t))} \geq c(m+1)^{-2/(\bar{s}+4)}.
\end{aligned}
$$

This gives the desired estimate and also shows the saturation claim for $s > \bar{s}+2$. □

6.5.2.3 From Hilbert space splittings to direct integrals of Hilbert spaces

As noted above, the construction of the vector-valued RKHS H of Y-valued functions over Ω induced by a feature map $\mathbf{R}$ and the construction of the associated operator P_ρ is a generalization of the Hilbert space splitting concept developed in Chapter 2. Here we present some material on the connection to direct integrals of Hilbert spaces and operators between them, which was developed as a tool within the theory of operator algebras. We refer to [177, Section IV.8] and [69] for more details and proofs of the following statements.

We simplify things by assuming that Ω is a separable complete metric space equipped with a σ-finite Borel measure ρ. Then, given a family of separable Hilbert spaces $(\mathbf{V};\mathbf{a}) := (V_\omega; a_\omega)_{\omega\in\Omega}$, its associated so-called *direct integral*

$$\tilde{V}_\rho := \int_\Omega^\oplus V_\omega \, d\rho(\omega)$$

is the space of all (equivalence classes of) functions $\tilde{v} : \omega \in \Omega \mapsto \tilde{v}(\omega) := v_\omega \in V_\omega$, so that the scalar-valued function

$$\omega \mapsto \|v_\omega\|^2_\omega = a_\omega(v_\omega, v_\omega)$$

is measurable and integrable, i.e.,

$$\|\tilde{v}\|^2_{\tilde{V}_\rho} := \int_\Omega \|v_\omega\|^2_\omega \, d\rho(\omega) < \infty.$$

Two functions $\tilde{v}$ and $\tilde{w}$ are equivalent if $v_\omega = w_\omega$ almost everywhere with respect to ρ. There is a technical condition, called the *measurability of the family* $(\mathbf{V};\mathbf{a})$, for the theory of direct integrals of Hilbert spaces to hold, which we will state a little later.

Obviously, the measurability of $\omega \mapsto \|v_\omega\|^2_\omega$ implies the measurability of the functions $\omega \mapsto a_\omega(v_\omega, w_\omega)$ and allows the definition of the scalar product on $\tilde{V}_\rho$ by

$$\tilde{a}(\tilde{v},\tilde{w}) := \int_\Omega a_\omega(v_\omega, w_\omega) \, d\rho(\omega).$$

Equipped with this scalar product, $\tilde{V}_\rho$ is itself a Hilbert space. The notation is the same as in Chapter 2, if one takes a discrete measure defined on $\Omega = \mathbb{N}$. A special case of the above definition is that of a *constant family*, where $(V_\omega; a_\omega)$ coincides with a fixed separable Hilbert space $(Y; (\cdot,\cdot)_Y)$ for all $\omega \in \Omega$. Then $\tilde{V}_\rho$ can be identified with the L^2 Bochner space $L^2_\rho(\Omega, Y)$. This setting was sufficient for the considerations in Section 6.5.

The above mentioned measurability of the family $(\mathbf{V}; \mathbf{a})$ is equivalent to the condition that the sets

$$\Omega_n := \{\omega \in \Omega : \dim(V_\omega) = n\}, \qquad n \in \mathbb{N},$$

are measurable. This implies that the set Ω_∞ is also measurable, where Ω_∞ contains those $\omega \in \Omega$ such that V_ω is an infinite-dimensional separable Hilbert space. This is the essence of the results in [69, Section II.1.4] or [177, Theorem 8.13]. In other words, up to isometry, direct integrals of measurable families $(\mathbf{V}; \mathbf{a})$ can be viewed as at most countable Hilbert sums of L^2 Bochner spaces associated with the sets Ω_n and Ω_∞. As a consequence, direct integrals themselves are separable Hilbert spaces, if Ω is a separable complete metric space and ρ is σ-finite.

Next, we consider a family

$$\mathbf{R} := \{R_\omega\}_{\omega \in \Omega}$$

of bounded linear operators $R_\omega : V_\omega \to V$, $\omega \in \Omega$, where $(V; a)$ is another separable Hilbert space. We call such a family *admissable* if the scalar-valued function

$$\omega \mapsto a(R_\omega v_\omega, v)$$

is measurable and integrable for all $\tilde{v} \in \tilde{V}_\rho$ and $v \in V$. Thus, such an admissible operator family $\mathbf{R}$ defines a bounded linear operator R_ρ from $\tilde{V}_\rho$ into V by the formula

$$a(R_\rho \tilde{v}, v) := \int_\Omega a(R_\omega v_\omega, v)\, d\rho(\omega), \qquad v \in V.$$

In fact, the formula on the right-hand side defines a linear functional on V for each $\tilde{v}$, which is bounded by the open mapping principle. By the Hahn-Banach theorem, this defines a unique element $R_\rho \tilde{v} \in V$, and the boundedness of R_ρ on $\tilde{V}_\rho$ again follows from the open mapping theorem. In the case of discrete measures on $\Omega = \mathbb{N}$, treated in Chapter 2, the admissibility requirement of $\mathbf{R}$ is another way of saying that the collection $(\mathbf{V}; \mathbf{a}; \mathbf{R}; \rho)$ is a space splitting associated with $(V; a)$, see Problem 6.14.

Once R_ρ is well-defined, its adjoint $R_\rho^* : V \to \tilde{V}_\rho$ is given by

$$(R_\rho^* v)_\omega = R_\omega^* v, \qquad \omega \in \Omega, \qquad v \in V,$$

which follows by comparing the formulas

$$\tilde{a}(R_\rho^* v, \tilde{v}) = \int_\Omega a_\omega((R_\rho^* v)_\omega, v_\omega)\, d\rho(\omega)$$

and

$$a(v, R_\rho \tilde{v}) = \int_\Omega a(R_\omega v_\omega, v)\, d\rho(\omega) = \int_\Omega a_\omega(R_\omega^* v, v_\omega)\, d\rho(\omega).$$

The Schwarz operator $P_\rho := R_\rho R_\rho^* \in \mathcal{B}_{pos}(V)$ is thus given by

$$\begin{aligned} a(P_\rho v, w) &= \tilde{a}(R_\rho^* v, R_\rho^* w) = \int_\Omega a_\omega((R_\rho^* v)_\omega, (R_\rho^* w)_\omega)\, d\rho(\omega) \\ &= \int_\Omega a_\omega(R_\omega^* v, R_\omega^* w)\, d\rho(\omega) = \int_\Omega a(P_\omega v, w)\, d\rho(\omega), \end{aligned}$$

where

$$P_\omega := R_\omega R_\omega^*.$$

Obviously, if ρ is a probability measure, then $P_\rho v = \mathbb{E}(P_\omega v)$. This was already used in Section 6.5.

The Gramian operator $\tilde{P}_\rho := R_\rho^* R_\rho : \tilde{V}_\rho \to \tilde{V}_\rho$ was not used above. It is given componentwise by

$$\begin{aligned} a_\omega((\tilde{P}_\rho \tilde{v})_\omega, w_\omega) &= a_\omega(R_\omega^* R_\rho \tilde{v}, w_\omega) = a(R_\rho \tilde{v}, R_\omega w_\omega) \\ &= \int_\Omega a(R_{\omega'} v_{\omega'}, R_\omega w_\omega)\, d\rho(\omega') = \int_\Omega a_{\omega'}(R_\omega^* R_{\omega'} v_{\omega'}, w_\omega)\, d\rho(\omega'), \end{aligned}$$

for each $\omega \in \Omega$.

Most of the theorems in Chapter 2 apply to this more general setting. The difficulty lies in verifying the measurability conditions, which are not relevant for discrete measures ρ on $\Omega = \mathbb{N}$. In the situation treated in Section 6.5, we had $V_\omega = Y$ and the measurability was guaranteed by the assumption (6.36).

6.6 Remarks on non-symmetric problems

In this section, let U and V be separable Hilbert spaces. We consider an operator equation

$$Bu = f \tag{6.63}$$

in V' for determining $u \in U$ with given $f \in V'$, where $B \in \mathcal{B}(U, V')$ is arbitrary. Solving (6.63) is equivalent to solving the variational problem

$$b(u, v) = f(v) \qquad \forall\, v \in V, \tag{6.64}$$

where

$$b(u, v) = \langle Bu, v \rangle_{V' \times V}, \qquad u \in U,\ v \in V,$$

is a bounded bilinear form on $U \times V$ generated by B. For simplicity, we will assume throughout this section that $f \in \operatorname{ran}(B)$ and $\ker(B) = \{0\}$. This guarantees the unique solvability of (6.63).

Any treatment of (6.63-6.64) (or the adjoint problem $B'v = g$ in U') using space splittings faces the obvious problem that the operators and bilinear forms are generally not symmetric, and U and V do not necessarily coincide. There are standard ways to symmetrize, which in one way or another are related to least-squares techniques. To see this, consider the quadratic minimization problem

$$J(u) := \|Bu - f\|_{V'}^2 = \|R_V^{-1}(Bu - f)\|_V^2 \to \min, \qquad u \in U, \tag{6.65}$$

which has the same solution as (6.64) under the above conditions. Here $R_V : V \to V'$ denotes the Riesz map. The necessary condition for a minimum in (6.65) can be transformed as follows:

$$\begin{aligned} 0 = (Bu - f, Bw)_{V'} &= \langle Bw, R_V^{-1}(Bu - f)\rangle_{V'\times V} \\ &= \langle B'R_V^{-1}(Bu - f), w\rangle_{U'\times U} = (B^*(Bu - f), w)_U \qquad \forall w \in U, \end{aligned}$$

where $B^* = R_U^{-1}B'R_V^{-1} : V' \to U$ is the Hilbert adjoint of $B : U \to V'$. So u must satisfy the operator equation

$$B'R_V^{-1}Bu = B'R_V^{-1}f \tag{6.66}$$

in U' governed by the symmetric operator $A := B'R_V^{-1}B : U \to U'$ or, equivalently, the normal equation

$$B^*Bu = B^*f$$

in U, where $B^*B \in \mathcal{B}_{pos}(U)$. Alternatively, one can solve the equation

$$BR_U^{-1}B'v = f, \tag{6.67}$$

in V' for $v \in V$, and recover the solution of the original problem by setting $u = R_U^{-1}B'v$. Obviously, this is the same as setting $u = B^*h$, where $h \in V'$ solves the operator equation

$$BB^*h = f.$$

The bilinear form $a : U \times U \to \mathbb{R}$ associated with the above A is given by

$$a(u, v) := \langle B'R_V^{-1}Bu, v\rangle_{U'\times U} = \langle Bv, R_V^{-1}Bu\rangle_{V'\times V} = (Bu, Bv)_{V'}.$$

Note that if $\mathrm{ran}(B)$ is not closed in V', this bilinear form does not lead to an equivalent scalar product on V, which is a prerequisite for using the theory of stable space splittings.

Finally, by adding the artificial constraint $B'v = 0$, the original problem can be rewritten as operator equation

$$\begin{aligned} Cv \ + \ Bu &= f \\ B'v \qquad &= 0 \end{aligned} \tag{6.68}$$

in $V' \times U'$ with solution $(v, u) \in V \times U$. The operator $C \in \mathcal{B}(V, V')$ should be such that the bilinear form on V induced by it is symmetric on V (and coercive on $\ker(B')$

if the latter is non-trivial). The simplest such operator is $C = R_V$, which results in $c(v, z) = (v, z)_V$. This ensures that (6.68) generates a mixed variational problem on $V \times U$. The reformulation (6.68) has recently received some attention in the context of so-called discontinuous Petrov-Galerkin schemes for certain non-symmetric PDE problems.

All of the above reformulations suffer from several drawbacks: The new formulations are closely related to the operator B^*B, and space splittings must be designed with respect to $A = B'R_V^{-1}B$, and not with respect to the original operator B. If B represents a PDE boundary value problem, this may unnecessarily complicate the solution process. Also, the spectral properties and especially the condition numbers may become significantly worse when passing from B to A. Finally, the Riesz mappings R_V^{-1} and R_U^{-1} play a role in the calculations with A or A', which can add another complication depending on the choice of the underlying scalar products in V and U.

Nevertheless, we will proceed with some elementary observations for the above symmetric reformulations of (6.63). In particular, we will show that, under appropriate inf-sup conditions on $b(\cdot, \cdot)$ and for given stable space splittings for $(V; (\cdot, \cdot)_V)$ or $(U; (\cdot, \cdot)_U)$, we automatically obtain stable space splittings whose associated Schwarz equations are equivalent to the original problem (6.66). Moreover, the condition numbers of the underlying Schwarz operators depend only on the condition numbers of the given space splittings and on the condition number of B^*B. This allows us to use the results of the analysis of Schwarz iteration schemes with finite Ω from the Chapters 4 and 5.

In addition to $\ker(B) = \{0\}$, we assume that the inf-sup condition (A.14) holds for $b : U \times V \longrightarrow \mathbb{R}$, which is equivalent to

$$\|Bu\|_{V'}^2 = (B^*Bu, u)_U \geq c_0^{-1}.$$

This implies that the bilinear form $a : U \times U \longrightarrow \mathbb{R}$ associated with A is coercive and thus defines an equivalent scalar product on U. Furthermore, the condition number of B^*B is well-defined.

In the following, we discuss three different methods to derive stable splittings for non-symmetric problems.

First, we present a space splitting motivated by the generalization of the Kaczmarz algorithm for solving (6.63) based on the reformulation (6.67). Let $(\mathbf{V}; \mathbf{a}; \mathbf{R}; \rho)$ be a stable space splitting for $(V; (\cdot, \cdot)_V)$ with condition number κ. Then $(\mathbf{V}; \mathbf{a}; \tilde{\mathbf{R}}; \rho)$, where the operators in $\tilde{\mathbf{R}}$ are given by $\tilde{R}_i = R_U^{-1}B'R_i : V_i \longrightarrow U$, is a stable space splitting for $(U; (\cdot, \cdot)_U)$. The Schwarz operator P_ρ is formally related to the solution of an operator equation involving the Riesz operator $R_U : U \longrightarrow U'$. It has the form

$$P_\rho = \sum_{i \in \Omega} \rho_i P_i, \qquad P_i = R_U^{-1}B'(R_i A_i^{-1} R_i')B, \quad i \in \Omega,$$

and satisfies

$$\kappa = \kappa(P_\rho) \leq \hat{\kappa} \cdot \kappa(B^*B), \tag{6.69}$$

where $\hat{\kappa}$ is the condition number of another Schwarz operator $\hat{P}_\rho$ defined next. The estimate (6.69) follows from the observation that

$$\hat{P}_\rho = \sum_{i\in\Omega} \rho_i \hat{P}_i, \qquad \hat{P}_i = R_i A_i^{-1} R_i' R_V, \quad i \in \Omega,$$

is the Schwarz operator of the space splitting for $(V; (\cdot,\cdot)_V)$ associated with the operator equation in V' governed by R_V. Hence

$$\frac{(P_\rho u, u)_U}{\|u\|_U^2} = \frac{(R_U^{-1} B' \hat{P}_\rho R_V^{-1} Bu, u)_U}{\|u\|_U^2} = \frac{\langle Bu, \hat{P}_\rho R_V^{-1} Bu, \rangle_{V'\times V}}{\|Bu\|_{V'}^2} \cdot \frac{(B^* Bu, u)_U}{\|u\|_U^2}$$

for all $u \in U$. So, since

$$\langle Bu, \hat{P}_\rho R_V^{-1} Bu, \rangle_{V'\times V} = (\hat{P}_\rho R_V^{-1} Bu, R_V^{-1} Bu), \qquad \|Bu\|_{V'}^2 = \|R_V^{-1} Bu\|_V^2,$$

and P_ρ, $\hat{P}_\rho$ and $B^* B$ are symmetric on the respective spaces, we get

$$\lambda_{\min}(P_\rho) \ge \lambda_{\min}(\hat{P}_\rho)\lambda_{\min}(B^* B), \qquad \lambda_{\max}(P_\rho) \le \lambda_{\max}(\hat{P}_\rho)\lambda_{\max}(B^* B).$$

This implies (6.69) with $\hat{\kappa} = \lambda_{\max}(\hat{P}_\rho)/\lambda_{\min}(\hat{P}_\rho)$.

Although the resulting Schwarz iterative schemes in U formally concern the trivial operator equation $R_U u = g$ with the right-hand side $g = R_U u \in U'$ given by the (unknown) solution u of (6.63) and are thus infeasible, when rewritten as iterations in V they represent Schwarz iterative schemes for (6.67) and are thus the analog of the Kaczmarz iteration for linear algebraic systems. To see this, consider the update step of the resulting Schwarz method for $R_U u = g$ in its simplest form, namely

$$u^{(m+1)} = u^{(m)} + \omega \rho_i R_U^{-1} B' R_i A_i^{-1} R_i' B e_u^{(m)}, \tag{6.70}$$

where $e_u^{(m)} = u - u^{(m)}$. The corresponding update step for a Schwarz method with respect to the space splitting $(\mathbf{V}; \mathbf{a}; \mathbf{R}; \rho)$ applied to (6.67) is

$$v^{(m+1)} = v^{(m)} + \omega \rho_i R_i A_i^{-1} R_i' B R_U^{-1} B' e_v^{(m)}, \tag{6.71}$$

where $e_v^{(m)} = v - v^{(m)}$ and v solves (6.67). Now, if one formally applies $R_U^{-1} B'$ to both sides of this update formula, sets $u^{(m)} = R_U^{-1} B' v^{(m)}$, and notes that due to $u = R_U^{-1} B' v$ one also has $R_U^{-1} B' e_v^{(m)} = e_u^{(m)}$, then the update formula (6.70) results. In practice, one would use (6.71). Note that $\ker(B') = \mathrm{ran}(R_V^{-1} B)^\perp$ can be non-trivial and components in $\ker(B')$ are not controlled in the iteration process (6.71). However, they have no influence on $u^{(m)}$, since they are annihilated after the application of B'.

As a second example in a similar spirit, if $(\mathbf{V}; \mathbf{a}; \mathbf{R}; \rho)$ is a stable space splitting for $(U; (\cdot,\cdot)_U)$ with condition number $\hat{\kappa}$, then it is a stable space splitting for $(U; (B\,\cdot, B\,\cdot)_{V'})$, and the condition number of the associated Schwarz operator

$$P_\rho = \sum_{i\in\Omega} \rho_i P_i, \qquad P_i := (R_i A_i^{-1} R_i') B' R_V^{-1} B, \quad i \in \Omega,$$

also admits the bound (6.69). This is because the bilinear form $a(u,u)$ associated with the reformulation (6.66) of (6.63) satisfies $a(u,u) = \|Bu\|_{V'}^2$, and, under our conditions on $b(\cdot,\cdot)$, induces an equivalent norm on U. The Schwarz operator

$$\hat{P}_\rho = \sum_{i\in\Omega} \rho_i P_i, \qquad P_i = R_i A_i^{-1} R_i' R_U, \quad i \in \Omega,$$

for the given stable space splitting for $(U; (\cdot,\cdot)_U)$ satisfies $P_\rho = \hat{P}_\rho B^* B$. The associated Schwarz iterative methods are Landweber-type iterations for the normal equation (6.66), and include the Riesz map R_V^{-1} besides the application of B and B'. This is in contrast to the previously considered Kaczmarz type iterations, where only R_U^{-1} appeared.

The third example is motivated by attempts to develop robust adaptive methods for convection-diffusion and transport problems, and has been propagated in various forms [47, 63, 155, 191]. We follow [47], where it is proposed to recast $Bu = f$ as symmetric variational problem in U, namely

$$\hat{b}(u,w) := (Bu, Bw)_{V'} = (f, Bw)_{V'}, \qquad \forall\, w \in U. \tag{6.72}$$

Here the scalar product in V' is induced by a properly chosen scalar product on V using the Riesz representation theorem, namely

$$(f,h)_{V'} := (R_V^{-1} f, R_V^{-1} h)_V \qquad \forall\, f,h \in V',$$

or, equivalently,

$$(R_V v, R_V z)_{V'} = (v,z)_V \qquad \forall\, v,z \in V.$$

The fact that $B : U \to V'$ is invertible guarantees that (6.72) and (6.64) have the same unique solution. Obviously, (6.72) is the variational formulation for the normal equation (6.66) of the least-squares problem (6.65). Now, in a second step, an operator-dependent Hilbert norm $\|\cdot\|_B$ on U is introduced, so that the symmetric bilinear form $\hat{b}$ in (6.72) becomes U-elliptic with ellipticity constants close to 1. Obviously,

$$\|u\|_B = \|Bu\|_{V'},$$

as advocated in [47], does this in an optimal way with both ellipticity constants equal to 1. In [155], for the special case $U = V$, the definition of $\|\cdot\|_B$ is obtained by the interpolation of two norms on U given by

$$\|u\|_{U,1}^2 = \|u\|_V^2 = b(u,u) = \langle B_{sym} u, u\rangle_{V'\times V}, \qquad \|u\|_{U,2}^2 = \|u\|_V^2 + \|B_{skew} u\|_{V'}^2,$$

where $B_{sym} := \frac{1}{2}(B + B')$ and $B_{skew} := \frac{1}{2}(B - B')$ denote the symmetric and skew-symmetric parts of B. Since

$$(Bu, Bw)_{V'} = b(u, R_V^{-1} Bw) = \langle Bw, R_V^{-1} Bu\rangle_{V'\times V},$$

and

$$(f, Bw)_{V'} = f(R_V^{-1}Bw) = \langle Bw, R_V^{-1}f\rangle_{V'\times V},$$

we arrive at the system (6.68) or, written as a mixed variational problem on $V \times U$, at

$$\begin{aligned} (v, z)_V + b(u, z) &= f(v') \qquad && \forall z \in V, \\ b(v, w) &= 0 && \forall w \in U \end{aligned}$$

by introducing the auxiliary variable $v = R_V^{-1}(f - Bu) \in V$ and setting $C = R_V$. This mixed formulation avoids the explicit use of the duality map R_V. However, it is an open problem how to preserve the excellent conditioning of the form $\hat{b}$ on a discrete level by an adequate space splitting approach. Working directly with the norm $\|u\|_B$ is the same as considering the space U with the scalar product generated by B^*B. A critical issue here is the choice of the Hilbert space norm $\|\cdot\|_V$ to start with.

In the rest of this section we formulate some results that extend the results of Chapter 6 on Schwarz approximation methods (infinite Ω) to non-symmetric B in the special situation of a dictionary splitting. To fit the approach and notation used in the previous sections of this chapter, and to avoid switching between Hilbert spaces and their duals, we now consider the problem of finding the solution $u \in U$ that satisfies the operator equation

$$Bu = v \tag{6.73}$$

with given operator $B \in \mathcal{B}(U, V)$ and right-hand side $v \in V$. To be consistent with the notation in (6.63) used so far in this section, we rename $R_V^{-1}B$ to B and set $v = R_V^{-1}f$. We again assume $\ker(B) = \{0\}$ and $v \in \operatorname{ran}(B)$, to ensure the unique solvability of (6.73).

The symmetrized versions of (6.73) are now

$$B^*Bu = B^*v$$

and

$$BB^*h = v, \qquad u = B^*h, \tag{6.74}$$

with $h \in V$, where the Hilbert adjoint $B^* \in \mathcal{B}(V, U)$ is defined by the equality $(Bu, w)_V = (u, B^*w)_U$ for all $u \in U$ and $w \in V$. To be consistent with this definition of Hilbert adjoints for operators from $\mathcal{B}(U, V)$ and $\mathcal{B}(V, U)$, we fix $(\cdot,\cdot)_U$ and $(\cdot,\cdot)_V$ as the designated scalar products in U and V, respectively.

We restrict ourselves to a Kaczmarz-type approximation method based on (6.74). For this, we take a countable collection $(\mathbf{V}; \mathbf{a}; \mathbf{R})$ associated with $(V; (\cdot,\cdot)_V)$ that satisfies the conditions formulated at the beginning of Section 6.1. In particular, the span of the ranges of the operators $R_i : V_i \to V$ is dense in V and their norms are uniformly bounded. Then the collection $(\mathbf{V}; \mathbf{a}; B^*\mathbf{R})$ related to $(U; (\cdot,\cdot)_U)$ also satisfies these conditions. For example, the counterpart of (6.1) is

$$\|B^*R_i\|_{V_i\to U}^2 \le \|B\|_{U\to V}^2 \sup_{i\in\Omega} \|R_i\|_{V_i\to V}^2 \le \|B\|_{U\to V}^2 \Lambda, \qquad i \in \Omega.$$

The density of the union of the ranges $\mathrm{ran}(B^*R_i)$ follows from the corresponding assumption on $\mathbf{R}$ and the density of $\mathrm{ran}(B^*)$ in U. The latter property holds because the closure of $\mathrm{ran}(B^*)$ in U is equal to $\ker(B)^\perp$ and $\ker(B) = \{0\}$ was assumed. The computation of the action of the operators

$$P_i := B^*R_iA_i^{-1}(B^*R_i)^* = B^*R_iA_i^{-1}(R_i^*B), \qquad i \in \Omega,$$

on errors $e^{(m)} = u - u^{(m)}$ will be necessary in the course of Schwarz approximation methods for solving $Bu = v$ with this space splitting at hand. But this is possible without knowing u, since the subproblem

$$a_i(R_i^*Be^{(m)}, v_i) = (Be^{(m)}, R_iv_i)_V = (v, R_iv_i)_V - (u^{(m)}, B^*R_iv_i)_U \quad \forall\, v_i \in V_i,$$

for the evaluation of $R_i^*Be^{(m)}$ depends on u only via the given right-hand side $v \in V$ of the operator equation. The formula is further simplified if we consider a collection $\mathbf{V}_D$ generated by a dictionary $D = \{\phi_i\}_{i\in\Omega}$ in V, which is the case closest to the classical setting for Kaczmarz methods in numerical linear algebra. In fact, we need the density of $\mathrm{span}(D)$ and the uniform boundedness of the dictionary elements in V. Then the formula for $P_ie^{(m)}$ is

$$P_ie^{(m)} = (Be^{(m)}, \phi_i)_VB^*\phi_i = ((v, \phi_i)_V - (u^{(m)}, B^*\phi_i)_U)B^*\phi_i. \tag{6.75}$$

Obviously, $B^*\phi_i$ is the analog of the transpose of the i-th row of B in the matrix case.

Thus, the convergence results of Sections 6.3 and 6.4 hold if one accepts the implicit regularity assumptions on u required for them. Let us start with the randomized Schwarz approximation method (6.10) with underlying probability distribution $\mathbf{p}$ and relaxation parameters $\alpha_m = (m+1)/(m+2)$. The operator $P_{\mathbf{p}}$, which is relevant for the analysis, is then

$$P_{\mathbf{p}} = \sum_{i\in\Omega} p_iP_i = \sum_{i\in\Omega} p_i(B^*\phi_i, \cdot)_UB^*\phi_i.$$

It follows from Lemma 2.10 that $P_{\mathbf{p}}$ is trace class, since $\dim(V_i) = 1$ and thus

$$\|B^*R_i\|_{HS}^2 = \|B^*\phi_i\|_U^2 \le \|B\|_{U\to V}^2 \sup_{i\in\Omega} \|\phi_i\|_V^2.$$

Furthermore, if B is compact and $\{\phi_i\}_{i\in\Omega} \subset V$, $\{\psi_i\}_{i\in\Omega} \subset U$ are the orthonormal systems in its singular value representation with corresponding singular values σ_i, see (A.8), then the operator $P_{\mathbf{p}}$ is given by

$$P_{\mathbf{p}} = \sum_{i\in\Omega} p_i\sigma_i^2(\psi_i, \cdot)_U\psi_i.$$

So it has ψ_k as eigenvectors and $\lambda_k = p_k\sigma_k^2 \to 0$ as eigenvalues, and the spaces $V_{P_{\mathbf{p}}}^s$ can be made explicit. In general this is not the case. As a corollary to Theorem 6.3 and its corollaries in Section 6.3, we get the following result for the dictionary case:

Corollary 6.5 *Under the stated conditions on $Bu = v$ and the dictionary D in V, consider the Schwarz approximation scheme (6.10) with parameters*

$$\alpha_m = \frac{m+1}{m+2}, \qquad \mu_m = \frac{(u - \alpha_m u^{(m)}, B^*\phi_{i_m})_U}{\big((v, \phi_{i_m})_V - (u^{(m)}, B^*\phi_{i_m})_U\big)\, \|B^*\phi_{i_m}\|_U^2},$$

with $P_{i_m}e^{(m)}$ computed as in (6.75) and i.i.d. indices $i_m \in \Omega$ randomly chosen according to $\mathbf{p}$. *Then we have $\mathbb{E}(\|u - u^{(m)}\|^2) \to 0$ for arbitrary $u^{(0)}$. Moreover, if $u \in V^s_{P_\mathbf{p}}$ for some $0 < s \le 1$, then*

$$\mathbb{E}(\|u - u^{(m)}\|^2) = \mathrm{O}((m+1)^{-s}), \qquad m \to \infty.$$

If $s = 1$, we have the more accurate estimate

$$\mathbb{E}(\|e^{(m)}\|_U^2) \le \frac{\|e^{(0)}\|_U^2}{(m+1)^2} + \frac{\|u\|_U^2 + \Lambda \|u\|^2_{V^1_{P_\mathbf{p}}}}{m+1}, \qquad m = 1, 2, \ldots,$$

where $\Lambda = \|B\|^2_{U \to V} \sup_{i \in \Omega} \|\phi_i\|_V^2$.

Note here that the formula for μ_m corresponds to the steepest descent rule.

A similar result can be formulated for the greedy version. To do so, we need to specify the definition of the classes $\mathcal{A}^p$: Simplifying to the dictionary case, we have

$$\mathcal{A}^p = \{u \in U : \|u\|^p_{\mathcal{A}^p} := \inf_{c_i : u = \sum_{i \in \Omega} c_i B^*\phi_i} \sum_{i \in \Omega} |c_i|^p < \infty\}, \quad 0 < p \le 2.$$

To make this consistent with the definition given in Section 6.4, and to avoid technical difficulties with the completion steps and factor space construction used there, we assume that $B^*\mathbf{V}_D$ is a Bessel system, i.e., the operator

$$P\mathbf{c} := \sum_{i \in \Omega} c_i B^*\phi_i,$$

which is well-defined for finitely supported coefficient sequences $\mathbf{c} = (c_i)_{i \in \Omega}$, extends to a bounded operator on $\ell^2(\Omega)$. Moreover, we simplify further by assuming $\ker(P) = \{0\}$. Then, by Lemma 6.1, we have

$$\mathcal{A}^p \subset U \qquad \text{for all } 0 < p \le 2.$$

In the special case of a compact operator B and $D = \{\phi_i\}_{i \in \Omega}$ defined from its singular value decomposition, the definition becomes even more explicit. We get

$$\mathcal{A}^p = \{u = \sum_{i \in \Omega} c_i \psi_i \in U : \|u\|^p_{\mathcal{A}^p} = \sum_{i \in \Omega} \frac{|c_i|^p}{\sigma_i^p} < \infty\}, \quad 0 < p \le 2.$$

With these preparations, we can now formulate the conclusions from Theorem 6.4 and its corollaries.

Corollary 6.6 *Under the stated conditions on $Bu = v$ and the dictionary D in V, consider the Schwarz approximation scheme (6.10) with parameters α_m, μ_m and $P_{i_m}e^{(m)}$ as in Corollary 6.5, but with i_m specified by the weak greedy rule*

$$\tilde{\Delta}_{i_m}^{(m)} \geq \beta \sup_{i\in\Omega} \tilde{\Delta}_i^{(m)}, \qquad \tilde{\Delta}_i^{(m)} := |R_i^* Be^{(m)}|^2 = |(v, \phi_i)_V - (u^{(m)}, B^*\phi_i)_U|^2,$$

with constant weakness parameter $0 < \beta \leq 1$. Then we have $\|u - u^{(m)}\|_U^2 \to 0$ for any $u^{(0)} \in U$. Moreover, if $u \in \mathcal{A}^p$ for some $1 \leq p < 2$, then

$$\|u - u^{(m)}\|^2 = \mathrm{O}((m+1)^{-(2/p-1)}), \qquad m \to \infty.$$

In particular, if $p = 1$ then we have the estimate

$$\mathbb{E}(\|e^{(m)}\|_U^2) \leq \frac{\|e^{(0)}\|_U^2}{(m+1)^2} + \frac{2\left(\|u\|_U^2 + \beta^{-1}\Lambda\|u\|_{\mathcal{A}^1}^2\right)}{m+1}, \qquad m = 1, 2, \ldots,$$

where $\Lambda = \|B\|_{U\to V}^2 \sup_{i\in\Omega} \|\phi_i\|_V^2$.

The weak greedy rule is a special instance of (4.63). Since this corollary is essentially a result about greedy approximation from dictionaries, the reader may want to consult [179] for further results and especially for counterexamples showing the impossibility of stronger convergence guarantees. Moreover, in complete analogy to the case of orthogonal splittings discussed in the introductory Section 6.1 of this chapter, more general statements hold in the case of compact B and the special choice of the dictionary $D = \{\phi_i\}_{i\in\Omega}$ obtained from its singular value decomposition. We refer to Problem 6.15.

Problems

6.1 Lower bounds for convergence in V_P^t.
Let P satisfy the conditions of Theorem 6.1. Consider any iterative method for solving the operator equation $Pu = g$, $g \in V_P^{s+2}$, such that the associated error sequence $e^{(m)} = u - u^{(m)}$ satisfies

$$e^{(m)} \in \mathcal{K}_m := e^{(0)} + \mathrm{span}(\{Pe^{(0)}, \ldots, P^m e^{(0)}\}), \qquad m = 1, 2, \ldots,$$

i.e., $e^{(m)}$ belongs to the m-th affine Krylov subspace generated by P from $e^{(0)}$. Prove that there are absolute constants $0 < c < C < \infty$ depending on $s - t$, such that

$$C(m+1)^{2(s-t)} \geq \varepsilon_m := \sup_{\|P\|\leq 1} \sup_{e^{(0)}\in V_P^s} \inf_{e^{(m)}\in\mathcal{K}_m} \frac{\|e^{(m)}\|_{V_P^t}^2}{\|e^{(0)}\|_{V_P^s}^2} \geq c(m+1)^{2(s-t)}$$

for any $m = 1, 2, \ldots$. This provides a generic lower bound on the convergence rate for Krylov subspace methods in the scale $V^t_{P_\rho}$.

Hints. For the solution of this problem and some history, see [100, Section 4-5]. Using eigenfunction expansions, write

$$e^{(m)} = p_m(P)e^{(0)} = \sum_{k=1}^{\infty} p_m(\lambda_k)c_k^{(0)}\psi_k,$$

and then estimate the V^t_P norm of $e^{(m)}$ by the V^s_P norm of $e^{(0)}$ following the proofs of Theorem 6.1 and 6.2. This should lead to the bound

$$\varepsilon_m \le \varepsilon'_m := \inf_{p_m:\, p_m(0)=1} \|\lambda^{s-t} p_m^2(\lambda)\|_{C([0,1])}.$$

Then solve the extremal problem for polynomials that defines ε'_m. Lower bounds for ε'_m can be obtained from the fact that polynomials p_m with $p_m(0) = 1$ satisfy $p(x) \ge 1/2$ for $x \in [0, c_0(m+1)^{-2}]$, where $c_0 > 0$ is independent of m (this follows from Markov's inequality for derivatives of polynomials). Upper bounds for ε'_m follow by considering polynomials p_m with sufficiently fast decay away from 0. A simple approach is to define such polynomials from powers of the trigonometric Dirichlet kernel. For $\phi \in [0, \pi/2]$ set

$$p_{k,r}(1-\cos\phi) := D_k(\phi)^r = \left(\frac{\sin((2k+1)\phi/2)}{(2k+1)\sin(\phi/2)}\right)^r, \qquad k, r \in \mathbb{N}.$$

Check that these $p_{k,r}$ are indeed polynomials of degree kr with $p_{k,r}(0) = 1$. They provide a matching upper bound for ε'_m if k, r are chosen such that $r-1 < 2(s-t) \le r$ and $kr \le m < (k+1)r$. See also the examples in [100].

Finally, to establish a lower bound on ε_m, construct an operator P with a sufficiently dense spectrum in $[(m+1)^{-2}, 1]$, choose an appropriate $e^{(0)}$, and recycle the arguments presented in the first part of Subsection 5.4.1 in conjunction with the reasoning leading to the lower bound on ε'_m.

6.2 More averaging helps: Analysis of (6.9).

a) Derive the recursion formulas to compute the quadruple of iterates

$$U^{(m)} := (u^{(m)}, u^{(m-1)}, (m+1)\bar{u}^{(m)}, \frac{1}{2}(m+1)(m+2)\hat{u}^{(m)}), \quad m = 1, 2, \ldots,$$

with the components defined in (6.6), (6.7), and (6.9) from the quadruple $U^{(m-1)}$, where $U^{(0)} = (u^{(0)}, u^{(0)}, u^{(0)}, u^{(0)})$. This implies a slight increase in memory and computational cost when using $\hat{u}^{(m)}$ instead of $\bar{u}^{(m)}$ as the final approximation.

b) Show that, under the assumptions of Theorem 6.1, the error $\hat{e}^{(m)} := u - \hat{u}^{(m)}$ of the iteration (6.9) satisfies a bound similar to (6.8), but with $\min(s-t, 1)$ replaced by $\min(s-t, 2)$.

Hint. For b), establish a formula for $\hat{c}_k^{(m)}$ using the results from the proof of Theorem 6.2.

6.3 A Nesterov-type iteration.
Consider the iteration

$$u^{m+1} = v^{(m)} + \zeta_m P e_v^{(m)}, \qquad m = 0, 1, \ldots,$$
$$v^{(m+1)} = (1 + \eta_m) u^{(m+1)} - \eta_m u^{(m)},$$

starting with $u^{(0)} = v^{(0)}$. Choose the parameters

$$\zeta_m = \lambda_{\max}(P)^{-1}, \quad \eta_m = \frac{\gamma_m}{\gamma_{m+1} + 1}, \quad \gamma_m = \frac{\beta_m}{\sqrt{\beta_m + 1} + 1},$$

where $\{\beta_m\}_{m \in \mathbb{Z}_+}$ is recursively given by $\sqrt{\beta_{m+1}} = \sqrt{\beta_m + 1} + 1$, $m = 0, 1, \ldots, \beta_0 = 0$. Following the proof of Theorem 4.12 and using the same notation, show that one has

$$\|e_u^{(m)}\|_{V_P^t}^2 \le \frac{\lambda_{\max}(P)}{\beta_m} \|e_u^{(0)}\|_{V_P^{t+1}}^2$$

and $\beta_m \ge (m+1)^2$ for $m = 1, 2, \ldots$.
Hints. As in the proof of Theorem 4.12, introduce the auxiliary sequence

$$w^{(m)} = (1 + \gamma_m) v^{(m)} - \gamma_m u^{(m)},$$

multiply the error recursion by $P^{t/2}$, and use

$$\| \cdot \|_{V_P^s}^2 = a(P^{-s} \cdot, \cdot) = \|P^{-s/2} \cdot \|^2.$$

By carefully repeating all the estimation steps, verify that

$$\|e_w^{(m+1)}\|_{V_P^{t+1}}^2 + \frac{\beta_{m+1}}{\lambda_{\max}(P)} \|e_u^{(m+1)}\|_{V_P^t}^2 \le \|e_w^{(m)}\|_{V_P^{t+1}}^2 + \frac{\beta_m}{\lambda_{\max}(P)} \|e_u^{(m)}\|_{V_P^t}^2,$$

$m = 0, 1, \ldots$. To simplify this task, set $\xi_m = 0$ and $\delta_m = 1$ in the derivations that lead to Theorem 4.12. Since $\gamma_0 = 0$ and $w^{(0)} = v^{(0)} = u^{(0)}$, the above statement follows. The quadratic growth of β_m is obvious, since $\beta_1 = 2$ and $\sqrt{\beta_{m+1}} > \sqrt{\beta_m} + 1$ by definition of this sequence.

6.4 Convergence in $V_{\mathbf{p}}^{-1}$.
In addition to the assumptions of Theorem 6.3, assume that

$$\mathbb{E}(a(P_{\mathbf{p}} P_i v, P_i v)) \le \Lambda' \|P_{\mathbf{p}} v\|^2, \qquad i \in \Omega, \quad v \in V.$$

Consider the randomized Schwarz iteration (6.10) with the given probability distribution $\mathbf{p}$ and derive an estimate for the expectation of the squared error in the $V_{P_{\mathbf{p}}}^{-1}$ norm, i.e., for

$$\mathbb{E}(\|e^{(m)}\|_{V_{P_{\mathbf{p}}}^{-1}}^2) = \mathbb{E}(a(P_{\mathbf{p}} e^{(m)}, e^{(m)})), \qquad m = 0, 1, \ldots,$$

if $u \in V = V_{P_{\mathbf{p}}}^0$.

Hint. Follow the proof of Theorem 6.3, but consider the expansion of

$$\|e^{(m+1)}\|^2_{V^{-1}_{P_\mathrm{p}}} = a(P_\mathrm{p} e^{(m+1)}, e^{(m+1)})$$

instead of $\|e^{(m+1)}\|^2$.

6.5 Proof of (6.19).
Consider the iteration (6.10) with the specifications assumed for Theorem 6.3.
a) Derive a recursion for the coefficients $c_k^{(m)}$ in the eigenexpansion

$$\mathbb{E}(e^{(m)}) := \sum_{k=1}^{\infty} c_k^{(m)} \psi_k, \qquad m = 1, 2, \ldots,$$

of the expected errors $\mathbb{E}(e^{(m)})$ in terms of the coefficients $c_k^{(0)}$ and d_k of the eigenexpansions of $e^{(0)}$ and u, respectively.
b) Based on the result of Part a), for given $m = 1, 2, \ldots$, choose the coefficients $c_k^{(0)}$ and d_k such that the quotient of the left-hand to the right-hand side in (6.19) is bounded away from zero, uniformly in m.
Hint. For Part b), use the explicit formulas for the $V^s_{P_\mathrm{p}}$ norms in terms of eigenexpansion coefficients in conjunction with

$$\mathbb{E}(\|e^{(m)}\|^2) \geq \|\mathbb{E}(e^{(m)})\|^2.$$

6.6 Analogs of Theorem 6.4 for other weak greedy rules.
a) Under the condition (6.26), establish the analog of the estimate (6.21) if the weak greedy rule (4.63) is replaced by either (4.57) or (4.60), but all other assumptions in Theorem 6.4 hold. In the case of (4.60), the condition (6.26) can be replaced by a normalization condition on the operator norm $\|R_i\|^2_{V_i \to V}$.
b) Provide versions of Theorem 6.4, where the steepest descent rule is replaced by a concrete step-size proposal for μ_m.
Hints. For a), two-sided estimates of $\Delta_i^{(m)}$ and $\bar{\Delta}_i^{(m)}$ by $\tilde{\Delta}_i^{(m)}$ are sufficient to derive (6.22), the rest of the argument is the same. For b), follow the hint in the proof of Theorem 6.4.

6.7 Properties of the minimization problems (6.38) and (6.41).
This problem uses the notation and assumptions introduced in Section 6.5, leading to the definition of $\mathbb{E}(R_\omega y)$ and P_ρ, which appear in the necessary condition (6.39) for a minimizer $u \in V$ for (6.38) to exist. In particular, let P_ρ be injective and compact, and denote by $\lambda_1 \geq \lambda_2 \geq \cdots$ its eigenvalues and $\psi_1, \psi_2, \ldots$ the corresponding eigenvectors forming a CONS in V.
a) Show that the quadratic functional $J(v)$ in (6.38) satisfies

$$J(v) = \|f_v - f_\mu\|_{L^2(\Omega,Y)} + \sigma^2,$$

where σ^2 is defined in (6.43). Moreover, show that $\mathbb{E}(R_\omega y) = \mathbb{E}(R_\omega f_\mu(\omega))$ by defining the latter expectation properly.

b) Show that the functions

$$\hat{\psi}_k(\omega) := \lambda_k^{-1/2} f_{\psi_k} = \lambda_k^{-1/2} R_\omega^* \psi_k, \qquad k = 1, 2, \ldots,$$

form a (not necessarily complete) orthonormal system in $L^2_\rho(\Omega, Y)$.
c) Using the result in Part b), show that $\mathbb{E}(R_\omega f_\mu(\omega)) \in \operatorname{ran}(P_\rho^{1/2})$.
d) Show that (6.41) possesses a unique solution u_r for any $r \geq 0$.
e) Prove: If $\mathbb{E}(R_\omega f_\mu(\omega)) \in \operatorname{ran}(P_\rho)$, then $u_r = u$ for $r \geq \|u\|_V$, where u solves (6.39) and is the unique global minimizer for (6.38).
f) Finally, show that if $\mathbb{E}(R_\omega f_\mu(\omega)) \notin \operatorname{ran}(P_\rho)$, then $\|u_r\|_V = r$ and $J(u_r)$ is strictly decreasing for $r \in [0, \infty)$, i.e., in this case a global minimum for (6.38) cannot exist.
Hints. We refer to [49, Chapter 1-2] for statements analogous to Part a) and d). Part b) is straightforward by inspection of $(\hat{\psi}_k, \hat{\psi}_i)_{L^2_\rho(\Omega,Y)}$. For Part c), recall that, by definition of $\mathbb{E}(R_\omega f_\mu(\omega))$, we have

$$\begin{aligned}(\mathbb{E}(R_\omega f_\mu(\omega)), \psi_k)_V &= \int_\Omega (R_\omega f_\mu(\omega), \psi_k)_V \, d\rho(\omega) \\ &= \int_\Omega (f_\mu(\omega), R_\omega^* \psi_k)_Y \, d\rho(\omega) = \lambda_k^{1/2} (f_\mu, \hat{\psi}_k)_{L^2_\rho(\Omega,Y)},\end{aligned}$$

$k = 1, 2, \ldots$. Thus,

$$\mathbb{E}(R_\omega f_\mu(\omega)) = P_\rho^{1/2} v$$

with

$$v = \sum_k (f_\mu, \hat{\psi}_k)_{L^2_\rho(\Omega,Y)} \psi_k \in V.$$

To obtain the statements in Parts d)-f), set

$$v = \sum_k c_k \psi_k, \qquad g := \mathbb{E}(R_\omega f_\mu(\omega)) = \sum_k g_k \psi_k,$$

and consider the Lagrangian for the constrained minimization problem $J(v) \longrightarrow \min$, $\|v\|_V = r$, in terms of the coefficients c_k, namely

$$F((c_k)_{k \in \mathbb{N}}, \lambda) := \sum_k \lambda_k c_k^2 - 2 \sum_k c_k g_k + \lambda\Big(\sum_k c_k^2 - r^2\Big) + \|f_\mu\|^2_{L^2_\rho(\Omega,Y)} + \sigma^2.$$

Then analyze its critical points (assume $\lambda \geq 0$).

6.8 Divergence in V of the online learning algorithm (6.44–6.45) when (6.38) has no global minimum.
Under the assumptions of the previous problem, show that

$$\|\mathbb{E}(u^m)\|_V \longrightarrow \infty, \qquad u^{(0)} \in V,$$

for the online algorithm (6.44–6.45), if the condition (6.40) is violated, i.e., if

$$g := \mathbb{E}(R_\omega y) = \mathbb{E}(R_\omega f_\mu) \notin \operatorname{ran}(P_\rho).$$

Hints. The expected trajectory $(U^{(m)} := \mathbb{E}(u^{(m)}))_{m\in\mathbb{Z}_+}$ satisfies the linear inhomogeneous recursion

$$U^{(m+1)} = \alpha_m((I-\mu_m P_\rho)U^{(m)} + \mu_m g), \qquad m = 0, 1, \ldots.$$

Using

$$\alpha_k = \frac{k+1}{k+2}, \qquad k = 0, \ldots, m-1,$$

this gives the explicit formula

$$(m+1)U^{(m)} = (\prod_{l=0}^{m-1}(I-\mu_l P_\rho))u^{(0)} + \underbrace{(\sum_{k=1}^{m} k\mu_{k-1}\prod_{l=k}^{m-1}(I-\mu_l P_\rho))g}_{G^{(m)}:=}, \qquad m = 1, 2, \ldots.$$

Argue that $\|U^{(m)}\|_V - (m+1)^{-1}\|G^{(m)}\|_V \to 0$ if $m\to\infty$, independent of $u^{(0)}$. To estimate $\|G^{(m)}\|_V^2$ from below, expand $G^{(m)}$ with respect to the CONS $(\psi_n)_{n\in\mathbb{N}}$ of eigenvectors of P_ρ, i.e.,

$$G^{(m)} = \sum_{n=1}^{\infty} \underbrace{(\sum_{k=1}^{m} k\mu_{k-1}\prod_{l=k}^{m-1}(1-\mu_l\lambda_n))}_{d^{(m)}(\lambda_n)} g_n\psi_n,$$

where $g_n = (g, \psi_n)$ are the coefficients of $g \notin \mathrm{ran}(P_\rho)$. It remains to estimate $d^{(m)}(\lambda)$ from below for the range $c_0/(m+1)^{1-t} \le \lambda \le \lambda_{\max}(P_\rho)$ with suitable constant $c_0 > 0$ using the properties of the step-size parameters μ_m from (6.45), in particular $0 < \lambda\mu_m \le 1/2$ can be used. To estimate the products $\prod_{l=k}^{m-1}(1-\mu_l\lambda)$, see the hints in Problems 6.12 and 6.13. This should yield a lower bound of

$$d^{(m)}(\lambda) \ge c_1\frac{m+1}{\lambda}, \qquad c_0(m+1)^{-(1-t)} \le \lambda \le \lambda_{\max}(P_\rho),$$

where the constants $c_0, c_1 > 0$ depend on t but not on λ and m. Substituted into the expression of norms in V in terms of coefficients with respect to $(\psi_n)_{n\in\mathbb{N}}$, this bound gives $(m+1)^{-2}\|G^{(m)}\|_V^2 \to \infty$ if $\sum_{n=1}^{\infty}\lambda_n^{-2}g_n^2 = \infty$, but the latter condition is equivalent to $g \notin \mathrm{ran}(P_\rho)$.

6.9 The case $s = 1$ in Theorem 6.5.
Prove that for $s = 1$ the convergence estimate in Theorem 6.5 can be improved to

$$\mathbb{E}(\|e^{(m)}\|^2) \le \frac{\|e^{(0)}\|^2 + m\|u\|^2}{(m+1)^2} + \frac{2A^2\Lambda\sigma^2}{(m+1)^{2t-1}} + \frac{\|u\|^2_{V^1_{P_\rho}}}{A(m+1)^{1-t}},$$

if μ_m satisfies (6.45). The best asymptotic error bound $O((m+1)^{-1/3})$ is obtained for $t = 2/3$ and $A^3 = \|u\|^2_{V^1_{P_\rho}}/(2\Lambda\sigma^2)$.

Hints. Start from (6.51) and set $h = u$ there. After taking full expectations, the error recursion $\varepsilon_{k+1}^2 \le \alpha_k^2(\varepsilon_k^2 + 2\Lambda\sigma^2\mu_k^2) + \bar{\alpha}_k^2(\|u\|^2 + \mu_k^{-1}\|u\|_{V_{P_\rho}^1}^2)$ can be used recursively for $k = m-1, \ldots, 0$, together with the definition of μ_m in (6.45) and the properties of the sequences α_m, which is similar to the proof of Theorem 6.5.

6.10 Arbitrary values $1/2 < t < 1$ **in Theorem 6.5**.
If the step-size sequence μ_m from (6.45) is defined with arbitrary $t \in (1/2, 1)$ in Theorem 6.5, while all other assumptions remain the same, then we have

$$\mathbb{E}(\|e^{(m)}\|^2) \le \frac{C_t^2}{(m+1)^{-\min(2t-1,(1-t)s)}}, \qquad m = 1, 2, \ldots.$$

The constant C_t^2 has a similar structure as the constant C^2 in (6.54) corresponding to the value $t = (s+1)/(s+2)$, but it also depends on t.
Hints. Note that the induction argument at the end of the proof holds for any $t \in (1/2, 1)$, one just has to make sure that

$$1 - 2r = a := \max(2 - 2t, -bs/2 + 1 - r, (1-s)b + t) > 0.$$

Fix t and go through the three possible cases

$$1 - 2r = a = 2 - 2t, \quad 1 - 2r = a = -bs/2 + 1 - r, \text{ and } 1 - 2r = a = (1-s)b + t,$$

while determining the free parameters r and b such that the remaining inequalities are satisfied and the value of r is maximized for given t. For $s = 1$, a more precise statement is possible using the simplification described in Problem 6.9.

6.11 Averaged online algorithm.
Prove an analog of Theorem 6.5 for the averaged sequence

$$\bar{u}^{(m)} := \frac{1}{m+1}\sum_{k=0}^{m} u^{(k)} = \frac{1}{m+1}u^{(m)} + \frac{m}{m+1}\bar{u}^{(m-1)},$$

where $u^{(m)}$ is obtained from the unregularized iteration (6.44) with $\alpha_m = 1$ and μ_m as in (6.45), $m = 0, 1, \ldots$.
Hint: $L^2(\Omega, Y)$ convergence in expectation of such averages has been studied in [67].

6.12 Properties of products Π_k^{m-1}.
Consider the products

$$\Pi_k^{m-1} := \prod_{l=k}^{m-1}(1 - \frac{a}{(l+1)^t}), \quad k = 0, \ldots, m-1, \qquad \Pi_m^{m-1} = 1,$$

where $0 < t < 1$ and $0 < a < 1$. Estimates for Π_k^{m-1} and quantities depending on them, such as

$$S_m := \sum_{k=0}^{m} \Pi_k^{m-1}, \quad S'_m := \sum_{k=0}^{m} \Pi_k^{m-1} s_k, \quad S''_m := \sum_{k=0}^{m-1} \Pi_{k+1}^{m-1} (k+1)^r,$$

were used in Subsection 6.5.2.2.

Prove the following inequalities, where the constants C may depend on $t \in (0,1)$ and $r \geq 0$, but not on $k = 0, \dots, m$, $m = 0, 1, \dots$, and $a \in (0,1)$:

$$\Pi_k^{m-1} \leq \exp(-a(1-t)^{-1}((m+1)^{1-t} - (k+1)^{1-t})), \tag{6.76}$$

$$S_m \leq C(m+1)\min(1, (a(m+1)^{1-t})^{-1}), \tag{6.77}$$

$$S'_m \leq C(m+1)^2 \min(1, (a(m+1)^{1-t})^{-2}), \tag{6.78}$$

$$S''_m \leq C(m+1)^{r+1} \min(1, (a(m+1)^{1-t})^{-1}). \tag{6.79}$$

Hints. The property (6.76) is a consequence of $\log(1-x) < -x$, $0 < x < 1$, applied in

$$\Pi_k^{m-1} = \exp\Big(\sum_{l=k}^{m-1} \log(1 - a(l+1)^{-t})\Big),$$

and of the inequality

$$\sum_{l=k}^{m-1} (l+1)^{-t} \geq \int_{k+1}^{m+1} x^{-t}\, dx.$$

The proof of the remaining statements benefits from the following construction: Introduce the index sets

$$I_n := \{0 \leq k \leq m : n \leq a(1-t)^{-1}((m+1)^{1-t} - (k+1)^{1-t}) < n+1\}, n = 0, 1, \dots.$$

To keep the notation short, the dependence of I_n on a and m is not explicitly stated. Obviously, $I_n = \emptyset$ for $n > a(1-t)^{-1}((m+1)^{1-t} - 1)$, and $\Pi_k^{m-1} \leq e^{-n}$, if $k \in I_n$. This gives

$$S_m \leq \sum_{n \geq 0} e^{-n} |I_n|,$$

where the cardinality of I_n can be bounded as follows: Since $k \in I_n$ is equivalent to

$$(m+1)^{1-t} - \frac{(1-t)(n+1)}{a} < (k+1)^{1-t} \leq (m+1)^{1-t} - \frac{(1-t)n}{a},$$

we have

$$|I_n| \leq (m+1)\left(\left(1 - \frac{(1-t)n}{a(m+1)^{1-t}}\right)^{1/(1-t)} - \left(1 - \frac{(1-t)(n+1)}{a(m+1)^{1-t}}\right)^{1/(1-t)}\right)$$

$$\leq (m+1)\frac{1}{a(m+1)^{1-t}}\left(1 - \frac{(1-t)n}{a(m+1)^{1-t}}\right)^{t/(1-t)} \leq \frac{m+1}{a(m+1)^{1-t}}.$$

Substitution gives $S_m \le e(m+1)(a(m+1)^{1-t})^{-1}/(e-1)$. Since $\Pi_k^{m-1} \le 1$, this is complemented by the trivial bound $S_m \le m+1$. Together, we have (6.77) with $C = e/(e-1)$. Note that (6.77) is the special case $r = 0$ of (6.79), which in turn follows from (6.77), since

$$S''_m \le \sum_{n\ge 0} e^{-n} \sum_{k\in I_n} (k+1)^r \le (m+1)^r \sum_{n\ge 0} e^{-n}|I_n| = (m+1)^r S_m.$$

Finally, (6.78) also follows from (6.77) if one substitutes

$$S_k \le C(k+1)\min(1,(a(k+1)^{1-t})^{-1}) \le C(m+1)\min(1,(a(m+1)^{1-t})^{-1})$$

into the formula for S'_m. Note that (6.78) can be extended to values $-1 < r < 0$ with little effort.

6.13 A matching lower bound for (6.60).
By repeating the arguments given in the proof of Theorem 6.6, establish the lower bound

$$\mathbb{E}((c_i^{(m)})^2) \ge c\Big(\min(1, A^{-2}\rho_i^{-2}(m+1)^{2t-2})c_i^2 \\ + A^2\min((m+1)^{1-2t}, A^{-1}\rho_i^{-1}(m+1)^{-t})\rho_i\sigma^2\Big), \qquad (6.80)$$

which is valid for $i, m \in \mathbb{N}$ with a constant $c > 0$ only depending on $t \in (0,1)$.
Hints. Denote

$$\tilde{\Pi}_k^{m-1} := (1-\rho_i\mu_{m-1}(2-\mu_{m-1}))\cdots(1-\rho_i\mu_k(2-\mu_k)), \quad 0 \le k \le m-1,$$

$\tilde{\Pi}_m^{m-1} = 1$, and observe that $\tilde{\Pi}_k^{m-1} \le \Pi_k^{m-1}$, where Π_k^{m-1} are the quantities defined in (6.59) with $a = A\rho_i$. Consequently, by repeating the recursive estimation of the expectations $\varepsilon_m = \mathbb{E}((c_i^{(m)})^2)$ in the proof of Theorem 6.6, one arrives at the lower estimate

$$\varepsilon_m \ge \frac{1}{(m+1)^2}\left(c_i^2 \sum_{k=0}^{m} \tilde{\Pi}_k^{m-1}\tilde{S}_k + \rho_i\sigma^2 \sum_{k=0}^{m-1}(k+1)^2\mu_k^2\tilde{\Pi}_{k+1}^{m-1}\right),$$

where

$$\tilde{S}_k = \sum_{l=0}^{k} \tilde{\Pi}_l^{k-1}$$

for $k = 0, \ldots, m-1$. To derive the lower bound (6.80), it remains to obtain lower bounds for $\tilde{\Pi}_{k+1}^{m-1}$, $\tilde{S}_k$, and related quantities. To do this, follow the hints for Problem 6.12, starting from a lower bound for $\tilde{\Pi}_{k+1}^{m-1}$ of the form

$$\tilde{\Pi}_k^{m-1} \ge \begin{cases} \exp(-C_t A\rho_i(m^{1-t}-k^{1-t})), & k > 0, \\ \exp(-C_t A\rho_i m^{1-t}), & k = 0, \end{cases}$$

where $C_t = C/(1-t)$ with some absolute constant C. To derive the latter, note that $0 < \mu_k \leq 1/2$ by (6.45) and thus $0 < \rho_i\mu_k(2-\mu_k) < 3/4$ and

$$\log(1-\rho_i\mu_k(2-\mu_k)) \geq -\frac{4}{3}\log(4)\rho_i\mu_k(2-\mu_k) \geq -\frac{8}{3}\log(4)\frac{A\rho_i}{(k+1)^t}.$$

Apply this to the expression for $\log(\tilde{\Pi}_k^{m-1})$ and use

$$\sum_{l=k+1}^{m} l^{-t} \leq \int_k^m x^{-t}\,dx = \frac{m^{1-t}-k^{1-t}}{1-t}, \qquad k \geq 1,$$

with an obvious modification for $k = 0$.

6.14 Admissibility of R for discrete measures.
The admissibility of an operator family $\mathbf{R}$ was defined in Subsection 6.5.2.3. Show that, for a discrete measure ρ on $\Omega = \mathbb{N}$, the admissibility of $\mathbf{R} = (R_i : V_i \to V)_{i\in\Omega}$ is equivalent to the collection $(\mathbf{V}; \mathbf{a}; \mathbf{R}; \rho)$ becoming a space splitting associated with $(V; a)$, see the definitions in Section 2.1.
Hint. Since measurability in the definition of admissibility is not an issue in this particular situation, you must show that the condition (2.2) implies the integrability condition

$$\sum_i \rho_i|a(R_i\tilde{v}_i, v)| < \infty$$

for all $\tilde{v} \in \tilde{V}_\rho$ and $v \in V$. The opposite direction was essentially established in Subsection 6.5.2.3.

6.15 Compact B and dictionaries D from its singular value decomposition.
Let $B : U \to V$ be compact with $\ker(B) = \{0\}$ and $\mathrm{ran}(B)$ dense in V. Let

$$Bu = \sum_i \sigma_i(u,\psi_i)_U\phi_i, \quad B^*v = \sum_i \sigma_i(v,\phi_i)_U\psi_i, \qquad u \in U,\ v \in V,$$

be its singular value decomposition, where the singular values $\sigma_i \searrow 0$ are monotonically ordered, and $\{\psi_i\}_{i\in\Omega}$ and $D := \{\phi_i\}_{i\in\Omega}$ are CONS in U and V, respectively. Provide stronger versions of the Corollaries 6.5 and 6.6.
Hint. Use the explicit descriptions of $P_{\mathbf{p}}$, $V^s_{P_{\mathbf{p}}}$, and $\mathcal{A}^p$ to reduce the estimates for the respective Schwarz approximation methods to estimates on sequence spaces, similar to the arguments in Section 6.1.

6.16 A non-symmetric toy problem.
Consider the one-dimensional convection-diffusion problem

$$-\epsilon u'' + u' = f, \qquad u \in U = V := H_0^1(G), \quad G := [0,1],$$

with a given right-hand side $f \in H^{-1}(G)$ and a constant value $\epsilon > 0$, see [62, 155, 156] for a detailed discussion of this model problem, including a reaction term. If we take

$$(u,v)_V := (u',v')_{L^2} = \int_0^1 u'(t)v'(t)\,dt$$

as the scalar product in V, then

$$b(u,v) := \epsilon(u,v)_V + (u',v)_{L^2} = \langle f,v\rangle_{V'\times V},$$

is the associated variational formulation.
a) Check that $B(\cdot,\cdot)$ satisfies the inf-sup condition and reveal the dependence of the constants on ϵ.
b) Check that, with the usual interpretation of distributional derivatives, the Riesz map is given by $R_V v = -v''$, $v \in V$, and give an explicit formula for R_V^{-1} using the Green function of the differential operator R_V acting on $H_0^1(G)$.
c) Find explicit representations for the operators B, B', and B^*B.
d) Check that the symmetric bilinear form associated with the operator equation (6.66) in $U = H_0^1(G)$ is bounded and coercive, but with a worse constant in the coercivity estimate. Write down the associated bilinear form and differential equation.
e) Is there a computationally efficient multiscale space splitting for the variational problem associated with B^*B, with condition numbers that are independent of ϵ, or at least only slightly dependent on it?
Hints. The constants in a) and b) degrade proportionally to ϵ and ϵ^2, when $\epsilon \to 0$. The formulas for B^*B and the associated symmetric bilinear form include integral operators and global averages, respectively. We do not know the answer to e). See [47] for a closer look at this problem.

Chapter 7
Applications to PDE solvers

Abstract In this chapter we describe how the theory of the previous chapters applies to the analysis of solvers for PDE discretizations. The focus is on finite element discretizations of a standard second-order elliptic boundary value problem. We discuss multilevel preconditioners for adaptive finite element spaces, non-overlapping and overlapping domain decomposition methods, and tensor product techniques.

7.1 Multilevel methods

In this section, we give selected examples of applications of the standard multiscale splittings discussed in Section 3.1 to the analysis of some concrete algorithms for solving second-order elliptic PDEs. The model equation is the scalar diffusion equation

$$-\nabla(w\nabla u) = f \tag{7.1}$$

in a polyhedral domain $G \subset \mathbb{R}^d$ with boundary $\Gamma = \partial G$, equipped for simplicity with homogeneous Dirichlet boundary conditions, where the diffusion coefficient $w \in L^\infty(G)$ is non-negative and uniformly bounded away from 0 a.e. in G. In its weak formulation, (7.1) is then a prototypical variational problem in the Sobolev space $H_0^1(G)$. First, we analyze subspace splittings of $H^s(G)$ with respect to finite element spaces. In particular, stable splittings for *adaptive* finite element spaces are considered. Besides the hierarchical basis method, we focus on multiscale splittings leading to additive BPX preconditioners and multiplicative multigrid solvers. Second, we establish the counterpart of Theorem 3.3 for the spaces $H_0^s(G)$, $1/2 < s < 3/2$, and the trace spaces $H^s(\Gamma)$, $0 < s < 1$. Third, we further generalize our results to the case of non-conforming finite elements. Finally, we consider the adaptive wavelet method. As a result, the Schwarz iterative methods presented in Chapter 4 yield optimal convergence rates for certain multilevel, multigrid, and wavelet methods for boundary value problems for elliptic PDEs, which are popular in simulations for engineering applications.

M. Griebel, P. Oswald, *Hilbert Space Splittings and Iterative Methods*, Springer Series in Computational Mathematics 62, https://doi.org/10.1007/978-3-031-74370-2_7

7.1.1 BPX preconditioning for adaptive finite element spaces

In this subsection, we modify the construction and analysis of finite element multiscale splittings for $H^s(G)$ from Subsection 3.1.3 in such a way that it leads to preconditioners of optimal complexity for discretizations with respect to finite element spaces obtained by so-called nested local refinement, which arise naturally in adaptive finite element methods.

7.1.1.1 Adaptive finite element spaces

We begin by constructing hierarchies of adaptive finite element subspaces of $H^s(G)$, $0 < s < 3/2$. Recall some notation from Subsection 3.1.3: Let $V_j = V(\mathcal{T}_j)$ be a nested sequence of Lagrange C^0 finite element spaces of fixed degree $p = 1, 2, \ldots$ associated with a sequence of uniformly regular and quasi-uniform (simplicial or rectangular) partitions $\mathcal{T}_j$ with maximal cell diameter $\mathrm{diam}(\mathcal{T}_j) \approx 2^{-j}$ obtained by dyadic refinement. The nodal basis functions spanning V_j are denoted by $N_{P,j}$ (the subscript j indicates the refinement level). Similarly, the associated dual functionals, which were used to define the quasi-interpolants Q_j for the proof of Theorem 3.3, are denoted by $\lambda_{P,j}$. Recall that their construction involved the family $\mathcal{B}_{r_0}$ of balls of radius $r \geq r_0$ contained in the standard cell Σ associated with the type of underlying partitions. For the sets of nodal points of $\mathcal{T}_j$ we use the notation $\mathcal{V}_j$.

Throughout this chapter, let J be a given fixed but arbitrary integer (in many places one can also allow $J = \infty$). For given subsets $\tilde{\mathcal{V}}_j \subset \mathcal{V}_j$, $j \leq J$, we set

$$\tilde{V}_j := \mathrm{span}(\{N_{P,j} : P \in \tilde{\mathcal{V}}_j\}), \quad \hat{V}_j := \mathrm{span}(\{\tilde{V}_k : k \leq j\}), \quad 0 \leq j \leq J. \quad (7.2)$$

The finite-dimensional spaces $\hat{V}_j \subset V_j$ and in particular $\hat{V}_J$ are thus obtained by selecting nodal basis functions from different levels of refinement and are our candidates for adaptive finite element spaces. This is in contrast to the traditional understanding of adaptive finite element methods, where such spaces are identified with $V(\mathcal{T})$ for an adaptively generated partition $\mathcal{T}$. The relationship between the two concepts will be discussed later.

Obviously, without further conditions on the subsets $\tilde{\mathcal{V}}_j$, this definition can lead to very exotic spaces $\hat{V}_J$. However, the following abstract condition allows meaningful statements about the uniform stability of splittings of $\hat{V}_J$ into a sequence of spaces $\tilde{V}_j$, $j \leq J$, generalizing Theorem 3.3. Let $\tilde{G}_j$, $j \leq J$, denote the common support of functions from $\tilde{V}_j$, $j \leq J$, (i.e., the union of the supports of all $N_{P,j}$ with $P \in \tilde{\mathcal{V}}_j$) and set $\hat{G}_j := \cup_{k=j}^{J} \tilde{G}_k$. Obviously, $\tilde{G}_j$ is a union of cells from $\mathcal{T}_j$ (cells of level j for short) and $\hat{G}_J = \tilde{G}_J$. We say that our construction obeys the *local refinement condition for given J* if, with some fixed $r_0 > 0$, for any $j = 0, \ldots, J-1$ and any nodal basis function $N_{P,j} \notin \tilde{V}_{j+1}$, there is at least one cell $C_i \in \mathcal{T}_j$ in the support of $N_{P,j}$ such that, when C_i is affine-linearly mapped to the standard cell Σ, the image of the subset $\Sigma_{i,j} := C_i \backslash \hat{G}_{j+1}$ contains a ball $B(x_0, r)$ of radius $r \geq r_0$. Note that $\Sigma_{i,j}$ depends on $C_i \in \mathcal{T}_j$ and the nodal point sets $\tilde{V}_k$, $k \geq j$, but not on the individual nodal point $P \in C_i$ satisfying the above assumption $N_{P,j} \notin \tilde{V}_{j+1}$. Moreover, if $C_i \not\subset \hat{G}_{j+1}$ for some cell $C_i \in \mathcal{T}_j$ then the condition $N_{P,j} \notin \tilde{V}_{j+1}$ holds for all nodal points $P \in C_i$. A simple example is shown in Figure 7.1.

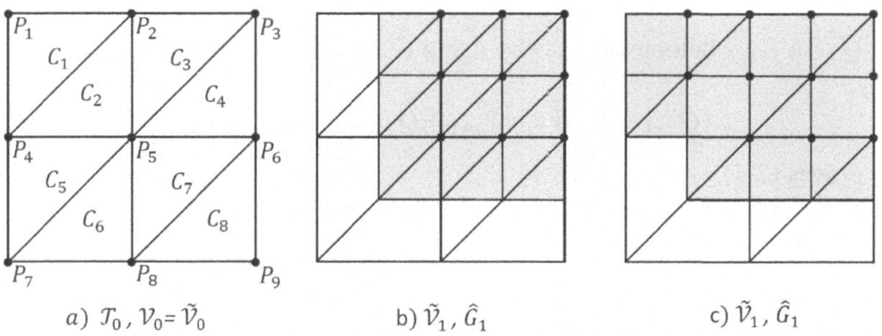

Fig. 7.1 Illustration of the local refinement condition. a) Initial triangulation $\mathcal{T}_0$ with $\mathcal{V}_0 = \tilde{\mathcal{V}}_0$. b) Locally refined mesh with $\tilde{\mathcal{V}}_1, \hat{G}_1$ that satisfies the local refinement condition for $J = 1$ with $N_{P_k} \notin \tilde{\mathcal{V}}_1$ for all $k \neq 3$ and $\Sigma_{i,1} \neq 0$ for all $i \neq 3, 4$. c) Locally refined mesh with $\tilde{\mathcal{V}}_1, \hat{G}_1$ that violates the local refinement condition because $N_{P_1} \in \tilde{\mathcal{V}}_1$ but $\operatorname{supp}(N_{P_1}) = C_1 \subset \hat{G}_1$.

This complicated-sounding condition ensures that, for the construction of the quasi-interpolants Q_j, $j = 0, \ldots, J-1$, in the proof of Theorem 3.3, we can always choose the dual functionals $\lambda_{P,j}$ such that

$$\lambda_{P,j}(v) = 0, \qquad v \in \hat{V}_{j+1,J} := \operatorname{span}(\{\tilde{V}_k : k := j+1, \ldots, J\}) \subset \hat{V}_J, \tag{7.3}$$

holds for all $P \in \mathcal{V}_j$ such that $N_{P,j} \notin \tilde{V}_{j+1}$. In fact, by the local refinement condition, for each such P we can take some C_i and and a ball $B(x_0, r)$ in the image of $\Sigma_{i,j}$ for the definition of q_{P,C_i} in (3.22) and set

$$\lambda_{P,j}(u) = \int_{C_i} u(x) q_{P,C_i}(x)\, dx.$$

For the remaining P, we pick any of the C_i containing P and use the same formula. This corresponds to setting $a_{P,i} = 1$ and $a_{P,i'} = 0$, $i' \neq i$, in (3.25). With this, because of $\operatorname{supp}(v) \subset \hat{G}_{j+1}$ for any $v \in \hat{V}_{j+1,J}$ and $\operatorname{supp}(q_{P,C_i}) \subset \Sigma_{i,j} \subset \bar{G} \backslash \hat{G}_{j+1}$ for all P with $N_{P,j} \notin \tilde{V}_{j+1}$ and its associated C_i, we arrive at (7.3).

The property (7.3) in turn implies that we have

$$(Q_{j+1} - Q_j)\hat{v}_J \in \tilde{V}_{j+1}, \qquad j = 0, \ldots, J-1, \tag{7.4}$$

for any function $\hat{v}_J$ in the adaptive finite element space $\hat{V}_J$. In fact, for $j = 0, \ldots, J-1$, any $\hat{v}_J \in \hat{V}_J$ can be written as

$$\hat{v}_J = v_j + \tilde{v}_{j+1} + w_{j+1,J},$$

where $v_j \in V_j$ is a linear combination of nodal basis functions $N_{P,j}$ not in $\tilde{V}_{j+1}$, $\tilde{v}_{j+1} \in \tilde{V}_{j+1}$, and $w_{j+1,J}$ belongs to $\hat{V}_{j+2,J}$ (note that for $j = J-1$ we have $w_{j+1,J} = w_{J,J} = 0$). Now by (7.3) we have $Q_j\hat{v}_J = Q_j v_j = v_j$, while

$$Q_{j+1}\hat{v}_J = Q_{j+1}(v_j + \tilde{v}_{j+1}) + Q_{j+1}w_{j+1,J} = v_j + \tilde{v}_{j+1} + Q_{j+1}w_{j+1,J},$$

where $Q_{j+1}w_{j+1,J}$ belongs to $\tilde{V}_{j+2}$ by using (7.3) with j replaced by $j+1$. Consequently,

$$(Q_{j+1} - Q_j)\hat{v}_J = \tilde{v}_{j+1} + Q_{j+1}w_{j+1,J} \in \tilde{V}_{j+1},$$

which proves (7.4).

With this at hand, we can formulate the main result of this section:

Theorem 7.1 *If the construction (7.2) satisfies the local refinement condition for J, then $(\tilde{V}_j; a_j)_{j\le J}$, where $a_j(\cdot,\cdot) = 2^{2js}(\cdot,\cdot)_{L^2}$ for $j = 0, 1, \dots, J$, is a space splitting for the adaptive finite element subspace $\hat{V}_J$, considered as a subspace of $H^s(G)$, $0 < s < 3/2$. Its stability constants depend on p and s, the initial partition $\mathcal{T}_0$, the regularity and quasi-uniformity constants of the dyadic refinement process leading to the partitions $\mathcal{T}_j$, $1 \le j \le J$, and the choice of $r_0 > 0$ in the local refinement condition, but are independent of J and other details of the local refinement.*

Proof Since $\tilde{V}_j \subset V_j$ and

$$\hat{V}_J = \mathrm{span}(\{\tilde{V}_j : j \le J\}) \subset V_J \subset H^s(G),$$

the considered space splittings for $(\hat{V}_J; (\cdot,\cdot)_{H^s(G)})$ can be interpreted as subsplittings of the space splitting of Theorem 3.3. Thus, according to the remarks in Subsection 2.3.3, to prove the statement we only need to verify the lower stability bound. To do this, consider the sequence of quasi-interpolants Q_j with the additional property (7.4), whose existence for $j < J$ is guaranteed by the local refinement condition (for $j \ge J$, the choice of Q_j is the same as in the proof of Theorem 3.3). This sequence is admissible in the proof of Theorem 3.3 and thus implies the norm equivalence

$$\|u\|_{H^s}^2 \approx \|Q_0u\|_{L^2}^2 + \sum_{j=1}^{\infty} 2^{2js}\|(Q_j - Q_{j-1})u\|_{L^2}^2, \qquad u \in H^s(G).$$

By (7.4) and $(Q_j - Q_{j-1})\hat{v}_J = 0$ for $j > J$ we get

$$\inf_{\tilde{v}_j\in\tilde{V}_j:\,\hat{v}_J=\sum_{j\le J}\tilde{v}_j} \sum_{j=0}^{J} 2^{2js}\|\tilde{v}_j\|_{L^2}^2$$

$$\le \|Q_0\hat{v}_J\|_{L^2}^2 + \sum_{j=1}^{J} 2^{2js}\|(Q_j - Q_{j-1})\hat{v}_J\|_{L^2}^2 \le C\|\hat{v}_J\|_{H^s}^2,$$

where C depends only on the properties used in the proof of Theorem 3.3, but not on the specifics of the construction of $\hat{V}_J$ and on J. This concludes the proof of Theorem 7.1. □

Using the L^2 stability of the nodal basis in $\tilde{V}_j \subset V_j$, we can reinterpret the result of Theorem 7.1 in the language of frames. In fact, for $0 < s < 3/2$, the system

$$\hat{\Phi}_J := \{\tilde{N}_{P,j} = d_{P,j} N_{P,j} : P \in \tilde{\mathcal{V}}_j, \; j = 0, \ldots, J\}, \tag{7.5}$$

consisting of nodal basis functions scaled by factors

$$d_{P,j} \approx 2^{j(d/2-s)}$$

and associated with the sets $\tilde{\mathcal{V}}_j$, is a frame for the adaptive finite element subspace $\hat{V}_J \subset H^s(G)$, if the assumptions of Theorem 7.1 hold. Furthermore, the frame bounds are independent of J. This follows from the use of the refinement and coarsening strategies for space splittings discussed in Subsection 2.3.2 and the definition of frames, since

$$\| \sum_{P \in \mathcal{V}_j} c_{P,j} N_{P,j} \|_{L^2}^2 \approx 2^{-jd} \sum_{P \in \mathcal{V}_j} c_{P,j}^2$$

for any set of coefficients $c_{P,j}$ and

$$\|N_{P,j}\|_{H^s}^2 \approx 2^{2js} \|N_{P,j}\|_{L^2}^2 \approx 2^{2j(s-d/2)}, \qquad 0 < s < 3/2,$$

with constants independent of j and $P \in \mathcal{V}_j$. We call $\hat{\Phi}_J$ a *BPX-frame* for $\hat{V}_J$, referring to the names of the authors of [22] who originally introduced a multilevel preconditioner based on $\hat{\Phi}_J$ for second-order elliptic problems ($s = 1$).

If we set $\tilde{\mathcal{V}}_j = \mathcal{V}_j$ for all $j \geq 0$, which corresponds to full dyadic refinement, we get $\tilde{V}_J = \hat{V}_J = V_J$ for all $J \geq 1$, and the local refinement condition is trivially satisfied. The message of Theorem 7.1 is then that the infinite space splitting of Theorem 3.3 inherits its stability properties to finite subsplittings for the subspaces $V_J \subset H^s(G)$, uniformly with respect to the discretization level $J = 1, 2, \ldots$. This leads to additive and multiplicative Schwarz methods of low complexity whose convergence properties do not degenerate with increasing J.

Examples of nodal point sets $\tilde{\mathcal{V}}_j$ for which the result of Theorem 7.1 on the stability of splittings for the resulting adaptive finite element space $\hat{V}_J$ does not hold are easy to find. For example, if we set $\tilde{\mathcal{V}}_J = \mathcal{V}_J$ and $\tilde{\mathcal{V}}_j = \emptyset$, $j < J$, then we have the well-known condition number behavior $\kappa_j \approx 2^{2js}$ for the nodal basis discretization of a $H^s(G)$-elliptic problem with respect to $\hat{V}_J = V_J$. Another example where $\tilde{\mathcal{V}}_j = \mathcal{V}_j \backslash \{P_0\}, 0 \leq j \geq J$, with fixed $P_0 \in \mathcal{V}_0$, is worked out in Problem 7.1.

7.1.1.2 Nested local refinement

Next, we consider the so-called nested local refinement, for which the local refinement condition formulated in the previous subsection is easy to verify and which is related to the practice of adaptive finite element methods. The typical adaptive finite element method builds the final discretization recursively by a sequence of

Solve - Estimate - Mark - Refine

steps. More specifically, given the current space $\hat{V}_j$ associated with the current partition $\hat{\mathcal{T}}_j$, a sufficiently accurate approximation $\hat{u}_j \in \hat{V}_j$ to the solution $u \in H^s(G)$ of the problem at hand is computed in a Solve step. It is desirable that the cost of the solver scales well with the dimension of $\hat{V}_j$. In the Estimate step, the current approximate solution $\hat{u}_j$ is used to obtain information about the local distribution of the error function

$$\hat{e}_j := u - \hat{u}_j,$$

typically in the form of so-called a posteriori error estimators built from local error indicators associated with the cells of $\hat{\mathcal{T}}_j$. Using some kind of thresholding, a certain number of cells in $\hat{\mathcal{T}}_j$ with large local error indicators are selected in the Mark step and subdivided in the subsequent Refine step to obtain the next refined partition $\hat{\mathcal{T}}_{j+1}$. The Refine step may include additional refinement to satisfy certain compatibility conditions of the partitions $\hat{\mathcal{T}}_j$ and to avoid so-called hanging nodes, for example.

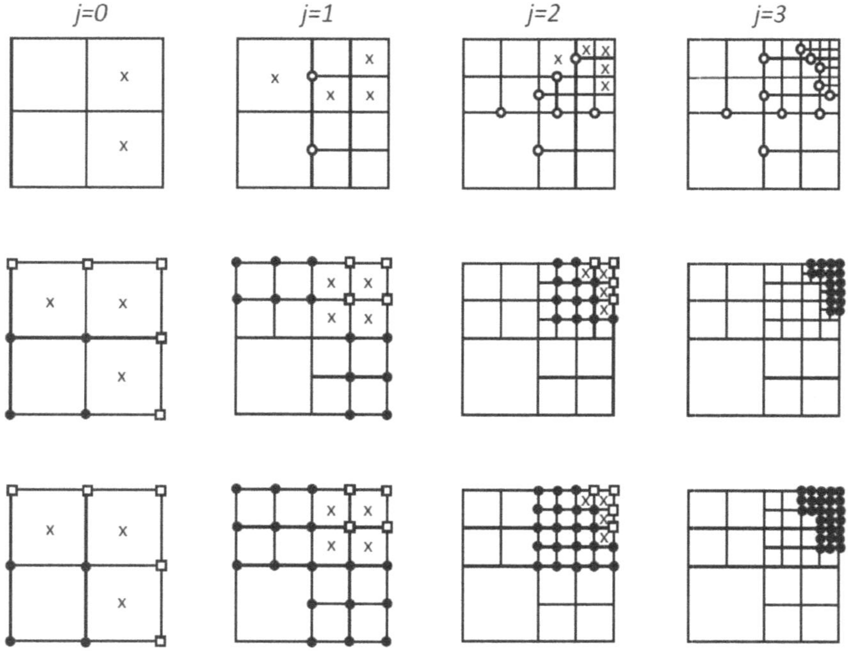

Fig. 7.2 Illustration of local refinement. Top row: Actual refinement process, where cells marked by × are subdivided in each step and hanging nodes marked by ∘ are created. Middle row: Nested local refinement corresponding to the previous example, with G_j marked by crosses and $\tilde{\mathcal{V}}_j$ consisting of nodal points "interior" to G_j (nodal points in $\tilde{\mathcal{V}}_j$ are indicated by • and nodal points in $\mathcal{V}_j \backslash \tilde{\mathcal{V}}_j$ are indicated by □). Bottom row: Same as middle row, but with $\tilde{\mathcal{V}}_j$ consisting of all nodal points in G_j.

For illustration, consider rectangular partitions in the plane, quadrisection as refinement, and bilinear finite elements ($p = 1$). Then the nodal point sets $\mathcal{V}_j$ are given by the vertices of $\mathcal{T}_j$. Figure 7.2 (first row) shows a sequence of rectangular partitions $\hat{\mathcal{T}}_j$, $j \leq 3$, which may have resulted from such a refinement process, where marked cells and hanging nodes are indicated by crosses and circles, respectively. To connect with our refinement paradigm based on a level-wise selection of basis functions, in the second row of Figure 7.2 we show a sequence of nodal point sets $\tilde{\mathcal{V}}_j$ that leads to a finite element space $\tilde{V}_3$ with similar approximation power as the $\hat{V}_3$ associated with the partition $\hat{\mathcal{T}}_3$ shown in the first row. Here, only nodal points interior to the region of marked cells of level j form the sets $\tilde{\mathcal{V}}_j$. This construction is equivalent to the construction of adaptive finite element spaces with hanging nodes. Note that in a realistic Solve - Estimate - Mark - Refine procedure, cells of level j can also be marked in later refinement steps, see the example in the first row. Alternatively, the sets $\tilde{\mathcal{V}}_j$ can be formed by all nodal points associated with marked cells of level j. This is shown in the third row of Figure 7.2. In both cases, the local refinement property holds: It is easy to check that the sets $\Sigma_{i,j}$ appearing in the construction of the quasi-interpolants used for proving Theorem 7.1 can be identified with cells of level j and level $j+1$ outside $\tilde{G}_j$ for the situation shown in the second row and in the third row of Figure 7.2, respectively. Note that we have $\tilde{G}_{j-1} \subset \tilde{G}_j$ which implies $\tilde{G}_j = \hat{G}_j$, $0 \leq j < J$.

More formally, the following result holds, which also explains the terminology of nested local refinement and covers the situation of adaptive finite element spaces with hanging nodes:

Lemma 7.1 *For the above setting of Lagrange C^0 finite element spaces V_j associated with the partition sequence $\mathcal{T}_j$ obtained by dyadic refinement, denote by $G_j \subset G$ the union of certain cells from $\mathcal{T}_j$ (think of cells of level j marked during a refinement process), and assume that the nestedness property*

$$G_j \subset G_{j-1}, \qquad j = 1, \ldots, J-1, \tag{7.6}$$

is satisfied. Let $\tilde{\mathcal{V}}_0 = \mathcal{V}_0$ and $\tilde{V}_0 = V_0$.
a) Let $\tilde{\mathcal{V}}_j$, $j = 1, 2, \ldots, J$, consist of all nodal points from $\mathcal{V}_j$ "interior" to G_{j-1}, i.e., all nodal points from $\mathcal{V}_j$ which are not in the closure of $G \backslash G_{j-1}$. Then the local refinement condition is satisfied for sufficiently small r_0.
b) If $\tilde{\mathcal{V}}_j = \mathcal{V}_j \cap G_{j-1}$, then the local refinement condition holds for sufficiently small r_0 if we additionally assume that

$$C \cap \tilde{G}_j \neq \emptyset \quad \text{for all cells } C \text{ of level } j-1 \text{ not in } G_{j-1}. \tag{7.7}$$

Proof By construction, the nestedness property (7.6), and the definition of $\tilde{\mathcal{V}}_j$ in Part a), we obviously have $\hat{G}_j = \tilde{G}_j = G_{j-1}$ for $j \geq 1$. Consequently, the support of any nodal basis $N_{P,j-1}$ which is not in $\tilde{V}_j$ must contain at least one cell $C_i \in \mathcal{T}_{j-1}$ of level $j-1$ which is not in $\hat{G}_j$, otherwise $\mathrm{supp}(N_{P,j-1}) \subset G_{j-1} = \tilde{G}_j$ implies $N_{P,j-1} \in \tilde{V}_j$. This cell can serve as $\Sigma_{i,j-1}$, which shows the statement of Part a).

From the example of G_j for $j = 2$ shown in Figure 7.3 on the left, it is clear that the local refinement property may not hold for the choice of $\tilde{\mathcal{V}}_j$ in Part b) without additional assumptions such as (7.7). We obviously have several $N_{P,1}$ not in $\hat{V}_2$, but $\tilde{G}_2 = \hat{G}_2 = \tilde{G}$ implies that it is impossible to find suitable non-empty sets $\Sigma_{i,1}$. However, since $\tilde{G}_j$ is the union of cells of level j, the additional assumption (7.7) implies that every cell C of level $j-1$ that is not in G_{j-1} must contain a cell C' of level j that is not in $\tilde{G}_j = \hat{G}_j$. Again, each $N_{P,j-1} \notin \tilde{V}_j$ must contain such a cell C (and thus also C') in its support. This C' can then serve as $\Sigma_{i,j-1}$, which establishes the nested local refinement condition for Part b). □

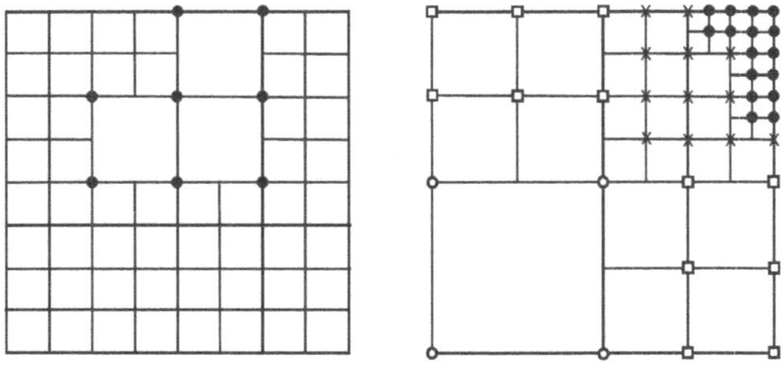

Fig. 7.3 Illustration of the violation of the local refinement property in Part b) of Lemma 7.1 (left) and of the construction of the basis $\tilde{\Phi}_J$ in $\hat{V}_J$ (right, the example corresponds to the situation depicted in the middle row of Figure 7.2 for $J = 3$).

7.1.1.3 Hierarchical basis splitting

Another choice for $\tilde{\mathcal{V}}_j$ that does not satisfy the local refinement condition is related to the so-called hierarchical basis (HB) method. Set

$$\tilde{\mathcal{V}}_j = \mathcal{V}_j \backslash \mathcal{V}_{j-1}, \quad j \geq 1, \qquad \tilde{\mathcal{V}}_0 = \mathcal{V}_0.$$

Obviously, we then have $\hat{V}_J = V_J$ for all J, and we get the direct sum decomposition

$$\hat{V}_J = V_J = \sum_{j=0}^{J} \tilde{V}_j, \qquad J \geq 0.$$

The associated system $\hat{\Phi}_J$ defined in (7.5) is an alternative basis for V_J, called the hierarchical basis. The counterpart of Theorem 7.1 for this special case is as follows.

Theorem 7.2 *The space splitting $(\tilde{V}_j; a_j)_{j\leq J}$ associated with the hierarchical basis construction above is a stable Riesz splitting for $(V_J; (\cdot,\cdot)_{H^s})$, $0 < s < 3/2$, whose condition number $\kappa_{HB,J}$ scales with J as*

$$\kappa_{HB,J} \approx \begin{cases} 1, & s > d/2, \\ J^2, & s = d/2, \\ 2^{2J(d/2-s)}, & s < d/2. \end{cases}$$

The constants in the two-sided estimate for $\kappa_{HB,J}$ depend on the parameters d, p, and s, on $\mathcal{T}_0$, and on the regularity and quasi-uniformity constants of the dyadic refinement process leading to the partitions $\mathcal{T}_j$, $j \geq 1$, but are independent of J.

Proof To prove Theorem 7.2, again we only have to deal with the lower stability estimate. Obviously, the unique hierarchical basis decomposition of $u_J \in V_J$ is given by

$$u_J = Q_0 u_J + \sum_{j=1}^{J} (Q_j - Q_{j-1}) u_J,$$

where $Q_j : C(\bar{G}) \to V_j$ temporarily denotes the nodal interpolation operators, i.e., applied to u_J, we have

$$Q_j u_J = \sum_{P \in \mathcal{V}_j} u_J(P) N_{P,j}, \qquad j = 0, 1, \ldots, J.$$

Let $u_J = \sum_{j=0}^{J} v_j$ be an arbitrary decomposition into $v_j \in V_j$, $j \leq J$. Then, using the L^2 stability of the nodal bases and the fact that $\mathcal{V}_j \subset \mathcal{V}_{j+1}$, we have

$$\begin{aligned} \|Q_0 u_J\|_{L^2}^2 &\leq C \sum_{P \in \mathcal{V}_0} |\sum_{j=0}^{J} v_j(P)|^2 \\ &\leq C \begin{cases} (J+1) \sum_{P \in \mathcal{V}_0} \sum_{j=0}^{J} |v_j(P)|^2, & \epsilon = 0, \\ \sum_{P \in \mathcal{V}_0} \sum_{j=0}^{J} 2^{2j\epsilon} |v_j(P)|^2, & \epsilon > 0, \end{cases} \\ &\leq C \begin{cases} (J+1) \sum_{j=0}^{J} 2^{jd} \|v_j\|_{L^2}^2, & \epsilon = 0, \\ \sum_{j=0}^{J} 2^{j(d+2\epsilon)} \|v_j\|_{L^2}^2, & \epsilon > 0. \end{cases} \end{aligned}$$

Since $(Q_k - Q_{k-1}) v_j = 0$ for $j < k$, we also have

$$\|(Q_k - Q_{k-1}) u_J\|_{L^2}^2 = \|(Q_k - Q_{k-1}) (\sum_{j=k}^{J} v_j)\|_{L^2}^2$$

$$\leq 2 \left(\|Q_k (\sum_{j=k}^{J} v_j)\|_{L^2}^2 + \|Q_{k-1} (\sum_{j=k}^{J} v_j)\|_{L^2}^2 \right) \leq C 2^{-kd} \sum_{P \in \mathcal{V}_k} |\sum_{j=k}^{J} v_j(P)|^2$$

$$\leq C \begin{cases} (J-k+1) \sum_{j=k}^{J} 2^{jd} \|v_j\|_{L^2}^2, & \epsilon = 0, \\ 2^{-2k\epsilon} \sum_{j=k}^{J} 2^{j(d+2\epsilon)} \|v_j\|_{L^2}^2, & \epsilon > 0. \end{cases}$$

In the case $s > d/2$ we fix a value $0 < \epsilon < s - d/2$. Substitution of the previous estimates then yields

$$
\begin{aligned}
\|Q_0 u_J\|_{L^2} &+ \sum_{k=1}^{J} 2^{2ks}\|(Q_k - Q_{k-1})u_J\|_{L^2}^2 \\
&\leq C \sum_{k=0}^{J} 2^{2k(s-d/2-\epsilon)} \sum_{j=k}^{J} 2^{2j(d/2+\epsilon)}\|v_j\|_{L^2}^2 \\
&= C \sum_{j=0}^{J} 2^{2j(d/2+\epsilon)}\|v_j\|_{L^2}^2 \sum_{k\leq j} 2^{2k(s-d/2-\epsilon)} \leq C \sum_{j=0}^{J} 2^{2js}\|v_j\|_{L^2}^2 .
\end{aligned}
$$

This gives the result for $3/2 > s > d/2$ if one takes the infimum with respect to all representations of u_J and uses the lower stability bound for the BPX-frame. If $s \leq d/2$, then the substitution of the estimates with $\epsilon = 0$ yields the additional forefactors J^2 for $s = d/2$, $d \leq 2$, and $2^{J(d/2-s)}$ for $0 < s < \min(d/2, 3/2)$, respectively. The construction of examples showing the necessity of these forefactors is outlined in Problem 7.2. This establishes Theorem 7.2. □

The example of the hierarchical basis splitting demonstrates, on the one hand, that the formulated local refinement condition is only a sufficient condition for the uniform stability of the associated multiscale splittings (consider the case $s > d/2$) and, on the other hand, that if it fails, the uniform stability of the associated multiscale splitting can be lost (consider the case $s \leq d/2$). The most important case is $p = 1$, $d = 2$, $s = 1$ (linear finite element discretizations of second-order elliptic problems in the plane), where the condition number growth is mild and the resulting solver is widely used. For example, it is implemented in the popular PLTMG software package [6, 7]. BPX and HB preconditioners are also used in other adaptive finite element codes [157].

7.1.1.4 Complexity of BPX preconditioning

Let us now give some details about the structure and complexity of the preconditioners resulting from the BPX-frame construction for adaptive finite element spaces. Suppose the linear system

$$\hat{A}_J \hat{x}_J = \hat{b}_J$$

is the discretization of an $H^s(G)$-elliptic problem, $0 < s < 3/2$, with respect to the system

$$\bar{\Phi}_J := \{N_{P,j} : P \in \bar{\mathcal{V}}_j,\ j = 0, \ldots, J\}, \tag{7.8}$$

in $\hat{V}_J$. The index sets $\bar{\mathcal{V}}_k$ in (7.8) are given by $\bar{\mathcal{V}}_J = \tilde{\mathcal{V}}_J$ and

$$\bar{\mathcal{V}}_j := \{P \in \tilde{\mathcal{V}}_j : N_{P,j} \notin \tilde{V}_{j+1}\} \subset \tilde{\mathcal{V}}_j, \qquad j < J.$$

The basis property of $\bar{\Phi}_J$ in $\hat{V}_J$ holds for nested local refinement (which we will assume from now on), its verification is left to the reader, see Problem 7.3 a). If the local refinement property does not hold, the basis property of $\bar{\Phi}_J$ in $\hat{V}_J$ may not be true, see Problem 7.3 c). The joint cardinality

$$\hat{n}'_J := \sum_{j=0}^{J} \bar{n}_j, \qquad \bar{n}_j = |\bar{\mathcal{V}}_j|, \quad j = 0, \ldots, J,$$

of the sets $\bar{\mathcal{V}}_j$ is equal to the cardinality of $\bar{\Phi}_J$ and the dimension of $\hat{V}_J$. It is shown in Problem 7.3 b) that the cardinality $\hat{n}_J$ of the (typically larger) BPX-frame $\hat{\Phi}_J$ is always comparable to $\hat{n}'_J$, i.e., we have $\hat{n}_J \leq C\hat{n}'_J$ with a constant C, which is independent of the choice of $\tilde{\mathcal{V}}_j$, $j \leq J$.

We will not discuss the assembly of $\hat{A}_J$ and $\hat{b}_J$ and the implementation of the corresponding matrix-vector products. For the discussion of the properties of the preconditioner $\hat{C}_J$ associated with the BPX-frame $\hat{\Phi}_J$ it is only relevant that we have

$$(\hat{A}_J\hat{y}_J, \hat{y}_J) \approx a(\hat{v}_J, \hat{v}_J) \approx \|\hat{v}_J\|^2_{H^s}, \qquad \hat{v}_J = \sum_{j=0}^{J} \sum_{P \in \tilde{\mathcal{V}}_j} y_{P,j} N_{P,j},$$

with constants that are independent of J. Here, $a(\cdot,\cdot)$ is the bilinear form of the underlying $H^s(G)$-elliptic problem and $\hat{y}_J$ is the vector containing the coefficients $y_{P,j}$ of $\hat{v}_J$ with respect to the basis $\hat{\Phi}_J$.

The space splitting corresponding to the BPX-frame $\hat{\Phi}_J$ consists of one-dimensional subspaces $V_{P,j}$ of $\hat{V}_j$ spanned by $N_{P,j}$ with $P \in \tilde{\mathcal{V}}_j$ for some $j \leq J$, and we use $N_{P,j}$ as the basis function in $V_{P,j}$. Thus, the preconditioning matrix

$$\hat{C}_J = \hat{R}_J \hat{D}_J^{-1} \hat{R}_J^T,$$

in the matrix representation

$$\hat{P}_J = \hat{C}_J \hat{A}_J$$

of the Schwarz operator, see Lemma 2.2, has a diagonal matrix $\hat{D}_J$ of size $\hat{n}_J \times \hat{n}_J$, which incorporates the scaling parameters for the nodal basis functions in $\hat{\Phi}_J$. By definition, the $\hat{n}'_J \times \hat{n}_J$ matrix $\hat{R}_J$ contains in each column the coefficients of the expression of $N_{P,k}$ with $\tilde{N}_{P,k} \in \hat{\Phi}_J$ with respect to the basis $\bar{\Phi}_J$ in $\hat{V}_J$ and is responsible for transforming the coefficient vector of a representation of $\hat{v}_J \in \hat{V}_J$ with respect to the BPX-frame $\hat{\Phi}_J$ into its coefficient vector with respect to the basis $\bar{\Phi}_J$. Although $\hat{R}_J$ can be quite dense, the matrix-vector multiplication with $\hat{R}_J$ and its transpose $\hat{R}_J^T$, if organized in the right order, can be done in $\mathrm{O}(\hat{n}'_J) = \mathrm{O}(\hat{n}_J)$ arithmetic operations. In fact, given any vector $\hat{x}_J$ of coefficients with respect to the BPX-frame $\hat{\Phi}_J$, we can decompose it into subvectors $\tilde{x}_j$ associated with the nodal point sets $\tilde{\mathcal{V}}_j$, $j = 0, \ldots, J$. Then, starting with $j = 0$, we store the subvector $\bar{x}_j$ of the current $\tilde{x}_j$ corresponding to the nodal points in $\bar{\mathcal{V}}_j \subset \tilde{\mathcal{V}}_j$ and update $\tilde{x}_{j+1}$ by incorporating the coefficients for the basis functions $N_{P,j}$ with $P \in \tilde{\mathcal{V}}_j \backslash \bar{\mathcal{V}}_j$ using their expressions in $\tilde{V}_{j+1}$. Then we replace j by $j+1$ and repeat. The concatenation of the subvectors $\bar{x}_j$, $j = 0, \ldots, J$, is $\hat{R}_J\hat{x}$, and the whole process takes $\mathrm{O}(\hat{n}_J)$ operations. Obviously, the products $\hat{R}_J^T\hat{x}$ can be computed with similar complexity if the order of execution is reversed.

An alternative representation of matrix-vector multiplication with $\hat{C}_J$, which is closer to the structure of V-cycle multigrid methods, uses the hierarchy of spaces $\hat{V}_j$ equipped with their bases $\bar{\Phi}_j$, $j = 1, \ldots, J$. Recall that $\hat{V}_j \subset \hat{V}_{j+1}$ and that $\bar{\Phi}_{j+1}$ is obtained by replacing nodal basis functions $N_{P,j}$ with $P \in \tilde{\mathcal{V}}_j \backslash \bar{\mathcal{V}}_j$ by the nodal basis in $\tilde{V}_{j+1}$. Consequently, the matrix representation I_j of the natural embedding $\hat{V}_j \subset \hat{V}_{j+1}$ with respect to these bases essentially acts only on the subvector corresponding to $N_{P,j}$ with $P \in \tilde{\mathcal{V}}_j \backslash \bar{\mathcal{V}}_j$. Denote the matrix corresponding to the natural embedding $\tilde{V}_j \subset \hat{V}_j$ by $\tilde{I}_j$. Obviously, the matrix representations corresponding to the operators $R_j : \tilde{V}_j \subset \hat{V}_J$ in the respective bases can then be expressed as

$$R_j = I_{J-1} \cdots I_{j+1} I_j \tilde{I}_j, \qquad j = 0, \ldots, J.$$

As a result, the matrix representation of the preconditioner $\hat{C}_J$ satisfies the recursion

$$\begin{aligned}
\hat{C}_j &= \sum_{k=0}^{j} R_k D_k^{-1} R_k^T = \sum_{k=0}^{j} I_{j-1} \cdots I_k \tilde{I}_k D_k^{-1} \tilde{I}_k^T I_k^T \cdots I_{j-1}^T \\
&= \tilde{I}_j D_j^{-1} \tilde{I}_j^T + I_{j-1} \left(\sum_{k=0}^{j-1} I_{j-2} \cdots I_k \tilde{I}_k D_k^{-1} \tilde{I}_k^T I_k^T \cdots I_{j-2}^T \right) I_{j-1}^T \\
&= \tilde{D}_j^{-1} + I_{j-1} \hat{C}_{j-1} I_{j-1}^T, \qquad j = 1, \ldots, J,
\end{aligned}$$

where $\hat{C}_0 = \tilde{D}_0^{-1}$. Here, $\tilde{D}_j^{-1} := \tilde{I}_j D_j^{-1} \tilde{I}_j^T$ and D_j denotes the diagonal matrices containing the scaling factors $d_{P,j}$ of the nodal basis functions $N_{P,j} \in \tilde{V}_j$, $j = 0, \ldots, J$, which appear in (7.5) and are needed for the frame property of $\hat{\Phi}_J$ to hold. Such recursions are typical for the matrix representations of V-cycle multigrid iterations, see the next subsection.

The solution of the preconditioned system $\hat{C}_J \hat{A}_J \hat{x}_J = \hat{C}_J \hat{b}_J$ by the Richardson iteration corresponds to the additive Schwarz method with the space splitting induced by the BPX-frame. In practice, the preconditioned cg-method with the preconditioner $\hat{C}_J$ is often preferred as the iterative solver for $\hat{A}_J \hat{x}_J = \hat{b}_J$. This avoids the search for appropriate damping parameters for the Richardson iteration to converge and further reduces the influence of the condition number of $\hat{P}_J$.

7.1.1.5 Multiplicative version and multigrid

Another possibility is to explore multiplicative Schwarz iterations by cycling through the splitting induced by the BPX-frame in a level-wise fashion. Here, the order of execution is important, which makes the use of greedy or stochastic orderings less attractive. To emphasize this and to interpret it in the language of multigrid methods, we rewrite the space splitting associated with the BPX-frame $\hat{\Phi}_J$ by equipping the subspaces $\tilde{V}_j$, $j \leq J$, with new bilinear forms $\tilde{a}_j(\cdot, \cdot)$ induced by the correspondingly scaled discrete $L^2(G)$ norms

$$\|\tilde{v}_j\|_j^2 := 2^{2j(s-d/2)} \sum_{P\in\tilde{\mathcal{V}}_j} c_{P,j}^2, \qquad \tilde{v}_j = \sum_{P\in\tilde{\mathcal{V}}_j} c_{P,j} N_{P,j}.$$

Then, due to the $L^2(G)$ stability of the nodal bases and under the conditions of Theorem 7.1, we get the uniform stability of the space splitting $(\tilde{V}_j; \tilde{a}_j)_{j=0,\dots,J}$ for $(\hat{V}_J; (\cdot,\cdot)_{H^s})$. Thus, according to Chapter 4 and in particular Corollary 4.1, for any fixed ordering of the subproblems associated with $(\tilde{V}_j; \tilde{a}_j)$, the multiplicative Schwarz method converges at a rate that depends only slightly on J, if the relaxation parameter ω is chosen properly. In fact, for $s = 1$ one can show strengthened Cauchy-Schwarz inequalities and obtain J-independent error rates using (4.31). Furthermore, if the order of execution of the subproblems is $j = J, J-1, \dots, 0$, the arithmetic complexity of one loop of this multiplicative Schwarz method can be shown to be $O(\hat{n}_J)$, i.e., it is proportional to the dimension of $\hat{V}_J$.

This multiplicative Schwarz method is then equivalent to a $V^{(1,0)}$ multigrid cycle with specific coarse-grid spaces and simple Jacobi-type smoothers. Similarly, the symmetric multiplicative Schwarz method with the order of execution $k = J, J-1, \dots, 0, 0, \dots, J-1, J$ can be interpreted as a $V^{(1,1)}$ multigrid cycle, see also [82, 83]. We only sketch the proof and refer to [21, 187] for full details. In addition to the previous subsection, we need further notation. We denote by

$$\hat{A}_j := I_j^T \cdots I_{J-1}^T \hat{A}_J I_{J-1} \cdots I_J, \qquad j = 0, \dots, J-1,$$

the so-called coarse-grid matrices defined by Galerkin projection from $\hat{A}_J$ and introduce so-called smoothers $\hat{S}_J$ given by

$$\hat{S}_j = I - \omega \tilde{D}_j^{-1} \hat{A}_j, \qquad j = 0, \dots, J.$$

Here and in the following formulas, I always denotes the identity matrix of appropriate size.

The error propagation matrix $\hat{M}_J$ of a $V^{(\nu,\mu)}$ multigrid cycle with these coarse-grid matrices and smoothers for the iterative solution of the system $\hat{A}_J \hat{x}_J = \hat{b}_J$ has the recursive structure

$$\hat{M}_j = \hat{S}_j^{\mu} (I - I_{j-1}(I - \hat{M}_{j-1}) \hat{A}_{j-1}^{-1} I_{j-1}^T \hat{A}_j) \hat{S}_j^{\nu}, \quad j = J, \dots, 1, \tag{7.9}$$

and $\hat{M}_0 = \hat{S}_0^{\mu+\nu}$. This is a partial case of [187, Theorem 2.4.1], where one can also see that only the matrix-vector multiplication with $\hat{A}_j$ but not with $\hat{A}_j^{-1}$ is needed in the actual implementation of the algorithm.

Now consider the case $\mu = 1$, $\nu = 0$ and introduce matrices $\hat{C}_j$ such that $\hat{M}_j = I - \hat{C}_j \hat{A}_j$, $j = 0, \dots, J$ (note that these $\hat{C}_j$ are different from those used in the previous subsection). Then the recursion (7.9) specializes to

$$\begin{aligned}\hat{M}_j &= I - \hat{C}_j \hat{A}_j = (I - I_{j-1} \hat{C}_{j-1} I_{j-1}^T \hat{A}_j)(I - \omega \tilde{D}_j^{-1} \hat{A}_j) \\ &= I - (\omega \tilde{D}_j^{-1} + I_{j-1} \hat{C}_{j-1} I_{j-1}^T - \omega I_{j-1} \hat{C}_{j-1} I_{j-1}^T \hat{A}_j \tilde{D}_j^{-1}) \hat{A}_j\end{aligned}$$

for $k = J, \ldots, 1$. Together with the formula for $\hat{M}_0$, this gives a recursion for $\hat{C}_J$ as

$$\hat{C}_j = \omega\tilde{D}_j^{-1} + I_{j-1}\hat{C}_{j-1}I_{j-1}^T - \omega I_{j-1}\hat{C}_{j-1}I_{j-1}^T\hat{A}_j\tilde{D}_j^{-1} \tag{7.10}$$

for $j = J, \ldots, 1$ and $\hat{C}_0 = \omega\tilde{D}_0^{-1}$ for $j = 0$.

It remains to compare with the error propagation matrix of the multiplicative Schwarz method based on the BPX-frame for $\hat{V}_J$ executed in a levelwise fashion in reverse order $j = J, \ldots, 0$ with the same relaxation parameter ω. Its error iteration is determined by the product matrix

$$(I - \omega R_0 D_0^{-1} R_0^T \hat{A}_J) \cdots (I - \omega R_J D_J^{-1} R_J^T \hat{A}_J),$$

while the residual iteration for

$$r^{(m)} = \hat{A}_J e^{(m)}$$

is governed by

$$(I - \omega\hat{A}_J R_0 D_0^{-1} R_0^T) \cdots (I - \omega\hat{A}_J R_J D_J^{-1} R_J^T).$$

The equivalence of the above $V^{(1,0)}$ multigrid cycle to the multiplicative Schwarz method then follows by establishing the identity

$$(I - \hat{A}_J\hat{C}_J) = (I - \omega\hat{A}_J R_0 D_0^{-1} R_0^T) \cdots (I - \omega\hat{A}_J R_J D_J^{-1} R_J^T). \tag{7.11}$$

To see this, set

$$\begin{aligned} K_j &:= \omega R_j D_j^{-1} R_j^T = \omega I_{J-1} \cdots I_j \tilde{D}_j^{-1} I_j^T \cdots I_{J-1}^T, \\ \hat{K}_j &:= I_{J-1} \cdots I_j \hat{C}_j I_j^T \cdots I_{J-1}^T, \qquad j = 0, \ldots, J, \end{aligned}$$

and check the identity

$$I - \hat{A}_J\hat{K}_j = (I - \hat{A}_J\hat{K}_{j-1})(I - \hat{A}_J K_j), \qquad j = 1, \ldots, J.$$

Since $\hat{K}_J = \hat{C}_J$ and $K_0 = \hat{K}_0$, this gives (7.11).

A final remark concerns the structure of the preconditioner associated with the additive Schwarz method. As shown in the previous subsection, its matrix representation satisfies the recursion

$$\hat{C}_j = \tilde{D}_j^{-1} + I_{j-1}\hat{C}_{j-1}I_{j-1}^T, \qquad \hat{C}_0 = \tilde{D}_0^{-1},$$

which is a concatenated version of (7.10). Up to the additive component $\tilde{D}_j^{-1}$, it also resembles the recursion for a $V^{(0,0)}$ multigrid cycle, since for $\mu = \nu = 0$ the recursion in (7.10) reduces to

$$\hat{C}_j = I_{j-1}\hat{C}_{j-1}I_{j-1}^T.$$

7.1.2 Splittings for $H^s(\Gamma)$ and $H_0^s(G)$

So far, we have relied exclusively on Theorem 3.3, which provides the basis for the analysis of multilevel solvers for finite element discretizations of $H^s(G)$-elliptic problems in polyhedral domains. This does not cover applications to second-order elliptic PDE problems with essential boundary conditions on $\Gamma = \partial G$. In particular, we are interested in space splittings for $H_0^1(G)$. It turns out that stability results for the spaces

$$H_0^s(G) = \{u \in H^s(G) : T_\Gamma u = 0\}, \qquad 1/2 < s < 3/2,$$

involving the subspaces

$$V_{j,0} = \{v_j \in V_j : v_j|_\Gamma = 0\} \subset V_j, \qquad j \geq 0,$$

follow easily from Theorem 3.3. The restriction $s > 1/2$ is natural since there is no meaningful trace operator T_Γ for all functions of $H^s(G)$ with $s \leq 1/2$. Recall that $T_\Gamma u = u|_\Gamma$ is the ordinary trace on Γ if $u \in C(\bar{G})$.

One approach to show this is to realize that the subspaces $V_{j,0}$ coincide with the adaptive finite element spaces $\hat{V}_j$ if we set $\tilde{\mathcal{V}} = \mathcal{V}_j \backslash \Gamma$, $j = 0, 1, \ldots$. In analogy to Theorem 7.1, one can prove a suitable lower stability estimate using the Scott-Zhang quasi-interpolants [162], for which the dual functionals $\lambda_{P,j}$ for nodal points $P \in \Gamma$ are defined by integration over $(d-1)$-dimensional boundary cells $C \cap \Gamma$, $C \in \mathcal{T}_j$. Here the constraint $s > 1/2$ is needed to compensate for the growth of the L^2 norms of the quasi-interpolants Q_j. We leave this to the reader, see Problem 3.5 b) and Problem 7.1 c) in the partial case $p = 1$ and $s = 1$.

We use a slightly different approach, which as a by-product also leads to finite element multiscale splittings for Sobolev spaces $H^{s-1/2}(\Gamma)$ and to associated bounded extension operators E_Γ from $H^{s-1/2}(\Gamma)$ to $H^s(G)$, which have their own applications to inhomogeneous boundary value problems as well as to boundary element and domain decomposition methods. For this purpose we introduce the restrictions $V_{j,\Gamma} = \{T_\Gamma v_j : v_j \in V_j\}$ of the finite element spaces V_j to Γ. Since $\mathcal{T}_j$ partitions the polyhedral domain $G \subset \mathbb{R}^d$, the boundary Γ is partitioned into $(d-1)$-dimensional boundary cells of cells in $\mathcal{T}_j$, and $V_{j,\Gamma}$ are the C^0 Lagrange finite element spaces with respect to the associated nested sequence of partitions

$$\mathcal{T}_{j,\Gamma} := \{C \cap \Gamma : C \in \mathcal{T}_j\}, \qquad j \geq 0,$$

of Γ. The natural nodal basis functions for $V_{j,\Gamma}$ are the restrictions of the nodal basis functions $N_{P,j}$ with $P \in \Gamma$ to Γ. Thus, having the characterization of Sobolev spaces in Theorem 3.3 in mind, we can formally introduce "Sobolev" spaces on Γ by setting

$$\tilde{H}^s(\Gamma) = \{u_\Gamma \in L^2(\Gamma) : \|u_\Gamma\|^2_{\tilde{H}^s(\Gamma)} := \inf_{\substack{v_{j,\Gamma} \in V_{j,\Gamma}: \\ u_\Gamma = \sum_{j \geq 0} v_{j,\Gamma}}} \sum_{j=0}^{\infty} 2^{2js} \|v_{j,\Gamma}\|^2_{L^2(\Gamma)} < \infty\},$$

where $s > 0$. Obviously, using any sequence of uniformly L^2 bounded projections $Q_{j,\Gamma} : L^2(\Gamma) \to V_{j,\Gamma}$, one can easily show that

$$\|u_\Gamma\|^2_{\tilde{H}^s(\Gamma)} \approx \|Q_{0,\Gamma}u_\Gamma\|^2_{L^2(\Gamma)} + \sum_{j=1}^{\infty} 2^{2js}\|(Q_{j,\Gamma} - Q_{j-1,\Gamma})u_\Gamma\|^2_{L^2(\Gamma)}$$

is an equivalent norm on $\tilde{H}^s(\Gamma)$.

Theorem 7.3 *Under the assumptions of Theorem 3.3, the following statements hold for* $1/2 < s < 3/2$*:*
a) We have $H^{s-1/2}(\Gamma) = \tilde{H}^{s-1/2}(\Gamma)$ *with equivalent norms. In particular, this implies that* $(V_{j,\Gamma}; 2^{2j(s-1/2)}(\cdot,\cdot)_{L^2(\Gamma)})_{j\ge 0}$ *is a stable space splitting of* $H^{s-1/2}(\Gamma)$.
b) Given an arbitrary sequence of uniformly $L^2(\Gamma)$ *bounded projections* $Q_{j,\Gamma} : L^2(\Gamma) \to V_{j,\Gamma}$, *the extension operator*

$$E_\Gamma := E_{0,\Gamma}Q_{0,\Gamma} + \sum_{j=1}^{\infty} E_{j,\Gamma}(Q_{j,\Gamma} - Q_{j-1,\Gamma})$$

is a bounded operator from $H^{s-1/2}(\Gamma)$ *into* $H^s(G)$. *Here,* $E_{j,\Gamma} : V_{j,\Gamma} \to V_j$ *is the "minimal" discrete extension operator of level j, given by* $E_{j,\Gamma}(N_{P,j}|_\Gamma) = N_{P,j}$, $P \in \mathcal{V}_j \cap \Gamma$. *Moreover, every* $u \in H^s(G)$ *possesses a unique decomposition* $u = v + E_\Gamma v_\Gamma$, *where* $v_\Gamma = T_\Gamma u \in H^{s-1/2}(\Gamma)$ *and* $v \in H^s_0(G)$.
c) The collection $(V_{j,0}; 2^{2js}(\cdot,\cdot)_{L^2(G)})_{j\ge 0}$ *represents a stable space splitting for* $H^s_0(G)$.

Proof Throughout this proof we assume that $1/2 < s < 3/2$. Elements in $V_{j,\Gamma}$ are denoted by $u_{j,\Gamma}$ or $v_{j,\Gamma}$ while the trace of $v_j \in V_j$ on Γ is denoted by $T_\Gamma v_j$ as before. We start by identifying $\tilde{H}^{s-1/2}(\Gamma)$ as the trace space of $H^s(G)$ up to equivalent norms. Recall that by the definition of the trace norm

$$\|u_\Gamma\|_{H^{s-1/2}(\Gamma)} = \inf_{u\in H^s(G):\, u_\Gamma = T_\Gamma u} \|u\|_{H^s(G)}$$

and the norm equivalence for $H^s(G)$ proved in Theorem 3.3, we have

$$\|u_\Gamma\|^2_{H^{s-1/2}(\Gamma)} \approx \inf_{v_j\in V_j:\, u_\Gamma = \sum_{j\ge 0} T_\Gamma v_j} \sum_{j\ge 0} 2^{2js}\|v_j\|^2_{L^2(G)}$$

for any $u_\Gamma \in H^{s-1/2}(\Gamma)$. By the L^2 stability of the respective nodal bases in V_j and $V_{j,\Gamma}$, and since the interior nodal values of v_j have no influence on the condition $u_\Gamma = \sum_{j\ge 0} T_\Gamma v_j$ under the infimum, we can replace $\|v_j\|^2_{L^2(G)}$ in the right-hand side by

$$2^{-jd} \sum_{P\in\mathcal{V}_j\cap\Gamma} v_j(P)^2 \approx 2^{-j}\|T_\Gamma v_j\|^2_{L^2(\Gamma)}.$$

After substitution, the claim in Part a) follows by changing the notation, while keeping in mind that $V_{j,\Gamma} = T_\Gamma V_j$.

For b), we use the L^2 stability of the corresponding nodal bases in $V_{j,\Gamma}$ and V_j to conclude that

$$\|E_{j,\Gamma}v_{j,\Gamma}\|^2_{L^2(G)} \approx 2^{-j}\|v_{j,\Gamma}\|^2_{L^2(\Gamma)}, \qquad v_{j,\Gamma} \in V_{j,\Gamma}, \quad j \geq 0.$$

Thus, taking an arbitrary decomposition

$$u_\Gamma = \sum_{j\geq 0} v_{j,\Gamma}$$

of $u_\Gamma \in H^{s-1/2}(\Gamma)$, by Theorem 3.3 and the assumed uniform $L^2(\Gamma)$ boundedness of the projections $Q_{j,\Gamma}$, we have

$$\begin{aligned}
\|E_\Gamma u_\Gamma\|^2_{H^s(G)} &\leq C\Bigg(\|E_{0,\Gamma}Q_{0,\Gamma}(\sum_{j\geq 0} v_{j,\Gamma})\|^2_{L^2(G)} \\
&\qquad + \sum_{k=1}^{\infty} 2^{2ks}\|E_{k,\Gamma}(Q_{k,\Gamma}-Q_{k-1,\Gamma})(\sum_{j\geq k} v_{j,\Gamma})\|^2_{L^2(G)}\Bigg) \\
&\leq C\left(\|Q_{0,\Gamma}(\sum_{j\geq 0} v_{j,\Gamma})\|^2_{L^2(\Gamma)} + \sum_{k=1}^{\infty} 2^{2ks-k}\|(Q_{k,\Gamma}-Q_{k-1,\Gamma})(\sum_{j\geq k} v_{j,\Gamma})\|^2_{L^2(\Gamma)}\right) \\
&\leq C\left(\sum_{j\geq 0} 2^{2\epsilon j}\|v_{j,\Gamma}\|^2_{L^2(\Gamma)} + \sum_{k=1}^{\infty} 2^{2ks-k}\cdot 2^{-2k\epsilon}\sum_{j\geq k} 2^{2\epsilon j}\|v_{j,\Gamma}\|^2_{L^2(\Gamma)}\right) \\
&\leq C\left(\sum_{j\geq 0} 2^{2\epsilon j}\|v_{j,\Gamma}\|^2_{L^2(\Gamma)} \sum_{k\leq j} 2^{2k(s-1/2-\epsilon)}\right) \leq C\left(\sum_{j\geq 0} 2^{2j(s-1/2)}\|v_{j,\Gamma}\|^2_{L^2(\Gamma)}\right),
\end{aligned}$$

where we chose ϵ so that $0 < \epsilon < s - 1/2$. After taking the infimum with respect to all representations of u_Γ, Part a) implies the boundedness of E_Γ. The uniqueness of the decomposition $u = v + E_\Gamma v_\Gamma$ into "interior" $v \in H^s_0(G)$ and "boundary" part $v_\Gamma \in H^{s-1/2}(\Gamma)$ follows by applying the trace operator T_Γ to both sides: We have $T_\Gamma u = T_\Gamma v + T_\Gamma E_\Gamma v_\Gamma = v_\Gamma$ and thus $v = u - E_\Gamma T_\Gamma u \in H^s(G)$. Remember that $T_\Gamma E_\Gamma$ is the identity operator on $H^{s-1/2}(\Gamma)$ and $E_\Gamma T_\Gamma$ is bounded on $H^s(G)$.

Finally, in Part c) we only have to establish the lower stability bound of the space splitting under consideration, i.e., for every $u \in H^s_0(G)$ we have to find a sequence $v_{j,0} \in V_{j,0}$, $j \geq 0$, such that

$$u = \sum_{j=0}^{\infty} v_{j,0} \;:\; \sum_{j=0}^{\infty} 2^{2js}\|v_{j,0}\|^2_{L^2(G)} \leq C\|u\|^2_{H^s(G)}. \tag{7.12}$$

By Theorem 3.3 and since $H^s_0(G) \subset H^s(G)$, there is a sequence $v_j \in V_j$, $j \geq 0$, such that

$$u = \sum_{j=0}^{\infty} v_j \ : \quad \sum_{j=0}^{\infty} 2^{2js} \|v_j\|^2_{L^2(G)} \le C \|u\|^2_{H^s(G)}. \tag{7.13}$$

Now set

$$u_j = \sum_{k=0}^{j} v_k, \qquad u_{j,0} = u_j - E_{j,\Gamma}(T_\Gamma u_j), \qquad j \ge 0.$$

By construction we have $u_{j,0} \in V_{j,0}$. It is not hard to see that $u_{j,0}$ converges to u in $H^s(G)$. Indeed,

$$u - u_{j,0} = E_{j,\Gamma}(T_\Gamma u_j) + \sum_{k>j} v_k$$

is an admissible decomposition in Theorem 3.3. This implies

$$\|u - u_{j,0}\|^2_{H^s(G)} \le C \left(2^{2js} \|E_{j,\Gamma}(T_\Gamma u_j)\|^2_{L^2(G)} + \sum_{k>j} 2^{2ks} \|v_k\|^2_{L^2(G)} \right).$$

To bound the term $\|E_{j,\Gamma}(u_j|_\Gamma)\|^2_{L^2(G)}$, we use the fact that $T_\Gamma u = 0$. In fact, this implies

$$T_\Gamma u_j = -T_\Gamma(u - u_j) = -\sum_{k>j} T_\Gamma v_k.$$

Using the properties of the extension operator $E_{j,\Gamma}$ at the discrete level j, we have

$$\begin{aligned} \|E_{j,\Gamma}(T_\Gamma u_j)\|^2_{L^2(G)} &\le C 2^{-j} \|T_\Gamma u_j\|^2_{L^2(\Gamma)} \le C 2^{-j(1+2\epsilon)} \sum_{k>j} 2^{2k\epsilon} \|T_\Gamma v_k\|^2_{L^2(\Gamma)} \\ &\le C 2^{-2j(1/2+\epsilon)} \sum_{k>j} 2^{2k(1/2+\epsilon)} \|v_k\|^2_{L^2(G)}, \end{aligned}$$

where we can take $\epsilon = s - 1/2 > 0$. This gets us

$$\|u - u_{j,0}\|^2_{H^s(G)} \le C \sum_{k>j} 2^{2ks} \|v_k\|^2_{L^2(G)},$$

and our assumption (7.13) gives the desired convergence statement.

To show (7.12), now set $v_{0,0} = u_{0,0} = v_0 - E_{0,\Gamma}(u_0|_\Gamma)$ and

$$v_{j,0} = u_{j,0} - u_{j-1,0} = v_j - E_{j,\Gamma}(T_\Gamma u_j) + E_{j-1,\Gamma}(T_\Gamma u_{j-1}), \qquad j \ge 1.$$

From the above we have

$$u = \sum_{j=0}^{\infty} v_{j,0}$$

with convergence in $H^s(G)$, and by considering the estimate for $\|E_{j,\Gamma} T_\Gamma u_j\|^2_{L^2(G)}$, now with some $0 < \epsilon < s - 1/2$, we also get

$$
\begin{aligned}
&\sum_{j=0}^{\infty} 2^{2js}\|v_{j,0}\|_{L^2(G)}^2 \\
&\quad \le 3\sum_{j=0}^{\infty} 2^{2js}(\|v_j\|_{L^2(G)}^2 + \|E_{j,\Gamma}T_\Gamma u_j\|_{L^2(G)}^2 + \|E_{j-1,\Gamma}T_\Gamma u_{j-1}\|_{L^2(G)}^2) \\
&\quad \le C\left(\sum_{j=0}^{\infty} 2^{2js}\|v_j\|_{L^2(G)}^2 + \sum_{j=0}^{\infty} 2^{2j(s-1/2-\epsilon)} \sum_{k\ge j} 2^{2k(1/2+\epsilon)}\|v_k\|_{L^2(G)}^2\right) \\
&\quad \le C\left(\sum_{j=0}^{\infty} 2^{2js}\|v_j\|_{L^2(G)}^2 + \sum_{k=0}^{\infty} 2^{2k(1/2+\epsilon)}\|v_k\|_{L^2(G)}^2 \sum_{j\le k} 2^{2j(s-1/2-\epsilon)}\right) \\
&\quad \le C\sum_{j=0}^{\infty} 2^{2js}\|v_j\|_{L^2(G)}^2 \le C\|u\|_{H^s(G)}^2.
\end{aligned}
$$

In the last estimation step, (7.12) was used. This concludes the proof of Theorem 7.2. □

Similar to Theorem 7.1, the statements of Theorem 7.2 can also be extended to nested local refinement and to the situation where zero traces are prescribed only on a part of Γ consisting of cells from the coarsest boundary partition $\mathcal{T}_{0,\Gamma}$. The extension operator E_Γ can be used to deal with inhomogeneous essential boundary conditions. For example, the Galerkin approximation $u_J \in V_J$ to the solution $u \in H^1(\Omega)$ of the Dirichlet problem of the diffusion equation (7.1) with boundary condition

$$T_\Gamma u = g \in H^{1/2}(\Gamma)$$

can be reduced to solving the variational problem

$$a(u_{J,0}, v_{J,0}) = \langle f, v_{J,0}\rangle_{H^{-1}\times H_0^1} - a(\tilde{E}_{J,\Gamma}g, v_{J,0}) \qquad \forall\, v_{J,0} \in V_{J,0}$$

on $V_{J,0}$ and setting $u_J = u_{J,0} + \tilde{E}_{J,\Gamma}g$ afterwards. Here, $a(\cdot,\cdot)$ is the H_0^1 elliptic bilinear form

$$a(u,v) := \int_G w(x)\nabla u(x)\cdot\nabla v(x)\,dx, \qquad u,v \in H_0^1(G),$$

generated by (7.1) and

$$\tilde{E}_{J,\Gamma} := \sum_{j=0}^{J} E_{j,\Gamma}(Q_{j,\Gamma} - Q_{j-1,\Gamma}), \qquad J = 0,1,\dots,$$

are the finite sections of E_Γ. Note that unlike the "minimal" discrete level J extension operators $E_{J,\Gamma}$, the operators $\tilde{E}_{J,\Gamma}$ are uniformly bounded operators from $H^{s-1/2}(\Gamma)$ to $V_j \subset H^s(G)$. When restricted to $V_{J,\Gamma}$, they have the usual properties of an extension operator. In particular, $T_\Gamma\tilde{E}_{J,\Gamma}$ is the identity on $V_{J,\Gamma}$.

7.1.3 Non-nested spaces: An example

In Subsection 3.2.2.2 we considered, as an example of the auxiliary space method, a stable space splitting for non-conforming linear finite element spaces $\hat{V}_{J,0}$, which allowed us to switch from discretizations of $H_0^1(G)$ elliptic problems with respect to $\hat{V}_{J,0}$ to discretizations with respect to the conforming linear finite element space $V_{J,0}$ on the same simplicial partition $\mathcal{T}_J$. In principle, together with other switching tricks explained in Subsection 3.2.2 and with the use of available efficient solvers for conforming finite element discretizations discussed earlier, this makes the investigation of multiscale splittings that directly use the sequence of non-nested non-conforming finite element spaces $\hat{V}_{j,0} := \hat{V}(\mathcal{T}_j)_0$, $j \le J$, less of a priority. However, since dealing with non-nested spaces and non-inherited bilinear forms in a multilevel setting is a topic of recurring interest and the subject has some history in the literature, we present an introductory treatment using this example.

Recall the setting and notation of Subsection 3.2.2.2. To create multiscale splittings for $(\hat{V}_{J,0}; \hat{a}_J)$, we now need to define operators $R_j : \hat{V}_{j,0} \to \hat{V}_{J,0}$ that differ from the natural embeddings. Given the need for efficient implementations, it is canonical to define them recursively by

$$R_j = R_{j+1} I_j = I_{J-1} \cdots I_j, \qquad j = 0, 1, \ldots, J-1,$$

whereas R_J is the identity operator on $\hat{V}_{J,0}$. The operators $I_j : \hat{V}_{j,0} \to \hat{V}_{j+1,0}$ are called *intergrid transfer operators* and must be chosen carefully. For efficiency, their matrix representations should be sparse and easy to implement. Moreover, in analogy to the conforming case, and since the nodal basis for $\hat{V}_j$ is L^2 stable, see (3.51), the family $(\hat{V}_j; 2^{2j}(\cdot,\cdot)_{L^2}; R_j)_{j=0,\ldots,J}$ is a plausible candidate for a good splitting with cheap solvers for the subproblems in the resulting Schwarz methods. It remains to verify the stability properties of such splittings depending on the choice of the intergrid transfer operators I_j.

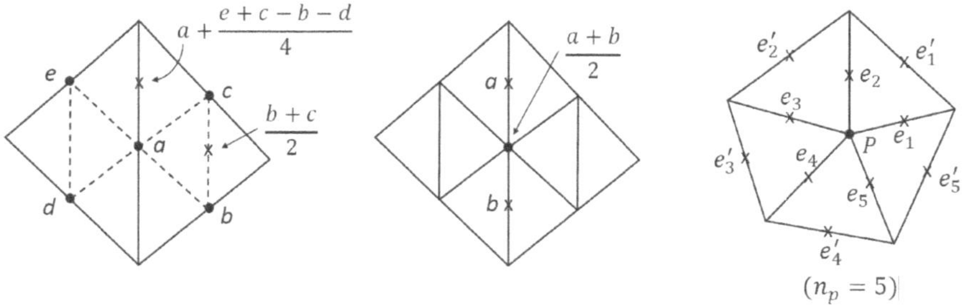

Fig. 7.4 Illustrations of the local rules defining intergrid transfer operators for non-conforming linear finite elements on triangles. Left: Natural $I_j : V_{j,0} \to V_{j+1,0}$, Center: Ortho-projections $Q_j : V_{j+1,0} \to V_{j,0}$, Right: Setting for the definition of the new I_j used in Theorem 7.4.

Historically, some natural candidates for I_j were examined first. Figure 7.4 shows on the left the local rules for defining the nodal values of $I_j v_j$ on $\mathcal{T}_{j+1}$ from the nodal values of v_j on $\mathcal{T}_j$ by edge averaging (this is natural since the values of the non-conforming linear finite element functions at each edge midpoint M_e are equal to the average along e of the linear functions in the adjacent triangles). For this particular set of intergrid transfer operators I_j, the uniform stability of the space splitting $(\hat{V}_j; 2^{2j}(\cdot,\cdot)_{L^2}; R_j)_{j=0,\dots,J}$ for $(\hat{V}_{J,0}; \hat{a}_J)$ was shown in [139], weaker $\mathrm{O}(J)$ condition number estimates and references to earlier work on the subject can be found in [133, 137]. They are based on a lower stability bound

$$|||\hat{u}_J|||^2 := \inf_{\hat{v}_j \in \hat{V}_{j,0}:\, \hat{u}_J = \sum_{j\le J} R_j \hat{v}_j} \sum_{j\le J} 2^{2j} \|\hat{v}_j\|_{L^2}^2 \le C \hat{a}_J(\hat{u}_J, \hat{u}_J),$$

which holds for all $\hat{u}_J \in \hat{V}_{J,0}$ with a constant C independent of J, see [133]. The proof uses the special decomposition

$$\hat{u}_J = R_0 S_0 \tilde{u}_J + \sum_{j=1}^{J} R_j (I - I_{j-1} Q_{j-1}) S_j \tilde{u}_J,$$

where I is the identity operator on the respective space, $Q_j : \hat{V}_{j+1,0} \to \hat{V}_{j,0}$ are the ortho-projections from $(\hat{V}_{j+1,0}; \hat{a}_{j+1})$ to $(\hat{V}_{j,0}; \hat{a}_j)$, $j = 0, \dots, J-1$, and $S_j = Q_j \cdots Q_{J-1}$ denotes their iterates (set $S_J = I$ for completeness). It turns out that the ortho-projections Q_j have a simple explicit expression in terms of local averages of nodal values along edges: The nodal value $(Q_j \hat{v}_{j+1})(M_e)$ for any edge e in $\mathcal{T}_j$ is equal to the average of the two nodal values of $\hat{v}_{j+1}$ associated with e, as shown in the center of Figure 7.4. This explicit form of Q_j demonstrates that $Q_j I_j$ is the identity on $\hat{V}_{j,0}$, which can be used to obtain estimates for the L^2 norm of

$$\hat{v}_j = (I - I_{j-1} Q_{j-1}) S_j \tilde{u}_J = S_j \tilde{u}_J - S_{j-1} \tilde{u}_J + S_{j-1} \tilde{u}_J - I_{j-1} S_{j-1} \tilde{u}_J$$

via the difference

$$\hat{a}_j(S_j \tilde{u}_J, S_j \tilde{u}_J) - \hat{a}_{j-1}(S_{j-1} \tilde{u}_J, S_{j-1} \tilde{u}_J) = \hat{a}_j(S_j \tilde{u}_J - Q_j S_j \tilde{u}_J, S_j \tilde{u}_J - Q_j S_j \tilde{u}_J)$$

using the ortho-projection property of the operators Q_j.

To prove the upper bound, one needs suitable Bernstein-type inequalities. For example, in [133, 138] such a Bernstein-type inequality involving the broken H_0^1 norm was given in the form

$$\hat{a}_J(R_j \hat{v}_j, R_j \hat{v}_j) \le C 2^{2j} \|\hat{v}_j\|_{L^2}^2, \qquad \hat{v}_j \in \hat{V}_{j,0}, \quad j = 0, \dots, J,$$

with a constant C independent of j and J, which led to the $\mathrm{O}(J)$ condition number estimate mentioned above. To establish the uniform boundedness of the condition numbers, an improved inverse inequality involving H^s norms with $s > 1$ was proved in [139] using methods from subdivision theory concerning the smoothness of refinable functions. These results are very technical and beyond the scope of this book.

Note that multigrid theory [17] with non-nested spaces and non-inherited bilinear forms (that is, coarse grid problems not obtained by Galerkin projection) gives optimal convergence results only for sophisticated V- and W-cycles, but leaves open the case of V-cycles with a few smoothing steps comparable to our additive and multiplicative Schwarz methods. Other families of intergrid transfer operators and other types of non-conforming finite elements are discussed in [138, 139], but the results are often incomplete.

We end with another proposal for I_j, which is less attractive from a computational point of view but provides optimal condition number estimates with little effort, by referring to Theorem 7.2. For given $\hat{v}_j \in \hat{V}_{j,0}$, we define

$$v_{j+1} := I_j \hat{v}_j \in V_{j+1,0} \subset \hat{V}_{j+1}$$

by

$$v_{j+1}(M_e) = \hat{v}_j(M_e) \tag{7.14}$$

for all edges e in $\mathcal{T}_j$ and by

$$v_{j+1}(P) = \begin{cases} 0, & P \in \mathcal{V}_j \cap \Gamma, \\ \frac{1}{n_P} \sum_{i=1}^{n_P} (2\hat{v}_j(M_{e_i}) - \hat{v}_j(M_{e_i'})), & P \in \mathcal{V}_j \backslash \Gamma, \end{cases} \tag{7.15}$$

where n_P is the valence of P, i.e., the number of triangles attached to P. The notation used in (7.15) is shown on the right in Figure 7.4. For the values $v_{j+1}(P)$ at the vertices $P \in \mathcal{V}_j$ interior to G, the formula simply says that $v_{j+1}(P)$ is equal to the average of the limit values at P of the restrictions of $\hat{v}_j$ to the triangles attached to P. This operator I_j, which was already mentioned in Subsection 3.2.2.2, has thus the important property that its restriction to $V_{j,0}$ is the natural embedding of $V_{j,0}$ in $V_{j+1,0}$. Consequently, any representation

$$\hat{u}_J = \sum_{j=0}^{J} R_j \hat{v}_j, \qquad \hat{v}_j \in \hat{V}_{j,0}, \quad j = 0, \dots, J, \tag{7.16}$$

can be rewritten as

$$\hat{u}_J = \hat{v}_J + \sum_{j=0}^{J-1} v_{j+1}, \qquad v_{j+1} = I_j \hat{v}_j \in V_{j+1,0}, \quad j = 0, \dots, J-1. \tag{7.17}$$

In other words, the representations (7.16), which appear in the $|||\cdot|||$ norm definition associated with the space splitting $(\hat{V}_{j,0}; 2^{2j}(\cdot,\cdot)_{L^2}; R_j)_{j=0,\dots,J}$, are in large part representations for the space splitting involving the conforming subspaces $V_{j,0}$. This observation is key to the proof of the following theorem.

Theorem 7.4 *With the set of intergrid transfer operators I_j defined by (7.14) and (7.15), the collection $(\hat{V}_{j,0}; 2^{2j}(\cdot,\cdot)_{L^2}; R_j)_{j=0,\dots,J}$ provides a stable space splitting for $(\hat{V}_J; \hat{a}_J)$ with constants and condition number estimates that are independent of J.*

Proof We start with the upper stability estimate. Consider an arbitrary representation (7.16). Set

$$u_J := \hat{u}_J - \hat{v}_J = \sum_{j=0}^{J-1} v_{j+1} \in V_{J,0}.$$

According to (7.17) and Theorem 7.2, we have

$$\hat{a}_J(u_J, u_J) = \|u_J\|_{H_0^1}^2 \leq C \sum_{j=0}^{J-1} 2^{2(j+1)} \|v_{j+1}\|_{L^2}^2 \leq C \sum_{j=0}^{J-1} 2^{2j} \|\hat{v}_j\|_{L^2}^2.$$

Here the inequality

$$\|v_{j+1}\|_{L^2}^2 = \|I_j \hat{v}_j\|_{L^2}^2 \leq C \|\hat{v}_j\|_{L^2}^2 \tag{7.18}$$

was used. We also have the Bernstein-type inequality

$$\hat{a}_J(\hat{v}_J, \hat{v}_J) \leq C 2^{2J} \|\hat{v}_J\|_{L^2}^2. \tag{7.19}$$

The verification of (7.18) and (7.19) is elementary and left to the reader as Problem 7.4. Putting it all together, we have

$$\hat{a}_J(\hat{u}_J, \hat{u}_J) \leq 2(\hat{a}_J(u_J, u_J) + \hat{a}_J(\hat{v}_J, \hat{v}_J)) \leq C \sum_{j=0}^{J} 2^{2j} \|\hat{v}_j\|_{L^2}^2.$$

By taking the infimum with respect to all possible representations (7.16) of $\hat{u}_J$, we get the upper bound

$$\hat{a}_J(\hat{u}_J, \hat{u}_J) \leq C \||\hat{u}_J\||^2.$$

To prove the lower stability bound, we define $u_{J-1} \in V_{J-1,0}$ by applying a suitable quasi-interpolant construction to $\hat{u}_J$. For example, set $u_{J-1}(P) = 0$ if $P \in \mathcal{V}_{J-1} \cap \Gamma$ and

$$u_{J-1}(P) = \frac{1}{n_P} \sum_{e:\, P \in e} \hat{u}(M_e), \qquad P \in \mathcal{V}_{J-1} \backslash \Gamma.$$

Thus, the values of u_{J-1} at interior nodal points $P \in \mathcal{V}_{J-1}$ are defined as averages of the midpoint values of $\hat{u}_J$ from edges e in $\mathcal{T}_J$ incident to P. Direct inspection shows

$$\hat{a}_J(u_{J-1}, u_{J-1}) \leq C \hat{a}_J(\hat{u}_J, \hat{u}_J) \tag{7.20}$$

and $\hat{w}_J := \hat{u}_J - u_{J-1} \in \hat{V}_{J,0}$ also satisfies

$$2^{2J} \|\hat{w}_J\|_{L^2}^2 \leq C \hat{a}_J(\hat{u}_J, \hat{u}_J). \tag{7.21}$$

The verification of these inequalities is left to the reader as Problem 7.5.

According to Theorem 7.3, the function $u_{J-1} \in V_{J-1,0} \subset \hat{V}_J$ admits a decomposition

$$u_{J-1} = \sum_{j=0}^{J-1} v_j, \qquad v_j \in V_{j,0} \subset \hat{V}_{j,0}, \tag{7.22}$$

such that

$$\sum_{j=0}^{J-1} 2^{2j}\|v_j\|_{L^2}^2 \le C\|u_{J-1}\|_{H^1}^2 \le C\hat{a}_J(u_{J-1},u_{J-1}) \le \hat{a}_J(u_J,u_J),$$

where the last estimation step follows from (7.20). Since by the construction of I_j we have $R_j v_j = v_j$, we conclude from the definition of $\hat{w}_J$, u_{J-1}, (7.22), and (7.21) that the decomposition

$$\hat{u}_J = \sum_{j=0}^{J-1} R_j v_j + \hat{w}_J$$

satisfies

$$|||\hat{u}_J|||^2 \le \sum_{j=0}^{J-1} 2^{2j}\|v_j\|_{L^2}^2 + 2^{2J}\|\hat{w}_J\|_{L^2}^2 \le C\hat{a}(\hat{u}_J,\hat{u}_J).$$

This shows the lower stability estimate and concludes the proof of Theorem 7.4. □

The implementation of the matrix-vector products associated with the intergrid transfer operators I_j used in Theorem 7.4 is more complex than for the traditional I_j mentioned above. Cost savings are achieved by replacing (7.15) with the simplified formula

$$v_{j+1}(P) = \begin{cases} 0, & P \in \mathcal{V}_j \cap \Gamma, \\ \hat{v}(M_{e_i}) + \hat{v}(M_{e_{i+1}}) - \hat{v}(M_{e_i'}), & P \in \mathcal{V}_j \backslash \Gamma, \end{cases}$$

where the index $i = i_P$ is arbitrarily chosen from the set $\{1,\dots,n_P\}$, compare the right side of Figure 7.4 for the notation. Obviously, the statement of Theorem 7.4 also holds for such I_j. The idea behind Theorem 7.4 can be adapted to other families of non-nested finite element spaces, see Problem 7.6 for an example.

7.1.4 Adaptive wavelet methods

In this subsection we refer to the material in Section 3.1 on multilevel Riesz splittings for Sobolev spaces obtained by wavelet constructions. In contrast to the BPX-type frame decompositions considered so far, which are often used for finite element discretizations, the Riesz basis property and the higher-order moment conditions available for wavelet systems offer additional advantages. On the one hand, the operator equation, when reformulated in wavelet coordinates, transforms into a linear system $\mathbf{A}\mathbf{x} = \mathbf{b}$ in $\ell^2(\Omega)$, where the condition number $\kappa(\mathbf{A}) := \|\mathbf{A}\|_{\ell^2}\|\mathbf{A}^{-1}\|_{\ell^2}$ is of moderate size and Ω denotes the index set of the wavelet system. On the other hand, under suitable moment and smoothness conditions on the wavelet system, this $\mathbf{A}$ can be approximated by finite-dimensional $\mathbf{A}_j$ in a controlled way with accuracies $\epsilon_j \to 0$, i.e., $\|\mathbf{A} - \mathbf{A}_j\|_{\ell^2} \le \epsilon_j$. This can be used to construct a true iterative scheme for solving the infinite linear system by accessing only finitely supported vectors in $\ell^2(\Omega)$. In particular, it opens the way to the design and analysis of adaptive discretization methods with optimal convergence properties for a whole class of operator equations in Sobolev spaces, including PDEs and boundary integral equations.

The analysis of adaptive wavelet methods, which began with the papers [45, 46], also had a major impact on the development of a similar rigorous theory of adaptive finite element methods, see [129, 130] for expositions. Our brief introduction to adaptive wavelet methods is based on the review article [173], we also refer to [42, 188].

To set up the notation, assume that the bilinear form a generated by $A : V \to V'$ is symmetric and V-elliptic, where V is an infinite-dimensional separable Hilbert space. In other words, $(V; a)$ is a Hilbert space and $a(v, v) \approx \|v\|_V^2$ for all $v \in V$. This is the setting of [173, Section 4-6] (for the discussion of the more general situation $B : V \to U'$, which is relevant for applications to PDE systems and space-time discretizations of parabolic equations, see [173]). Let

$$\Psi = \{\psi_\lambda : \lambda \in \Omega\}$$

be the wavelet system under consideration. We assume that Ψ is a Riesz basis for V, i.e., $\mathrm{span}(\Psi)$ is dense in V and

$$\alpha\|\mathbf{c}\|_{\ell^2}^2 \le \|\sum_{\lambda\in\Omega} c_\lambda\psi_\lambda\|_V^2 \le \beta\|\mathbf{c}\|_{\ell^2}^2$$

for all finitely supported $\mathbf{c} \in \ell^2(\Omega)$. Here $0 < \alpha \le \beta < \infty$ denote the lower and upper stability constants of the underlying Riesz splitting, and $\kappa_\Psi := \alpha/\beta$ is its condition number. Then the operator equation $Au = f$ with given right-hand side $f \in V'$ and unique solution $u \in V$ can be transformed into the linear system

$$\mathbf{Ax} = \mathbf{b} \tag{7.23}$$

in $\ell^2(\Omega)$, where $\mathbf{x} \in \ell^2(\Omega)$ denotes the vector of the unknown wavelet coefficients of u, i.e.,

$$u = \sum_{\lambda\in\Lambda} x_\lambda\psi_\lambda,$$

while the entries of the matrix $\mathbf{A}$ and the right-hand side $\mathbf{b}$ are given by

$$\mathbf{A}_{\lambda',\lambda} = \langle A\psi_{\lambda'}, \psi_\lambda\rangle_{V'\times V}, \qquad b_\lambda = \langle f, \psi_\lambda\rangle_{V'\times V}, \qquad \lambda', \lambda \in \Omega.$$

Obviously, $\mathbf{A}$ is symmetric, positive definite and invertible as operator acting on $\ell^2(\Omega)$. Since

$$(\mathbf{Ax}, \mathbf{x})_{\ell^2} = \sum_{\lambda'}\left(\sum_{\lambda}\langle A\psi_{\lambda'}, \psi_\lambda\rangle_{V'\times V}\, x_\lambda\right)x_{\lambda'} = \langle Au, u\rangle_{V'\times V},$$

the spectrum of $\mathbf{A}$ is contained in the interval

$$\sigma(\mathbf{A}) = [\|\mathbf{A}^{-1}\|_{\ell^2}^{-1}, \|\mathbf{A}\|_{\ell^2}] \subset [\alpha\|A^{-1}\|_{V'\to V}^{-1}, \beta\|A\|_{V\to V'}].$$

In other words, if $\kappa(A) := \|A\|_{V\to V'}\|A^{-1}\|_{V'\to V}^{-1}$ and κ_Ψ are of moderate size then $\kappa(\mathbf{A}) \le \kappa(A)\kappa_\Psi$ is also of moderate size.

The Riesz basis property of Ψ with respect to V has two important consequences: First, for any subset $\Lambda \subset \Omega$, the coefficient vector $\mathbf{x}_\Lambda \in \ell^2(\Lambda)$ of the Galerkin solution u_Λ of the equation $Au = f$ with respect to the subspace $V_\Lambda \subset V$ spanned by the wavelets ψ_λ with $\lambda \in \Lambda$ solves a linear system

$$\mathbf{A}_\Lambda \mathbf{x}_\Lambda = \mathbf{b}_\Lambda, \qquad \kappa(\mathbf{A}_\Lambda) \leq \kappa(\mathbf{A}) \tag{7.24}$$

from (7.23) with submatrix $\mathbf{A}_\Lambda$ of $\mathbf{A}$ and subvector $\mathbf{b}_\Lambda$ of $\mathbf{b}$. We also have

$$\|\mathbf{A}^{-1}\|_{\ell^2}^{-1} \leq \|\mathbf{A}_\Lambda^{-1}\|_{\ell^2}^{-1} \leq \|\mathbf{A}_\Lambda\|_{\ell^2} \leq \|\mathbf{A}\|_{\ell^2}, \tag{7.25}$$

independent of Λ. Thus, if standard iterative methods are used to solve (7.24), then the error reduction per iteration step is guaranteed by a constant factor < 1, independent of Λ. Here and in the following, ℓ^2 norms and scalar products are taken with respect to the underlying index sets Λ or Ω as needed, but we do not show this in every formula. Also, linear combinations of vectors and matrix-vector products involving different support sets $\Lambda, \Lambda' \subset \Omega$ are always interpreted as if the vectors were expanded by zeros to $\Lambda \cup \Lambda'$.

Second, the norms associated with $v \in V$ can be estimated by the vector $\mathbf{Ay}$, where $\mathbf{y}$ is the coefficient vector of v. For example, using the energy norm $\|\cdot\|_\mathbf{A}$ induced by $\mathbf{A}$ on $\ell^2(\Omega)$, we have

$$\|\mathbf{A}\|_{\ell^2}^{-1}\|\mathbf{Ay}\|_{\ell^2}^2 \leq \|\mathbf{y}\|_\mathbf{A}^2 := (\mathbf{Ay}, \mathbf{y})_{\ell^2} \leq \|\mathbf{A}^{-1}\|_{\ell^2}\|\mathbf{Ay}\|_{\ell^2}^2 \tag{7.26}$$

for each $\mathbf{y} \in \ell^2(\Omega)$. Furthermore, if $\Lambda' \subset \Lambda \subset \Omega$, then the energy norm error of any vector $\mathbf{y}_{\Lambda'} \in \ell^2(\Lambda')$ with respect to the solution $\mathbf{x}_\Lambda$ of the Galerkin system (7.24) can be estimated from below by the $\ell^2(\Lambda)$ norm of the residual $\mathbf{b}_\Lambda - \mathbf{A}_\Lambda \mathbf{y}_{\Lambda'}$. Indeed,

$$\|\mathbf{x}_\Lambda - \mathbf{y}_{\Lambda'}\|_\mathbf{A}^2 \geq \|\mathbf{A}\|_{\ell^2}^{-1}\|\mathbf{A}_\Lambda(\mathbf{x}_\Lambda - \mathbf{y}_{\Lambda'})\|_{\ell^2}^2 = \|\mathbf{A}\|_{\ell^2}^{-1}\|\mathbf{b}_\Lambda - \mathbf{A}_\Lambda \mathbf{y}_{\Lambda'}\|_{\ell^2}^2. \tag{7.27}$$

Since we have

$$\|\mathbf{x} - \mathbf{y}_{\Lambda'}\|_\mathbf{A}^2 = \|\mathbf{x} - \mathbf{x}_\Lambda\|_\mathbf{A}^2 + \|\mathbf{x}_\Lambda - \mathbf{y}_{\Lambda'}\|_\mathbf{A}^2 \geq \|\mathbf{x}_\Lambda - \mathbf{y}_{\Lambda'}\|_\mathbf{A}^2 \tag{7.28}$$

by Galerkin orthogonality, we can also use it to obtain residual based estimates for the error with respect to the solution $\mathbf{x}$ of (7.23). Since residual information is more readily available during the execution of an approximation method, this allows for effective error control. Both (7.27) and (7.28) will be instrumental in the convergence proof for the adaptive wavelet method described next.

Roughly speaking, the adaptive wavelet method aims to solve (7.24) for an approximation $\mathbf{x}_\epsilon$ with error $\|\mathbf{x} - \mathbf{x}_\epsilon\|_{\ell^2} \leq \epsilon$ such that $\mathbf{x}_\epsilon$ contains as few non-zero entries as possible and can be computed with a total number of elementary operations proportional to the support size of $\mathbf{x}_\epsilon$. In a sense, it is a special version of a greedy Schwarz approximation method as discussed in Section 6.4, with additional guarantees regarding the required computational work. We start with an abstract version called the exact adaptive wavelet Galerkin method (exact-AWGM) in [173], which ignores the work estimates, before elaborating on its economic realization.

7.1.4.1 Abstract adaptive wavelet method

The optimality criterion specifying what is meant by "as few as possible non-zero entries" is motivated by concepts from nonlinear approximation [64], as briefly mentioned at the end of Section 6.4. For our situation of a Riesz splitting induced by a wavelet system Ψ in V, the definition (6.32) is

$$\sigma_N(u) = \inf_{\Lambda\subset\Omega:\ |\Lambda|\le N}\ \inf_{v_\Lambda\in V_\Lambda} \|u - v_\Lambda\|_V, \qquad N = 1, 2, \ldots.$$

To avoid switching between the V and $\ell^2(\Omega)$ norms, from now on we will use the analogous definition in terms of wavelet coordinates. With $\mathbf{x} \in \ell^2(\Omega)$ being the wavelet coordinate vector associated with $u \in V$, we set

$$\sigma_N(\mathbf{x}) = \inf_{\Lambda\subset\Omega:\ |\Lambda|\le N}\ \inf_{\mathbf{y}_\Lambda\in\ell^2(\Lambda)} \|\mathbf{x} - \mathbf{y}_\Lambda\|_{\ell^2} = (\sum_{k=N+1}^{\infty} (x_k^*)^2)^{1/2}, \qquad N \in \mathbb{N},$$

where x_k^*, $k \in \mathbb{N}$, denotes the entries of the non-increasing rearrangement of the sequence $(|x_\lambda|)_{\lambda\in\Omega}$. In other words, when working with a Riesz basis, best N-term approximation of u means finding its N largest coefficients x_λ in absolute value. To prove meaningful statements for realistic algorithms, we introduce approximation classes $\mathcal{A}^s$, $s > 0$, in terms of the decay of the best N-term approximation as follows: We say that $\mathbf{x} \in \mathcal{A}^s$ if

$$\|\mathbf{x}\|_{\mathcal{A}^s} := \sup_{N\in\mathbb{N}} (N+1)^s \sigma_N(\mathbf{x}) < \infty. \tag{7.29}$$

These classes are slightly different from the classes $\mathcal{A}^p$, $0 < p < 2$, defined in Section 6.4 (note that elements from such an $\mathcal{A}^p$ would satisfy (7.29) with $s = 1/p - 1/2$).

If $\mathbf{x} \in \mathcal{A}^s$ solves the equation (7.23), then quasi-optimality of an adaptive wavelet method [173] means that the method finds, for given $\epsilon > 0$, a finite subset $\Lambda_\epsilon \subset \Omega$ and an approximation $\mathbf{x}_\epsilon \in \ell^2(\Lambda_\epsilon)$ with

$$\|\mathbf{x} - \mathbf{x}_\epsilon\|_{\ell^2} \le \epsilon, \qquad |\Lambda_\epsilon| \le C_1 \epsilon^{-1/s} \|\mathbf{x}\|_{\mathcal{A}^s}^{1/s}. \tag{7.30}$$

Also, the number of operations to find Λ_ϵ and $\mathbf{x}_\epsilon$ should be bounded by $C_2|\Lambda_\epsilon|$ and the constants C_1 and C_2 should be independent of $\mathbf{x}$. Obviously, replacing $\|\mathbf{x} - \mathbf{x}_\epsilon\|_{\ell^2}$ with other norms such as $\|\mathbf{x} - \mathbf{x}_\epsilon\|_{\mathbf{A}}$ or $\|u - u_\epsilon\|_V$ will not change the quasi-optimality, it will only affect the values of C_1 and C_2. It is easy to see from (7.30) and the definition (7.29) that, because of

$$|\Lambda_\epsilon|^s \sigma_{|\Lambda_\epsilon|}(\mathbf{x}) \le |\Lambda_\epsilon|^s \|\mathbf{x} - \mathbf{x}_\epsilon\|_{\ell^2} \le C_1^s \|\mathbf{x}\|_{\mathcal{A}^s} = C_1^s \sup_{N\ge 1} (N+1)^s \sigma_N(\mathbf{x}),$$

one should not expect values $C_1 < 1$. Obviously, $C_2 \geq 1$ should also be assumed. The next theorem (see [173, Proposition 4]) shows that, neglecting the required bound on the number of operations, there exists an idealized iterative scheme satisfying the condition (7.30).

Theorem 7.5 *Consider the system (7.23) with solution* $\mathbf{x}$. *Set* $\Lambda_0 = \emptyset$ *and* $\mathbf{x}^{(0)} = \mathbf{0}$. *Given a fixed parameter* $0 < \mu < \kappa(\mathbf{A})^{-1}$, *choose the* smallest *set* $\Lambda_{m+1} \supset \Lambda_m$ *for* $m = 0, 1, \dots$, *so that*

$$\|\mathbf{b}_{\Lambda_{m+1}} - \mathbf{A}_{\Lambda_{m+1}}\mathbf{x}^{(m)}\|_{\ell^2}^2 \geq \mu\|\mathbf{b} - \mathbf{A}\mathbf{x}^{(m)}\|_{\ell^2}^2. \tag{7.31}$$

Next solve (7.24), where $\Lambda = \Lambda_{m+1}$, *and denote the solution by* $\mathbf{x}^{(m+1)}$.
Then we have

$$\|\mathbf{x} - \mathbf{x}^{(m)}\|_{\mathbf{A}}^2 \leq (1 - \mu/\kappa(\mathbf{A}))^m \|\mathbf{x}\|_{\mathbf{A}}^2, \qquad m = 0, 1, \dots, \tag{7.32}$$

and if $\mathbf{x} \in \mathcal{A}^s$, *we get*

$$|\Lambda_m| \leq C\|\mathbf{x} - \mathbf{x}^{(m-1)}\|_{\mathbf{A}}^{-1/s} \|\mathbf{x}\|_{\mathcal{A}^s}^{1/s}, \qquad m = 1, 2, \dots, \tag{7.33}$$

where C *is a constant independent of* $\mathbf{x}$.

Proof To show (7.32), we set $\Lambda' = \Lambda_m \subset \Lambda = \Lambda_{m+1}$ and $\mathbf{x}_\Lambda = \mathbf{x}^{(m+1)}$, $\mathbf{y}_{\Lambda'} = \mathbf{x}^{(m)}$ in (7.27) and additionally use (7.28). This gives

$$\begin{aligned}
\|\mathbf{x} - \mathbf{x}^{(m+1)}\|_{\mathbf{A}}^2 &= \|\mathbf{x} - \mathbf{x}^{(m)}\|_{\mathbf{A}}^2 - \|\mathbf{x}^{(m+1)} - \mathbf{x}^{(m)}\|_{\mathbf{A}}^2 \\
&\leq \|\mathbf{x} - \mathbf{x}^{(m)}\|_{\mathbf{A}}^2 - \|\mathbf{A}\|_{\ell^2}^{-1} \|\mathbf{b}_{\Lambda_{m+1}} - \mathbf{A}_{\Lambda_{m+1}}\mathbf{x}^{(m)}\|_{\ell^2}^2 \\
&\leq \|\mathbf{x} - \mathbf{x}^{(m)}\|_{\mathbf{A}}^2 - \mu\|\mathbf{A}\|_{\ell^2}^{-1} \|\mathbf{A}(\mathbf{x} - \mathbf{x}^{(m)})\|_{\ell^2}^2 \\
&\leq \|\mathbf{x} - \mathbf{x}^{(m)}\|_{\mathbf{A}}^2 - \mu\|\mathbf{A}\|_{\ell^2}^{-1} \|\mathbf{A}^{-1}\|_{\ell^2}^{-1} \|\mathbf{x} - \mathbf{x}^{(m)}\|_{\mathbf{A}}^2,
\end{aligned}$$

where $m = 0, 1, \dots$. As a result we have

$$\|\mathbf{x} - \mathbf{x}^{(m+1)}\|_{\mathbf{A}}^2 \leq (1 - \mu/\kappa(\mathbf{A}))\|\mathbf{x} - \mathbf{x}^{(m)}\|_{\mathbf{A}}^2, \qquad m = 0, 1, \dots, \tag{7.34}$$

and we get the estimate (7.32) by recursion.

To show (7.33), we derive a bound on $|\Lambda_{m+1} \backslash \Lambda_m|$ as follows: Denote by $\mathbf{x}_N$ the subvector of $\mathbf{x}$ containing its N largest entries in absolute value (i.e., $\sigma_N(\mathbf{x}) = \|\mathbf{x} - \mathbf{x}_N\|_{\ell^2}$) and by Λ_N, $|\Lambda_N| \leq N$, its support set. Let N_0 be the *smallest* N such that

$$\|\mathbf{x} - \mathbf{x}_N\|_{\mathbf{A}}^2 \leq (1 - \mu\kappa(\mathbf{A}))\|\mathbf{x} - \mathbf{x}^{(m)}\|_{\mathbf{A}}^2.$$

Now consider the choice $\Lambda = \Lambda_{N_0} \cup \Lambda_m \supset \Lambda' = \Lambda_m$ and $\mathbf{y}_{\Lambda'} = \mathbf{x}^{(m)}$ in (7.27). Since

$$(1 - \mu\kappa(\mathbf{A}))\|\mathbf{x} - \mathbf{x}^{(m)}\|_{\mathbf{A}}^2 \geq \|\mathbf{x} - \mathbf{x}_{N_0}\|_{\mathbf{A}}^2 \geq \|\mathbf{x} - \mathbf{x}_\Lambda\|_{\mathbf{A}}^2 = \|\mathbf{x} - \mathbf{x}^{(m)}\|_{\mathbf{A}}^2 - \|\mathbf{x}_\Lambda - \mathbf{x}^{(m)}\|_{\mathbf{A}}^2,$$

we get by definition of N_0, the minimization property of the Galerkin projection $\mathbf{x}_\Lambda$, and (7.28) the bound

$$\|\mathbf{x}_\Lambda - \mathbf{x}^{(m)}\|_{\mathbf{A}}^2 \ge \mu\kappa(\mathbf{A})\|\mathbf{x} - \mathbf{x}^{(m)}\|_{\mathbf{A}}^2.$$

With this and with (7.26) and (7.25), we have

$$\begin{aligned}\|\mathbf{b}_\Lambda - \mathbf{A}_\Lambda\mathbf{x}^{(m)}\|_{\ell^2}^2 &= \|\mathbf{A}_\Lambda(\mathbf{x}_\Lambda - \mathbf{x}^{(m)})\|_{\ell^2}^2 \ge \|\mathbf{A}_\Lambda^{-1}\|_{\ell^2}^{-1}\|\mathbf{x}_\Lambda - \mathbf{x}^{(m)}\|_{\mathbf{A}}^2 \\ &\ge \|\mathbf{A}^{-1}\|_{\ell^2}^{-1}\mu\kappa(\mathbf{A})\|\mathbf{x} - \mathbf{x}^{(m)}\|_{\mathbf{A}}^2 \ge \mu\|\mathbf{b} - \mathbf{A}\mathbf{x}^{(m)}\|_{\ell^2}^2.\end{aligned}$$

This is, $\Lambda = \Lambda_{N_0} \cup \Lambda_m$ satisfies the same inequality as Λ_{m+1}. But by its definition in Theorem 7.5, Λ_{m+1} is the smallest index set that satisfies this inequality. Consequently,

$$|\Lambda_{m+1}\backslash\Lambda_m| = |\Lambda_{m+1}| - |\Lambda_m| \le |\Lambda| - |\Lambda_m| \le N_0.$$

Since

$$N_0 \le N_1 := \min\{N : \sigma_N(\mathbf{x})^2 = \|\mathbf{x} - \mathbf{x}_N\|_{\ell^2}^2 \le \|\mathbf{A}\|_{\ell^2}^{-1}(1 - \mu\kappa(\mathbf{A}))\|\mathbf{x} - \mathbf{x}^{(m)}\|_{\mathbf{A}}^2\},$$

we see by definition of $\mathcal{A}^s$ that

$$\|\mathbf{x}\|_{\mathcal{A}^s}^2 \ge N_1^{2s}\sigma_{N_1-1}(\mathbf{x})^2 \ge N_1^{2s}\|\mathbf{A}\|_{\ell^2}^{-1}(1 - \mu\kappa(\mathbf{A}))\|\mathbf{x} - \mathbf{x}^{(m)}\|_{\mathbf{A}}^2.$$

This implies

$$|\Lambda_{m+1}\backslash\Lambda_m| \le N_0 \le N_1 \le \|\mathbf{x}\|_{\mathcal{A}^s}^{1/s}(\|\mathbf{A}\|_{\ell^2}^{-1}(1 - \mu\kappa(\mathbf{A}))\|\mathbf{x} - \mathbf{x}^{(m)}\|_{\mathbf{A}}^2)^{-1/(2s)}. \quad (7.35)$$

It remains to add these bounds and to consider the already proven exponential decay of the norms $\|\mathbf{x} - \mathbf{x}^{(m)}\|_{\mathbf{A}}^2$. To shorten the notation, temporarily set $\rho := 1 - \mu/\kappa(\mathbf{A})$ and $C_0 := (\|\mathbf{A}\|_{\ell^2}^{-1}(1 - \mu\kappa(\mathbf{A})))^{-1/(2s)}$. From (7.34) we have

$$\|\mathbf{x} - \mathbf{x}^{(m-1)}\|_{\mathbf{A}}^2/\|\mathbf{x} - \mathbf{x}^{(k)}\|_{\mathbf{A}}^2 \le \rho^{m-1-k}, \qquad k = 0, \dots, m-1,$$

which is used together with (7.35) for the final estimate of $|\Lambda_m|$. We get

$$\begin{aligned}|\Lambda_m| &= \sum_{k=0}^{m-1}|\Lambda_{k+1}\backslash\Lambda_k| \le C_0\|\mathbf{x}\|_{\mathcal{A}^s}^{1/s}\sum_{k=0}^{m-1}\|\mathbf{x} - \mathbf{x}^{(k)}\|_{\mathbf{A}}^{-1/s} \\ &\le C_0\|\mathbf{x}\|_{\mathcal{A}^s}^{1/s}\|\mathbf{x} - \mathbf{x}^{(m-1)}\|_{\mathbf{A}}^{-1/s}\sum_{k=0}^{m-1}\rho^{(m-1-k)/2s} \\ &\le \frac{(\|\mathbf{A}\|_{\ell^2}^{-1}(1 - \mu\kappa(\mathbf{A})))^{-1/(2s)}}{1 - \rho^{1/2s}}\|\mathbf{x}\|_{\mathcal{A}^s}^{1/s}\|\mathbf{x} - \mathbf{x}^{(m-1)}\|_{\mathbf{A}}^{-1/s}.\end{aligned}$$

This shows (7.33) with a constant depending on $\kappa(A)$, μ, and s. □

To see that the result of Theorem 7.5 guarantees the quasi-optimality of the abstract iterative scheme (neglecting the computational cost), one has to switch from the $\|\cdot\|_{\mathbf{A}}$ norms to the equivalent $\|\cdot\|_{\ell^2}$ norms and choose, for given $0 < \epsilon < \|\mathbf{x}\|_{\ell^2}$, the first $m > 0$ such that $\|\mathbf{x} - \mathbf{x}^{(m)}\|_{\ell^2} \le \epsilon$ (for $\epsilon \ge \|\mathbf{x}\|_{\ell^2}$, $\mathbf{x}_\epsilon = \mathbf{0}$ suffices and (7.30) trivially holds). Such an m must exist because of (7.32). We denote it by m_ϵ and set $\mathbf{x}_\epsilon = \mathbf{x}^{(m_\epsilon)}$, $\Lambda_\epsilon = \Lambda_{m_\epsilon}$. Since

$$\|\mathbf{x} - \mathbf{x}^{(m_\epsilon - 1)}\|_{\mathbf{A}} \ge \|\mathbf{A}^{-1}\|^{-1/2}\|\mathbf{x} - \mathbf{x}^{(m_\epsilon - 1)}\|_{\ell^2} > \|\mathbf{A}^{-1}\|^{-1/2}\epsilon,$$

(7.33) implies the second estimate in (7.30) by our choice of m_ϵ.

The parameter μ is the only free parameter in the abstract iterative scheme of Theorem 7.5. Choosing it close to $\kappa(\mathbf{A})^{-1}$ gives a smaller value of ρ and implies a faster convergence of the iterates $\mathbf{x}^{(m)}$. At the same time, this requires larger index sets Λ_{m+1} in (7.31). If we choose μ close to zero, then it becomes easier to satisfy (7.31) with small index sets $\Lambda_{m+1} \supset \Lambda_m$ and thus with slow growth of the cardinalities $|\Lambda_m|$. However, to satisfy the error criterion for given ϵ, larger values of m_ϵ and more iteration steps are required.

7.1.4.2 Practical adaptive wavelet method

The abstract iteration scheme of Theorem 7.4 cannot be implemented with a number of operations close to the size of $\Lambda_\epsilon = \Lambda_{m_\epsilon}$ as required by the quasi-optimality condition. The main obstacles are the exact solution of the linear systems (7.24) and the stringent update rule (7.31) which require access to the infinite system $\mathbf{A}$ and the right-hand side $\mathbf{b}$. In particular, already the computation of all entries of $\mathbf{A}_\Lambda$ cannot be done with $O(|\Lambda|)$ operations. Also, in applications to boundary integral equations $\mathbf{A}_\Lambda$ is even dense. In this subsection, we provide a list of modifications needed to overcome these obstacles. The necessary changes concern the approximate computation of the right-hand side $\mathbf{b}$ in (7.23), the approximate computation of matrix-vector products $\mathbf{A}\mathbf{y}_\Lambda$, the iterative solution of (7.24), which uses the previous two routines, and an economical replacement of the $\Lambda_m \to \Lambda_{m+1}$ rule. For full details and complete proofs, the reader is referred to [173] and the original references cited therein.

Approximate computation of the right-hand side $\mathbf{b}$. This routine is called **RHS** in [173] and is considered the easiest task, since $f \in V'$ is explicitly available. We say that $f \in V'$ admits the routine $\mathbf{RHS}(\epsilon) := \mathbf{b}_\epsilon$ if, for any $\epsilon > 0$, we can compute $\mathbf{b}_\epsilon$ with finite support Λ_ϵ such that

$$\|\mathbf{b} - \mathbf{b}_\epsilon\|_{\ell^2} \le \epsilon, \qquad |\Lambda_\epsilon| \le c_1 \min(\{N : \sigma_N(\mathbf{b}) \le \epsilon\}). \tag{7.36}$$

In addition, finding $\mathbf{b}_\epsilon$ should require $\le c_2(|\Lambda_\epsilon| + 1)$ operations. Here c_1, c_2 are absolute constants, independent of f and ϵ.

Roughly speaking, the condition (7.36) says that we should be able to find an approximation $\mathbf{b}_\epsilon$ to the best N-term approximation $\mathbf{b}_N$ (i.e., the N largest entries in absolute value) with $\le c_1 N$ entries, and we can do this with a number of operations proportional to its support length. In general, the availability of $\mathbf{RHS}(\epsilon)$ is an assumption and can only be substantiated under further assumptions. For example, the

$\bar{s}$-admissibility of $\mathbf{A}$ introduced next implies (7.36) for $0 < s \le \bar{s}$, see Corollary 3.1 in [173]. In practice, one computes an approximation $\mathbf{b}_{\epsilon'}$ with very small $\epsilon' > 0$ and large support by determining enough coefficients b_λ with high accuracy, then reordering the entries of $\mathbf{b}_{\epsilon'}$ by decreasing absolute value and implementing **RHS**(ϵ) for $\epsilon \ge 2\epsilon'$ in an obvious way. The cost of this preprocessing step is often negligible. It also leads to simplifications in other routines, e.g., in the iterative scheme for solving (7.24).

Approximate matrix-vector products with $\mathbf{A}$. The routine **APPLY** in [173] is based on certain compressibility assumptions about $\mathbf{A}$. We say that $\mathbf{A}$ is $\bar{s}$-admissible for some fixed $\bar{s} > 0$ if, for any $\epsilon > 0$, any $\mathbf{y} \in \ell^2(\Lambda')$ with finite support set $\Lambda' \subset \Omega$ and any $0 < s \le \bar{s}$, we can compute an approximation $\mathbf{z}_\epsilon \in \ell^2(\Lambda)$ to $\mathbf{Ay}$ with finite $\Lambda \subset \Omega$ that satisfies

$$\|\mathbf{Ay} - \mathbf{z}_\epsilon\|_{\ell^2} \le \epsilon, \qquad |\Lambda| \le c_3 \epsilon^{-1/s} \|\mathbf{y}\|_{\mathcal{A}^s}^{1/s}.$$

Additionally, finding $\mathbf{z}_\epsilon$ requires $\le c_4(\epsilon^{-1/s}\|\mathbf{y}\|_{\mathcal{A}^s}^{1/s}|\Lambda'| + 1)$ operations. Here c_3, c_4 are absolute constants independent of $\mathbf{y}$ and ϵ. We set $\mathbf{APPLY}(\mathbf{y}, \epsilon) := \mathbf{z}_\epsilon$.

In [173, Section 3], the existence of such a routine **APPLY** and thus the $\bar{s}$-admissibility of $\mathbf{A}$ is shown under the condition that $\mathbf{A}$ is s^*-computable ($s^* > \bar{s}$). This means that $\mathbf{A}$ can be replaced (with error $e_j \to 0$ in the $\ell^2(\Omega)$ operator norm) by sparse $\mathbf{A}_j$ containing $\le n_j$ non-zero entries in each column so that

$$n_{j+1}/n_j \le c_5, \qquad e_j n_j^s \le c_6, \quad 0 < s < s^*, \qquad j = 1, 2, \ldots.$$

For $j = 0$, one sets $\mathbf{A}_0 = \mathbf{0}$ and $n_0 = 0$.

The construction of **APPLY** and the proof of its properties is technically demanding, we refer to [173, Section 3] and the original articles. The same is true for the verification of the s^*-computability condition, which depends heavily on A, and the properties of Ψ. In particular, it depends on the polynomial reproduction order, the order of vanishing moments, the smoothness, and the local structure of the wavelets $\psi_\lambda \in \Psi$. The best studied case are Ψ consisting of tensor products of univariate biorthogonal spline wavelets (or similar multivariate constructions). The s^*-computability of PDE operators A is discussed in [173, Section 5 and 7], for boundary integral operators, see [101, Section 4].

Approximate solution of (7.24). In the abstract adaptive algorithm, the solution of (7.24) is required for $\Lambda = \Lambda_{m+1} \supset \Lambda_m$, with $\mathbf{x}^{(m)} \in \ell^2(\Lambda^{(m)}) \subset \ell^2(\Lambda)$ available from the previous step and satisfying (7.31). The corresponding routine **GALERKIN** depends on two parameters $\delta, \epsilon > 0$ and makes a similar assumption, namely that we have a starting vector $\mathbf{y}_\Lambda \in \ell^2(\Lambda)$ (the counterpart of $\mathbf{x}^{(m)}$) such that

$$\|\mathbf{b}_\Lambda - \mathbf{A}_\Lambda \mathbf{y}_\Lambda\|_{\ell^2} \le \delta. \tag{7.37}$$

Furthermore, fix $\alpha > 0$, $0 < \rho < 1$ and $K = 1, 2, \ldots$ so that

$$\|I - \alpha\mathbf{A}\|_{\ell^2} \le \rho < 1, \qquad 2\rho^K \le \frac{\epsilon}{\delta}, \qquad a := \frac{\delta\rho}{2\alpha K\|\mathbf{A}\|_{\ell^2}}. \tag{7.38}$$

With this, set $\mathbf{z}^{(0)} = \mathbf{y}_\Lambda$, repeat

$$\mathbf{z}^{(k)} = \mathbf{z}^{(k-1)} + \alpha(\mathbf{RHS}(a\rho^k) - \mathbf{APPLY}(\mathbf{z}^{(k-1)}, a\rho^k))|_\Lambda \text{ for } k = 1, \dots, K,$$

and set $\mathbf{GALERKIN}(\Lambda, \mathbf{y}_\Lambda, \delta, \epsilon) := \mathbf{z}^{(K)}$. In other words, run K steps of an inexact Richardson iteration for (7.24), where the residual update is computed with increasing accuracy (entries outside Λ are neglected, this is indicated in the update formula by $|_\Lambda$).

The following result is Proposition 4.4 in [173]:

Lemma 7.2 *Under the conditions given above for the execution of the routine* $\mathbf{GALERKIN}(\Lambda, \mathbf{y}_\Lambda, \delta, \epsilon)$*, we have*

$$\|\mathbf{b}_\Lambda - \mathbf{A}_\Lambda\mathbf{z}^{(K)}\|_{\ell^2} \le \epsilon.$$

Moreover, if $\mathbf{A}$ *is* $\bar{s}$*-admissible and* $\mathbf{x} \in \mathcal{A}^s$ *for some* $0 < s \le \bar{s}$*, then the number of operations needed to compute* $\mathbf{z}^{(K)}$ *is bounded by*

$$c_7(\delta^{-1/s}(\|\mathbf{x}\|_{\mathcal{A}^s}^{1/s} + \|\mathbf{y}_\Lambda\|_{\mathcal{A}^s}^{1/s}) + |\Lambda| + 1),$$

where c_7 *depends on* δ/ϵ *but is otherwise independent of the input.*

Proof Define for $k = 1, \dots, K$ the differences

$$\mathbf{y}^{(k-1)} := (\mathbf{b}_\Lambda - \mathbf{A}_\Lambda\mathbf{z}^{(k-1)}) - (\mathbf{RHS}(a\rho^k) - \mathbf{APPLY}(\mathbf{z}^{(k-1)}, a\rho^k))|_\Lambda.$$

Then, by the properties of **RHS** and **APPLY**, we have

$$\begin{aligned}\|\mathbf{y}^{(k-1)}\|_{\ell^2} &\le \|\mathbf{b}_\Lambda - \mathbf{RHS}(a\rho^k)|_\Lambda\|_{\ell^2} + \|\mathbf{A}_\Lambda\mathbf{z}^{(k-1)} - \mathbf{APPLY}(\mathbf{z}^{(k-1)}, a\rho^k)|_\Lambda\|_{\ell^2} \\ &\le 2a\rho^k\end{aligned}$$

for $k = 1, \dots, K$. Since

$$\begin{aligned}\mathbf{x}_\Lambda - \mathbf{z}^{(K)} &= (I - \alpha\mathbf{A}_\Lambda)(\mathbf{x}_\Lambda - \mathbf{z}^{(K-1)}) + \alpha\mathbf{y}^{(K-1)} = \cdots \\ &= (I - \alpha\mathbf{A}_\Lambda)^K(\mathbf{x}_\Lambda - \mathbf{y}_\Lambda) + \alpha\sum_{k=0}^{K-1}(I - \alpha\mathbf{A}_\Lambda)^k\mathbf{y}^{(K-1-k)},\end{aligned}$$

and thus

$$\begin{aligned}\mathbf{b}_\Lambda - \mathbf{A}_\Lambda\mathbf{z}^{(K)} &= \mathbf{A}_\Lambda(\mathbf{x}_\Lambda - \mathbf{z}^{(K)}) \\ &= (I - \alpha\mathbf{A}_\Lambda)^K(\mathbf{b}_\Lambda - \mathbf{A}_\Lambda\mathbf{y}_\Lambda) + \alpha\sum_{k=0}^{K-1}\mathbf{A}_\Lambda(I - \alpha\mathbf{A}_\Lambda)^k\mathbf{y}^{(K-1-k)},\end{aligned}$$

we get

$$\|\mathbf{b}_\Lambda - \mathbf{A}_\Lambda \mathbf{z}^{(K)}\|_{\ell^2} \le \rho^K \|\mathbf{b}_\Lambda - \mathbf{A}_\Lambda \mathbf{y}_\Lambda\|_{\ell^2} + \alpha \|\mathbf{A}_\Lambda\|_{\ell^2} \sum_{k=0}^{K-1} \rho^k \|\mathbf{y}^{(K-1-k)}\|_{\ell^2}$$

$$\le \rho^K \delta + 2a\alpha \|\mathbf{A}_\Lambda\|_{\ell^2} \sum_{k=0}^{K-1} \rho^{K-1} = \rho^K (\delta + 2a\alpha K \|\mathbf{A}_\Lambda\|_{\ell^2}/\rho) \le \epsilon,$$

where the definition of K and a in (7.38) was used in the last step. The estimate for the number of operations follows from the corresponding estimates for the routines **RHS** and **APPLY** combined with the assumed $\bar{s}$-admissibility (see Proposition 3.1 and Corollary 3.1 in [173]) and the fact that $K \le \log_2(\delta/\epsilon) + 1$. This concludes the proof. □

Let us make two remarks here: The parameter α is internal to the routine, and one is tempted to choose it to minimize ρ and thus K. This in turn requires good upper and lower bounds for $\|\mathbf{A}\|_{\ell^2}$ and $\|\mathbf{A}^{-1}\|_{\ell^2}^{-1}$, respectively. In the proof of the above result, starting from the definition of a, the possibly unknown value of $\|\mathbf{A}\|_{\ell^2}$ can be replaced by any suitable upper bound.

Practical rule of expanding index sets. Recall that the expansion rule $\Lambda_m \longrightarrow \Lambda_{m+1}$ in the abstract adaptive wavelet method was based on satisfying (7.31). This means, given the residual vector $\mathbf{r}_m = \mathbf{b} - \mathbf{A}\mathbf{x}_{\Lambda_m}$ of the current approximation,

$$\Lambda_{m+1} \supset \Lambda_m$$

is the set of smallest size such that the square of the ℓ^2 norm of $\mathbf{r}_m|_{\Lambda_{m+1}} = b_{\Lambda_{m+1}} - \mathbf{A}_{\Lambda_{m+1}}\mathbf{x}_{\Lambda_m}$ is $\ge \mu\|\mathbf{r}_m\|_{\ell^2}^2$ for some fixed $0 < \mu < \kappa(\mathbf{A})^{-1}$. Note that $\mathbf{r}_m|_{\Lambda_m}$ is the zero vector, since $\mathbf{x}_{\Lambda_m}$ is the Galerkin solution. In the practical adaptive wavelet algorithm residuals are only approximated, the iterates come from approximate Galerkin solutions, and the search for a minimally extended index set must also be relaxed. The routine **EXPAND** of [173] starts with a given subset $\Lambda' \subset \Omega$ (the support set of the current approximation) and a finitely supported vector $\mathbf{r}$ (an approximation of the current residual) with support set contained in $\Lambda_{\mathbf{r}} \supset \Lambda'$ as input and outputs a new index set $\Lambda \supset \Lambda'$ such that on the one hand

$$\|\mathbf{r}|_\Lambda\|_{\ell^2}^2 \ge \mu \|\mathbf{r}\|_{\ell^2}^2, \tag{7.39}$$

and on the other hand

$$|\Lambda \backslash \Lambda'| \le c_7 \min(\{|\hat{\Lambda} \backslash \Lambda'| : \ \hat{\Lambda} \supset \Lambda', \ \|\mathbf{r}|_{\hat{\Lambda}}\|_{\ell^2}^2 \ge \mu \|\mathbf{r}\|_{\ell^2}^2\}). \tag{7.40}$$

The parameter $0 < \mu < 1$ is fixed and we set $\mathbf{EXPAND}(\Lambda', \mathbf{r}, \mu) := \Lambda$. The way to satisfy (7.39) and (7.40) is to approximately search for as few as possible of the largest entries of $\mathbf{r}$ with index in $\Lambda_{\mathbf{r}} \backslash \Lambda'$ such that (7.39) holds, and add the corresponding indices to Λ' to get Λ. It turns out that this can be done with $\le |\Lambda'| + |\Lambda_{\mathbf{r}}| + 1$ operations. The details can be found in [173].

Now that we have the routines **RHS**, **APPLY**, **GALERKIN** and **EXPAND**, we can formulate the final adaptive wavelet method called **AWGM** in [173]. This requires proper tuning of the parameters of each routine. In particular, as the algorithm progresses, the error tolerances of the routines involved should decrease in a carefully monitored way to guarantee the error bound and, at the same time, to control the support size of the final approximation and the number of operations to find it.

The routine **AWGM** has 4 global parameters: $\theta > 0$, $0 < \gamma < \kappa(\mathbf{A})^{-3/2}$, $(\gamma\kappa(\mathbf{A}))^2 < \mu < \kappa(A)^{-1}$ and $0 < \delta < \mu^{1/2} - \gamma\kappa(\mathbf{A})$. The parameters μ and γ are used in the calls to **EXPAND** and **GALERKIN**, respectively, while the parameters θ, δ enter an inner loop that detects the current residual and residual error. More specifically, the routine **AWGM** starts with $\mathbf{y}^{(0)} = \mathbf{0}$, $\Lambda_0 = \emptyset$ and a tolerance ϵ_{-1}, where $\epsilon_{-1} \approx \|\mathbf{b}\|_{\ell^2}$ is desirable (we choose the notation $\mathbf{y}^{(m)}$ for the successive approximations to emphasize that they are different from the Galerkin solutions $\mathbf{x}^{(m)} = \mathbf{x}_{\Lambda_m}$). Then, for $m = 0, 1, \ldots,$ until termination, we first compute in an inner loop an approximation $\mathbf{r}^{(m)}$ to the current residual $\mathbf{b} - \mathbf{A}\mathbf{y}^{(m)}$ and determine ϵ_m as follows: Starting with $\zeta = \theta\epsilon_{m-1}$, calculate

$$\mathbf{r}^{(m)} = \mathbf{RHS}(\zeta/2) - \mathbf{APPLY}(\mathbf{y}^{(m)}, \zeta/2)$$

and set

$$\epsilon_m := \|\mathbf{r}^{(m)}\|_{\ell^2} + \zeta.$$

If $\epsilon_m \le \epsilon$, then set $\mathbf{y}_\epsilon := \mathbf{y}^{(m)}$ and $\Lambda_\epsilon := \Lambda_m$ and exit. Otherwise, if $\zeta > \delta\|\mathbf{r}^{(m)}\|_{\ell^2}$, then set $\zeta := \zeta/2$ and determine the new $\mathbf{r}^{(m)}$ and ϵ_m using the same formulas above. This process is repeated until it terminates or until $\zeta \le \delta\|\mathbf{r}^{(m)}\|$ holds for the first time. In the latter case, we find the new, expanded index set

$$\Lambda_{m+1} := \mathbf{EXPAND}(\Lambda_m, \mathbf{r}^{(m)}, \mu)$$

and the new approximation

$$\mathbf{y}^{(m+1)} := \mathbf{GALERKIN}(\Lambda_{m+1}, \mathbf{y}^{(m)}, \epsilon_m, \gamma\|\mathbf{r}^{(m)}\|_{\ell^2}).$$

Then we set $m := m + 1$ and repeat.

The following main result is quoted from [173, Theorem 4.1].

Theorem 7.6 *The adaptive wavelet algorithm described above terminates after finitely many steps with a final approximation $\mathbf{y}_\epsilon$ such that*

$$\|\mathbf{b} - \mathbf{A}\mathbf{y}_\epsilon\|_{\ell^2} \le \epsilon. \tag{7.41}$$

If $\mathbf{x} \in \mathcal{A}^s$ for some $s > 0$, then the size of the support set Λ_ϵ is bounded by

$$|\Lambda_\epsilon| \le c_8\epsilon^{-1/s}\|\mathbf{x}\|_{\mathcal{A}^s}^{1/s}.$$

Finally, if $\epsilon_{-1} \approx \|\mathbf{b}\|_{\ell^2} > \epsilon$ and $\mathbf{A}$ is $\bar{s}$-admissible, then the number of operations required to compute $\mathbf{y}_\epsilon$ and Λ_ϵ is bounded by $c_9\epsilon^{-1/s}\|\mathbf{x}\|_{\mathcal{A}^s}^{1/s}$.

Proof We only give the proof of termination after finally many steps and show the estimate (7.41) following the arguments for Proposition 4.2 in [173]. Admittedly, compared to Theorem 7.5, the estimate for the operation count is the main novelty of the result, but we refrain from reproducing its very technical proof here.

First, if, for some m, $r^{(m)}$ is computed in the inner loop with some ζ, we have

$$\begin{aligned}\|\mathbf{b} - \mathbf{A}\mathbf{y}^{(m)}\|_{\ell^2} &\le \|\mathbf{r}^{(m)}\|_{\ell^2} + \|\mathbf{b} - \mathbf{RHS}(\zeta/2)\|_{\ell^2} + \|\mathbf{A}\mathbf{y}^{(m)} - \mathbf{APPLY}(\mathbf{y}^{(m)}\zeta/2)\|_{\ell^2}\\ &\le \|\mathbf{r}^{(m)}\|_{\ell^2} + \zeta\end{aligned}$$

by the properties of **RHS** and **APPLY**, and similarly

$$\|\mathbf{r}^{(m)}\|_{\ell^2} \le \|\mathbf{b} - \mathbf{A}\mathbf{y}^{(m)}\|_{\ell^2} + \zeta.$$

This has two consequences. On the one hand, if the whole process terminates in the inner loop for this m, then we actually have $\|\mathbf{b}-\mathbf{A}\mathbf{y}_\epsilon\|_{\ell^2} \le \epsilon$. On the other hand, if the process does not terminate during the execution of this loop, then the exit criterion of the inner loop is definitely met after a finite number of steps. In fact, for $\zeta \to 0$, we have $\|\mathbf{r}^{(m)}\|_{\ell^2} \to \|\mathbf{b} - \mathbf{A}\mathbf{y}^{(m)}\|_{\ell^2}$ and thus $\delta\|\mathbf{r}^{(m)}\|_{\ell^2} - \zeta \to \delta\|\mathbf{b} - \mathbf{A}\mathbf{y}^{(m)}\|_{\ell^2} > 0$ (the case $\|\mathbf{b} - \mathbf{A}\mathbf{y}^{(m)}\|_{\ell^2} = 0$ can be excluded, since then $\|\mathbf{r}^{(m)}\|_{\ell^2} + \zeta \to 0$, which implies exit after finitely many steps). Also, at the exit of the inner loop, we have

$$\begin{aligned}\|\mathbf{b} - \mathbf{A}\mathbf{y}^{(m)}\|_{\ell^2} &\le \epsilon_m \le (1+\delta)\|\mathbf{r}^{(m)}\|_{\ell^2},\\ \|\mathbf{b} - \mathbf{A}\mathbf{y}^{(m)} - \mathbf{r}^{(m)}\|_{\ell^2} &\le \delta\|\mathbf{r}^{(m)}\|_{\ell^2}.\end{aligned} \tag{7.42}$$

Consequently, if the process does not terminate for this m, then **EXPAND** returns Λ_{m+1} so that

$$\|\mathbf{r}^{(m)}|_{\Lambda_{m+1}}\|_{\ell^2} \ge \mu^{1/2}\|\mathbf{r}^{(m)}\|_{\ell^2}, \tag{7.43}$$

see (7.39). Furthermore, the implicit condition (7.37) for executing the procedure **GALERKIN**$(\Lambda_{m+1}, \mathbf{y}^{(m)}, \epsilon_m, \gamma\|\mathbf{r}^{(m)}\|_{\ell^2})$ holds because

$$\|\mathbf{b}_{\Lambda_{m+1}} - \mathbf{A}_{\Lambda_{m+1}}\mathbf{y}^{(m)}\|_{\ell^2} \le \|\mathbf{b} - \mathbf{A}\mathbf{y}^{(m)}\|_{\ell^2} \le \epsilon_m$$

according to (7.42). So we can compute $\mathbf{y}^{(m+1)}$, which satisfies

$$\|\mathbf{b}_{\Lambda_{m+1}} - \mathbf{A}_{\Lambda_{m+1}}\mathbf{y}^{(m+1)}\|_{\ell^2} \le \gamma\|\mathbf{r}^{(m)}\|_{\ell^2}. \tag{7.44}$$

From this we can deduce an error reduction

$$\|\mathbf{x} - \mathbf{y}^{(m+1)}\|_{\mathbf{A}} \le \rho\|\mathbf{x} - \mathbf{y}^{(m)}\|_{\mathbf{A}} \tag{7.45}$$

for this step by some constant factor $\rho < 1$ depending on μ, γ, and δ. To see this, denote by $\mathbf{x}_{\Lambda_{m+1}}$ the Galerkin solution of (7.24) with $\Lambda = \Lambda_{m+1}$. Then, by Galerkin orthogonality, we have

$$\|\mathbf{x} - \mathbf{y}^{(m+1)}\|_{\mathbf{A}}^2 = \|\mathbf{x} - \mathbf{x}_{\Lambda_{m+1}}\|_{\mathbf{A}}^2 + \|\mathbf{x}_{\Lambda_{m+1}} - \mathbf{y}^{(m+1)}\|_{\mathbf{A}}^2.$$

For the second term we have

$$\begin{aligned}\|\mathbf{x}_{\Lambda_{m+1}} - \mathbf{y}^{(m+1)}\|_{\mathbf{A}}^2 &\le \|\mathbf{A}^{-1}\|_{\ell^2} \|\mathbf{b}_{\Lambda_{m+1}} - \mathbf{A}_{\Lambda_{m+1}} \mathbf{y}^{(m+1)}\|_{\ell^2}^2 \\ &\le \|\mathbf{A}^{-1}\|_{\ell^2} \gamma^2 \|\mathbf{r}^{(m)}\|_{\ell^2} \\ &\le \|\mathbf{A}^{-1}\|_{\ell^2} \frac{\gamma^2}{(1-\delta)^2} \|\mathbf{b} - \mathbf{A}\mathbf{y}^{(m)}\|_{\ell^2}^2 \\ &\le \kappa(\mathbf{A}) \frac{\gamma^2}{(1-\delta)^2} \|\mathbf{x} - \mathbf{y}^{(m)}\|_{\mathbf{A}}^2\end{aligned}$$

using (7.44) and the second estimate in (7.42). For the first term, we argue as follows: From the second inequality in (7.42) we have

$$\begin{aligned}\|\mathbf{r}^{(m)}|_{\Lambda_{m+1}}\|_{\ell^2} &\le \|(\mathbf{b} - \mathbf{A}\mathbf{y}^{(m)})|_{\Lambda_{m+1}}\|_{\ell^2} + \|(\mathbf{b} - \mathbf{A}\mathbf{y}^{(m)}) - \mathbf{r}^{(m)}\|_{\ell^2} \\ &\le \|\mathbf{b}_{\Lambda_{m+1}} - \mathbf{A}_{\Lambda_{m+1}} \mathbf{y}^{(m)}\|_{\ell^2} + \delta \|\mathbf{r}^{(m)}\|_{\ell^2},\end{aligned}$$

and, consequently, by (7.43) and the first inequality in (7.42), we get

$$\|\mathbf{b}_{\Lambda_{m+1}} - \mathbf{A}_{\Lambda_{m+1}} \mathbf{y}^{(m)}\|_{\ell^2} \ge (\mu^{1/2} - \delta)\|\mathbf{r}^{(m)}\|_{\ell^2} \ge \frac{\mu^{1/2} - \delta}{1+\delta} \|\mathbf{b} - \mathbf{A}\mathbf{y}^{(m)}\|_{\ell^2}.$$

So, assuming that

$$\tilde{\mu} := \left(\frac{\mu^{1/2} - \delta}{1+\delta}\right)^2 < \kappa(\mathbf{A})^{-1},$$

we can repeat the argument in Theorem 7.5 for estimating the term $\|\mathbf{x} - \mathbf{x}^{(m+1)}\|_{\mathbf{A}}^2$ and get

$$\|\mathbf{x} - \mathbf{x}_{\Lambda_{m+1}}\|_{\mathbf{A}}^2 \le (1 - \tilde{\mu}/\kappa(\mathbf{A}))\|\mathbf{x} - \mathbf{y}^{(m)}\|_{\mathbf{A}}^2.$$

Putting it all together we get

$$\|\mathbf{x} - \mathbf{y}^{(m+1)}\|_{\mathbf{A}}^2 \le (1 - \tilde{\mu}\kappa(\mathbf{A})^{-1} + \kappa(\mathbf{A})\frac{\gamma^2}{(1-\delta)^2})\|\mathbf{x} - \mathbf{y}^{(m)}\|_{\mathbf{A}}^2.$$

This gives (7.45) with some $\rho < 1$ when

$$\tilde{\mu}\kappa(\mathbf{A})^{-1} > \kappa(\mathbf{A})\frac{\gamma^2}{(1-\delta)^2},$$

which, according to the assumptions about γ and μ, holds for

$$0 < \delta < \delta_{\max} = \mu^{1/2} - \gamma\kappa(\mathbf{A}).$$

The decay estimate (7.45) together with (7.42) implies that the termination criterion $\epsilon_m \le \epsilon$ must be satisfied for some finite m. If, on the contrary, this does not happen at the exit of each inner loop, then we would have

$$\epsilon < \epsilon_m = \|\mathbf{r}^{(m)}\|_{\ell^2} + \zeta \le (1+\delta^{-1})\|\mathbf{r}^{(m)}\|_{\ell^2} \le \frac{1+\delta}{\delta(1-\delta)}\|\mathbf{b} - \mathbf{A}\mathbf{y}^{(m)}\|_{\ell^2}$$
$$\le \frac{1+\delta}{\delta(1-\delta)}\|\mathbf{A}\|_{\ell^2}^{1/2}\|\mathbf{x} - \mathbf{y}^{(m)}\|_{\mathbf{A}},$$

which is impossible for $m \to \infty$ according to (7.45). Together with (7.41) this shows the finite termination of **AWGM** after finitely many steps. □

We conclude with a few remarks. First, the modern theory of adaptive finite element methods [129, 130, 171] has followed a similar path, with some necessary changes. Instead of working with wavelet Riesz discretizations, their construction is based on BPX-type frame decompositions $\Phi = \{\phi_\lambda\}_{\lambda\in\Omega}$. Since the scaled nodal basis functions $\phi_\lambda = \tilde{N}_{P,j}$ from which these systems are composed do not satisfy moment conditions, compressibility arguments such as s^*-computability cannot be applied. This is the reason why the extensions of Theorem 7.6 to frames considered in [173, Section 6] do not automatically apply to BPX-type frames Φ. Thus, quasi-optimality of an adaptive finite element method can only be expected if it produces subspaces

$$V_{\Lambda_m} = \text{span}(\{\phi_\lambda : \lambda \in \Lambda_m\})$$

with finite $\Lambda_m \subset \Omega$, for which the associated Galerkin systems (7.24) have uniformly bounded condition numbers after preconditioning and admit an $O(|\Lambda_m|)$ operation count for the involved matrix-vector products. This in turn is only possible for PDE problems and index sets Λ_m corresponding to nested refinement (see Section 7.1 for $H^1(G)$ problems). The other important change concerns the routine **EXPAND**. Since economical residual computations via frame coefficients are not available, the determination of a new, expanded index set $\Lambda_{m+1} \supset \Lambda_m$ is now based on local a posteriori error estimators and a marking strategy that refines all cells with large local error indicators such that a fixed portion (determined by the parameter μ) of the current error is captured.

Second, the adaptive wavelet method finds a subspace V_{Λ_ϵ} of dimension $|\Lambda_\epsilon|$ and an approximate solution u_ϵ with error $\le \epsilon$ involving a number of operations roughly bounded by $O(|\Lambda_\epsilon|)$, independent of ϵ. Moreover, $|\Lambda_\epsilon|$ is quasi-optimal, i.e., it has the smallest possible size up to constants to reach an error level $\le \epsilon$ at all. This is in contrast to many standard iterative methods for solving systems such as (7.24), which at best require a number of operations of $O(|\Lambda| \log(1/\epsilon))$ to achieve a final error of $O(\epsilon)$. In other words, even if we knew such a quasi-optimal Λ_ϵ, a standard iterative solver would have a slightly higher cost than the adaptive wavelet method when $\epsilon \to 0$, at least asymptotically. The improvement was achieved by performing a fixed, ϵ-independent number of inexact iteration steps for a sequence of subproblems

$$\mathbf{A}_{\Lambda_m}\mathbf{x}_{\Lambda_m} = \mathbf{b}_{\Lambda_m}$$

of smaller size and thus smaller associated number of operations. Similar improvements are possible in some other situations, such as for discretizations associated with multiscale decompositions. Full multigrid methods [187] are a prominent example.

7.2 Domain decomposition methods

Besides multiscale splitting, there is another paradigm for decomposing discretization spaces for elliptic PDE problems on domains $G \subset \mathbb{R}^d$ into subproblems, namely *domain decomposition*. In one version, the domain G is partitioned into a family of pairwise disjoint subdomains G_i, and in each step the algorithm solves the problem restricted to G_i plus a correction corresponding to transmission conditions associated with the system of boundaries between adjacent subdomains, called the interface $\tilde{\Gamma}$. In another version, the domain decomposition is just a locally finite cover of G obtained by extending each G_i to a larger region G_i' with sufficient overlap between adjacent subdomains, in which case the transmission conditions are implicitly taken care of. Applied to the linear systems arising in discretization schemes, this results in non-overlapping and overlapping block-coordinate decompositions, where the blocks are now chosen according to a decomposition of the physical domain G. The partitioning of a domain is an interesting problem in itself, governed by load-balancing considerations (the local subproblems should have approximately the same runtime complexity) and by keeping communication costs to a minimum (usually achieved by keeping the inner boundaries and overlap regions between neighboring subdomains small). The idea has gained great popularity with the advent of massively parallel and distributed computing, although its roots go far back to the Schwarz alternating method [161] for proving the existence of solutions to the Dirichlet problem on complicated domains inaccessible by Fourier methods. We refer to [180] for an introduction to the theory with emphasis on three-dimensional problems and to https://www.ddm.org for a repository of the proceedings of a conference series devoted to the theory, implementation, and applications of domain decomposition methods.

Rather than strive for maximum generality, our goal in this section is to introduce some space splittings underlying the basic domain decomposition methods in a model situation where the problem at hand arises from a *linear finite element discretization* of the scalar diffusion equation (7.1) with homogeneous Dirichlet data $T_\Gamma u = 0$. We further assume that the finite element discretization spaces under consideration are subject to multiscale refinement as assumed in Theorem 3.3. In other words, the considered spaces are linear finite element subspaces $V_{J,0}$ of $H_0^1(G)$ for which uniformly stable BPX-frame splittings exist. It turns out that by regrouping the BPX-frame, new splittings and preconditioners typical for domain decomposition methods can be derived, together with bounds on their stability constants.

7.2.1 Non-overlapping domain decomposition

Consider a variational problem with a symmetric H_0^1 elliptic bilinear form $a(\cdot,\cdot)$ on the finite element subspace $V_{J,0}$ associated with the partition $\mathcal{T}_J$, where J is fixed but arbitrary. Theorem 7.3 c) implies that the BPX-frame $\Phi_{J,0}$ consisting of all appropriately scaled nodal basis functions associated with nodal points $P \notin \Gamma$ and

$j \leq J$ has a bounded condition number, uniformly in J. For simplicity, consider as subdomains G_i, $i = 1, \ldots, N$, the cells of the partition $\mathcal{T}_{j_0}$, where $0 < j_0 < J$. This gives us $N \approx 2^{j_0 d}$ for the number of subdomains and $\dim(V_{G_i}) \approx 2^{(J-j_0)d}$ for the size of the local subspace

$$V_{G_i} := \{v_J \in V_{J,0} : \operatorname{supp}(v_{J,0}) \subset G_i\}, \qquad i = 1, \ldots, N,$$

associated with G_i. Figure 7.5 (left) shows this situation for the decomposition of a square G in $\mathbb{R}^2$ into square subdomains G_i. The sets $\Gamma_{ik} := G_i \cap G_k$ are not empty if G_i and G_k are neighbors. In general, they are lower-dimensional cells and can be ordered by dimensionality. For example, if $d = 3$, we typically have faces, edges, and vertices in $\mathcal{T}_{j_0}$ among the Γ_{ik}. Their union is the interface $\tilde{\Gamma}$.

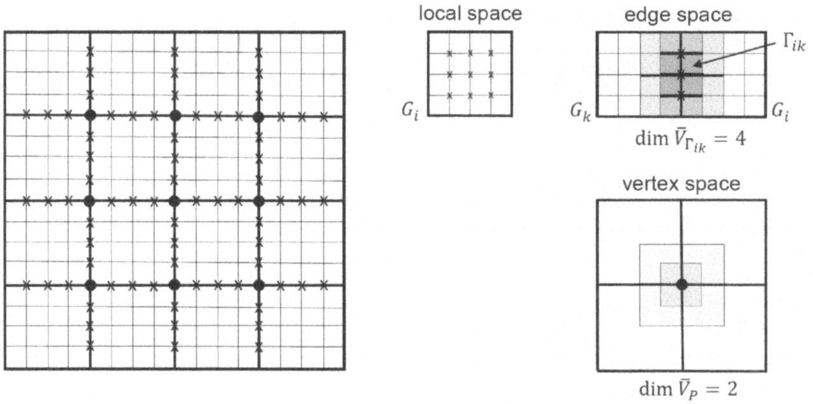

Fig. 7.5 Illustration of the construction of non-overlapping domain decomposition splittings. Left: Subdomains G_i and interface $\tilde{\Gamma}$ for uniformly refined rectangular partitions ($j_0 = 1$, $J = 3$). Right: Local, edge and vertex spaces associated with G_i and Γ_{ik}, respectively.

The BPX-frame $\Phi_{J,0}$ associated with the H_0^1 elliptic problem on $V_{J,0}$ consists of all scaled nodal basis functions

$$\tilde{N}_{P,j} = d_{P,j} N_{P,j}, \quad d_{P,j} \approx 2^{j(d/2-1)}, \quad P \in \mathcal{V}_j \backslash \Gamma, \quad j \leq J.$$

We will partition $\Phi_{J,0}$ as follows: For each G_i, we denote by Φ_{G_i} the collection of scaled nodal basis functions $\tilde{N}_{P,j} \in \Phi_{J,0}$ with support in G_i, i.e., $N_{j,P}$ with P inside G_i and $j_0 < j \leq J$. Scaled nodal basis functions $\tilde{N}_{P,j}$ with $P \in \tilde{\Gamma}$ and $j_0 < j \leq J$ are collected in $\bar{\Phi}_{\tilde{\Gamma}}$. So the BPX-frame $\Phi_{J,0}$ is partitioned according to

$$\Phi_{J,0} = \Phi_{G_1} \cup \cdots \cup \Phi_{G_N} \cup \Phi_{j_0,0} \cup \bar{\Phi}_{\tilde{\Gamma}}.$$

The subspace of $V_{J,0}$ spanned by $\bar{\Phi}_{\tilde{\Gamma}}$ is denoted by $\bar{V}_{\tilde{\Gamma}}$. Elements of $\bar{V}_{\tilde{\Gamma}}$ and V_{G_i} are denoted by $\bar{v}_{\tilde{\Gamma}}$ and v_{G_i}, respectively.

Now consider the $|||\cdot|||$ norms associated with the BPX-frame and its components introduced above. Recall that for a frame $\Phi = (\phi_i)$ in a Hilbert space, the triple bar norm is given by

$$|||u|||_{\Phi}^2 := \inf_{u=\sum_i c_i \phi_i} \sum_i c_i^2.$$

This notation will be used below. As consequence of Theorem 7.3 c) we have

$$\begin{aligned} a(u_{J,0}, u_{J,0}) &\approx \|u_{J,0}\|_{H_0^1}^2 \approx |||u_{J,0}|||_{\Phi_{J,0}}^2 \\ &= \inf_{u_{J,0}=\sum_i v_{G_i}+v_{j_0,0}+\bar{v}_{\bar{\Gamma}}} \big(\sum_i |||v_{G_i}|||_{\Phi_{G_i}}^2 + |||v_{j_0,0}|||_{\Phi_{j_0,0}}^2 + |||\bar{v}_{\bar{\Gamma}}|||_{\bar{\Phi}_{\bar{\Gamma}}}^2\big). \end{aligned}$$

For the terms related to $V_{j_0,0}$ and V_{G_j}, again with reference to Theorem 7.3 c) and in the latter case by an additional domain dilation argument (or by reference to Problem 7.7), the triple bar norms can be replaced by

$$|||v_{j_0,0}|||_{\Phi_{j_0,0}}^2 \approx \|v_{j_0,0}\|_{H_0^1}^2 \approx a(v_{j_0,0}, v_{j_0,0})$$

and

$$|||v_{G_i}|||_{\Phi_{G_i}}^2 \approx \|v_{G_i}\|_{H_0^1(G_i)}^2 \approx a(v_{G_i}, v_{G_i}), \qquad i = 1, \ldots, N.$$

All constants in these norm equivalences are independent of $0 < j_0 < J$. Substitution yields

$$a(u_{J,0}, u_{J,0}) \approx \inf_{u_{J,0}=\sum_i v_{G_i}+v_{j_0,0}+\bar{v}_{\bar{\Gamma}}} \big(\sum_i a(v_{G_i}, v_{G_i}) + a(v_{j_0,0}, v_{j_0,0}) + |||\bar{v}_{\bar{\Gamma}}|||_{\bar{\Phi}_{\bar{\Gamma}}}^2\big). \tag{7.46}$$

First, we formulate an intermediate result that is fundamental to the theoretical analysis of many versions of non-overlapping domain decomposition methods.

Theorem 7.7 *Let the domain partition $\{G_i\}$ be given by the cells of $\mathcal{T}_{j_0}$ as described above, and let $a(\cdot,\cdot)$ be a symmetric $H_0^1(G)$ elliptic bilinear form. Then the collection*

$$\mathbf{V}_{DD,0} = (V_{G_i},\ i = 1, \ldots, N,\ \bar{V}_{\bar{\Gamma}},\ V_{j_0,0})$$

is a stable subspace splitting for $(V_{J,0}; a)$ with stability constants independent of j_0 and J.

Proof The statement is equivalent to showing the norm equivalence

$$\begin{aligned} &a(u_{J,0}, u_{J,0}) \\ &\approx \inf_{u_{J,0}=\sum_i v_{G_i}+\bar{v}_{\bar{\Gamma}}+v_{j_0,0}} \left(a(\sum_i v_{G_i}, \sum_i v_{G_i}) + a(\bar{v}_{\bar{\Gamma}}, \bar{v}_{\bar{\Gamma}}) + a(v_{j_0,0}, v_{j_0,0}) \right) \end{aligned} \tag{7.47}$$

with constants uniform in $u_{J,0} \in V_{J,0}$ and $j_0 < J$. The upper estimate in (7.47) is straightforward, since for any decomposition of $u_{J,0}$, we have

$$a(u_{J,0}, u_{J,0}) \leq 3\left(a(\sum_i v_{G_i}, \sum_i v_{G_i}) + a(\bar{v}_{\bar{\Gamma}}, \bar{v}_{\bar{\Gamma}}) + a(v_{j_0,0}, v_{j_0,0})\right)$$

by the subadditivity of the norm induced by $a(\cdot,\cdot)$. Moreover, due to the obvious properties of the H_0^1 norm, the first term can be bounded according to the disjoint supports of the v_{G_i} as

$$a(\sum_i v_{G_i}, \sum_i v_{G_i}) \leq C\|\sum_i v_{G_i}\|^2_{H_0^1(G)} = C\sum_i \|v_{G_i}\|^2_{H_0^1(G_i)} \leq C\sum_i a(v_{G_i}, v_{G_i}).$$

After substitution, all that remains is to take the infimum with respect to all possible decompositions

$$u_{J,0} = \sum_i v_{G_i} + \bar{v}_{\bar{\Gamma}} + v_{j_0,0}.$$

This approach to the upper stability bound, which uses finite overlap properties of the supports of the terms in the underlying decompositions, can be invoked for all of the subspace splittings discussed in this subsection. Therefore, we will often be brief on their derivation and concentrate on the more challenging lower stability bounds.

To prove the latter in our case, we construct a special decomposition of $u_{J,0} \in V_{J,0}$ by using the L^2 ortho-projections

$$u_{j,0} := Q_{V_{j,0}} u_{J,0}, \quad j \leq J,$$

namely

$$u_{J,0} = u_{j_0,0} + \underbrace{E_{\bar{\Gamma}} u_{J,0}}_{\bar{w}_{\bar{\Gamma}}:=} + \underbrace{(u_{J,0} - u_{j_0,0} - E_{\bar{\Gamma}} u_{J,0})}_{w_{J,0}:=}.$$

Here the linear operator $E_{\bar{\Gamma}}$ depends on j_0 and is given by

$$E_{\bar{\Gamma}} u_{J,0} = \sum_{j=j_0+1}^{J} E_{j,\bar{\Gamma}} T_{\bar{\Gamma}}(u_{j,0} - u_{j-1,0}), \tag{7.48}$$

where the "minimal" discrete extension operators $E_{j,\bar{\Gamma}}$ are defined in complete analogy to the $E_{j,\Gamma}$ in Theorem 7.3 b). Obviously,

$$\bar{w}_{\bar{\Gamma}} := E_{\bar{\Gamma}} u_{J,0}$$

belongs to $\bar{V}_{\bar{\Gamma}}$. Furthermore, we have

$$T_{\bar{\Gamma}} E_{\bar{\Gamma}} u_{J,0} = T_{\bar{\Gamma}}(u_{J,0} - u_{j_0,0}),$$

which implies $T_{\bar{\Gamma}} w_{J,0} = 0$ and $w_{J,0} = \sum_i u_{G_i}$ for some $u_{G_i} \in V_{G_i}$, $i = 1, \ldots, N$. Finally,

$$\|E_{\bar{\Gamma}} u_{J,0}\|_{H_0^1} \leq C\|u_{J,0} - u_{j_0,0}\|_{H_0^1}, \qquad u_{J,0} \in V_{J,0}, \tag{7.49}$$

holds with a constant independent of j_0 and J. In fact, by Theorem 7.3 c) and the properties of the L^2 orthoprojections $Q_{V_{j,0}}$, we have on the one hand

$$a(u_{J,0}, u_{J,0}) \approx \|u_{J,0}\|^2_{H_0^1} \approx \|u_{0,0}\|^2_{L^2} + \sum_{j=1}^{J} 2^{2j} \|u_{j,0} - u_{j-1,0}\|^2_{L^2} \tag{7.50}$$

and on the other hand

$$\|w_{\tilde{\Gamma}}\|^2_{H_0^1} \leq C \sum_{j=j_0+1}^{J} 2^{2j} \|E_{j,\tilde{\Gamma}} T_{\tilde{\Gamma}}(u_{j,0} - u_{j-1,0})\|^2_{L^2}.$$

But by construction of the $E_{j,\tilde{\Gamma}}$ we also have

$$\|E_{j,\tilde{\Gamma}} T_{\tilde{\Gamma}} v_{j,0}\|^2_{L^2} \approx 2^{-j} \|T_{\tilde{\Gamma}} v_{j,0}\|^2_{L^2(\tilde{\Gamma})} \leq C\|v_{j,0}\|^2_{L^2}$$

for any $v_{j,0} \in V_{j,0}$ and $j_0 < j \leq J$, which allows us to continue with

$$\|w_{\tilde{\Gamma}}\|^2_{H_0^1} \leq C \sum_{j=j_0+1}^{J} 2^{2j} \|u_{j,0} - u_{j-1,0}\|^2_{L^2} \leq C\|u_{J,0} - u_{j_0,0}\|^2_{H_0^1}.$$

This proves (7.49) and gives the bound

$$a(\bar{w}_{\tilde{\Gamma}}, \bar{w}_{\tilde{\Gamma}}) \leq C\|\bar{w}_{\tilde{\Gamma}}\|^2_{H_0^1} \leq C\|u_{J,0} - u_{j_0,0}\|^2_{H_0^1} \leq Ca(u_{J,0}, u_{J,0})$$

for the second term $\bar{w}_{\tilde{\Gamma}}$ of the decomposition. As already mentioned, the last term $w_{J,0}$ is the sum of certain $u_{G_i} \in V_{G_i}$ with disjoint support. Thus

$$\begin{aligned}\sum_i a(u_{G_i}, u_{G_i}) &\leq Ca(w_{J,0}, w_{J,0}) = Ca(u_{J,0} - u_{j_0,0} - \bar{w}_{\tilde{\Gamma}}, u_{J,0} - u_{j_0,0} - \bar{w}_{\tilde{\Gamma}}) \\ &\leq C(a(u_{J,0} - u_{j_0,0}, u_{J,0} - u_{j_0,0}) + a(\bar{w}_{\tilde{\Gamma}}, \bar{w}_{\tilde{\Gamma}})).\end{aligned}$$

Finally, since $Q_{V_{j,0}} u_{j_0,0} = Q_{V_{j,0}} u_{J,0} = u_{j,0}$ for $j \leq j_0$, we have from (7.50)

$$\begin{aligned}&a(u_{j_0,0}, u_{j_0,0}) + a(u_{J,0} - u_{j_0,0}, u_{J,0} - u_{j_0,0}) \\ &\qquad \approx \|u_{0,0}\|^2_{L^2} + \sum_{j=1}^{J} 2^{2j} \|u_{j,0} - u_{j-1,0}\|^2_{L^2} \approx a(u_{J,0}, u_{J,0}),\end{aligned}$$

Combining the last three bounds, we get

$$\sum_i a(u_{G_i}, u_{G_i}) + a(\bar{w}_{\tilde{\Gamma}}, \bar{w}_{\tilde{\Gamma}}) + a(u_{j_0,0}, u_{j_0,0}) \leq Ca(u_{J,0}, u_{J,0}),$$

where C is independent of j_0 and J. This gives the lower stability bound. □

Theorem 7.7 essentially says that the H_0^1 elliptic problem on $V_{J,0}$ can be stably split into many independent so-called *local problems* on V_{G_i}, $i = 1, \ldots, N$, a so-called *coarse problem* on $V_{j_0,0}$, and a so-called *interface problem* on $V_{\tilde{\Gamma}}$. Due to their derivation from a partitioning of the BPX-frame $\Phi_{J,0}$, each of them comes with a BPX-type frame, which would lead to a multilevel preconditioner for each of the subproblems. However, this splitting does not immediately translate into a practical iteration scheme suitable for parallel implementations. While the local problems can be processed completely independently and in parallel, the main obstacles are the problems associated with the subspaces $V_{j_0,0}$ and $\bar{V}_{\tilde{\Gamma}}$, which overlap with the local problems and lead to large communication costs.

Instead of the splitting $\mathbf{V}_{DD,0}$, we can also consider the subspace splitting

$$\mathbf{V}_{DD} := (V_{G_i},\, i = 1, \ldots, N,\, \tilde{V}_{\tilde{\Gamma}})$$

for $(V_{J,0}; a)$ by merging the coarse and interface spaces into a slightly larger interface space $\tilde{V}_{\tilde{\Gamma}} := \mathrm{span}(V_{j_0,0}, \bar{V}_{\tilde{\Gamma}})$, thus formally eliminating the coarse space. A few changes in the proof of Theorem 7.7 show that $\mathbf{V}_{DD}$ is also uniformly stable (in the proof of the lower stability estimate, just set $\tilde{v}_{\tilde{\Gamma}} = u_{j,0} + E_{\tilde{\Gamma}} u_{J,0} \in \tilde{V}_{\tilde{\Gamma}}$ and use the triangle inequality). However, this does not cure the issues with the global interface problem metioned above.

As an attempt to reduce these difficulties, the space $\bar{V}_{\tilde{\Gamma}}$ underlying the interface problem can be decomposed into local components $\bar{V}_{\Gamma_{ik}}$. If Γ_{ik} is zero-dimensional (i.e., if it consists of a single common vertex point $P \notin \Gamma$ of G_i and G_k), then $\bar{V}_P := \bar{V}_{\Gamma_{ik}}$ is spanned by all $\tilde{N}_{P,j}$ with that P and $j_0 < j \le J$. In all other cases, $\bar{V}_{\Gamma_{ik}}$ is spanned by all $\tilde{N}_{P,j} \in \bar{\Phi}_{\tilde{\Gamma}}$ such that P belongs to the interior of Γ_{ik} and $j_0 < j \le J$. For the typical partition sequences, the Γ_{ik} are common vertices, edges, faces, $\cdots$ of neighboring cells G_i and G_k from $\mathcal{T}_{j_0}$, and the spaces $\bar{V}_{\Gamma_{ik}}$ are called vertex, edge, face, $\cdots$ spaces, respectively. This process of decomposing the interface $\tilde{\Gamma}$ into lower-dimensional cells is commonly called *iterative substructuring*. For $d = 2$ and rectangular partitions, the nodal sets and support regions associated with the vertex and edge spaces $\bar{V}_{\Gamma_{ik}}$ are shown in Figure 7.5 (right).

Since in our notation we can have $\Gamma_{ik} = \Gamma_{lm}$ for different index pairs (i, k) and (l, m), in the sequel it is silently assumed that for each entity (vertex, edge, face, $\cdots$) only one space $\bar{V}_{\Gamma_{ik}}$ is used. We have the following consequence of the norm equivalence (7.46) for the resulting subspace splitting. It looks much more suitable for parallel implementations, but still does not avoid the global problem associated with $V_{j_0,0}$. This result is behind the theoretical results for many *primal iterative substructuring methods*, see [180, Chapter 4-5] or [197].

Theorem 7.8 *Under the conditions stated in Theorem 7.7, the refined subspace collection*

$$\mathbf{V}_{DD,1} := (V_{G_i},\, i = 1, \ldots, N,\, \bar{V}_{\Gamma_{ik}},\, \Gamma_{ik} \neq \emptyset,\, V_{j_0,0}). \tag{7.51}$$

provides a subspace splitting for $(V_{J,0}; a)$ with uniformly bounded stability constants and condition number.

Proof The upper stability bound is obvious, since the maximum number of v_{G_i} and $\bar{v}_{\Gamma_{ik}}$ whose supports contain any single point $x \in G$ depends only on the geometry of $\mathcal{T}_{j_0}$. Consider the triple bar norm in (7.46) for arbitrary $\bar{v}_{\bar{\Gamma}} \in \bar{V}_{\bar{\Gamma}}$. Since $\bar{\Phi}_{\Gamma_{ik}} \cap \bar{\Phi}_{\Gamma_{lm}} = \emptyset$ if $\Gamma_{ik} \neq \Gamma_{lm}$ and multiple occurrences of $\bar{V}_{\Gamma_{ik}}$ in the subspace collection $\mathbf{V}_{DD,1}$ are not allowed, it is clear that

$$|||\bar{v}_{\bar{\Gamma}}|||^2_{\bar{\Phi}_{\bar{\Gamma}}} = \inf_{\bar{v}_{\bar{\Gamma}} = \sum_{\Gamma_{ik}} \bar{v}_{\Gamma_{ik}}} |||\bar{v}_{\Gamma_{ik}}|||^2_{\bar{\Phi}_{\Gamma_{ik}}}.$$

On the other hand, since the upper stability bound holds for the infinite BPX-frame $\Phi_{J,0}$, it is also applicable to any subsystem (the constant can only improve). Applying this to all $\bar{\Phi}_{\Gamma_{ik}}$ separately, we have

$$\inf_{\bar{v}_{\bar{\Gamma}} = \sum_{\Gamma_{ik}} \bar{v}_{\Gamma_{ik}}} a(\bar{v}_{\Gamma_{ik}}, \bar{v}_{\Gamma_{ik}}) \leq C \inf_{\bar{v}_{\bar{\Gamma}} = \sum_{\Gamma_{ik}} \bar{v}_{\Gamma_{ik}}} |||\bar{v}_{\Gamma_{ik}}|||^2_{\bar{\Phi}_{\Gamma_{ik}}}.$$

The lower stability bound for the refined subspace splitting $\mathbf{V}_{DD,1}$ then follows by substitution. □

From a parallelization point of view, the $\bar{V}_{\Gamma_{ik}}$ problems are not completely independent: Besides communication with the coarse space $V_{j_0,0}$, they also involve (via the associated natural injection operators and their adjoints) communication with some of their neighbors. For example, to solve the problem on V_{G_i}, one must exchange information with all $\bar{V}_{\Gamma_{ik}}$ problems such that $\Gamma_{ik} \neq \emptyset$. Moreover, the problems on $\bar{V}_{\Gamma_{ik}}$ associated with vertices, edges, faces, $\cdots$ require information exchange with all their neighboring V_{G_i} (and through them with other $\bar{V}_{\Gamma_{lm}}$). For instance, if $d = 2$, the vertex problem at P needs to exchange information with the local V_{G_i} problems of all its neighboring G_i, the number of which is equal to the valence of P, while the edge problems need to communicate with only two neighboring local problems. This is similar for $d = 3$, where the local communication costs are even higher, with the vertex spaces being the biggest bottleneck. The other obvious bottleneck is the coarse problem associated with $V_{j_0,0}$.

To remedy this situation, there have been proposals to design preconditioners by omitting either the coarse problem or the vertex problems. Omitting both at the same time is problematic, since then the span of the remaining subspaces contains only finite element functions that vanish at the vertices of the G_i, i.e., in our setting, at the vertices of the cells in $\mathcal{T}_{j_0}$. Such strategies come with a penalty, since the uniform stability of the new splittings with respect to $j_0 < J$ is lost. This is shown by the following result:

Theorem 7.9 *Under the conditions of Theorem 7.7, the collection*

$$\mathbf{V}_{DD,2} := (V_{G_i},\ i = 1, \ldots, N,\ \bar{V}_{\Gamma_{ik}},\ \Gamma_{ik} \neq \emptyset).$$

(i.e., $\mathbf{V}_{DD,1}$ *without the coarse space component* $V_{j_0,0}$*) provides a subspace splitting for* $(V_{J,0}; a)$*, where the condition number* $\kappa_{DD,2}$ *scales as*

$$\kappa_{DD,2} \approx 2^{2j_0}$$

with constants independent of $j_0 < J$. *Furthermore, if the collection* $\mathbf{V}_{DD,3}$ *is obtained from* $\mathbf{V}_{DD,1}$ *by dropping all vertex spaces* $\bar{V}_P$ *with* $P \in \mathcal{V}_{j_0} \backslash \Gamma$, *then*

$$\kappa_{DD,3} \leq C \begin{cases} (J - j_0)^2, & d = 2, \\ 2^{(J-j_0)(d-2)}, & d \geq 3, \end{cases}$$

again with constants independent of $j_0 < J$.

Proof Since the subspace collections $\mathbf{V}_{DD,k}$, $k = 2, 3$, represent subsplittings of $\mathbf{V}_{DD,1}$, compare the language introduced in Subsection 2.3.3, we only have to worry about the lower stability estimate. For the splitting $\mathbf{V}_{DD,2}$, the result is straightforward, since this space splitting is obtained by partitioning the shortened BPX-frame

$$\Phi_{J,j_0+1,0} = \{\tilde{N}_{P,j} : P \in \mathcal{V}_{j,0}, j = j_0 + 1, \dots, J\}.$$

By the same reasoning that led to Theorem 7.8, we now have the norm equivalence

$$|||u_{J,0}|||^2_{\Phi_{J,j_0+1,0}} \approx |||u_{J,0}|||^2_{\mathbf{V}_{DD,2}},$$

where

$$|||u_{J,0}|||^2_{\mathbf{V}_{DD,2}} := \inf_{u_{J,0}=\sum_i v_{G_i}+\sum_{\Gamma_{ik}} \bar{v}_{\Gamma_{ik}}} \sum_i a(v_{G_i}, v_{G_i}) + \sum_{\Gamma_{ik}} a(\bar{v}_{\Gamma_{ik}}, \bar{v}_{\Gamma_{ik}}).$$

However, we have

$$\begin{aligned} |||u_{J,0}|||^2_{\Phi_{J,j_0+1,0}} &\approx \inf_{u_{J,0}=\sum_{j=j_0+1}^J v_{j,0}} \sum_{j=j_0+1}^{J} 2^{2j} \|v_{j,0}\|^2_{L^2} \\ &\approx 2^{2(j_0+1)} \|Q_{V_{j_0+1,0}} u_{J,0}\|^2_{L^2} + \sum_{j=j_0+2}^{J} 2^{2j} \|(Q_{V_{j,0}} - Q_{V_{j-1,0}}) u_{J,0}\|^2_{L^2} \\ &\leq C 2^{2(j_0+1)} \left(\|Q_{V_{0,0}} u_{J,0}\|^2_{L^2} + \sum_{j=1}^{J} 2^{2j} \|(Q_{V_{j,0}} - Q_{V_{j-1,0}}) u_{J,0}\|^2_{L^2} \right) \\ &\leq C 2^{2j_0} a(u_{J,0}, u_{J,0}) \end{aligned}$$

by the properties of the BPX-frame in $V_{J,0}$ and the L^2 ortho-projections $Q_{V_{j,0}}$.

The estimate cannot be improved. To see this, choose any $u_{J,0} = u_{0,0} \in V_{0,0}$. So, the lower stability constant of the subspace splitting $\mathbf{V}_{DD,2}$ satisfies $A \approx 2^{-2j_0}$. This implies $\kappa_{DD,2} \approx 2^{2j_0}$, since $B \approx 1$ for the upper stability constant (consider any $u_{J,0}$ in the L^2 orthogonal complement of $V_{j_0,0}$).

To prove the necessary lower stability bound for $\mathbf{V}_{DD,3}$, one can use the result of Problem 7.9. Take there $\mathbf{W} = (\bar{V}_P)_{P \in \mathcal{V}_{j_0} \backslash \Gamma}$ and $\mathbf{V} = \mathbf{V}_{DD,3}$. Then, by Theorem 7.8, the stability constants of the subspace splitting $(\mathbf{V}; \mathbf{W}) = \mathbf{V}_{DD,1}$ for $(V_{J,0}, a)$ are ≈ 1, independent of $j_0 < J$. Now consider an arbitrary

$$w = \sum_{P \in \mathcal{V}_{j_0} \setminus \Gamma} \bar{u}_P,$$

and note that we have

$$|||w|||^2_{\mathbf{W}} = \sum_P a(\bar{u}_P, \bar{u}_P)$$

due to the linear independence of the $\bar{V}_P$. Each $\bar{u}_P$ has a unique representation

$$\bar{u}_P = \sum_{j=j_0+1}^{J} c_{P,j} N_{P,j} = \sum_{j=j_0+1}^{J} \frac{c_{P,j}}{\tilde{N}_{P,j}(P)} \tilde{N}_{P,j},$$

such that

$$a(\bar{u}_P, \bar{u}_P) \approx \sum_{j=j_0+1}^{J} 2^{j(2-d)} c^2_{P,j}.$$

We decompose $\bar{u}_P$ with respect to $\mathbf{V}_{DD,3}$ as follows: We start with

$$\bar{u}_P = \bar{u}_P(P) N_{P,j_0} + \underbrace{\sum_{j=j_0+1}^{J} c_{P,j} N_{P,j} - (\sum_{j=j_0+1}^{J} c_{P,j}) N_{P,j_0}}_{\bar{u}_{P,j_0+1} :=}.$$

To shorten the notation, set $s_k := \sum_{j=k}^{J} c_{P,j}$ and

$$\bar{u}_{P,k} := \sum_{j=k}^{J} c_{P,j} N_{P,j} - s_k N_{P,k-1}, \qquad k = j_0 + 1, \ldots, J - 1.$$

By expressing $N_{P,k-1}$ in the nodal basis of $V_{k,0}$, i.e.,

$$N_{P,k-1} = N_{P,k} + \sum_{P' \in \mathcal{V}_k :\, P' \neq P} a_{P',k} N_{P',k},$$

one checks by recursion for $k = j_0 + 1, \ldots, J - 1$ that

$$\begin{aligned} \bar{u}_{P,j_0+1} &= \bar{u}_{P,j_0+2} - s_{j_0+1} \sum_{P' \in \mathcal{V}_{j_0+1} :\, P' \neq P} a_{P',j_0+1} N_{P',j_0+1} \\ &= \cdots = - \sum_{j=j_0+1}^{J-1} s_j \sum_{P' \in \mathcal{V}_j :\, P' \neq P} a_{P',j} N_{P',j}. \end{aligned}$$

The latter sum can be rewritten as

$$\bar{u}_{P,j_0+1} = \sum_{G_i :\, P \in G_i} v_{G_i,P} + \sum_{\Gamma_{ik} :\, P \in \Gamma_{ik} \neq \{P\}} \bar{v}_{\Gamma_{ik},P},$$

where

$$v_{G_i,P} = \sum_{j=j_0+1}^{J-1} \Big(\sum_{\tilde{N}_{P',j} \in \Phi_{G_i}} s_j a_{P',j} N_{P',j} \Big)$$

and

$$\bar{v}_{\Gamma_{ik},P} = \sum_{j=j_0+1}^{J-1} \Big(\sum_{\tilde{N}_{P',j} \in \bar{\Phi}_{\Gamma_{ik}}} s_j a_{P',j} N_{P',j} \Big).$$

It remains to note that the coefficients $a_{P',j}$ are uniformly bounded and the associated $\tilde{N}_{P',j}$ form a certain subsystem of the BPX-frame $\Phi_{J,0}$. By the upper stability estimate of the latter we get

$$a(v_{G_i,P}, v_{G_i,P}) \leq C \sum_{j=j_0+1}^{J-1} s_j^2 2^{j(2-d)},$$

and since

$$s_j^2 = \Big(\sum_{k=j}^{J} c_{P,k}\Big)^2 \leq \begin{cases} (J-j-1) \sum_{k=j}^{J} c_{P,k}^2, & d=2, \\ C 2^{J(d-2)} \sum_{k=j}^{J} 2^{k(2-d)} c_{P,k}^2, & d>2, \end{cases}$$

we get after substitution

$$\begin{aligned} a(v_{G_i,P}, v_{G_i,P}) &\leq C \begin{cases} (J-j_0)^2 \sum_{k=j_0+1}^{J} c_{P,k}^2, & d=2, \\ 2^{(J-j_0)(d-2)} \sum_{k=j_0+1}^{J} 2^{k(2-d)} c_{P,k}^2, & d>2, \end{cases} \\ &\leq C \begin{cases} (J-j_0)^2 a(\bar{u}_P, \bar{u}_P), & d=2, \\ 2^{(J-j_0)(d-2)} a(\bar{u}_P, \bar{u}_P), & d>2. \end{cases} \end{aligned}$$

Such upper bounds also hold for all $\bar{v}_{\Gamma_{ik}}$ with $P \in \Gamma_{ik}$ by similar reasoning. Since the number of terms v_{G_i} and $\bar{v}_{\Gamma_{ik}}$ associated with a given P is uniformly bounded, we finally have

$$\begin{aligned} \sum_{G_i:\, P \in G_i} a(v_{G_i,P}, v_{G_i,P}) &+ \sum_{\Gamma_{ik}:\, P \in \Gamma_{ik} \neq \{P\}} a(\bar{v}_{\Gamma_{ik},P}, \bar{v}_{\Gamma_{ik},P}) \\ &\leq C \begin{cases} (J-j_0)^2 a(\bar{u}_P, \bar{u}_P), & d=2, \\ 2^{(J-j_0)(d-2)} a(\bar{u}_P, \bar{u}_P), & d>2, \end{cases} \end{aligned} \tag{7.52}$$

After adding up the representations for all $\bar{u}_P$, we get

$$w = u_{j_0,0} + \sum_{P \in \mathcal{V}_{j_0} \setminus \Gamma} \sum_{G_i:\, P \in G_i} v_{G_i,P} + \sum_{\Gamma_{ik}:\, P \in \Gamma_{ik} \neq \{P\}} \bar{v}_{\Gamma_{ik},P},$$

where

$$u_{j_0,0} = \sum_{P \in \mathcal{V}_{j_0} \setminus \Gamma} \bar{u}_P(P) N_{P,j_0} \in V_{j_0,0}.$$

Using the above bound for s_{j_0+1} for each P, we have

$$
\begin{aligned}
a(u_{j_0,0}, u_{j_0,0}) &\leq C 2^{j_0(2-d)} \sum_{P \in \mathcal{V}_{j_0} \setminus \Gamma} \Big(\sum_{j=j_0+1}^{J} c_{P,j} \Big)^2 \\
&\leq C \sum_{P \in \mathcal{V}_{j_0} \setminus \Gamma} a(\bar{u}_P, \bar{u}_P) \begin{cases} (J - j_0), & d = 2, \\ 2^{(J-j_0)(d-2)}, & d > 2. \end{cases}
\end{aligned}
$$

Together with (7.52) and the fact that the number of occurrences of components associated with the same subdomain G_i or the same edge, face, $\cdots$ Γ_{ik} is uniformly bounded by a constant C_0, we conclude that

$$
\begin{aligned}
|||w|||_{DD,3}^2 &\leq a(u_{j_0,0}, u_{j_0,0}) + C_0 \sum_{P \in \mathcal{V}_{j_0} \setminus \Gamma} \sum_{G_i : P \in G_i} a(v_{G_i,P}, v_{G_i,P}) \\
&\quad + \sum_{\Gamma_{ik} : P \in \Gamma_{ik} \neq \{P\}} a(\bar{v}_{\Gamma_{ik},P}, \bar{v}_{\Gamma_{ik},P}) \\
&\leq C \sum_{P \in \mathcal{V}_{j_0} \setminus \Gamma} a(\bar{u}_P, \bar{u}_P) \begin{cases} (J - j_0)^2, & d = 2, \\ 2^{(J-j_0)(d-2)}, & d > 2, \end{cases} \\
&\leq C |||w|||_{\mathbf{W}}^2 \begin{cases} (J - j_0)^2, & d = 2, \\ 2^{(J-j_0)(d-2)}, & d > 2. \end{cases}
\end{aligned}
$$

Thus, the statement of Problem 7.9 yields a lower stability constant of $\approx (J - j_0)^{-2}$ for $d = 2$ and $\approx 2^{(J-j_0)(2-d)}$ if $d > 2$. This finally gives the result for $\mathbf{V}_{DD,3}$ in Theorem 7.9. □

We conclude with comments on the connection to the theory of primal iterative substructuring algorithms as presented in [180] or [197]. First, this theory serves a larger class of elliptic problems. For scalar elliptic problems, particular emphasis is placed on situations where the diffusion coefficient is approximately piecewise constant with respect to the domain partition $\{G_i\}$, i.e., where

$$
a(u, u) \approx a_w(u, u) := \sum_i w_i \int_{G_i} \|\nabla u\|_2^2 \, dx
$$

with moderate ellipticity constants but arbitrary weights w_i. In our setting, where all stability bounds depend on the ellipticity constants with respect to the standard H_0^1 norm, this would lead to strong dependencies of all results on the ratio $(\max_i w_i)/(\min_i w_i)$. Another aspect is that our approach relies on the existence of a BPX-frame, which in turn depends on a refinement structure. Although such an assumption is often satisfied in adaptive finite element methods, and we believe that the above statements can be extended to adaptive finite element spaces as discussed in Subsection 7.1.1, there are situations where refinement structures are not present, but the theory of iterative substructuring algorithms can still be developed.

There is another important methodological difference. We considered the interface problem on $\tilde{V}_{\tilde{\Gamma}}$ and its decomposition into problems on $V_{j_0,0}$ and on many $\bar{V}_{\Gamma_{ik}}$ associated with the substructures Γ_{ik} as problems on subspaces of $V_{J,0}$. Instead, the traditional substructuring methodology replaces it by the so-called *Schur complement problem* for finding the trace

$$u_{\tilde{\Gamma}} := T_{\tilde{\Gamma}} u_{J,0}$$

on the interface $\tilde{\Gamma}$ and decomposing it in a similar way into a coarse problem and vertex, edge, face, $\cdots$ problems. To describe some details, we restrict ourselves to the bilinear form $a(u,u) = a_w(u,u)$ on a linear finite element space $V_{J,0}$ (this is the setting in [180, 197]). Let $V_{\tilde{\Gamma}} := \{v_{\tilde{\Gamma}} : \; v_{\tilde{\Gamma}} = T_{\tilde{\Gamma}} v_{J,0} \text{ for some } v_{J,0} \in V_{J,0}\}$ denote the trace space of $V_{J,0}$ with respect to the interface $\tilde{\Gamma}$. Then the Schur complement form on $V_{\tilde{\Gamma}} \times V_{\tilde{\Gamma}}$ is given by

$$s_{\tilde{\Gamma}}(v_{\tilde{\Gamma}}, v_{\tilde{\Gamma}}) := \inf_{v_{J,0} \in V_{J,0}:\, v_{\tilde{\Gamma}} := T_{\tilde{\Gamma}} v_{J,0}} a(v_{J,0}, v_{J,0}).$$

The solution $v_{J,0}$ of the minimization problem on the right is called the *harmonic extension* of $v_{\tilde{\Gamma}}$ and is denoted by $E_{\tilde{\Gamma}} v_{\tilde{\Gamma}}$. Obviously, by definition we have

$$s_{\tilde{\Gamma}}(v_{\tilde{\Gamma}}, v_{\tilde{\Gamma}}) = a(E_{\tilde{\Gamma}} v_{\tilde{\Gamma}}, E_{\tilde{\Gamma}} v_{\tilde{\Gamma}}),$$

which allows us to identify $(V_{\tilde{\Gamma}}; s_{\tilde{\Gamma}})$ with the subspace $E_{\tilde{\Gamma}} V_{\tilde{\Gamma}}$ of $(V_{J,0}; a)$. If we denote by $u_{G_i} \in V_{G_i}$ the a-orthogonal projections of a given $u_{J,0} \in V_{J,0}$ and by $u_{\tilde{\Gamma}} = T_{\tilde{\Gamma}} u_{J,0}$ its trace on $\tilde{\Gamma}$, then we have an a-orthogonal decomposition

$$u_{J,0} = \sum_{i=1}^{N} u_{G_i} + E_{\tilde{\Gamma}} u_{\tilde{\Gamma}},$$

where

$$a(u_{J,0}, u_{J,0}) = \sum_{i=1}^{N} a(u_{G_i}, u_{G_i}) + s_{\tilde{\Gamma}}(u_{\tilde{\Gamma}}, u_{\tilde{\Gamma}}).$$

To rephrase this in our language of space splittings, the collection

$$\mathbf{V}'_{DD} = (V_{G_i}, i = 1, \ldots, N, \; E_{\tilde{\Gamma}} V_{\tilde{\Gamma}})$$

is an orthogonal subspace splitting for $(V; a)$. Furthermore, if we replace the space of harmonic extensions $E_{\tilde{\Gamma}} V_{\tilde{\Gamma}}$ by $(V_{\tilde{\Gamma}}; s_{\tilde{\Gamma}})$ complemented by $E_{\tilde{\Gamma}}$ as embedding operator $R_{\tilde{\Gamma}} : V_{\tilde{\Gamma}} \longrightarrow V_{J,0}$, the new space splitting has optimal stability constants $A = B = 1$. The latter splitting is the starting point for the analysis of iterative substructuring methods [180].

Since the assembly of the matrix representation $S_{\tilde{\Gamma}}$ of the bilinear form $s_{\tilde{\Gamma}}(\cdot,\cdot)$ and its exact inversion are more costly than the inversion of the linear system of the original finite element problem on $V_{J,0}$, a direct use of this space splitting is not justified. The standard workaround is to consider local Schur complement forms $s_{\Gamma_{ik}}(\cdot,\cdot)$ and local harmonic extension operators $E_{\Gamma_{ik}}$ associated with the vertices, edges, faces, $\cdots$ Γ_{ik} and a coarse grid problem on $(V_{j_0}; a)$. This is done as follows: Let G_{ik} denote the union of all cells in $\mathcal{T}_{j_0}$ that share Γ_{ik}, introduce the local finite element space $V_{G_{ik}} = V_{J,0} \cap H_0^1(G_{ik})$ and let $V_{\Gamma_{ik}}$ be the local trace space of $V_{G_{ik}}$ and $T_{\Gamma_{ik}} : V_{G_{ik}} \to V_{\Gamma_{ik}}$ the associated trace map. Then define

$$s_{\Gamma_{ik}}(v_{\Gamma_{ik}}, v_{\Gamma_{ik}}) = \inf_{v_{\Gamma_{ik}} = T_{\Gamma_{ik}} v_{G_{ik}}} a(v_{G_{ik}}, v_{G_{ik}}) \tag{7.53}$$

and denote by $E_{\Gamma_{ik}} v_{\Gamma_{ik}}$ the minimizer in $V_{G_{ik}}$ of the optimization problem on the right-hand side of (7.53). It can be shown (see Problem 7.10) that the resulting refined subspace collection

$$\mathbf{V}'_{DD,1} := (V_{G_i}, i = 1, \ldots, N;\ E_{\Gamma_{ik}} V_{\Gamma_{ik}}, \Gamma_{ik} \neq \emptyset, V_{j_0,0}) \tag{7.54}$$

provides a stable subspace splitting with stability constants uniformly bounded with respect to $j_0 < J$. Furthermore, the subspaces $E_{\Gamma_{ik}} V_{\Gamma_{ik}}$ of local harmonic extensions can be replaced by $(V_{\Gamma_{ik}}; s_{\Gamma_{ik}})$ with embedding $R_{\Gamma_{ik}} = E_{\Gamma_{ik}}$ (extended by zero to all of G). This notation is deliberately chosen to emphasize the analogy with the subspace splitting (7.51) treated in Theorem 7.8. It is possible to reinterpret (7.54) in the sense of (7.51) and vice versa, see Problem 7.11. The difference is in the size of the local subspaces $V_{G_{ik}}$ and $\bar{V}_{\Gamma_{ik}}$ in $V_{J,0}$ associated with Γ_{ik}.

If one opts for the version using $(V_{\Gamma_{ik}}; s_{\Gamma_{ik}})$ spaces, the remaining task is the solution of variational problems with the bilinear forms $s_{\Gamma_{ik}}(\cdot,\cdot)$ and the computation of harmonic extensions. As in most theoretical work on iterative substructuring methods, we assume for simplicity that the latter is done exactly and focus on the local Schur complement problems. Since the vertex problems involve the determination of a single value, there is nothing to solve. Therefore, let $\dim(\Gamma_{ik}) > 0$. The definition of $s_{\Gamma_{ik}}(\cdot,\cdot)$ is equivalent to the definition of a discrete trace norm on $V_{\Gamma_{ik}} = T_{\Gamma_{ik}} V_{G_{ik}}$ onto lower-dimensional manifolds. In Subsection 7.1.2, we already treated such a problem for $H^s(G)$ with polyhedral domain $G \subset \mathbb{R}^d$ and considered the trace to its $(d-1)$-dimensional boundary Γ, where the boundary space was identified with $H^{s-1/2}(\Gamma)$. We now have spaces $H_0^1(G_{ik}) \cap V_{J,0}$ and traces to inner polyhedral manifolds Γ_{ik} of dimension $d-1, \ldots, 1, 0$. Thus, for $(d-1)$-dimensional Γ_{ik} (edges for $d = 2$, faces for $d = 3$), we expect the Schur complement form $s_{\Gamma_{ik}}$ to be spectrally equivalent to the scalar product in $H^{1/2}(\Gamma_{ik}) \cap V_{\Gamma_{ik},0}$, while lower-dimensional traces are not well-defined for general functions from $H_0^1(G_{ik})$. Note at this point that the subsystem

$$\Phi_{G_{ik}} := \{\tilde{N}_{P,j} :\ \mathrm{supp}(N_{P,j}) \subset G_{ik},\ j = j_0 + 1, \ldots, J\}$$

of the BPX-frame $\Phi_{J,0}$ is a frame in $V_{G_{ik}}$ due to Problem 7.7. Following the proof of Theorem 7.3, one checks that

$$s_{\Gamma_{ik}}(v_{\Gamma_{ik}}, v_{\Gamma_{ik}}) \approx \inf_{v_{\Gamma_{ik}} = \sum_{j=j_0+1}^{J} v_{j,\Gamma_{ik}}} \sum_{j=j_0+1}^{J} 2^{(2-r_{ik})j} \|v_{j,\Gamma_{ik}}\|_{L^2(\Gamma_{ik})}^2,$$

where $r_{ik} := d - \dim(\Gamma_{ik})$ is the dimension gap of the substructure Γ_{ik} and the $v_{j,\Gamma_{ik}}$ belong to the subspaces

$$V_{j,\Gamma_{ik}} := V_{\Gamma_{ik}} \cap T_{\Gamma_{ik}} V_{j,0}$$

of $V_{\Gamma_{ik}}$, $j = j_0 + 1, \ldots, J$.

Thus, if we take the nodal basis in $V_{J,0}$ to represent its elements and consequently the $L^2(\Gamma_{ik})$ stable nodal basis in $V_{\Gamma_{ik}}$ to represent $v_{\Gamma_{ik}}$, then the condition $\kappa_{\Gamma_{ik}}$ of the Schur complement problem is determined by the ratio of the best lower and upper bounds for the quotient

$$\frac{s_{\Gamma_{ik}}(v_{\Gamma_{ik}}, v_{\Gamma_{ik}})}{\|v_{\Gamma_{ik}}\|^2_{L^2(\Gamma_{ik})}} \approx \frac{\inf_{v_{\Gamma_{ik}}=\sum_{j=j_0+1}^{J} v_{j,\Gamma_{ik}}} \sum_{j=j_0+1}^{J} 2^{(2-r_{ik})j} \|v_{j,\Gamma_{ik}}\|^2_{L^2(\Gamma_{ik})}}{\|v_{\Gamma_{ik}}\|^2_{L^2(\Gamma_{ik})}}.$$

This immediately gives us

$$\kappa_{\Gamma_{ik}} \approx \begin{cases} 2^{J-j_0}, & r_{ik} = 1, \\ J - j_0, & r_{ik} = 2, \\ 1, & 2 < r_{ik} < d. \end{cases} \tag{7.55}$$

The message is: If O(1) conditioning of the described space splitting independent of $j_0 < J$ is the goal, and if exact inversion is not possible, then preconditioning of the local Schur complement problems governed by $s_{\Gamma_{ik}}(\cdot,\cdot)$ is only necessary for $r_{ik} = 1, 2$.

Let us briefly point out the connection between our statements in this subsection and the results discussed in [180, Chapter 4-5]. There, the theory of iterative substructuring methods is developed in terms of linear finite element subspaces $V_0(\mathcal{T}') \subset V_0(\mathcal{T})$ of $H_0^1(G)$ over simplicial partitions $\mathcal{T}' < \mathcal{T}$ of mesh-sizes

$$H := \operatorname{diam}(\mathcal{T}') < h := \operatorname{diam}(\mathcal{T}).$$

Furthermore, the cells of $\mathcal{T}'$ are taken as subdomains G_i and $V_0(\mathcal{T}')$ serves as the coarse space component. In the notation of this subsection we have $H \approx 2^{-j_0}$ and $h = 2^{-J}$.

For $d = 3$, Theorem 7.8 is related to Algorithms 5.5 and 5.7 in [180], which concern versions of Schur complement solvers with a condition number behavior close to O(1). The appearance of the additional factors $(1 + \log(H/h))^2$ is due to the use of the nodal interpolation operator on $V_0(\mathcal{T})$ and the replacement of the Schur complement forms for the edges by $L^2(\Gamma_{ik})$ scalar products (compare (7.55) for $d = 3$ and $r_{ik} = 2$ and note that in our case $\log(H/h) \approx J - j_0$).

Algorithm 5.2 in [180] corresponds to the splitting $\mathbf{V}_{DD,3}$, where the vertex spaces $\bar{V}_P$ are dropped from $\mathbf{V}_{DD,1}$. Theorem 5.3 in [180] claims for Algorithm 5.2 a condition number behavior of $O((1 + \log(H/h))^2)$ for $d = 2$ and $O((H/h)(1 + \log(H/h))^2)$ for $d = 3$. This agrees with the result in Theorem 7.3. However, unlike our result, Theorem 5.3 in [180] also guarantees that the constants in the condition number estimate are independent of the size of the coefficients w_i that characterize the diffusion coefficient.

7.2.2 Overlapping domain decomposition

Overlapping domain decomposition methods use decompositions into local problems with respect to enlarged regions $G_i' \supset G_i$ and a coarse problem (dropping it has the same negative effect as in the non-overlapping case). With sufficient overlap between the G_i', there is no need for a separate interface problem. We keep the restrictions and notation of the previous subsection. Recall that we assume the multiscale setting of Theorem 7.3 and consider $(V_{J,0}; a)$, where the bilinear form $a(\cdot,\cdot)$ is equivalent to the $H_0^1(G)$ scalar product. As before, the domain partition $\{G_i\}$ is given by the cells in $\mathcal{T}_{j_0}$, which we now replace by larger regions G_i' by attaching one layer of cells from $\mathcal{T}_{j_1}$ to G_i, where $j_0 \le j_1 \le J$ is fixed, i.e.,

$$G_i' = \cup\{C \in \mathcal{T}_{j_1} : G_i \cap C \neq \emptyset\}, \qquad i = 1, \ldots, N.$$

The overlap region (the union of all intersections $G_i' \cap G_k'$) is thus a layer of cells in $\mathcal{T}_{j_1}$ attached to the interface $\tilde{\Gamma}$. Its width $\delta \approx 2^{-j_1}$ characterizes the amount of overlap. Figure 7.6 shows the situation for $d = 2$.

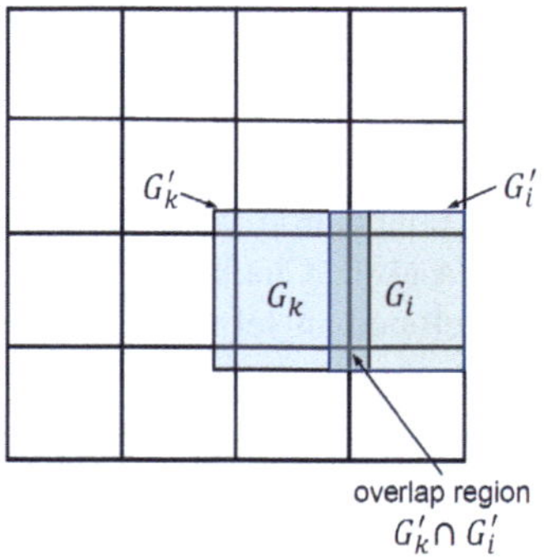

Fig. 7.6 Illustration of the construction of an overlapping domain decomposition.

Next, define

$$V_{G_i'} = \{v_{j,0} \in V_{J,0} : \operatorname{supp}(v_{j,0}) \subset G_i'\}.$$

These are now the local subspaces which, together with the coarse space $V_{j_0,0}$, are the standard components of an overlapping domain decomposition preconditioner.

Theorem 7.10 *Under the above assumptions, the collection*

$$\mathbf{V} = (V_{G_i'}, i = 1, \ldots, N, V_{j_0,0})$$

provides a stable subspace splitting for $(V_{J,0}; a)$ *with condition number* $\kappa_{\mathbf{V}} \le C2^{j_1 - j_0}$*, where* C *is independent of* $0 < j_0 \le j_1 < J$.

Proof The upper stability bound follows from the finite overlap property of $\mathcal{T}_{j_0}$, which is due to the regularity and quasi-uniformity assumptions on the refinement process. To see this, let N_o denote the maximum number of G_i' such that the intersection of their interiors is non-empty. Then for any decomposition $u_{J,0} = v_{j_0,0} + \sum_i v_{G_i'}$, where $v_{j_0,0} \in V_{j_0,0}$ and $v_{G_i'} \in V_{G_i'}$, $i = 1, \dots, N$, no more than $N_o + 1$ terms can vanish at any point in G. So we have

$$a(u_{J,0}, u_{J,0}) \approx \|u_{J,0}\|^2_{H_0^1} \le C(N_o + 1)(\|v_{j_0,0}\|^2_{H_0^1} + \sum_i \|v_{G_i'}\|^2_{H_0^1})$$
$$\le C(N_o + 1)(a(v_{j_0,0}, v_{J,0}) + \sum_i a(v_{G_i'}, v_{G_i'})).$$

To show the lower stability bound, we fix a decomposition with the property

$$u_{J,0} = \sum_{j=0}^{J} v_{j,0} : \qquad \sum_{j=0}^{J} 2^{2j} \|v_{j,0}\|^2_{L^2} \le C\|u_{J,0}\|^2_{H_0^1} \le Ca(u_{J,0}, u_{J,0}). \tag{7.56}$$

According to Theorem 7.3 c), such a decomposition exists. By the same theorem we also have

$$u_{j_0,0} = \sum_{j=0}^{j_0} v_{j,0} : \qquad a(u_{j_0,0}, u_{j_0,0}) \le C \sum_{j=0}^{j_0} 2^{2j} \|v_{j,0}\|^2_{L^2}. \tag{7.57}$$

Each of the $v_{j,0}$ with $j > j_0$ is now decomposed with respect to the $V_{G_i'}$, for which we constantly use the L^2 stability of the nodal basis in $V_{j,0}$. For $j = j_1, \dots, J$, the decomposition is straightforward: Just note that, by the construction of the extended regions G_i', each $N_{P,j}$ has a support that is completely contained in at least one G_i'. Thus, by arbitrarily assigning the $N_{P,j}$ to one of these G_i' and splitting the nodal basis representation of $v_{j,0}$ accordingly, we obtain

$$v_{j,0} = \sum_{i=1}^{N} u_{j,G_i'} : \qquad \sum_i \|u_{j,G_i'}\|^2_{L^2} \le C\|v_{j,0}\|^2_{L^2}, \tag{7.58}$$

where $u_{j,G_i'} \in V_{G_i'}$, $i = 1, \dots, N$ and $j = j_1, \dots, J$.

For the remaining levels $j = j_0+1, \dots, j_1-1$, we have to be more careful, because the nodal basis functions $N_{P,j}$ with $P \in \tilde{\Gamma}$ do not belong to any of the $V_{G_i'}$. The following trick helps: For each $j = j_0 + 1, \dots, j_1 - 1$, consider the decomposition

$$v_{j,0} = \underbrace{(v_{j,0} - E_{j,\tilde{\Gamma}} v_{j,\tilde{\Gamma}})}_{w_j :=} + \sum_{l=j}^{j_1-1} \underbrace{(E_{l,\tilde{\Gamma}} - E_{l+1,\tilde{\Gamma}}) v_{j,\tilde{\Gamma}}}_{w_{l+1,j} :=} + E_{j_1,0} v_{j,\tilde{\Gamma}},$$

where $v_{j,\tilde{\Gamma}} = T_{\tilde{\Gamma}} v_{j,0}$ and the "minimal" discrete extension operators $E_{l,\tilde{\Gamma}}$ of level $l = j_1, \dots, J$ are the same as used before in the previous subsection, see (7.48). For arbitrary $v_{j,0} \in V_{j,0}$ and their traces $v_{j,\tilde{\Gamma}} = T_{\tilde{\Gamma}} v_{j,0}$ on $\tilde{\Gamma}$, we have

$$\|E_{l,\tilde{\Gamma}} v_{j,\tilde{\Gamma}}\|_{L^2}^2 \approx 2^{-l} \|v_{j,\tilde{\Gamma}}\|_{L^2(\tilde{\Gamma})}^2 \leq C 2^{j-l} \|v_{j,0}\|_{L^2}^2; \quad l = j, \ldots, J, \tag{7.59}$$

by their definition and the properties of the L^2 norms with respect to the nodal bases in $V_{\tilde{\Gamma}}$, $V_{j,0}$ and $V_{j_1,0}$, Adding up the above representations, we get

$$\begin{aligned}\sum_{j=j_0+1}^{j_1-1} v_{j,0} &= \sum_{j=j_0+1}^{j_1-1} (w_j + \sum_{l=j}^{j_1-1} w_{l+1,j} + E_{j_1,0} v_{j,\tilde{\Gamma}}) \\ &= w_{j_0+1} + \sum_{l=j_0+2}^{j_1-1} (w_l + \sum_{j=j_0+1}^{l-1} w_{l,j}) + \sum_{j=j_0+1}^{j_1-1} (w_{j_1,l} + E_{j_1,0} v_{j,\tilde{\Gamma}}) = \sum_{l=j_0+1}^{j_1} w_{l,0},\end{aligned}$$

with $w_{l,0} \in V_{l,0}$, where

$$\begin{aligned} w_{j_0+1,0} &= w_{j_0+1}, \\ w_{l,0} &= w_l + \sum_{j=j_0+1}^{l-1} w_{l,j} = w_l + (E_{l-1,\tilde{\Gamma}} - E_{l,\tilde{\Gamma}}) \sum_{j=j_0+1}^{l-1} v_{j,\tilde{\Gamma}}\end{aligned}$$

for $j = j_0 + 2, \ldots, j_1 + 1$, and similarly

$$w_{j_1,0} = \sum_{j=j_0+1}^{j_1-1} (w_{j_1,l} + E_{j_1,0} v_{j,\tilde{\Gamma}}) = E_{j_1,0}(\sum_{j=j_0+1}^{j_1-1} v_{j,\tilde{\Gamma}}).$$

By construction, these $w_{l,0}$, $l = j_0+1, \ldots, j_1$, have a zero trace on $\tilde{\Gamma}$, and their nodal basis representations can therefore be split, similar to (7.58), as

$$w_{l,0} = \sum_{i=1}^{N} w_{l,G_i'} : \qquad \sum_i \|w_{l,G_i'}\|_{L^2}^2 \leq C \|w_{l,0}\|_{L^2}^2, \quad w_{l,G_i'} \in V_{G_i'}, \tag{7.60}$$

where $w_{l,G_i'} \in V_{G_i'}$ for $i = 1, \ldots, N$ and $l = j_0 + 1, \ldots, j_1$.

We are now ready to prove a lower stability bound. The decomposition we use is

$$u_{J,0} = u_{j_0,0} + \sum_{i=1}^{N} \underbrace{\left(\sum_{l=j_0+1}^{j_1} w_{l,G_i'} + \sum_{l=j_1}^{J} u_{l,G_i'} \right)}_{\tilde{u}_{G_i'} :=},$$

where, according to the upper stability bound in Theorem 7.3 c), we have

$$a(\tilde{u}_{G_i'}, \tilde{u}_{G_i'}) \leq C \left(\sum_{l=j_0+1}^{j_1} 2^{2l} \|w_{l,G_i'}\|_{L^2(G_i')}^2 + \sum_{l=j_1}^{J} 2^{2j} \|u_{l,G_i'}\|_{L^2(G_i')}^2 \right), \quad i = 1, \ldots, N.$$

Together with (7.57), (7.58) and (7.60), we get after summation

$$a(u_{j_0,0}, u_{j_0,0}) + \sum_{i=1}^{N} a(\tilde{u}_{G_i'}, \tilde{u}_{G_i'})$$
$$\leq C\left(\sum_{j=0}^{j_0} 2^{2j} \|v_{j,0}\|_{L^2}^2 + \sum_{l=j_0+1}^{j_1} 2^{2l} \|w_{l,0}\|_{L^2}^2 + \sum_{j=j_1}^{J} 2^{2j} \|v_{j,0}\|_{L^2}^2\right).$$

It remains to deal with the second sum. By definition (see the lines above (7.60)), the $w_{l,0}$ are sums of w_l and terms involving "minimal" discrete extensions of the levels l and $l-1$ of the traces $v_{j,\tilde{\Gamma}}$, $j < l$. For the former we have

$$\|w_l\|_{L^2}^2 = \|v_{l,0} - E_{l,\tilde{\Gamma}} v_{l,0}\|_{L^2}^2 \leq C\|v_{l,0}\|_{L^2}^2,$$

since the nodal basis representation of $v_{l,0} - E_{l,\tilde{\Gamma}} v_{l,0}$ is obtained from that of $v_{l,0}$ by setting the coefficients associated with $N_{P,l}$, $P \in \tilde{\Gamma}$, to zero. This immediately gives

$$\|w_{j_0+1,0}\|_{L^2}^2 \leq C\|v_{j_0+1,0}\|_{L^2}^2.$$

For $l = j_0 + 2, \ldots, j_1 - 1$, we can additionally use (7.59), which results in

$$\|w_{l,0}\|_{L^2}^2 \leq 2(\|w_l\|_{L^2}^2 + \|(E_{l-1,\tilde{\Gamma}} - E_{l,\tilde{\Gamma}}) \sum_{j=j_0+1}^{l-1} v_{j,\tilde{\Gamma}}\|_{L^2}^2)$$
$$\leq C(\|v_{l,0}\|_{L^2}^2 + 2^{-l}\| \sum_{j=j_0+1}^{l-1} v_{j,\tilde{\Gamma}}\|_{L^2(\tilde{\Gamma})}^2).$$

The sum on the right can be bounded by

$$\| \sum_{j=j_0+1}^{l-1} v_{j,\tilde{\Gamma}}\|_{L^2(\tilde{\Gamma})}^2 = \| \sum_{j=j_0+1}^{l-1} 2^{-j/2}(2^{j/2} v_{j,\tilde{\Gamma}})\|_{L^2(\tilde{\Gamma})}^2$$
$$\leq C2^{-j_0} \sum_{j=j_0+1}^{l-1} 2^j \|v_{j,\tilde{\Gamma}}\|_{L^2(\tilde{\Gamma})}^2 \leq C2^{-j_0} \sum_{j=j_0+1}^{l-1} 2^{2j} \|v_{j,0}\|_{L^2}^2.$$

After substitution we get

$$\|w_{l,0}\|_{L^2}^2 \leq C(\|v_{l,0}\|_{L^2}^2 + 2^{-l-j_0} \sum_{j=j_0+1}^{l-1} 2^{2j} \|v_{j,0}\|_{L^2}^2)$$

for $l = j_0 + 2, \ldots, j_1 - 1$. The analogous estimate holds for $l = j_1$, but without the term $\|v_{j_1,0}\|_{L^2}^2$.

With this, we are ready to finish. We have

$$\sum_{l=j_0+1}^{j_1} 2^{2l}\|w_{l,0}\|_{L^2}^2 \le C\left(\sum_{l=j_0+1}^{j_1-1} 2^{2l}\|v_{l,0}\|_{L^2}^2 + 2^{-j_0}\sum_{l=j_0+2}^{j_1} 2^l \sum_{j=j_0+1}^{l-1} 2^{2j}\|v_{j,0}\|_{L^2}^2\right)$$
$$\le C2^{j_1-j_0}\sum_{j=j_0+1}^{j_1-1} 2^{2j}\|v_{j,0}\|_{L^2}^2,$$

and after a final substitution, together with (7.56), we get the desired lower stability bound

$$a(u_{j_0,0}, u_{j_0,0}) + \sum_{i=1}^{N} a(\tilde{u}_{G_i'}, \tilde{u}_{G_i'}) \le C2^{j_1-j_0} a(u_{J,0}, u_{J,0}).$$

The bound for the condition number $\kappa_{\mathbf{V}}$ follows by comparing the upper and lower stability bounds. It cannot be improved, which is intuitively clear, but the construction of counterexamples depends on the finite element type, is technically cumbersome, and is not given here. □

Theorem 7.10 is a particular instance of Theorem 3.13 in [180], which is proved for linear finite element spaces $V_0(\mathcal{T}) \subset H_0^1(G)$ over regular, not necessarily quasi-uniform simplicial partitions $\mathcal{T}$ for $d = 2$ and $d = 3$. It concerns a similar decomposition into locally overlapping spaces $V_{G_i'}$ and a coarse space $V_0(\mathcal{T}')$ over a simplicial partition $\mathcal{T}'$ whose cells have a size comparable to the diameters H_i of the shape-regular regions G_i'. It is not assumed that $\mathcal{T}' \prec \mathcal{T}$, and finite element functions from $V_0(\mathcal{T}')$ are simply mapped to $V_0(\mathcal{T})$ by nodal interpolation. The G_i' must be unions of cells from $\mathcal{T}$, and the overlap parameter for G_i' is defined by

$$\delta_i := \inf_{x\in G_i'} \max_{k=1,\dots,N:\, x\in G_k'} \operatorname{dist}(x, \partial G_k'\backslash\Gamma),$$

$i = 1,\dots,N$. The quantity $H/\delta := \max_{i=1,\dots,N} H_i/\delta_i$ enters the bound for the condition number

$$\kappa_{\mathbf{V}} \le C(1 + H/\delta),$$

which is given in [180, Theorem 3.13]. In our particular case, we have $H_i \approx 2^{-j_0}$ and $\delta_i \approx 2^{-j_1}$, so that $H/\delta \approx 2^{j_1-j_0}$. Thus, our Theorem 7.10 is consistent with the result of [180].

We conclude with a remark on the iterative solvers of domain decomposition type, which are based on the above subspace splittings of $V_{J,0}$. In contrast to the multilevel preconditioners and the multiplicative Schwarz implementations briefly discussed in Subsections 7.1.1.4 and 7.1.1.5, where the order of execution is important, now the order of execution of the subproblems can be arbitrary and in particular random, since the application of the natural injection operators R_i from the subspaces V_{G_i} and $V_{G_i'}$ into $V_{J,0}$ and their adjoints is trivial. Also, strengthened Cauchy-Schwarz inequalities are readily available, so there is no logarithmic penalty to be expected for multiplicative Schwarz versions. We refer to [90] for some experiments with a randomized domain decomposition algorithm motivated by fault mitigation in large parallel computer networks.

7.3 Tensor product decomposition methods

In higher dimensions, tensor product techniques are often appropriate. First, in many applications with dimension $d > 3$, the domains $G \subset \mathbb{R}^d$ are product domains anyway. Second, the handling of discretization spaces obtained by arbitrary spatial refinement underlying the constructions described in Chapter 3 and in this chapter is already difficult for $d = 3$ and becomes prohibitive for larger d. Moreover, the dimension of the standard discretization spaces V_j grows with $\approx 2^{dj}$, leading to the well-known *curse of dimension*. This reduces their practical relevance to quite small values of j, which are usually not suitable to obtain a sufficiently small approximation error. Furthermore, the convergence rate does not generally improve with increasing dimension d, but rather remains the same up to d-dependent constants.

In this section, we use two examples to show how space splittings for Sobolev spaces on product domains can be obtained from their counterparts defined on intervals. The first example concerns the construction of robust finite element BPX-type and prewavelet preconditioners for the anisotropic diffusion equation

$$\sum_{i=1}^{d} w_i \partial^2_{x_i,x_i} u + w_0 u = f \tag{7.61}$$

with constant coefficients and homogeneous Dirichlet or Neumann boundary conditions. Here we assume $w_0 \geq 0$ and $w_i > 0$, $i = 1, \ldots, d$. For simplicity, we take the unit cube $G = (0,1)^d$ as a model for a bounded product domain in $\mathbb{R}^d$. Note that the weights w_i in (7.61) play a different role than the w_i in the scalar diffusion equation considered in the previous section. The second example concerns the construction of economic splittings for sparse grid discretizations, which are a workhorse for many simulations in higher dimensions, mainly due to the much smaller growth of the dimension of the underlying discretization spaces with respect to d and resolution level j.

7.3.1 Tensor product Sobolev spaces and robust splittings

We begin with the following observations: For the parameters $s_i \geq 0$, $i = 1, \ldots, d$, we set

$$H^{\mathbf{s}}_{mix}(G) := H^{s_1}((0,1)) \otimes \cdots \otimes H^{s_d}((0,1)), \qquad \mathbf{s} := (s_1, \ldots, s_d).$$

These spaces are called Sobolev spaces of functions of dominating mixed derivative. For example, when equipped with the classical Sobolev norm $\|\cdot\|_{H^1,1}$ (see Subsection A.5.1), the tensor product of two $H^1((0,1))$ spaces has the norm

$$\|u\|^2_{H^{(1,1)}_{mix}} = \|u\|^2_{L^2} + \|\partial_{x_1} u\|^2_{L^2} + \|\partial_{x_2} u\|^2_{L^2} + \|\partial^2_{x_1,x_2} u\|^2_{L^2}.$$

Here the term containing the mixed second-order derivative has highest order and cannot be bounded by the other terms, which explains the name dominating mixed derivative.

Let us introduce the *weighted intersection* $(H; b_{\mathbf{c}}) := \cap_i c_i (H_i; b_i)$ of given Hilbert spaces $(H_i; b_i)$, where $c_i > 0$ are scaling constants, by setting

$$H := \cap_i H_i, \qquad b_{\mathbf{c}}(u, u) := \sum_i c_i b_i(u, u). \tag{7.62}$$

Note that for the arguments below it is sufficient to assume in (7.62) that the bilinear forms $b_i(\cdot, \cdot)$ induce only semi-norms on H_i. It turns out that Sobolev spaces on product domains G can be expressed as weighted intersections of Sobolev spaces of functions with dominating mixed derivative. We demonstrate this using an example motivated by an application to the anisotropic diffusion equation (7.61).

To do this, let $\mathbf{e}_i$ be the i-th unit coordinate vector in $\mathbb{Z}^d$, $i = 1, \ldots, d$, and define $H_{mix}^{\mathbf{e}_i}$ as the tensor product of $d - 1$ $L^2((0, 1))$ spaces and (at the i-th position) one $H^1((0, 1))$ space with norm

$$\|u\|^2_{H_{mix}^{\mathbf{e}_i}} = \|u\|^2_{L^2} + \|\partial_{x_i} u\|^2_{L^2}.$$

So the intersection $H := \cap_{i=1}^d H_{mix}^{\mathbf{e}_i}$ coincides with $H^1(G)$ and produces an equivalent norm

$$\|u\|^2_H = d\|u\|^2_{L^2} + \sum_{i=1}^d \|\partial_{x_i} u\|^2_{L^2} = d\|u\|^2_{L^2} + \|\nabla u\|^2_{L^2}.$$

More importantly, the same procedure gives the norm

$$\|u\|^2_U = \bar{w}_0 \|u\|^2_{L^2} + \sum_{i=1}^d w_i \|\partial_{x_i} u\|^2_{L^2}, \qquad \bar{w}_0 := \sum_{i=0}^d w_i,$$

for the weighted intersection

$$U := w_0 L^2(G) \cap w_1 H_{mix}^{\mathbf{e}_1} \cap \cdots \cap w_d H_{mix}^{\mathbf{e}_d},$$

which, up to the lower order L^2 term, is identical to the energy norm induced by the Neumann problem for the anisotropic diffusion problem (7.61).

Similar statements hold for $H_0^1(G)$. Here it is convenient to take the equivalent norm $\|u\|_{H_0^1} := \|\nabla u\|_{L^2}$ for both $H_0^1((0, 1))$ and $H_0^1(G)$ spaces. Then, using the same notation but replacing $H^1((0, 1))$ in the definition of $H_{mix}^{\mathbf{e}_i}$ with $H_0^1((0, 1))$ and denoting the result by $H_{mix,0}^{\mathbf{e}_i}$ we have

$$\|u\|^2_{H_0} = \sum_{i=1}^d \|u^{(\mathbf{e}_i)}\|^2_{L^2} = \|\nabla u\|^2_{L^2} = \|u\|^2_{H_0^1},$$

if we set $H_0 := \cap_{i=1}^d H_{mix,0}^{\mathbf{e}_i} = H_0^1(G)$. We also have

$$\|u\|_{U_0}^2 = w_0\|u\|_{L^2}^2 + \sum_{i=1}^{d} w_i\|\partial_{x_i}u\|_{L^2}^2, \tag{7.63}$$

if we use the weighted intersection

$$U_0 := w_0 L^2(G) \cap w_1 H^{\mathbf{e}_1}_{mix,0} \cap \cdots \cap w_d H^{\mathbf{e}_d}_{mix,0}. \tag{7.64}$$

The bilinear form induced by the weighted intersection norm (7.63) on U_0 coincides with the bilinear form $a(\cdot,\cdot)$ induced by (7.61) on $H_0^1(G)$. Thus, if we produce splittings for the weighted intersection U_0 with w_i-independent condition number estimates, we arrive at a robust preconditioner construction for solving (7.61).

To achieve this for certain multiscale splittings, we use Lemma 2.7 and Lemma 2.9. For now, we do not specify the concrete nature (finite elements, wavelets,...) of the nested, dense sequence of subspaces

$$\{0\} \neq V_0 \subset V_1 \subset \cdots V_j \subset \cdots,$$

of $H_0^1((0,1))$ (or later of $H^1((0,1))$). We simply assume that $(V_j; 2^{2j}(\cdot,\cdot)_{L^2})_{j\geq 0}$ provides a stable space splitting for $(H_0^1((0,1)), (\cdot,\cdot)_{H_0^1})$ with condition number κ, and that

$$W_0 := V_0, \qquad W_j = V_j \ominus_{L^2} V_{j-1}, \qquad j \geq 1,$$

is the sequence of associated L^2 orthogonal complement spaces. By Lemma 2.9, the space splitting $(W_j; 2^{2j}(\cdot,\cdot)_{L^2})_{j\geq 0}$ is then also stable for $(H_0^1((0,1)), (\cdot,\cdot)_{H_0^1})$, with the same condition number κ. This holds because, in the notation of Lemma 2.9, we compute $\theta_j = (4/3)2^{-2j}$ from $\rho_j = 2^{-2j}$. Furthermore, $(W_j)_{j\geq 0}$ is an orthogonal subspace splitting for $L^2((0,1))$ with condition number 1. By Lemma 2.7, this gives stable splittings of the form

$$(W_{\mathbf{j}}; 2^{2j_i}(\cdot,\cdot)_{L^2(G)})_{\mathbf{j}\geq\mathbf{0}}, \quad W_{\mathbf{j}} := W_{j_1} \otimes \cdots \otimes W_{j_d}, \quad \mathbf{j} := (j_1,\ldots,j_d) \in \mathbb{Z}_+^d,$$

for each of the $H^{\mathbf{e}_i}_{mix,0}$, $i = 1,\ldots,d$. Their condition numbers are all equal to κ. Also, $(W_{\mathbf{j}})_{\mathbf{j}\geq\mathbf{0}}$ is an orthogonal subspace splitting for $L^2(G)$, with condition number 1.

Thus, adding these splittings according to the definition of the intersection norm, we get the following result:

Lemma 7.3 *If $(V_j; 2^{2j}(\cdot,\cdot)_{L^2})_{j\geq 0}$ is a space splitting for $(H_0^1((0,1)); (\cdot,\cdot)_{H_0^1})$ with condition number κ as given above, then the Riesz splitting*

$$(W_{\mathbf{j}}; (\pi^{-2}w_0 + \sum_{i=1}^{d} w_i 2^{2j_i})(\cdot,\cdot)_{L^2(G)})_{\mathbf{j}\geq\mathbf{0}} \tag{7.65}$$

is a stable space splitting for $U_0 = H_0^1(G)$ equipped with the bilinear form induced by the anisotropic diffusion equation (7.61). Its condition number is κ and independent of the coefficients w_i, $i = 0,\ldots,d$.

Proof The only tricky point is the appearance of the scaling factor π^{-2} for the L^2 term. Let $\lambda_{\min}$ and $\lambda_{\max}$ be the best constants in the norm equivalence

$$\lambda_{\max}^{-1}\|u\|_{H_0^1}^2 \le |||u|||_{\mathbf{W}}^2 := \sum_{j=0}^{\infty} 2^{2j}\|Q_{W_j}u\|_{L^2}^2 \le \lambda_{\min}^{-1}\|u\|_{H_0^1}^2, \qquad u \in H_0^1((0,1)),$$

associated with the space splitting $(W_j; 2^{2j}(\cdot,\cdot)_{L^2})_{j\ge 0}$. From the above we have

$$\kappa = \lambda_{\max}/\lambda_{\min}.$$

So by Lemma 2.7 we also have

$$\lambda_{\max}^{-1}\|u\|_{H^{\mathbf{e}_i}_{mix,0}}^2 \le \sum_{\mathbf{j}\ge\mathbf{0}} 2^{2j_i}\|Q_{W_\mathbf{j}}u\|_{L^2}^2 \le \lambda_{\min}^{-1}\|u\|_{H^{\mathbf{e}_i}_{mix,0}}^2, \qquad u \in H^{\mathbf{e}_i}_{mix,0},$$

$i = 1, \ldots, d$. For the L^2 term, we have the equality

$$\|u\|_{L^2}^2 = \sum_{\mathbf{j}\ge\mathbf{0}} \|Q_{W_\mathbf{j}}u\|_{L^2}^2, \qquad u \in L^2(G).$$

After these inequalities are added to a two-sided inequality for the weighted intersection norm, the desired result, i.e., the final condition number estimate by κ independent of the w_i, follows if

$$\lambda_{\max}^{-1} \le 1 \le \lambda_{\min}^{-1}.$$

In general, however, this may not hold. What we can be sure of is that

$$\lambda_{\max}^{-1} \le \pi^{-2} \le \lambda_{\min}^{-1}. \tag{7.66}$$

In fact, the Poincaré inequality on $(0, 1)$ has the form

$$\|u\|_{L^2}^2 \le \pi^{-2}\|u\|_{H_0^1}^2, \qquad u \in H_0^1((0,1)),$$

with equality for

$$u_0(x) = \sin(\pi x).$$

With this u_0 we have

$$\pi^{-2}\|u_0\|_{H_0^1}^2 = \|u_0\|_{L^2}^2 \le |||u_0|||_{\mathbf{W}}^2 \le \lambda_{\min}^{-1}\|u_0\|_{H_0^1}^2,$$

which gives $\pi^{-2} \le \lambda_{\min}^{-1}$. Next, take an arbitrary $0 \ne v_0 \in V_0 = W_0$. Then

$$\lambda_{\max}^{-1}\|v_0\|_{H_0^1}^2 \le |||v_0|||_{\mathbf{W}}^2 = \|v_0\|_{L^2}^2 \le \pi^{-2}\|v_0\|_{H_0^1},$$

where the Poincaré inequality was used in the last step. This completes the proof of (7.66).

With (7.66) at hand, we have for the U_0 norm in (7.64) the bound

$$\begin{aligned}
\|u\|_{U_0}^2 &= \sum_{i=1}^{d} w_i \|u\|_{H^{\mathbf{e}_i}_{mix,0}}^2 + w_0\pi^{-2}(\pi^2\|u\|_{L^2}^2) \\
&\le \lambda_{\max}\left(\sum_{i=1}^{d} w_i \sum_{\mathbf{j}\ge\mathbf{0}} 2^{2j_i}\|Q_{W_\mathbf{j}}u\|_{L^2}^2 + w_0\pi^{-2}\sum_{\mathbf{j}\ge\mathbf{0}}\|Q_{W_\mathbf{j}}u\|_{L^2}^2\right) \\
&\le \lambda_{\max}\left(\sum_{\mathbf{j}\ge\mathbf{0}}(\sum_{i=1}^{d} w_i 2^{2j_i} + w_0\pi^{-2})\|Q_{W_\mathbf{j}}u\|_{L^2}^2\right).
\end{aligned}$$

A similar argument leads to the lower bound, and this completes the proof of Lemma 7.3. □

The practical use of this splitting depends on suitable L^2 stable bases in the L^2 orthogonal complement spaces $W_j \subset H_0^1((0,1))$. In the linear finite element case considered in [87], piecewise linear prewavelets, as introduced in Subsection 3.1.4.1, were used. Since they result, after proper scaling, only in semi-orthogonal Riesz bases, the discretization of the subproblems on the subspaces $W_\mathbf{j}$ leads to rapidly growing condition numbers with increasing d. In [68], also devoted to (7.61) and equipped with mixtures of Dirichlet and Neumann boundary conditions, the authors used fully orthogonal piecewise polynomial multiwavelets, where the discretization of the L^2 problems on the subspaces $W_\mathbf{j}$ becomes trivial. However, the construction and implementation of these multiwavelets is not easy, and the constants in the required norm equivalences for $H_0^1((0,1))$ and $H^1((0,1))$ are relatively large. The main focus in [68] is the development of an adaptive wavelet solver for (7.61), analogous to Subsection 7.1.4.

So it is worthwhile to return to splittings involving the spaces

$$V_\mathbf{j} = V_{j_1} \otimes \cdots \otimes V_{j_d}, \quad \mathbf{j} \ge \mathbf{0},$$

which, at least in the context of finite elements, allow much simpler basis constructions than the $W_\mathbf{j}$. This was done in [87] for $d = 2$. To explain the idea, we give up some optimality in the final constants. First, let $l_i \in \mathbb{Z}$ be the *smallest* integers such that

$$w_0\pi^{-2} \le 4^{l_0}, \qquad w_i \le 4^{l_i}, \quad i = 1, \ldots, d.$$

If we replace the forefactors in (7.65) with the new factors $(4^{l_d} + \sum_{i=1}^{d} 4^{j_i+l_i})$, then the modified splitting has a condition number of $\le 4\kappa$ in the worst case. Now set

$$\ell := \max_{i=0,\ldots,d} l_i, \qquad \mathbf{l}_k = (\ell - l_1 + k, \ldots, \ell - l_d + k) \ge \mathbf{0}, \quad k = 0, 1, \ldots. \quad (7.67)$$

Since for $\mathbf{0} \le \mathbf{j} \le \mathbf{l}_0$ we obviously have

$$4^{\ell} \le 4^{l_0} + \sum_{i=1}^{d} 4^{j_i+l_i} \le (d+1)4^{\ell},$$

we conclude that

$$4^{\ell}\|Q_{V_{\mathbf{l}_0}}u\|_{L^2}^2 \le \sum_{\mathbf{0}\le\mathbf{j}\le\mathbf{l}_0} (4^{l_0} + \sum_{i=1}^{d} 4^{j_i+l_i})\|Q_{W_{\mathbf{j}}}u\|_{L^2}^2 \le (d+1)4^{\ell}\|Q_{V_{\mathbf{l}_0}}u\|_{L^2}^2.$$

For $k = 1, 2, \dots$ we similarly find

$$4^{l_0} + d4^{\ell+k-1} \le 4^{l_0} + \sum_{i=1}^{d} 4^{j_i+l_i} \le 4^{l_0} + d4^{\ell+k}, \qquad \mathbf{l}_{k-1} < \mathbf{j} \le \mathbf{l}_k,$$

and get

$$\begin{aligned}\frac{d}{4}4^{\ell+k}\|(Q_{V_{\mathbf{l}_k}} - Q_{V_{\mathbf{l}_{k-1}}})u\|_{L^2}^2 &\le \sum_{\mathbf{l}_{k-1}<\mathbf{j}\le\mathbf{l}_k} (4^{l_0} + \sum_{i=1}^{d} 4^{j_i+l_d})\|Q_{W_{\mathbf{j}}}u\|_{L^2}^2 \\ &\le (d+1)4^{\ell+k}\|(Q_{V_{\mathbf{l}_k}} - Q_{V_{\mathbf{l}_{k-1}}})u\|_{L^2}^2.\end{aligned}$$

Thus, summation over all $k = 0, 1, \dots$ and another use of Lemma 2.9 gives the following result:

Theorem 7.11 *If $(V_j; 2^{2j}(\cdot,\cdot)_{L^2})_{j\ge0}$ is a space splitting for $(H_0^1((0,1)); (\cdot,\cdot)_{H_0^1})$ with condition number κ as specified above, then*

$$(V_{\mathbf{l}_k}; 2^{2(\ell+k)}(\cdot,\cdot)_{L^2(G)})_{k\ge0}$$

is a stable space splitting for $U_0 = H_0^1(G)$ equipped with the bilinear form induced by the anisotropic diffusion equation (7.61). Its condition number is $\le 4(d+1)\kappa$ and is independent of the coefficients w_i, $i = 0, \dots, d$. The integer $\ell \in \mathbb{Z}$ and the multi-indices $\mathbf{l}_k \in \mathbb{Z}_+^d$ are defined in (7.67) and depend on the coefficients w_i via the integers l_i.

The factor $(d + 1)$ in the condition number estimate is due to the rough treatment of the terms with $\mathbf{0} \le \mathbf{j} \le \mathbf{l}_0$ and can probably be removed. Clearly, Theorem 7.11 concerns a redundant space splitting and leads to another BPX-frame consisting of properly scaled nodal basis functions from the involved subspaces $V_{\mathbf{l}_k}$. This opens the way for the construction of adaptive finite element subspaces that inherit good condition number estimates. For this purpose, the quasi-interpolant techniques and arguments from Subsections 7.1.1 and 7.1.2 can be reused. If the initial sequence V_j consists of linear finite element subspaces of $H_0^1((0,1))$ on uniform partitions of step-size 2^{-j-1} (this guarantees our condition $V_0 \neq 0$), then the $V_{\mathbf{l}_k}$ are multilinear finite element spaces on anisotropic d-dimensional rectangular partitions of step-size $2^{l_i-\ell-k}$ in i-th coordinate direction.

If a specific solution space $V_{\mathbf{J}}$ with fixed $\mathbf{J} = (J_1, \ldots, J_d)$ is given, then the argument for the infinite space splitting derived in Theorem 7.11 needs only a few changes. In fact, as a consequence of Lemma 7.3, we can start from the Riesz subsplitting

$$(W_{\mathbf{j}}; (\pi^2 w_0 + \sum_{i=1}^{d} w_i 2^{2j_i})(\cdot,\cdot)_{L^2(G)})_{\mathbf{0}\le\mathbf{j}\le\mathbf{J}}$$

with the condition $\le \kappa$ for $V_{\mathbf{J}}$ considered as subspace of the energy space generated by (7.61) and repeat the same replacement steps. The only difference is that the spaces $W_{\mathbf{j}}$ with $\mathbf{j}$ not subordinated to $\mathbf{J}$ are neglected. This leads to the slightly modified, finite stable space splitting

$$(V_{\tilde{\mathbf{I}}_k}; 2^{2(\ell+k)}(\cdot,\cdot)_{L^2(G)})_{0\le k\le k_{\mathbf{J}}}, \qquad \tilde{\mathbf{I}}_k := \min(\mathbf{I}_k, \mathbf{J}),$$

for $V_{\mathbf{J}}$, for which the condition number bound of Theorem 7.11 remains in place. Here, $k_{\mathbf{J}}$ denotes the smallest k such that $\mathbf{I}_k \ge \mathbf{J}$. Similar consequences hold for sparse grid spaces (another class of discretization spaces spanned by certain $V_{\mathbf{j}}$ or $W_{\mathbf{j}}$), which we discuss in the next subsection.

Let us make a few comments about the changes required when zero Neumann conditions (or mixtures of zero Neumann and Dirichlet boundary conditions as in [68]) complement the anisotropic diffusion equation (7.61). Then we need multiscale space splittings for one or several of the spaces

$$H^1((0,1)) \supset \left\{ \begin{matrix} \{u \in H^1((0,1)) : \; u(0) = 0\} \\ \{u \in H^1((0,1)) : \; u(1) = 0\} \end{matrix} \right\} \supset H_0^1((0,1)).$$

When using wavelet or finite element constructions, this means changes for nodal basis functions, boundary wavelets and prewavelets associated with $x = 0$ or $x = 1$, respectively. As $w_0 \longrightarrow 0$, we expect problems, since for $w_0 = 0$ there is a necessary condition

$$\int_G f(x)\, dx = 0$$

for solvability, and solutions of the Neumann problem for (7.61) are unique only up to constants.

For the sake of clarity, we will treat only the pure Neumann problem, set $w_0 = 0$, and consider dimension $d = 2$. The reader can easily extend this to the other cases. To save our tensor product argument, we will assume that $(V_j, 2^{2j}(\cdot,\cdot)_{L^2})_{j\ge 0}$ is a stable space splitting with condition number κ for $H^1((0,1))$, where V_0 contains the one-dimenional space W_{-1} of constant functions on $(0,1)$, and is L^2 orthogonally decomposed into W_{-1} and

$$W_0 := V_0 \ominus_{L^2} W_{-1}.$$

As a consequence, $(W_j, 2^{2j}(\cdot,\cdot)_{L^2})_{j\ge -1}$ is an L^2 orthogonal space splitting for $H^1((0,1))$ with the same κ. It is preferable to introduce

$$\tilde{H}^1((0,1)) = H^1((0,1)) \ominus_{L^2} W_{-1}, \qquad \|u\|_{\tilde{H}^1}^2 := \|u'\|_{L^2}^2.$$

Let $\bar{u} = \int_0^1 u(x)\,dx$. Then we have the one-dimensional Poincaré-Sobolev inequality

$$\|u - \bar{u}\|_{L^2}^2 \le \pi^{-2}\|u\|_{\tilde{H}^1}^2$$

with the extremal function $u_0(x) = \cos(\pi x)$. This shows that the $\tilde{H}^1$ norm and the original H^1 norm are equivalent on $\tilde{H}^1((0,1))$, i.e.,

$$\|u\|_{\tilde{H}^1}^2 \le \|u\|_{H^1}^2 \le (1+\pi^{-2})\|u\|_{\tilde{H}^1}^2, \qquad u \in \tilde{H}^1((0,1)),$$

and the space splitting $(W_j, 2^{2j}(\cdot,\cdot)_{L^2})_{j\ge 0}$ for $\tilde{H}^1((0,1))$ is stable with condition number

$$\kappa' \le (1+\pi^{-2})\kappa.$$

Again, one can check that the associated stability constants satisfy $\lambda'_{\min} \le \pi^2 \le \lambda'_{\max}$ (this fact would be needed if $w_0 \ne 0$).

Now we carefully repeat the above replacement steps. First, we see by Lemma 2.7 that

$$\tilde{H}_{mix}^{(1,0)} := \tilde{H}^1((0,1)) \otimes L^2((0,1))$$

has norm $\|\partial_{x_1} u\|_2^2$ and admits a space splitting

$$(W_{j_1,j_2}, 2^{2j_1}(\cdot,\cdot)_{L^2})_{j_1\ge 0, j_2 \ge -1}$$

with the same condition number κ'. A similar result holds for $\tilde{H}_{mix}^{(0,1)}$. So the space

$$\tilde{H}^1(G) := H^1(G) \ominus_{L^2} W_{-1,-1},$$

where $W_{-1,-1} = W_{-1} \otimes W_{-1} = \{const.\}$, can be equipped with the intersection norm

$$a(u,u) := \|u\|^2_{w_1\tilde{H}_{mix}^{(1,0)} \cap w_2 \tilde{H}_{mix}^{(0,1)}},$$

which coincides with the energy norm induced by the homogeneous Neumann problem for (7.61). Here we have tacitly extended the norms and space splittings by zero from $\tilde{H}_{mix}^{(0,1)}$ and $\tilde{H}_{mix}^{(1,0)}$ to all of $\tilde{H}^1(G)$. Furthermore, $(\tilde{H}^1(G); a)$ admits the stable space splitting

$$\begin{aligned}(W_{j_1,j_2}; (w_1 2^{2j_1} + w_2 2^{2j_2})(\cdot,\cdot)_{L^2})_{j_1,j_2\ge 0} \\ \cup (W_{j_1,-1}; w_1 2^{2j_1}(\cdot,\cdot)_{L^2})_{j_1\ge 0} \cup (W_{-1,j_2}; w_2 2^{2j_2}(\cdot,\cdot)_{L^2})_{j_2\ge 0}\end{aligned} \tag{7.68}$$

with condition number κ'. Thus, in contrast to $H_0^1(G)$ (i.e., the case of zero Dirichlet conditions), some new space components associated with essentially univariate functions have appeared. In fact, the elements of $W_{j_1,-1}$ and W_{-1,j_2} are only functions of x_1 and x_2, respectively.

Again, it is desirable to replace the splitting into W_{j_1,j_2} spaces by a splitting using the more covenient V_{j_1,j_2} spaces. For $d = 2$ we can simply assume that $w_1 \ge w_2$ and determine $\ell \ge 0$ as the integer such that $4^\ell \approx w_1/w_2$ holds best. Then

$$(W_{j_1,j_2}; (4^{\ell+j_1} + 4^{j_2})(\cdot,\cdot)_{L^2})_{j_1,j_2\ge 0}$$
$$\cup (W_{j_1,-1}; 4^{\ell+j_1}(\cdot,\cdot)_{L^2})_{j_1\ge 0} \cup (W_{-1,j_2}; 4^{j_2}(\cdot,\cdot)^2_{L})_{j_2\ge 0} \tag{7.69}$$

is a stable space splitting for $(\tilde{H}^1(G); a)$ with condition number $\le 2\kappa'$.

Now we apply the coarsening strategy to this splitting by combining spaces W_{j_1,j_2} with scaling factors $\approx 4^k$ in (7.69) into larger spaces $\tilde{W}_k$. To do this, set

$$\tilde{W}_k := W_{-1,k},$$

for $k = 0, \ldots, \ell - 1$,

$$\tilde{W}_\ell := W_{0,-1} \oplus_{L^2} \cdots \oplus_{L^2} W_{0,\ell} \oplus_{L^2} W_{-1,\ell},$$

and

$$\tilde{W}_k := W_{k-\ell,k-1} \oplus_{L^2} W_{k-\ell-1,k} \oplus_{L^2} W_{k-\ell,k} \oplus_{L^2} W_{-1,k}$$

for $k > \ell$. Checking the forefactors in (7.69) associated with the W_{j_1,j_2} that form $\tilde{W}_k$, we immediately see that

$$(\tilde{W}_k; 2^{2k}(\cdot,\cdot)_{L^2})_{k\ge 0} \tag{7.70}$$

remains a stable space splitting for $(\tilde{H}^1(G); a)$ with condition number $\le 4\kappa'$. The space splittings (7.68-7.70) give rise to asymptotically optimal finite element prewavelet and wavelet preconditioners for the homogeneous Neumann problem, see Problem 7.11 for the linear finite element case.

The last step is to realize that, by setting

$$\tilde{V}_{j_1,j_2} := V_{j_1,j_2} \ominus_{L^2} \{const.\},$$

we have

$$\tilde{W}_0 = \tilde{V}_{-1,0}, \qquad \tilde{W}_k = \tilde{V}_{-1,k} \ominus_{L^2} \tilde{V}_{-1,k-1}, \quad k = 1, \ldots, \ell - 1,$$

and

$$\tilde{W}_k = \tilde{V}_{k-\ell,k} \ominus_{L^2} \tilde{V}_{k-\ell-1,k-1}, \quad k = \ell, \ell + 1, \ldots.$$

Applying Lemma 2.9 again leads us to the final stable space splitting

$$(\tilde{V}_{-1,k}; 2^{2k}(\cdot,\cdot)_{L^2})_{0\le k<\ell} \cup (\tilde{V}_{k-\ell,k}; 2^{2k}(\cdot,\cdot)_{L^2})_{k\ge\ell} \tag{7.71}$$

for $(\tilde{H}^1(G); a)$, with condition number $\le 2\kappa' \le 4\kappa$, which is independent of the weights ω_i in (7.61). Stable space splittings for $(\tilde{V}_{J_1,J_2}; a)$, $J_1, J_2 \ge 0$, with similar properties are immediate, just replace all $\tilde{V}_{j_1,j_2}$ appearing in (7.71) by $\tilde{V}_{\min(j_1,J_1),\min(j_2,J_2)}$. Note that the space splitting (7.71) and its subsplittings are with respect to the spaces $\tilde{V}_{j_1,j_2} = V_{j_1,j_2} \ominus_{L^2} \{const.\}$ and not V_{j_1,j_2}. However, after replacing the L^2 scalar product on $\tilde{V}_{j_1,j_2} \subset V_{j_1,j_2}$ by the discrete L^2 scalar product resulting from the L^2 stable nodal basis in V_{j_1,j_2}, the solution of the local problems on $\tilde{V}_{j_1,j_2}$ can be computed by a simple rank-one modification of the solution on V_{j_1,j_2}.

7.3.2 Sparse grid spaces

On tensor product domains, there is another popular construction of discretization spaces for PDE problems, namely sparse grid spaces, which were originally introduced in the finite element context in [203]. In the Fourier series context, they are called hyperbolic cross spaces, see Subsection 3.1.2. In numerical integration and approximation, similar constructions have long been used. For details and historical remarks, we refer to the review article [29] .

We will restrict ourselves to finite element sparse grid subspaces of $H_0^1(G)$, where

$$G = (0, 1)^d$$

is again the d-dimensional cube, and reuse the simplified notation of the previous subsection. We start from the stable splitting of $H_0^1(G)$ into L^2 orthogonal spaces

$$(W_{\mathbf{j}}; (\sum_{i=1}^{d} 2^{2j_i})(\cdot, \cdot)_{L^2})_{\mathbf{j}\geq\mathbf{0}}, \tag{7.72}$$

obtained from a univariate stable space splitting $(V_j; 2^{2j}(\cdot, \cdot)_{L^2})_{j\geq 0}$ with $V_0 \neq \{0\}$ of $H_0^1((0, 1))$ by the tensor product construction outlined in the previous subsection. The condition number κ of (7.72) is inherited from this univariate splitting. This corresponds to setting the weights in (7.61) to $w_0 = 0$ and $w_i = 1$, $i = 1, \ldots, d$, which means that we restrict ourselves to the homogeneous Dirichlet problem for the Laplacian on $G = (0, 1)^d$. The reader may extend the results to the general case, see [86, Theorem 5] and Problem 7.12 for $d = 2$. Obviously, tensor products of simultaneous Riesz bases, and in particular biorthogonal wavelet bases in $H_0^1((0, 1))$ and $L^2((0, 1))$, as described in Subsection 3.1.4, would lead to similar constructions and results.

Let $\Lambda \subset \mathbb{Z}_+^d$ be a finite *monotone* index set, i.e., $\mathbf{j} \in \Lambda$ implies $\mathbf{k} \in \Lambda$ for all $\mathbf{0} \leq \mathbf{k} \leq \mathbf{j}$. For further use, we denote the (non-monotone) subset of extremal indices in Λ by

$$\Lambda' := \{\mathbf{j} \in \Lambda : \ \mathbf{k} > \mathbf{j} \ \Rightarrow \mathbf{k} \notin \Lambda\} \subset \Lambda.$$

We define

$$S_\Lambda = \mathrm{span}((W_{\mathbf{j}}, \mathbf{j} \in \Lambda)) = \mathrm{span}((V_{\mathbf{j}}, \mathbf{j} \in \Lambda')),$$

and call it the generalized sparse grid space in $H_0^1(G)$ associated with $(V_j)_{j\geq 0}$ and Λ. If $\dim(V_j) \approx 2^j$ grows geometrically, then it is not hard to check that

$$\dim(S_\Lambda) \approx \sum_{\mathbf{j}\in\Lambda'} \dim(V_{\mathbf{j}}).$$

Here are three relevant examples:

Full grid space. First, set $\Lambda_{\mathbf{J}} = \{\mathbf{j} : \ \mathbf{0} \leq \mathbf{j} \leq \mathbf{J}\}$, where $\mathbf{J} \in \mathbb{Z}_+^d$. Then $\Lambda'_{\mathbf{J}} = \{\mathbf{J}\}$ and $S_{\Lambda_{\mathbf{J}}} = V_{\mathbf{J}}$ recovers the usual anisotropic full grid spaces.

Sparse grid space. Second, set

$$\Lambda_J = \{\mathbf{j} \in \mathbb{Z}_+^d : |\mathbf{j}|_1 := j_1 + \cdots + j_d \le J\}.$$

Then $\Lambda'_J = \{\mathbf{j} \in \mathbb{Z}_+^d : |\mathbf{j}|_1 := j_1 + \cdots + j_d = J\}$ and

$$S_J := S_{\Lambda_J} = \text{span}((V_\mathbf{j}, |\mathbf{j}|_1 = J))$$

is called the standard sparse grid space of level J associated with $(V_j)_{j\ge 0}$. Since the cardinality of the set Λ'_J is

$$|\Lambda'_J| = \binom{d+J-1}{d-1} \approx J^{d-1},$$

and $\dim(W_\mathbf{j}) \approx \dim(V_\mathbf{j}) \approx 2^{|\mathbf{j}|_1}$, we have $\dim(S_J) \approx J^{d-1}2^J$ with constants that depend on d but not on J.

Energy-optimized sparse grid space. Third, let

$$\Lambda_{J,\alpha} := \{\mathbf{j} \in \mathbb{Z}_+^d : |\mathbf{j}|_1 \le \alpha J + (1-\alpha)|\mathbf{j}|_\infty\},$$

where $|\mathbf{j}|_\infty = \max_{i=1,\dots,d} j_d$ and $\alpha \ge 0$ is a parameter. The associated spaces

$$S_{J,a} := S_{\Lambda_{J,a}}$$

are called energy-optimized sparse grid spaces and have been studied in [85] using a slightly different but equivalent definition and notation. For $\alpha = 1$ we have

$$S_{J,1} = S_J,$$

while for $\alpha \longrightarrow \infty$ the limit space is

$$S_{J,\infty} := \text{span}((V_{(J-1,\dots,J-1)}, V_{(J,0,\dots,0)}, \dots, V_{(0,\dots,0,J)}))$$

and close to the full grid space $V_{J,\dots,J}$. For $\alpha = 0$, the endpoint space is

$$S_{J,0} := \text{span}((V_{(J,0,\dots,0)}, \dots, V_{(0,\dots,0,J)})).$$

The most interesting case is $0 < \alpha < 1$, where one can prove that

$$\dim(S_{J,\alpha}) \approx 2^J, \quad J \ge 0,$$

with constants depending on d and α but independent of J. Since, under suitable regularity assumptions, the subspaces $S_{J,\alpha}$ still yield approximation rates of the form $\mathrm{O}(2^{-\beta J})$ in standard Sobolev norms with some $\beta > 0$, which are comparable to the approximation rates of full grid spaces $V_{(J,\dots,J)}$, the slow dimension growth for $\alpha \le 1$ is attractive. We refer to [29, 85] for more details.

In the following, we focus on obtaining uniformly stable space splittings for S_Λ, considered as a subspace of $H_0^1(G)$. Obviously, since L^2 stable finite element prewavelet implementations are available, we can rely on the L^2 orthogonal subsplittings

$$(W_{\mathbf{j}}; (\sum_{i=1}^{d} 2^{2j_i})(\cdot,\cdot)_{L^2})_{\mathbf{j}\in\Lambda} \tag{7.73}$$

of the Riesz splitting (7.72) with condition number $\leq \kappa$ and its further discretization into a Riesz basis splitting using the restriction of the multivariate prewavelet basis.

In the past, bases in generalized sparse grid spaces have been conveniently introduced by tensor products of univariate hierarchical bases. For linear finite elements, if we again identify V_j with the space $V_{j+1,0}$ of piecewise linear functions v_j on the uniform partition $\mathcal{T}_{j+1}$ of step-size 2^{-j-1} on $(0,1)$ such that $v_j(0) = v_j(1) = 0$, then the univariate hierarchical basis

$$\mathcal{H} := \cup_{j\geq 0}\mathcal{H}_j, \qquad \mathcal{H}_j = \{h_{i,j} := N_{(2i-1)2^{-j-1},j+1}, i = 1,\ldots,2^j\}$$

is given by the nodal basis functions $N_{P,j+1}$ associated with the nodal points P in $\mathcal{V}_{j+1}\backslash\mathcal{V}_j$. This system coincides with the so-called Faber system, introduced in 1910 as the first example of a Schauder basis in the subspace

$$C_0([0,1]) = \{u \in C([0,1]) : \; u(0) = u(1) = 0\}$$

of the Banach space of continuous functions on the interval $[0,1]$. It happens to be a complete orthogonal system in $(H_0^1((0,1)); (\cdot,\cdot)_{H_0^1})$. But unlike the L^2 semi-orthogonal prewavelet systems, it is not a Riesz basis in any $L_p((0,1))$.

We will use $\mathcal{H}$ to create algebraic bases $\mathcal{H}_\Lambda$ in S_Λ as follows:

$$\mathcal{H}_\Lambda := \cup_{\mathbf{j}\in\Lambda}\mathcal{H}_{\mathbf{j}}, \quad \mathcal{H}_{\mathbf{j}} := \{h_{\mathbf{i},\mathbf{j}}(x) := \prod_{r=1}^{d} h_{i_r,j_r}(x_r),\; i_r = 1,\ldots,2^{j_r},\; r = 1,\ldots,d\}.$$

The basis property is obvious. We mention that an analog of Theorem 7.2 is proved in [140], which estimates the condition number $\kappa_{HB,\Lambda}$ of the Galerkin matrix of an $H_0^1(G)$ elliptic problem on S_Λ with respect to the properly scaled hierarchical basis $\mathcal{H}_\Lambda$. In particular, it was found that

$$\kappa_{HB,\Lambda_{J,\alpha}} \approx J^{d-1}2^{J\alpha(d-1)/(\alpha+(d-1))}, \quad J \geq 1, \tag{7.74}$$

with constants depending on $d > 1$ and $\alpha > 0$. Note that in the isotropic full grid case associated with $\Lambda_{(J,\ldots,J)}$, the tensor product hierarchical basis $\mathcal{H}_{\Lambda_{(J,\ldots,J)}}$ is quite different from the hierarchical basis underlying Theorem 7.2. In fact, each $h_{\mathbf{i},\mathbf{j}}$ in $\mathcal{H}_{\Lambda_{(J,\ldots,J)}}$ is the product of nodal basis functions of different levels $j_1+1,\ldots,j_d+1$ ($j_i \leq J$) in each coordinate direction and has support in d-dimensional rectangles of side-lengths $2^{-j_1},\ldots,2^{-j_d}$. In contrast, the hierarchical basis functions in Theorem 7.2 are nodal basis functions $N_{P,j+1}$ with $P \in \mathcal{V}_{j+1}\backslash\mathcal{V}_j$, $j \leq J$, i.e., they are products of univariate nodal basis functions of the same level j and with support in d-dimensional cubes of side-length 2^{-j}. This explains the different asymptotic behavior of the condition numbers, namely

$$\kappa_{HB,\Lambda_{(J,\ldots,J)}} \approx J^{d-1}2^{J(d-1)}, \qquad \kappa_{HB,J} \approx \begin{cases} J^2, & d=2, \\ 2^{J(d-2)}, & d>2, \end{cases} \qquad J \geq 1.$$

The exponential growth of the condition numbers in (7.74) for $d \geq 2$ makes the use of hierarchical basis preconditioners for generalized sparse grid discretizations less attractive and motivates the search for better alternatives.

We therefore turn to preconditioners for $(S_\Lambda; (\cdot,\cdot)_{H_0^1})$ discretizations based on splittings into spaces of the form $(V_{\mathbf{j}}; 2^{2|\mathbf{j}|_\infty}(\cdot,\cdot)_{L^2})$, where the indices $\mathbf{j} \in \Lambda$ are properly chosen so that the condition number κ_Λ of the resulting splittings is as small as possible. The advantage of such space splittings is that, after switching to the corresponding nodal basis representations in the involved $V_{\mathbf{j}}$, they lead to BPX-type preconditioners with condition number $\approx \kappa_\Lambda$. If elements of S_Λ are represented with respect to $\mathcal{H}_\Lambda$, then the natural embedding operators $R_{\mathbf{j}} : V_{\mathbf{j}} \longrightarrow S_\Lambda$ require a basis change from the nodal basis in $V_{\mathbf{j}}$ to the tensor product hierarchical basis $\mathcal{H}_{\Lambda_{\mathbf{j}}}$ in $V_{\mathbf{j}}$, followed by an extension-by-zero step to obtain the coefficients with respect to $\mathcal{H}_\Lambda$. Thus, while the subproblem solvers, the embedding operators $R_{\mathbf{j}}$ and their adjoints have efficient implementations, the question remains whether such splittings exhibit moderate condition number growth.

For this, we proceed as before: First we combine the $W_{\mathbf{j}}$ spaces in the space splitting (7.73) with similar forefactors $2^{2j_1} + \cdots + 2^{2j_d} \approx 2^{2|\mathbf{j}|_\infty}$ into a new finite sequence of L^2 orthogonal subspaces

$$W_{k,\Lambda} := \bigoplus \{W_{\mathbf{j}} : \mathbf{j} \in \Lambda,\ |\mathbf{j}|_\infty = k\}, \qquad k = 0, 1, \ldots, k_\Lambda,$$

where

$$k_\Lambda := \max_{\mathbf{j}\in\Lambda} |\mathbf{j}|_\infty .$$

Clearly, since $2^{2k} \leq 2^{2j_1} + \cdots + 2^{2j_d} \leq d2^{2k}$ for all $W_{\mathbf{j}} \subset W_{k,\Lambda}$, this leads to a new space splitting

$$(W_{k,\Lambda}; 2^{2k}(\cdot,\cdot)_{L^2})_{0\leq k\leq k_\Lambda} \tag{7.75}$$

for $(S_\Lambda; (\cdot,\cdot)_{H_0^1})$ with condition number $\leq d\kappa$. Now, let Λ_k'' denote the *minimal* index set such that

$$W_{k,\Lambda} \subset \operatorname{span}((V_{\mathbf{j}} : \mathbf{j} \in \Lambda_k'')).$$

Obviously, $\mathbf{j} \in \Lambda_k''$ implies $|\mathbf{j}|_\infty = k$ and $\mathbf{j} \in \Lambda$. Indices $\mathbf{j}$ for which there is another $\mathbf{j}' > \mathbf{j}$ with these necessary properties do not belong to Λ_k''. If $(k, \ldots, k)$ belongs to Λ, then Λ_k'' consists of that single index.

For $d = 2$, Figure 7.7 schematically shows, for some of the $S_{J,\alpha}$ with $J = 12$, the associated index sets $\Lambda_{J,\alpha}$ together with the "active" indices $\mathbf{j} \in \Lambda_k''$, $k = 0, \ldots, J$, involved in the splitting (7.76). It turns out that the cardinality of the index sets Λ_k'' is important, as can be seen from the proof of the following result:

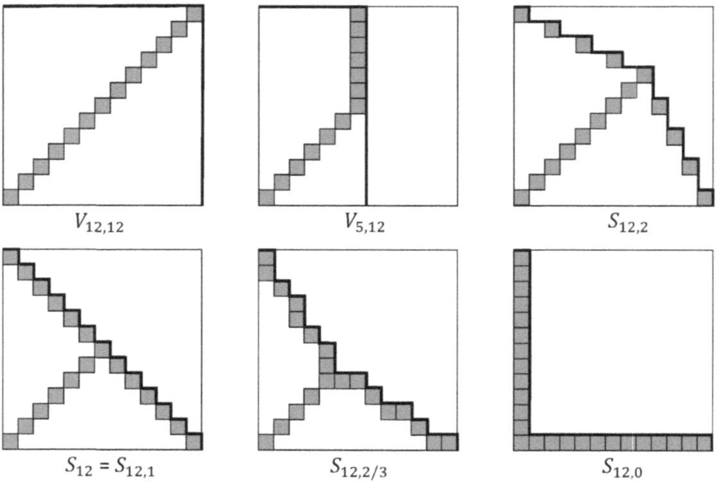

Fig. 7.7 Index sets Λ and $\cup_{k=0}^{k_\Lambda}\Lambda_k''$ in (7.76) associated with some $V_{\mathbf{J}}$ and energy-optimized sparse grid spaces $S_{J,\alpha}$ for $d = 2$, $|\mathbf{J}|_\infty = J = 12$, and various α.

Theorem 7.12 *With the given tensor product splitting (7.72), the space splitting*

$$((V_{\mathbf{j}}; 2^{2k}(\cdot,\cdot)_{L^2}))_{\mathbf{j}\in\Lambda_k'',\, k=0,\dots,k_\Lambda} \tag{7.76}$$

provides a stable space splitting for $(S_\Lambda; (\cdot,\cdot)_{H_0^1})$ *with condition number*

$$\kappa_\Lambda \le \frac{4}{3}(\max_{k=0,\dots,k_\Lambda} |\Lambda_k''|)d\kappa.$$

We have an upper bound

$$\max_{k=0,\dots,k_\Lambda} |\Lambda_k''| \le d(k_\Lambda + 1)^{d-2},$$

independent of Λ. *In particular, for the standard* ($\alpha = 1$) *and energy-optimized sparse grid spaces* $S_{J,\alpha}$ *of level* J, *we obtain a condition number behavior of* $\kappa_{\Lambda_{J,\alpha}} \le C(J+1)^{d-2}$, $J \ge 0$, *with constants depending on* α *and* $d \ge 2$, *but not on* J.

Proof The condition number estimate for the space splitting (7.76) can be seen as follows: First, for each $\mathbf{j} \in \Lambda_k''$, the subspace $W_{\mathbf{j}}$ is L^2 orthogonal to all other $V_{\mathbf{j}'}$ with $\mathbf{j}' \in \Lambda_k''$ (otherwise we would have $V_{\mathbf{j}} \subset V_{\mathbf{j}'}$, which contradicts the minimality of Λ_k''). So, considering the decomposition

$$u_\Lambda = \sum_{k=0}^{k_\Lambda} \sum_{\mathbf{j}\in\Lambda_k''} Q_{W_{\mathbf{j}}} u_\Lambda, \qquad u_\Lambda \in S_\Lambda,$$

we get an upper bound for the norm $|||\cdot|||_{\mathbf{V}}$ by

$$|||u_\Lambda|||_{\mathbf{V}}^2 := \inf_{u_\Lambda=\sum_k \sum_{\mathbf{j}\in\Lambda_k''} v_{\mathbf{j}}} \sum_{k=0}^{k_\Lambda} \sum_{\mathbf{j}\in\Lambda_k''} 2^{2k}\|v_{\mathbf{j}}\|_{L^2}^2 \le \sum_{\mathbf{j}\in\Lambda} 2^{2|\mathbf{j}|_\infty}\|Q_{W_{\mathbf{j}}}u_\Lambda\|_{L^2}^2 =: |||u_\Lambda|||_{\mathbf{W}}^2.$$

For the lower bound, we take any decomposition

$$u_\Lambda = \sum_{k=0}^{k_\Lambda} \sum_{\mathbf{j}\in\Lambda_k''} v_{\mathbf{j}}$$

for $u_\Lambda \in S_\Lambda$. Denote by $w_{\mathbf{l},\mathbf{j}} = Q_{W_{\mathbf{l}}} v_{\mathbf{j}}$, $\mathbf{l} \le \mathbf{j}$, the L^2 orthoprojection of $v_{\mathbf{j}}$ onto $W_{\mathbf{l}}$. Then

$$u_\Lambda = \sum_{\mathbf{l}\in\Lambda} \underbrace{\Big(\sum_{k\ge|\mathbf{l}|_\infty} \sum_{\mathbf{j}\in\Lambda_k'':\mathbf{j}\ge\mathbf{l}} w_{\mathbf{l},\mathbf{j}}\Big)}_{Q_{W_{\mathbf{l}}}u_\Lambda=}$$

is the unique decomposition of u_Λ in terms of the Riesz splitting (7.73). Furthermore, since

$$\begin{aligned} 2^{2|\mathbf{l}|_\infty}\|Q_{W_{\mathbf{l}}}u_\Lambda\|_{L^2}^2 &\le \Big(\sum_{k=|\mathbf{l}|_\infty}^{k_\Lambda} 4^{|\mathbf{l}|_\infty-k}\Big)\Big(\sum_{k=|\mathbf{l}|_\infty}^{k_\Lambda} 2^{2k}\Big\|\sum_{\mathbf{j}\in\Lambda_k'':\mathbf{j}\ge\mathbf{l}} w_{\mathbf{l},\mathbf{j}}\Big\|_{L^2}^2\Big) \\ &\le \frac{4}{3}\sum_{k=|\mathbf{l}|_\infty}^{k_\Lambda} \sum_{\mathbf{j}\in\Lambda_k'':\mathbf{j}\ge\mathbf{l}} 2^{2k}|\Lambda_k''|\|w_{\mathbf{l},\mathbf{j}}\|_{L^2}^2, \end{aligned}$$

we get

$$\begin{aligned} |||u_\Lambda|||_{\mathbf{W}}^2 &\le \frac{4}{3}\big(\max_{k=0,\ldots,k_\Lambda}|\Lambda_k''|\big)\sum_{\mathbf{l}\in\Lambda}\sum_{k=|\mathbf{l}|_\infty}^{k_\Lambda}\sum_{\mathbf{j}\in\Lambda_k'':\mathbf{j}\ge\mathbf{l}} 2^{2k}\|w_{\mathbf{l},\mathbf{j}}\|_{L^2}^2 \\ &\le \frac{4}{3}\big(\max_{k=0,\ldots,k_\Lambda}|\Lambda_k''|\big)\sum_{k=0}^{k_\Lambda} 2^{2k}\sum_{\mathbf{j}\in\Lambda_k''}\sum_{\mathbf{l}\le\mathbf{j}}\|w_{\mathbf{l},\mathbf{j}}\|_{L^2}^2 \\ &= \frac{4}{3}\big(\max_{k=0,\ldots,k_\Lambda}|\Lambda_k''|\big)\sum_{k=0}^{k_\Lambda} 2^{2k}\sum_{\mathbf{j}\in\Lambda_k''}\|v_{\mathbf{j}}\|_{L^2}^2. \end{aligned}$$

After taking the infimum with respect to all possible representations of u_Λ, we get the lower bound

$$|||u_\Lambda|||_{\mathbf{W}}^2 \le \frac{4}{3}\big(\max_{k=0,\ldots,k_\Lambda}|\Lambda_k''|\big)|||u_\Lambda|||_{\mathbf{V}}^2.$$

Together with the upper bound above and the condition number estimate for the space splitting (7.75), this proves the bound for the condition number κ_Λ.

Bounds for $|\Lambda''_k|$ can be obtained by the following reasoning: For any $\mathbf{j} \in \Lambda''_k$ we have $j_i = k$ for some i and $j_{i'} \le k$ for the remaining $i' \ne i$. First, consider the set $\Lambda_{k,1}$ of all $\mathbf{j}' \in \mathbb{Z}_+^{d-1}$ such that $(k, \mathbf{j}') \in \Lambda$ and $|\mathbf{j}'|_\infty \le k$. Obviously this is a monotone set in $\mathbb{Z}_+^{d-1}$ and $(k, \mathbf{j}') \in \Lambda''_k$ if and only if $\mathbf{j}'$ is an extremal index in $\Lambda_{k,1}$. By the result of Problem 7.13, the number of $\mathbf{j} \in \Lambda''_k$ of the form $(k, \mathbf{j}')$ is therefore bounded by $(k+1)^{d-2}$. Consequently, $|\Lambda''_k| \le d(k+1)^{d-2}$, since the same reasoning must be applied to the similarly defined sets $\Lambda_{k,i}$, $i = 2, \ldots, d$. This gives the estimate for $\max_{k=0,\ldots,k_\Lambda} |\Lambda''_k|$. It is easy to see from the examples in Figure 7.7 that the resulting bound $|\Lambda''_k| \le 2$ is sharp for $d = 2$.

It remains to show that, for $\Lambda = \Lambda_{J,\alpha}$, the above estimate for $|\Lambda''_k|$ cannot be improved, up to a multiplicative constant. Without loss of generality, let $d > 2$ and assume that J is sufficiently large. Then, for the index set of the energy-optimized sparse grid spaces of level J, we can prove the lower bound

$$\max_{k=0,\ldots,J} |\Lambda''_k| \ge c(J+1)^{d-2}, \qquad J \ge 1, \tag{7.77}$$

with a constant $c > 0$ independent of J. Since for $\Lambda = \Lambda_{J,\alpha}$ we have $k_\Lambda = J$, which in this case matches the upper bound stated in Theorem 7.12 for general Λ.

To establish (7.77), consider again the set

$$\Lambda_{k,1} = \{\mathbf{j}' \in \mathbb{Z}_+^{d-1} : (k, \mathbf{j}') \in \Lambda_{J,\alpha},\ |\mathbf{j}|_\infty \le k\}$$

and recall that each extremal index $\mathbf{j}'$ in $\Lambda_{k,1}$ corresponds to a unique index $(k, \mathbf{j}')$ in Λ''_k. So it remains to count the extremal indices in $\Lambda_{k,1}$ to get a lower bound for $|\Lambda''_k|$. The condition $\mathbf{j} := (k, \mathbf{j}') \in \Lambda_{J,\alpha}$ can be rewritten as

$$|\mathbf{j}|_1 = k + |\mathbf{j}'|_1 \le \alpha(J-k) + k \quad \Leftrightarrow \quad |\mathbf{j}'|_1 \le \alpha(J-k).$$

Let m be the largest integer such that $m \le \alpha(J-k)$. Then the set of extremal points in $\Lambda_{k,1}$ is given by

$$(\Lambda_{k,1})' = \{\mathbf{j}' \in \mathbb{Z}_+^{d-1} : |\mathbf{j}'|_1 = m,\ |\mathbf{j}'|_\infty \le k\}.$$

Now, for sufficiently large J, choose the smallest k such that $m \le k$. For such k, the condition $|\mathbf{j}'|_1 \le m$ automatically implies $|\mathbf{j}'|_\infty \le m \le k$ such that $(\Lambda_{k,1})'$ coincides with the set Λ'_m of extremal indices of the $(d-1)$-dimensional standard sparse grid index set of level m. Thus, for these choices of m and k, we have

$$|\Lambda''_k| \ge |(\Lambda_{k,1})'| = |\Lambda'_m| = \binom{m+d-2}{d-2} \ge cm^{d-2}.$$

The argument for (7.77) concludes by showing that $m \ge cJ$. In fact, by definition of m and k, we have

$$k - 1 < m \le \alpha(J-k) < (m+1) \le k+1 \quad \Leftrightarrow \quad \frac{\alpha J - 1}{\alpha + 1} < k < \frac{\alpha J + 1}{\alpha + 1}.$$

So if $J \geq 2/\alpha$ then

$$m \geq k > \frac{\alpha J - 1}{\alpha + 1} \geq \frac{(\alpha - \alpha/2)J}{\alpha + 1} = \frac{\alpha}{2(\alpha + 1)} J.$$

This finally confirms (7.77) and finishes the proof of Theorem 7.12. □

It turns out that for $d \geq 3$ there is no chance to get a uniform stability result with **V** space splittings that is valid for all $(S_\Lambda; (\cdot,\cdot)_{H_0^1})$. Consequently, the BPX-type preconditioners for sparse grid discretizations of $H_0^1(G)$ elliptic problems resulting from these splittings cannot be asymptotically optimal for $J \longrightarrow \infty$ and general Λ. Indeed, it was shown in [84] that, for *any* space splitting of the form $(V_{\mathbf{j}}; \rho_{\mathbf{j}}(\cdot,\cdot)_{L^2})$ with $\rho_{\mathbf{j}} \geq 0$ for the standard sparse grid spaces $(S_J; (\cdot,\cdot)_{H_0^1})$, we have a lower bound for the condition number of $\kappa_{\Lambda_J} \geq c(J+1)^{d-2}$. There it was suggested to also allow for some negative scaling factors $\rho_{\mathbf{j}} < 0$ to remedy the situation. However, such extensions cannot be handled within the space splitting theory treated in this book.

Problems

7.1 BPX-frame: Examples.
Consider linear finite elements with respect to dyadically refined triangulations $\mathcal{T}_j$, $j \geq 0$, on a polygonal domain $G \subset \mathbb{R}^2$. Let P_0 be a fixed nodal point in $\mathcal{T}_0$. Set $s = 1$.
a) Define the adaptive finite element spaces $\hat{V}_J$ by setting $\tilde{\mathcal{V}}_j = \{P_0\}$, $j = 0, 1, \ldots, J$. Show that the BPX-frame $\hat{\Phi}_J$ associated with the subspaces $\hat{V}_J \subset H^1(G)$ is uniformly stable, $J \geq 1$. More precisely, show that it is even a Riesz basis in $\hat{\Phi}_J$.
b) Define the adaptive finite element spaces $\hat{V}_J$ by setting $\tilde{\mathcal{V}}_j = \mathcal{V}_j \backslash \{P_0\}$, $j = 0, 1, \ldots, J$. Show that the associated BPX-frame for the subspaces $\hat{V}_J \subset H^1(G)$ is not uniformly stable, $J \geq 1$, and that its condition number scales as $\approx J^2$, $J \longrightarrow \infty$.
c) Define the finite element spaces $V_{J,0} = \hat{V}_J$ by setting $\tilde{\mathcal{V}}_j = \mathcal{V}_j \backslash \Gamma$, $j = 0, 1, \ldots, J$, where Γ is the boundary of G. Show that the associated BPX-frame for the subspaces $V_{J,0} \subset H_0^1(G)$ is uniformly stable, $J \geq 1$.
Hints. For b), compare with the proof of Theorem 7.2 and Problem 7.2 for $s = 1$, $d = 2$. Part c) is a partial case of finite element subsplittings for $H_0^1(G)$.

7.2 Counterexamples to Theorem 7.2.
Show that the estimates in the proof of Theorem 7.2 cannot be improved for $s \leq d/2$, i.e., show that, for some positive constant c, it holds that

$$\sup_{u_J \in V_J : \|u_J\|_{H^s} \leq 1} \|Q_0 u_J\|_{L^2} + \sum_{j=1}^{J} 2^{2js} \|(Q_j - Q_{j-1}) u_J\|_{L^2}^2$$
$$\geq c \begin{cases} J^2, & s = d/2,\ d \leq 2, \\ 2^{2J(d/2-s)}, & 0 < s < \min(d/2, 3/2), \end{cases}$$

where $Q_j : C(\bar{G}) \longrightarrow V_j$ denote the nodal interpolation operators.

7.3 Basis and dimension of $\hat{V}_J$.
Consider the construction of $\hat{V}_J$ for nodal point sets $\tilde{\mathcal{V}}_j$, $j \le J$, satisfying the local refinement condition of Subsection 7.1.1.1 for J and fixed $r_0 > 0$.
a) Prove that $\bar{\Phi}_J$, defined in (7.8), is an algebraic basis in $\hat{V}_J$.
b) Show the inequality $|\hat{\Phi}_J| \le C|\bar{\Phi}_J| = C\dim(\hat{V}_J)$, where C is independent of $\hat{V}_J$, but may depend on the type of finite element spaces and dyadic refinement used.
c) Without the local refinement condition, the statement in a) may not hold. Give an example.
Hints. a) It is obvious that $\bar{\Phi}_J$ spans $\hat{V}_J$. If the local refinement condition holds, then the dual functionals $\lambda_{P,j}$ constructed for the proof of Theorem 7.1 can be used to prove linear independence. For b), note that $\tilde{n}_J := |\tilde{\mathcal{V}}_J| = |\bar{\mathcal{V}}_J|$ and $\tilde{n}_j := |\tilde{\mathcal{V}}_j| \le |\bar{\mathcal{V}}_j| + c|\tilde{\mathcal{V}}_{j+1}|$, $j < J$, with some constant $c < 1$ depending on the type of finite element spaces and dyadic refinement. Then use this recursively to bound $|\hat{\Phi}_J| = \tilde{n}_0 + \cdots + \tilde{n}_J$. For c), consider linear finite elements for $d = 2$, take a triangle C as G and set $\mathcal{T}_0 = \{C\}$. Let $\mathcal{T}_1$ be obtained by quadrisection. Set $\tilde{\mathcal{V}}_0 = \mathcal{V}_0$ and $\tilde{\mathcal{V}}_1 = \mathcal{V}_1 \backslash \{M_e\}$, where M_e is the midpoint of an arbitrarily chosen edge e of C. Then the cardinality of $\bar{\Phi}_1$ is 7, while $\hat{V}_1 = V_1$ has dimension 6.

7.4 Auxiliary inequalities for the upper stability estimate in Theorem 7.4.
Prove the inequalities (7.18) and (7.19).
Hints: Use the L^2 stability of the nodal bases in $V_{k+1,0}$ and $\hat{V}_{k,0}$, respectively, to establish (7.18). For the Bernstein-type inequality (7.19) in $\hat{V}_{J,0}$, additionally use (3.52).

7.5 Auxiliary inequalities for the lower stability estimate in Theorem 7.4.
Prove the inequalities (7.20) and (7.21).
Hints: Adapt the considerations in Subsection 3.2.2.2, where analogous inequalities are established. Again, the expressions (3.51) and (3.52) for L^2 and broken H_0^1 norms of elements of $\hat{V}_{J,0}$ are needed.

7.6 Other examples of multiscale splittings with non-nested spaces.
Consider the sequences $V_{j,0}$ of bilinear and trilinear finite element subspaces of $H_0^1(G)$, where $G \subset \mathbb{R}^d$ is a rectangular domain for $d = 2$ and $d = 3$, respectively, and the underlying partition sequence $(\mathcal{T}_j)_{j\ge 0}$ consists of d-dimensional rectangles and is obtained by uniform dyadic refinement. The corresponding non-conforming Wilson finite element spaces $\hat{V}_{j,0} \supset V_{j,0}$ consist of all piecewise polynomial functions on $\mathcal{T}_j$ that are continuous at the vertices $P \in \mathcal{V}_j$, vanish at Γ, and are locally on each rectangle $C \in \mathcal{T}_j$ an enrichment of $V_{j,0}$ with the quadratic monomials x_i^2. Prove an analog of Theorem 7.4.
Hints: The Wilson rectangle ($d = 2$) and the Wilson brick ($d = 3$) are described in [41] and have historically been used for applications in plane elasticity when $d = 2$. The standard "nodal" basis for representing functions from $\hat{V}_{j,0}$ is given by the nodal basis functions of $V_{j,0}$ and, for any d-dimensional rectangle

$$C = \prod_{i=1}^{d} [a_i, b_i] \in \mathcal{T}_j,$$

by the additional quadratic bubble functions

$$N_{C,i}(x) = (b_i - x_i)(x_i - a_i), \quad x \in C, \qquad N_{C,i}(x) = 0, \quad x \notin C,$$

$i = 1, \ldots, d$. The discrete $H_0^1(G)$ bilinear forms on $\hat{V}_{j,0}$ are given by

$$\hat{a}_j(\hat{v}_{j,0}, \hat{v}_{j,0}) = \sum_{C \in \mathcal{T}_j} \|\nabla \hat{v}_{j,0}\|^2_{L^2(C)}, \qquad \hat{v}_{j,0} \in \hat{V}_{j,0},$$

equivalent discrete norms involving differences along edges in $\mathcal{T}_{j+1}$ can be easily defined. An intergrid transfer operator $I_j : \hat{V}_{j,0} \to V_{j+1,0} \subset \hat{V}_{j,0}$ with all necessary properties can be defined by interpolation of function values (where $\hat{v}_{j,0}$ is continuous) and by interpolation of averages (at edge midpoints and face centers, where $\hat{v}_{j,0}$ has jumps). Note that a simpler way to develop a multiscale splitting for $\hat{V}_{j,0}$ is to use the auxiliary space method, as described in Subsection 3.2.2.2 of Chapter 3.

7.7 The spaces in the refined subspace splitting (7.51) in Theorem 7.8 are adaptive finite element spaces.
Let the notation and assumptions of Subsection 7.2.1 hold. Show that the subspaces V_{G_i}, V_{G_ik} and $\bar{V}_{\Gamma_{ik}}$ are special instances of adaptive finite element spaces to which Theorem 7.1 applies, see Subsection 7.1.1. Conclude that the properly scaled subsystems Φ_{G_i}, $\Phi_{G_{ik}}$ and $\bar{\Phi}_{\Gamma_{ik}}$ of the BPX-frame for $V_{J,0}$ are frames in $(V_{G_i}; a)$, $(V_{G_{ik}}; a)$ and $(\bar{V}_{\Gamma_{ik}}; a)$ with uniform frame bounds, respectively.
Hints: Identify the nodal sets $\tilde{\mathcal{V}}_j$, $j = j_0 + 1, \ldots, J$, associated with the systems Φ_{G_i}, $\Phi_{G_{ik}}$ and $\bar{\Phi}_{\Gamma_{ik}}$, additionally set $\tilde{\mathcal{V}}_j = \emptyset$, $j = 0, \ldots, j_0$, and verify the local refinement condition for them.

7.8 Modifications for $p > 1$.
The considerations in Subsection 7.2.1 can be carried out for C^0 Lagrange finite element spaces $V_{J,0}$ of arbitrary polynomial degree $p > 1$ with a few modifications. For fixed $0 < j_0 < J$, consider a modified BPX-frame $\Phi_{j_0,J,0}$ for $V_{J,0}$ constructed as follows: For $j = j_0 + 1, \ldots, J$, take all $\tilde{N}_{P,j}$ from $V_{j,0}$ as before. For $j \le j_0$, include all scaled *linear finite element* nodal basis functions $\tilde{N}_{P,j}$ associated with the interior vertices P in $\mathcal{T}_j$. These functions will form the new subset $\bar{\Phi}_{j_0,0}$ (the BPX frame for the linear finite element subspace of level j_0, which we will temporarily denote by $\bar{V}_{j_0,0}$). For $j = j_0$, also include in $\Phi_{j_0,J,0}$ all scaled nodal basis functions $\tilde{N}_{P,j_0}$ from $V_{j_0,0}$ with nodal points not associated with vertices in $\mathcal{T}_{j_0}$. These functions are additionally included in the subsystems Φ_{G_i} (if P is interior to G_i) and $\bar{\Phi}_{\Gamma_{ik}}$ (if P is interior to Γ_{ik}). In the latter case this leads to slightly larger subspaces $\bar{V}_{\Gamma_{ik}}$. The vertex systems $\bar{\Phi}_P$ and vertex spaces $\bar{V}_P$ are not affected. With these changes, and considering $\bar{V}_{j_0,0}$ as coarse space, show the analogues of Theorems 7.7-7.9.
Hints: The frame property of $\Phi_{j_0,J,0}$ (with constants independent of $j_0 < J$) follows if one proves Theorem 7.3 c) for the nested sequence

$$\bar{V}_{0,0} \subset \cdots \subset \bar{V}_{j_0-1,0} \subset V_{j_0,0} \subset \cdots \subset V_{J,0} \subset \cdots$$

(i.e., if one switches for $j < j_0$ to the smaller subspaces of linear finite elements) and uses the uniform L^2 stability of the nodal bases in these spaces. Here, an additional argument is needed only for $j = j_0$, since in the construction of $\Phi_{j_0,J,0}$ the nodal basis in $V_{j_0,0}$ was modified by replacing the $\tilde{N}_{P,j_0}$ associated with vertices in $\mathcal{T}_{j_0}$ with the corresponding scaled nodal basis functions from the linear finite element space $\bar{V}_{j_0,0}$.

7.9 A useful result on the lower stability constant when deleting components from a subspace splitting.
Let $\mathbf{V} = (V_i)_{i\in\Omega_V}$ and $\mathbf{W} = (W_j)_{j\in\Omega_W}$ be subspace collections in a Hilbert space $(V;a)$ and suppose that their concatenation $(\mathbf{V},\mathbf{W})$ is a stable subspace splitting for $(V;a)$ with stability constants A and B, i.e.,

$$B^{-1}a(u,u) \le |||u|||^2 := \inf_{u=\sum_i v_i+\sum_j w_j} \sum_i a(v_i,v_i) + \sum_j a(w_j,w_j) \le A^{-1}a(u,u)$$

for all $u \in V$. Suppose the span of $\mathbf{V}$ is still dense in V. Show the following statement: $\mathbf{V}$ is a stable subspace splitting for $(V;a)$ such that

$$\bar{B}^{-1}a(u,u) \le |||u|||_{\mathbf{V}}^2 := \inf_{u=\sum_i v_i} \sum_i a(v_i,v_i) \le \bar{A}^{-1}a(u,u), \qquad u \in V,$$

provided that

$$|||w|||_{\mathbf{V}}^2 \le A_0|||w|||_{\mathbf{W}}^2, \qquad w \in W := \mathrm{span}(\mathbf{W}).$$

The stability constants of $\mathbf{V}$ satisfy

$$\frac{A}{2A_0} \le \bar{A} \le \bar{B} \le B.$$

If $\bar{B} \approx B \approx A \approx 1$, then the condition number $\kappa_{\mathbf{V}}$ of the subspace splitting $\mathbf{V}$ behaves as $\kappa_{\mathbf{V}} \approx A_0^{-1}$ if A_0 is the largest possible value in the assumed inequality for w.
Hints: The bound $\bar{B} \le B$ is always true for subsplittings. For the lower stability bound, just note that

$$|||u|||_{\mathbf{V}}^2 \le 2(|||v|||_{\mathbf{V}}^2 + |||w|||_{\mathbf{V}}^2), \qquad u = v + w,$$

for any $w \in W$, take the infimum with respect to $u = v + w$, and replace $|||w|||_{\mathbf{V}}^2$ with its upper bound $A_0|||w|||_{\mathbf{V}}^2$. Then compare with the definition of $|||u|||^2$.

7.10 Equivalent formulation of Theorem 7.8 using local Schur complement problems.
a) Prove that Theorem 7.8 admits an equivalent formulation if, in the construction of the subspace splitting $\mathbf{V}_{DD,1}$, the subspaces $\bar{V}_{\Gamma_{ik}}$ are replaced by the local Schur complement spaces $(V_{\Gamma_{ik}}; s_{\Gamma_{ik}})$, which are equipped with extension operators $E_{\Gamma_{ik}} : V_{\Gamma_{ik}} \to \bar{V}_{\Gamma_{ik}}$ such that

$$a(E_{\Gamma_{ik}} v_{\Gamma_{ik}}, E_{\Gamma_{ik}} v_{\Gamma_{ik}}) \le C s_{\Gamma_{ik}}(v_{\Gamma_{ik}}, v_{\Gamma_{ik}}), \qquad v_{\Gamma_{ik}} \in V_{\Gamma_{ik}},$$

with a constant C independent of Γ_{ik} and $j_0 < J$. In particular, the harmonic extension satisfies this condition with $C = 1$.
b) Prove that the statement of Theorem 7.8 remains valid if the spaces $\bar{V}_{\Gamma_{ik}}$ are replaced by the larger spaces $V_{G_{ik}}$. Moreover, similar to a), it admits an equivalent formulation using local Schur complement spaces.
Hints: For a), the upper stability bound poses no difficulty, since $E_{\Gamma_{ik}} V_{\Gamma_{ik}} \subset \bar{V}_{\Gamma_{ik}}$. For the lower stability bound, take any decomposition

$$u = u_{j_0,0} + \sum_i u_{G_i} + \sum_{\Gamma_{ik}} \bar{u}_{\Gamma_{ik}}$$

so that

$$a(u_{j_0,0}, u_{j_0,0}) + \sum_i a(u_{G_i}, u_{G_i}) + \sum_{\Gamma_{ik}} a(\bar{u}_{\Gamma_{ik}}, \bar{u}_{\Gamma_{ik}}) = A^{-1} a(u, u),$$

where A is the lower stability constant for the splitting considered in Theorem 7.8. Set $u_{\Gamma_{ik}} := T_{\Gamma_{ik}} \bar{u}_{\Gamma_{ik}}$ and change the above decomposition to

$$u = u_{j_0,0} + \underbrace{(\sum_i u_{G_i} + \sum_{\Gamma_{ik}} (\bar{u}_{\Gamma_{ik}} - E_{\Gamma_{ik}} u_{\Gamma_{ik}}))}_{\sum_i v_{G_i} :=} + \sum_{\Gamma_{ik}} E_{\Gamma_{ik}} u_{\Gamma_{ik}}.$$

Then estimate the expression

$$a(u_{j_0,0}, u_{j_0,0}) + \sum_i a(v_{G_i}, v_{G_i}) + \sum_{\Gamma_{ik}} s_{\Gamma_{ik}}(u_{\Gamma_{ik}}, u_{\Gamma_{ik}})$$

using the definition of $s_{\Gamma_{ik}}(\cdot,\cdot)$, the properties of the extension operators, and the finite overlap properties of the local terms involved. For example, you can use

$$\sum_i a(v_{G_i}, v_{G_i}) \approx a(\sum_i v_{G_i}, \sum_i v_{G_i}).$$

For b), use

$$V_{G_{ik}} \supset \bar{V}_{\Gamma_{ik}}$$

for the lower stability bound and the finite overlap property for the upper stability estimate. The equivalent formulation with local Schur complement spaces can be established as suggested for a).

7.11 Linear finite element prewavelets for $H^1((0,1))$.
The functions ψ_i, $i = 0, \ldots, 4$, are shown in Figure 7.8 (for the "boundary" prewavelet ψ_3, compare also Figure 3.7), while the "inner" prewavelet ψ (not shown in Figure 7.8) is the same as in the $H_0^1((0,1))$ case, see Figure 3.5 b).

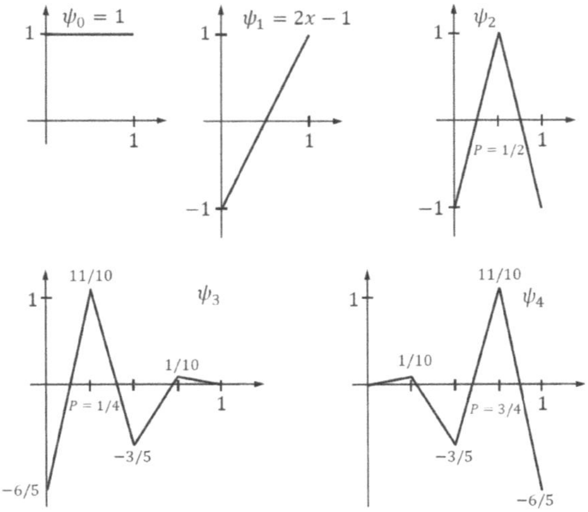

Fig. 7.8 Linear finite element prewavelets for $H^1((0,1))$. Upper row: L^2 orthogonal functions spanning V_0. Lower row: Left and right boundary prewavelets.

a) Show that the collection $\mathbf{W} = (W_j)_{j\geq -1}$ given by

$$W_{-1} = \mathrm{span}(\psi_0), \quad W_0 = \mathrm{span}(\psi_1, \psi_2), \quad W_1 = \mathrm{span}(\psi_3, \psi_4)$$

and

$$W_j = \mathrm{span}(\{\psi_3(2^{j-1}\cdot), \psi(2^j \cdot - i),\ i = 0, \ldots, 2^{j-1} - 1, \psi_4(1 - 2^{j-1}(1 - \cdot)\}),$$

$j \geq 2$, is an L^2 orthogonal Riesz splitting for $L^2((0,1))$.
b) Verify that the functions spanning W_j form, after appropriate scaling, uniformly stable Riesz bases for $(W_j; (\cdot,\cdot)_{L^2})$, $j \geq 2$.
Hints. In a), the orthogonality $W_j \perp_{L^2} W_k$ for $j \neq k$ follows from checking the orthogonality $W_j \perp_{L^2} V_j$, $j \geq 1$, and observing that $V_{j+1} \supset W_j$, where V_j is the space of linear splines with respect to the uniform partition of $[0,1]$ of step-size 2^{-j}. For b), the L^2 Riesz basis property for each W_j, $j \geq 2$, can be obtained by constructing an L^2 biorthogonal system (e.g., of piecewise constant or piecewise linear functions). Alternatively, adapt the argument in Subsection 3.1.4.2, which leads to Theorem 3.5. To get uniform Riesz bounds, the L^2 normalization of each prewavelet function is sufficient. That $W_j = V_{j+1} \ominus V_j$ then follows by a dimension counting argument which in turn implies the completeness of $\mathbf{W}$ in $L^2((0,1))$.

7.12 Robust splittings for sparse grid spaces ($d = 2$).
Let $d = 2$ and consider bilinear finite element spaces. Derive $\mathbf{W}$ and $\mathbf{V}$ space splittings for $(S_\Lambda; a)$ with condition number estimates independent of w_i, $i = 0, 1, 2$, and the monotone index set $\Lambda \subset \mathbb{Z}^2_+$, where $a(\cdot,\cdot)$ is the bilinear form generated by the homogeneous Dirichlet problem for (7.61).

Hints: Start from the Riesz splitting (7.73), now associated with (7.61) and the weights w_i properly incorporated. The **V** space splitting can be obtained by following the proof of Theorem 7.11, with modifications as in the proof of Theorem 7.12. In particular, the spaces $V_{\mathbf{1}+k}$ in the splitting of Theorem 7.11 must be replaced by several $V_{\mathbf{j}}$, taking the structure of Λ into account.

7.13 The number of extremal points in Λ.
Given an arbitrary monotone index set $\emptyset \neq \Lambda \subset \{0, \ldots, k\}^d$, show that the cardinality of the set Λ' of its extremal points is bounded by

$$|\Lambda'| \leq (k+1)^{d-1}.$$

Hints: For $d = 1$ and $d = 2$, the best possible bounds are $|\Lambda'| = 1$ and $|\Lambda'| \leq k + 1$, respectively (for $d = 2$, the bound is achieved for the index set associated with the standard sparse grid space of level k). For general d and k, use induction by considering the $k + 1$ different $(d - 1)$-dimensional index sets

$$\Lambda_j = \{\mathbf{j}' \in \mathbb{Z}_+^{d-1} : (j, \mathbf{j}') \in \Lambda\}, \quad j = 0, \ldots, k,$$

and by realizing that for extremal indices in Λ' of the form $(j, \mathbf{j}')$, the subindex $\mathbf{j}'$ must be extremal for Λ_j. Then apply the bound for dimension $d-1$, which is justified since Λ_j is again monotone and a subset of $\{0, \ldots, k\}^{d-1}$. Note that for $d > 2$ the bound can be slightly improved by a constant factor < 1. Since $|\Lambda'| \geq c(k + 1)^{d-1}$ for the index set of the standard d-dimensional sparse grid space of level k, the exponential $d - 1$ in the estimate for $|\Lambda'|$ cannot be improved.

Correction to: Hilbert Space Splittings and Iterative Methods

Michael Griebel • Peter Oswald

Correction to:
M. Griebel, P. Oswald, *Hilbert Space Splittings and Iterative Methods*, Springer Series in Computational Mathematics 62, https://doi.org/10.1007/978-3-031-74370-2

The original version of the book was published with several errors which were noticed by the authors after the book was published. Authors provided the list of errors and is available at the end of the book.

The following is a list of corrections of typographical, notational, mathematical, and other errors and mistakes in the text that were discovered after the manuscript was submitted. Readers are encouraged to contact the authors if they find other inconsistencies and errors, or if they have suggestions for improvement. The solutions to the problems in the book are available from the authors on request.

- p.36, line -1: It should be p_i, not $\mathbf{p}_i$, in the formula.
- p.64, line -7: The problem heading should read "Stable subspace splittings ...".
- p.106, Fig. 3.5 a): The notation attached to the graph should be ϕ, not ψ.
- p.111, line -7 ff: From the formula (3.34) on page 111 to the formulation of Theorem 3.5 on page 113, the previously introduced notation $N_{P,j}$ for finite element nodal basis functions of level j changes to $N_{j,P}$, which is a notational inconsistency.
- p.160, line 3: It should be $\tilde{a}_\rho$, not $\tilde{a}$, in the denominator of the formula.
- p.167, lines 9 and 10: Drop the subscript 2 in the norms.
- p.179, line -12: Replace "With this choice of I_m, we have" with "With this choice of I_m, for the iteration (4.1) with constant step-size parameter $\omega_m = \lambda_{\max}^{-1}$, we have".
- p.184, line -1: It should be $e_w^{(m)}$, not $e_w^{(m+1)}$.
- p.185, line 8: It should be $a(P_\rho^{-1}(\delta_m e_w^{(m)} + (1-\delta_m)e_v^{(m)}), d_v^{(m)})$ or, alternatively, $a(\delta_m e_w^{(m)} + (1-\delta_m)e_v^{(m)}, P_\rho^{-1} d_v^{(m)})$.
- p.185, line -7: Obsolete (repeated from the previous line).
- p.192, line 1: Replace r_m with r_m'.

The updated version of this book can be found at
https://doi.org/10.1007/978-3-031-74370-2

M. Griebel, P. Oswald, *Hilbert Space Splittings and Iterative Methods*, Springer Series in Computational Mathematics 62, https://doi.org/10.1007/978-3-031-74370-2_8

- p.197, Problem 4.3: Delete Part b) here. This has already been established in Part b) of Problem 2.1.
- p.197, line 14: It should be $W_i^\perp = \mathrm{ran}(R_i)$, not W_i. This is because $\cos(\angle(W_i, W_j)) = \cos(\angle(W_i^\perp, W_j^\perp))$ only if $W_i \cap W_j = \{0\}$.
- p.197, line -16, formula (4.108), and p.198, line 1: It should be the operator norm $\|M_{\rho,\omega}\|_V$ instead of $\|M_{\rho,\omega}\|$.
- p.198, line 4: Replace for clarity with

$$u_1 = Q_{W_i^\perp} u, \quad u_2 = Q_{U_{i+1}} Q_{W_i} u, \quad u_3 = Q_{U_{i+1}^\perp} Q_{W_i} u.$$

- p.198, line 7: Replace $\|u\|^2$ with $\|u_2 + u_3\|^2$ or, better, by $(\|u_2\|^2 + \|u_3\|^2)$.
- p.198, lines -11 and -10: Replace the sentence "Find a complementing ..." with "For the upper bound, note that $\bar{M}_{\rho,\omega} \in \mathcal{B}_{pos}(V)$."
- p.198, line -9: It should be $I - \omega C_j A$ instead of $I - C_j A$ in the product of operators.
- p.199, lines 15 and 16: Replace the formula containing $r(M_{\eta,\xi})$ in line 15 with

$$\sqrt{\eta} < r(M_{\eta,\xi}) = \begin{cases} p_{\max} + \sqrt{p_{\max}^2 - \eta}, \\ \max(p_{\min} + \sqrt{p_{\min}^2 - \eta}, p_{\max} + \sqrt{p_{\max}^2 - \eta}), \\ p_{\min} + \sqrt{p_{\min}^2 - \eta}, \end{cases}$$

for the three cases

$$\begin{array}{l} 0 < \xi \le (1 - \sqrt{\eta})^2/\lambda_{\min}, \\ (1 - \sqrt{\eta})^2/\lambda_{\min} < \xi < (1 + \sqrt{\eta})^2/\lambda_{\max}, \\ \xi \ge (1 + \sqrt{\eta})^2/\lambda_{\max}, \end{array}$$

respectively, and define $p_{\min} = -(1 + \eta - \zeta\lambda)/2$ in line 16.
- p.200, lines -1 and -4: Replace β_m with δ_m three times.
- p.210, line 14: Change the norm in the denominator to $\|A^+ D^{1/2} I_{\mathbf{p}}^{-1/2} y\|_2$.
- p.211, line 8: Remove the obsolete I_ρ from the formula.
- p.213, line -1: It should be $i = 1, \ldots, n'$.
- p.215, line 1: In the matrix that appears in the formula (5.19), replace γ^{-1} with $\gamma^{-1/2}$ (two times) and γ^{-2} with γ^{-1}.
- p.228 ff: Starting with the formula in line -10, the notation for the norm $\|\cdot\|_{\tilde{V},1-s}$ is shortened to $\|\cdot\|_{1-s}$ in many places. This affects p.228 (lines -10 and -3), p.229 (lines 9,-3,-2), p.230 (lines 2,8,10,11,12), p.231 (lines 8,9,15) and p.249 (lines 8,13,18). Replace with $\|\cdot\|_{\tilde{V},1-s}$ for consistency.
- p.237, line 19: The lower bound in (5.57) should be corrected to

$$\frac{\mathbb{E}(\|e_x^{(m)}\|^2)}{\|e_x^{(0)}\|^2} \ge 1 - \frac{Cm}{n(n+1)}, \qquad m = 1, \ldots, N.$$

In Problem 5.8 this is established with $C = 2\sqrt{e}$.
- p.246, lines -3 to -1: Ignore. We have no direct counterexample for this bound, but we cannot prove it either. Our initial sketch of a proof was flawed.

- p.247, line 6: After "... Schwarz iteration" insert the reference "(5.3)".
- p.247, line 7: Add the sentence "Furthermore, assume $i_0 = 1$."
- p.247, lines 9 and 14: The formulae in these lines should be corrected to

$$\|e^{(m+1)}\|^2 = c^{2m}\|e^{(1)}\|^2 = \frac{c^{2m}}{1-c^2}(r_1^{(0)} - cr_2^{(0)})^2 \leq c^{2m}\|e^{(0)}\|^2$$

and

$$\mathbb{E}(\|e^{(m+1)}\|^2) = \left(\frac{1+c^2}{2}\right)^m \mathbb{E}(\|e^{(1)}\|^2) \leq \frac{1+c}{2}\left(\frac{1+c^2}{2}\right)^m \|e^{(0)}\|^2,$$

respectively.
- p.249, line 2: The expression on the right-hand side of the inequality should include the term $|||\tilde{e}_u^{(0)}|||_{\tilde{V},1-s}$, which for $J \in \mathcal{F}_L^{1,1}$ is not necessarily bounded from above by a multiple of $\Delta\tilde{J}_u^{(0)}$, as erroneously stated at the end of the hints to the problem. But, for example,

$$\Delta\tilde{J}_u^{(m)} \leq C\frac{\Delta\tilde{J}_u^{(0)} + |||\tilde{e}_u^{(0)}|||_{\tilde{V},1-s}}{m^2}, \qquad m = 1, 2, \ldots,$$

is a correct (though not optimal in terms of constants) replacement.
- p.249, lines -12 to -10: Replace with

$$(\alpha'_{m+1})^{-1} = 1 + \gamma_{m+1} = 1/2 + \sqrt{(\alpha'_m)^{-2} + 1/4} \geq 1/2 + (\alpha'_m)^{-2}.$$

Iterating this gives the result with a constant C depending on the quotient $n(\tilde{S}'_s)^2/\tilde{S}_s$ and the choice of $(\alpha'_0)^{-1} = 1 + \gamma_0$.
- p.249, line -5: Replace the subscript ℓ^2 with 2.
- p.250, line 2: Should be corrected to $N - \frac{j+1}{2}$.
- p.250, line 12: To match the notation in the text, it should read $e_x^{(m)}, e_y^{(m)}, e_z^{(m)} \in X_j$ instead of $e^{(m)} \in X_j$.
- p.250, line -11: In the formula, correct to

$$\ldots = N - \frac{2sr}{s+r} \geq N - \frac{s+r}{2}, \ldots$$

- p.251, line 8: Replace $2 \leq \ldots$ with $2 < \ldots$
- p.251, lines 10 to 15: The hint for b) is correct in principle. But the proof is much simpler if one aims at proving

$$\mathbb{E}'(\|e^{(m+1)}\|^2) \geq q\|e^{(m)}\|^2, \quad m \geq 0,$$

with some constant $q > 1$ depending on ω and α.
- p.259, line 9: An index shift is required: Replace $\bar{u}^{(m+1)}$ and $\bar{u}^{(m)}$ with $\bar{u}^{(m)}$ and $\bar{u}^{(m-1)}$, respectively. This is consistent with the following considerations.
- p.260, lines -9 and -8: The subscripts of the norms should read V_P^t and V_P^s, respectively.

- p.269, line -1 and p.270, line 10: It should read $a(e^{(m)}, u)$, not $a(e, u)$.
- p.288, line -6: It should be $\bar{\varepsilon}_{m+1} = \alpha_m(1 - \rho_i\mu_m)\bar{\varepsilon}_m + \bar{\alpha}_m c_i$.
- p.289, line 3: The last term should be $\bar{\alpha}_{m-1}c_i$, not $\bar{\alpha}_{m-2}c_i$.
- p.289, line 6: The specification $S_0 = 1$ is missing.
- p.303, line -2: Add the natural assumption $t < s$ (or $s - t > 0$).
- p.303, line -1: The exponents in the two-sided estimate should be $-2(s-t)$ or $2(t-s)$, not $2(s-t)$. Strictly speaking, the hints for the problem on p.304 are more appropriate for the related quantity

$$\bar{\varepsilon}_m := \sup_{\|P\|\le 1} \inf_{p_m:\, p_m(0)=1} \sup_{e^{(0)}\in V_P^s} \frac{\|p_m(P)e^{(0)}\|^2_{V_P^t}}{\|e^{(0)}\|^2_{V_P^s}},$$

  which describes the best possible asymptotic convergence rate for *linear* Krylov subspace methods, while ε_m ($\le \bar{\varepsilon}_m$) is related to optimal best approximation from $\mathcal{K}_m$ and also covers nonlinear Krylov subspace methods. Currently we only have a proof of the lower bound for $\bar{\varepsilon}_m$ but not for ε_m.
- p.305. line 3: It should be $u^{(m+1)}$, not u^{m+1}.
- p.305, line 6: The denominator in the definition of γ_m should be $\sqrt{\beta_m + 1/4} + 1/2$.
- p.305, line 7: The recursion should be $\sqrt{\beta_{m+1}} = \sqrt{\beta_m + 1/4} + 1/2$.
- p.305, line 10: It should be $\beta_m \ge (m+1)^2/4$, not $\beta_m \ge (m+1)^2$.
- p.306, line 12: Correct the text to read "uniformly bounded away from zero for infinitely many m, depending on the spectrum of $P_{\mathbf{p}}$."
- p.308, lines -3 and -1 and p.309, line 2: σ should be replaced with σ_H.
- p.310, line 4: It should be $m = 1, 2, \ldots$, the quantities $S_0 = S_0' = 1$ and $S_0'' = 0$ do not need to be estimated.
- p.310, line -2: In this inequality a term +1 is missing (this is because the number of integers in an interval $[a, b)$ belongs to the interval $(b - a - 1, b - a + 1)$ and in the worst case can be arbitrarily close to the endpoints of this interval). This +1 can be compensated by an additional constant in the following estimates. Note that the estimation of $|I_n|$ is simplified if one first observes that $|I_0| \ge |I_n|$ for all $n \ge 1$.
- p.311, line 8: Reference should be to (6.79).
- p.312, line 15: In the formula it should read v_i, not $\tilde{v}_i$.
- p.312, line -4: For consistency it should be $G := (0, 1)$.
- p.313, line 5: It should be $b(\cdot,\cdot)$, not $B(\cdot,\cdot)$.
- p.313, line -4: It should read "The constants in a) and d)", not "The constants in a) and b)".
- p.313, line -2: In the meantime, we have realized that there are several (finite element and wavelet) multiscale splittings that provide answers for Part e) of the problem.
- p.334, lines -4 and -5, p.335, line 7: It should read $\hat{V}_{j,0}$, not $\hat{V}_j$ (analogously on p.336, lines 11 and -2 and on p.337, line -4).
- p.337, line -11: It should be $\hat{u}_J(M_e)$ in the formula, not $\hat{u}(M_e)$.

- p.355, line -6: The := sign can be replaced with = since $\bar{w}_{\bar{\Gamma}}$ was already defined in line -11.
- p.373, line -4: For notational consistency, replace the subscript $L^2(G)$ with L^2.
- p.387, Problem 7.1 b): We are currently not sure if the statement is correct. An upper $O(J^2)$ bound for the condition number is relatively easy to show, nontrivial lower bounds are still missing.
- p.387, lines -13 and -10: Correct to "subspace $\hat{V}_J \subset$ ... instead of "subspaces $\hat{V}_J \subset$ ....
- p.388, line 20: It better should read $V_{j+1,0}$ and $\hat{V}_{j,0}$, instead of $V_{k+1,0}$ and $\hat{V}_{k,0}$.
- p.389, line 14: It should be $V_{G_{ik}}$, not $V_{G_i k}$.
- p.389, Problem 7.8: We have solved this problem in a sightly different way, namely starting from the ladder

$$\bar{V}_{0,0} \subset \ldots \subset \bar{V}_{j_0,0} \subset V_{j_0+1,0} \subset \ldots \subset V_{J,0}, \qquad 0 \le j_0 < J.$$

  Then the formal definitions of V_{G_i}, $\bar{V}_{\Gamma_{ik}}$ and the vertex spaces $\bar{V}_P$ as spans of scaled nodal basis functions $\tilde{N}_{P',j}$, $j_0 < j \le J$, for nodal points P' interior to G_i, interior to Γ_{ik} and with $P' = P$, respectively, remain the same and the proofs become clearer. The differences with the original formulation of Problem 7.8 are minimal.
- p.390, lines -14 and -12: It must read

$$\frac{A}{2\max(1, A_0)} \le \bar{A} \le \bar{B} \le B$$

  and $\kappa_V \approx (\max(1, A_0))^{-1}$, respectively.
- p.392, Fig. 7.8: For better compatibility with Figure 3.5 b), the function values shown for ψ_3 should be multiplied by $5/6$, i.e., they should be -1, $11/12$, $-1/2$, $1/12$ (similar changes for ψ_4).
- p.392, line -1: Replace the word "exponential" with "exponent".

Appendix A
Hilbert space basics

This appendix is a service for the reader who needs a quick reminder. It establishes notation and language and briefly reviews well-known facts, mostly about separable Hilbert spaces and bounded linear operators acting on them, that are used throughout the book. No proofs are given. Familiarity with set-theoretic notation and concepts from real and functional analysis, measure theory, and linear algebra is assumed. Most of the material can be found in accessible and more systematic and detailed form in [3, 150, 192], for more specialized topics we also refer to [13, 26]. We have chosen to restrict our exposition to real Hilbert spaces which is, with a few exceptions and notational simplifications, the same as in the complex case. In the final Section A.5 we introduce Sobolev spaces and state some of their properties, which are used in Chapters 3 and 7.

A.1 Spaces: Basic notation, definitions and properties

A.1.1 Metric spaces

Let X be an arbitrary non-empty set. The pair $(X; d_X)$ is called a *metric space* if the map $d_X : X \times X \to \mathbb{R}_+$ (called a *metric* or distance function on X) satisfies

$$0 \le d_X(x, y) = d_X(y, x) \le d_X(x, z) + d_X(z, y), \qquad x, y, z \in X,$$

and if $d_X(x, y) = 0$ implies $x = y$. The topology on X induced by the metric d_X is generated by the set of open balls $B_{X,r}(x) = \{y \in X : d_X(x, y) < r\}, x \in X, r > 0$. Obviously, any subset $Y \subset X$ can be transformed into a metric space $(Y; d_X|_Y)$ by restricting d_X to $Y \times Y$. In the following, when the metric is fixed, we often omit the letter d_X in the notation and speak of the metric space X and its subset Y equipped with the metric on X.

M. Griebel, P. Oswald, *Hilbert Space Splittings and Iterative Methods*, Springer Series in Computational Mathematics 62, https://doi.org/10.1007/978-3-031-74370-2

Notions like convergence, completeness, compactness, continuity of maps between metric spaces, etc. can be defined using sequences. Here is a list of relevant definitions and properties:

- A sequence $\tilde{x} := (x_n)_{n\in\mathbb{N}}$ of elements $x_n \in X$ *converges in* X if there exists an $x \in X$ such that $d_X(x_n, x) \to 0$ for $n \to \infty$. This x is unique, and is called the limit of $\tilde{x}$. The shorthand notation is $x_n \to x$ in X.
- A subset $Y \subset X$ is called *dense* in X if for each $x \in X$ there exists a sequence $\tilde{y} := (y_n)_{n\in\mathbb{N}}$ of elements $y_n \in Y$ that converges to x in X.
- The metric space $(X; d_X)$ is called *separable* if it contains an at most countable dense subset.
- A subset $Y \subset X$ is *closed* in X if $y_n \to x$ in X for some sequence of elements $y_n \in Y$ implies $x \in Y$. For example, if Y is dense and closed in X, then $Y = X$.
- We call $\tilde{x}$ a *Cauchy sequence* in X if $d_X(x_n, x_m) \to 0$ for $n, m \to \infty$. If every Cauchy sequence converges in X, then $(X; d_X)$ is called *complete*. If $(X; d_X)$ is not complete, then it can be completed: There exists a complete metric space $(\bar{X}; d_{\bar{X}})$, called the *completion of* $(X; d_X)$, such that $X \subset \bar{X}$ is dense in $(\bar{X}; d_{\bar{X}})$, and d_X is equal to the restriction of $d_{\bar{X}}$ to X, i.e.,

$$d_X(x, y) = d_{\bar{X}}(x, y), \qquad x, y \in X.$$

  The completion is unique up to isomorphism. In particular, if $(X; d_X)$ is complete and Y is a dense subset X, then $(X; d_X)$ coincides with the completion of $(Y; d_X|_Y)$.
- A metric space $(X; d_X)$ is called *(sequentially) compact* if every sequence $\tilde{x}$ contains a subsequence converging in X. If $(X; d_X)$ is separable, then sequential compactness implies the usual notion of compactness (every open cover of X contains a finite subcover). A subset $Y \subset X$ is compact if the induced metric space $(Y; d_X|_Y)$ is compact.

A.1.2 Normed linear spaces

In most of this book we only treat *normed linear spaces* X over $\mathbb{R}$, where X is a linear space and the metric $d_X(x, y) := \|x - y\|_X$ is induced by a *norm* $\|\cdot\|_X : X \to \mathbb{R}_+$ characterized by the properties

$$0 \le \|x + y\|_X \le \|x\|_X + \|y\|_X, \qquad \|\lambda x\|_X = |\lambda|\|x\|_X, \tag{A.1}$$

which are satisfied for all $x, y \in X$ and $\lambda \in \mathbb{R}$, and

$$\|x\|_X = 0 \quad \Leftrightarrow \quad x = 0, \qquad x \in X.$$

If only (A.1) is satisfied, we call $\|\cdot\|_X$ a *semi-norm*.

The above definitions and statements for metric spaces carry over to normed linear spaces. For example, any linear set $Y \subset X$ becomes a normed linear space by itself if the norm on Y is inherited from X. Complete normed spaces are called *Banach spaces*. The completion of a subset $Y \subset X$ (considered as a metric space with the metric inherited from the norm in X) of a Banach space X can be identified with its closure

$$\mathrm{clos}_X(Y) = \{x \in X : \ y_n \longrightarrow x \text{ for some sequence } (y_n)_{n\in\mathbb{N}} \text{ with } y_n \in Y\}. \quad (A.2)$$

If the context is clear and it does not conflict with other notations, we also use $\bar{Y}$ for the closure of Y.

For normed linear spaces, there are additional aspects due to the linear structure on X: We encounter linear combinations and series, linear subsets and subspaces, linear functionals and operators, just to name a few. The main facts and notational conventions we rely on are listed next.

- *Linear sets* (or linear manifolds) in a linear space X are non-empty subsets that are algebraically closed under addition and scalar multiplication. The *span* of an arbitrary subset $Y \neq \emptyset$ denoted by $\mathrm{span}(Y)$ is the smallest linear set in X that contains Y. Equivalently, $\mathrm{span}(Y)$ is the set of all finite linear combinations of elements from Y. The span of a collection of subsets $(Y_\omega)_{\omega\in\Omega}$ of X is simply defined as the span of their union $Y = \cup_{\omega\in\Omega} Y_\omega$. Furthermore, Y is a linear set in X if and only if $\mathrm{span}(Y) = Y$.
- A family $Y = (y_\omega)_{\omega\in\Omega} \subset X$ of elements in X is called *linearly independent* if each $x \in \mathrm{span}(Y)$ has a unique expression by a finite linear combination of the $y_\omega \in Y$. Equivalently, and more traditionally, linear independence is defined by the property that

$$\sum_{i=1}^{n} \lambda_i y_{\omega_i} = 0 \qquad \Rightarrow \qquad \lambda_1 = \cdots = \lambda_n = 0$$

  for any choice of $n \in \mathbb{N}$ different $\omega_i \in \Omega$ and $\lambda_i \in \mathbb{R}$. If Y is a finite set, then obviously n does not exceed the cardinality of Y. More generally, a family $(Y_\omega)_{\omega\in\Omega}$ of linear sets $Y_\omega \subset X$ is linearly independent if

$$\sum_{i=1}^{n} y_{\omega_i} = 0 \qquad \Rightarrow \qquad y_{\omega_1} = \cdots = y_{\omega_n} = 0$$

  for any selection of $y_{\omega_i} \in Y_{\omega_i}$ with different ω_i, $i = 1, \ldots, n$.
- An *infinite series* $\sum_{i=1}^{\infty} x_i$ with terms x_i from X is called convergent if the sequence of its partial sums converges in X to some $x \in X$, called the *sum*, namely

$$s_n := \sum_{i=1}^{n} x_i \longrightarrow x, \qquad n \longrightarrow \infty.$$

  A notion we encounter more often is *unconditional convergence*, which requires that the rearranged series $\sum_{i=1}^{\infty} x_{\sigma_i}$ converges for any permutation σ of $\mathbb{N}$ (and then necessarily to the same sum). A simple sufficient condition for unconditional convergence in Banach spaces is as follows:

Lemma A.1 *Let X be a Banach space. If $x_i \in X$, $i \in \mathbb{N}$, satisfies*

$$\sum_{i=1}^{\infty} \|x_i\|_X < \infty,$$

then the series $\sum_{i\in\mathbb{N}} x_i$ converges unconditionally in X, and its sum x satisfies

$$\|x\|_X \le \sum_{i=1}^{\infty} \|x_i\|_X < \infty.$$

These definitions carry over to series with arbitrary countable index sets Ω other than $\mathbb{N}$. Series with respect to uncountable index sets Ω can only converge if they contain at most countably many non-zero terms.

- A closed linear set in a normed linear space X is called a *subspace*. Note that this is not the universal convention in the literature, where linear sets and subspaces are often called subspaces (or linear manifolds) and closed subspaces, respectively. Given a subset $Y \subset X$, the closure of its span in X is the smallest subspace containing Y. Linear sets and their closure to subspaces play a central role in our considerations in Chapter 2.

Most spaces of interest in numerical analysis are separable Banach spaces. Here are two examples of Banach spaces encountered in this book:

- *Sequence spaces*. For any at most countable index set Ω, define the spaces of real sequences
$$\ell^p(\Omega) := \{x = (x_i)_{i\in\Omega} : \ \|x\|_{\ell^p} < \infty\}$$
with norm
$$\|x\|_{\ell^p} := \begin{cases} (\sum_{i\in\Omega} |x_i|^p)^{1/p}, & 1 \le p < \infty, \\ \sup_{i\in\Omega} |x_i|, & p = \infty. \end{cases}$$
Addition and scalar multiplication for sequences are defined componentwise. For $1 \le p < \infty$ the spaces $\ell^p(\Omega)$ are separable. We also consider weighted versions of sequence spaces, and we consider L^p spaces of functions defined on certain measure spaces. When $\Omega = \{1, \dots, n\}$ is finite, we often identify $\ell^p(\Omega)$ with $\mathbb{R}^n$ equipped with the p-norm and shorten the notation for the norm to $\|x\|_p$, $1 \le p \le \infty$ (for $\alpha \in \mathbb{Z}^n$ we also use $|\alpha|_p$).
- *Continuous functions*. Let (Ω, d_Ω) be a compact metric space and Y a Banach space. Then the set $C(\Omega, Y)$ of all continuous functions $f : \Omega \to Y$ equipped with the norm
$$\|f\|_{C(\Omega,Y)} := \max_{\omega\in\Omega} \|f(\omega)\|_Y,$$
is a separable Banach space. For functions, addition and scalar multiplication are defined pointwise. If $Y = \mathbb{R}$, we simply write $C(\Omega)$.

A.1.3 Linear operators

A map $T : X \to Y$ from one linear space X to another linear space Y is called a *linear operator* if its *domain* of definition $\mathrm{dom}(T)$ is a linear set in X and if T preserves the linear structure, namely

$$T(\lambda x + x') = \lambda Tx + Tx', \qquad x, x' \in \mathrm{dom}(T), \quad \lambda \in \mathbb{R}.$$

The *range* of T is defined as

$$\mathrm{ran}(T) := \{y \in Y : \ y = Tx \text{ for some } x \in \mathrm{dom}(T)\},$$

and is a linear set in Y. We call T *injective* (or one-to-one) if $Tx = Tx'$ implies $x = x'$ for all $x, x' \in \mathrm{dom}(T)$, and *surjective* (or onto Y) if $\mathrm{ran}(T) = Y$. The *null-space* (or kernel) of a linear operator T is another linear set in X, given by

$$\ker(T) := \{x \in \mathrm{dom}(T) : \ Tx = 0\}.$$

The injectivity of T is equivalent to the requirement that $\ker(T) = \{0\}$. We use the word *bijection* for linear operators T with domain $\mathrm{dom}(T) = X$ that are injective and surjective at the same time.

In this book we deal almost exclusively with bounded linear operators acting between Banach spaces X and Y, where $T : X \to Y$ is called *bounded* if there is a constant $C < \infty$ such that

$$\|Tx\|_Y \le C\|x\|_X, \qquad x \in \mathrm{dom}(T). \tag{A.3}$$

The *operator norm* of T (with respect to $\mathrm{dom}(T)$) is defined as the infimum of the admissible constants C in (A.3), namely

$$\|T\|_{X\to Y} := \inf_{(A.3)} C = \sup_{0\neq x\in\mathrm{dom}(T)} \frac{\|Tx\|_Y}{\|x\|_X} = \sup_{x\in\mathrm{dom}(T):\ \|x\|_X=1} \|Tx\|_Y. \tag{A.4}$$

If $T : X \to X$, we shorten the notation for the operator norm (A.4) to $\|T\|_X$.

Boundedness for linear operators is equivalent to their *continuity*: If $x_n \to x$ in X for x_n and x belongs to $\mathrm{dom}(T)$, then $Tx_n \to Tx$ in Y. Similarly, the map $x \in \mathrm{dom}(T) \to \|Tx\|_Y \in \mathbb{R}$ is continuous. Furthermore, the null-space $\ker(T)$ of a bounded operator T is automatically a subspace of X if $\mathrm{dom}(T)$ is a subspace of X.

We often rely on the *extension by continuity* principle for densely defined bounded operators between Banach spaces: An operator $T : X \to Y$ is *densely defined* on the Banach space X if $\mathrm{dom}(T)$ is dense in X.

Lemma A.2 *If X and Y are Banach spaces and the bounded linear operator T : $X \to Y$ is densely defined on X, then there is a unique bounded linear operator $\bar{T} : X \to Y$ with domain* $\mathrm{dom}(\bar{T}) = X$ *such that*

$$\bar{T}x = Tx, \qquad x \in \mathrm{dom}(T), \qquad \|\bar{T}\|_{X\to Y} = \|T\|_{X\to Y}.$$

The action of $\bar{T}$ is given by the limit process

$$Tx_n \to \bar{T}x \quad \text{in } Y,$$

where $x_n \to x$ in X for sequences from $\mathrm{dom}(T)$. The correctness of this definition and the remaining statements of Lemma A.2 follow because T transforms any Cauchy sequence $(x_n)_{n\in\mathbb{N}}$ in $\mathrm{dom}(T)$ to a Cauchy sequence $(Tx_n)_{n\in\mathbb{N}}$ in Y due to (A.3), and since completion and closure of subsets are the same in Banach spaces, see (A.2).

Finally, we define the inverse of a bounded linear operator $T : X \to Y$ with $\mathrm{dom}(T) = X$: We call a linear operator $S : Y \to X$ the *inverse* of T and denote it by $T^{-1} = S$ if S is bounded and ST is the identity operator in X (left inverse) and TS is the identity operator in Y (right inverse). The following theorem gives two convenient criteria for the invertibility of operators between Banach spaces.

Theorem A.1 *Let $T : X \to Y$ be a bounded linear operator between the Banach spaces X and Y.*
a) If $\mathrm{dom(T)} = X$*, then T is invertible if and only if it is injective, i.e.,* $\ker(T) = \{0\}$*, and surjective, i.e.,* $\mathrm{ran}(T) = Y$*.*
b) If T is densely defined on X, then its extension by continuity $\bar{T}$ is invertible if and only if $\mathrm{ran}(T)$ *is dense in Y and there is a constant $c > 0$ such that*

$$\|Tx\|_Y \geq c\|x\|_X, \qquad x \in \mathrm{dom(T)}.$$

In this case,

$$\|\bar{T}^{-1}\|_{Y\to X} \leq c^{-1}.$$

The non-trivial statement in Part a) of Theorem A.1, namely the fact that the bijectivity of T automatically implies the boundedness of T^{-1}, is a consequence of the famous open mapping principle for Banach spaces. The following criterion for the *equivalence of norms* on the same Banach space follows by applying Theorem A.1 to the identity operator on X: If $\|\cdot\|_{X,1}$ and $\|\cdot\|_{X,2}$ are two norms defined on X, both of which make X a Banach space, and

$$\|x\|_{X,1} \leq C\|x\|_{X,2}, \qquad x \in X,$$

for some constant $C < \infty$, then the two norms are equivalent, i.e., there is also a constant $c > 0$ such that

$$c\|x\|_{X,2} \leq \|x\|_{X,1} \leq C\|x\|_{X,2}, \qquad x \in X.$$

The equivalence of different norms on Hilbert spaces is a recurring theme in this book.

Whenever there is no confusion, we use the same letter T for the original T and its extension by continuity $\bar{T}$, and assume that the domain of T is X, even if T was originally only densely defined. Extension by continuity can also be applied without the density assumption on $\mathrm{dom}(T)$, in which case $\mathrm{dom}(\bar{T}) = \mathrm{clos}_X(\mathrm{dom}(T))$ is a subspace of X.

A.1.4 Dual spaces and adjoint operators

By definition, *linear functionals on* X are linear operators mapping a Banach space X into the scalar field $\mathbb{R}$. The *dual space* X' of a Banach space X is the set of all bounded linear functionals on X with $\mathrm{dom}(T) = X$, equipped with the norm

$$\|f\|_{X'} := \|f\|_{X\to\mathbb{R}} = \sup_{0\neq x\in X} \frac{|f(x)|}{\|x\|_X} = \sup_{x\in X:\, \|x\|_X=1} |f(x)|.$$

X' is also a Banach space. The evaluation of a functional $f \in X'$ at x is sometimes written as the evaluation of a bilinear form defined on $X' \times X$, called the *duality pairing*, namely

$$\langle f, x\rangle_{X'\times X} := f(x), \qquad f \in X', \quad x \in X.$$

Since $|\langle f, x\rangle_{X'\times X}| = |f(x)| \leq \|f\|_{X'}\|x\|_X$, the duality pairing is bounded and thus continuous on $X' \times X$ by the definition of the norm in X'. If we take $X = \ell^p(\Omega)$ with at most countable Ω as an example, $1 < p < \infty$, then the dual space is isomorphic to $\ell^q(\Omega)$, where $1 < q < \infty$ is given by $1 = 1/p + 1/q$, in the sense that each $f \in \ell^p(\Omega)'$ is generated by a unique $y \in \ell^q(\Omega)$, i.e.,

$$f(x) = \sum_{i\in\Omega} y_i x_i, \qquad x \in \ell^p(\Omega); \qquad \|f\|_{\ell^p(\Omega)'} = \|y\|_{\ell^q(\Omega)}.$$

In this case, the duality pairing can be identified with the usual scalar product of two sequences. For $\ell^\infty(\Omega)$ and for spaces of continuous functions, the characterization of their dual spaces is more complicated, but not needed for our purposes.

If $T : X \to Y$ is a bounded linear operator between Banach spaces, then it has a unique *adjoint*

$$T' : Y' \to X'$$

given by the formula

$$\langle T'f, x\rangle_{X'\times X} = (T'f)(x) = f(Tx) = \langle f, Tx\rangle_{Y'\times Y}, \qquad x \in X, \quad f \in Y'.$$

The linearity of T' is obvious, its boundedness and the formula

$$\|T'\|_{Y'\to X'} = \|T\|_{X\to Y}$$

follow from the inequalities

$$|(T'f)(x)| = |f(Tx)| \leq \|f\|_{Y'}\|Tx\|_Y \leq \|f\|_{Y'}\|T\|_{X\to Y}\|x\|_X,$$

which are tight on the set of all possible pairs $f \in Y'$ and $x \in X$. The adjoint generalizes the notion of transpose (or adjoint) for matrices. Note that in Hilbert spaces we use a second, slightly different notion of adjoint, denoted T^*, see Section A.2.

A.1.5 Hilbert spaces

Hilbert spaces are a special class of Banach spaces, in which the norm is induced from an inner or scalar product.

A linear space H over the scalar field $\mathbb{R}$ is called a *real* inner product or *pre-Hilbert space* if it is equipped with a symmetric and positive definite bilinear form $(\cdot,\cdot)_H : H \times H \to \mathbb{R}$, called a *scalar product* on H. More explicitly, the scalar product must satisfy the properties

$$(x,y)_H = (y,x)_H, \qquad (\lambda x + y, z)_H = \lambda(x,z)_H + (y,z)_H, \tag{A.5}$$

for all $x, y, z \in H$, $\lambda \in \mathbb{R}$, and

$$(x,x)_H > 0, \qquad 0 \neq x \in H. \tag{A.6}$$

The norm on H induced by the scalar product is defined by $\|x\|_H := (x,x)_H^{1/2}$, and H becomes a normed linear space. If the strict inequality $>$ is replaced by $\geq$ in (A.6), then in general $\|\cdot\|_H$ is only a semi-norm on H. A pre-Hilbert space H is called a *Hilbert space* if it is complete.

Besides $H = \mathbb{R}^n$, which becomes a Hilbert space when equipped with the Euclidean scalar product, a standard example is the sequence space $\ell^2(\Omega)$ with scalar product

$$(x,y)_{\ell^2(\Omega)} = \sum_{i\in\Omega} y_i x_i, \qquad x = (x_i)_{i\in\Omega},\ y = (y_i)_{i\in\Omega} \in \ell^2(\Omega).$$

The scalar product $(\cdot,\cdot)_H$ in a pre-Hilbert space can be recovered from the norm $\|\cdot\|$ by the parallelogram identity

$$(x,y)_H = \frac{1}{4}(\|x+y\|_H^2 - \|x-y\|_H^2), \qquad x, y \in H.$$

Combined with the usual completion procedure, this fact allows us to define Hilbert spaces H by specifying the norm $\|\cdot\|_H$ instead of the scalar product on a dense set in H. For this however, (A.5) must be checked for the resulting bilinear form $(\cdot,\cdot)_H$. Another basic property of Hilbert spaces is the Cauchy-Schwarz inequality

$$|(x,y)_H| \leq \|x\|_H \|y\|_H, \qquad x, y \in H.$$

The scalar product allows us to define orthogonality, which is a crucial concept in Hilbert spaces. Two elements $x, y \in H$ are called *orthogonal* if $(x,y)_H = 0$ (for short $x \perp y$). Similarly, two subsets $X, Y \subset H$ are orthogonal (for short $X \perp Y$) if $x \perp y$ for all $x \in X$ and $y \in Y$. If X is a subset in H, then its *orthogonal complement*

$$X^\perp = \operatorname{span}(X)^\perp := \{y \in H : \ y \perp x \text{ for all } x \in X\}$$

is automatically a subspace of H. From this definition it follows that

$$X^{\perp\perp} := (X^\perp)^\perp = \mathrm{clos}(\mathrm{span}(X)).$$

If W is a subspace of H, then W and its complement $W^\perp$ form an *orthogonal subspace decomposition* of H, denoted by

$$H = W \oplus W^\perp,$$

in the sense that each $x \in H$ can be uniquely decomposed as

$$x = w + w^\perp, \quad \|x\|_H^2 = \|w\|_H^2 + \|w^\perp\|_H^2, \qquad w \in W,\ w^\perp \in W^\perp \quad (w \perp w^\perp).$$

The maps $x \longrightarrow w$ and $x \longrightarrow w^\perp$ are called *ortho-projections* onto W and $W^\perp$, and are denoted by Q_W and $Q_{W^\perp} = I - Q_W$, respectively. Alternatively, $Q_W x$ can be defined as the element $w \in W$ that is closest to x in norm distance, i.e.,

$$Q_W x = \mathrm{argmin}_{w \in W} \|x - w\|_H^2,$$

or as the unique $w \in W$ such that

$$(w, w')_H = (x, w')_H \qquad \text{for all } w' \in W.$$

Since W and $W^\perp$ as subspaces of H can be orthogonally decomposed into subspaces on their own, we obtain orthogonal subspace decompositions into many orthogonal subspaces. The fundamental result in this direction is the Fischer-Riesz theorem, which guarantees the existence of complete orthonormal systems (CONS) for any Hilbert space H.

The formal definition of a CONS is as follows: Let $\Omega \neq \emptyset$ be an arbitrary index set. A system $\mathcal{E} := \{e_\omega\}_{\omega\in\Omega}$ of non-zero elements $e_\omega \in H$ is called *orthogonal* if $e_\omega \perp e_{\omega'}$ whenever $\omega \neq \omega'$. It is called *orthonormal* if additionally $\|e_\omega\|_H = 1$ for all $\omega \in \Omega$. Also $\mathcal{E}$ is called complete if it is maximal, i.e., it cannot be enlarged to another orthogonal system $\mathcal{E}' \supset \mathcal{E}$ by adding an element $e' \neq 0$ orthogonal to all e_ω in $\mathcal{E}$.

The completeness of an orthogonal system $\mathcal{E}$ is also characterized by the property $\mathrm{clos}_H(\mathrm{span}(\mathcal{E})) = H$, i.e., by the density of the span of $\mathcal{E}$ in H.

Theorem A.2 *a) Every Hilbert space H possesses a CONS $\mathcal{E}$.*
b) The index sets Ω associated with different CONS for the same Hilbert space H have the same cardinality, called the dimension *of H. In particular, H is separable if and only if Ω is finite (then H is finite-dimensional) or countable (then it is separable infinite-dimensional).*
c) If $\mathcal{E}$ is a CONS in H, then every $x \in H$ possesses a unique series expansion with respect to $\mathcal{E}$, given by

$$x = \sum_{\omega\in\Omega} (x, e_\omega)_H e_\omega, \qquad \|x\|_H^2 = \sum_{\omega\in\Omega} (x, e_\omega)_H^2. \tag{A.7}$$

In (A.7), at most countably many coefficients (x, e_ω) are non-zero, and convergence is unconditional in H.

This theorem shows that up to isomorphism there is only one separable infinite-dimensional Hilbert space H, namely $\ell^2(\mathbb{N})$. The isomorphism between H and $\ell^2(\mathbb{N})$ is given by the map

$$x \longrightarrow ((x, e_i)_H)_{i\in\Omega},$$

where the index set of the CONS $\mathcal{E}$ is identified with $\mathbb{N}$ by bijection. Finite-dimensional Hilbert spaces are isomorphic to the Euclidean space $\mathbb{R}^n$, where n is the dimension. Finally, note that every CONS $\mathcal{E}$ for H generates an orthogonal subspace decomposition

$$H = \bigoplus_{i\in\Omega} W_i$$

into one-dimensional subspaces W_i spanned by $e_i \in \mathcal{E}$, $i \in \Omega$.

The other crucial property of Hilbert spaces that distinguishes them from Banach spaces is their *self-duality*. This follows from the Riesz representation theorem.

Theorem A.3 *Every $x \in H$ generates a bounded linear functional $f_x \in H'$ of the same norm $\|f_x\|_{H'} = \|x\|_H$ given by the formula*

$$f_x(y) := (x, y)_H \qquad y \in H.$$

Conversely, for every $f \in H'$ there is a unique $x \in H$ such that

$$f = f_x.$$

Theorem A.3 shows that the *duality map* $R_H : H \longrightarrow H'$ given by

$$R_H x = f_x$$

realizes an isomorphism between the Hilbert space H and its dual H'. Its inverse $R_H^{-1} : H' \longrightarrow H$ allows us to identify each $f \in H'$ with an element $x_f \in H$ such that

$$f(y) = (x_f, y)_H \quad \text{for all } y \in H.$$

This self-duality of Hilbert spaces leads to a second definition of *adjoint* (the so-called Hilbert space adjoint) of a densely defined bounded operator $T : H_1 \longrightarrow H_2$ between Hilbert spaces H_1 and H_2, which is different from the (Banach space) adjoint T': The new adjoint $T^* : H_2 \longrightarrow H_1$ is given by

$$T^* = R_{H_1}^{-1} T' R_{H_2},$$

and allows us to interpret T' as an operator between the original Hilbert spaces, and not their duals. A more explicit, equivalent definition of T^* is to define $z := T^* y$ for given $y \in H_2$ as the unique $z \in H_1$ such that

$$(z, x)_{H_1} = (Tx, y) \qquad \forall\, x \in \operatorname{dom}(T).$$

This also shows that $\operatorname{dom}(T^*) = H_2$. Note at this point that $\|T^*\|_{H_2\to H_1} = \|T'\|_{H_2'\to H_1'}$ $(= \|T\|_{H_1\to H_2})$, since the duality maps are isometries.

A.2 Operators between Hilbert spaces

A.2.1 $\mathcal{B}(H_1, H_2)$

From now on we restrict ourselves to bounded linear operators $T : H_1 \to H_2$, which act between two Hilbert spaces with $\mathrm{dom}(T) = H_1$. With the extension by continuity principle discussed in Subsection A.1.3 at hand, this is not a severe restriction. The space of all such operators is denoted by $\mathcal{B}(H_1, H_2)$ and becomes a Banach space by itself when equipped with the operator norm $\|\cdot\|_{H_1 \to H_2}$. Here the linear structure is defined pointwise, i.e., $(\lambda T + S)x := \lambda Tx + Sx$ for arbitrary $x \in H_1, T, S \in \mathcal{B}(H_1, H_2)$, and $\lambda \in \mathbb{R}$. If $H_1 = H_2 = H$, we shorten the notation to $\mathcal{B}(H) := \mathcal{B}(H, H)$. The space $\mathcal{B}(H)$ is a *Banach algebra with unit*, where the unit is the identity operator I on H and the multiplication operation is given by the composition $(TS)x := T(Sx)$ for all $x \in H$ and $T, S \in \mathcal{B}(H)$. Here is a short list of general properties of operators in $\mathcal{B}(H_1, H_2)$ and $\mathcal{B}(H)$ that are used frequently throughout the book.

- The inequalities

  $$\|T + S\|_{H_1 \to H_2} \le \|T\|_{H_1 \to H_2} + \|S\|_{H_1 \to H_2}, \qquad \|\lambda T\|_{H_1 \to H_2} = |\lambda| \|T\|_{H_1 \to H_2}$$

  valid for $T, S \in \mathcal{B}(H_1, H_2)$, $\lambda \in \mathbb{R}$, and

  $$\|TS\|_{H_0 \to H_2} \le \|T\|_{H_1 \to H_2} \|S\|_{H_0 \to H_1}, \qquad T \in \mathcal{B}(H_1, H_2), \quad S \in \mathcal{B}(H_0, H_1),$$

  follow directly from the formula (A.4) for the operator norm.
- New operators are often defined by limiting processes. Besides convergence in the operator norm, also called *uniform convergence* and denoted by $T_n \to T$, there are other useful notions of convergence for operators. We call the sequence $(T_n)_{n \in \mathbb{N}}$ of bounded linear operators $T_n \in \mathcal{B}(H_1, H_2)$ *strongly convergent* and denote it by $T_n \xrightarrow{s} T$ if the sequence $(T_n x)_{n \in \mathbb{N}}$ converges for every $x \in H_1$ to some $y =: Tx$ in H_2. Strong convergence is conveniently used to derive new bounded linear operators, since automatically $T \in \mathcal{B}(H_1, H_2)$ and

  $$\|T\|_{H_1 \to H_2} \le \liminf_{n \to \infty} \|T_n\|_{H_1 \to H_2}.$$

  This follows from the uniform boundedness principle, see [192, Section 4.5]. However, strong convergence does not imply uniform convergence. Similar conclusions can be drawn if $(T_n x, y)_{H_2} \to (Tx, y)_{H_2}$ as $n \to \infty$ for all $x \in H_1$ and $y \in H_2$. This is called *weak convergence*.
- The *spectrum* $\sigma(T)$ of $T \in \mathcal{B}(H)$ is the set of all complex numbers λ such that the operator $\lambda I - T$ is *not* invertible on H. The complement set $\rho(T) := \mathbb{C} \backslash \sigma(T)$ is called the *residual set* of T. To define it formally, we need to consider the complexification of H, which can be identified with the linear space

  $$H_c = \{x + iy : x, y \in H\}$$

over $\mathbb{C}$ (where i is the complex unit). The addition in H_c is inherited componentwise from H, and the scalar multiplication with complex numbers $\lambda = a + ib$ is defined by $\lambda(x + iy) = (ax - by) + i(ay + bx), a, b \in \mathbb{R}$. Equipped with the complex-valued bilinear form $(x + iy, x' + iy')_{H_c} := (x, x')_H + (y, y')_H + i(-(x, y')_H + (y, x')_H)$ (here taken conjugate linear in the second variable), where $x, y, x', y' \in H$, the linear space H_c becomes a complex Hilbert space with norm $\|x + iy\|_{H_c} = (\|x\|_H^2 + \|y\|_H^2)^{1/2}$, whose restriction to elements of the form $x + i0$ coincides with H. The operator T extends to an operator T_c, which acts on H_c by the formula $T_c(x + iy) = Tx + iTy$. Thus, $\lambda I - T_c$ is a well-defined bounded linear operator on H_c for any $\lambda \in \mathbb{C}$, and $\sigma(T) := \sigma(T_c)$ and $\rho(T) := \rho(T_c)$ are properly defined by checking the invertibility of $\lambda I - T_c$ on H_c.
For the purpose of this book, it is sufficient to know the following facts: The spectrum of $T \in \mathcal{B}(H)$ is a non-empty compact subset of $\mathbb{C}$ contained in a disk of radius $\|T\|_H$ around the origin, i.e.,

$$|\lambda| \leq \|T\|_H, \qquad \lambda \in \sigma(T).$$

A sometimes tighter confinement of the spectrum is given by the *numerical range* of T_c, namely

$$\sigma(T) + \sigma(T^*) \subset \{(T_c(x + iy), x + iy)_{H_c} : \|x + iy\|_{H_c}^2 = \|x\|_H^2 + \|y\|_H^2 = 1\}.$$

If $\lambda \in \sigma(T)$ then $\bar{\lambda} \in \sigma(T^*)$ and vice versa. The *point spectrum* $\sigma_p(T) \subset \sigma(T)$ of T is defined as the set of all *eigenvalues* of T, i.e., all $\lambda \in \mathbb{C}$ such that $Tz = \lambda z$ for some non-zero $z = x + iy \in H_c$. There are important subsets of $\mathcal{B}(H)$, namely compact and symmetric operators, for which either $\sigma(T)\backslash\{0\} = \sigma_p(T)$ or $\sigma(T) \subset \mathbb{R}$ holds. Note that complexification can be avoided in the latter case.

- For the convergence theory of iterative methods in a Hilbert space H, the *spectral radius*

$$r(T) := \max_{\lambda \in \sigma(T)} |\lambda|$$

of an operator $T \in \mathcal{B}(H)$ plays an important role. Alternatively one has

$$r(T) = \lim_{n \to \infty} \|T^n\|_H^{1/n}.$$

If $r(T) < 1$, then starting with any $e^{(0)} \in H$, the fixpoint iteration scheme

$$e^{(m+1)} = Te^{(m)}, \qquad m = 0, 1, \ldots,$$

converges to zero in H, and for any $\epsilon > 0$ there is a constant C_ϵ such that

$$\|e^{(m)}\| \leq C_\epsilon (r(T) + \epsilon)^m \|e^{(0)}\|, \qquad m = 1, 2, \ldots.$$

In other words, $\limsup_{m \to \infty} (\|e^{(m)}\| / \|e^{(0)}\|)^{1/m} \leq r(T)$. Typically, $e^{(m)}$ stands for the error after the m-th step of a stationary linear iteration scheme with iteration operator T, and the estimate says that the average error reduction per step is at least asymptotically determined by $r(T)$.

A.2.2 Special classes of operators in $\mathcal{B}(H)$

We start with a list of definitions. Recall that H is a real Hilbert space.

An operator

$$T \in \mathcal{B}(H)$$

is called

- *normal* if $TT^* = T^*T$,
- *symmetric* (or self-adjoint) if $T = T^*$,
- *orthogonal* if T is invertible and $T^{-1} = T^*$,
- a *projection* if $T^2 = T$,
- an *ortho-projection* if it is a projection and symmetric,
- *positive* if $(Tx, x)_H \geq 0$ for all $x \in H$,
- *positive definite* (or strictly positive) if $(Tx, x)_H > 0$ for all $0 \neq x \in H$,
- *coercive* (or strongly positive) if there exists a constant $c_0 > 0$ such that

$$(Tx, x)_H \geq c_0 \|x\|_H^2, \qquad 0 \neq x \in H.$$

By Theorem A.1, the coercivity implies the invertibility of T when considered as operator from H to the subspace $\mathrm{ran}(T)$ of H. For operators in $\mathcal{B}(H)$, the terms self-adjoint and symmetric are the same, and we use symmetric throughout the book. The set of all symmetric operators in $\mathcal{B}(H)$ is denoted by $\mathcal{B}_{sym}(H)$. Two standard ways to artificially construct symmetric operators are as follows. If $T \in \mathcal{B}(H)$, then the operator

$$\frac{1}{2}(T + T^*)$$

is symmetric and is called the *symmetric part* of T. If $T \in \mathcal{B}(H_1, H_2)$, then $T^*T \in \mathcal{B}_{sym}(H_1)$ and $TT^* \in \mathcal{B}_{sym}(H_2)$. Such operators arise in the context of least-squares methods for operator equations. The definition of orthogonality extends to operators $T \in \mathcal{B}(H_1, H_2)$ as is, it is equivalent to requiring $\mathrm{ran}(T) = H_2$ and the isometry property

$$(Tx, Ty)_{H_2} = (x, y)_{H_1}, \qquad x, y \in H_1,$$

to hold. The terminology for positive operators is not entirely consistent in the literature. In the case of real Hilbert spaces, the corresponding notations are usually applied only to symmetric operators. For complex Hilbert spaces, positivity of $T \in \mathcal{B}(H)$ automatically implies symmetry. We use $\mathcal{B}_{pos}(H)$ for the subset of all positive symmetric operators of $\mathcal{B}(H)$. If $T \in \mathcal{B}(H_1, H_2)$, then $T^*T \in \mathcal{B}_{pos}(H_1)$ and $TT^* \in \mathcal{B}_{pos}(H_2)$. These operators become positive definite if T and T^* are injective.

Obviously, symmetric and orthogonal operators (and thus ortho-projections) are normal. For the class of normal operators, and in particular for symmetric operators, a powerful spectral theorem holds (see [192, Chapter 7] or [150, Chapter VII] for the functional calculus for symmetric operators). We need only the following consequences for symmetric operators $T \in \mathcal{B}_{sym}(H)$, which are covered by [150, Chapter VI.3-4]:

- The spectrum of T is a compact subset of the real line: $\sigma(T) \subset [-\|T\|_H, \|T\|_H]$. If $T \in \mathcal{B}_{pos}(H)$, then $\sigma(T) \subset [0, \|T\|_H]$, and if T is coercive, then $\sigma(T) \subset [c_0, \|T\|_H]$, where $c_0 > 0$ is the constant in the definition of coercivity.
  More precisely, if $[\lambda_{\min}(T), \lambda_{\max}(T)]$ denotes the *smallest* interval containing $\sigma(T)$, then
  $$\lambda_{\min}(T) = \min_{\lambda\in\sigma(T)} \lambda = \inf_{x\in H:\, \|x\|_H=1} (Tx, x)_H,$$
  $$\lambda_{\max}(T) = \max_{\lambda\in\sigma(T)} \lambda = \sup_{x\in H:\, \|x\|_H=1} (Tx, x)_H = \|T\|_H.$$
  The numbers $\lambda_{\min}(T)$ and $\lambda_{\max}(T)$ are called the *lower* and *upper spectral bounds* of T. The operator T is positive if and only if $\lambda_{\min}(T) \geq 0$, and it is coercive if and only if $\lambda_{\min}(T) > 0$. In the latter case, we define its *condition number* by
  $$\kappa(T) := \frac{\lambda_{\max}(T)}{\lambda_{\min}(T)}.$$
- If $T \in \mathcal{B}_{pos}(H)$, then there exists a unique positive symmetric operator $S \in \mathcal{B}_{pos}(H)$ such that $S^2 = T$. This S is called the *square root* of T, and is denoted by $\sqrt{T}$ or $T^{1/2}$. Its spectrum is given by $\sigma(\sqrt{T}) = \{\lambda \in \mathbb{R}_+ : \lambda^2 \in \sigma(T)\}$. Thus, $\|\sqrt{T}\|_H^2 = \|T\|_H$. If $S \in \mathcal{B}(H)$ commutes with T, then it also commutes with $\sqrt{T}$.
- The map $|\cdot| : \mathcal{B}(H) \to \mathcal{B}_{pos}(H)$, given by $|T| = \sqrt{T^*T}$, associates with T a positive symmetric operator, called the *absolute value* of T. Its spectrum is given by $\sigma(|T|) = \{|\lambda| : \lambda \in \sigma(T)\}$. So $\| \, |T| \, \|_H = \|T\|_H$. If $T \in \mathcal{B}_{pos}(H)$, then $|T| = T$. If $S \in \mathcal{B}(H)$ commutes with T, then it also commutes with $|T|$.

The last two statements are special instances of the so-called functional calculus, which allows the construction of operators $\phi(T)$ for any continuous function $\phi : \sigma(T) \to \mathbb{R}$ and any symmetric operator T. We do not go into detail. In this book we only need the formula

$$\|\phi(T)\|_H = \max_{\lambda\in\sigma(T)} |\phi(\lambda)|, \qquad T \in \mathcal{B}_{sym}(H),$$

for the norm of $\phi(T)$. The construction of $\phi(T)$ for the special case of compact $T \in \mathcal{B}_{sym}(H)$ is outlined below.

A.2.3 Compact operators on separable Hilbert spaces

Throughout this section, let H, H_1, H_2 be separable Hilbert spaces. We call $T \in \mathcal{B}(H_1, H_2)$ *compact* if, for any bounded sequence $(x_n)_{n\in\mathbb{N}}$ of $x_n \in H_1$, the sequence $(Tx_n)_{n\in\mathbb{N}}$ contains a subsequence converging in H_2. The set of all compact operators $T \in \mathcal{B}(H_1, H_2)$ forms a subspace of $\mathcal{B}(H_1, H_2)$. Finite-rank operators are compact and form a dense subset in the subspace of all compact operators. Recall that a linear operator $T \in \mathcal{B}(H_1, H_2)$ is *finite-rank* if its range $\operatorname{ran}(T)$ is a finite-dimensional subspace of H_2. If $T \in \mathcal{B}(H_1, H)$ and $S \in \mathcal{B}(H, H_2)$ and at least one of these operators is compact, then $ST \in \mathcal{B}(H_1, H_2)$ is compact. Also $T \in \mathcal{B}(H_1, H_2)$ is compact if and only if $T^* \in \mathcal{B}(H_2, H_1)$ is compact.

Compact operators in $\mathcal{B}(H_1, H_2)$ have a simple structure: There are (not necessarily complete) orthonormal systems $\Psi := (\psi_k)_{k\in\Omega}$ in H_1 and $\Phi := (\phi_k)_{k\in\Omega}$ in H_2 with the same, at most countable index set $\Omega = \mathbb{N}$ or $\Omega = \{1, \ldots, n\}$ and positive numbers

$$\sigma_k \in (0, \|T\|_{H_1\to H_2}], \quad k \in \Omega,$$

called *singular values of* T, such that

$$Tx = \sum_{k\in\Omega} \sigma_k (x, \psi_k)_{H_1} \phi_k. \tag{A.8}$$

If $\Omega = \mathbb{N}$, then $\sigma_k \to 0$ for $k \to \infty$. By setting $x = \psi_m$ in (A.8), we see that $\phi_m = \sigma_m^{-1} T\psi_m$, $m \in \Omega$, and the subspace generated by Φ coincides with the closure of $\operatorname{ran}(T)$. Thus, if necessary, Φ can be extended to a CONS in H_2 by adding a CONS of $\ker(T^*) = \operatorname{ran}(T)^\perp$ to it. Similarly, Ψ can be turned into a CONS for H_1 by adding a CONS for $\ker(T)$.

The formula (A.8) is a straightforward generalization of the singular value decomposition for matrices. In fact, we have

$$T = U_\Phi \Sigma U_\Psi^*.$$

Here, $\Sigma : \ell^2(\Omega) \to \ell^2(\Omega)$ is the diagonal multiplication operator given by the (positive) singular values, and $U_\Psi : \ell^2(\Omega) \to H_1$ is given by

$$U_\Psi c = \sum_{k\in\Omega} c_k \psi_k, \qquad c = (c_k)_{k\in\Omega} \in \ell^2(\Omega).$$

A similar definition holds for U_Φ where ψ_k is replaced by ϕ_k. Since $U_\Psi^* x = ((x, \psi_k)_{H_1})_{k\in\Omega}$, the above identity for T is equivalent to (A.8). Furthermore, $U_\Psi^* U_\Psi$ and $U_\Phi^* U_\Phi$ are the identity operators on the subspace $\ker(T^*)^\perp$ generated by Ψ in H_1 and the subspace $\ker(T)^\perp$ generated by Φ in H_2, respectively.

If Ω is countable, we have $\lim_{k\to\infty} \sigma_k = 0$. In this case it is easy to show that T is the limit in $\mathcal{B}(H_1, H_2)$ of the finite-rank operators

$$S_n x = \sum_{k=1}^{n} \sigma_k (x, \psi_k)_{H_1} \phi_k, \qquad n \to \infty.$$

This approximability by finite-rank operators in the operator norm is a very powerful and desirable property from the point of view of developing numerical discretization methods. It also characterizes compact operators: $T \in \mathcal{B}(H_1, H_2)$ is compact if and only if it is the limit of a sequence of finite-rank operators in $\mathcal{B}(H_1, H_2)$. To get fast convergence $S_n \to T$, reordering the index set so that $\sigma_1 \geq \sigma_2 \geq \cdots$ is usually recommended.

The representation formula (A.8) follows from the Hilbert-Schmidt theorem for compact symmetric operators: If $T \in \mathcal{B}_{sym}(H)$ is compact, then there exists a (not necessarily complete) orthonormal system of eigenvectors $(\psi_k)_{k\in\Omega}$ in H, and associated non-zero eigenvalues $\lambda_k \in \mathbb{R}$, $k \in \Omega$, so that

$$Tx = \sum_{k\in\Omega} \lambda_k (x, \psi_k)_H \psi_k. \tag{A.9}$$

If $\Omega = \mathbb{N}$, then necessarily $\lambda_k \to 0$ for $k \to \infty$. The spectrum of such an operator is a pure point spectrum consisting of the positive eigenvalues λ_k, complemented by $\lambda = 0$ if $\lambda = 0$ is not an eigenvalue. The only case where $\lambda = 0$ does not belong to $\sigma(T)$ is when H is finite-dimensional, and T is symmetric positive definite. To get (A.8), set $H = H_1$ and apply (A.9) to the operator $T^*T \in \mathcal{B}_{pos}(H) \subset \mathcal{B}_{sym}(H)$. Since T^*T is positive, one has $\lambda_k > 0$. It remains to set $\sigma_k = \sqrt{\lambda_k}$ and $\phi_k = \sigma_k^{-1} T\psi_k$. In particular, the singular values of a compact $T \in \mathcal{B}(H)$ are exactly the positive eigenvalues of $|T|$, since in this case

$$|T| = \sqrt{T^*T} = \sum_{k\in\Omega} \sqrt{\lambda_k} (x, \psi_k)_H \psi_k.$$

See [150, Section VI.5] for more details.

The *Schatten classes* $\mathcal{S}_q(H_1, H_2)$ $(1 \le q < \infty)$ are classes of compact operators in $\mathcal{B}(H_1, H_2)$ defined in terms of singular values: $T \in \mathcal{B}(H_1, H_2)$ belongs to $\mathcal{S}_q(H_1, H_2)$ if T is compact and its singular values satisfy

$$\|T\|_{\mathcal{S}_q} := \Big(\sum_{k\in\Omega} \sigma_k^q\Big)^{1/q} < \infty.$$

Equipped with this norm, $\mathcal{S}_q(H_1, H_2)$ becomes a Banach space, and even a Hilbert space if $q = 2$.

We are mainly interested in the cases $q = 2$ (Hilbert-Schmidt operators) and $q = 1$ (trace class operators), for which alternative definitions are also available. In fact, an operator $T \in \mathcal{B}(H_1, H_2)$ is called a *Hilbert-Schmidt operator* if

$$\|T\|_{\mathrm{HS}}^2 := \sum_{k\in\Omega} \|T\psi_k\|_{H_2}^2 < \infty \tag{A.10}$$

for some CONS $(\psi_k)_{k\in\Omega}$ in H_1. The right-hand side in the definition of $\|T\|_{\mathrm{HS}}^2$ does not depend on the particular choice of the CONS in H_1. It is not hard to see that Hilbert-Schmidt operators must be compact, since they are the limit of finite-rank operators, and we have

$$T_n = \sum_{k=1}^{n} (\cdot, \psi_k) T\psi_k \to T$$

in the operator norm. It follows by computing the HS norm of T in (A.10) using (A.8) that the set of Hilbert-Schmidt operators $T \in \mathcal{B}(H_1, H_2)$ coincides with $\mathcal{S}_2(H_1, H_2)$, with equality of the associated norms $\|T\|_{\mathrm{HS}} = \|T\|_{\mathcal{S}_2}$.

The definition of trace class operators is given only for the case $H_1 = H_2 = H$, as used in this book. The *trace* of a positive symmetric operator $T \in \mathcal{B}_{pos}(H)$ is defined as

$$\mathrm{Tr}(T) := \sum_{k\in\Omega} (T\psi_k, \psi_k)_H,$$

where the right-hand side does not depend on the choice of the CONS Ψ in H. An operator $T \in \mathcal{B}(H)$ is called *trace class* if

$$\|T\|_{\mathrm{Tr}} = \mathrm{Tr}(|T|) := \sum_{k\in\Omega} (|T|\psi_k, \psi_k)_H < \infty.$$

Remember that $|T| = \sqrt{T^*T}$ is a positive symmetric operator by construction.

Since $\|T\|_{\mathrm{HS}} \leq \|T\|_{\mathrm{Tr}}$, trace class operators are Hilbert-Schmidt and thus compact. It can then be recovered from (A.9) and the definition of $|T|$ that the set of trace class operators $T \in \mathcal{B}(H)$ is equal to the Schatten class

$$\mathcal{S}_1(H) := \mathcal{S}_1(H, H),$$

with equality $\|T\|_{\mathrm{Tr}} = \|T\|_{\mathcal{S}_1}$ of the norms.

These facts, as well as other properties of Hilbert-Schmidt and trace class operators, can be found in [150, Section VI.6]. In particular, operators $T \in \mathcal{B}(H)$ of the form $T = S^*S$, where $S \in \mathcal{B}(H, H_1)$ for some separable Hilbert space H_1, are trace class if and only if S is a Hilbert-Schmidt operator. Moreover, in this case

$$\|T\|_{\mathrm{Tr}} = \|S\|_{\mathrm{HS}}^2.$$

We conclude this subsection with the definition of the operator $\phi(T)$ for arbitrary continuous functions $\phi : \sigma(T) \to \mathbb{R}$ if $T \in \mathcal{B}_{sym}(H)$ is compact. Given the representation (A.9) for T, we set

$$\phi(T)x = \sum_{k\in\Omega} \phi(\lambda_k)(\cdot, \psi_k)\psi_k + \phi(0)x_0,$$

where x_0 is the ortho-projection of x onto $\ker(T)$. The additional term involving x_0 is only present if $\lambda = 0$ is an eigenvalue of T and if $\phi(0) \neq 0$. This definition is consistent with the continuous functional calculus for normal operators, and recovers the already given definitions of the square root, the absolute value, and the integer powers T^n. Furthermore, for $s > 0$ the function $\phi(t) = t^s$ is defined and continuous for $t \geq 0$. Thus, the powers T^s of a compact operator $T \in \mathcal{B}_{pos}(H)$ are given by

$$T^s x = \sum_{k\in\Omega} \lambda_k^s (x, \psi_k)_H \psi_k. \tag{A.11}$$

In certain situations it makes sense to formally define powers T^s also for $s \leq 0$ by the formula (A.11), see Section 2.4.2 for an example.

A.3 Linear equations and variational problems

For this section, see [15], [16], and [26]. Let U, V be Hilbert spaces, and let

$$a : U \times V \to \mathbb{R}$$

denote a *bounded bilinear form*, i.e., $a(\cdot, v)$ is an element of U' for any fixed $v \in V$, $a(u, \cdot)$ is an element of V' for any fixed $u \in U$, and, for some positive constant C_0, we have

$$|a(u, v)| \leq C_0 \|u\|_U \|v\|_V, \qquad u \in U, \quad v \in V.$$

Then, finding $u \in U$ for given $f \in V'$ such that

$$a(u, v) = f(v) \qquad \forall\, v \in V, \tag{A.12}$$

is called a *variational problem on* U with *test space* V. The space U is called the *trial space*. Often, the trial space and the test space are the same ($U = V$), in which case we use the letter V for both spaces, and simply speak of a *variational problem on* V.

According to our assumptions about a and the Riesz representation theorem, the bilinear form a generates an operator $A \in \mathcal{B}(U, V')$ with norm $\|A\|_{U \to V'} \leq C_0$ given by

$$a(u, v) = \langle Au, v \rangle_{V' \times V}, \qquad u \in U, \quad v \in V,$$

and (A.12) is equivalent to finding solutions $u \in U$ of the linear equation

$$Au = f \tag{A.13}$$

for given $f \in V'$. Conversely, the linear equation (A.13) generates an equivalent variational problem (A.12). Obviously, the existence of solutions is guaranteed if $f \in \operatorname{ran}(A)$, while the uniqueness requires the injectivity of A or, equivalently, that $\ker(A) = \{0\}$ holds. The problems (A.12) and (A.13) are called *well-posed* if A is invertible, i.e., if A is injective and surjective at the same time.

Theorem A.4 *Let $a : U \times V \to \mathbb{R}$ be a bounded bilinear form, where U, V are Hilbert spaces. Then the variational problem (A.12) is well-posed if there exist constants $c_0, c_0' > 0$ such that*

$$\inf_{0 \neq u \in U} \sup_{0 \neq v \in V} \frac{a(u, v)}{\|u\|_U \|v\|_V} \geq c_0 > 0, \tag{A.14}$$

and

$$\inf_{0 \neq v \in V} \sup_{0 \neq u \in U} \frac{a(u, v)}{\|u\|_U \|v\|_V} \geq c_0' > 0. \tag{A.15}$$

If this is the case, then $\|A^{-1}\|_{V' \to U} \leq c_0^{-1}$, and $\|(A')^{-1}\|_{U' \to V} \leq (c_0')^{-1}$ by symmetry.

This important result, attributed to [123], follows directly from Part b) of Theorem A.1, since (A.14) is equivalent to

$$\|Au\|_{V'} \geq c_0\|u\|_U$$

by definition of the norm in V', while the second condition (A.15) is equivalent to

$$\|A'v\|_{U'} \geq c_0'\|v\|_V$$

and ensures that ran(A) is dense in V'. Conditions of the form (A.14) and (A.15) are called inf-sup conditions, and they play a central role in the numerical analysis of variational problems. If one is only interested in proving the bounded invertibility of A, one can replace the second inf-sup condition (A.15) by verifying that for every $0 \neq v \in V$ there is at least one $u \in U$ such that $a(u, v) \neq 0$. This also implies the density of ran(A) in V' and is often easier to prove than (A.15).

Now consider the case $U = V$. Then a simpler sufficient condition for well-posedness is coercivity: A bounded bilinear form $a : V \times V \longrightarrow \mathbb{R}$ is called *coercive* if

$$a(v, v) \geq c_0\|v\|_V^2, \qquad v \in V, \tag{A.16}$$

for some $c_0 > 0$. Obviously, (A.16) implies both conditions in Theorem A.4. A bounded and coercive bilinear form on V is also called V-elliptic, where c_0, C_0 are the associated ellipticity constants.

Even more can be said for symmetric variational problems on V: A bounded bilinear form $a : V \times V \longrightarrow \mathbb{R}$ is called *symmetric* if

$$a(u, v) = a(v, u), \qquad u, v \in V, \tag{A.17}$$

or, equivalently, if $A = A' : V \longrightarrow V'$. Operators A for which the associated bilinear form satisfies (A.17) and (A.16) will also be called symmetric and coercive, respectively.

Theorem A.5 *Let the bounded bilinear form $a : V \times V \longrightarrow \mathbb{R}$ on the Hilbert space V be coercive, i.e., it satisfies (A.16). Then the variational problem (A.12) is well-posed, and*

$$\|A^{-1}\|_{V' \to V} \leq c_0^{-1}.$$

If, in addition, a is symmetric, i.e., if it satisfies (A.17), then the problems (A.12) and (A.13), respectively, are equivalent to solving the quadratic optimization problem

$$J(u) := \frac{1}{2}a(u, u) - \langle f, u\rangle_{V' \times V} \longrightarrow \min \tag{A.18}$$

on V. Furthermore, $a(\cdot, \cdot)$ and $\|u\| := a(u, u)^{1/2}$ define a new scalar product and equivalent norm on V, and we have

$$c_0\|u\|_V^2 \leq a(u, u) = \|u\|^2 \leq C_0\|u\|_V^2, \qquad v \in V.$$

The well-posedness result is the famous Lax-Milgram theorem. Most of the core material of this book is related to designing iterative solvers for variational problems under the conditions of Theorem A.5. The connection between variational and optimization problems on Hilbert spaces is more general: Any unconstrained strongly convex differentiable minimization problem

$$J(u) \longrightarrow \min$$

on V with smooth $J : V \longrightarrow \mathbb{R}$ leads to a, in general, nonlinear variational problem induced by the (Fréchet or Gâteaux) derivative $\nabla J : V \longrightarrow V'$: Find $u \in V$ such that

$$\langle \nabla J(u), v\rangle_{V' \times V} = 0.$$

If $J(u)$ is a quadratic functional on V defined as in (A.18), but without the symmetry assumption on the bilinear form a, then the strong convexity of J is equivalent to the coercivity of a, and

$$\nabla J(u) = \frac{1}{2}(A + A')u - f.$$

Thus, unconstrained strongly convex quadratic minimization problems on V always lead to variational problems with symmetric and coercive forms, the theory of which is covered by Theorem A.5.

Arbitrary linear equations $Tu = v$ with given $T \in \mathcal{B}(U, V)$ and $v \in V$, where U, V are fixed Hilbert spaces, can be transformed into variational problems in several ways. First, one can consider the variational problem for $u \in U$ given by

$$a(u, w) := (Tu, w)_V = (v, w)_V \qquad \forall\, w \in V.$$

If $U \neq V$, then for the existence and uniqueness of solutions we can only rely on Theorem A.4.

Alternatively, as is typical for least-squares formulations of linear equations, one can replace $Tu = v$ by the linear equation

$$T^*Tu = T^*v \tag{A.19}$$

in U, or by a formulation

$$TT^*w = v, \quad u = T^*w, \tag{A.20}$$

as a linear equation in V. Equation (A.19) is the necessary optimality condition for the least-squares minimization problem

$$J(u) := \frac{1}{2}\|Tu - v\|_V^2 \to \min$$

on U, while (A.20) is related to solving the minimization problem

$$J(w) := \frac{1}{2}\|T^*w\|_U^2 - (y, w)_V \to \min$$

on V.

There is another important class of variational problems, so-called *mixed variational problems*, see [16, Section 4.2] for a detailed account. They are posed on a pair of Hilbert spaces V, W and governed by two bounded bilinear forms

$$a : V \times V \to \mathbb{R} \quad \text{and} \quad b : V \times W \to \mathbb{R}.$$

Given $f \in V'$ and $g \in W'$, one looks for solution pairs $(u, p) \in V \times W$ of the system

$$\begin{aligned} a(u, v) + b(v, p) &= f(v) \qquad \forall\, v \in V, \\ b(u, q) \qquad\qquad &= g(q) \qquad \forall\, q \in W. \end{aligned} \tag{A.21}$$

If $A : V \to V'$ and $B : V \to W'$ are the bounded linear operators generated by the bilinear forms a and b, respectively, then (A.21) is equivalent to the system of operator equations

$$Au + B'p = f, \qquad Bu = g.$$

If a is symmetric, then (A.21) represents the optimality conditions of the constrained quadratic optimization problem

$$J(u) := \frac{1}{2}a(u, u) - f(u) \to \min, \qquad Bu = g.$$

The system in (A.21) can also be written as a single variational problem on the Hilbert space

$$V \oplus W = \{(u, p) \in V \times W : \|(u, p)\|_{V \oplus W} := (\|u\|_V^2 + \|p\|_W^2)^{1/2}\},$$

and its well-posedness can be studied using either Theorem A.4 or Theorem A.5. Alternatively, if $g \in \operatorname{ran}(B) \subset W'$, then the general solution of the equation $Bu = g$ is satisfied for $u = u_0 + z$, where $Bu_0 = g$ and $z \in Z := \ker(B)$ is arbitrary. So to satisfy the first equation in (A.21), we must have

$$a(z, z') = f(z') - a(u_0, z'), \qquad \forall\, z' \in Z$$

for the unknown $z \in Z$. In other words, a mixed variational problem on $V \otimes W$ can be reduced to a variational problem with the bilinear form a on the subspace $Z = \ker(B)$ of V. This approach is equivalent to eliminating the constraint $Bu = g$ and is not usually pursued in practice, since Z may have a more complicated structure than V. However, it can be used to show that the well-posedness of (A.21) is guaranteed if the inf-sup conditions (A.14) and (A.15) hold on Z (i.e., with $u, v \in V = U$ replaced by $z, z' \in Z$) and are complemented by an inf-sup condition for the bilinear form b, namely

$$\inf_{q \in W} \sup_{v \in V} \frac{b(v, q)}{\|v\|_V \|q\|_W} \geq \beta > 0.$$

The last condition ensures that B is surjective.

A.4 Constructions on Hilbert spaces

A.4.1 Hilbert sums

Direct sums of Hilbert spaces (for short *Hilbert sums*) are explored throughout the book. The direct sum

$$\tilde{H} := \sum_{\omega\in\Omega}^{\oplus} H_\omega$$

of a collection $\mathbf{H} := (H_\omega)_{\omega\in\Omega}$ of (non-trivial) Hilbert spaces H_ω consists of all families $\tilde{x} := (x_\omega)_{\omega\in\Omega}$, where $x_\omega \in H_\omega$, $\omega \in \Omega$, such that

$$\|\tilde{x}\|_{\tilde{H}} := \Big(\sum_{\omega\in\Omega} \|x_\omega\|^2_{H_\omega}\Big)^{1/2} < \infty.$$

The Hilbert sum $\tilde{H}$ of $\mathbf{H}$ is itself a Hilbert space. It can be viewed as a generalization of the space $\ell^2(\Omega)$ of real-valued sequences with index set Ω to "variable" Hilbert space-valued functions with domain Ω. Since every $\tilde{x} \in \tilde{H}$ can have at most countably many non-zero coordinates x_ω, the set $\tilde{H}_0$ of all finitely supported families $\tilde{x}$ is dense in $\tilde{H}$.

The space $\tilde{H}$ is separable (this is the case we are most interested in) if and only if Ω is at most countable, and every H_ω is separable. For convenience, in our further discussion of Hilbert sums we assume that $\mathbf{H} = (H_i)_{i\in\Omega}$ is a collection of at most countably many separable Hilbert spaces with index set $\Omega = \mathbb{N}$ or $\Omega = \{1, \ldots, n\}$ for some $n \in \mathbb{N}$. Given a sequence $\rho := (\rho_i)_{i\in\Omega}$ of positive real numbers $\rho_i > 0$, we also consider the *weighted Hilbert sum* $\tilde{H}_\rho$ of the collection $\mathbf{H}$ consisting of sequences $\tilde{x} = (x_i)_{i\in\Omega}$ with $x_i \in H_i$ for all $i \in \Omega$ and finite norm

$$\|\tilde{x}\|_{\tilde{H}_\rho} := \Big(\sum_{i\in\Omega} \rho_i \|x_i\|^2_{H_i}\Big)^{1/2} < \infty.$$

The main theme in our treatment of iterative methods for solving operator equations in a given Hilbert space H is to relate H to a weighted Hilbert sum $\tilde{H}_\rho$ of a suitable collection $\mathbf{H} = (H_i)_{i\in\Omega}$ of Hilbert spaces, and to transform the original problem on H into an equivalent operator equation on $\tilde{H}_\rho$. In this way, iterative schemes on H can be built from components that solve operator equations on the spaces H_i rather than on H itself. The Fourier method for solving $Tx = y$ with compact $T \in \mathcal{B}_{sym}(H)$ and $y \in \mathrm{ran}(T)$ can be seen as an extreme (but not necessarily numerically advantageous) instance of this approach. Here we would choose the collection of one-dimensional subspaces H_i spanned by the eigenvectors ψ_i as the family $\mathbf{H}$, and solving the problem on H_i boils down to computing the Fourier coefficient $(x, \psi_i)_H = \lambda_i^{-1}(y, \psi_i)_H$ if we set $\rho_i = 1$.

$$R\tilde{w} = \sum_{i\in\Omega} w_i, \qquad \tilde{w} = (w_i)_{i\in\Omega} \in \tilde{W},$$

The definition of Hilbert sums has been generalized to families $\mathbf{H}$ defined on a measure space $(\Omega, \mathcal{F}, \rho)$. Roughly speaking, the so-called *direct integral of* $\mathbf{H}$, denoted by

$$\tilde{H}_\rho := \int_\Omega^\oplus H_\omega \, d\rho(\omega),$$

is the set of all (equivalence classes of) measurable functions $\tilde{x} : \omega \in \Omega \to \tilde{x}(\omega) \in H_\omega$ such that

$$\|\tilde{x}\|_{\tilde{H}_\rho}^2 := \int_\Omega \|x(\omega)\|_{H_\omega}^2 \, d\rho(\omega) < \infty.$$

The notation is consistent with the previously introduced weighted Hilbert sum for $\mathbf{H}$ with at most countable Ω (just take the discrete measure on Ω generated by the ρ_i). The exact definition requires the specification of the meaning of measurability, and is often given under certain restrictions on the measure space. Here, positive Borel measures on compact separable metric spaces Ω are sufficient. We will not go into the details (see the monographs [69, Sections II.1-2], [177, Section IV.8] and Subsection 6.5.2.3). For our purposes in Section 6.5, a simplified treatment based on the Bochner space of square-integrable Hilbert space-valued functions will suffice.

A.4.2 Bochner integration and $L^2_\rho(\Omega, H)$

The few facts from the integration theory of Hilbert space-valued functions that we need can be found in [51, 200]. Consider an arbitrary measure space $(\Omega, \mathcal{F}, \rho)$ and a Hilbert space H. As usual, Ω is a set, $\mathcal{F}$ is a σ-algebra on Ω, and ρ is a measure on $(\Omega, \mathcal{F})$. A Hilbert space-valued function $\tilde{x} : \Omega \to H$ is called *simple* if it takes finitely many non-zero values $x_i \in H$ on measurable subsets Ω_i of finite measure, and is zero otherwise. More precisely, $\tilde{x}$ is simple if, for some $n \in \mathbb{Z}_+$, we have $\tilde{x}(\omega) = x_i \in H$ for $\omega \in \Omega_i$, and $\tilde{x}(\omega) = 0$ for $\omega \in \Omega \setminus \cup_{i=1}^n \Omega_i$, where the Ω_i belong to $\mathcal{F}$, are pairwise disjoint, and have finite measure, $i = 1, \ldots, n$. For such a simple function, its Bochner integral is defined as

$$\int_\Omega \tilde{x}(\omega) \, d\rho(\omega) := \sum_{i=1}^n \rho(\Omega_i) x_i \in H.$$

A function $\tilde{x} : \Omega \to H$ is called *Bochner measurable* if it is the pointwise ρ-almost everywhere (a.e.) limit of a sequence $(\tilde{x}^{(k)})_{k \in \mathbb{N}}$ of simple functions, i.e., if there holds $\|\tilde{x}^{(k)}(\omega) - \tilde{x}(\omega)\|_H \to 0$ for $\omega \in \Omega \setminus N$, where $\rho(N) = 0$. A function $\tilde{x}$ is *Bochner integrable* if a sequence of simple functions can be chosen such that

$$\int_\Omega \|\tilde{x}^{(k)}(\omega) - \tilde{x}(\omega)\|_H \, d\rho(\omega) \to 0.$$

Obviously, integrability requires measurability. The *Bochner integral* of $\tilde{x}$ over $(\Omega, \mathcal{F}, \rho)$ is then defined by

$$\int_{\Omega} \tilde{x}(\omega)\, d\rho(\omega) := \lim_{k\to\infty} \int_{\Omega} \tilde{x}^{(k)}(\omega)\, d\rho(\omega).$$

The set of all (equivalence classes of) Bochner integrable functions is denoted by $L^1_\rho(\Omega, H)$. Then a theorem of Bochner states that $\tilde{x} \in L^1_\rho(\Omega, H)$ if and only if $\tilde{x}$ is Bochner measurable, and

$$\|\tilde{x}\|_{L^1_\rho(\Omega,H)} := \int_{\Omega} \|\tilde{x}(\omega)\|_H\, d\rho(\omega) < \infty.$$

With this norm, $L^1_\rho(\Omega, H)$ becomes a Banach space. We note that when Bochner integration is applied to a particular $\tilde{x}$, it is often difficult to verify the measurability of $\tilde{x}$. In this respect, a theorem of Pettis can be helpful: If H is separable, then Bochner measurability follows from *weak Bochner measurability*, which requires that the scalar function $\omega \to (\tilde{x}(\omega), x)_H$ is measurable for every $x \in H$. This avoids the explicit use of simple functions. Up to this point, the statements hold even if H is replaced by an arbitrary (separable) Banach space.

Similarly, we can define the space $L^2_\rho(\Omega, H)$ as the set of all (equivalence classes of) Bochner measurable functions $\tilde{x} : \Omega \to H$ such that

$$\int_{\Omega} \|\tilde{x}^{(k)}(\omega) - \tilde{x}(\omega)\|^2_H\, d\rho(\omega) \to 0$$

for some sequence of simple functions, or, equivalently, as the set of all Bochner measurable $\tilde{x}$ with finite integral

$$\|\tilde{x}\|_{L^2_\rho(\Omega,H)} := \left(\int_{\Omega} \|\tilde{x}(\omega)\|^2_H\, d\rho(\omega) \right)^{1/2} < \infty.$$

Then, equipped with this norm and the scalar product

$$(\tilde{x}, \tilde{y})_{L^2_\rho(\Omega,H)} := \int_{\Omega} (\tilde{x}(\omega), \tilde{y}(\omega))_H\, d\rho(\omega), \qquad \tilde{x}, \tilde{y} \in L^2_\rho(\Omega, H),$$

$L^2_\rho(\Omega, H)$ becomes a Hilbert space.

A.4.3 Tensor products of Hilbert spaces

Another useful tool, especially in the analysis of algorithms for problems in function spaces over domains in high dimensions, is tensorization. For an introduction, see [192]. To avoid too many indices, let us consider the *tensor product* $(V \otimes U; a \otimes b)$ of two separable Hilbert spaces $(V; a)$ and $(U; b)$, which is defined as the completion of the linear span of the set of elementary tensors $v \otimes u$. The completion is done with respect to a new scalar product $a \otimes b$, which is defined for elementary tensors by

$$a \otimes b(v \otimes u, v' \otimes u') = a(v, v')b(u, u'), \qquad v, v' \in V, \quad u, u' \in U,$$

and is extended by linearity to their span. Equivalently, the tensor product can be defined by taking arbitrary CONS $\{e_i\}_{i\in\mathbb{N}}$ and $\{f_j\}_{j\in\mathbb{N}}$ in V and U, respectively, and by postulating that $\{e_i \otimes f_j\}_{(i,j)\in\mathbb{N}^2}$ is a complete orthonormal system in $V \otimes W$. Thus, $V \otimes U$ is defined by isometry with $\ell^2(\mathbb{N} \times \mathbb{N})$, i.e.,

$$V \otimes U = \{z = \sum_{i,j\in\mathbb{N}} c_{ij} e_i \otimes f_j : \|z\|_{V\times U} = (\sum_{i,j\in\mathbb{N}} c_{ij}^2)^{1/2} < \infty\}.$$

For our setting, this definition is more convenient. It does not matter which CONS are chosen in V and U. A third equivalent definition identifies elementary tensors $v \otimes u$ with rank-one operators in $T = T_{v\otimes u} \in \mathcal{B}(V', U)$ by the formula

$$T_{v\otimes u}(g) = g(v)u, \qquad g \in V',$$

and $V \otimes U$ with the Hilbert space $\mathcal{S}_2(V', U)$ of Hilbert-Schmidt operators with scalar product

$$(T, S)_{HS} := \sum_{n\in\mathbb{N}} (T\psi'_n, S\psi'_n)_U,$$

where $(\psi'_n)_{n\in\mathbb{N}}$ is an arbitrary CONS in V'.

The interpretation of elementary tensors $v \otimes u$ with respect to the originally given spaces V and U depends on their nature. If V and U are Hilbert spaces of real-valued functions over certain domains G_V and G_U, then $V \otimes U$ is often identified with a function space over the product domain $G_V \times G_U$ by setting

$$(v \otimes u)(x, y) = v(x)u(y), \quad x \in G_V, \ y \in G_U, \qquad v \in V, \ u \in U.$$

The tensor product $T \otimes S : V_1 \otimes U_1 \to V_2 \otimes U_2$ of two operators $T \in \mathcal{B}(V_1, V_2)$ and $S \in \mathcal{B}(U_1, U_2)$ is first defined for elementary tensors by

$$(T \otimes S)(v_1 \otimes u_1) = (Tv_1) \otimes (Su_1), \qquad v_1 \in V_1, \quad u_1 \in U_1,$$

and then extended to all $V_1 \otimes U_1$ by linearity and completion by continuity. Equivalently, with CONS $(e_i^k)_{i\in\mathbb{N}}$ and $(f_i^k)_{i\in\mathbb{N}}$ fixed in V_k and U_k with $k = 1, 2$, the matrix representation of $T \otimes S$ with respect to the resulting tensor product CONS is given by the Kronecker product of the matrix representations of T and S. Thus,

$$\|T \otimes S\|_{V_1\otimes U_1 \to V_2\otimes U_2} = \|T\|_{V_1\to V_2} \|S\|_{U_1\to U_2}.$$

Also, if $T \in \mathcal{B}_{sym}(V)$ and $S \in \mathcal{B}_{sym}(U)$, then $T \otimes S \in \mathcal{B}_{sym}(V \otimes U)$. For the spectra we have

$$\sigma(T \otimes S) = \{ts : \ t \in \sigma(T), \ s \in \sigma(S)\}.$$

This follows from the properties of the Kronecker product.

A.5 Sobolev spaces on domains

A.5.1 Definition

Sobolev spaces of functions on open sets $G \subset \mathbb{R}^d$ with (generalized) derivatives in $L^p(G)$ are indispensable in the modern theory of linear and nonlinear partial differential equations (PDEs). We consider here only the Hilbert space case $p = 2$, where Sobolev spaces are denoted by $H^s(G)$ with smoothness parameter $s \in \mathbb{R}$. There are several ways to define Sobolev spaces. The traditional way is to first introduce the spaces $H^m(G)$ of integer smoothness $m \in \mathbb{Z}_+$ as sets of all functions $f \in L^2(G)$ with generalized derivatives $f^{(\alpha)} \in L^2(G)$ of order $|\alpha| \leq m$. For non-integer $s \in \mathbb{R}_+$, $H^s(G)$ is then defined by complex interpolation, while for $s < 0$ duality arguments are used. This approach, together with the main results on embedding, trace and extension theorems, is presented in, e.g., [1, 30, 92]. Alternatively, one can start with Sobolev spaces on $G = \mathbb{R}^d$ (or on compact d-dimensional smooth manifolds without boundary), defined in terms of Fourier-analytic or atomic decompositions, and then specify spaces on domains $G \subset \mathbb{R}^d$ by restriction. We refer to [2, 183, 184], the special case of the periodic cube $G = T^d$ is outlined in Subsection 3.1.2. Both approaches lead, up to equivalent norms, to the same spaces if the boundary $\Gamma := \partial G$ of G satisfies some mild regularity conditions, which are also a prerequisite for the validity of embedding and trace theorems.

Since one of the main applications in this book concerns iterative methods for finite element discretizations of elliptic boundary value problems on polyhedral domains, we focus on Sobolev spaces on bounded Lipschitz domains G. More precisely, we assume that $G \subset \mathbb{R}^d$ is open and bounded, and that its boundary Γ admits a finite cover of rotated open d-dimensional rectangular domains U_i, $i = 1, \ldots, n$, together with Lipschitz-continuous functions $f_i : \mathbb{R}^{d-1} \to \mathbb{R}$ such that

$$G \cap U_i = \{y \in U_i : 0 < y_d < f_i(y_1, \ldots, y_{d-1})\}$$

in a rotated Cartesian coordinate system adjusted to U_i. Figure A.1 illustrates this definition. Most (but not all) bounded polyhedral domains are bounded Lipschitz domains. An exception is, for example, a slit domain in $\mathbb{R}^2$, which requires special treatment. Other, even weaker conditions on G, such as the (uniform) cone or the segment condition, and similar definitions for unbounded domains are discussed in [1, Section 4.4].

To avoid using the language of distribution theory, in the formulas below we implicitly assume that $f : G \to \mathbb{R}$ is the restriction to G of a compactly supported infinitely many times differentiable function on $\mathbb{R}^d$, i.e.,

$$f = g|_G, \qquad g \in C_0^\infty(\mathbb{R}^d).$$

We set

$$\|f\|_{H^s} := \inf_{g \in C_0^\infty(\mathbb{R}^d):\, f = g|_G} \|g\|_{H^s(\mathbb{R}^d)}, \tag{A.22}$$

where

$$\|g\|^2_{H^s(\mathbb{R}^d)} := \int_{\mathbb{R}^d} (1 + |\theta|^2)^s |\mathcal{F}g(\theta)|^2 \, d\theta, \qquad s \in \mathbb{R}. \tag{A.23}$$

The Sobolev space $H^s(G)$ can then be defined as the completion of $C^\infty(\bar{G})$ with respect to the norm (A.22). The spaces $H^s(\mathbb{R}^d)$ are similarly defined by completion of $C_0^\infty(\mathbb{R}^d)$ with respect to the norm (A.23). Here and in many other places in this book, we omit the domain G in the notation for function space norms such as $H^s(G)$, if G is fixed and there is no confusion with other domains. The above definition of Sobolev spaces on domains is called *extrinsic* (or by restriction). The advantage of the extrinsic definition is that many properties of $H^s(G)$ can now be derived canonically from properties of $H^s(\mathbb{R}^d)$. The latter spaces allow the use of powerful tools from harmonic analysis and are somewhat easier to handle.

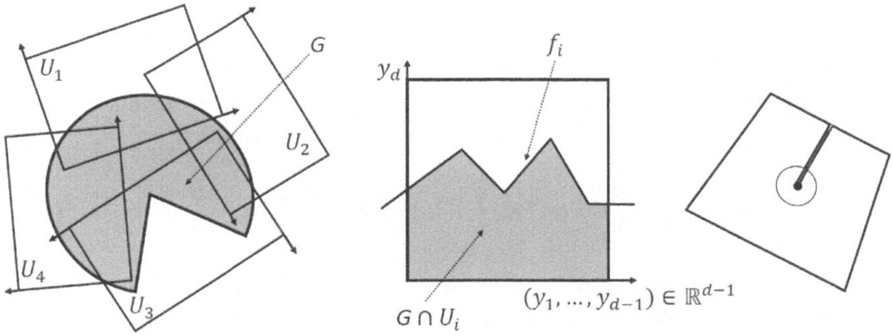

Fig. A.1 Left and center: Illustration of the definition of Lipschitz domains, Right: Example of a slit domain (the tip of the slit cannot be covered by a U_i with Lipschitz-continuous f_i).

Applications to PDE problems require intrinsic definitions in terms of differentiability properties of functions. For example, the treatment of $2m$-th order elliptic problems by variational methods leads to energy spaces governed by norms of the form

$$\|f\|^2_{H^m,1} := \sum_{|\alpha| \le m} \|f^{(\alpha)}\|^2_{L^2}$$

or

$$\|f\|^2_{H^m,2} := \|f\|^2_{L^2(G)} + |f|^2_{H^m}, \qquad |f|^2_{H^m} := \sum_{|\alpha|=m} \|f^{(\alpha)}\|^2_{L^2}.$$

Let us make a few remarks about the notation used in this book for derivatives of functions of d variables. Partial derivative operators are denoted by ∂_{x_i}, $\partial^m_{x_i} = (\partial_{x_i})^m$, $\partial^2_{x_i,x_j} := \partial_{x_i}\partial_{x_j}$ and so on. For example,

$$\nabla := (\partial_{x_1}, \ldots, \partial_{x_d})^T \quad \text{and} \quad \Delta := \sum_{i=1}^{d} \partial^2_{x_i,x_i}$$

stand for the gradient and Laplace operator, respectively. We also use multi-index notation: For $\alpha \in \mathbb{Z}_+^d$ we set

$$f^{(\alpha)} := (\partial_{x_1})^{\alpha_1} \cdots (\partial_{x_d})^{\alpha_d} f$$

where $|\alpha| = |\alpha|_1 = \sum_{i=1}^d \alpha_i$ denotes the order of this derivative. Again, the required definition of generalized derivatives can be avoided by defining the spaces by completion of $C^\infty(\bar{G})$ with respect to the above norms, see e.g. [25, Theorem 1.3.4]. For bounded Lipschitz domains, these norms are equivalent to the norm defined in (A.22) for $s = m$.

The spaces $H^s(G)$ with non-integer $0 < s < m$ are then introduced by real or complex interpolation. We use the real method [9, 11], for which we need the K-functional for the pair $(L^2(G); H^m(G))$ given by

$$K(f, t; L^2(G), H^m(G)) := \inf_{f=f_1+f_2} \|f_1\|_{L^2} + t|f_2|_{H^m}, \quad t > 0.$$

Note that in the definition of K-functionals for this pair, the semi-norm $|f_2|_{H^m}$ can be replaced by the norm $\|f_2\|_{H^m}$, which leads to equivalent results. Then, for $0 < s < m$, we define $H^s(G)$ as the set of all $f \in L^2(G)$ for which

$$\|f\|_{H^s,3}^2 := \|f\|_{L^2}^2 + \int_0^\infty (t^{-s} K(f, t^m; L^2(G), H^m(G)))^2 \frac{dt}{t} < \infty. \qquad \text{(A.24)}$$

The space $H^s(G)$ does not depend on the choice of the integer $m > s$. The corresponding norms are equivalent for different admissible m and are (for bounded Lischitz domains) also equivalent to the extrinsic definition. In particular, we have

$$\|f\|_{H^m}^2 \approx \|f\|_{H^m,1}^2 \approx \|f\|_{H^m,2}^2 \approx \|f\|_{H^m,3}^2, \qquad f \in H^m(G),$$

$m = 1, 2, \ldots$, for bounded Lipschitz domains G.

There are other more explicit definitions of $H^s(G)$, $s > 0$, in terms of differences. We mention only the one using the m-th order L^2 modulus of continuity

$$w_m(t, f)_{L^2} = \sup_{h \in \mathbb{R}^d:\, \|h\|_2 \le t} \|\Delta_h^m f\|_{L^2(G_{m,h})}^2, \qquad t > 0, \quad f \in L^2(G),$$

where $\Delta_h^m := (\Delta_h)^m$ with $\Delta_h f(x) = f(x+h) - f(x)$ denotes the m-th order forward difference operator and

$$G_{m,h} := \{x \in G : \ x + \xi h \in G,\ 0 \le \xi \le m\}.$$

For a class of domains $G \subset \mathbb{R}^d$ including bounded Lipschitz domains, it was proved in [9, 105] that

$$K(f, t^m; L^2(G), H^m(G)) \approx w_m(t, f)_{L^2}, \qquad t > 0.$$

Thus, replacing the K-functional $K(f, t^m; L^2(G), H^m(G))$ in (A.24) by $w_m(t, f)_{L^2}$ and discretizing the integral leads to another equivalent norm for $H^s(G)$, $0 < s < m$:

$$\|f\|_{H^m,4}^2 := \|f\|_{L^2}^2 + \sum_{j=0}^{\infty} 2^{2sj} w_m(2^{-j}, f)_{L^2}^2.$$

The modulus $w_m(t, f)_{L^2}$ has useful properties that allow easy proofs in many cases. Obviously, $w_m(t, f)_{L^2} \searrow 0$ if $t \searrow 0$, and only its behavior for $t \le \mathrm{diam}(G)/m$ matters, since it is constant for $t \ge \mathrm{diam}(G)/m$. We also have

$$w_m(t, f)_{L^2} \le 2w_{m-1}(t, f)_{L^2} \le \cdots \le 2^m \|f\|_{L^2},$$

$$w_m(t, f)_{L^2} \le Ctw_{m-1}(t, \nabla f)_{L^2}, \qquad f \in H^1(G),$$

and

$$w_m(t, f)_{L^2}^2 \approx t^{-d} \int_{\|h\|_2 \le t} \|\Delta_h^m f\|_{L^2(G_{m,h})}^2 \, dh, \qquad 0 < t \le \mathrm{diam}(G)/m.$$

The proofs of these properties are based on elementary identities for difference operators.

For $s < 0$, besides using the extrinsic definition (A.22), one could alternatively identify $H^s(G)$ with the dual of $H^{-s}(G)$, thus avoiding the detour via spaces on $\mathbb{R}^d$. For $s \le -1/2$ this leads to slightly different spaces. Since we refer to Sobolev spaces of negative smoothness only in a few places in this book, we do not go into detail and refer to [1, 2]. Below we mention a partial result, namely the definition of the space $H^{-1}(G)$ as the dual of the space $H_0^1(G)$, which is equivalent to the extrinsic definition (A.22) for $s = -1$.

A.5.2 Some properties of Sobolev spaces

We list some properties of Sobolev spaces that are occasionally used in Chapters 3 and 7. Throughout this subsection, $G \subset \mathbb{R}^d$ is assumed to be a bounded Lipschitz domain. First, functions in $H^s(G)$ can be identified with continuous functions if $s > d/2$ and the associated embedding $H^s(G) \subset C(\bar{G})$ is even compact. In particular, we have

$$\|f\|_{L^\infty} \le C\|f\|_{H^s}, \qquad f \in H^s(G), \quad s > d/2, \tag{A.25}$$

with some constant that depends only on s and G. This is a special case of the Sobolev embedding theorem, see for example [30]. For $s \le d/2$ the bound (A.25) fails, which also explains the non-optimality of space splittings for finite element subspaces of $H^s(G)$ based on nodal interpolation operators for such s, see Subsection 7.1.3 in Chapter 7. Moreover, differential operators of order k (such as ∂_{x_i} $(k = 1)$ or the Laplace operator Δ $(k = 2)$) act boundedly from $H^s(G)$ to $H^{s-k}(G)$.

Next, we cite a result on L^2 approximation by multivariate polynomials, the so-called Bramble-Hilbert lemma [19, 25]. It is crucial for the study of local approximation schemes such as splines, finite element functions, and wavelets.

Lemma A.3 *Let $G \subset \mathbb{R}^d$ be a bounded Lipschitz domain and let $m = 1, 2, \ldots$. Then, for any $f \in H^m(G)$, there exists a polynomial q of total degree $\leq m - 1$ such that*

$$|f - q|_{H^k} \leq C(\mathrm{diam}(G))^{m-k}|f|_{H^m}, \qquad k = 0, \ldots, m,$$

where the constant C depends only on m and G. If G is star-shaped with respect to a ball of radius $\geq \gamma\,\mathrm{diam}(G)$, then the constant C in the estimate depends on G only via the value of γ and the dimension d.

A domain G is star-shaped with respect to a ball B if, for any $x \in G$, the closed convex hull of the set $B \cup \{x\}$ is contained in G. Lemma A.3 is used in the proof of Theorem 3.3, see Subsection 3.1.3.

The trace $T_\Gamma f := f|_\Gamma$ of a continuous function $f \in C(\bar{G})$ on the boundary Γ of a bounded Lipschitz domain G is naturally defined. Thus, by the above mentioned embedding result for $H^s(G)$ into $C(\bar{G})$, the operator T_Γ is also well-defined and bounded on $H^s(G)$, $s > d/2$. Obviously, there is no meaningful extension of T_Γ to $H^0(G) = L^2(G)$. An abstract definition of the *trace space* $X(\Gamma)$ associated with a Banach space $X(G)$ is as follows: Suppose the linear set $X(G) \cap C(\bar{G})$ is dense in $X(G)$. If a function $g \in C(\Gamma)$ is such that $g = T_\Gamma f$ for some $f \in X(G) \cap C(\bar{G})$, then

$$\|g\|_{X(\Gamma)} := \inf_{f \in X(G) \cap C(\bar{G}):\, g = T_\Gamma f} \|f\|_{X(G)}$$

defines a semi-norm on a certain linear set in $C(\Gamma)$. If it is a norm, i.e., $\|g\|_{X(\Gamma)} = 0$ implies $g = 0$, then the trace space $X(\Gamma)$ is defined as the completion of this linear set with respect to the norm $\|\cdot\|_{X(\Gamma)}$. With this definition it was shown that the trace space for $H^s(G)$, $1/2 < s < 3/2$, coincides with $H^{s-1/2}(\Gamma)$ up to equivalent norms. We refer for example to [2, Theorem 9.2.1] or [92]. The latter space can be defined by the sum of Sobolev norms over $(d-1)$-dimensional domains associated with the finite cover of Γ given by the sets U_i and the local parameterizations f_i, see [1, 172]. For domains with smoother boundary functions, the result holds for larger values of s. For polyhedral Lipschitz domains equipped with sequences of dyadically refined partitions $\mathcal{T}_j$, a characterization of the trace space for $H^s(G)$, $1/2 < s < 3/2$, is given in Theorem 7.3 in Subsection 7.1.2. It uses approximation spaces associated with the sequence of Lagrange C^0 finite element spaces over the partitions $\mathcal{T}_j|_\Gamma$ induced from $\mathcal{T}_j$ by restriction to Γ.

Finally, to meet the needs of PDE applications, spaces with boundary value restrictions are needed. Canonically, $H_0^s(G)$ denotes the closure of $C_0^\infty(G)$ with respect to the $\|\cdot\|_{H^s}$ norm given by (A.22). For $s > 1/2$ this is a proper subspace of $H^s(G)$. For bounded Lipschitz domains and values $1/2 < s < 3/2$, it has a more explicit equivalent definition, namely

$$H_0^s(G) = \{f \in H^s(G) : T_\Gamma f = 0\}.$$

This is used in Chapter 7. In addition, we have an equivalent description of Sobolev spaces of negative smoothness, which avoids the detour via $\mathbb{R}^d$ in the extrinsic definition, namely

$$H^{-s}(G) = (H_0^s(G))', \qquad s > 0,$$

with equivalent norms. Here, duality with respect to $(\cdot,\cdot)_{L^2(G)}$ can be used to interpret this fact. For $|s| < 1/2$ there is no difference between $H^s(G)$ and $H_0^s(G)$ while for $s > 1/2$ the dual space $(H_0^s(G))'$ is strictly larger than $(H^s(G))'$. We do not go into detail, since duality and negative smoothness are not used prominently in this book, with a few exceptions in Subsection 3.1.1 of Chapter 3.

A.5.3 Sobolev spaces and boundary value problems for PDEs

The space splitting concept and the analysis of iterative methods based on it have been motivated to a large extent by the development of efficient numerical methods for solving elliptic PDE problems. In Chapter 3 and Chapter 7 of this book, we have included illustrative examples of subspace splittings for Sobolev spaces related to standard discretization schemes for scalar second-order elliptic PDE problems. Here we briefly give some background on this problem class. More details can be found in [2, Section 11].

Let $G \subset \mathbb{R}^d$ be a bounded Lipschitz domain and Γ be its boundary. Since Γ is represented by a finite number of Lipschitz functions f_i, which are a.e. differentiable in their respective coordinate systems, the unit normal vector $n = n(x) := (n_1(x), \ldots, n_d(x))^T$ is well defined a.e. on Γ. As a preparation for transforming PDEs into variational problems on Sobolev spaces, we mention the integration by parts formula

$$(\partial_{x_j} u, v)_{L^2(G)} + (u, \partial_{x_j} v)_{L^2(G)} = (n_j T_\Gamma u, T_\Gamma v)_{L^2(\Gamma)}, \qquad j = 1, \ldots, d, \quad \text{(A.26)}$$

valid for $u, v \in H^1(G)$, see [2, Proposition 9.3.3].

A symmetric and second-order strongly elliptic PDE is given in divergence form by

$$(Lu)(x) := -\nabla^T \mathbf{a}(x) \nabla u(x) + c(x)u(x) = f(x), \qquad x \in G.$$

Here,

$$\mathbf{a}(x) = (a_{ij}(x))_{i,j=1}^d$$

is a symmetric, uniformly positive definite and bounded real-valued $d \times d$ matrix function with entries $a_{ij} \in C(\bar{G})$, i.e.,

$$\mathbf{a}(x) = \mathbf{a}(x)^T, \qquad \alpha\|\xi\|_2^2 \le \xi^T \mathbf{a}(x)\xi \le \beta\|\xi\|_2^2, \quad \xi \in \mathbb{R}^d,$$

holds for a.e. $x \in G$ with constants $0 < \alpha \le \beta$, and $c(x) \ge 0$ belongs to $L^\infty(G)$. The matrix $\mathbf{a}(x)$ is often called a diffusion tensor, special cases are $\mathbf{a}(x) = w(x)I$ (scalar diffusion) and diagonal matrices $\mathbf{a}(x) = \mathrm{diag}(w_i(x))$ (anisotropic diffusion), and α, β are the ellipticity constants of the principal part of the operator L. The following arguments extend to piecewise continuous $\mathbf{a}(x)$. Obviously, if $\mathbf{a}(x) = I$ and $c(x) = 0$, we get the Laplace operator $L = -\Delta$. We use some special cases for illustration in Chapters 3 and 7.

To obtain variational formulations for the equation $Lu = f$, we need Green's formula

$$(Lu, v)_{L^2(G)} = \underbrace{\int_G \nabla u(x)^T \mathbf{a}(x) \nabla v(x) + c(x)u(x)v(x)\,dx}_{a_L(u,v):=} - \int_\Gamma \underbrace{\left(n(x)^T \mathbf{a}(x) T_\Gamma \nabla u(x)\right)}_{T_{\Gamma,L,n}u(x):=} T_\Gamma v(x)\,dx, \tag{A.27}$$

which holds for $u \in H^2(G)$, $v \in H^1(G)$ according to (A.26). In fact, $u \in H^2$ implies $\nabla u(x) \in H^1(G)^d$ and $T_\Gamma \nabla u(x) \in H^{1/2}(\Gamma)^d$ is well-defined. Furthermore, we have $f = Lu \in L^2(G)$. If we assume that

$$u \in H^2(G) \cap H_0^1(G) = \{u \in H^2(G) : T_\Gamma u = 0\} \quad \text{and} \quad v \in H_0^1(G),$$

then (A.27) simplifies to

$$(f, v)_{L^2(G)} = (Lu, v)_{L^2(G)} = a_L(u, v), \qquad v \in H_0^1(G).$$

The bilinear form $a_L(u, v)$ is symmetric, bounded and coercive on $V := H_0^1$. Thus it uniquely defines a symmetric, boundedly invertible linear operator

$$A_L : V \to V' = H^{-1}(G).$$

The solution $u \in V = H_0^1(G)$ of the associated operator equation $A_L u = f$ with $f \in H^{-1}(G)$ or, equivalently, the solution of the variational problem

$$a_L(u, v) = \langle f, v \rangle_{V' \times V} \qquad \forall\, v \in V \tag{A.28}$$

on $V = H_0^1(G)$ is then interpreted as the weak or generalized solution of the homogeneous Dirichlet problem

$$Lu = f, \qquad T_\Gamma u = 0,$$

with right-hand side $f \in H^{-1}(G)$.

The inhomogeneous Dirichlet problem, where the boundary condition is replaced by $T_\Gamma u = g \in H^{1/2}(\Gamma)$, is reduced to solving (A.28) by first finding some extension $u_g \in H^1(G)$ of the boundary data g, i.e., $T_\Gamma u_g = g$, and then solving (A.28) with a modified right-hand side for $\tilde{u} = u - u_g \in V$, i.e.,

$$a_L(\tilde{u}, v) = \langle f, v \rangle_{V' \times V} - a_L(u_g, v) \qquad \forall\, v \in V.$$

The construction of extension operators is a topic of independent interest, examples related to finite element discretizations are discussed in Section 7.1.

The definition of the Neumann problem for $Lu = f$ and its weak formulation require some further comments. Since the trace of the gradient $\nabla u \in L^2(G)^d$ is not well-defined for $u \in V := H^1(G)$, one proceeds as follows: Assuming for the moment that u and $f = Lu$ are fixed, we rewrite (A.27) in the form

$$(T_{\Gamma,L,n}u, T_\Gamma v)_{L^2(\Gamma)} = a_L(u,v) - (f,v)_{L^2(G)}.$$

If $u \in H^1(G)$ and if f belongs to the dual space of $H^1(G)$ with respect to the duality pairing $(\cdot,\cdot)_{L^2}$, then the right-hand side is a continuous linear functional on $H^1(G)$ with argument v. By definition of $T_\Gamma v$, the left-hand side then represents a continuous linear functional with argument $w = T_\Gamma v$ on $H^{1/2}(\Gamma)$ and thus uniquely defines $T_{\Gamma,L,n}u$ as an element of $H^{1/2}(\Gamma)'$. In this sense, the weak solution $u \in V := H^1(G)$ of the (inhomogeneous) Neumann problem

$$Lu = f, \qquad T_{\Gamma,L,n}u = g \in H^{1/2}(\Gamma)',$$

with right-hand side $f \in V'$ is now defined as the solution of

$$a_L(u,v) = \langle f, v\rangle_{V'\times V} + \langle g, T_\Gamma v\rangle_{(H^{1/2})'\times H^{1/2}} \qquad \forall\, v \in V. \tag{A.29}$$

As before, there is an associated operator $A_{L,n}$, now acting from $H^1(G)$ to $H^1(G)'$. But there is an obstacle: If $c(x) = 0$, then the bilinear form $a_L(\cdot,\cdot)$ is not coercive on $H^1(G)$, and solutions of (A.29), if they exist, are not unique. This is because constant functions solve the problem for $f = 0$, $g = 0$ and belong to the null space of $A_{L,n}$. Moreover, the problem (A.29) does not admit a solution for every pair $(f,g) \in V' \times H^{1/2}(\Gamma)'$. A necessary (and sufficient) condition for solvability is

$$\langle f, 1\rangle_{V'\times V} + \langle g, 1\rangle_{(H^{1/2})'\times H^{1/2}} = 0. \tag{A.30}$$

For small $c(x)$, the lower ellipticity constant (i.e., the lower bound on the ratio $a_L(u,u)/\|u\|^2_{H^1}$) can become very small, and in the case $c(x) \to 0$ the Neumann problem should be considered as a singularly perturbed problem. The obvious remedy is to assume the solvability condition (A.30) and replace $V = H^1(G)$ by the factor space $V = H^1(G)/\{const.\}$. An example of this procedure can be found in Section 7.3.

There are many other types of boundary value problems associated with such L. Since in this book we use PDE problems only to illustrate the application of the space splitting concept for iterative methods, we restrict ourselves to the situation of the homogeneous Dirichlet and Neumann problems described above. Similarly, there is a powerful elliptic regularity theory concerning the mapping properties in Sobolev spaces $H^s(G)$ of the solution operator of boundary value problems associated with L, which we do not need in this book. See e.g. [182] for domains with smooth boundary Γ, [2] for Lipschitz domains, and [60, 92] for some types of polyhedral domains.

References

1. Adams, R., Fournier, J.: *Sobolev Spaces*. 2nd edition, Academic Press, Amsterdam (2003)
2. Agranovich, M.: *Sobolev Spaces, Their Generalizations and Elliptic Problems in Smooth and Lipschitz Domains*. Springer, Cham (2015)
3. Akhiezer, N., Glazman, I.: *Theory of Linear Operators in Hilbert Space*. Dover Books on Mathematics, Dover Publications, Mineola (1993)
4. Arnold, D.: *Finite Element Exterior Calculus*. SIAM, Philadelphia (2018)
5. Axelsson, O.: *Iterative Solution Methods*. Cambridge University Press, Cambridge (1994)
6. Bank, R.: *PLTMG - A Software Package for Solving Elliptic Partial Differential Equations*. SIAM, Philadelphia (1990)
7. Bank, R.: Hierarchical bases and the finite element method. *Acta Numerica* **5**, 1–43 (1996)
8. Barron, A., Cohen, A., Dahmen, W., DeVore, R.: Approximation and learning by greedy algorithms. *Ann. Statistics* **3**, 64–94 (2008)
9. Bennett, C., Sharpley, R.: *Interpolation of Operators*. Acad. Press, New York (1988)
10. Benveniste, A., Metivier, M., Priouret, P.: *Adaptive Algorithms and Stochastic Approximations*. Springer, Berlin (1990)
11. Bergh, J., Löfström, J.: *Interpolation Spaces. An Introduction*. Springer, Berlin (1976)
12. Beylkin, G., Coifman, R., Rokhlin, V.: The fast wavelet transform and numerical algorithms. *Comm. Pure Appl. Math.* **44**, 141–183 (1991)
13. Birman, M., Solomjak, M.: *Spectral Theory of Self-Adjoint Operators in Hilbert Space*. Reidel, Dordrecht (1987)
14. Bohn, B., Garcke, J., Griebel, M.: *Algorithmic Mathematics in Machine Learning*. Data Science Book Series, SIAM, Philadelphia (2024)
15. Bochev, P., Gunzburger, M.: *Least-Squares Finite Element Methods*. Applied Mathematical Sciences, Vol. 166, Springer, New York (2009)
16. Boffi, D., Brezzi, F., Fortin, M.: *Mixed Finite Element Methods and Applications*. Springer, Heidelberg (2013)
17. Bramble, J.: *Multigrid Methods*. Pitman Research Notes in Mathematical Sciences 294, Longman Scientific and Technical, Essex (1993)
18. Bramble, J.: Multigrid Methods in Finite Elements. In *Multiscale Problems and Methods in Numerical Simulations*. Lecture Notes in Mathematics, Vol. 1825, Canuto, C., ed., pp. 97–151, Springer, Berlin, Heidelberg (2003)
19. Bramble, J., Hilbert, S.: Estimation of linear functionals on Sobolev spaces with application to Fourier transforms and spline interpolation. *SIAM J. Numer. Anal.* **7**, 112–124 (1970)
20. Bramble, J., Pasciak, J., Wang, J., Xu, J.: Convergence estimates for product iterative methods with applications to domain decomposition. *Math. Comp.* **57**, 1–21 (1991)
21. Bramble, J., Pasciak, J., Wang, J., Xu, J.: Convergence estimates for multigrid algorithms without regularity assumptions. *Math. Comp.* **57**, 23–45 (1991)

M. Griebel, P. Oswald, *Hilbert Space Splittings and Iterative Methods*, Springer Series in Computational Mathematics 62, https://doi.org/10.1007/978-3-031-74370-2

22. Bramble, J., Pasciak, J., Xu, J.: Parallel multilevel preconditioners. *Math. Comp.* **54**, 1–22 (1990)
23. Brandt, A.: Multi-grid techniques: 1984 guide with applications to fluid dynamics. *GMD-Studien.* no 85, St. Augustin (1984)
24. Brenner, S.: Preconditioning complicated finite elements by simple finite elements. *SIAM J. Sci. Comput.* **17**, 1269–1274 (1996)
25. Brenner, S., Scott, L.: *The Mathematical Theory of Finite Elements*. Springer, Berlin (2002)
26. Brezis, H.: *Functional Analysis, Sobolev Spaces, and Partial Differential Equations*. Universitext, Springer, New York (2011)
27. Brezzi, F., Fortin, M.: *Mixed and Hybrid Finite Element Methods*. Springer, New York (1991)
28. Brigham, E.: *The Fast Fourier Transform*. Prentice-Hall, Englewood Cliffs (1974)
29. Bungartz, H.-J., Griebel, M.: Sparse grids. *Acta Numerica* **13**, 147–269 (2004)
30. Burenkov, V.: *Sobolev Spaces on Domains*. Teubner-Texte zur Mathematik, Stuttgart (1998)
31. Carmeli, C., De Vito, E., Toigo, A.: Vector valued reproducing kernel Hilbert spaces of integrable functions and Mercer theorem. *Anal. Appl.* **4:4**, 377–408 (2006)
32. Carmeli, C., De Vito, E., Toigo, A., Umanita, V.: Vector valued reproducing kernel Hilbert spaces and universality. *Anal. Appl.* **8:1**, 19–61 (2010)
33. Carnicer, J., Dahmen, W., Peña, J.: Local decomposition of refinable spaces. *Appl. Comput. Harmon. Anal.* **3**, 127–153 (1996)
34. Casazza, P., Edidin, D., Kalra, D., Paulsen, V.: Projections and the Kadison-Singer problem. *Oper. Matrices* **1:3**, 391–408 (2007)
35. Casazza, P., Kutyniok, G.: Frames of subspaces. In *Wavelets, Frames and Operator Theory*, Contemp. Math. **345**, 87–113 (2004)
36. Casazza, P., Kutyniok, G., Li, S.: Fusion frames and distributed processing. *Appl. Comput. Harmon. Anal.* **25**, 114–132 (2008)
37. Cegielski, A.: *Iterative Methods for Fixed Point Problems in Hilbert Spaces*. Lecture Notes in Mathematics, Vol. 2057, Springer, Berlin, Heidelberg (2012)
38. Chen, L.: Deriving the X-Z identity from auxiliary space method. In *Domain Decomposition Methods in Science and Engineering XIX*. Lecture Notes in Computational Science and Engineering, Vol. 78, Huang Y., Kornhuber R., Widlund, O., Xu, J., eds., pp. 309–316, Springer, Berlin, Heidelberg (2011)
39. Christensen, O.: *An Introduction to Frames and Riesz Bases*. Birkhäuser, Boston (2003)
40. Chui, C., Wang, J.-Z.: On compactly supported spline wavelets and a duality principle. *Trans. Amer. Math. Soc.* **360:2**, 903–915 (1992)
41. Ciarlet, P.: *The Finite Element Method for Elliptic Problems*. SIAM Classics in Appl. Math., Vol. 40, SIAM, Philadelphia (2002)
42. Cohen, A.: *Numerical Analysis of Wavelet Methods*. Studies in Mathematics and its Applications, Elsevier, Amsterdam (2003)
43. Cohen, A., Daubechies, I.: A new technique to estimate the regularity of refinable functions. *Revista Mathematica Iberoamericana* **12:2**, 527–591 (1996)
44. Cohen, A., Daubechies, I., Feauveau, J.-C.: Biorthogonal bases of compactly wavelets. *Communications on Pure and Applied Mathematics* **45**, 485–560 (1992)
45. Cohen, A., Dahmen, W., DeVore, R.: Adaptive wavelet methods for elliptic operator equations - convergence rates. *Math. Comp.* **70**, 27–75 (2001)
46. Cohen, A., Dahmen, W., DeVore, R.: Adaptive wavelet methods II: Beyond the elliptic case. *Found. Comput. Math.* **2:3**, 203–245 (2002)
47. Cohen, A., Dahmen, W., Welper, G.: Adaptivity and variational stabilization for convection-diffusion equations. *ESAIM: Math. Model. Numer. Anal.* **46**, 1247–1273 (2012)
48. Cooley, J., Tukey, J.: An algorithm for the machine calculation of complex Fourier series. *Math. Comp.* **19:90**, 297–301 (1969)
49. Cucker, F., Zhou, D.-X.: *Learning Theory: An Approximation Theory Viewpoint*. Cambridge University Press, Cambridge (2007)
50. Cucker, F., Smale, S.: On the mathematical foundations of learning. *Bull. Amer. Math. Soc.* **39:1**, 1–49 (2002)

51. Da Prato, G., Zabczyk, J.: *Stochastic Equations in Infinite Dimensions*. Cambridge University Press, Cambridge (1992)
52. Dahlke, S., Friedrich, U., Keding, P., Raasch, T., Sieber, A.: Adaptive quarkonial domain decomposition methods for elliptic partial differential equations. *IMA J. Numer. Anal.* **41:4**, 2608–2638 (2021)
53. Dahmen, W.: Stability of multiscale transformations. *J. Fourier Anal. Appl.* **2**, 341–361 (1996)
54. Dahmen, W.: Wavelet and multiscale methods for operator equations. *Acta Numerica* **6**, 55–228 (1997)
55. Dahmen, W., Kunoth, A.: Multilevel preconditioning. *Numerische Mathematik* **63**, 315–344 (1992)
56. Dahmen, W., Kunoth, A., Urban, K.: Biorthogonal spline wavelets on the interval: Stability and moment conditions. *Appl. Comput. Harmon. Anal.* **6**, 132–196 (1999)
57. Dahmen, W., Stevenson, R.: Element-by-element construction of wavelets satisfying stability and moment conditions. *SIAM J. Numer. Anal.* **37**, 319–352 (1999)
58. Daubechies, I.: Orthonormal bases of compactly supported wavelets. *Comm. Pure Appl. Math.* **41**, 909–996 (1988)
59. Daubechies, I.: *Ten Lectures on Wavelets*. SIAM, Philadelphia (1992)
60. Dauge, M.: *Elliptic Boundary Value Problems in Corner Domains*. Lecture Notes in Mathematics, Vol. 1341, Springer, Berlin, Heidelberg (1988)
61. De Boor, C.: *A Practical Guide to Splines*. Springer, New York (1978)
62. Demkowicz, L., Gopalakrishnan, J.: A class of discontinuous Petrov-Galerkin methods. Part II: Optimal test functions. *Numer. Meth. PDE* **27:1**, 70–105 (2011)
63. Demkowicz, L., Gopalakrishnan, J.: Discontinuous Petrov-Galerkin (DPG) method. In *Encyclopedia of Computational Mechanics*, Volume 2: Fundamentals, Part 2, Stein, E., Hughes, T., de Borst, R., eds., pp. 777–792, Wiley, Chichester (2017)
64. DeVore, R.: Nonlinear approximation. *Acta Numerica* **7**, 51–150 (1998)
65. Deutsch, F.: *Best Approximation in Inner Product Spaces*. Springer, New York (2001)
66. Deutsch, F., Hundal, H.: The rate of convergence for the method of alternating projections II. *J. Math. Anal. Appl.* **205**, 381–405 (1997)
67. Dieuleveut, A., Bach, F.: Nonparametric stochastic approximation with large step-sizes. *Ann. Statistics* **44:4**, 1363–1399 (2016)
68. Dijkema, T., Schwab, C., Stevenson, R.: An adaptive wavelet method for solving high-dimensional elliptic PDEs. *Constr. Approx.* **30:3**, 423–455 (2009)
69. Dixmier, J.: *Von Neumann Algebras*. North-Holland, Amsterdam (1981)
70. Duflo, M.: *Random Iterative Systems*. Springer, Berlin (1997)
71. Engl, H., Hanke, M., Neubauer, A.: *Regularization of Inverse Problems*. Kluwer Acad. Publ., Dordrecht (2000)
72. Eijkhout, V., Vassilevski, P.: The role of the strengthened Cauchy-Buniakowskii-Schwarz inequality in multilevel methods. *SIAM Rev.* **33**, 405–419 (1991)
73. Faustmann, M., Melenk, M., Parvizi, M.: On the stability of Scott-Zhang type operators and application to multilevel preconditioning in fractional diffusion. *M2AN* **55**, 595–625 (2021)
74. Fercoq, O., Qu, Z.: Adaptive restart of accelerated gradient methods under local quadratic growth condition. *IMA J. Numer. Anal.* **39:4**, 2069–2095 (2019)
75. Fercoq, O., Richtárik, P.: Optimization in high dimensions via accelerated, parallel, and proximal coordinate descent. *SIAM Rev.* **58:4**, 739–771 (2016)
76. Fornasier, M.: Quasi-orthogonal decompositions of structured frames. *J. Math. Anal. Appl.* **289**, 180–199 (2004)
77. Galantai, A.: *Projectors and Projection Methods*. Springer, New York (2004)
78. Gander, M., Wanner, G.: The origins of the alternating Schwarz method. In *Domain Decomposition Methods in Science and Engineering XXI*, Lecture Notes in Computational Science and Engineering, Vol. 98, Erhel, J., Gander, M., Halpern, L., Pichot, G., Sassi, T., Widlund, O., eds., pp. 487–496, Springer, Cham (2014)
79. Gorbunov, E. Hanzely, F., Richtárik, P.: A unified theory of SGD: Variance reduction, sampling, quantization and coordinate descent. In *Proceedings of the Twenty Third International Conference on Artificial Intelligence and Statistics*, PMLR **108**, pp. 680–690 (2020)

80. Golub, G., Van Loan, C.: *Matrix Computations*. 4th edition, Johns Hopkins University Press, Baltimore (2013)
81. Gower, R., Richtárik, P.: Randomized iterative methods for linear systems. *SIAM J. Matr. Anal. Appl.* **36:4**, 1660–1690 (2015)
82. Griebel, M.: *Multilevelmethoden als Iterationsverfahren über Erzeugendensystemen*. Teubner Skripten zur Numerik, Teubner, Stuttgart (1994)
83. Griebel, M.: Multilevel algorithms considered as iterative methods on semidefinite systems. *SIAM Int. J. Sci. Stat. Comput.* **15:3**, 547–565 (1994)
84. Griebel, M., Hullmann, A., Oswald, P.: Optimal scalings for sparse grid discretizations. *Numer. Lin. Alg. Appl.* **22:1**, 76–100 (2014)
85. Griebel, M., Knapek, S.: Optimized general sparse grid approximation spaces for operator equations. *Math. Comp.* **78**, 2223–2257 (2009)
86. Griebel, M., Oswald, P.: On the abstract theory of additive and multiplicative Schwarz algorithms. *Numerische Mathematik* **70**, 163–180 (1995)
87. Griebel, M., Oswald, P.: Tensor product type subspace splittings and multilevel iterative methods for anisotropic problems. *Adv. Comput. Meth.* **4**, 171–2006 (1995)
88. Griebel, M., Oswald, P.: Schwarz iterative methods: Infinite space splittings. *Constr. Approx.* **44:1**, 121–139 (2016)
89. Griebel, M., Oswald, P.: Stochastic subspace correction in Hilbert space. *Constr. Approx.* **48:3**, 501–521 (2018)
90. Griebel, M., Oswald, P.: Stochastic subspace correction methods and fault tolerance. *Math. Comp.* **89**, 279–312 (2020)
91. Griebel, M., Oswald, P.: Convergence analysis of online algorithms for vector-valued kernel regression. *Constr. Approx.* (submitted). *INS Preprint* No. 2302, University of Bonn, arXiv:2309.07779 (2023)
92. Grisvard, P.: *Elliptic Problems in Non-Smooth Domains*. Pitman, Boston (1985)
93. Gröchenig, K.-H.: *Foundations of Time-Frequency Analysis*. Birkhäuser, Boston (2001)
94. Gu, M.: Subspace iteration randomization and singular value problems. *SIAM J. Sci. Comput.* **37:3**, 1139–1173 (2015)
95. Hackbusch, W.: *Iterative Solution of Large Sparse Systems of Equations*. Applied Mathematical Sciences, Vol. 95, 2nd edition, Springer, Cham (2016)
96. Hackbusch, W.: *Multigrid Methods and Applications*. Springer Series in Computational Mathematics, Vol. 4, Springer, Berlin (1985)
97. Halko, N., Martinsson, P., Tropp, J.: Finding structure in randomness: Probabilistic algorithms for constructing approximate matrix decompositions. *SIAM Rev.* **53:2**, 217–288 (2011)
98. Han, B., Michelle, M.: Wavelets on intervals derived from arbitrary compactly supported biorthogonal multiwavelets. *Appl. Comput. Harmon. Anal.* **53**, 270–331 (2021)
99. Han, B.: *Framelets and Wavelets: Algorithms, Analysis, and Applications*. Applied and Numerical Harmonic Analysis. Birkhäuser, Cham (2017)
100. Hanke, M.: Accelerated Landweber iterations for the solution of ill-posed equations. *Numerische Mathematik* **60**, 341–373 (1991)
101. Harbrecht, H., Schneider, R.: Rapid solution of boundary integral equations by wavelet Galerkin schemes. In *Multiscale, Nonlinear and Adaptive Approximation*, DeVore, R., Kunoth, A., eds., pp. 249–294, Springer, Berlin, Heidelberg (2009)
102. Hernandez, E., Weiss, G.: *A First Course on Wavelets*. CRC Press, Boca Raton (1996)
103. Hsiao, G., Wendland, W.: *Boundary Integral Equations*. Applied Mathematical Sciences, Vol. 164, 2nd edition, Springer, Cham (2021)
104. Hytönen, T., van Neerven, J., Veraar, M., Weis, L.: *Analysis in Banach Spaces I: Martingales and Littlewood-Paley Theory*. Springer, Cham (2016)
105. Johnen, H., Scherer, K.: On the equivalence of the K-functional and the moduli of continuity and some applications. In *Constructive Theory of Functions of Several Varibles*. Lecture Notes in Mathematics, Vol. 571, Schempp, W., Zeller, K., eds., pp. 119–140, Springer, Berlin (1976)

106. Karkulik, M., Melenk, M., Rieder, A.: Stable decompositions of hp-BEM spaces and an optimal Schwarz preconditioner for the hypersingular integral operator in 3D. *M2AN* **54**, 145–180 (2020)
107. Kelley, C.: *Iterative Methods for Optimization*. Frontiers in Applied Mathematics, SIAM, Philadelphia (1999)
108. Klusowski, J., Siegel, J.: Sharp convergence rates for matching pursuit. *IEEE Trans. Information Theory* (submitted), arxiv:2307.07679v2 (2023)
109. Knuth, D.: *The Art of Computer Programming, Vol. 3: Sorting and Searching*. 2nd edition, Addison-Wesley, Redwood City (1998)
110. Korotov, S., Plaza, A., Suárez, J.: Longest-edge n-section algorithms: Properties and open problems. *J. Comput. Appl. Math.* **293**, 139–146 (2016)
111. Kushner, H., Yin, G.: *Stochastic Approximation and Recursive Algorithms and Applications*. 2nd edition, Springer, New York (2003)
112. Lan, G.: *First-order and Stochastic Optimization Methods for Machine Learning*. Springer Series in the Data Sciences. Springer, Cham (2020)
113. Lee, Y., Sidford, A.: Efficient accelerated coordinate descent methods and faster algorithms for solving linear systems. In *Foundations of Computer Science (FOCS)*, 2013 IEEE 54th Annual Symposium, pp. 147–156, IEEE (2013)
114. Leventhal, D., Lewis, A.: Randomized methods for linear constraints: Convergence rates and conditioning. *Math. Oper. Res.* **35:3**, 641–654 (2010)
115. Liesen, J., Strakos, Z.: *Krylov Subspace Methods, Principles and Analysis*. Numerical Mathematics and Scientific Computing, Oxford University Press, Oxford (2013)
116. Lorentz, R., Oswald, P.: Criteria for hierarchical bases in Sobolev spaces. *Appl. Comput. Harmon. Anal.* **8:1**, 32–85 (2000)
117. Mahoney, M.: Randomized algorithms for matrices and data. *Foundations and Trends in Machine Learning* **3:2**, 123–224 (2011)
118. Marcus, A., Spielman, D., Srivastava, N.: Interlacing families II: Mixed characteristic polynomials and the Kadison-Singer problem. *Annals of Math.* **182:1**, 327–350 (2015)
119. Martinsson, P.-G., Tropp, J.: Randomized numerical linear algebra: Foundations and algorithms. *Acta Numerica* **29**, 403–572 (2020)
120. Matsokin, A., Nepomnyaschikh, S.: The Schwarz alternation method in a subspace. *Izv. Vyssh. Uchebn. Zaved. Mat.* **85**, 61–66 (1985)
121. Micchelli, C., Pontil, M.: On learning vector-valued functions. *Neural. Comput.* **17:1**, 177–204 (2005)
122. Meyer, Y.: *Wavelets and Operators*. Cambridge University Press, Cambridge (1992)
123. Nečas, J.: *Les Méthodes Directes en Theorie des Équations Élliptiques*. Masson, Paris (1967)
124. Needell, D., Tropp, J.: Paved with good intentions: Analysis of a randomized block Kaczmarz method. *Lin. Alg. Appl.* **441**, 199–221 (2014)
125. Nepomnyaschikh, S.: Decomposition and fictitious domain methods for elliptic boundary value problems. In *Proceedings of the 5th Conf. Domain Decomposition methods for PDE*, Keyes, D., Chan, R., Meurant, G., Scroggs, J., Voigt, R., eds., pp. 62–72, SIAM, Philadelphia (1992)
126. Nesterov, Y.: A method for solving a convex programming problem with convergence rate $O(1/k^2)$. *Sov. Math. Dokl.* **27:2**, 372–376 (1983)
127. Nesterov, Y.: Efficiency of coordinate descent methods on huge-scale optimization problems. *SIAM J. Optim.* **22:2**, 341–362 (2012)
128. Nesterov, Y.: *Lectures on Convex Optimization*. 2nd edition, Springer, New York (2018)
129. Nochetto, R., Siebert, K., Veeser, A.: Theory of adaptive finite element methods: An introduction. In *Multiscale, Nonlinear and Adaptive Approximation*, DeVore, R., Kunoth, A., eds., pp. 409–542, Springer, Berlin, Heidelberg (2009)
130. Nochetto, R., Veeser, A.: Primer of adaptive finite element methods. In *Multiscale and Adaptivity: Modeling, Numerics and Applications*, Lecture Notes in Mathematics, Vol. 2040, C.I.M.E. Summer School, Cetraro, Italy, 2009, Naldi, R., Russo, G., eds., pp. 125–225, Springer, Heidelberg (2011)

131. Novikov, I., Protasov, V., Skopina, M.: *Wavelet Theory*. Transl. Math. Monogr., Vol. 239, AMS, Providence (2011)
132. O'Donoghue, B., Candes, E.: Adaptive restart for accelerated gradient schemes. *Found. Comput. Math.* **15:3**, 715–732 (2015)
133. Oswald, P.: On a hierarchical basis multilevel method with non-confroming P1 elements. *Numerische Mathematik* **62**, 189–212 (1992)
134. Oswald, P.: On a BPX-preconditioner for P1 elements. *Computing* **51**, 125–133 (1993)
135. Oswald, P.: *Multilevel Finite Element Approximation: Theory and Applications*. Teubner Skripte zur Numerik, Teubner, Stuttgart (1994)
136. Oswald, P.: On the convergence rate of SOR: A worst case estimate. *Computing* **52**, 245–255 (1994)
137. Oswald, P.: Preconditioners for nocnconforming discretizations. *Math. Comp.* **65**, 923–941 (1996)
138. Oswald, P.: Intergrid transfer operators and multilevel preconditioners for non-conforming discretizations. *Appl. Numer. Math.* **23**, 139–158 (1997)
139. Oswald, P.: Optimality of multilevel preconditioning for non-conforming P1 finite elements. *Numerische Mathematik* **111**, 267–291 (2008)
140. Oswald, P.: Estimates for generalized sparse grid hierarchical basis preconditioners. *INS Preprint* No. 1803, University of Bonn, arXiv:1803.01782 (2018)
141. Oswald, P., Zhou, W.: Convergence analysis for Kaczmarz-type methods in a Hilbert space framework. *Lin. Alg. Appl.* **478**, 131–161 (2015)
142. Oswald P., Zhou, W.: Random reordering in SOR-type methods. *Numerische Mathematik* **135**, 1207–1220 (2017), Correction: Random reordering in SOR-type methods. **154**, 521–525 (2023)
143. Pages, G.: *Numerical Probability. An Introduction with Applications to Finance*. Springer, New York (2018)
144. Petra, S., Popa, C.: Single projection Kaczmarz extended algorithms. *Numer. Algorithms* **73**, 791–806 (2016)
145. Pierra, G.: Decomposition through formalization in a product space. *Mathematical Programming* **28**, 96–115 (1984)
146. Pisier, G.: Grothendieck's theorem, past and present. *Bull. Amer. Math. Soc.* **49**, 237–323 (2012)
147. Polyak, B.: Some methods of speeding up the convergence of iteration methods. *USSR Comput. Math. and Math. Physics* **4:5**, 1–17 (1964)
148. Popa, C.: Least-squares solution of overdetermined inconsistent linear systems using Kaczmarz relaxation. *Int. J. Comp. Math.* **55:1-2**, 79–89 (1995)
149. Ravichandran, M., Srivastava, N.: Asymptotically optimal multi-paving. *International Mathematics Research Notices* **14**, 10908–10940 (2021)
150. Reed, M., Simon, B.: *Methods of Modern Mathematical Physics I, Functional Analysis*. Revised and enlarged edition, Acad. Press, San Diego (1980)
151. Rokhlin, V., Tygert, M.: A fast randomized algorithm for overdetermined linear least-squares regression. *Proc. Natl. Acad. Sci. USA* **105**, 13212–13217 (2008)
152. Roulet, V., d'Aspremont, A.: Sharpness, restart and acceleration. *SIAM J. Optim.* **30:1**, 262–289 (2020)
153. Saad, Y.: *Iterative Methods for Sparse Linear Systems*. 2nd edition, SIAM, Philadelphia (2003)
154. Saad, Y.: Iterative methods for linear systems of equations: A brief historical journey. *Contemporary Mathematics* **754**, 75 Years of Mathematics of Computation, 197–215 (2020)
155. Sangalli, G.: A uniform analysis of non-symmetric and coercive linear operators. *SIAM J. Math. Anal.* **36**, 2033–2048 (2005)
156. Sangalli, G.: Robust a-posteriori estimator for advection-diffusion-reaction problems. *Math. Comp.* **77:261**, 41–70 (2008)
157. Schmidt, A., Siebert, K.: *Design of Adaptive Finite Element Software: The Finite Element Toolbox ALBERTA*. Lecture Notes in Computational Science and Engineering, Vol. 42, Springer, Berlin, Heidelberg (2005)

158. Schölkopf, B., Smola, A.: *Learning with Kernels*. MIT Press, Cambridge (2002)
159. Schumaker, L.: *Spline Functions*. Wiley, New York (1981)
160. Schwab, C.: *p- and hp-Finite Element Methods*. Numerical Mathematics and Scientific Computation, Oxford University Press, Oxford (1988)
161. Schwarz, H.: Über einen Grenzübergang durch alternierendes Verfahren. *Vierteljahrsschrift der Naturforschenden Gesellschaft in Zürich* **15**, 272–286 (1870), also in Schwarz, H.: Gesammelte Mathematische Abhandlungen, Vol. 2., pp. 133–143, Springer, Berlin (1890)
162. Scott, R., Zhang, S.: Finite element interpolation of non-smooth functions satisfying boundary conditions. *Math. Comp.* **54**, 483–493 (1990)
163. Sedgewick, R., Wayne, K.: *Algorithms*. 4th edition, Addison-Wesley, Upper Saddle River (2011)
164. Seidel, L.: Über ein Verfahren, die Gleichungen, auf welche die Methode der kleinsten Quadrate führt, sowie lineare Gleichungen überhaupt, durch successive Annäherung aufzulösen. *Abhandlungen der Mathematisch-Physikalischen Klasse der Königlich Bayerischen Akademie der Wissenschaften* (in German) **11:3**, 81–108 (1874)
165. Shiryayev, A.: *Probability*. Springer, New York (1984)
166. Sil'nichenko, A.: Rate of convergence of greedy algorithms. *Math. Notes* **76:3**, 582–586 (2004)
167. Southwell, R.: *Relaxation Methods in Engineering Science - A Treatise in Approximate Computation*. Oxford University Press, Oxford (1940)
168. Spielman, D., Srivastava, N.: An elementary proof of the restricted invertibility theorem. *Israel Journal of Mathematics* **190:1**, 83–91 (2012)
169. Steinbach, O.: *Numerical Approximation Methods for Elliptic Boundary Value Problems*. Springer, New York (2008)
170. Steinwart, I., Christmann, A.: *Support Vector Machines*. Springer, New York (2008)
171. Stevenson, R.: Optimality of a standard adaptive finite element method. *Found. Comput. Math.* **7**, 245–269 (2007)
172. Stevenson, R.: The completion of locally refined simplicial partitions created by bisection. *Math. Comp.* **77**, 227–241 (2008)
173. Stevenson, R.: Adaptive wavelet methods for solving operator equations: An overview. In *Multiscale, Nonlinear and Adaptive Approximation*, DeVore, R., Kunoth, A., eds., pp. 543–597, Springer, Berlin, Heidelberg (2009)
174. Strohmer, T., Vershynin, R.: A randomized Kaczmarz algorithm with exponential convergence. *J. Four. Anal. Appl.* **15:1**, 262–278 (2009)
175. Sun, W.: G-frames and G-Riesz bases. *J. Math. Anal. Appl.* **322**, 437–452 (2006)
176. Sweldens, W.: The lifting scheme: A custom-design construction of biorthogonal wavelets. *Appl. Comput. Harmon. Anal.* **3**, 186–200 (1996)
177. Takesaki, M.: *Theory of Operator Algebras I*. Springer, New York (1979)
178. Temlyakov, V.: Greedy approximation. *Acta Numerica* **17**, 235–409 (2008)
179. Temlyakov, V.: *Greedy Approximation*. Cambridge University Press, Cambridge (2012)
180. Toselli, A., Widlund, O.: *Domain Decomposition Methods - Algorithms and Theory*. Springer, Berlin (2005)
181. Trefethen, L., Bau III, D.: *Numerical Linear Algebra*. SIAM, Philadelphia (1997)
182. Triebel, H.: *Interpolation Theory, Function Spaces, Differential Operators*. North Holland, Amsterdam (1978)
183. Triebel, H.: *The Structure of Functions*. Birkhäuser, Basel (2001)
184. Triebel, H.: *Theory of Function Spaces III*. Birkhäuser, Basel (2006)
185. Triebel, H.: *Function Spaces and Wavelets on Domains*. European Math. Soc. Publishing House, Zürich (2008).
186. Tropp, J.: Column subset selection, matrix factorization, and eigenvalue optimization. In *Proceedings of the Twentieth Annual ACM-SIAM Symposium on Discrete Algorithms (SODA)*, Mathieu, C., ed., pp. 978–986, SIAM, Philadelphia (2009)
187. Trottenberg, U., Oosterle C., Schüller A.: *Multigrid*. Acad. Press, San Diego (2001)
188. Urban, K.: *Wavelet Methods for Elliptic Partial Differential Equations*. Oxford University Press, Oxford (2009)

189. Vapnik, V.: *Statistical Learning Theory*. Wiley, New York (1998)
190. Vassilevski, P.: *Multilevel Block Factorization Preconditioners: Matrix-Based Analysis and Algorithms for Solving Finite Element Equations*. Springer, New York (2010)
191. Verfürth, R.: A posteriori error estimators for convection-diffusion equations. *Numerische Mathematik* **80:4**, 641–663 (1998)
192. Weidmann, J.: *Linear Operators in Hilbert Spaces*. Grad. Texts Math., Vol. 68, Springer, Berlin (1980)
193. Woodruff, D.: Sketching as a tool for numerical linear algebra. *Foundations and Trends in Theoretical Computer Science* **10:1-2**, 1–157 (2014)
194. Xu, J.: Iterative methods by space decomposition and subspace correction. *SIAM Rev.* **34:4**, 581–613 (1992)
195. Xu, J.: The auxiliary space method and optimal multigrid preconditioning for unstructured grids. *Computing* **56**, 215–235 (1996)
196. Xu, J., Chen, L., Nochetto, R.: Optimal multilevel methods for H(grad), H(curl), and H(div) systems on graded and unstructured grids. In *Multiscale, Nonlinear and Adaptive Approximation*, DeVore, R., Kunoth, A., eds., pp. 599–659, Springer, Berlin, Heidelberg (2009)
197. Xu, J., Zou, J.: Some non-overlapping domain decomposition algorithms. *SIAM Rev.* **40**, 379–412 (1998)
198. Xu, J., Zikatanov, L.: The method of alternating projections and the method of subspace corrections in Hilbert space. *J. Amer. Math. Soc.* **15:3**, 573–597 (2002)
199. Xu, J., Zikatanov, L.: Algebraic multigrid methods. *Acta Numerica* **26**, 591–721 (2017)
200. Yosida, K.: *Functional Analysis*. 6th edition, Springer, Berlin (1980)
201. Youssef, P.: A note on column subset selection. *International Mathematics Research Notices* **23**, 6431–6447 (2014)
202. Yserentant, H.: Old and new convergence proofs for multigrid methods. *Acta Numerica* **2**, 285–326 (1993)
203. Zenger, C.: Sparse grids. In *Parallel Algorithms for Partial Differential Equations*, Hackbusch, W., ed., NNFM, Vol. 31, Vieweg, Braunschweig (1991)
204. Zouzias, A., Freris, N.: Randomized extended Kaczmarz for solving least squares. *SIAM J. Matrix Anal. Appl.* **34:2**, 773–793 (2013)

Index

M. Griebel, P. Oswald, *Hilbert Space Splittings and Iterative Methods*, Springer Series in Computational Mathematics 62, https://doi.org/10.1007/978-3-031-74370-2

The manufacturer's authorised representative in the EU is Springer Nature Customer Service Centre GmbH, Europaplatz 3, 69115 Heidelberg, Germany. If you have any concerns regarding our products, please contact ProductSafety@springernature.com

Printed and bound by CPI Group (UK) Ltd, Croydon, CR0 4YY
30/09/2026
02225598-0013